S0-AZP-650

HARCOURT

Math

CALIFORNIA EDITION

Harcourt School Publishers
Orlando • Boston • Dallas • Chicago • San Diego
www.harcourtschool.com

Grateful acknowledgment is made to the National Audubon Society for permission to reprint map from *Great Backyard Bird Count 2000* by BirdSource (http://www.birdsource.org). Copyright 2000.

Printed in the United States of America

ISBN 0-15-315516-7

3 4 5 6 7 8 9 10 032 2004 2003 2002 2001

Mathematics Advisor

David Singer
Professor of Mathematics
Case Western Reserve University
Cleveland, Ohio

Senior Author

Evan M. Maletsky
Professor of Mathematics
Montclair State University
Upper Montclair, New Jersey

Authors

Angela Giglio Andrews
Math Teacher, Scott School
Naperville District #203
Naperville, Illinois

Grace M. Burton
Chair, Department of Curricular Studies
Professor, School of Education
University of North Carolina at Wilmington
Wilmington, North Carolina

Howard C. Johnson
Dean of the Graduate School
Associate Vice Chancellor for Academic Affairs
Professor, Mathematics and Mathematics Education
Syracuse University
Syracuse, New York

Lynda A. Luckie
Administrator/Math Specialist
Gwinnett County Public Schools
Lawrenceville, Georgia

Joyce C. McLeod
Visiting Professor
Rollins College
Winter Park, Florida

Vicki Newman
Classroom Teacher
McGaugh Elementary School
Los Alamitos Unified School District
Seal Beach, California

Janet K. Scheer
Executive Director
Create A Vision
Foster City, California

Karen A. Schultz
College of Education
Georgia State University
Atlanta, Georgia

Program Consultants and Specialists

Janet S. Abbott
Mathematics Consultant
California

Arax Miller
Curriculum Coordinator and English Department Chairperson
Chamlian School
Glendale, California

Lois Harrison-Jones
Education and Management Consultant
Dallas, Texas

Rebecca Valbuena
Language Development Specialist
Stanton Elementary School
Glendora, California

Use Whole Numbers and Decimals

1 PLACE VALUE OF WHOLE NUMBERS

2 PLACE VALUE OF DECIMALS 16

Daily Review and Practice
Quick Review
Mixed Review and Test Prep
Intervention
Troubleshooting, pp. H2–H5
Extra Practice
pp. H32–H35

Technology Resources

Harcourt Math Newsroom Video:
Chapter 3, p. 36

E-Lab:
Chapter 2, p. 21
Chapter 4, p. 52

Multimedia Glossary:
The Learning Site at **www.harcourtschool.com/glossary/math**

UNIT 2

CHAPTERS 5–8

Algebra, Data, and Graphing

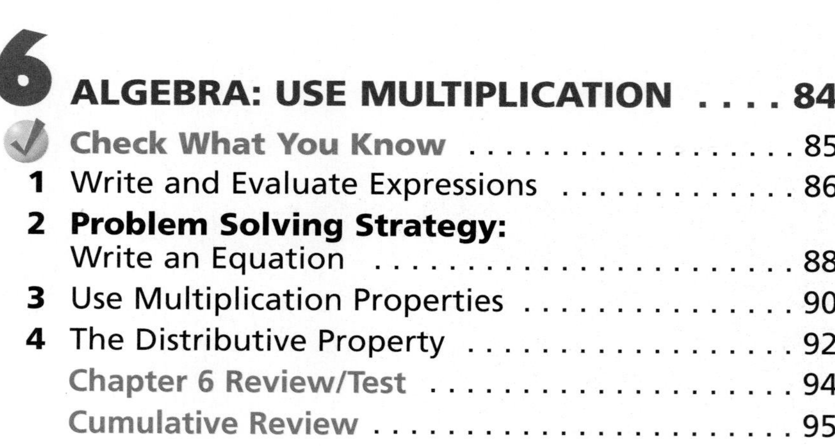

Daily Review and Practice
Quick Review
Mixed Review and Test Prep

Intervention
Troubleshooting, pp. H6–H10

Extra Practice
pp. H36–H39

Technology Resources

Harcourt Math Newsroom Video: *Chapter 8, p. 119*

E-Lab:
Chapter 5, p. 76
Chapter 6, p. 92
Chapter 7, p. 104
Chapter 8, p. 131

Multimedia Glossary: *The Learning Site* at **www.harcourtschool.com/glossary/math**

UNIT WRAPUP

UNIT 3

CHAPTERS 9–10

Multiply Whole Numbers and Decimals

Daily Review and Practice
- Quick Review
- Mixed Review and Test Prep

Intervention
- Troubleshooting, pp. H11–H12, H14

Extra Practice
- pp. H40–H41

Technology Resources

Harcourt Math Newsroom Video:
Chapter 9, p. 145

E-Lab:
Chapter 10, p. 159

Mighty Math Calculating Crew:
Chapter 9, p. 149

Multimedia Glossary:
The Learning Site at **www.harcourtschool.com/glossary/math**

UNIT 4

CHAPTERS 11–14

Divide Whole Numbers and Decimals

Daily Review and Practice
- Quick Review
- Mixed Review and Test Prep

Intervention
- Troubleshooting, pp. H13–H17

Extra Practice
- pp. H42–H45

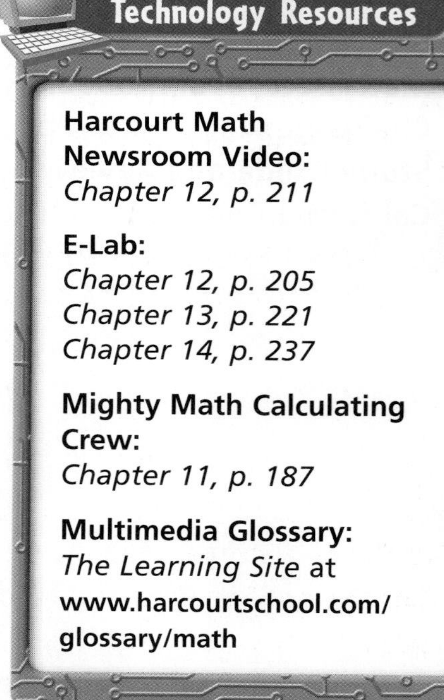

Technology Resources

Harcourt Math Newsroom Video:
Chapter 12, p. 211

E-Lab:
Chapter 12, p. 205
Chapter 13, p. 221
Chapter 14, p. 237

Mighty Math Calculating Crew:
Chapter 11, p. 187

Multimedia Glossary:
The Learning Site at **www.harcourtschool.com/glossary/math**

UNIT WRAPUP

UNIT 5
CHAPTERS 15–18

Fractions, Ratio, and Percent

15 NUMBER THEORY 252

16 FRACTION CONCEPTS 278

Daily Review and Practice
Quick Review
Mixed Review and Test Prep
Intervention
Troubleshooting, pp. H18–H21
Extra Practice
pp. H46–H49

Technology Resources

Harcourt Math Newsroom Video:
Chapter 17, p. 308

E-Lab:
Chapter 15, pp. 254, 257, 265
Chapter 17, p. 303
Chapter 18, p. 317

Mighty Math Calculating Crew:
Chapter 16, p. 287

Mighty Math Number Heroes:
Chapter 18, p. 320

Multimedia Glossary:
The Learning Site at
www.harcourtschool.com/glossary/math

UNIT 6
CHAPTERS 19–22

Operations with Fractions

UNIT 7

CHAPTERS 23–26

Algebra and Geometry

Daily Review and Practice
  Quick Review
  Mixed Review and Test Prep
Intervention
  Troubleshooting, pp. H10, H25–H28
Extra Practice
  pp. H54–H57

Technology Resources

Harcourt Math Newsroom Video:
Chapter 26, p. 479

E-Lab:
Chapter 23, p. 427
Chapter 26, pp. 484, 491

Mighty Math Calculating Crew:
Chapter 23, p. 423

Mighty Math Astro Algebra:
Chapter 24, p. 443

Mighty Math Number Heroes:
Chapter 25, p. 471

Multimedia Glossary:
The Learning Site at
www.harcourtschool.com/glossary/math

UNIT WRAPUP

UNIT 8

CHAPTERS 27–29

Measurement

27 CUSTOMARY AND METRIC SYSTEMS 502

28 PERIMETER AND AREA 522

Daily Review and Practice
Quick Review
Mixed Review and Test Prep
Intervention
Troubleshooting, pp. H28–H31
Extra Practice
pp. H58–H60

Technology Resources

Harcourt Math Newsroom Video:
Chapter 27, p. 507

E-Lab:
Chapter 27, p. 505
Chapter 27, p. 515
Chapter 28, p. 527
Chapter 29, p. 554

Multimedia Glossary:
The Learning Site at
www.harcourtschool.com/glossary/math

Probability

Daily Review and Practice
- Quick Review
- Mixed Review and Test Prep

Intervention
- Troubleshooting, p. H31

Extra Practice
- p. H61

UNIT WRAPUP

Technology Resources

E-Lab:
Chapter 30, p. 571

Mighty Math Number Heroes:
Chapter 30, p. 576

Multimedia Glossary:
The Learning Site at **www.harcourtschool.com/glossary/math**

STUDENT HANDBOOK

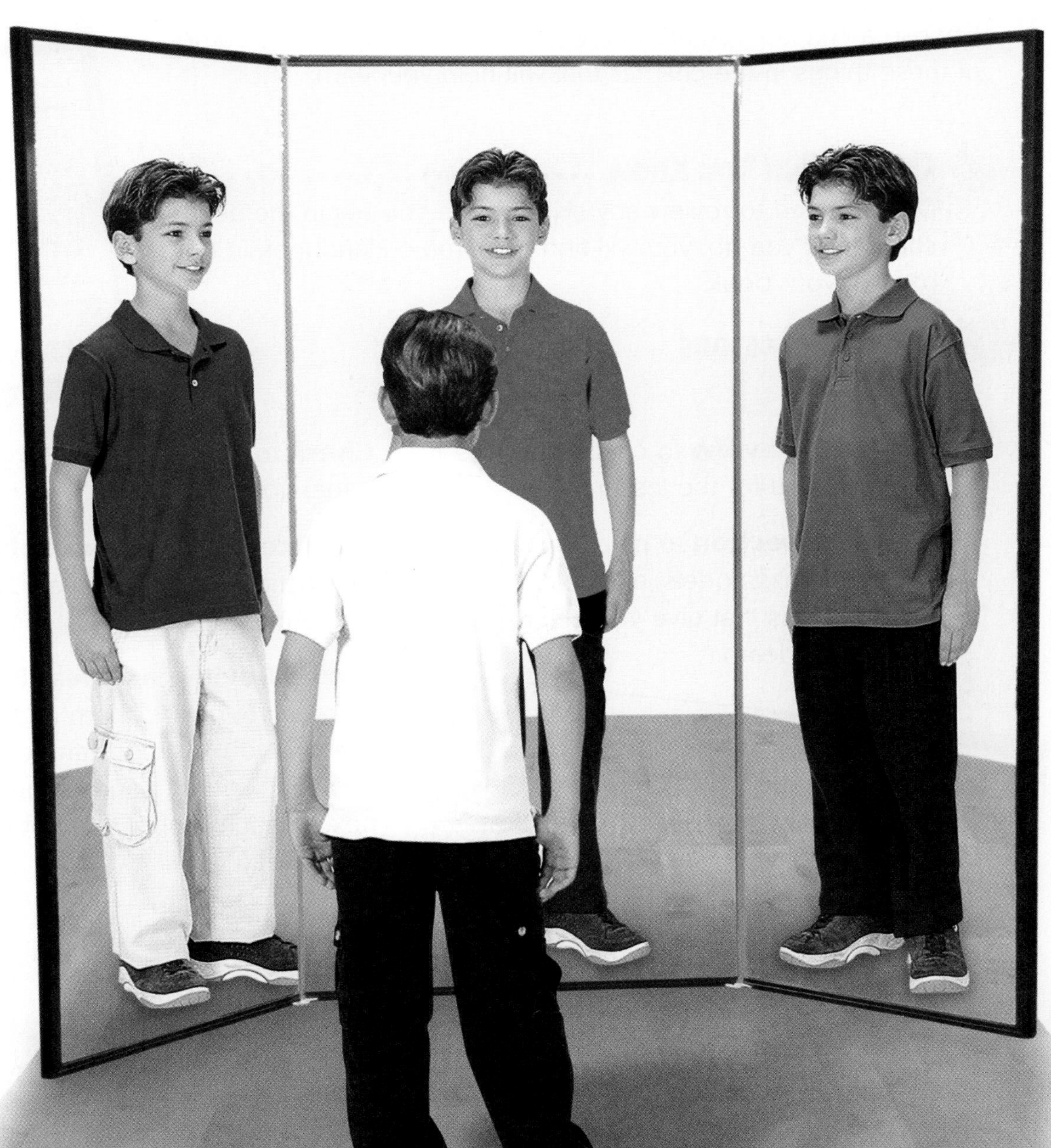

Welcome!

The authors of *Harcourt Math* want you to enjoy learning math and to feel confident that you can do it. We invite you to share your math book with family members. Take them on a guided tour through your book!

The Guided Tour

Choose a chapter you are interested in. Show your family some of these things in the chapter that will help you learn.

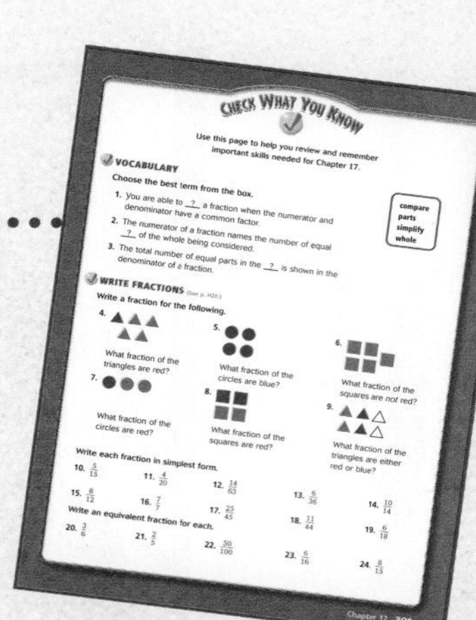

CHECK WHAT YOU KNOW

Use this page to help you review and remember important skills needed for Chapter 17.

VOCABULARY

Choose the best term from the box.

compare
parts
simplify
whole

1. You are able to __?__ a fraction when the numerator and denominator have a common factor.
2. The numerator of a fraction names the number of equal __?__ of the whole being considered.
3. The total number of equal parts in the __?__ is shown in the denominator of a fraction.

WRITE FRACTIONS (See p. H20.)

Write a fraction for the following.

4. What fraction of the triangles are red?
5. What fraction of the circles are blue?
6. What fraction of the squares are *not* red?
7. What fraction of the circles are red?
8. What fraction of the squares are red?
9. What fraction of the triangles are either red or blue?

Write each fraction in simplest form.

10. $\frac{5}{15}$ 11. $\frac{4}{20}$ 12. $\frac{14}{63}$ 13. $\frac{6}{36}$ 14. $\frac{10}{14}$
15. $\frac{8}{12}$ 16. $\frac{7}{7}$ 17. $\frac{25}{45}$ 18. $\frac{11}{44}$ 19. $\frac{6}{18}$

Write an equivalent fraction for each.

20. $\frac{3}{6}$ 21. $\frac{2}{5}$ 22. $\frac{50}{100}$ 23. $\frac{6}{16}$ 24. $\frac{8}{15}$

Chapter 17 301

Check What You Know

Do you need to review any skills before you begin the next chapter? If you do, you will find help in the Handbook in the back of your book.

The Math Lessons

- **Quick Review** to check the skills you need for the lesson.
- **Learn section** to help you study problems, models, examples, and questions that give you different ways to learn.
- **Check** to make sure you understood the lesson.
- **Practice and Problem Solving** to practice what you have just learned.
- **Mixed Review and Test Prep** to keep your skills sharp and to prepare you for important tests. Look back at the pages shown next to each problem to get help if you need it.

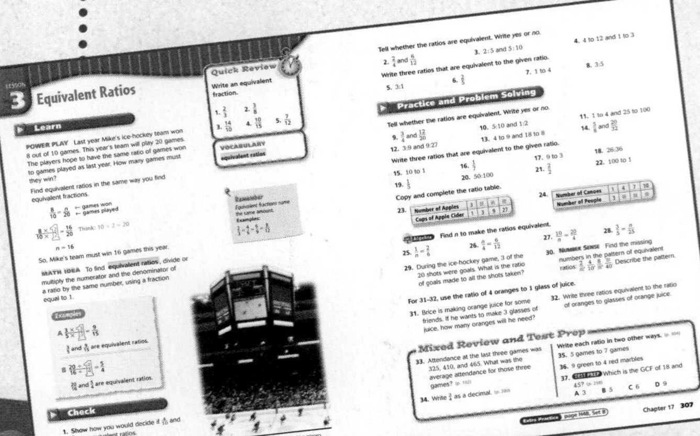

LESSON 3 Equivalent Ratios

Learn

POWER PLAY Last year Mike's ice-hockey team won 8 out of 10 games. This year's team will play 20 games. The players hope to have the same ratio of games won to games played as last year. How many games must they win?

Find equivalent ratios in the same way you find equivalent fractions.

$\frac{8}{10} = \frac{n}{20}$ ← games won ← games played

$\frac{8 \times 2}{10 \times 2} = \frac{16}{20}$ Think: $10 \times 2 = 20$

$n = 16$

So, Mike's team must win 16 games this year.

MATH IDEA To find **equivalent ratios**, divide or multiply the numerator and the denominator of a ratio by the same number, using a fraction equal to 1.

Examples

A $\frac{3 \times 3}{5 \times 3} = \frac{9}{15}$ — $\frac{3}{5}$ and $\frac{9}{15}$ are equivalent ratios.

B $\frac{20 \div 4}{16 \div 4} = \frac{5}{4}$ — $\frac{20}{16}$ and $\frac{5}{4}$ are equivalent ratios.

Quick Review

Write an equivalent fraction.

1. $\frac{2}{3}$ 2. $\frac{3}{8}$ 3. $\frac{14}{10}$ 4. $\frac{10}{15}$ 5. $\frac{7}{12}$

VOCABULARY

equivalent ratios

Remember

Equivalent fractions name the same amount.

Examples: $\frac{2}{3} = \frac{4}{6} = \frac{6}{9} = \frac{8}{12}$

Check

1. Show how you would decide if $\frac{4}{10}$ and $\frac{2}{5}$ are equivalent ratios.

Tell whether the ratios are equivalent. Write *yes* or *no*.

2. $\frac{2}{4}$ and $\frac{6}{12}$ 3. 2:5 and 5:10 4. 4 to 12 and 1 to 3

Write three ratios that are equivalent to the given ratio.

5. 3:1 6. $\frac{2}{5}$ 7. 1 to 4 8. 3:5

Practice and Problem Solving

Tell whether the ratios are equivalent. Write *yes* or *no*.

9. $\frac{3}{4}$ and $\frac{12}{20}$ 10. 5:10 and 1:2 11. 1 to 4 and 25 to 100
12. 3:9 and 9:27 13. 4 to 9 and 18 to 8 14. $\frac{5}{8}$ and $\frac{20}{32}$

Write three ratios that are equivalent to the given ratio.

15. 10 to 1 16. $\frac{1}{7}$ 17. 9 to 3 18. 26:36
19. $\frac{1}{5}$ 20. 50:100 21. $\frac{2}{2}$ 22. 100 to 1

Copy and complete the ratio table.

23.

Number of Apples	3	■	■	■
Cups of Apple Cider	1	3	9	27

24.

Number of Canoes	1	4	7	10
Number of People	3	■	■	■

Algebra Find n to make the ratios equivalent.

25. $\frac{1}{n} = \frac{2}{6}$ 26. $\frac{n}{4} = \frac{6}{12}$ 27. $\frac{10}{n} = \frac{20}{4}$ 28. $\frac{3}{5} = \frac{n}{25}$

29. During the ice-hockey game, 3 of the 20 shots were goals. What is the ratio of goals made to all the shots taken?

30. NUMBER SENSE Find the missing numbers in the pattern of equivalent ratios: $\frac{2}{■}, \frac{4}{10}, \frac{8}{■}, \frac{■}{40}$. Describe the pattern.

For 31–32, use the ratio of 4 oranges to 1 glass of juice.

31. Brice is making orange juice for some friends. If he wants to make 3 glasses of juice, how many oranges will he need?

32. Write three ratios equivalent to the ratio of oranges to glasses of orange juice.

Mixed Review and Test Prep

33. Attendance at the last three games was 325, 410, and 465. What was the average attendance for those three games? (p. 102)

34. Write $\frac{3}{4}$ as a decimal. (p. 280)

Write each ratio in two other ways. (p. 304)

35. 5 games to 7 games

36. 9 green to 4 red marbles

37. TEST PREP Which is the GCF of 18 and 45? (p. 258)
A 3 B 5 C 6 D 9

Extra Practice page H48, Set B

306 — Chapter 17 307

Student Handbook

Now show your family the **Student Handbook** in the back of your book. The sections will help you in many different ways.

 **Troubleshooting** will help you review and remember skills from last year.

 **Extra Practice** can be used to make sure that you are ready to move on to the next lesson.

 **Sharpen Your Test-Taking Skills** will help you feel confident that you can do well on a test.

 **Basic Facts Tests** will check whether you have memorized all of the basic facts and will show you which facts you still need to practice.

 **California Standards** tell what you are expected to learn this year. Each lesson's standards are listed at the bottom of the lesson page.

Invite your family members to

- talk with you about what you are learning.
- help you correct errors you have made on completed work.
- help you set a time and find a quiet place to do math homework.
- help you memorize the addition, subtraction, multiplication, and division facts.
- solve problems as you play together, shop together, and do household chores.
- visit **The Learning Site** at **www.harcourtschool.com**
- have **Fun with Math!**

Have a great year!

The Authors

Be a Good Problem Solver!

You need to organize your thinking. You can use problem solving steps to stay on track.

Use these problem solving steps. They can help you think through a problem.

the problem.

What are you asked to find?	Restate the question in your own words.
What information will you use?	Look for numbers. Find how they are related.
Is there information you will not use? If so, what?	Decide whether you need all the information you are given.

Plan a strategy to solve.

What strategy can you use to solve the problem?	Think about some problem solving strategies you can use. Then choose one.

Solve the problem.

How can you use the strategy to solve the problem?	Follow your plan. Show your solution.

Check your answer.

Look back at the problem. Does the answer make sense? Explain.	Be sure you answered the question that is asked.
What other strategy could you use?	Solving the problem by another method is a good way to check your work.

PROBLEM SOLVING STRATEGIES

- **Draw a Diagram or Picture**
- Make a Model or Act it Out
- Make an Organized List
- Find a Pattern
- Make a Table or Graph
- Predict and Test
- Work Backward
- Solve a Simpler Problem
- Write an Equation
- Use Logical Reasoning

Try It

Here's how you can use the problem solving steps to solve a problem.

Draw a Diagram

PROBLEM Carey and Joan are sisters. Carey is 4 years older than Joan. The sum of their ages is 16. How old is each sister?

the problem.
I need to find the age of each sister. One is 4 years older than the other. Their ages together equal 16.

a strategy to solve.
I could *draw a diagram*. I could draw boxes to stand for their ages. Then I can find out how to make the boxes match the information in the problem.

the problem.
I'll let one box be Joan's age and a little larger box be Carey's age, because she is 4 years older. My diagram looks like this:

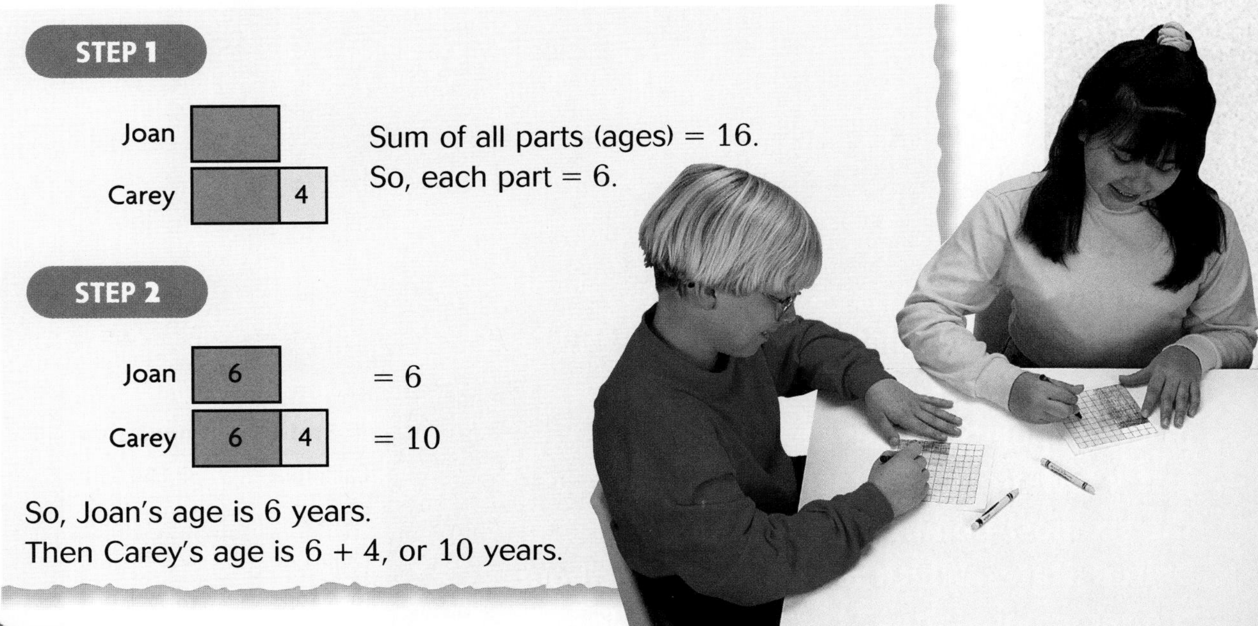

So, Joan's age is 6 years.
Then Carey's age is 6 + 4, or 10 years.

Check your answer.
Carey is 4 years older than Joan. The sum of their ages is 6 + 10, or 16 years. This matches the problem.

Place Value of Whole Numbers

Satellite image of North America at night

One person is added to the United States population every 17 seconds. Much of the population lives in just a few areas. The table lists the six most-populated states in the United States. Which two states have about the same population? List the states in order from the state with the greatest population to the state with the least population.

MOST-POPULATED STATES

State	Population
California	32,666,550
Florida	14,915,980
Illinois	12,045,326
New York	18,175,301
Pennsylvania	12,001,451
Texas	19,759,614

Use this page to help you review and remember important skills needed for Chapter 1.

VOCABULARY

Choose the best term from the box.

expanded form
digits
period
standard form

1. Each group of three digits separated by a comma in a large number is called a __?__.
2. The number 500 + 30 + 4 is written in __?__.
3. The symbols 0, 1, 2, 3, 4, 5, 6, 7, 8, and 9 are __?__.

PLACE VALUE (See p. H2.)

Write the value of the digit 6 in each number.

4. 62,980
5. 368
6. 456,709
7. 906

Write the value of the blue digit.

8. 560
9. 3,072
10. 47,092
11. 302,561
12. 6,257
13. 13,348
14. 97,812
15. 120,465

READ AND WRITE WHOLE NUMBERS (See p. H2.)

Write the number in standard form.

16. sixteen thousand, forty
17. 50,000 + 4,000 + 200 + 4
18. eight hundred seven thousand, forty-two
19. 400,000 + 20,000 + 3,000 + 50 + 6

Write the number in expanded form.

20. 387
21. 2,412
22. 43,671
23. 35,902
24. 60,234
25. 12,084
26. 213,087
27. 692,530

Write the number in word form.

28. 409
29. 2,010
30. 5,102
31. 73,249
32. 14,013
33. 89,738
34. 311,952
35. 616,870

LESSON 1

HANDS ON
Understand Millions

Quick Review

Write the number that is 100 more than the given number.

1. 0 **2.** 25
3. 309 **4.** 999
5. 1,009

MATERIALS
grid paper, crayons, tape

Explore

There are at least one million different species of insects, including beetles, butterflies, ants, and bees. Here's an activity to help you understand the size of one million.

STEP 1

Use a 10 × 10 sheet of grid paper. Color a dot in each box on the grid paper. Let each dot represent one insect.

STEP 2

How many insects does the grid paper show? Write the total at the bottom of the grid paper.

STEP 3

Tape your grid paper to another student's grid paper. Write the total number of insects shown by the 2 sheets of grid paper.

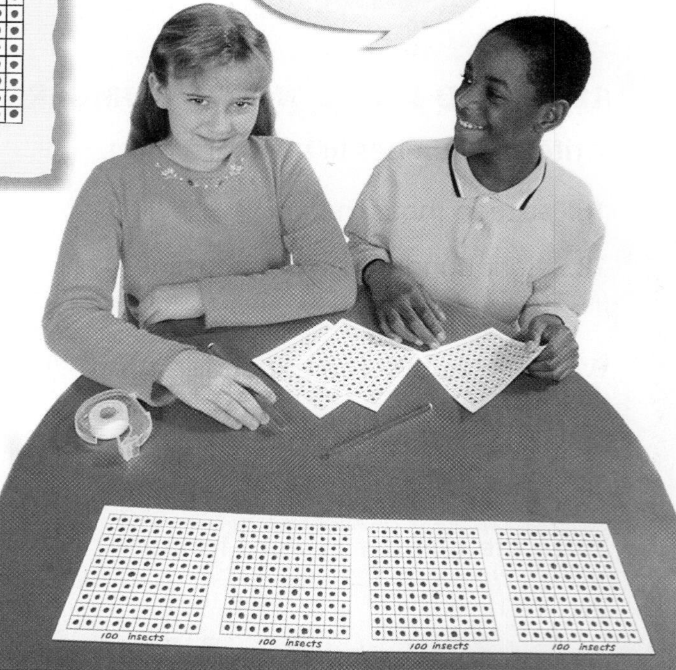

Try It

How many 10 × 10 sheets of grid paper will you need to show

a. 500 insects?
b. 700 insects?

Tape 10 sheets of grid paper together. Label the total number of insects shown.

- Will you need more than or fewer than 100 sheets to show 1 million insects? Explain.

CALIFORNIA STANDARDS **NS 1.0** Students compute with very large and very small numbers, positive integers, decimals, and fractions and understand the relationship between decimals, fractions, and percents. They understand the relative magnitudes of numbers. **MR 1.1** Analyze problems by identifying relationships, distinguishing relevant from irrelevant information, sequencing and prioritizing information, and observing patterns. *also* **NS 1.1, MR 1.2, MR 2.0, MR 2.2, MR 2.3, MR 2.4, MR 3.0, MR 3.2, MR 3.3**

Connect

You can look for a pattern to find how many sheets of grid paper you will need to show 1 million insects.

Copy and complete the table.

Sheets of Grid Paper	Total Insects
1	100
10	1,000
■	10,000
■	100,000
■	1,000,000

Talk About It

- What pattern did you use to complete the table above?

REASONING In what other ways could you model 1 million?

MATH IDEA Picturing what a model of a million looks like helps you better understand the relative size of numbers.

Honey bee swarm ▲

Practice

1. How many groups of 100 are in 1,000? 100,000? 1,000,000?
2. If you earned $1,000 a week, how long would it take to earn a million dollars?
3. Time how long it takes you to count to 100. Now, estimate how long it would take you to count to 1,000,000.
4. **REASONING** If one million pennies are equal to $10,000, what is the value of one million dimes?
5. **REASONING** If you could write a number every second, how many hours would it take to write all the numbers from 1 to 1,000,000?

Mixed Review and Test Prep

USE DATA **Use the graph for 6–7.**

6. Which school has the most students? the least?
7. Which school has the most male students?
8. 9×8
9. $30 \div 6$
10. Which is the sum 80,444 + 812,045?

   A 792,489
   B 892,489
   C 892,589
   D 902,489

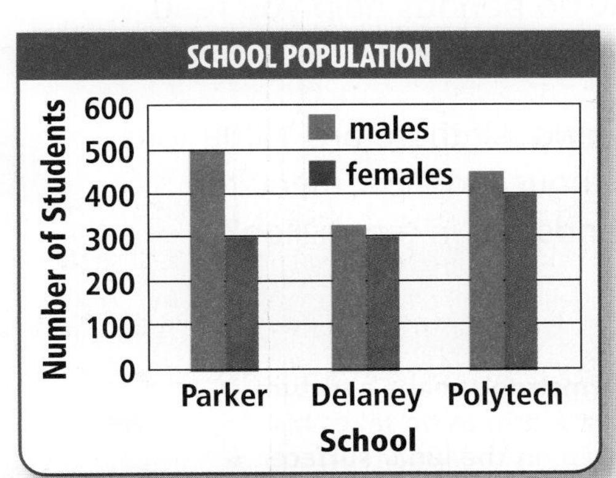

LESSON 2

Millions and Billions

Quick Review

Multiply each number by 10.

1. 20 **2.** 380
3. 44 **4.** 5,870
5. 116,900

VOCABULARY

billion

Learn

MOONWALK The first moon landing was one of the most-watched events in television history. An estimated 726,300,000 people watched Neil Armstrong take the first step on the moon.

You can show this number on a place-value chart.

Starting from the right, each group of 3 digits forms a period. Commas separate the periods.

"Coming to you *live* from the moon."

MILLIONS			THOUSANDS			ONES		
Hundreds	Tens	Ones	Hundreds	Tens	Ones	Hundreds	Tens	Ones
7	2	6,	3	0	0,	0	0	0

Standard Form: 726,300,000

Expanded Form: 700,000,000 + 20,000,000 + 6,000,000 + 300,000

Word Form: seven hundred twenty-six million, three hundred thousand

You can find the value of a digit in a number by multiplying the digit by its place value.

Find the value of the blue digit in 2,345,890.

$8 \times 100 = 800$

- How do periods help you read a number?

REASONING If there are 1,000 tens in ten thousand, how many ten thousands are in ten million?

Neil Armstrong snaps a picture of Buzz Aldrin as he takes his first step on the lunar surface. ▶

CALIFORNIA STANDARDS NS 1.0 Students compute with very large and very small numbers, positive integers, decimals, and fractions and understand the relationship between decimals, fractions, and percents. They understand the relative magnitudes of numbers. **NS 1.1** Estimate, round, and manipulate very large (e.g., millions) and very small numbers. *also* **MR 1.0, MR 1.1, MR 1.2, MR 2.0, MR 2.2, MR 2.3, MR 2.4, MR 3.0, MR 3.2, MR 3.3**

Billions

Scientists have no evidence of life on the moon. Earth is the only planet known to support life. At the beginning of the twenty-first century, there were about 6,009,421,000 people on Earth.

A billion has one more period than a million. One **billion** is 1,000 million.

BILLIONS			MILLIONS			THOUSANDS			ONES		
Hundreds	Tens	Ones	Hundreds	Tens	Ones	Hundreds	Tens	Ones	Hundreds	Tens	Ones
		6,	0	0	9,	4	2	1,	0	0	0

Standard Form: 6,009,421,000

Expanded Form: 6,000,000,000 + 9,000,000 + 400,000 + 20,000 + 1,000

Word Form: six billion, nine million, four hundred twenty-one thousand

REASONING If there are 1,000 millions in 1 billion, how many 10 millions are in 10 billion?

MATH IDEA The value of a digit in a number is the product of that digit and its place value.

Check

1. **Explain** how to find the value of a digit in a number.

Write the value of the blue digit.

2. 46,785,039
3. 39,806,527
4. 5,148,713,002
5. 268,432,116

Write each number in standard form.

6. five million, three hundred two thousand, fourteen
7. twelve billion, six hundred eleven thousand, one hundred seven
8. thirty-three million, two hundred seven thousand, twelve
9. twenty billion, five hundred thirty million, two hundred

For 10–13, use the number 4,302,698,051.

10. Write the name of the period that has the digits 698.
11. Write the place-value position of the digit 4.
12. Write the digit in the ten-millions place.
13. Write the name of the period that has the digits 302.

LESSON CONTINUES

Practice and Problem Solving

Write the value of the blue digit.

14. 35,427,231 **15.** 780,904,652 **16.** 413,916,102 **17.** 19,413,572

18. 8,102,673,124 **19.** 14,956,630,210 **20.** 9,124,432,212 **21.** 424,984,127

Write each number in standard form.

22. twenty-one million, eleven thousand, two hundred twelve

23. fifty-three billion, two million, one hundred sixteen thousand, seven

In which number does the digit 7 have the greater place value? Explain.

24. **a.** 17,854
b. 105,079

25. **a.** 7,089,000,000
b. 7,089,000

26. **a.** 750
b. 750,000

For 27–28, copy and complete.

27. The number 2,984,052,681 represents two billion, __?__ hundred eighty-four million, fifty-two thousand, __?__ hundred eighty-one.

28. Algebra 5,000,000,000 + ■ + 70,000,000 + 1,000,000 + 500,000 + 60,000 + ■ + 50 + 1 = 5,371,560,851

Write each number in two different ways.

29. two million, three hundred six thousand, fifteen

30. 1,000,000,000 +10,000,000 + 10,000 + 100 + 1

31. 65,200,108 **32.** 207,111,006 **33.** 1,480,200,965

USE DATA For 34–35, use the table.

34. Which planet is an average of ninety-two million, nine hundred thousand miles from the sun?

35. About how many billion miles is the average distance of Neptune from the sun?

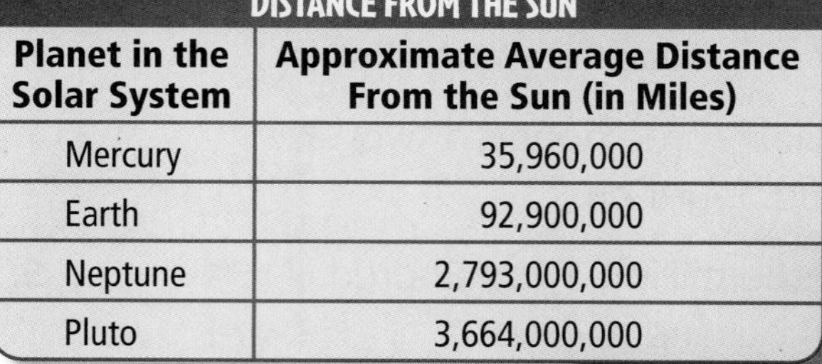

DISTANCE FROM THE SUN

Planet in the Solar System	Approximate Average Distance From the Sun (in Miles)
Mercury	35,960,000
Earth	92,900,000
Neptune	2,793,000,000
Pluto	3,664,000,000

36. What's the Error? Mary wrote the number 5,67,890. Explain her error. Write it correctly in standard form.

37. Write About It Write a rule for changing a number from standard form to expanded form. Show 3 examples.

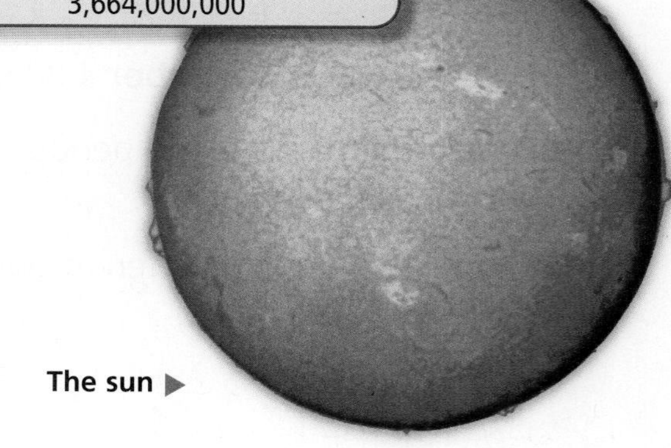

The sun ▶

38. **REASONING** Write the greatest 10-digit whole number possible without repeating a digit. What is the value of the digit 6?

39. **REASONING** Write the least 9-digit number possible without repeating a digit. What is the value of the digit 2?

Mixed Review and Test Prep

Write the value of the blue digit. (p. 4)

40. 73,215

41. 678,621

42. Write the factors of 24.

Copy and complete.

43. $6 \times 2 = 4 + ■$

44. $50 - 34 = 4 \times ■$

45. How many sheets of 10×10 grid paper do you need to model 1 billion? (p. 2)

There are 100 dimes in a roll.

46. How much money is each roll worth? (p. 2)

47. How many rolls of dimes do you need for $1,000,000? (p. 2)

Choose the letter of the next three numbers in the pattern.

48. **TEST PREP** 9, 12, 15, 18

A 19, 20, 21
B 21, 24, 27
C 20, 23, 26
D 23, 25, 27

49. **TEST PREP** 1, 2, 4, 7, 11

F 16, 22, 29
G 14, 18, 23
H 16, 24, 31
J 13, 15, 17

LINKUP to Reading

STRATEGY • USE GRAPHIC AIDS
Analyzing each part of a graphic aid, such as a graph, diagram, or table, will help you understand the data that are presented.

On October 12, 1999, the world's 6 billionth person was born. The United Nations designated that day as the "Day of 6 Billion."

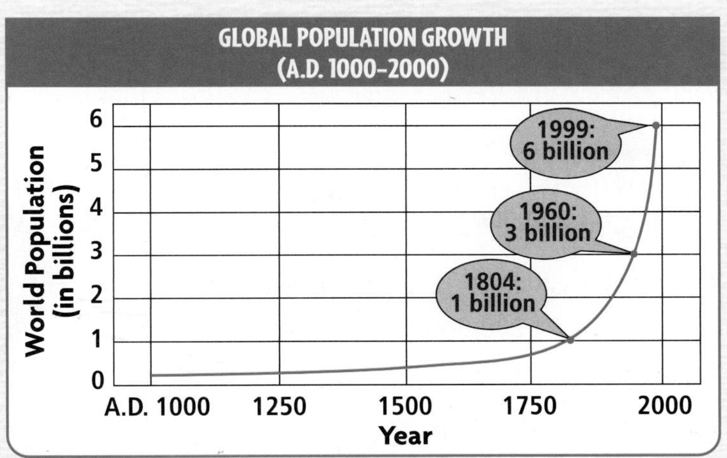

1. Has the world population increased or decreased from A.D. 1000 to 2000?

2. The Census Bureau estimated the U.S. population to be two hundred seventy-four million, nine hundred forty-three thousand in the year 2000. Write that number in standard form.

LESSON

3 Compare Numbers

Quick Review

Write the value of the blue digit.

1. 60 **2.** 4,260
3. 5,209 **4.** 2,854
5. 11,463

Learn

POPULAR POOCHES The beagle and the poodle are two popular dog breeds in the United States. In a recent year, there were 56,946 registered beagles and 56,803 registered poodles. Which breed had more registered dogs?

You can see a relationship between 56,803 and 56,946 on a number line.

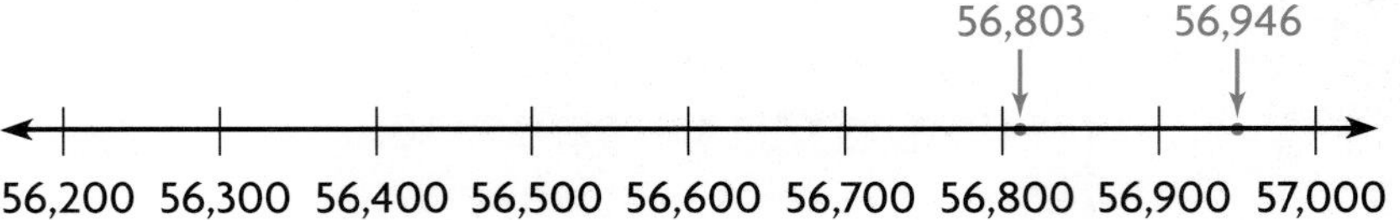

Remember
As you move from left to right on a number line, the numbers increase. As you move right to left, the numbers decrease.

Compare the digits in each place-value position. Check each place until the digits are different.

STEP 1

Compare the ten thousands.
56,946
↓ same
56,803

STEP 2

Compare the thousands.
56,946
↓ same
56,803

STEP 3

Compare the hundreds.
56,946
↓ 9 > 8
56,803

So, 56,946 > 56,803.

Since 56,946 > 56,803, there were more beagles than poodles.

Examples Compare.

A 84,200,000
↓ 2 < 9
84,900,000

So, 84,200,000 < 84,900,000.

B 6,024,850
↓ 6 > 0
143,850

There are no millions in 143,850.

So, 6,024,850 > 143,850.

CALIFORNIA STANDARDS NS 1.0 Students compute with very large and very small numbers, positive integers, decimals, and fractions and understand the relationship between decimals, fractions, and percents. They understand the relative magnitudes of numbers. **NS 1.1** Estimate, round, and manipulate very large (e.g., millions) and very small numbers. *also* **MR 1.0, MR 1.1, MR 2.0, MR 2.3, MR 2.4, MR 3.0, MR 3.2, MR 3.3**

Check

1. **Explain** whether the number seven billion is greater than or less than seven million.

Start at the left. Name the first place-value position where the digits differ. Name the greater number.

2. 11,733; 11,418 **3.** 8,007; 87,000 **4.** 205,763; 205,796

Compare. Write <, >, or = for each ●.

5. 180,225 ● 12,651 **6.** 3,154,270 ● 3,154,830 **7.** 198,335 ● 198,335

Practice and Problem Solving

Start at the left. Name the first place-value position where the digits differ. Name the greater number.

8. 351,604; 351,408 **9.** 6,712; 61,365 **10.** 680,742; 680,789

Compare. Write <, >, or = for each ●.

11. 48,922 ● 49,800 **12.** 15,650 ● 5,650 **13.** 187,418 ● 178,418

14. 451,300 ● 452,030 **15.** 1,800,250 ● 1,800,250 **16.** 20,767,300 ● 2,767,071

17. 509,782,650 ● 509,562,710 **18.** 1,098,254,701 ● 1,082,276,535

Find the missing digit.

19. 1,988,678 < 1,98■,678 **20.** 192,717,568 > 19■,717,568

21. **REASONING** Pat sits two rows behind Joshua. Mary sits in the third row, which is the row in front of Joshua. In which row does Pat sit?

22. Last year, 456,985 people attended animal shows. This year 456,345 people attended. Which year had a greater attendance?

23. **Write About It** Explain how to compare 1,879,987 and 1,979,987.

Mixed Review and Test Prep

24. 36 ÷ 3 **25.** 9 × 6

26. Write 3,000,000,000 + 60,000,000 + 8,000,000 + 20,000 + 6,000 + 9 in standard form. (p. 4)

27. 9,011 − 1,235

28. **TEST PREP** How many dimes are in $55? (p. 2)

A 55 **C** 500

B 5,050 **D** 550

Extra Practice page H32, Set B

LESSON

Order Numbers

Quick Review

Compare. Write <, >, or = for each ●.

1. 15 ● 18
2. 65 ● 56
3. 131 ● 131
4. 443 ● 334
5. 1,098 ● 1,908

Learn

GREATEST PLACES Mammoth Cave, a national park in Kentucky, covers 52,419 acres. Bryce Canyon in Utah covers 35,835 acres, and Mesa Verde in Colorado covers 52,122 acres.

One Way

You can use a number line to order the sizes of the parks.

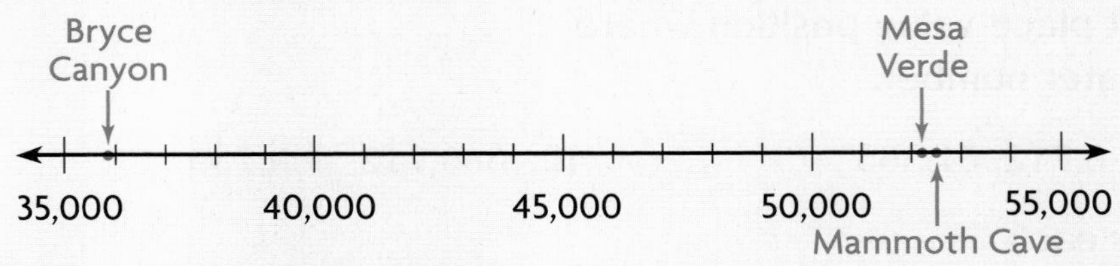

Another Way

Use place value to order 52,419; 35,835; and 52,122 from greatest to least.

STEP 1

Compare the ten thousands.

52,419
↓
52,122
↓
35,835 5 > 3

So, 35,835 is the least of the three numbers.

STEP 2

Compare the other two numbers by comparing the thousands.

52,419
↓ same
52,122

STEP 3

Compare the hundreds.

52,419
↓ 4 > 1
52,122

Order the numbers.

So, 52,419 > 52,122 > 35,835.

In order from greatest to least areas, the parks are Mammoth Cave, Mesa Verde, Bryce Canyon.

- Order the areas of the parks from least to greatest.

MATH IDEA To compare and order numbers, start at the left and compare the digits in each place-value position. Then use these comparisons to order the numbers.

CALIFORNIA STANDARDS NS 1.0 Students compute with very large and very small numbers, positive integers, decimals, and fractions and understand the relationship between decimals, fractions, and percents. They understand the relative magnitudes of numbers. **NS 1.1** Estimate, round, and manipulate very large (e.g., millions) and very small numbers. *also* **MR 1.0, MR 1.1, MR 2.0, MR 2.3, MR 2.4, MR 3.0, MR 3.2, MR 3.3**

Check

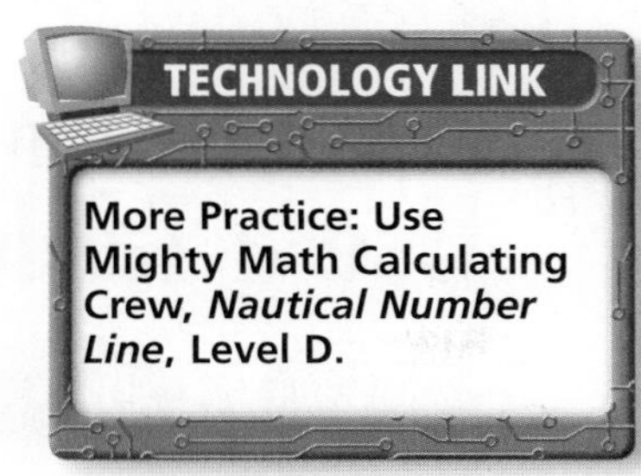

TECHNOLOGY LINK
More Practice: Use Mighty Math Calculating Crew, *Nautical Number Line*, Level D.

1. **Describe** the order of all the parks' areas from greatest to least if Carlsbad Caverns, which covers 46,755 acres, and Acadia, which covers 41,231 acres, are added to the list.

Order the numbers from greatest to least.

2. 13,978; 13,792; 17,379
3. 980,435; 980,526; 980,472

Order the numbers from least to greatest.

4. 78,940; 78,914; 78,801
5. 346,720; 342,820; 347,200

Practice and Problem Solving

Order from greatest to least.

6. 26,295; 216,925; 219,625
7. 235,289; 236,287; 236,178
8. 78,935; 77,590; 178,286
9. 234,650,200; 234,850,100; 38,950,500

Order from least to greatest.

10. 895,000; 8,595,000; 859,000
11. 4,210,632; 2,410,781; 4,120,681
12. 49,086; 49,680; 48,690
13. 335,219; 336,007; 336,278

Complete. Use the greatest possible digits.

14. $14■,599 > 148,600 > 1■5,431$
15. $5,■41,600 < 5,8■1,600 < 5,941,600$

REASONING Kim lives 455 miles from the canyon. Juan lives 110 miles farther from the canyon than Kim. Darren lives 150 miles closer to the canyon than Juan. Use this information for 16 and 17.

16. How far does each person live from the canyon?
17. Who lives closest to the canyon? Who lives farthest away?
18. **What's the Question?** In a set of three numbers, 345,798 is the least and 355,798 is the greatest. The answer is 350,798.

Mixed Review and Test Prep

19. 9×3
20. 8×4

Write the number that is 1,000 more.

21. 4,099
22. 45,230
23. **TEST PREP** Which number is one million, seventeen thousand, four hundred sixty-five written in standard form? (p. 4)

A 1,170,465
B 1,017,645
C 1,017,465
D 1,017,346

LESSON 5

Problem Solving Skill
Use a Table

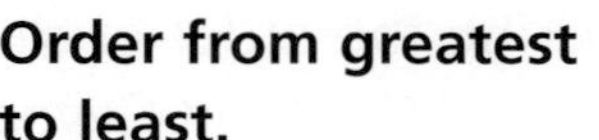

Quick Review

Order from greatest to least.

1. 68; 72; 59; 69
2. 298; 289; 299
3. 1,086; 1,098; 1,068
4. 27,119; 26,120; 27,120
5. 609,095; 609,509; 609,059

SIZE IT UP Rhode Island is the smallest state in the United States. Delaware is the second smallest state. What is the difference between the areas of Delaware and Rhode Island?

AREA AND POPULATION OF STATES

Name	Area (sq mi)	Population
California	158,869	32,268,301
Delaware	2,396	731,581
Michigan	58,527	9,773,892
Nevada	110,567	1,676,809
New Jersey	8,215	8,052,849
Rhode Island	1,231	987,429

Find the areas of Delaware and Rhode Island. Then subtract to find the difference.

Area of Delaware	2,396	square miles
Area of Rhode Island	−1,231	square miles
	1,165	square miles

So, the difference in areas is 1,165 square miles.

Talk About It

- How would you change the table to show the state populations in order from greatest to least?
- Explain how reading the data from a table helps you solve problems.

CALIFORNIA STANDARDS MR 2.3 Use a variety of methods, such as words, numbers, symbols, charts, graphs, tables, diagrams, and models, to explain mathematical reasoning. **NS 1.1** Estimate, round, and manipulate very large (e.g., millions) and very small numbers. *also* **NS 1.0, MR 1.0, MR 1.1, MR 2.0, MR 2.4, MR 3.0, MR 3.2, MR 3.3**

Problem Solving Practice

USE DATA For 1–4, use the table.

CAPITAL CITIES		
City	**State**	**Population**
Atlanta	Georgia	394,017
Juneau	Alaska	26,751
Phoenix	Arizona	983,392
Montpelier	Vermont	8,247
Santa Fe	New Mexico	55,859

1. Which city has the greatest population? the least?

2. How much greater is the population of Santa Fe than the population of Juneau?

3. **Algebra** Which equation shows the combined population of Atlanta and Phoenix if n represents the total population?

   A $394{,}017 + 983{,}392 = n$
   B $n + 983{,}392 = 394{,}017$
   C $n = 983{,}392 - 394{,}017$
   D $2{,}394{,}017 + n = 983{,}392$

4. Which state's capital city has a population of one million when rounded to the nearest million?

   F Georgia
   G Alaska
   H Arizona
   J New Mexico

Mixed Applications

5. The clock shows the time Mike's 1-hour flight landed in Atlanta. It took him 30 minutes to drive to the airport, and he was 45 minutes early for his flight. At what time did he leave his house?

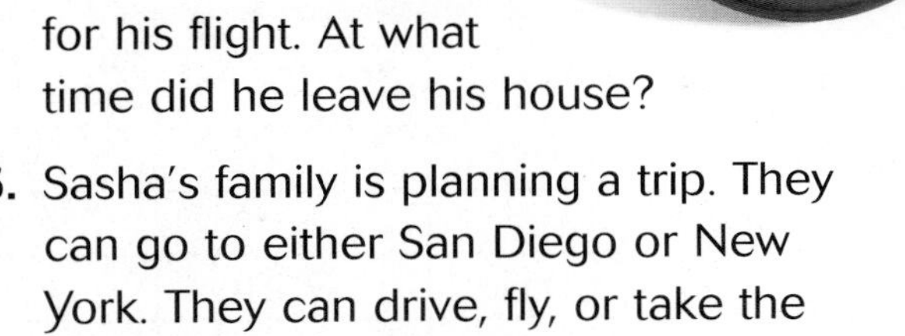

6. Sasha's family is planning a trip. They can go to either San Diego or New York. They can drive, fly, or take the train to their destination. Make a list of their choices.

7. Bill said, "The greater the state's area, the greater the population." How can you use the table on page 12 to decide whether Bill's statement is correct?

8. **Algebra** The Smiths drove a total of 540 miles in 2 days. They drove twice as far on Tuesday as they did on Monday. How many miles did they drive on Monday? on Tuesday?

9. **Write a problem** using the Capital Cities table above. Then solve the problem.

Review/Test

CHECK VOCABULARY AND CONCEPTS

Choose the best term from the box.

million
billion
ten thousand
thousand

1. A thousand millions is equal to one __?__. (p. 5)
2. A thousand thousands is equal to one __?__. (p. 4)
3. A million is made up of 100 groups of __?__. (p. 3)

CHECK SKILLS

Write the value of the blue digit. (pp. 4–7)

4. 5,323,022
5. 7,914,605,004
6. 27,140,652

Write each number in two different ways. (pp. 4–7)

7. 2,030,909
8. three billion, two hundred three million, forty-two thousand, five

Start at the left. Name the first place-value position where the digits differ. Name the greater number. (pp. 8–9)

9. 7,532; 7,523
10. 260,789; 269,243
11. 19,287; 319,310

Compare. Write <, >, or = for each ●. (pp. 8–9)

12. 104,690 ● 140,690
13. 3,250 ● 13,250
14. 9,782,650 ● 9,782,650

Order from least to greatest. (pp. 10–11)

15. 9,519; 10,003; 9,195
16. 14,326; 14,623; 13,426
17. 1,502,369; 1,501,369; 1,507,369

CHECK PROBLEM SOLVING

USE DATA For 18–19, use the table. (pp. 12–13)

18. Order the lakes from the least area to the greatest area.
19. List the lakes in order from the least maximum depth to the greatest maximum depth.
20. In the future, a spaceship might be built that could travel 100,000 miles a second. How long would it take to travel 1 million miles? 10 million miles? 1 billion miles? (pp. 2–3)

AREAS OF THE GREAT LAKES

Lake	Area (sq mi)	Maximum Depth (ft)
Michigan	22,300	923
Erie	9,910	210
Ontario	7,550	802
Superior	31,700	1,330
Huron	23,000	750

Cumulative Review

Look for important words.
See item **6.**
The important word is **not. Not** tells you to find a value other than one billion.
Also see problem **2**, p. H62.

For 1–9, choose the best answer.

1. $\begin{array}{r} 3{,}974 \\ +1{,}638 \\ \hline \end{array}$

   A 5,612 **C** 4,502
   B 4,612 **D** Not here

2. How is 2,000,000 + 20,000 + 200 + 2 written in standard form?

   F 2,020,202 **H** 2,200,220
   G 2,222 **J** 220,202

3. Which names the least number?

   A 4,521,206
   B three million, seven hundred fifty-one thousand, nine hundred five
   C 3,790,358
   D 4,000,000 + 3,000 + 700 + 20

4. Which number has a 4 in the ten-thousands place?

   F 10,400 **H** 3,742,619
   G 14,307 **J** 8,405,361

5. Which names a number greater than 906,572?

   A 799,999
   B nine hundred sixty-three thousand, four hundred one
   C 900,000 + 900 + 80 + 5
   D 905,672

6. Which of these is *not* equivalent to one billion?

   F 100 ten millions
   G 1,000 millions
   H 1,000,000 thousands
   J 10,000 ten thousands

7. Which symbol makes this true?
   1,064,615 ● 1,041,578

   A =
   B <
   C >
   D +

For 8–9, use the table.

SOME LANGUAGES IN THE UNITED STATES	
Language	**Number of Speakers**
French	1,702,000
Greek	388,000
Chinese	1,249,000
Spanish	17,339,000

8. Which language is spoken by the greatest number of people?

   F French **H** Chinese
   G Greek **J** Spanish

9. Which shows the languages listed in order from the least number of people to the greatest number of people who speak the language?

   A Spanish, French, Chinese, Greek
   B Greek, Chinese, Spanish, French
   C Greek, Chinese, French, Spanish
   D French, Greek, Spanish, Chinese

CHAPTER 2 Place Value of Decimals

Pollen was first described in ancient writings. Because some types of pollen were used as medicines, pollen has been called the world's first health food. Pollen is unique because it contains nearly every known nutrient required for a balanced human diet. Look at the table. Which pollen grain is the smallest?

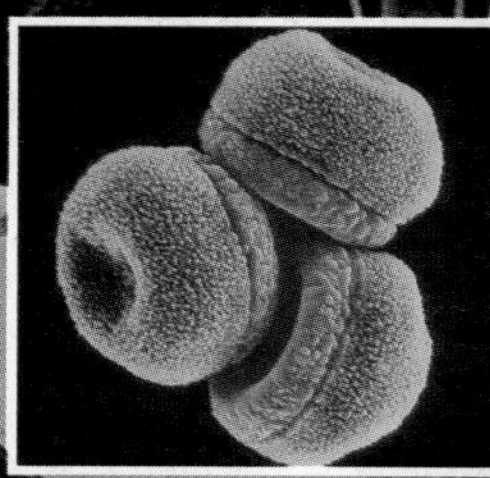
Pollen of giant water lilies

POLLEN GRAIN SIZES

Pollen Type	Grain Size
Giant water lily	0.02 cm
Red clover	0.0002 cm
Daisy	0.0025 cm
Snapdragon	0.0015 cm

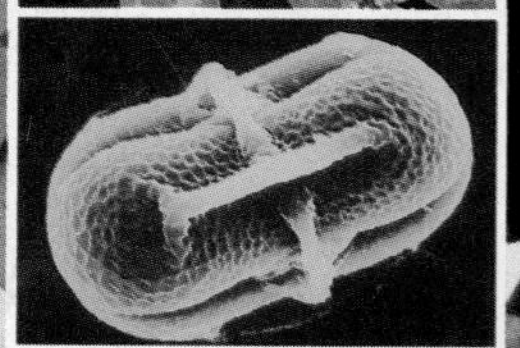
Red clover and pollen

Daisies and pollen

Snapdragon and pollen

Use this page to help you review and remember important skills needed for Chapter 2.

VOCABULARY

Choose the best term from the box.

whole number
decimal point
equivalent
decimal

1. A number with one or more digits to the right of the decimal point is a __?__.
2. In a decimal number, a __?__ separates the ones and the tenths places.
3. Numbers having the same value are __?__.

PLACE VALUE (See p. H2.)

Copy and complete the chart for 302,584.

	THOUSANDS			ONES		
	Hundreds	■	Ones	Hundreds	Tens	■
4.	3	0	■	5	■	4
5.	■	$0 \times 10{,}000$	$2 \times 1{,}000$	■	8×10	4×1
6.	■	0	2,000	■	80	4

Write the value of the blue digit.

7. 4,132
8. 418
9. 12,300
10. 108,543
11. 512,570
12. 2,632,105
13. 27,492,503
14. 1,204,672,314
15. 891,563
16. 13,065,814
17. 92,516,782
18. 24,073,546,752

READ AND WRITE DECIMALS (See p. H3.)

Write as a decimal.

19. 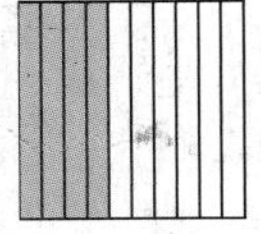

20. 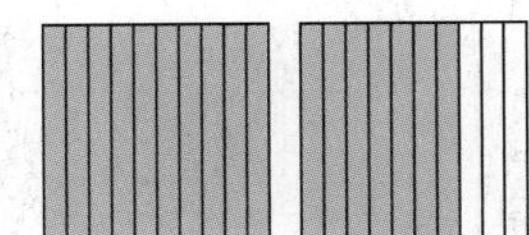

21. 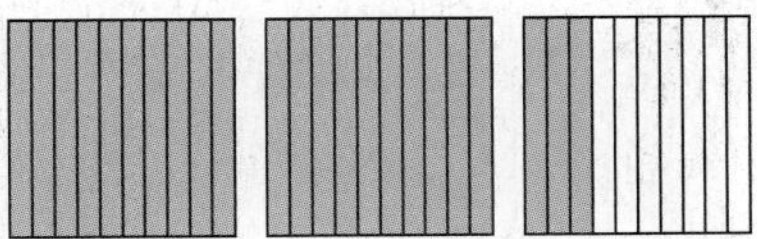

Write the number in standard form.

22. fourteen and three tenths
23. $7{,}000 + 4 + 0.1$
24. $3{,}000 + 20 + 7 + 0.2$
25. sixteen and four tenths

LESSON 

Tenths and Hundredths

Quick Review

Name the place value of the digit 8 in each number.

1. 1,845
2. 408
3. 8,209
4. 284
5. 80,672

VOCABULARY

tenth hundredth

Learn

WATCH IT GROW Decimal numbers name wholes and parts of a whole.

One whole is shaded.	One tenth is shaded.	One hundredth is shaded.
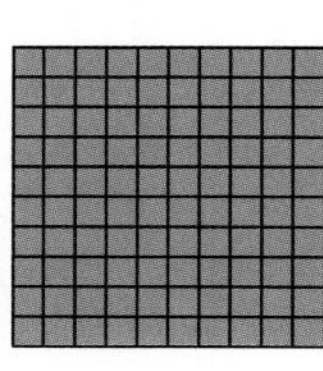	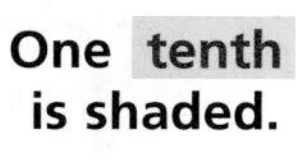 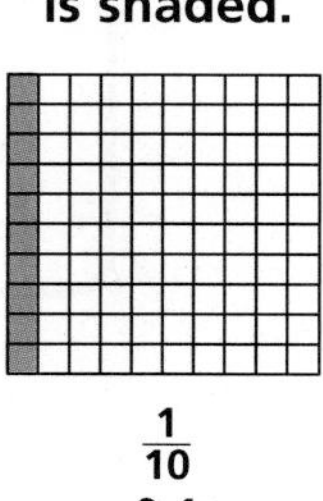	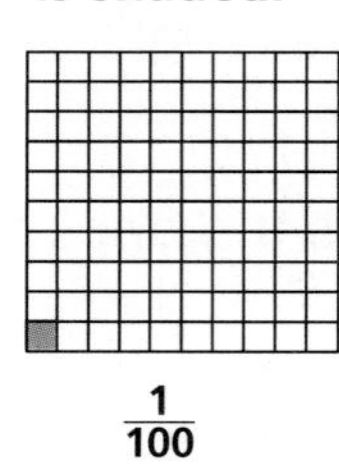
1	$\frac{1}{10}$	$\frac{1}{100}$
1.0	0.1	0.01
one	one tenth	one hundredth

You can use a place-value chart to find the value of each digit in the decimal 1.75.

Ones	Tenths	Hundredths
1	7	5
1 × 1 = 1.0	7 × 0.1 = 0.7	5 × 0.01 = 0.05

Standard Form: 1.75

Expanded Form: $1 + 0.7 + 0.05$

Word Form: one and seventy-five hundredths

- How are $1\frac{75}{100}$ and 1.75 related?

▲ Scientists use decimals when measuring plant growth. Bamboo is the fastest-growing land plant. It grows about 25 centimeters a day, or 1.75 meters a week.

Check

1. **Write** the next number in this pattern: $10, 1, \frac{1}{10}, \frac{1}{100}$. Describe the pattern.

Write as a decimal and a fraction or mixed number.

2. 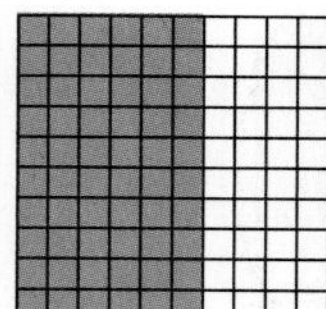

3. 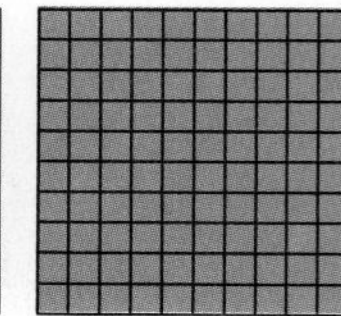

4. 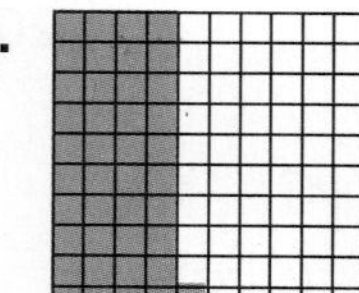

5. $3 + 0.57$

6. one and six tenths

7. thirty-two hundredths

CALIFORNIA STANDARDS NS 1.0 Students compute with very large and very small numbers, positive integers, decimals, and fractions and understand the relationship between decimals, fractions, and percents. They understand the relative magnitudes of numbers. **MR 2.3** Use a variety of methods, such as words, numbers, symbols, charts, graphs, tables, diagrams, and models, to explain mathematical reasoning. *also* **MR 1.1, MR 2.0, MR 2.4, MR 3.3**

Practice and Problem Solving

Write as a decimal and a fraction or mixed number.

8. 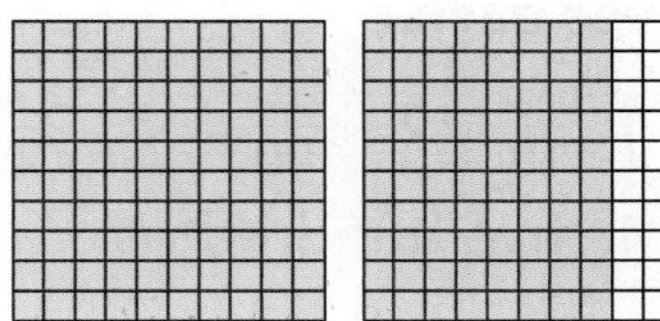

9. 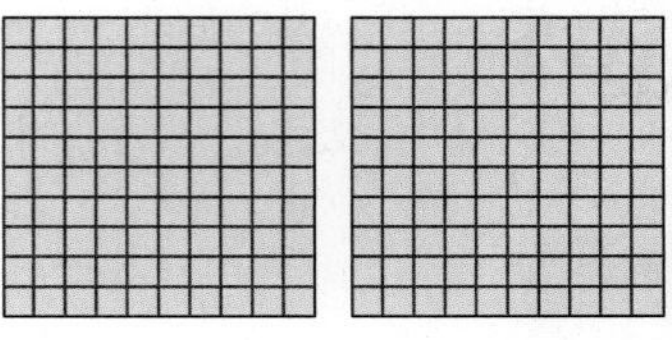

10. 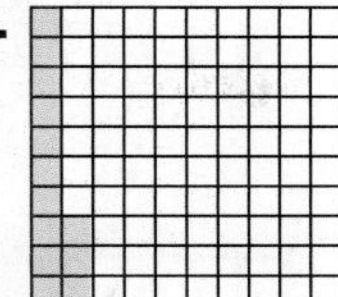

11. $4 + 0.7$

12. $17 + 0.70 + 0.01$

13. $0.30 + 0.04$

14. three and six tenths

15. eighteen hundredths

16. one and six hundredths

17. one and four tenths

18. four and eight hundredths

19. forty-five hundredths

Write the missing decimal in each pattern. Describe the pattern.

20. 0.2, 0.4, 0.6, ■, 1.0

21. 0.75, 0.80, 0.85, ■, 0.95

22. 2.84, 2.81, 2.78, 2.75, ■

23. 7.03, 7.08, 6.88, 6.93, 6.73, ■, 6.58

USE DATA **For 24–31, use the circle graph.**

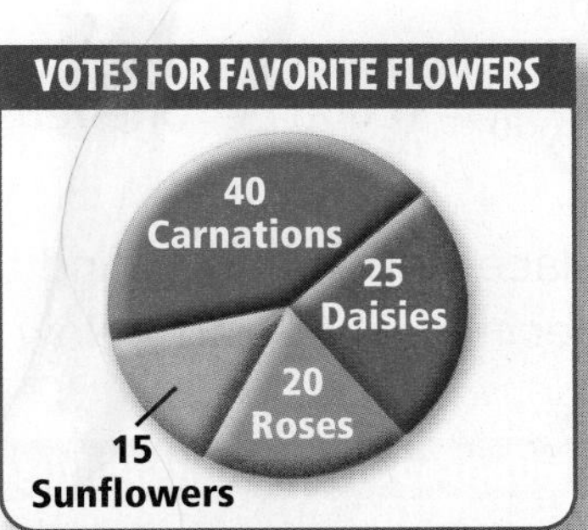

24. How many people voted?

25. What fraction of the people voted for daisies?

26. Write a decimal to show what part of the votes were for carnations.

27. What fraction of the votes were for carnations or roses?

28. Which flower received the most votes? the fewest votes?

29. **Write a problem** that can be solved using the circle graph.

30. **What if** 0.35 of the same number of people voted for roses? How many people would have voted for roses?

31. Mary says more than half of the voters chose daisies or sunflowers. Is Mary correct? Explain.

Mixed Review and Test Prep

32. Order 3,019,531; 3,019,643; and 3,019,639 from greatest to least. (p. 10)

33. The time is 5:25. What time will it be in 1 hour and 35 minutes?

34. $700 + 850$

35. $1,031 - 103$

36. **TEST PREP** How many hours are in a week?

A 168
B 148
C 60
D 24

LESSON 2

Thousandths and Ten-Thousandths

Quick Review

Write the decimal for each.

1. $\frac{8}{10}$ 2. $\frac{24}{100}$ 3. $2\frac{67}{100}$
4. five hundredths
5. six and fourteen hundredths

VOCABULARY

thousandth ten-thousandth

Learn

ZOOM IN If you divide one whole by 1,000, you get one **thousandth**. If one square on a hundredth model were magnified and one column of it were shaded, you could see one thousandth.

one

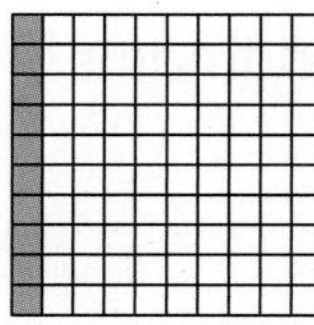

one tenth

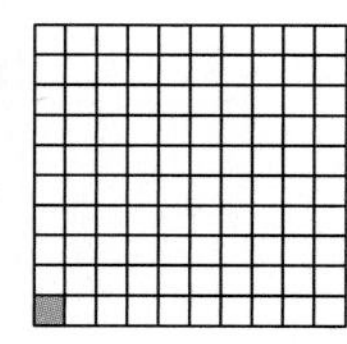

one hundredth

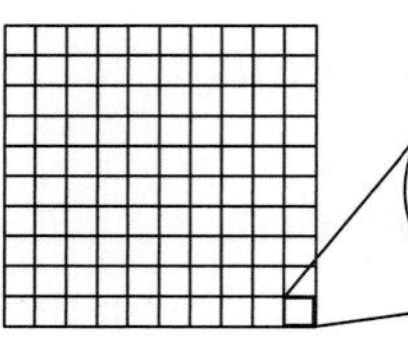

one thousandth

If you divide one whole by 10,000, you get one **ten-thousandth**.

$1 \div 10{,}000 = \frac{1}{10{,}000} = 0.0001$ = one *ten-thousandth*

You can use a place-value chart to find the value of each digit in a decimal. The chart below shows 2.7835.

Ones	Tenths	Hundredths	Thousandths	Ten-thousandths
2	7	8	3	5
2 × 1	7 × 0.1	8 × 0.01	3 × 0.001	5 × 0.0001
2.0	0.7	0.08	0.003	0.0005

Standard Form: 2.7835
Expanded Form: 2 + 0.7 + 0.08 + 0.003 + 0.0005
Word Form: two and seven thousand, eight hundred thirty-five ten-thousandths

- How many thousandths are in one hundredth? in one tenth? in one?

MATH IDEA In a decimal number, each successive place to the right of the decimal point is one tenth of the place before it.

▲ Scientists sometimes use thousandths to measure liquids. One milliliter is one thousandth of a liter.

CALIFORNIA STANDARDS NS 1.0 Students compute with very large and very small numbers, positive integers, decimals, and fractions and understand the relationship between decimals, fractions, and percents. They understand the relative magnitudes of numbers. **MR 2.4** Express the solution clearly and logically by using appropriate mathematical notation and terms and clear language; support solutions with evidence in both verbal and symbolic work. *also* **NS 1.1, MR 1.0, MR 1.1, MR 2.0, MR 2.3**

Check

1. How many ten-thousandths are in one thousandth? in one hundredth? in one tenth? in one?

Write each decimal in expanded form, in word form, and as a fraction.

2. 1.405 3. 0.134 4. 1.045 5. 2.107 6. 3.0003

Practice and Problem Solving

Write each decimal in expanded form, in word form, and as a fraction.

7. 1.067 8. 0.1234 9. 12.065 10. 3.206 11. 1.582

Write in standard form.

12. eight thousandths
13. fifty-four ten-thousandths
14. five hundredths
15. one and sixty-two thousandths

TECHNOLOGY LINK

More Practice Use E-Lab, *Thousandths*.

www.harcourtschool.com/elab2002

Write in expanded form.

16. 2.9 17. 0.43 18. 1.026 19. 7.1478

Write in word form.

20. 4.12 21. 11.03 22. 3.246 23. 0.5757

REASONING In 24–27, the decimal point in some of the numbers has been placed incorrectly. Write the correct decimal number for each.

24. The car usually travels on the highway at 5.05 miles per hour.
25. When Wanda was sick, her temperature was 10.15 degrees.
26. In the fishing tournament, only fish 9 inches or longer can be kept. Joe threw back his fish because it measured 85 inches.
27. The high jump was won by Jake Carr, whose jump measured 0.49 feet high.
28. A dozen oranges costs $0.98. How many oranges can you buy for $0.49?
29. **REASONING** What decimal is exactly halfway between 0.002 and 0.003?

Mixed Review and Test Prep

30. Write 3.5 as a mixed number. (p. 18)
31. Order 1,250,869; 1,205,896; and 1,205,869 from least to greatest. (p. 10)
32. 30×40
33. 10×20
34. **TEST PREP** Which tells how many sides a hexagon has?
    - **A** 3 sides
    - **B** 4 sides
    - **C** 6 sides
    - **D** 8 sides

Extra Practice page H33, Set B

LESSON 3

Equivalent Decimals

Learn

SAME BUT DIFFERENT Newborn giraffes are about 1.8 m tall, and adult giraffes can be from 4.25 m to 5.5 m tall. Write an equivalent decimal for 1.8.

Equivalent decimals are different names for the same number or amount. These are some different ways to express the decimal 1.8.

one and eight tenths   1 + 0.8   1.80   1.800

To determine if two decimals are equivalent, line up the decimal points and compare the digits in the same place-value positions.

Examples

A Equivalent Decimals
0.4
0.40

B Not Equivalent Decimals
0.8
0.08

MATH IDEA Equivalent decimals can be formed by placing zeros to the right of the last digit after the decimal point. The zeros do not change the value of the decimal.

Quick Review

Write in expanded form.

1. 3.7 **2.** 0.72
3. 4.908 **4.** 8.365
5. 17.0002

VOCABULARY

equivalent decimals

▲ **Giraffes are the tallest animals living in Africa.**

Check

1. Describe how you can determine if two decimals are equivalent.

Write *equivalent* or *not equivalent* to describe each pair of decimals.

2. 0.09 and 0.009 **3.** 3.8 and 3.80 **4.** 0.60 and 0.600 **5.** 7.2 and 7.02

Write an equivalent decimal for each number.

6. 5.3 **7.** 0.034 **8.** 0.1230 **9.** 9.030

CALIFORNIA STANDARDS NS 1.0 Students compute with very large and very small numbers, positive integers, decimals, and fractions and understand the relationship between decimals, fractions, and percents. They understand the relative magnitudes of numbers. **NS 1.1** Estimate, round, and manipulate very large and very small (e.g. thousandths) numbers. *also* **MR 1.0, MR 2.0, MR 2.4, MR 3.2**

Practice and Problem Solving

Write *equivalent* or *not equivalent* to describe each pair of decimals.

10. 1.02 and 1.20
11. 1.28 and 1.280
12. 3.007 and 3.07
13. 7.02 and 7.020
14. 4.09 and 4.099
15. 4.008 and 4.08

Write an equivalent decimal for each number.

16. 0.03
17. 4.630
18. 0.2
19. 5.600
20. 0.83
21. 5.550
22. 7.10
23. 0.900
24. 0.103
25. 2.4

Write the two decimals that are equivalent.

26. 0.0502, 0.00502, 0.05020

27. 0.017, 0.01700, 0.00170

28. 1.00050, 1.0050, 1.005

29. 8.01, 8.0010, 8.01000

USE DATA For 30–31, use the table.

MAMMALS WITH THE LONGEST TAILS

Mammal	Tail Length in Meters
Asian elephant	1.50
Leopard	1.40
African elephant	1.3
African buffalo	1.10
Giraffe	1.1

30. Which two animals have the same tail length?

31. Write an equivalent decimal for the length of a leopard's tail.

32. A 0.5-pound block of cheddar cheese costs $1.89. The Swiss cheese costs $2.98 per pound. Which cheese is less expensive per pound? Explain.

33. The cash register showed change of $2.50. Miko said this is two and one half dollars. Is Miko correct? Explain.

34. **What's the Error?** Jeb's batting average is .309, and Tom's is .390. Tom says they have the same average. Describe Tom's error.

35. **REASONING** Explain how to use a number line to show that 0.4 is not equivalent to 0.04.

Mixed Review and Test Prep

36. 70 + 80 + 90

37. 320 + 110 + 440

38. Order from greatest to least: 60,692; 605,962; 60,962 (p. 10)

39. True or false: 909,909 > 909,099? (p. 8)

40. **TEST PREP** Which shows three million, four hundred fifty-six thousand, four hundred thirty-two in standard form? (p. 4)

A 3,456,423
B 3,456,432
C 3,000,000 + 400,000 + 50,000 + 6,000 + 400 + 30 + 2
D 3,465,432

LESSON 4

Compare and Order Decimals

Quick Review

Compare.

1. 34 ● 43
2. 250 ● 205
3. 600 ● 600
4. 450 ● 4,500
5. 26,983 ● 29,683

Learn

MORE OR LESS At the gymnastics meet, Mindy scored 8.69 on the floor exercise, and Stephanie scored 8.85. Whose score was higher?

You can use a number line to compare decimals.

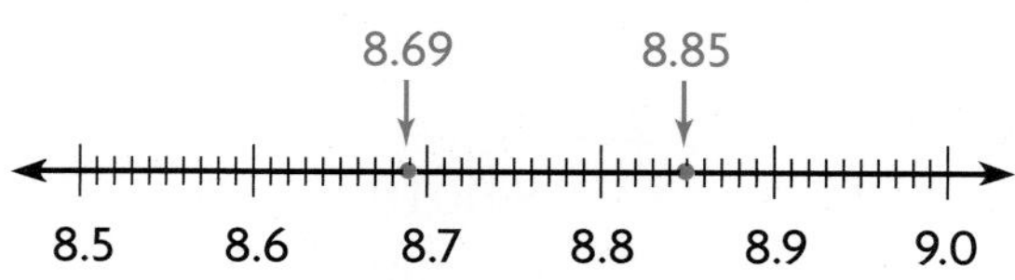

8.85 is to the right of 8.69. So, 8.85 is greater than 8.69. Since 8.85 > 8.69, Stephanie's score is higher.

You can also use place value to compare decimals.

Compare 3.452 and 3.456. First line up the decimal points. Then look for the first place where the digits are different.

STEP 1 Begin at the left. Compare the ones.

3.452
↓ same
3.456

STEP 2 Compare the tenths.

3.452
↓ same
3.456

STEP 3 Compare the hundredths.

3.452
↓ same
3.456

STEP 4 Compare the thousandths.

3.452
↓ 6 > 2, or 2 < 6
3.456

So, 3.456 > 3.452, or 3.452 < 3.456.

Order 2.4853, 2.4844, and 2.4862 from least to greatest.

STEP 1 Begin at the left. Compare the ones and tenths.

2.4853
↓↓
2.4844 same
↓↓
2.4862

STEP 2 Compare the hundredths.

2.4853
↓
2.4844 same
↓
2.4862

STEP 3 Compare the thousandths.

2.4853
↓
2.4844 4 < 5 < 6
↓
2.4862

STEP 4 Order the numbers.

2.4844 < 2.4853 < 2.4862

- How would you order the numbers from greatest to least?

CALIFORNIA STANDARDS **NS 1.5** Identify and represent on a number line decimals, fractions, mixed numbers, and positive and negative integers. **NS 1.1** Estimate, round, and manipulate very large and very small (e.g. thousandths) numbers. *also* **NS 1.0, MR 1.0, MR 2.0, MR 2.4, MR 3.2**

Check

1. **Explain** how the order is changed if a fourth number, 2.4856, is included in the list of numbers in the second example.

Write <, >, or = for each ●. Use the number line.

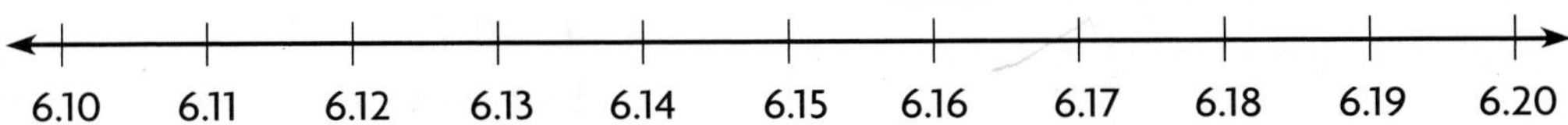

2. 6.152 ● 6.125
3. 6.14 ● 6.140
4. 6.114 ● 6.118
5. 6.109 ● 6.190
6. 6.170 ● 6.175
7. 6.176 ● 6.167

Practice and Problem Solving

Write <, >, or = for each ●.

8. 0.65 ● 0.63
9. 1.4 ● 1.400
10. 229.035 ● 229.406
11. 132.94 ● 132.48
12. 156.93 ● 156.98
13. 99.989 ● 99.998
14. 0.905 ● 0.905
15. 63.938 ● 63.939
16. 476.069 ● 476.096

Order from least to greatest.

17. 6.58, 6.38, 6.29, 7.08
18. 13.393, 13.309, 13.339, 13.039
19. 4.102, 4.105, 4.118, 4.110
20. 15.259, 15.389, 15.291, 15.301

USE DATA **For 21–23, use the menu.**

21. Write the salads in order from the least expensive to the most expensive.
22. Which two items each cost more than yogurt but less than onion soup?
23. **NUMBER SENSE** Lisa has $5 in her pocket. What two soups can she buy?
24. **Write About It** Explain how to compare 1.23 and 1.27.

Power Lunch Menu
Bean Soup $2.25
Onion Soup $2.95
Tomato Soup $2.75
Taco Salad $1.89
Garden Salad $1.80
Fruit Salad $1.85
Yogurt $1.95

Mixed Review and Test Prep

25. True or false: 2.05 = 2.50? (p. 22)
26. Write 3.025 in word form. (p. 20)
27. 2.98 − 1.80
28. 1.15 + 4.25
29. **TEST PREP** Which decimal is not equivalent to five hundred ten thousandths? (p. 22)
   - **A** 0.501
   - **B** 0.5100
   - **C** 0.510
   - **D** 0.51

Extra Practice page H33, Set D

LESSON 5

Problem Solving Skill

Draw Conclusions

Quick Review

1. 7×4
2. $18 \div 6$
3. $25 \div 5$
4. $120 \div 12$
5. $24.08 \div 10$

THINK FOR YOURSELF Sometimes you may need to analyze data to draw conclusions.

The table and the bar graph below are about an experiment involving five of the same type of plant. You can use the data to draw conclusions.

PLANT FOOD DROPS USED

Plant	Number of Plant Food Drops
A	1
B	5
C	9
D	13
E	17

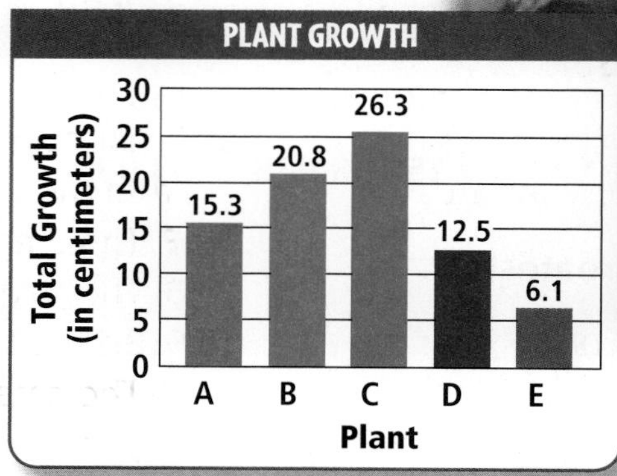

▲ This Pacific giant kelp can grow 18 inches in one day.

Analyze	Conclusion
• Look at the graph. Which plant grew the tallest?	Plant C grew the tallest.
• Look at the table. How many drops of plant food did the tallest plant receive?	Plant C received 9 drops.
• What relationship do you notice between the number of drops of plant food and the plant's growth?	Up to 9 drops, more plant food helped the plants grow taller. With more than 9 drops, growth decreased.

- What conclusion can you draw about Plants A, B, and C? about Plant E?
- **What if** the total growth for Plant D was 35.89 cm, and for Plant E was 38 cm? What conclusion could you draw about all five plants?

CALIFORNIA STANDARDS MR 2.0 Students use strategies, skills, and concepts in finding solutions. **MR 2.3** Use a variety of methods such as words, numbers, symbols, charts, graphs, tables, diagrams, and models to explain mathematical reasoning. **NS 1.1** Estimate, round, and manipulate very large and very small (e.g. thousandths) numbers. *also* **NS 1.0, MR 1.0, MR 2.4, MR 3.2**

Problem Solving Practice

For 1–4, can the conclusion be drawn from the information given? Write *yes* or *no*. Explain your choice.

Mrs. Carson measured the heights of her fifth graders. The shortest student is 50 inches tall, and the tallest student is 64 inches tall. There are 25 students in Mrs. Carson's class.

1. All of the students are over 5 feet tall.
2. The tallest student in the class is over 5 feet tall.
3. All of the students are shorter than Mrs. Carson.
4. The difference between the tallest student and the shortest student in the class is 14 inches.

Lena thinks that the small box of cereal is a better buy than the large box. The large box has 10 servings, and the small box has 8 servings. The small box costs $3.20, and the large box costs $3.90.

5. Which expression describes the price per serving for the small box?
   A $3.20 × 8
   B $3.20 ÷ 20
   C $3.20 ÷ 8
   D $3.20 × 20
6. What conclusion can you draw from the data?
   F The small box costs less per serving.
   G The large box costs less per serving.
   H Both boxes cost the same per serving.
   J The small box costs twice as much per serving.

Mixed Applications

USE DATA For 7–11, use the table.

7. Does the weight of the animal determine its speed? Explain.
8. **What's the Error?** Don says that the more an animal eats, the more it sleeps. Explain the error in Don's conclusion.
9. Which animal eats the greatest amount of fruits and vegetables? How many kilograms does it eat each day?
10. How many hours longer does the bear sleep than the elephant? than the giraffe?
11. **What's the Question?** Use the data from the table. The answer is 2 times as many kilograms.

ANIMALS

	Elephant	Bear	Giraffe
Weight	5,450 kg	725 kg	1,180 kg
Speed	51 kph	48 kph	51 kph
Sleep per day	180 min	8 hours	240 min
Fruits and vegetables eaten per day	1,800 g	0.9 kg	100 g

1 kilogram (kg) = 1,000 grams (g)
1 hour = 60 minutes (min)
kph = kilometers per hour

CHAPTER 2

Review/Test

CHECK VOCABULARY AND CONCEPTS

Choose the best term from the box.

tenth
hundredth
decimal
thousandth
ten-thousandths

1. A decimal or fraction that names one part of ten equal parts is one _?_. (p. 18)
2. If you divide one whole by 1,000, you get one _?_. (p. 20)
3. A decimal or fraction that names one part of 100 equal parts is one _?_. (p. 18)
4. The decimal 0.0016 is sixteen _?_. (p. 20)

CHECK SKILLS

Write as a decimal and a fraction or mixed number. (pp. 18–21)

5. twenty-one and twenty-one hundredths
6. seven and six tenths
7. three hundred forty-nine thousandths

Write each decimal in expanded form and in word form. (pp. 18–21)

8. 17.02
9. 1.002
10. 13.201
11. 4.076

Write *equivalent* or *not equivalent* to describe each pair of decimals. (pp. 22–23)

12. 0.650 and 0.65
13. 9.502 and 9.52
14. 3.0040 and 3.040
15. 10.01 and 10.010

Order from least to greatest. (pp. 24–25)

16. 0.056; 0.56; 0.059
17. 5.98; 5.908; 5.809
18. 6.969; 9.696; 6.696

CHECK PROBLEM SOLVING

Solve. (pp. 26–27)

19. Jim worked 5 hours last week and earned $31.25. Sara earned $43.75 and worked 7 hours. Latasha earned $56.25 and worked 9 hours. Draw a conclusion about this information.
20. Kerry, Stu, Beth, and Julio are in line. Kerry is first in line. Julio is between Kerry and Stu. Beth is after Stu. Beth is last in line. Draw a conclusion about where Julio stands in line.

Cumulative Review

Understand the problem.
See item **10**.

Remember that the value of a digit depends on its place-value position in the number. You need to find the value of a 9 in the hundredths place.

Also see problem **1**, p. H62.

For 1–11, choose the best answer.

1. $\begin{array}{r} 3{,}124 \\ -2{,}986 \\ \hline \end{array}$

   A 138   C 1,242
   B 1,138   D Not here

2. How is 8.0019 written in words?

   F eight and nineteen
   G eight and nineteen tenths
   H eight and nineteen thousandths
   J eight and nineteen ten-thousandths

3. Which fraction is equivalent to 0.9?

   A $\frac{1}{9}$ C $\frac{9}{100}$
   B $\frac{9}{10}$ D $\frac{9}{1{,}000}$

4. Which shows the numbers in order from greatest to least?

   F 4.1998, 4.1975, 4.1996
   G 4.1996, 4.1975, 4.1998
   H 4.1996, 4.1998, 4.1975
   J 4.1998, 4.1996, 4.1975

5. How many thousands are in one million?

   A 10   C 1,000
   B 100   D 10,000

6. Which number is equivalent to 3.02?

   F 3.002   H 3.20
   G 3.020   J 30.2

7. How is 8 + 0.4 + 0.02 written in standard form?

   A 8.0402   C 8.42
   B 8.402   D Not here

For 8–9, use the table.

RADIO STORE PRICE LIST	
Radio headset	$9.05
Portable radio	$8.99
Clock radio	$9.98
AM radio	$8.89

8. Which costs less than $8.90?

   F Radio headset
   G Portable radio
   H Clock radio
   J AM radio

9. Which costs more than the portable radio but less than the clock radio?

   A Radio headset   C Clock radio
   B Portable radio   D AM radio

10. What is the value of the 9 in 3.596?

    F 0.009   H 0.9
    G 0.09   J 900

11. Which digit makes this number sentence true?

    2,465,123 > 2,4■5,123

    A 5   C 8
    B 6   D 9

CHAPTER 3

Add and Subtract Whole Numbers

An animal species that is in danger of dying out is described as being endangered. The U.S. Department of the Interior keeps track of which animals are endangered. There are more than 1,000 species of animals endangered today in the United States. Look at the table. How many more endangered species are mammals than are fish, insects, and reptiles?

ENDANGERED SPECIES

Group	Number of Species
Birds	274
Fish	119
Insects	41
Mammals	331
Reptiles	114

Gray wolf

Use this page to help you review and remember important skills needed for Chapter 3.

VOCABULARY

Choose the best term from the box.

1. The total when you add is the ? .
2. The amount left when you subtract is the ? .
3. The numbers that are added are the ? .

addends
sum
difference
product

ROUNDING (See p. H3.)

Round each number to the nearest ten.

4. 328 **5.** 81 **6.** 427 **7.** 365

8. 625 **9.** 704 **10.** 450 **11.** 4,712

Round each number to the nearest hundred.

12. 923 **13.** 6,350 **14.** 1,057 **15.** 8,204

16. 768 **17.** 4,942 **18.** 3,680 **19.** 9,458

MENTAL MATH: FIND A RULE (See p. H4.)

Write a rule. Use mental math.

20.

Input	Output
7	10
17	20
27	30
37	40

21.

Input	Output
12	20
22	30
32	40
42	50

22.

Input	Output
65	55
55	45
45	35
35	25

23.

Input	Output
100	65
150	115
200	165
250	215

SUBTRACTION ACROSS ZEROS (See p. H4.)

Subtract. Regroup when necessary.

24. $30 - 25$ **25.** $70 - 36$ **26.** $200 - 156$ **27.** $700 - 293$ **28.** $400 - 351$

29. $50 - 37$ **30.** $560 - 208$ **31.** $600 - 248$ **32.** $700 - 235$ **33.** $7{,}300 - 163$

LESSON 1

Round Whole Numbers

Learn

CROWD PLEASERS The programs for this year's women's college basketball championships need to be ordered. The programs come in boxes of 10,000 each. Last year, 17,976 people attended the championships. How many boxes of programs should be ordered?

To be sure enough programs are ordered, round 17,976 up to the next ten thousand. You can use a number line to round.

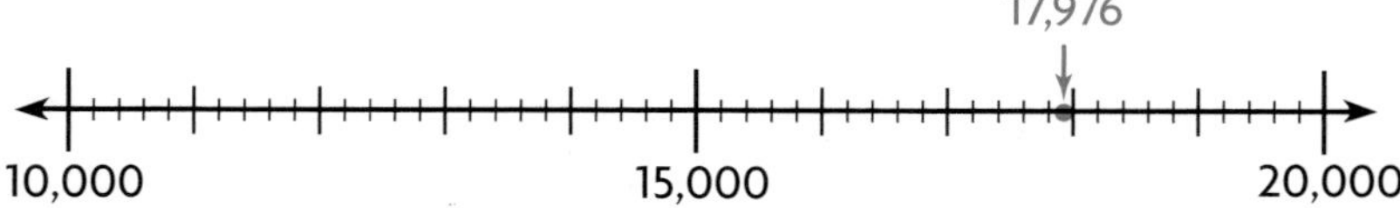

17,976 rounded to the next ten thousand is 20,000.

So, 2 boxes of 10,000 programs should be ordered.

Another way to round is to use the rounding rules.

Round 149,987 to the nearest hundred thousand.

STEP 1

Decide the place to which you want to round.

↓
149,987

STEP 2

If the digit to the right is less than 5, round down.
If the digit to the right is 5 or greater, round up.

149,987
↓ $4 < 5$
100,000 Round down.

Quick Review

What is the value of each 5?

1. 3,582,017
2. 9,023,456
3. 5,034,768
4. 2,057,316
5. 17,063,521

▲ 1982 was the first year the NCAA held championships for women's basketball. About 10,000 people attended.

Examples Round each number to the place of the blue digit.

A

1,265,483
↓ $2 < 5$
1,000,000 Round down.

1,265,483 rounded to the nearest million is 1,000,000.

B

1,265,483
↓ $6 > 5$
1,300,000 Round up.

1,265,483 rounded to the nearest hundred thousand is 1,300,000.

C

1,265,483
↓ $5 = 5$
1,270,000 Round up.

1,265,483 rounded to the nearest ten thousand is 1,270,000.

CALIFORNIA STANDARDS NS 1.1 Estimate, round, and manipulate very large and very small numbers. *also* **MR 1.0, MR 2.0, MR 2.3, MR 2.4, MR 2.5**

Check

1. **Tell** whether 12,735,489 is closer to 12,740,000 or to 12,730,000.

Round each number to the place of the blue digit.

2. 2,681
3. 178,365
4. 1,532,300
5. 33,689
6. 6,023,490

Round 1,654,508 to the place named.

7. thousands
8. ten thousands
9. hundred thousands
10. hundreds
11. tens
12. millions

Practice and Problem Solving

Round each number to the place of the blue digit.

13. 78,210
14. 350,962
15. 5,811,326
16. 606,310
17. 890,352

Round 2,908,365 to the place named.

18. thousands
19. ten thousands
20. hundred thousands
21. hundreds
22. tens
23. millions

Name the place to which each number was rounded.

24. 191,562 to 190,000
25. 4,736,810 to 5,000,000
26. 80,154 to 80,200

USE DATA **For 27–29, use the table.**

NCAA Women's Basketball Championship Game Attendance

Year	Attendance
1992	12,072
1993	16,141
1994	11,966
1995	18,038
1996	23,291
1997	16,714

27. Which 2 years had the same attendance, rounded to the nearest ten thousand?
28. How many people in all attended the NCAA Women's Championships in 1995, 1996, and 1997?
29. **What's the Error?** Sean says that, rounded to the nearest thousand, the attendance in 1997 was the same as in 1993. Describe Sean's error. Explain how to correct it.

Mixed Review and Test Prep

30. 68 + 25 = ■
31. 71 − 29 = ■
32. Use mental math to find the sum 15 + 8 + 2 + 5.
33. How many sides does a quadrilateral have?
34. **TEST PREP** In which number does the digit 2 have the least value? (p. 20)

    A 2.564
    B 4.256
    C 5.264
    D 6.524

Extra Practice page H34, Set A

LESSON

2 Estimate Sums and Differences

Quick Review

Round to the nearest thousand.

1. 457,986
2. 2,954
3. 326,198
4. 2,057,516
5. 9,051,687

Learn

SUPER SCOOP The world's largest ice-cream sundae used 715,040 ounces of ice cream with 163,603 ounces of syrup and toppings. About how many ounces did the giant sundae weigh?

Estimate the sum by rounding.

715,040 →	700,000	Round to the nearest hundred thousand.
+163,603 →	+200,000	
	900,000	

The sundae weighed about 900,000 ounces!

The sundae was covered with 163,603 ounces of syrup and other toppings. The other toppings weighed 8,595 ounces. About how many ounces of syrup were used on the sundae?

Estimate the difference by rounding.

163,603 →	164,000	Round to the nearest thousand.
− 8,595 →	− 9,000	
	155,000	

The sundae had about 155,000 ounces of syrup!

- **What if** the other toppings had weighed 1,096 ounces less? About how many ounces of syrup would have been used?

MATH IDEA You can estimate sums and differences by rounding the numbers in the problem before performing the operation.

CALIFORNIA STANDARDS NS 1.0 Students compute with very large and very small numbers, positive integers, decimals, and fractions and understand the relationship between decimals, fractions, and percents. They understand the relative magnitudes of numbers. **NS 1.1** Estimate, round, and manipulate very large and very small numbers. *also* **MR 1.0, MR 2.0, MR 2.3, MR 3.2**

Check

1. **Describe** how you can use mental math when estimating sums and differences.

Estimate by rounding.

2. $\begin{array}{r} 435{,}476 \\ -241{,}131 \\ \hline \end{array}$

3. $\begin{array}{r} 845{,}008 \\ +124{,}895 \\ \hline \end{array}$

4. $\begin{array}{r} 752{,}401 \\ -491{,}922 \\ \hline \end{array}$

5. $\begin{array}{r} 4{,}801{,}421 \\ -1{,}632{,}970 \\ \hline \end{array}$

Practice and Problem Solving

Estimate by rounding.

6. $\begin{array}{r} 482{,}631 \\ -222{,}965 \\ \hline \end{array}$

7. $\begin{array}{r} 502{,}963 \\ -132{,}631 \\ \hline \end{array}$

8. $\begin{array}{r} 493{,}582 \\ +182{,}785 \\ \hline \end{array}$

9. $\begin{array}{r} 5{,}842{,}110 \\ +1{,}216{,}850 \\ \hline \end{array}$

10. 397,352 + 187,590
11. 512,824 − 495,008
12. 89,405 + 321,945

Estimate to compare. Write < or > for each ●.

13. 69,210 + 24,391 ● 68,258 + 45,924
14. 74,361 + 24,391 ● 91,308 − 25,924
15. 82,356 − 14,638 ● 86,551 − 13,725
16. 23,689 + 89,204 ● 86,931 + 29,563

USE DATA For 17–20, use the table.

WORLD'S LARGEST DESSERTS	
Dessert	**Weight (in pounds)**
Cherry Pie	37,740
Iced Cake	128,238
Apple Pie	30,115
Popsicle	12,346
Lollipop	2,220

17. About how many pounds do the pies weigh altogether?
18. About how much more does the cake weigh than the popsicle?
19. About how much less is the weight of the apple pie than of the cherry pie?
20. **What if** there were 16,209 pounds of icing on the cake? About how much of the total weight would be just cake?
21. **What's the Question?** Ally and three friends are in line at the checkout counter. Ally is right behind Eric and in front of Lynn. Lee is last. Eric is the answer.

Mixed Review and Test Prep

22. 4,954 + 2,608 = ■
23. Which is an equivalent decimal for 52.70: 52.7 or 52.700? (p. 22)
24. Compare 7.064 and 7.644 using <, >, or =. (p. 24)
25. Round 1,435,987 to the nearest hundred thousand. (p. 32)
26. **TEST PREP** Which is 5,567,398 rounded to the nearest million? (p. 32)
    - **A** 5,000,000
    - **B** 5,600,000
    - **C** 6,000,000
    - **D** 6,500,000

Extra Practice page H34, Set B

LESSON 3

Add and Subtract Whole Numbers

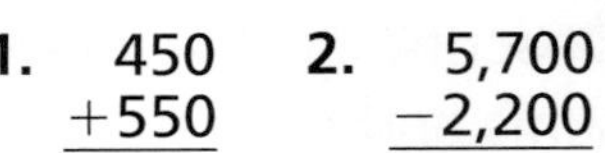

Quick Review

1. 450 + 550
2. 5,700 − 2,200
3. 4,200 − 600
4. 4,000 + 8,000 + 6,000
5. 73 + 27 + 42 + 58

Learn

MAKE A DIFFERENCE! At The Wildlife Center, volunteers are helping scientists place black-footed ferrets back into their natural habitats. This year's volunteers have logged in 2,385 hours of service. Last year they had 1,098 hours. How many hours did they have for both years?

Add. $2,385 + 1,098$

Estimate. $2,000 + 1,000 = 3,000$

$$\begin{array}{r} {\scriptstyle 1\,1\ \ } \\ 2,385 \\ +1,098 \\ \hline 3,483 \end{array}$$

Start with the ones. Regroup as needed.

▲ A black-footed ferret

So, the volunteers logged in 3,483 hours for both years. This is close to the estimate, so the answer is reasonable.

How many more hours were logged in this year than last year?

Subtract. $2,385 - 1,098$

Estimate. $2,000 - 1,000 = 1,000$

$$\begin{array}{r} {\scriptstyle 17\ \ } \\ {\scriptstyle 2\ 7\ 15} \\ 2,385 \\ -1,098 \\ \hline 1,287 \end{array}$$

Start with the ones. Regroup as needed.

So, the volunteers logged in 1,287 more hours this year than last year. This is close to the estimate, so the answer is reasonable.

- How do you use place value when you add and subtract?

TECHNOLOGY LINK

To learn more about addition and subtraction, watch the **Harcourt Math Newsroom Video** ***Black-footed Ferrets.***

CALIFORNIA STANDARDS NS 1.0 Students compute with very large and very small numbers, positive integers, decimals, and fractions and understand the relationship between decimals, fractions, and percents. They understand the relative magnitudes of numbers. *also* **MR 1.0, MR 2.0, MR 2.1, MR 2.3, MR 3.0, MR 3.1, MR 3.2, MR 3.3**

Check

1. **Explain** how you know which place value to regroup first when adding.

Find the sum or difference. Estimate to check.

2. $\begin{array}{r} 6{,}317 \\ +8{,}903 \\ \hline \end{array}$

3. $\begin{array}{r} 12{,}911 \\ -10{,}260 \\ \hline \end{array}$

4. $\begin{array}{r} 11{,}004 \\ +\ 8{,}986 \\ \hline \end{array}$

5. $\begin{array}{r} 67{,}411 \\ -29{,}897 \\ \hline \end{array}$

6. $\begin{array}{r} 98{,}623 \\ -71{,}128 \\ \hline \end{array}$

Practice and Problem Solving

Find the sum or difference. Estimate to check.

7. $\begin{array}{r} 5{,}752 \\ -1{,}842 \\ \hline \end{array}$

8. $\begin{array}{r} 8{,}875 \\ +4{,}908 \\ \hline \end{array}$

9. $\begin{array}{r} 27{,}032 \\ -19{,}986 \\ \hline \end{array}$

10. $\begin{array}{r} 17{,}620 \\ -\ 9{,}003 \\ \hline \end{array}$

11. $\begin{array}{r} 30{,}605 \\ +18{,}746 \\ \hline \end{array}$

12. $234 + 125 + 681$
13. $6{,}010 - 498$
14. $2{,}943 + 7{,}894$
15. $1{,}298 + 786 + 412$
16. $11{,}832 + 28{,}778$
17. $61{,}933 - 12{,}040$
18. $208 + 816 + 7{,}049$
19. $17{,}001 - 8{,}542$
20. $20{,}123 - 14{,}897$

USE DATA For 21–23, use the circle graphs.

21. How many species of birds are *not* threatened and *not* extinct? How many species of reptiles? How many species of mammals?

22. How many threatened and extinct species are there in all three classes of animals?

23. **Write a problem** using the information in the circle graphs.

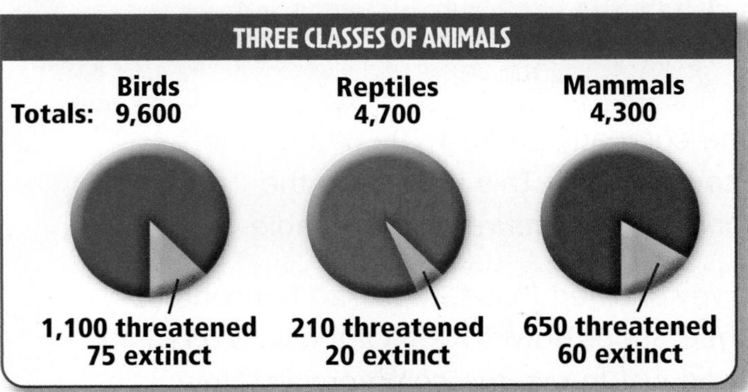

Mixed Review and Test Prep

24. Round 5,567,398 to the nearest million. (p. 32)

Write an equivalent decimal for each. (p. 22)

25. 7.50
26. 4.06

27. Order 2,430,717; 2,340,717; and 2,470,717 from least to greatest. (p. 10)

28. **TEST PREP** In which does the digit 9 have the greatest value? (p. 20)

A 1.953
B 2.193
C 5.359
D 9.153

Extra Practice page H34, Set C

LESSON 

Add and Subtract Greater Numbers

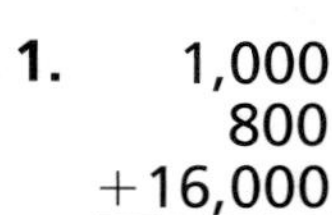

Quick Review

1. 1,000 + 800 + 16,000
2. 800 − 500
3. 11,900 − 9,300
4. 151,000 + 27,000 + 419,000
5. 220,000 + 110,000 + 440,000

▶ Learn

AD TIME ADDS UP! During Super Bowl XXXII, a 30-second commercial cost $1,599,905. A 1-minute commercial cost $3,199,987. If a company bought a 1-minute commercial and a 30-second commercial, how much did the company pay?

You add greater numbers the same way you add small numbers.

Add. 3,199,987 + 1,599,905

Estimate. 3,000,000 + 2,000,000 = 5,000,000

```
   111  1
 $3,199,987   Start with the ones.
+ 1,599,905   Regroup as needed.
 $4,799,892
```

So, the company paid $4,799,892 for the two commercials. This is close to the estimate, so the answer is reasonable.

A survey showed that 43,613,920 households watched Super Bowl XXXII. Of those, 3,052,740 watched just to see the commercials. How many tuned in to see the game?

Subtract. 43,613,920 − 3,052,740

Estimate. 44,000,000 − 3,000,000 = 41,000,000

```
     5 11   8 12
 43,6 1 3,9 2 0   Start with the ones.
− 3,052,740       Regroup as needed.
 40,561,180
```

So, 40,561,180 households tuned in to see the game. This is close to the estimate, so the answer is reasonable.

CALIFORNIA STANDARDS NS 1.0 Students compute with very large and very small numbers, positive integers, decimals, and fractions and understand the relationship between decimals, fractions, and percents. They understand the relative magnitudes of numbers. **NS 1.1** Estimate, round, and manipulate very large and very small numbers. *also* **MR 1.0, MR 2.0, MR 2.1, MR 2.2, MR 2.3, MR 3.0, MR 3.1, MR 3.2, MR 3.3**

Check

1. **Explain** how you know when to regroup in an addition problem.

Find the sum or difference. Estimate to check.

2. 3,456,219 + 5,462,793

3. 7,379,250 − 1,538,260

4. 23,432,213 + 13,513,568

5. 6,113,752 − 1,456,998

Practice and Problem Solving

Find the sum or difference. Estimate to check.

6. 18,245,620 − 5,750,003

7. 2,461,305 + 7,011,347

8. 5,388,214 − 843,412

9. 3,849,672 − 3,618,770

10. 2,870,034 + 6,088,197
11. 5,047,205 − 3,235,675
12. 3,556,014 + 879,843
13. 7,840,272 − 7,035,810
14. 8,373,514 + 632,753
15. 2,876,456 − 1,073,869

Find the missing digits.

16. 7,2■5 + 6,■72 = 1■,357

17. 8,47■ − 4,3■8 = ■,145

18. 1,369,36■ − 526,■23 = 84■,239

19. 4,382,1■7 + 4■8,287 = 4,■20,404

USE DATA **For 20–21, use the table.**

20. How many more households watched Super Bowl XX compared to Super Bowl XVI?

21. What is the order of the Super Bowls from the one with the highest ratings to the one with the lowest ratings?

22. Look back at page 38. How much would a company pay for three 30-second commercials on Super Bowl XXXII?

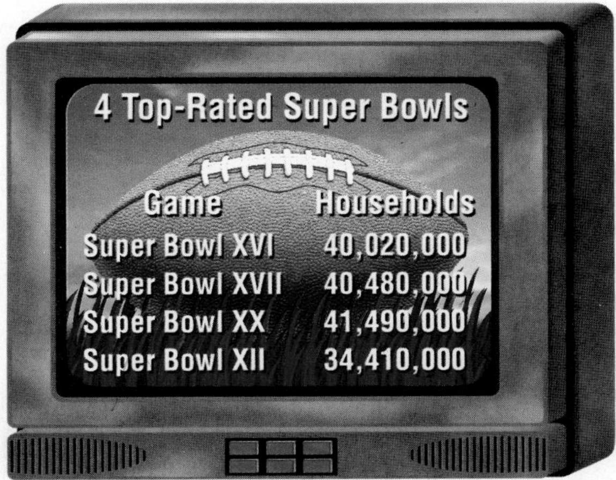

4 Top-Rated Super Bowls

Game	Households
Super Bowl XVI	40,020,000
Super Bowl XVII	40,480,000
Super Bowl XX	41,490,000
Super Bowl XII	34,410,000

Mixed Review and Test Prep

Round each to the nearest hundred thousand. (p. 32)

23. 685,321

24. 1,095,302

25. 28 − ■ = 17

26. Write 45,032,005 in word form. (p. 4)

27. **TEST PREP** In which does the digit 3 have the least value? (p. 20)

A 3.158 **B** 5.139 **C** 5.319 **D** 9.153

Extra Practice page H34, Set D

LESSON 5

Problem Solving Strategy

Use Logical Reasoning

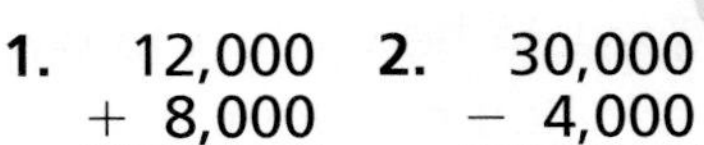

Quick Review

1. 12,000 + 8,000
2. 30,000 − 4,000
3. 430,000 − 230,000
4. 156,000 + 8,000
5. 1,500,000 + 3,500,000

PROBLEM Paco, Rob, Shawna, and Milly won the first four awards at Field Day. Rob won fewer than 5,000 points. Shawna won the fewest number of points. Paco won 355 more points than Shawna. How many points did they each win?

FIELD DAY
Points Scored
1ST Award 6,050 pts.
2ND Award 4,230 pts.
3RD Award 1,820 pts.
4TH Award 1,465 pts.

- What do you need to find out?
- What information will you use?

- What strategy can you use?

You can *use logical reasoning* to organize the information in a table.

- How can you solve the problem?

Make a table to show the information you know. Since each person won one of the awards, there can only be one *yes* in each row and column of the table.

Shawna won the fewest number of points. Put *yes* in the 1,465 column for Shawna and *no* in the remaining columns.

Paco won 355 + 1,465 points. Put *yes* in the 1,820 column for Paco.

Rob must have won 4,230 points, since his award was fewer than 5,000 points.

So, Milly won the remaining award, with 6,050 points.

	6,050	4,230	1,820	1,465
Paco	No	No	Yes	No
Rob	No	Yes	No	No
Shawna	No	No	No	Yes
Milly	Yes	No	No	No

- Look back at the problem. Does the answer make sense for the problem? Explain.

CALIFORNIA STANDARDS **MR 1.1** Analyze problems by identifying relationships, distinguishing relevant from irrelevant information, sequencing and prioritizing information, and observing patterns. **MR 2.3** Use a variety of methods, such as words, numbers, symbols, charts, graphs, tables, diagrams, and models, to explain mathematical reasoning. *also* **NS 1.0, NS 1.1, MR 1.0, MR 2.0, MR 2.4, MR 2.6, MR 3.0, MR 3.1, MR 3.2, MR 3.3**

Problem Solving Practice

Draw a Diagram or Picture
Make a Model or Act It Out
Make an Organized List
Find a Pattern
Make a Table or Graph
Predict and Test
Work Backward
Solve a Simpler Problem
Write an Equation
Use Logical Reasoning

Use logical reasoning to solve.

1. **What if** the problem did not say that Shawna won the fewest number of points? Could you still solve the problem? Explain.
2. Tom, Andy, and Neil mixed up their clothes at the beach. Each one had someone else's hat and sandals. Andy had Tom's hat and Neil's sandals. Whose hat and sandals did each person have?

Ari, Sue, and Jack are hauling truckloads of fruit. The loads weigh 87,500; 76,750; and 91,250 pounds. Ari's load weighs more than Sue's, but their loads weigh the same when rounded to the nearest ten thousand. Jack's load is lighter than Ari's.

3. Which list of names is in order from greatest to least truckload weight?

   A Jill, Jack, Ari
   B Ari, Sue, Jack
   C Sue, Jack, Ari
   D Jack, Ari, Sue

4. Which information is *not* necessary to solve problem 3?

   F Ari's load weighs more than Sue's.
   G Ari's and Sue's weights rounded to the nearest ten thousand are equal.
   H Jack's load weighs less than Ari's.
   J Weights: 87,500, 76,750, 91,250

Mixed Strategy Practice

5. Decimals A, B, C, D, and E are shown at the right. C is the least of the decimals. D is greater than A. C and E have the same thousandths digit. D is less than B. Which decimal goes with which letter?

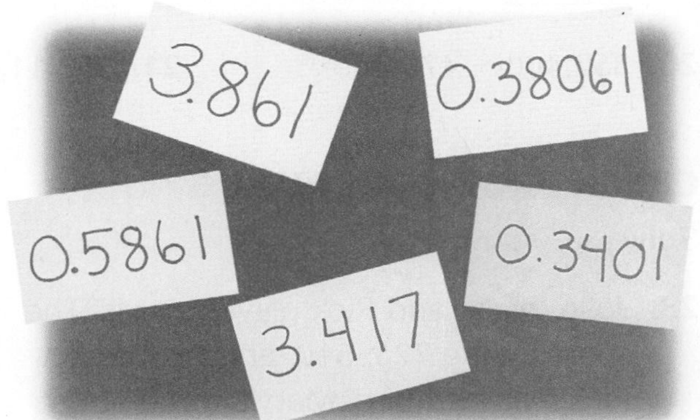

6. **REASONING** You have two sand-filled egg timers, one for 9 minutes and one for 2 minutes. How can you use them to time a 5-minute egg?
7. Lightning can be as hot as 50,000°F. The surface of the sun is about 11,000°F. How much hotter can lightning be than the surface of the sun?
8. The East Plant of the Citrus Company processed 1,675,000 pounds of oranges for juice last year. The West Plant processed 857,500 pounds. How many pounds of oranges did the Citrus Company process last year?

Review/Test

CHECK VOCABULARY AND CONCEPTS

Choose the best term from the box.

reasonable
round
estimate

1. To check your answer for reasonableness, you can __?__ a sum or difference first. (p. 34)
2. You can __?__ numbers by using the rounding rules. (p. 32)

CHECK SKILLS

Round each number to the place of the blue digit. (pp. 32–33)

3. 6,384 4. 1,962 5. 708,350 6. 614,223 7. 817,510 8. 2,630,658

Name the place to which each number was rounded. (pp. 32–33)

9. 54,386 to 54,000 10. 36,547 to 36,500 11. 4,287,540 to 4,300,000

Estimate by rounding. (pp. 34–35)

12. $\begin{array}{r} 507,294 \\ +\ 45,602 \\ \hline \end{array}$ 13. $\begin{array}{r} 740,329 \\ -184,053 \\ \hline \end{array}$ 14. $\begin{array}{r} 129,054 \\ +562,081 \\ \hline \end{array}$ 15. $\begin{array}{r} 8,682,505 \\ -2,321,008 \\ \hline \end{array}$

Find the sum or difference. (pp. 36–39)

16. $\begin{array}{r} 5,060 \\ +2,054 \\ \hline \end{array}$ 17. $\begin{array}{r} 8,450,729 \\ -5,563,072 \\ \hline \end{array}$ 18. $\begin{array}{r} 6,074,851 \\ +5,733,509 \\ \hline \end{array}$ 19. $\begin{array}{r} 8,382,900 \\ -3,985,631 \\ \hline \end{array}$

20. 352,400 + 475,355 21. 752,113 − 402,784 22. 843,017 + 674,335

CHECK PROBLEM SOLVING

Solve. (pp. 40–41)

23. José, Mary, and Rick played cards. The scores were 725, 510, and 225 points. Rick was *not* third. Mary scored about 300 points more than José. Who got which score?

24. Ann, Linda, and Hugh have pets. One of them has a turtle, one has a cat, and the other has a goldfish. Ann's pet has 4 feet, but cannot climb trees. Hugh is allergic to fur. Who has which pet?

25. Which of these numbers are equal when rounded to the nearest ten thousand: 568,235; 562,361; and 565,921?

Cumulative Review

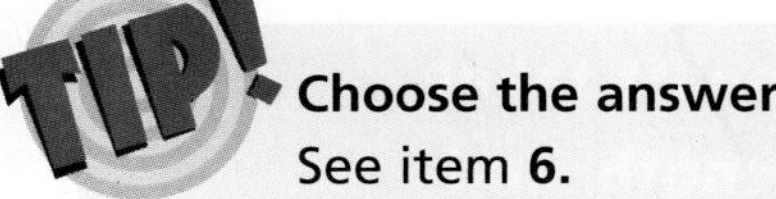

Choose the answer.
See item **6**.

If your answer does not match any of the choices, check your computation. If your computation is correct, mark the letter next to "Not here."

Also see problem **6**, p. H64.

For 1–9, choose the best answer.

1. How is one million, one hundred thousand, two hundred and five tenths written in standard form?

   A 1,100,200.5 **C** 1,120,000.5
   B 1,100,205 **D** Not here

For 2–3, use the bar graph.

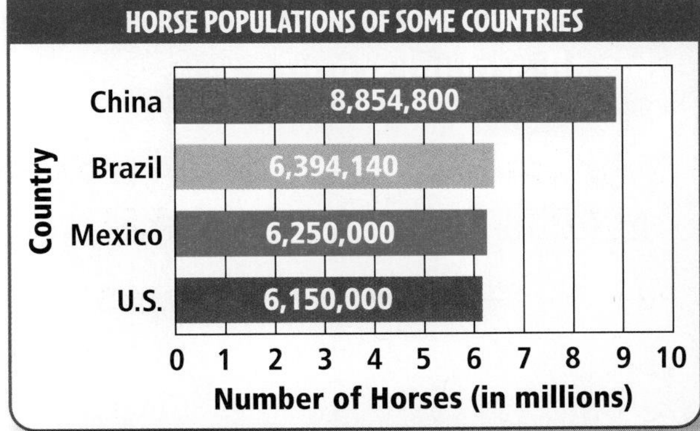

2. How many more horses are there in China than in the United States?

   F 1,704,800 **H** 2,704,800
   G 2,604,800 **J** 15,004,800

3. What is the horse population of Brazil rounded to the nearest hundred thousand?

   A 6,300,000 **C** 6,400,000
   B 6,390,000 **D** 6,490,000

4. A pizza parlor offers sausage, ham, mushrooms, and pineapple as toppings. Ari, Rob, Darius, and Max each ordered a different topping. Darius ordered mushrooms. Rob did not order a meat topping. Max does not like sausage. What topping did Ari order?

   F ham **H** mushrooms
   G sausage **J** pineapple

5. Which number is less than forty million, sixty-five thousand, two?

   A 40,650,002 **C** 40,065,002
   B 40,560,002 **D** 40,056,002

6. $\begin{array}{r} 866{,}022 \\ +233{,}053 \\ \hline \end{array}$

   F 199,075 **H** 632,969
   G 109,907 **J** Not here

7. In which pair are the numbers *not* equivalent?

   A 0.25 and 0.250
   B 0.035 and 0.0350
   C 0.02 and 0.002
   D 0.0056 and 0.005600

8. What is 1,717,854 rounded to the nearest ten thousand?

   F 1,710,000 **H** 1,719,000
   G 1,718,000 **J** 1,720,000

9. What is 5,908,231 rounded to the nearest hundred thousand?

   A 5,800,000 **C** 5,970,000
   B 5,900,000 **D** 6,000,000

Add and Subtract Decimals

To become a gymnast takes coordination, strength, and determination. In men's gymnastics, the gymnast begins each routine with 8.60 points plus 1.40 points in possible bonus points. A perfect routine is awarded a score of 10.00. Judges make deductions for any missing elements that are required in the exercise and for any flaws in execution. The all-around score is the total of all the events' scores. Below are Chris Young's scores for the 1999 U.S. Nationals meet. What was his all-around score?

Chris Young performs a Healy Twirl at the 1999 U.S. Nationals.

CHRIS YOUNG'S GYMNASTICS SCORES

Event	Floor Exercise	Pommel Horse	Still Rings	Vault	Parallel Bars	Horizontal Bar	All-Around
Score	9.25	9.20	8.10	9.50	8.85	8.65	■

Use this page to help you review and remember important skills needed for Chapter 4.

VOCABULARY

Choose the best term from the box.

digit hundredth thousandth tenth

1. If you divide one whole by 10, you get $\frac{1}{10}$, or one __?__.
2. If you divide one whole by 100, you get $\frac{1}{100}$, or one __?__.
3. If you divide one whole by 1,000, you get $\frac{1}{1,000}$, or one __?__.

ROUNDING (See p. H3.)

Round each number to the nearest hundred.

4. 57
5. 70
6. 109
7. 259
8. 179
9. 449
10. 371
11. 987
12. 835
13. 2,089
14. 1,465
15. 99.9
16. 649.9
17. 2,233.3
18. 1,315.5

MENTAL MATH STRATEGIES: DECIMALS (See p. H5.)

Copy and complete. Use mental math.

19. $0.2 + 0.8 = \blacksquare$
$0.7 + 0.3 = \blacksquare$
$0.5 + 0.5 = \blacksquare$
$0.4 + 0.6 = \blacksquare$
20. $0.6 + 0.4 = \blacksquare$
$0.2 + 1.8 = \blacksquare$
$1.1 + 1.9 = \blacksquare$
$1.3 + 2.7 = \blacksquare$
21. $1.0 - 0.2 = \blacksquare$
$1.0 - 0.4 = \blacksquare$
$1.0 - 0.6 = \blacksquare$
$1.0 - 0.8 = \blacksquare$

ADD AND SUBTRACT DECIMALS (See p. H5.)

Add or subtract. Regroup when necessary.

22. $\begin{array}{r} 2.0 \\ -0.5 \\ \hline \end{array}$
23. $\begin{array}{r} 7.5 \\ +3.2 \\ \hline \end{array}$
24. $\begin{array}{r} 20.0 \\ -\ 5.5 \\ \hline \end{array}$
25. $\begin{array}{r} 6.7 \\ -2.3 \\ \hline \end{array}$
26. $\begin{array}{r} 4.0 \\ +3.5 \\ \hline \end{array}$
27. $\begin{array}{r} 4.1 \\ -2.7 \\ \hline \end{array}$
28. $\begin{array}{r} 6.5 \\ +7.8 \\ \hline \end{array}$
29. $\begin{array}{r} 30.0 \\ -19.6 \\ \hline \end{array}$
30. $\begin{array}{r} 8.3 \\ +11.9 \\ \hline \end{array}$
31. $\begin{array}{r} 10.0 \\ -\ 9.8 \\ \hline \end{array}$
32. $2.4 + 7.9$
33. $13.1 - 2.2$

LESSON 1

Round Decimals

Quick Review

Round each number to the place of the blue digit.

1. 572
2. 1,194
3. 25,357
4. 175,506
5. 415,723

Learn

WINGING IT The Pygmy Blue is the smallest butterfly in the U.S. Its average width is 0.563 inch. What is the length rounded to the nearest hundredth of an inch?

You can use a number line to round 0.563 to the nearest hundredth.

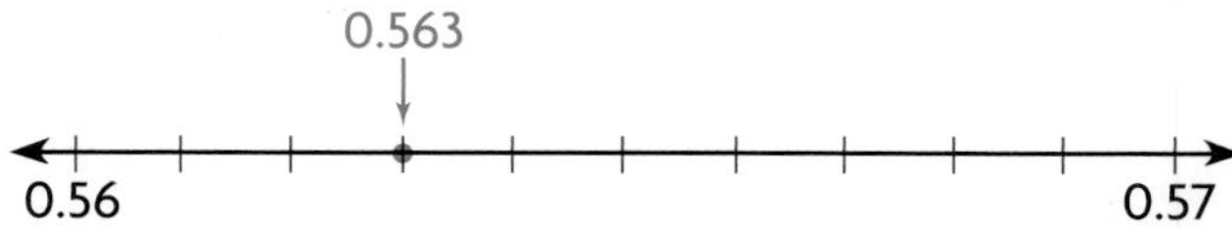

So, 0.563 inch rounded to the nearest hundredth of an inch is 0.56 inch.

Remember
Rounding rules:
- Find the place to which you want to round.
- If the digit to the right is less than 5, round down.
- If the digit to the right is 5 or greater, round up.

Examples

Round 0.4537 to the place of the blue digit.

A
0.4537 $7 > 5$
↓
0.454 Round up.

B
0.4537 $3 < 5$
↓
0.45 Round down.

C
0.4537 $5 = 5$
↓
0.5 Round up.

Check

1. **Explain** how you can use the rounding rules to round $8.49 to the nearest dollar and to the nearest tenth of a dollar.

Round each number to the place of the blue digit.

2. 0.783
3. 0.256
4. 0.1732
5. 0.685
6. 0.66

Round 0.6285 to the place named.

7. tenths
8. hundredths
9. thousandths
10. ones

CALIFORNIA STANDARDS NS 1.1 Estimate, round, and manipulate very large and very small (e.g., thousandths) numbers. **🔑NS 1.5** Identify and represent on a number line decimals, fractions, mixed numbers, and positive and negative integers. *also* **NS 1.0, MR 1.0, MR 2.0, MR 2.3, MR 2.4, MR 2.5**

Practice and Problem Solving

Round each number to the place of the blue digit.

11. 3.193 **12.** 29.423 **13.** 2.0475 **14.** 0.86 **15.** 1.234

Round 2.5438 to the place named.

16. tenths **17.** hundredths **18.** thousandths **19.** ones

Name the place to which each number was rounded.

20. 0.562 to 0.56 **21.** 6.8354 to 6.835 **22.** 80.154 to 80.2

23. 1.7592 to 1.759 **24.** 5.9273 to 6 **25.** 2.3625 to 2.36

Round to the nearest tenth of a dollar and to the nearest dollar.

26. \$10.56 **27.** \$2.73 **28.** \$0.98 **29.** \$5.45

30. \$4.02 **31.** \$32.36 **32.** \$9.49 **33.** \$10.91

Round each number to the nearest hundredth.

34. nine hundred thirty-four thousandths

35. 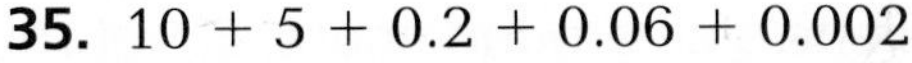 $10 + 5 + 0.2 + 0.06 + 0.002$

36. seven and eighty-three thousandths

37. $2 + 0.8 + 0.09 + 0.005 + 0.0003$

USE DATA For 38–40, use the graph.

38. Round the width of the Dwarf Blue to the nearest tenth of an inch.

39. Which two butterflies are the same width when each width is rounded to the nearest hundredth of an inch?

40. Which butterfly is 0.88 inch wide when its width is rounded to the nearest hundredth of an inch?

41. Write About It Round 0.9999 to the nearest thousandth. Explain what happens.

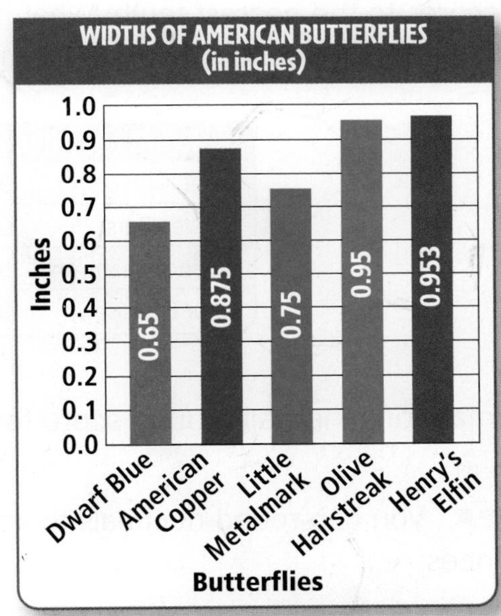

Mixed Review and Test Prep

42. 106,857 − 98,214 (p. 38)

43. 368,591 + 489,721 (p. 38)

44. Order 1.65, 1.56, and 1.6 from greatest to least. (p. 24)

45. $256 + 711 + 304$ (p. 36)

46. TEST PREP Which number is thirty million, three thousand written in standard form? (p. 4)

A 3,003,000
B 3,303,000
C 30,003,000
D 30,300,000

LESSON 2

Estimate Decimal Sums and Differences

Quick Review

Write the value of the blue digit.

1. 3.76 **2.** 6.408
3. 8.591 **4.** 2.099
5. 25.203

Learn

TO THE POINT Janet bought a ticket to a gymnastics meet. The ticket cost $8.79 and Janet paid with a $10 bill. About how much change did she receive?

Estimate by rounding to the nearest whole number, or dollar. Then subtract.

$$\begin{array}{rcr} \$10.00 & \rightarrow & \$10 \\ -\ 8.79 & \rightarrow & -\ 9 \\ \hline & & \$\ 1 \end{array}$$

So, Janet received about $1.00 in change.

To estimate Ilene's all-around score, Janet rounds her individual scores to the nearest tenth. What is the estimate of Ilene's all-around score for the meet?

$$\begin{array}{rcr} 9.20 & \rightarrow & 9.2 \\ 9.575 & \rightarrow & 9.6 \\ 9.40 & \rightarrow & 9.4 \\ +9.575 & \rightarrow & +9.6 \\ \hline & & 37.8 \end{array}$$

Event	Ilene's Score
Vault	9.20
Uneven Bars	9.575
Balance Beam	9.40
Floor Exercise	9.575

▲ Ilene Cohen at the 1999 High School Nationals

So the estimate of Ilene's all-around score for the meet is 37.8.

MATH IDEA You can round decimals to estimate sums and differences.

Check

1. **Explain** how to estimate 32.56 + 18.35.

Estimate the sum or difference to the nearest whole number or dollar.

2. $\begin{array}{r} 0.22 \\ +3.51 \\ \hline \end{array}$ **3.** $\begin{array}{r} \$1.78 \\ -\ 1.04 \\ \hline \end{array}$ **4.** $\begin{array}{r} 4.87 \\ +2.94 \\ \hline \end{array}$ **5.** $\begin{array}{r} \$12.89 \\ -\ 8.78 \\ \hline \end{array}$ **6.** $\begin{array}{r} 5.681 \\ +3.025 \\ \hline \end{array}$

CALIFORNIA STANDARDS NS 1.1 Estimate, round, and manipulate very large and very small (e.g., thousandths) numbers. **NS 2.0** Students perform calculations and solve problems involving addition, subtraction, and simple multiplication and division of fractions and decimals. *also* **NS 1.0, NS 2.1, MR 1.0, MR 2.0, MR 2.3, MR 2.5, MR 3.0**

Practice and Problem Solving

Estimate the sum or difference to the nearest whole number or dollar.

7. $\$7.92 + 5.39$
8. $7.36 - 6.41$
9. $\$30.23 + 13.65$
10. $10.362 - 0.985$
11. $2.978 + 1.351$
12. $40.67 - 19.08$
13. $\$11.98 + 2.59$
14. $2.704 + 1.818$
15. $\$100.00 - 79.65$
16. $7.153 + 4.099$

Estimate the sum or difference to the nearest tenth.

17. $2.39 + 8.06$
18. $0.702 - 0.397$
19. $14.782 + 8.11$
20. $50.111 - 6.75$
21. $8.741 - 1.42$
22. $14.89 - 7.34$
23. $3.872 + 12.94$
24. $37.054 + 27.92$

Estimate to compare. Write $<$ or $>$ for each ●.

25. $8.14 - 4.89$ ● $7.45 - 2.37$
26. $3.82 + 5.46$ ● $6.45 + 2.09$
27. $7.925 + 5.392$ ● $6.401 + 8.396$
28. $9.269 - 1.423$ ● $5.857 - 2.419$

USE DATA For 29–30, use the table.

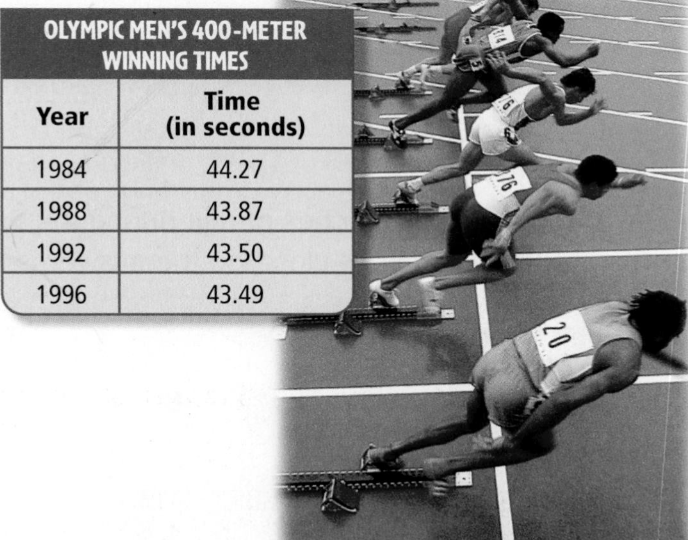

OLYMPIC MEN'S 400-METER WINNING TIMES	
Year	Time (in seconds)
1984	44.27
1988	43.87
1992	43.50
1996	43.49

29. Estimate the difference between the fastest and slowest race times to the nearest tenth.
30. The world record for the 400-meter dash is 43.29 seconds. About how much faster is that than 1996's time?
31. **REASONING** There are 1.609344 kilometers in a mile. Estimate to the nearest hundredth how many kilometers there are in 3 miles.
32. **Write About It** How is estimating decimals like estimating whole numbers?

Mixed Review and Test Prep

33. $21{,}095 + 409{,}739$ (p. 38)
34. $5{,}012 - 987$ (p. 38)

Write an equivalent decimal for each. (p. 22)

35. 0.105
36. 2.720

37. **TEST PREP** In which number does the digit 8 have the greatest value? (p. 2)

A 91,358
B 53,891
C 38,195
D 19,583

LESSON

3 Add and Subtract Decimals

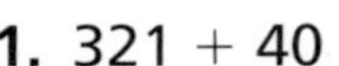

Quick Review

1. 321 + 40
2. 64 − 17
3. 116 + 203
4. 969 − 333
5. 1,075 + 762 + 956

Learn

SPLISH SPLASH Tyler's science class recorded the weather for the first week in March. Tyler's job was to measure the rainfall. It rained on 2 days during that week. On Tuesday, Tyler recorded 1.95 inches of rain, and on Friday another 0.85 inch. How much rain did Tyler record during the first week in March?

Add. 1.95 + 0.85

Estimate. 2 + 1 = 3

STEP 1

Line up the decimal points to align place-value positions. Add the hundredths.

$$\begin{array}{r} \scriptstyle 1 \\ 1.95 \\ +0.85 \\ \hline 0 \end{array}$$

STEP 2

Add the tenths.

$$\begin{array}{r} \scriptstyle 1\ 1 \\ 1.95 \\ +0.85 \\ \hline 80 \end{array}$$

STEP 3

Add the ones. Place the decimal point in the sum.

$$\begin{array}{r} \scriptstyle 1\ 1 \\ 1.95 \\ +0.85 \\ \hline 2.80 \end{array}$$

Tyler recorded 2.80 inches of rain during the first week of March. The estimate is close, so the answer is reasonable.

Examples

A

Line up the decimal points.

$$\begin{array}{r} \scriptstyle 1\ 1 \\ 2.69 \\ +3.83 \\ \hline 6.52 \end{array}$$

Place the decimal point in the sum.

B

$$\begin{array}{r} \scriptstyle 1\ 1 \\ 13.76 \\ +\ 8.50 \\ \hline 22.26 \end{array}$$

← Place a zero for an equivalent decimal.

- Why is it helpful to show an equivalent decimal when adding?

CALIFORNIA STANDARDS NS 1.0 Students compute with very large and very small numbers, positive integers, decimals, and fractions and understand the relationship between decimals, fractions, and percents. They understand the relative magnitudes of numbers. **O━ NS 2.1** Add, subtract, multiply, and divide with decimals; add with negative integers; subtract positive integers from negative integers; and verify the reasonableness of the results. *also* **NS 1.1, NS 2.0, MR 1.0, MR 2.0, MR 2.1, MR 2.3, MR 2.4, MR 3.1, MR 3.2, MR 3.3**

Subtract Decimals

Tyler recorded 2.75 inches of rain during the first week in April, and then he measured 4.92 inches of rain for the entire month of April. How much rain fell in April after the first week?

Subtract. $4.92 - 2.75$

Estimate. $5 - 3 = 2$

STEP 1

Line up the decimal points to align place-value positions. Subtract the hundredths. Regroup if needed.

$$\begin{array}{r} \scriptstyle 8\,12 \\ 4.9\!\!\!/2\!\!\!/ \\ -2.75 \\ \hline 7 \end{array}$$

STEP 2

Subtract the tenths. Regroup if needed.

$$\begin{array}{r} \scriptstyle 8\,12 \\ 4.9\!\!\!/2\!\!\!/ \\ -2.75 \\ \hline 17 \end{array}$$

STEP 3

Subtract the ones.

$$\begin{array}{r} \scriptstyle 8\,12 \\ 4.9\!\!\!/2\!\!\!/ \\ -2.75 \\ \hline 2.17 \end{array}$$

Place the decimal point.

So, 2.17 inches of rain fell in April after the first week. The estimate is close, so the answer is reasonable.

Examples

A 5.12 − 1.08

$$\begin{array}{r} \scriptstyle 0\,12 \\ 5.1\!\!\!/2\!\!\!/ \\ -1.08 \\ \hline 4.04 \end{array}$$

B 24.23 − 11.6

$$\begin{array}{r} \scriptstyle 3\,12 \\ 24\!\!\!/.2\!\!\!/3 \\ -11.60 \\ \hline 12.63 \end{array}$$

← Place a zero to show an equivalent decimal.

- Where do you place the decimal point in the answer?

MATH IDEA You can add and subtract decimals the same way you add and subtract whole numbers if you line up the decimal points first.

Check

1. **Explain** why you can place a zero to the right of a decimal point without changing the value of the number.

Find the sum. Estimate to check.

2. $\begin{array}{r} 0.3 \\ +0.9 \\ \hline \end{array}$

3. $\begin{array}{r} 2.07 \\ 1.15 \\ +0.62 \\ \hline \end{array}$

4. $\begin{array}{r} 11.74 \\ 5.12 \\ 6.03 \\ +\ 1.54 \\ \hline \end{array}$

5. $\begin{array}{r} 5.08 \\ +4.18 \\ \hline \end{array}$

6. $\begin{array}{r} 12.1 \\ +\ 9.01 \\ \hline \end{array}$

LESSON CONTINUES

Find the difference. Estimate to check.

7. $\begin{array}{r} 0.8 \\ -0.3 \\ \hline \end{array}$
8. $\begin{array}{r} 1.23 \\ -0.47 \\ \hline \end{array}$
9. $\begin{array}{r} 4.06 \\ -2.85 \\ \hline \end{array}$
10. $\begin{array}{r} 10.2 \\ -\ 8.67 \\ \hline \end{array}$
11. $\begin{array}{r} 23.05 \\ -12.8 \\ \hline \end{array}$

12. $4.3 - 3.6$
13. $1.6 - 0.8$
14. $7.2 - 5.69$
15. $17.13 - 10.9$

Practice and Problem Solving

Find the sum or difference. Estimate to check.

16. $\begin{array}{r} 2.1 \\ +1.7 \\ \hline \end{array}$
17. $\begin{array}{r} 13.59 \\ 3.41 \\ +\ 2.54 \\ \hline \end{array}$
18. $\begin{array}{r} 2.45 \\ +0.63 \\ \hline \end{array}$
19. $\begin{array}{r} 8.007 \\ 3.985 \\ +1.125 \\ \hline \end{array}$
20. $\begin{array}{r} 3.17 \\ +0.456 \\ \hline \end{array}$

21. $\begin{array}{r} 5.6 \\ 7.8 \\ +1.7 \\ \hline \end{array}$
22. $\begin{array}{r} 40.8 \\ 25.17 \\ +16.3 \\ \hline \end{array}$
23. $\begin{array}{r} 6.008 \\ +1.883 \\ \hline \end{array}$
24. $\begin{array}{r} 45.903 \\ 9.374 \\ +51.28 \\ \hline \end{array}$
25. $\begin{array}{r} 13.046 \\ +\ 0.298 \\ \hline \end{array}$

26. $\begin{array}{r} 1.5 \\ -0.7 \\ \hline \end{array}$
27. $\begin{array}{r} 4.12 \\ -1.85 \\ \hline \end{array}$
28. $\begin{array}{r} 7.09 \\ -5.64 \\ \hline \end{array}$
29. $\begin{array}{r} 16.1 \\ -\ 7.34 \\ \hline \end{array}$
30. $\begin{array}{r} 43.18 \\ -29.9 \\ \hline \end{array}$

31. $\begin{array}{r} 20.1 \\ -\ 8.6 \\ \hline \end{array}$
32. $\begin{array}{r} 3.15 \\ -1.99 \\ \hline \end{array}$
33. $\begin{array}{r} 5.07 \\ -0.68 \\ \hline \end{array}$
34. $\begin{array}{r} 22.6 \\ -\ 9.45 \\ \hline \end{array}$
35. $\begin{array}{r} 21.06 \\ -\ 9.33 \\ \hline \end{array}$

36. $1.84 + 1.92$
37. $1.34 + 4.61$
38. $2.05 + 0.97$
39. $1.2 - 0.6$
40. $5.3 - 2.68$
41. $19.07 - 12.6$

Algebra Find a pattern. Write a rule. Use your rule to write the next two numbers in the pattern.

42. 2.3, 2.5, 2.4, 2.6, 2.5, ■, ■
43. 0.452, 0.552, 0.542, 0.642, 0.632, ■, ■
44. 4.06, 3.96, 4.01, 3.91, 3.96, ■, ■
45. 1.725, 1.75, 2.75, 2.775, 3.775, ■, ■

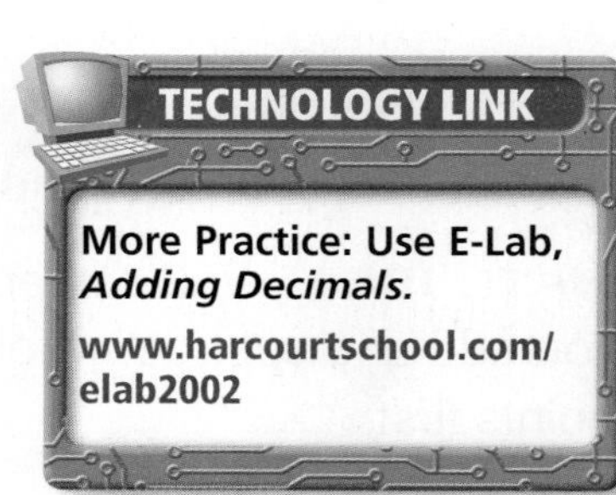

TECHNOLOGY LINK
More Practice: Use E-Lab, *Adding Decimals.*
www.harcourtschool.com/elab2002

46. **What's the Error?** Gail says that $3.2 + 1.45 = 1.77$. Describe and correct Gail's error.

47. **GEOMETRY** Bob is enclosing a rectangular field with fencing. The field is 120 feet long and 50 feet wide. How much fencing does he need?

48. **REASONING** Without adding, how do you know that the sum of 36.179 and 8.63 will have a 9 in the thousandths place?

49. **What if** it only rained 1.34 inches on Monday and 0.25 inches on Saturday in one week? How much rain would have been recorded in that week?

50. Steve spent $9.45 on a ticket to the game and $3.75 for snacks. He had $7.35 left at the end of the day. How much did he have at the beginning of the day?

Mixed Review and Test Prep

51. Write an equivalent decimal for 7.58. (p. 22)

Write the value of the blue digit.

52. 34,097,829 (p. 4)

53. 561.087 (p. 18)

Complete.

54. $3 \times 6 = 4 + ■$

55. $63 - 43 = 2 \times ■$

56. Write the factors of 25.

57. $847 + 916$ (p. 36)

58. $1{,}240 - 623$ (p. 36)

Choose the letter of the next three numbers in the pattern. (p. 18)

59. TEST PREP 1.8, 2.3, 2.0, 2.5, 2.2

- **A** 1.9, 2.4, 2.1
- **B** 2.5, 2.2, 2.7
- **C** 2.7, 2.4, 2.9
- **D** 2.7, 3.2, 3.7

LINKUP to Science

Scientists use special instruments to collect data about weather conditions. The graph below shows the U.S. cities with the least average amount of rainfall per year.

USE DATA **For 1–3, use the graph.**

1. How much less was the mean annual precipitation in Las Vegas, Nevada, than in Bakersfield, California?
2. How much greater was the mean annual precipitation in Phoenix, Arizona, than in Bishop, California?
3. How much greater was the mean annual precipitation in the fifth driest city than in the driest city?

TOP 5 DRIEST U.S. CITIES (mean annual precipitation in inches)

Yuma, Arizona	2.65
Las Vegas, Nevada	4.19
Bishop, California	5.61
Bakersfield, California	5.72
Phoenix, Arizona	7.11

Key: Each cactus = 1 inch of rain.

Extra Practice page H35, Set C

LESSON 4

Zeros in Subtraction

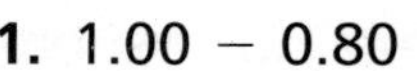

Quick Review

1. 1.00 − 0.80
2. 3.55 − 3.45
3. 4.70 − 2.0
4. 80 − 5.09
5. Round $4.75 to the nearest dollar.

Learn

PIT STOP Kristen saved $60 from her allowance. She wants to buy a remote control race car that costs $45.98 including sales tax. How much will Kristen have left after she purchases the race car?

Subtract. $60.00 − $45.98

Estimate. $60 − $46 = $14

Then find the exact answer.

```
     9 9
   5 10 10 10
  $6 0.0 0     • Place zeros to show an equivalent
                 decimal. Line up the decimal points.
 − 4 5.9 8     • Subtract.
  $1 4.0 2     • Place a decimal point in the difference.
```

Kristen will have $14.02 left. The estimate is close, so the answer is reasonable.

MATH IDEA Sometimes, in order to subtract decimals, you need to write equivalent decimals by placing zeros at the right.

Examples

A 2.9 − 0.74

```
   8 10
  2.9 0  ← Place a zero to show
 −0.7 4    an equivalent decimal.
  2.1 6
```

B 1.6 − 0.342

```
      9
    5 10 10
  1.6 0 0
 −0.3 4 2
  1.2 5 8
```

C 10.478 − 9.53

```
  9  14
  10.4 7 8
 − 9.5 3 0
   0.9 4 8
```

CALIFORNIA STANDARDS NS 1.0 Students compute with very large and very small numbers, positive integers, decimals, and fractions and understand the relationship between decimals, fractions, and percents. They understand the relative magnitudes of numbers. **NS 2.1** Add, subtract, multiply, and divide with decimals; add with negative integers; subtract positive integers; and verify the reasonableness of the results. *also* **NS 1.1, NS 2.0, MR 1.0, MR 2.0, MR 2.1, MR 2.2, MR 2.3, MR 3.0, MR 3.1, MR 3.2, MR 3.3**

Check

1. **Explain** how to subtract 0.25 from 3.

Find the difference.

2. $\begin{array}{r} 3.9 \\ -3.25 \\ \hline \end{array}$
3. $\begin{array}{r} 2.08 \\ -0.74 \\ \hline \end{array}$
4. $\begin{array}{r} 3.007 \\ -0.456 \\ \hline \end{array}$
5. $\begin{array}{r} 302.3 \\ -\ \ 19.635 \\ \hline \end{array}$
6. $\begin{array}{r} 250 \\ -195.32 \\ \hline \end{array}$

7. $3.25 - 0.856$
8. $5.1 - 4.932$
9. $0.6 - 0.099$

Practice and Problem Solving

Find the difference.

10. $\begin{array}{r} 1.3 \\ -0.73 \\ \hline \end{array}$
11. $\begin{array}{r} 3.05 \\ -2.6 \\ \hline \end{array}$
12. $\begin{array}{r} 4.2 \\ -1.02 \\ \hline \end{array}$
13. $\begin{array}{r} 132.5 \\ -\ \ 41.646 \\ \hline \end{array}$
14. $\begin{array}{r} 720.654 \\ -118.702 \\ \hline \end{array}$

15. $2.45 - 1.39$
16. $1.4 - 0.422$
17. $\$2.10 - \$1.20$
18. $5.95 - 1.994$
19. $4.7 - 1.616$
20. $\$5.31 - \$2.26$

Algebra **Find the value of the expression.**

21. $5.01 - n$ if $n = 1.52$
22. $n + (0.5 - 0.2)$ if $n = 2$
23. $(3.6 + n) - 1.1$ if $n = 1.3$
24. $10.25 + n$ if $n = 6.75$

USE DATA **For 25–27, use the table.**

INDIANAPOLIS 500 WINNING TIMES	
Year	**Average Speed (mph)**
1996	147.956
1995	153.616
1994	160.872
1993	157.207

25. How much faster was 1995's winner than 1996's winner?
26. Find the difference between the speeds of the 1994 and 1993 winners.
27. Find the difference between the fastest and slowest speeds.
28. **Write a problem** about the amount of change someone would receive from $10 when buying three items for lunch.
29. **What's the Question?** Jeff is 1.325 meters tall and Maria is 1.245 meters tall. The answer is 0.08 meter taller.

Mixed Review and Test Prep

30. Round 1.3609 to the nearest thousandth. (p. 46)
31. $1,569,201 - 523,641$ (p. 38)
32. $7.865 + 0.603$ (p. 50)
33. Write 0.1078 in word form. (p. 20)
34. **TEST PREP** Which 7 has the least value? (p. 20)

 A 7.026 **B** 2.067
 C 6.270 **D** 0.726

Extra Practice page H35, Set D

LESSON 5

Problem Solving Skill

Estimate or Find Exact Answer

Quick Review

Round 714,549 to the nearest:

1. ten
2. hundred
3. ten thousand
4. thousand
5. hundred thousand

TO THE PENNY? Enrique has $12 to buy supplies for his new kitten. He needs to buy a toy for $3.98, a collar for $3.79, and treats for $2.49. Does he have enough money to pay for all three items? How much change should he receive?

Sometimes you need an *exact answer* to a problem. Sometimes an *estimate* will do.

First Question

Does Enrique have enough money to pay for all three items?

Enrique can estimate by rounding each price up.

$3.98	→	$4.00	Round each up.
3.79	→	4.00	
+ 2.49	→	+ 3.00	
		$11.00	

11 < 12, so $12.00 will be enough to pay for all three items.

Second Question

How much change should Enrique receive?

Once you pay for something, you need to know what your exact change should be. To find Enrique's change, you need an exact answer.

$3.98	$12.00
3.79	− 10.26
+ 2.49	$1.74 change
$10.26	

Since the exact cost of the three items is $10.26, Enrique's exact change should be $1.74.

Talk About It

- Why did Enrique round up to the next higher dollar when estimating the total cost?
- What words in a question might be a clue that an estimate will do?

CALIFORNIA STANDARDS MR 1.0 Students make decisions about how to approach problems. **MR 2.5** Indicate the relative advantages of exact and approximate solutions to problems and give answers to a specified degree of accuracy. *also* **NS 1.0, NS 1.1, NS 2.0, NS 2.1, MR 2.0, MR 2.4, MR 2.6, MR 3.0, MR 3.1, MR 3.2, MR 3.3**

Problem Solving Practice

Decide whether you need an exact answer or an estimate. Then solve.

1. André has 15 meters of wood. He uses 7 meters for a bookcase and 2.5 meters for shelves. How much wood does he have left?

2. Barbara needs 10 gallons of water to fill her aquarium. How many times will she have to fill a 2.5-gallon bucket to fill the aquarium?

3. Don has $20 for groceries. He wants to buy a turkey for $13.89, potatoes for $2.79, and cranberry sauce for $1.98. Does he have enough money to pay for all three items?

4. Tony's party starts in 3 hours. He figures it will take about 1 hour to cook, 40 minutes to clean, and 30 minutes to decorate. Will he have time to shower before the guests arrive?

Mike received $40 for his birthday. He wants to buy headphones for $12.98, a microphone for $11.99, and a CD for $12.95.

5. Which question about Mike's shopping requires an exact answer?

   A Does he have enough money to buy all three items?

   B About how much will he pay?

   C How much change should he get?

   D Which two items cost about $26?

6. Which equation represents Mike's change?

   F $\$40 - \$38 = \$2$

   G $\$12.98 + \$11.99 + \$12.95 = \$37.92$

   H $\$13 + \$12 + \$13 = \$38$

   J $\$40 - \$37.92 = \$2.08$

Mixed Applications

For 7–10, use the items shown at the right.

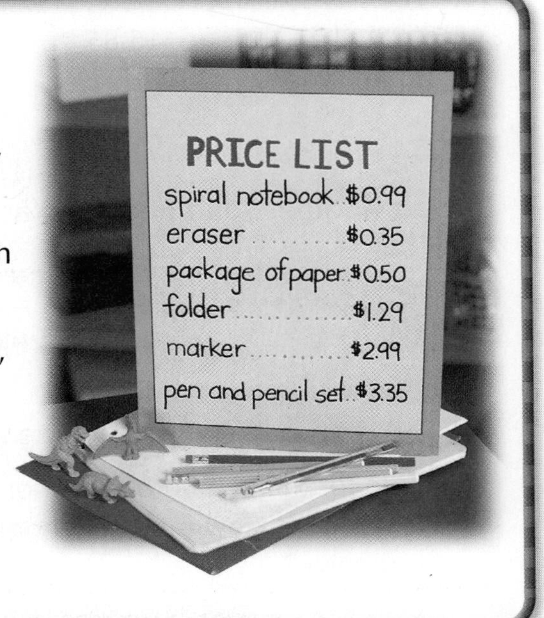

PRICE LIST
spiral notebook $0.99
eraser $0.35
package of paper $0.50
folder $1.29
marker $2.99
pen and pencil set $3.35

7. James has $10 to buy 1 folder, 1 pen and pencil set, and 2 spiral notebooks. Does he have enough money for all the items?

8. Jan buys three markers and an eraser. If she pays with a $20 bill, how much change will she receive?

9. Cindy has $2.50. If she can get only one of each item, what combination of supplies can she afford?

10. Tanya bought a folder and an eraser. The answer is $3.36. What's the question?

11. Describe situations that require an exact sum or difference and those where an estimate will do.

CHAPTER 4

Review/Test

CHECK VOCABULARY AND CONCEPTS

Choose the best term from the box.

decimal point
estimate
exact answer

1. When you round the numbers in a problem before you compute, you get an __?__ of the answer. (p. 48)
2. A mark that separates the whole-number places from the decimal places in a decimal number is the __?__. (p. 50)

CHECK SKILLS

Round each number to the place of the blue digit. (pp. 46–47)

3. 47.28
4. 219.375
5. 43.623
6. 4,305.62
7. 4.3819

Round 745.1583 to the place named. (pp. 46–47)

8. ones
9. tenths
10. hundredths
11. thousandths

Estimate the sum or difference to the nearest whole number or dollar. (pp. 48–49)

12. $\begin{array}{r} 5.72 \\ +8.04 \\ \hline \end{array}$
13. $\begin{array}{r} 17.19 \\ -\ 5.2 \\ \hline \end{array}$
14. $\begin{array}{r} \$25.23 \\ +\ 17.16 \\ \hline \end{array}$
15. $\begin{array}{r} \$312.753 \\ -\ 199.821 \\ \hline \end{array}$
16. $\begin{array}{r} 64.502 \\ +34.771 \\ \hline \end{array}$

Find the sum or difference. (pp. 50–55)

17. $\begin{array}{r} 54.90 \\ +82.04 \\ \hline \end{array}$
18. $\begin{array}{r} 32.9 \\ -15.85 \\ \hline \end{array}$
19. $\begin{array}{r} 6.062 \\ +17.581 \\ \hline \end{array}$
20. $\begin{array}{r} 23.009 \\ -\ 5.3 \\ \hline \end{array}$
21. $\begin{array}{r} 0.94 \\ +6.9531 \\ \hline \end{array}$

CHECK PROBLEM SOLVING

Decide whether you need an exact answer or an estimate. Then solve. (pp. 56–57)

22. Chet had \$20. He paid \$8.95 for admission and \$5.25 for lunch. Will he be able to buy a hat for \$10?
23. Lynn's soccer practice starts at 4:45. She finishes around 6:00. How long is her soccer practice?
24. Tom wants to buy a pattern for \$3.95, fabric for \$8.99, and buttons for \$5.75. He has \$20. How much change should he receive?
25. Ronda needs 16.5 pounds of soil for a science project. If she has one 12-pound bag, how much more soil does she need?

Cumulative Review

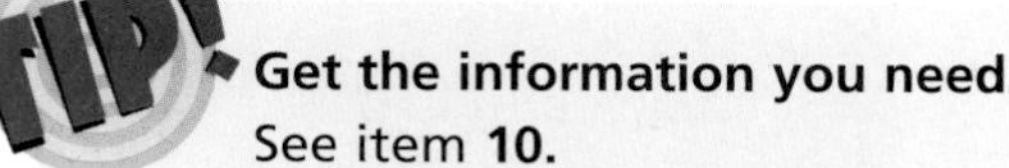

Get the information you need.
See item **10.**
Write the number in standard form. Identify the thousandths place and look at the digit to the right to decide how to round the thousandths digit.
Also see problem **3,** p. H63.

For 1–10, choose the best answer.

1. How is 50,000 + 6,000 + 300 + 40 written in standard form?

A 506,340 **C** 56,304
B 56,340 **D** 56.340

2. What is the value of the 5 in 18.7652?

F 5,000 **H** 0.05
G 0.5 **J** 0.005

3. Raymond had \$15.75. He spent \$4.85 for baseball cards and \$7.35 for comic books. How much money did he have then?

A \$13.25 **C** \$3.55
B \$8.40 **D** Not here

4. What is 1.2345 rounded to the nearest thousandth?

F 1.2 **H** 1.234
G 1.23 **J** 1.235

5.
$$\begin{array}{r} 71.58 \\ +46.19 \\ \hline \end{array}$$

A 117.77 **C** 117.58
B 117.67 **D** Not here

6. Which symbol makes this a true number sentence?

5.539 ● 5.53

F < **H** =
G > **J** ×

7. Which number is equivalent to $\frac{42}{100}$?

A 4.2 **C** 0.42
B 4.02 **D** 0.042

For 8–9, use the table.

LONGEST LONG JUMPS		
Year	**Distance in Meters**	**Athlete**
1991	8.87	Lewis
1991	8.95	Powell
1987	8.86	Emmiyan
1968	8.90	Beamon

8. What is the difference between the longest and the shortest jumps listed?

F 0.03 m **H** 0.3 m
G 0.09 m **J** 0.90 m

9. What is the difference between the two jumps listed for 1991?

A 0.012 m **C** 0.12 m
B 0.08 m **D** 0.8 m

10. What is one and seven hundred fifty-nine ten-thousandths rounded to the nearest thousandth?

F 1.076 **H** 1.075
G 1.0759 **J** 1.07

UNIT 1

MATH DETECTIVE

Digital Fill-In

REASONING In each case, use the clues to make a number.

Case 1

Use the digits 1, 2, 5, 7, 9 to fill in the boxes below to form a number.

Each digit is to be used only once. 6 ■ 8, ■ ■ 0, ■ 3 ■

Clues The number

- rounds to 700,000,000 when rounded to the nearest hundred million.
- is even.
- has more hundreds than ten thousands.
- is greater than 618,790,632.

Case 2

Use the digits 2, 3, 6, 7 to fill in the boxes below to form a number.

Each digit is to be used only once. 4, ■ ■ 5. ■ 8 ■

Clues The number

- rounds to 5,000 when rounded to the nearest thousand.
- has twice as many tens as thousandths.

Think It Over!

- In Case 1, could you have arranged the given digits differently and still have met all the given clues? Explain.
- In Case 2, if the number rounded to 4,000, what would your answer be?
- **STRETCH YOUR THINKING** Complete the middle number so the three numbers are in order from least to greatest. Is there more than one way to complete the number?
  1.05, 1.0 ■ ■, 1.06.

UNIT 1

Challenge

Roman Numerals

About 2,500 years ago, the Romans wrote numbers by using letters. Today we call those *Roman numerals*. The ordinary numbers we use are called *decimal numbers*.

I	V	X	L	C	D	M	$\overline{V}$
↓	↓	↓	↓	↓	↓	↓	↓
1	5	10	50	100	500	1,000	5,000

To change Roman numerals to decimal numbers, follow these rules:

> If the values of the letters from left to right decrease or stay the same, add.
>
> If the value of the letter to the left is less, subtract. You cannot have more than three of any symbol in a row.

Examples Change each Roman numeral to a decimal number.

A Roman numeral: XXVI

X	X	V	I	
↓	↓	↓	↓	
10 +	10 +	5 +	1	= 26

Decimal number: 26

B Roman numeral: CCLXIV

Think: $1 < 5$, so subtract.

C	C	L	X	IV	
↓	↓	↓	↓	↓	
100 +	100 +	50 +	10 +	(5 − 1)	= 264

Decimal number: 264

Talk About It

- How do you know whether to add or subtract to change a Roman numeral to a decimal number?
- Where do you see Roman numerals today?

Try It

Write each Roman numeral as a decimal number.

1. III **2.** DCIV **3.** CXLV **4.** DCCIII **5.** $\overline{V}$MMIV **6.** MCCXII

Write each decimal number as a Roman numeral.

7. 11 **8.** 79 **9.** 452 **10.** 341 **11.** 1,927 **12.** 6,205

Study Guide and Review

VOCABULARY

Choose the best term from the box.

1. A decimal or fraction that names one part of ten equal parts is one __?__. (p. 18)
2. The number 12,865,105 is 13,000,000 when rounded to the nearest __?__. (p. 4)

million
tenth
billion

STUDY AND SOLVE

Chapter 1

Identify place value of whole numbers.

Write the value of the blue digit.
4,395,752,023 — The 9 is in the ten millions place.
90,000,000

Write the value of the blue digit. (pp. 2–7)

3. 34,902
4. 12,045,748
5. 8,157,993,030
6. 13,987,456
7. 501,653
8. 7,961,614

Compare and order whole numbers.

Compare 217,932,654 and 217,805,308.
217,932,654 • Start at the left.
↓ • Compare digits.
217,805,308 $2 = 2$, $1 = 1$, $7 = 7$, $9 > 8$
So, $217{,}932{,}654 > 217{,}805{,}308$.

Compare. Write $<$, $>$, or $=$ for each ●. (pp. 8–11)

9. 256,087 ● 256,807
10. 4,083,147 ● 4,803,610
11. Order 39,325; 38,865; and 39,789 from greatest to least.

Chapter 2

Identify, read, and write decimals.

Write in word form.
4.012 — The 2 is in the thousandths place.
four and twelve thousandths

Write each decimal in expanded form, in word form, and as a fraction. (pp. 18–21)

12. 1.25
13. 0.357
14. 11.04
15. 0.2689
16. 1.0763
17. 4.2038

Compare and order decimals.

Compare 5.6275 and 5.6837.
5.6275 • Line up the decimal points.
↓ • Compare digits from the left. $5 = 5$, $6 = 6$, $2 < 8$
5.6837
So, $5.6275 < 5.6837$.

Write $<$, $>$, or $=$ for each ●. (pp. 24–25)

18. 2.041 ● 2.410
19. 4.110 ● 4.11
20. 0.043 ● 0.034
21. 1.5439 ● 1.532
22. Order 11.897, 11.987, and 11.917 from least to greatest.

Chapter 3

Estimate sums and differences of whole numbers.

Estimate by rounding.

514,752 →	500,000	to the nearest
+173,091 →	+200,000	hundred
	700,000	thousand

Estimate by rounding. (pp. 34–35)

23. 2,389,241 + 8,542,334

24. 687,511 − 325,874

25. 5,641,833 + 1,655,100

Add and subtract whole numbers.

3,648,054	• Start with the ones.
−1,724,321	• Regroup as needed.
1,923,733	

Find the sum or difference. (pp. 36–39)

26. 10,225 − 8,541

27. 9,514,687 + 3,408,994

Chapter 4

Estimate sums and differences of decimals.

Estimate by rounding.

4.76 →	5	to the nearest whole
−1.34 →	−1	number
	4	

Estimate the sum or difference to the nearest whole number. (pp. 48–49)

28. 3.98 + 8.64

29. 10.684 − 5.621

30. 2.671 + 4.494

31. 8.76 − 3.52

32. 0.57 + 0.82

Add and subtract decimals.

6.851	• Start with the ones.
+4.307	• Regroup as needed.
11.158	

Find the sum or difference. (pp. 50–55)

33. 9.681 − 4.320

34. 25.871 + 9.51

35. 12.67 − 8.38

PROBLEM SOLVING PRACTICE

Solve. (pp. 40–41; 56–57)

36. Yesterday was not Tuesday or Thursday. Tomorrow will not be Tuesday or Friday. Today is a weekday. What day is it?

37. Tim has $10.00 and wants to buy four items for $2.79, $3.49, $0.89, and $1.99. Does he need an estimate or an exact answer to decide if he will have enough money? Explain.

California Connections

POPULATION

California has the greatest population of any state in the United States. In 1990, the population of California was 29,558,000. It has been estimated that there will be about 50 million people living in California by 2025.

Los Angeles is the most populated city in California and the second most populated city in the United States. ▼

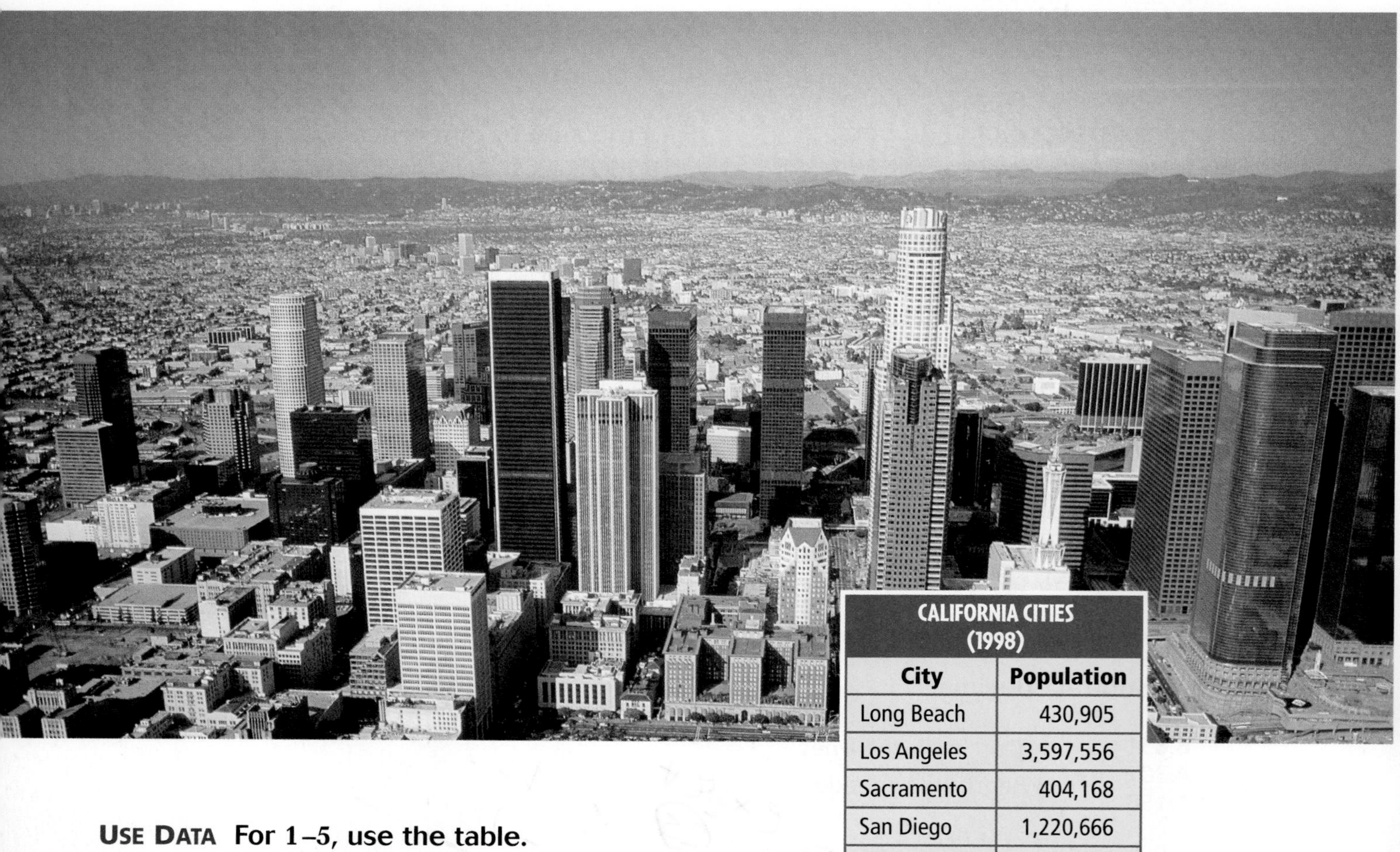

CALIFORNIA CITIES (1998)	
City	**Population**
Long Beach	430,905
Los Angeles	3,597,556
Sacramento	404,168
San Diego	1,220,666
San Francisco	745,774
San Jose	861,284

USE DATA For 1–5, use the table.

1. Which city has the greatest population? the least population?
2. How much greater is the population of San Jose than the population of San Francisco?
3. If the populations are rounded to the nearest hundred thousand, which two cities have the same population? Explain.
4. What is the total population of Los Angeles and San Diego? Explain how you would check to make sure your answer is reasonable.
5. **STRETCH YOUR THINKING** Find the population of the city or town where you live and compare it to the population for one of the cities in the table.

PRECIPITATION

Most of the people in California live along the coast, where the climate is mild. But even along the coast the climate varies a great deal. Rainfall is far greater in northern California than in southern California.

AVERAGE RAINFALL (in inches)

	Jan	Feb	Mar	Apr	May	Jun	Jul	Aug	Sep	Oct	Nov	Dec
Crescent City	13.08	10.76	11.53	5.97	3.78	1.51	0.38	0.97	2.12	5.80	13.34	14.08
San Diego	1.80	1.53	1.77	0.79	0.19	0.07	0.02	0.10	0.24	0.37	1.45	1.57

USE DATA For 1–6, use the table.

1. Write an equivalent decimal for the average rainfall in San Diego in January.
2. Write the average rainfall in Crescent City in December using a fraction.
3. In San Diego four months have an average rainfall greater than 1.50 in. List these months in order from greatest to least amounts of rainfall.
4. Use $<$, $>$, or $=$ to compare the average rainfall in July for Crescent City and San Diego. Explain how a number line shows the answer.
5. In January, February, and March, how much greater is the total average rainfall in Crescent City than in San Diego?
6. In which city is the total average rainfall in September and October about 8 inches? Explain how you know.

Battery Point Lighthouse in Crescent City is on a tiny offshore island that you can walk to at low tide.▼

CHAPTER 5

Algebra: Use Addition

During the Great Backyard Bird Count, bird enthusiasts across North America count the birds in their backyards, parks, or other natural areas for four days. In 2000, a total of 419 species and 4,760,263 individual birds were counted. The map shows the average number of belted kingfishers at different places. The greatest count in one location was 70 kingfishers in Ochopee, Florida. Write an equation to find how many were counted in the rest of Florida.

During the count, a total of 359 belted kingfishers were sighted across Florida.

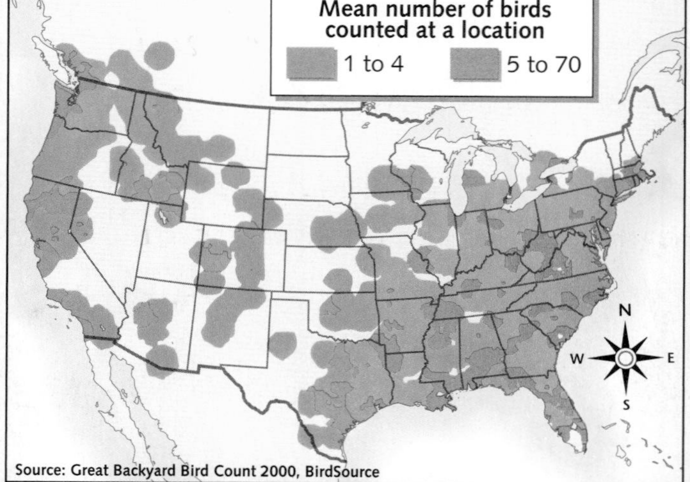

CHECK WHAT YOU KNOW

Use this page to help you review and remember important skills needed for Chapter 5.

VOCABULARY

Choose the best term from the box.

difference
parentheses
missing factor
missing addend

1. In $12 - 8 = n$, n represents the ? .
2. In $3 + n = 5$, n represents the ? .
3. In $6 + (8 - 2)$, the ? tell you which operation to do first.

ADDITION AND SUBTRACTION (See p. H6.)

Write the value of n.

4. $3 + 11 = n$
5. $9 + 13 = n$
6. $14 + 7 = n$
7. $9 + 8 = n$
8. $5 + 10 = n$
9. $16 + 8 = n$
10. $5 + 13 = n$
11. $12 + 6 = n$
12. $n = 6 + 8$
13. $18 - 9 = n$
14. $22 - 9 = n$
15. $23 - 10 = n$
16. $11 - 6 = n$
17. $8 - n = 5$
18. $10 - n = 4$
19. $n = 25 + 11$
20. $n = 17 + 13$
21. $n = 41 - 11$

ADDITION PROPERTIES (See p. H6.)

Write the letter of the addition property used in each equation.

A Order Property **B** Grouping Property **C** Zero Property

22. $9 + 8 = 8 + 9$
23. $24 + 0 = 24$
24. $(5 + 6) + 3 = 5 + (6 + 3)$
25. $0 + 12 = 12$
26. $23 + 9 = 9 + 23$
27. $7 + (1 + 4) = (7 + 1) + 4$

MENTAL MATH: FUNCTION TABLES (See p. H7.)

Copy and complete the table. Use mental math.

28.

Rule: Add 7.					
Input	4	8	12	16	20
Output	11	15	■	■	■

29.

Rule: Subtract 8.					
Input	16	21	26	31	36
Output	8	13	■	■	■

LESSON

1 Expressions and Variables

Learn

GO FISH An **expression** is a mathematical phrase that combines numbers, operation signs, and sometimes variables. It does *not* have an equal sign. These are some expressions:

$15 + 4 \qquad 9 - 7 \qquad (4 + 6) - 11 \qquad 3 \times 2$

You can write an expression and find its value.

Matt had 12 fish in his tank. Then he added 3 more fish. How many fish are in his tank now?

12 fish in tank	**plus**	**3 more fish**
↓	↓	↓
12	+	3

↓

15

So, Matt has 15 fish in the tank now.

An expression *may* have a variable. A **variable** is a letter or symbol that stands for a number. These are some expressions with a variable:

$4 + n \qquad f - 5 \qquad 6 \times a \qquad 3 + x - 2$

You can use a variable in an expression to model a situation.

Julie had 16 fish in her tank. She gave some fish to her sister. Write an expression to show this.

16 fish in tank	**minus**	**some fish**
↓	↓	↓
16	−	f

- What can you use to represent a number that you do not know in an expression?

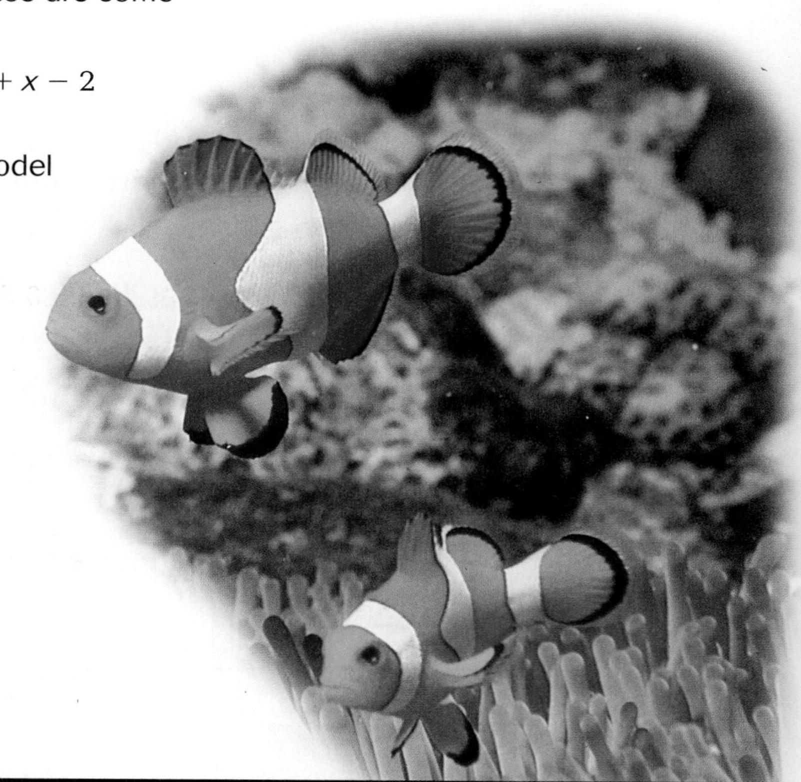

Quick Review

Write the words for each.

1. $3 + 5$
2. $6 - 2$
3. $10 + 2 + 8$
4. $(2 + 7) - 4$
5. $9 - (5 + 1)$

VOCABULARY

expression

variable

CALIFORNIA STANDARDS AF 1.0 Students use variables in simple expressions, compute the value of the expression for specific values of the variable, and plot and interpret the results. **🔑AF 1.2** Use a letter to represent an unknown number; write and evaluate simple algebraic expressions in one variable by substitution. *also* **MR 2.3, MR 2.4, MR 3.1, MR 3.2, MR 3.3**

Find the Value

MATH IDEA You can find the value of an expression by replacing the variable with a number and using that number to compute the result.

Examples

A Sara had 15 apples in a basket. Some were eaten.

Let a = the number of apples eaten.

apples in a basket		apples eaten
↓		↓
15	−	a

Expression: $15 - a$

If 10 apples were eaten, how many were left?

Let $a = 10$.

$15 - a$
↓ ↓
$15 - 10 = 5$

So, 5 apples were left.

B There are 11 people joining the swim team. Let p = people already on the swim team.

on swim team		joining swim team
↓		↓
p	+	11

Expression: $p + 11$

If 23 people were already on the team, how many people are on the swim team now?

Let $p = 23$.

$p + 11$
↓ ↓
$23 + 11 = 34$

So, there are 34 people on the swim team now.

C Tom had 18 baseball cards. He traded some of his cards for 3 of Tammy's cards.

Let c = number of cards Tom traded.

cards Tom had		cards Tom traded		cards Tom received
↓		↓		↓
18	−	c	+	3

Expression: $18 - c + 3$

If Tom traded 6 cards, how many cards does Tom have now?

Let $c = 6$.

$18 - c + 3$
↓ ↓ ↓
$18 - 6 + 3 = 15$

So, Tom has 15 baseball cards.

Check

1. **REASONING** **Explain** why 6 is an unreasonable value for n in this expression: Lyle has some fish. He buys 3 more fish and then sells 10 fish. Expression: $(n + 3) - 10$.

Write an expression. Find the value.

2. Sam had 10 dimes. He gave Gary 4.
3. eleven added to five
4. Kim had 24 cookies, ate 6, and then baked 12 more.

Write an expression with a variable. Explain what the variable represents.

5. Rona had 96 pencils and then sold some.
6. Tim had 17 cards. Joe gave him more.
7. Lee gave me 12 cookies. I ate 6 and then baked more.

LESSON CONTINUES

Practice and Problem Solving

Write an expression. Find the value.

8. Eleven cars were in the race. Seven more cars joined the race.

9. Len had 17 shells. He gave 5 shells to Joan.

10. The gardener grew 10 rosebushes. Then he sold 2 of them.

11. Sam had 23 books. Ann gave him 7 more books, and he gave 13 to Ellen.

12. There were 21 cups on the shelf. Three fell off, and six were added.

Write an expression with a variable. Explain what the variable represents.

13. Don had 37 toys but lost some.

14. Holly bought some books. Jon gave her 9 more.

15. I had 14 pretzels. I ate some and gave 3 to Mia.

16. Dina had 73 pens. She sold some.

17. Julio had some CDs. He gave 2 to Eric and bought 6 new CDs.

18. Allie had 26 beanbag toys. She gave 9 away and got more for her birthday.

Find the value of the expression.

19. $\$50 + n$ if n is \$20

20. $n + 3{,}600$ if n is 400

21. $n - 35$ if n is 100

22. $2{,}000 - n$ if n is 1,000

23. $\$10 + (\$4 - n)$ if n is \$2

24. $(22 + n) - 4$ if n is 8

For 25–26 choose the expression for each situation.

25. The temperature dropped 10 degrees and then went up 4 degrees.

A $(n + 10) + 4$
B $(n - 10) + 4$
C $(10 - n) + 4$
D $(n + 10) - 4$

26. At the bus stop, 5 of the 23 passengers got off the bus and other people got on.

F $(n + 23) + 5$
G $(n - 23) + 5$
H $(23 - n) + 5$
J $(23 - 5) + n$

Use the expression to complete each table.

27.

n	$n + 6$
4	10
5	11
6	■
7	■

28.

n	$n - 3$
8	5
12	■
16	■
20	17

29.

n	$20 - n$
5	■
8	12
10	■
15	■

30. Mary has tuna salad and lemonade for lunch. The tuna salad costs $\$2.29$. If n represents the cost of a lemonade, what expression represents the cost of Mary's lunch? If the lemonade costs $\$1.25$, how much does her lunch cost?

31. Sharon is 2 years younger than Mario. If a represents Mario's age, what expression represents Sharon's age? Mario is 12 years old. How old is Sharon?

TECHNOLOGY LINK

More Practice: Use Mighty Math Astro Algebra, Cargo Bay, *Dr. X Xpresses Herself,* Levels Red A and Green B.

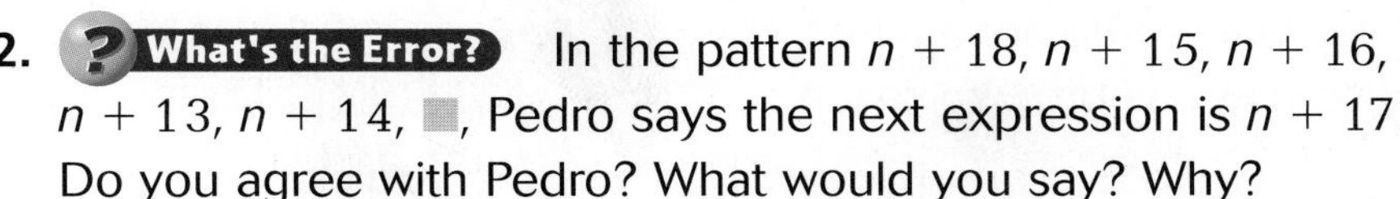

32. **What's the Error?** In the pattern $n + 18$, $n + 15$, $n + 16$, $n + 13$, $n + 14$, ■, Pedro says the next expression is $n + 17$. Do you agree with Pedro? What would you say? Why?

33. **MEASUREMENT** In 1932, the Indy 500 auto race was won at a speed of 104.144 mph. The following year, the winning speed was 0.018 mph faster. In 1934, the winning speed was 0.701 mph faster than in 1933. What was the winning speed in 1934?

34. **Write About It** Write an expression for this problem and explain what the variable represents: Chris had 7 trading cards. He traded some to Cindy, and she gave him 5 more.

Mixed Review and Test Prep

35. Use mental math to find the sum $30 + 40 + 500 + 200$. (p. 36)

36. Write an equivalent decimal for 9.09. (p. 22)

37. Write the greatest 7-digit whole number that is possible without repeating digits. (p. 4)

38. **TEST PREP** Which number is less than $100{,}000{,}000$? (p. 4)

A $3{,}000{,}000{,}000$ **C** $300{,}000{,}000$
B $1{,}000{,}000{,}000$ **D** $30{,}000{,}000$

39. **TEST PREP** Which number is between 0.56 and 0.57? (p. 20)

F 0.55 **H** 0.565
G 0.555 **J** 0.575

Thinker's Corner

JUMPING TO CONCLUSIONS Sam found a mysterious document. He was able to identify some of the numbers and two of the operation signs. He also concluded that [fish] and [octopus] are two different operations and that [starfish], [turtle], and [seahorse] are three different numbers from 1–10.

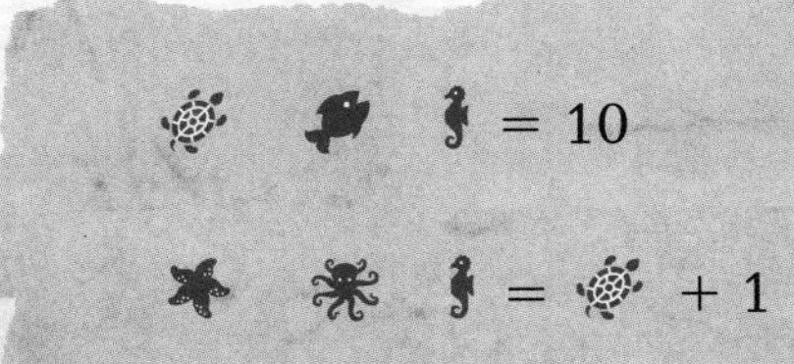

1. Work with a partner to find the numbers and the operations.

2. Use what you found to make your own puzzle with your partner.

Extra Practice page H36, Set A

LESSON 2

Write Equations

Quick Review

Write each in words.

1. $2 + 5 = 7$
2. $8 - 3 = 5$
3. $11 + 2 = 13$
4. $12 = 21 - 9$
5. $6 + 2 + 5 = 13$

VOCABULARY

equation

Learn

HISSSS! The reticulated python is the longest snake in the world. At birth it is 2 feet long, and some adults are 29 feet long. How much do these pythons grow to reach adult length?

Write an equation with a variable to model the problem. An **equation** is a number sentence that uses the equal sign to show that two amounts are equal. Use the variable *f* for the number of feet the python grows.

number of feet at birth	+	number of feet of growth	=	number of feet as an adult
↓		↓		↓
2	+	*f*	=	29

So, the equation is $2 + f = 29$.

▲ **The largest python is the reticulated python. One found in Indonesia in 1912 was 33 feet long.**

Examples

A Tate had 12 CDs and got more for her birthday. Now she has 21 CDs. How many CDs did Tate get?

12 CDs	+	birthday CDs	=	21 CDs
↓		↓		↓
12	+	*c*	=	21

B After Paolo paid $7 for a movie, he had $5. How much money did Paolo have to start with?

money to start with	–	$7	=	$5
↓		↓		↓
m	–	7	=	5

MATH IDEA You can write an equation to represent two amounts that are equal.

Check

1. **Tell** how an equation is different from an expression.

Write an equation. Explain what the variable represents.

2. Erica had 24 crayons. After she put some on the table, she had 12 left in the box. How many crayons did she put on the table?
3. Jim had some pencils. After Lee gave him 5 new pencils, he had 12. How many pencils did Jim have to start with?

CALIFORNIA STANDARDS AF 1.1 Use information taken from a graph or equation to answer questions about a problem situation. **O━ AF 1.2** Use a letter to represent an unknown number: write and evaluate simple algebraic expressions in one variable by substitution. *also* **AF 1.0, MR 1.0, MR 1.1, MR 2.3, MR 2.4, MR 3.0**

Practice and Problem Solving

Write an equation. Explain what the variable represents.

4. Ashley has 16 beads. She needs 24 beads to make a bracelet. How many more beads does she need?

5. A cheetah runs 70 mph, and a lion runs 50 mph. What is the difference in their speeds?

6. Bob paid \$3 for breakfast and \$5 for lunch and then bought dinner. He spent \$20 in all. How much did he pay for dinner?

7. After a number is subtracted from 22 and 5 is added, the result is 10. What number is subtracted?

8. After the temperature rose 15 degrees, it was 95 degrees outside. What was the temperature to start with?

9. After 7 people quit the tennis team, there were 30 players. How many players were on the team to start with?

Write a problem for the equation. State what the variable *n* represents.

10. $31 - n = 20$

11. $n + 5 = 13$

12. $(6 + n) - 4 = 17$

USE DATA For 13–15, use the bar graph to write an equation with a variable.

13. What is the difference between the length of an anaconda and the length of a cobra?

14. When a rattlesnake and another snake are laid end to end, their total length is 17 feet. How long is the other snake?

15. An anaconda grows 25 feet to reach its adult length. How long are anacondas at birth?

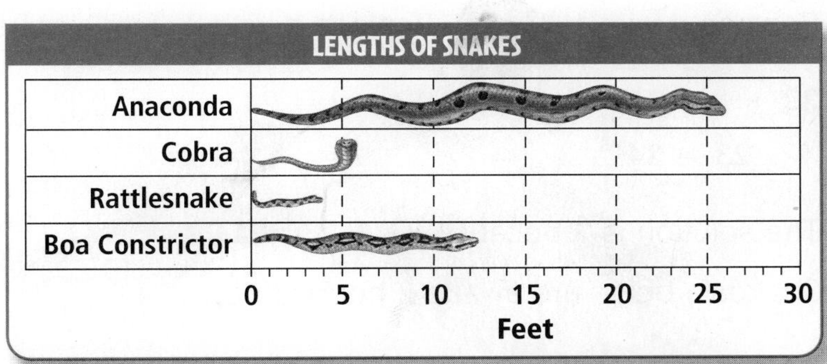

16. **Geometry** A square's side is *n* feet long, and its perimeter is 20 feet. Write an equation for the perimeter of this square.

17. Tran arrived at the mall at 1:15. She spent 45 minutes in the gift shop, 35 minutes in the shoe store, and 1 hour and 10 minutes in the music store. When did she leave the mall?

Mixed Review and Test Prep

18. Write an expression for this sentence: The temperature rose 5 degrees. (p. 68)

19. 7×4

20. 6×9

21. Use mental math to find the sum $6 + 13 + 7 + 24$. (p. 36)

22. **TEST PREP** In which number does the digit 6 have the greatest value? (p. 4)

A 56,953
B 2,560,333
C 36,531,228
D 4,523,635,200

LESSON 3

Solve Equations

Learn

ZZZZZ . . . Koala bears sleep most of the day because of their low-energy diet of leaves. In fact, of the 24 hours in a day, koalas sleep 22 hours. How many hours a day are koala bears awake?

Write an equation to model this situation.

hours asleep	+	**hours awake**	=	**hours in a day**
↓		↓		↓
22	+	h	=	24

If an equation contains a variable, you solve the equation by finding a value for the variable that makes the equation true. That value is the **solution**.

Which of the numbers 1, 2, or 5 is the solution of the equation?

Replace n with 1.	Replace n with 2.	Replace n with 5.
$22 + 1 \stackrel{?}{=} 24$	$22 + 2 \stackrel{?}{=} 24$	$22 + 5 \stackrel{?}{=} 24$
$23 = 24$ false	$24 = 24$ true	$27 = 24$ false

The solution is 2 because $22 + 2 = 24$.

So, koala bears are awake 2 hours a day.

Examples

A Is 4, 5, or 7 the solution of $3 + n = 10$?

$3 + 4 = 10$ Replace n with 4.
$7 = 10$ false

$3 + 5 = 10$ Replace n with 5.
$8 = 10$ false

$3 + 7 = 10$ Replace n with 7.
$10 = 10$ true

So, $n = 7$.

B Is 10, 11, or 13 the solution of $n - 6 = 7$?

$10 - 6 = 7$ Replace n with 10.
$4 = 7$ false

$11 - 6 = 7$ Replace n with 11.
$5 = 7$ false

$13 - 6 = 7$ Replace n with 13.
$7 = 7$ true

So, $n = 13$.

Quick Review

Find the value of each expression if n equals 5.

1. $n + 7$
2. $2 \times n$
3. $25 \div n$
4. $n - 2$
5. $(11 + n) - 12$

VOCABULARY

solution

Koala bears eat about 2.5 pounds of eucalyptus leaves each day, which is almost one tenth of their total body weight. ▼

CALIFORNIA STANDARDS AF 1.1 Use information taken from a graph or equation to answer questions about a problem situation. O–– AF 1.2 Use a letter to represent an unknown number: write and evaluate simple algebraic expressions in one variable by substitution. *also* AF 1.0, MR 1.0, MR 1.1, MR 2.4, MR 3.2, MR 3.3

Mental Math

Jason and his family traveled in Australia for 2 weeks. In his journal, Jason kept a count of all the koalas he saw. In the first week, he saw 13 koalas. During the entire trip, he counted 22 koalas. How many koalas did Jason see in the second week of the trip?

Write an equation to solve the situation.

koalas seen first week	+	**koalas seen second week**	=	**total koalas seen**
↓		↓		↓
13	+	k	=	22

You can solve the equation using mental math.

$13 + k = 22$ **Think:** 13 plus what number equals 22?

$k = 9$

Check: $13 + 9 = 22$ Replace k with 9.

$22 = 22$ √ The equation checks. The value of k is 9.

Examples

A $5 + n = 12$ **Think:** 5 plus what number equals 12?

$n = 7$

Check: $5 + 7 = 12$

$12 = 12$ √

B $7 = 17 - n$ **Think:** 17 minus what number equals 7?

$10 = n$

Check: $7 = 17 - 10$

$7 = 7$ √

MATH IDEA The value of the variable that makes the equation true is the solution of the equation.

Check

1. **Tell** how you know when the value of a variable in an equation is correct?

Which of the numbers 6, 8, or 11 is the solution of the equation?

2. $9 + n = 17$
3. $n - 2 = 9$
4. $n + 7 = 13$
5. $11 - n = 3$

Which of the numbers 3, 5, or 15 is the solution of the equation?

6. $n + 10 = 15$
7. $4 + n = 19$
8. $21 = n + 18$
9. $16 - n = 11$

Use mental math to solve each equation. Check your solution.

10. $8 + n = 15$
11. $25 - n = 5$
12. $\$60 - n = \$52$
13. $n + 10 = 72$
14. $8 = n + 4$
15. $n - \$7 = \$30$
16. $36 = n - 10$
17. $n + 6 = 13$

LESSON CONTINUES

Practice and Problem Solving

TECHNOLOGY LINK

More Practice: Use E-Lab, ***Modeling Addition and Subtraction Equations.***

www.harcourtschool.com/elab2002

Write which of the numbers 5, 7, or 12 is the solution of the equation.

18. $4 + n = 11$ **19.** $n + 6 = 18$ **20.** $53 - n = 48$

21. $2 = n - 10$ **22.** $n = 2 + 5$ **23.** $45 - 38 = n$

24. $16 = a + 9$ **25.** $b = 23 - 18$ **26.** $37 - 25 = c$

Use mental math to solve each equation. Check your solution.

27. $11 = n - 5$ **28.** $n - 3 = 20$ **29.** $n + 9 = 26$ **30.** $32 = 10 + n$

31. $n + 8 = 14$ **32.** $15 = n - 9$ **33.** $n + 35 = 48$ **34.** $(n - 24) + 3 = 7$

Solve the equation. Check your solution.

35. $25 + n = 31$ **36.** $n - 17 = 8$ **37.** $26 - n = 13$ **38.** $7 + (2 + n) = 18$

39. $30 = 17 + a$ **40.** $b = 12 - 9$ **41.** $51 - 18 = c$ **42.** $d = 10 + 43$

43. $x + 9 = 26$ **44.** $29 - y = 7$ **45.** $72 + 54 = z$ **46.** $44 = 52 - n$

REASONING For 47–50, each symbol represents one number. Find the value of each symbol.

47. ❤ + 3 = 9; 4 + ★ = ❤

48. ▲ + 6 = 10; ▲ − ♣ = 1

49. 12 − ♦ = 7; ♦ + ■ = 6

50. 5 = 12 − ●; ● + ▶ = 15

USE DATA For 51–54, use the bar graph to write the equation. Then solve.

51. How many hours a day is a cat awake?

52. Which animal sleeps 4 times as many hours as the horse?

53. How many more hours a day does a lion sleep than a cow?

54. Which animal sleeps 5 hours more per day than a horse?

55. **What's the Question?** Bob has 9 pets: 2 dogs, 4 cats, and some birds. Tim uses the equation $6 + n = 9$. The answer is 3 birds.

56. **REASONING** Jamaal and Pete are waiters. In tips, they have 113 one-dollar bills, 13 ten-dollar bills, and 9 one-dollar bills. How can they split the tips so that each one has the same amount of money?

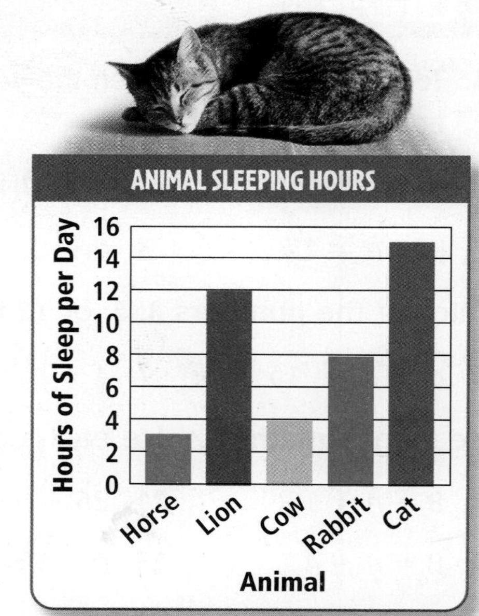

Mixed Review and Test Prep

57. Write an expression for this situation: Tom has seven more CDs than Rick. (p. 68)

58. TEST PREP Which is 1.0395 rounded to the nearest hundredth? (p. 46)

A 1.03 **C** 1.04
B 1.039 **D** 1.14

59. $2.08 + 0.9$ (p. 50) **60.** $1.07 - 0.59$ (p. 50)

61. TEST PREP Maria is 16. She is 2 years older than Joan. How old is Joan? (p. 69)

F 14 yr old **H** 16 yr old
G 15 yr old **J** 18 yr old

LINKUP to Reading

STRATEGY • CHOOSE RELEVANT INFORMATION
Sometimes you must decide which information in a word problem is relevant, or needed to solve the problem.

At the veterinary clinic, Ben fed 14 dogs, 3 birds, and some cats. One of the cats was recovering from surgery and couldn't eat yet. Three of the dogs were getting routine shots. There were 23 animals at the clinic. How many cats did Ben feed?

Read each fact and decide whether it is *relevant* or *not relevant* to solving the problem.

- Ben fed the animals at the clinic. relevant
- There were 14 dogs, 3 birds, and some cats. relevant
- One cat couldn't eat. relevant
- Three dogs were getting routine shots. not relevant
- There were 23 animals at the clinic. relevant

Solve the problem.
Find the number of cats.

$14 + 3 + n = 23$
$n = 6$ Think of the missing addend. There were 6 cats.

$6 - 1 = 5$ So, Ben fed 5 cats.

Solve. Write any information that is *not* relevant.

1. Lori got 10 baseball cards from Bill and gave Lyn 8 cards. After she bought 7 new cards and repaired 2 old ones, she had 30 cards. How many cards did Lori have to start with?

2. John was putting away books from a cart. He put 5 books on the science shelf and 8 on the mystery shelf. He put away 3 fiction books from a table. He had 4 books left on the cart. How many books did John have to start with?

Extra Practice page H36, Set C

LESSON

Use Addition Properties

Learn

FAIR PRICE? Claire and her brother Tom went to the fair. Claire paid \$5 for a hat. For lunch she bought a drink for \$1 and a sandwich for \$4. Tom bought a drink for \$1 and a salad. He also paid \$4 for a video game. They both spent the same amount of money. How much was Tom's salad?

You can write an equation to model this problem.
Let s = the amount Tom spent on a salad.

hat + **lunch** = **lunch** + **game**

$5 + (1 + 4) = (s + 1) + 4$

Solve for s.

$5 + (1 + 4) = (s + 1) + 4$ Think about the Grouping Property. s must equal 5.

$5 + (1 + 4) = (5 + 1) + 4$

So, Tom spent \$5 for his salad.

The Order Property is also called the **Commutative Property**. The Grouping Property is also called the **Associative Property**.

MATH IDEA You can use the properties of addition to help you solve problems.

Examples Find the value of n.

A $(12 + 8) + 3 = 12 + (n + 3)$

Using the Associative Property, n must be 8.

B $17 + 4 = n + 17$

Using the Commutative Property, n must be 4.

C $n + 0 = 6$

Using the Zero Property, n must be 6.

- How are the Commutative and the Associative Properties different?

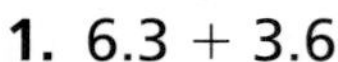

Quick Review

1. $6.3 + 3.6$
2. $1.05 - 0.01$
3. $950 + 2,050$
4. $(23 + 5) - 14$
5. $(6 + 8) + (4 - 1)$

VOCABULARY

Commutative Property

Associative Property

Remember

You can use mental math and the properties of addition to solve problems.

Order Property

$12 + 8 = 8 + 12$

Grouping Property

$(5 + 6) + 7 = 5 + (6 + 7)$

Zero Property

$37 + 0 = 37$

CALIFORNIA STANDARDS **AF 1.0** Students use variables in simple expressions, compute the value of the expression for specific values of the variable, and plot and interpret the results. **AF 1.2** Use a letter to represent an unknown number: write and evaluate simple algebraic expressions in one variable by substitution. *also* **AF 1.1, MR 1.1, MR 2.2, MR 3.0, MR 3.2**

Check

1. **Tell** which property says that $4 + 6$ and $6 + 4$ have the same sum.

Name the property used in each equation.

2. $12 + 3 = 3 + 12$
3. $3 + (5 + 7) = (3 + 5) + 7$
4. $5.9 + 0 = 5.9$

Find the value of *n*. Identify the addition property used.

5. $n + 15 = 15 + 3$
6. $(16 + 5) + 4 = 16 + (n + 4)$
7. $n + 1.5 = 1.5$

Practice and Problem Solving

Name the addition property used in each equation.

8. $239 + 0 = 239$
9. $0.5 + 6 = 6 + 0.5$
10. $(8 + 2) + 5 = 8 + (2 + 5)$

Find the value of *n*. Identify the addition property used.

11. $7 + 18 = n + 7$
12. $256 + (15 + n) = (256 + 15) + 2$
13. $0 + n = 5{,}681$
14. $26 + (n + 7) = (26 + 3) + 7$
15. $10 + (15 + 5) = (10 + n) + 5$
16. $4.7 + n = 2 + 4.7$

Algebra **Name the addition property used in each equation.**

17. $(a + b) + c = a + (b + c)$
18. $a + 0 = a$
19. $a + b = b + a$
20. ▲ + (◯ + ★) = (▲ + ◯) + ★
21. **LOGIC** Erica sold more subscriptions than Chris but fewer than Bob. Candy sold the fewest subscriptions. Who sold the most?
22. Sam and Joe went to the fair. Sam bought ride tickets for \$6 and some peanuts. Joe paid \$3 for popcorn and \$6 for tickets. They both spent the same amount. How much did Sam spend on peanuts?
23. **REASONING** Does the Commutative Property work for subtraction? Explain.

Mixed Review and Test Prep

24. How many groups of 100,000 are in one billion? (p. 5)
25. $23.908 + 6.07$ (p. 50)
26. Order the numbers 50,555; 55,505; and 50,005 from least to greatest. (p. 11)
27. Round 1.376 to the nearest tenth. (p. 46)
28. **TEST PREP** Which decimal is greater than eight tenths but less than 1? (p. 24)
    - **A** 0.798
    - **B** 0.819
    - **C** 0.098
    - **D** 1.978

Extra Practice page H36, Set D

LESSON 5

Problem Solving Skill
Use a Formula

Quick Review

Find the value if $n = 8$.

1. $n + 5$
2. $14 - n$
3. $n + n + n$
4. $(n + 7) + 10$
5. $4 + n + 6 + n$

BOW WOW BORDER Julio is building a pen in the backyard for his dog. He made a drawing of the pen and measured each side. How can Julio find the number of feet of fencing to buy?

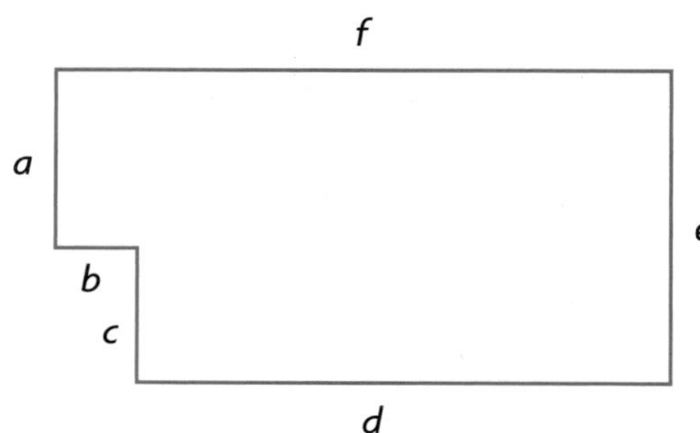

Julio's Measurements

Side a = 6.5 feet
Side b = 3 feet
Side c = 4.5 feet
Side d = 15.5 feet
Side e = 11 feet
Side f = 18.5 feet

You can use a formula to find the perimeter of the pen. The formula will tell you how many feet of fencing Julio needs to buy. Let a to f represent the lengths of the sides.

$$P = a + b + c + d + e + f$$

$$\downarrow \quad \downarrow \quad \downarrow \quad \downarrow \quad \downarrow \quad \downarrow$$

$$P = 6.5 + 3 + 4.5 + 15.5 + 11 + 18.5$$

$$P = 59$$

So, Julio needs to buy 59 feet of fencing.

Examples

A

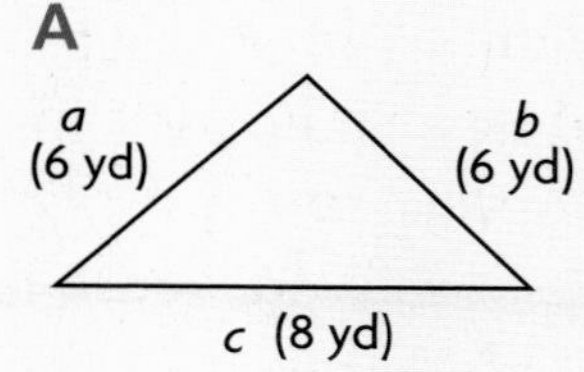

Find the perimeter.
$P = a + b + c$
$P = 6 + 6 + 8$
$P = 20$

The perimeter is 20 yd.

B

d
c (6.5 ft)
a (7.5 ft)
b (7 ft)

Find side d if the perimeter is 26.5 ft.
$P = a + b + c + d$
$26.5 = 7.5 + 7 + 6.5 + d$
$d = 5.5$

Side d is 5.5 ft.

Talk About It

- Is a formula an equation? Explain.

CALIFORNIA STANDARDS **AF 1.2** Use a letter to represent an unknown number: write and evaluate simple algebraic expressions in one variable by substitution. **MR 2.3** Use a variety of methods, such as words, numbers, symbols, charts, graphs, tables, diagrams, and models, to explain mathematical reasoning. *also* **AF 1.0, AF 1.1, MR 1.0, MR 2.0, MR 2.3, MR 3.2, MR 3.3**

Problem Solving Practice

Use a formula to solve.

1. Mathew is making a garden that is rectangular. The length of the garden is 12 feet and the width is 10 feet. How much fencing will Mathew need to enclose the garden?

2. Elena runs around the park to train for the marathon. She runs 2 miles east, 2.5 miles south, 2 miles west, and then she runs north back to where she began. The perimeter of the park is 9 miles. How far north does she run?

3. Which shows how to find the amount of fencing John needs for his backyard?

   40 ft
   30 ft

   A $P = 120 \text{ ft} \div 40 \text{ ft}$
   B $A = 30 \text{ ft} \times 40 \text{ ft}$
   C $P = 40 \text{ ft} + 30 \text{ ft} + 30 \text{ ft} + 40 \text{ ft}$
   D $P = 40 \text{ ft} + 30 \text{ ft} - 50 \text{ ft}$

4. Which equation might you use to find the length of a missing side of a pentagon?

   F $200 = a + 60 + 40 + 60$
   G $125 = a + 30 + 20 + 30 + 20$
   H $6 \times s = 150$
   J $70 = 16 + 12 + 47$

Mixed Applications

5. Use the digits 4, 7, 9, and 0 to write a number less than 10 with 0 in the hundredths place, 4 in the thousandths place, and 9 in the tenths place.

6. Kara's family drove 325 miles on Monday and completed the 500-mile trip on Tuesday. How far did they drive on Tuesday? Write an equation and solve.

7. **REASONING** Andie, Cassie, and Emma play on the basketball team. In one year, Andie scored 236 points. Cassie scored 28 fewer points. Emma scored 12 points more than Cassie. How many points did Emma score?

8. Classes at the middle school begin at 8:30 A.M. Each of 6 class periods is 50 minutes long, lunch period is 25 minutes, and there is a 7-minute break between periods. At what time is school dismissed?

9. **Write a problem** about a garden and the amount of fencing needed to enclose it. Use a formula to solve.

10. Sammy and his family drove for 3 days. They drove twice as far on Tuesday as they did on Monday. On Wednesday they drove 100 miles more than on Tuesday. They drove 218 miles on Monday. How many miles in all did they drive?

Review/Test

CHECK VOCABULARY AND CONCEPTS

Choose the best term from the box.

equation
equal sign
expression
variable

1. A part of a number sentence that combines numbers and operation signs is a(n) _?_. (p. 68)
2. A number sentence that uses the equal sign to show that two amounts are equal is a(n) _?_. (p. 72)
3. A letter or a symbol that stands for a number is called a(n) _?_. (p. 68)

CHECK SKILLS

Write an expression and explain what the variable represents. (pp. 68–71)

4. The school has 16 computers. More will be added next year.
5. There were 53 passengers on the train. Then some got off.

Find the value of the expression. (pp. 68–71)

6. $17 - n$ if n is 5
7. $n + 92$ if n is 0
8. $13 + n + 4$ if n is 10

Write an equation. Explain what the variable represents. (pp. 72–73)

9. Tony had 17 pencils. After he gave some to Marla, he had 9. How many pencils did he give to Marla?
10. Julie has 12 CDs. After buying new ones, she had 18 CDs. How many did she buy?

Use mental math to solve. Check your solution. (pp. 74–77)

11. $40 = n + 31$
12. $(n + 1) - 4 = 7$
13. $17 + 5 = 5 + n$
14. $4 = n - 18$
15. $(n - 13) + 4 = 5$
16. $18 - n = 9 + 3$

Find the value of *n*. Identify the addition property used. (pp. 78–79)

17. $n + 12 = 12 + 7$
18. $(n + 6) + 3 = 4 + (6 + 3)$

CHECK PROBLEM SOLVING

Solve. (pp. 80–81)

19. Melissa is making a rectangular garden. The length of the garden is 18 feet, and the width is 12 feet. How much fencing will Melissa need to enclose the garden?
20. Richy knows that the sidewalk around a four-sided yard is 180 feet. Three of the sides are 45 feet, 35 feet, and 60 feet. What is the length of the fourth side?

Cumulative Review

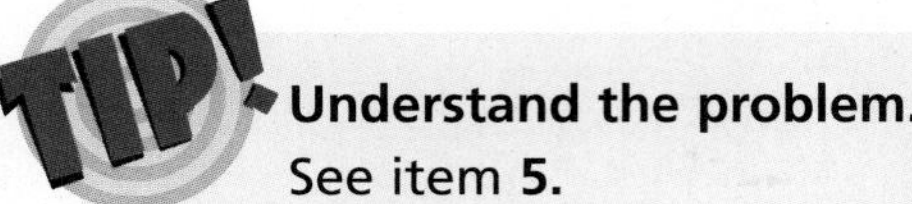

Understand the problem.
See item **5.**

The equation tells you that $n - 5$ equals $7 + 12$, which is 19. Since 5 is subtracted from n, you need to find a number that is 5 more than 19.

Also see problem **1,** p. H62.

For 1–9, choose the best answer.

1. Which situation is modeled by the equation $10 - a = 6$?

A Some number minus ten is six.
B Ten minus six is some number.
C Ten minus some number is six.
D Six minus some number is ten.

---

2. Which expression illustrates the Commutative Property of Addition?

F $16 + 7 = 16 + 7 + 0$
G $16 + 7 = 7 + 16$
H $16 + 0 = 16$
J $(16 + 7) + 2 = 16 + (7 + 2)$

---

3. What is the value of $21 - (6 + n)$ if $n = 10$?

A 5 **C** 15
B 11 **D** 16

---

4. Which is *not* equivalent to 75.1?

F 75.10 **H** $65.1 + 10.0$
G $76.0 - 0.9$ **J** $75.0 + 0.01$

---

5. What is the value of n for $n - 5 = 7 + 12$?

A 24 **C** 14
B 19 **D** Not here

---

6. At the station, 86 people got off the train and 35 stayed on. Which equation can be used to find p, the number of people on the train before some got off?

F $86 - 35 = p$ **H** $p - 86 = 35$
G $86 - p = 35$ **J** $35 + p = 86$

For 7–8, use the bar graph.

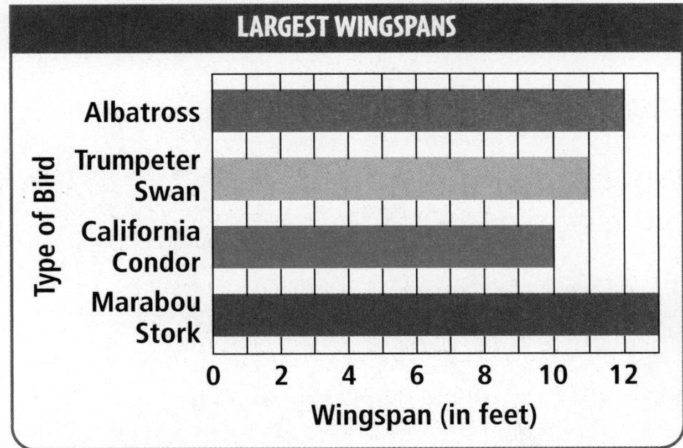

7. Which equation can be used to find w, the difference in the wingspans of the albatross and the trumpeter swan?

A $w - 12 = 11$ **C** $12 - w = 11$
B $12 - 11 = w$ **D** $11 - w = 12$

8. Which bird's wingspan is 3 feet longer than the California condor's?

F Marabou stork
G Trumpeter swan
H Albatross
J Not here

---

9. A small soda costs 60¢ and a large one costs 90¢. Which expression could be used to find the total cost of s small sodas and l large sodas?

A $(60 + 90) \times (s + l)$
B $60l + 90s$
C $(60 + s) \times (90 + l)$
D $60s + 90l$

Chapter 6 Algebra: Use Multiplication

Domino toppling is an activity in which dominoes are set up so that a pattern is formed when they fall over. Below is a table of three famous domino designs. The National Engineers' Week setup used *n* times the number of dominoes used in the Shadowbox setup. Write and solve an equation to find *n*.

Domino design for the Children's Museum

DOMINO TOPPLE SETUPS

Domino Design	Children's Museum	Shadowbox	National Engineers' Week
Number of Dominoes	1,900	3,000	30,000

CHECK WHAT YOU KNOW

Use this page to help you review and remember important skills needed for Chapter 6.

VOCABULARY

Choose the best term from the box.

expression
factor
Order Property
product

1. A mathematical phrase that combines numbers, operation signs, and sometimes variables is called an __?__.
2. The answer to a multiplication problem is the __?__.
3. The property that states two numbers can be multiplied in any order and the product remains the same is the __?__.

MULTIPLICATION FACTS THROUGH 12 (See p. H7.)

Find the value of *n*.

4. $3 \times 12 = n$
5. $2 \times 6 = n$
6. $8 \times 5 = n$
7. $n = 4 \times 10$
8. $6 \times 7 = n$
9. $n = 9 \times 11$
10. $n = 5 \times 9$
11. $n = 7 \times 8$
12. $10 \times 12 = n$

USE PARENTHESES (See p. H8.)

Choose the expression that shows the given value.

13. 32
    a. $2 + (10 \times 3)$
    b. $(2 + 10) \times 3$
14. 0
    a. $15 - (3 \times 5)$
    b. $(15 - 3) \times 5$
15. 24
    a. $4 + (8 \times 2)$
    b. $(4 + 8) \times 2$
16. 50
    a. $6 \times (9 - 4)$
    b. $(6 \times 9) - 4$
17. 34
    a. $3 \times (11 + 1)$
    b. $(3 \times 11) + 1$
18. 12
    a. $12 + (0 \times 4)$
    b. $(12 + 0) \times 4$

MULTIPLICATION PROPERTIES (See p. H8.)

Write the letter of the multiplication property used in each equation.

a. Order Property
b. Grouping Property
c. Property of One
d. Zero Property

19. $8 \times 5 = 5 \times 8$
20. $(4 \times 2) \times 5 = 4 \times (2 \times 5)$
21. $(8 \times 2) \times 5 = 8 \times (2 \times 5)$
22. $25 \times 1 = 25$
23. $78 \times 0 = 0$
24. $4 \times 11 = 11 \times 4$

LESSON 1

Write and Evaluate Expressions

Quick Review

1. $100 - 1$
2. $25 + 10$
3. 46×10
4. $200 \div 10$
5. $200 - 19$

VOCABULARY

evaluate

Learn

ROW BY ROW Kirstin has 4 rows of dominoes. She has the same number of dominoes in each row.

You can write an expression to model the total number of dominoes even if the number in each row is unknown.

Let d = number of dominoes in each row.

4 rows	**times**	**number in each row**
↓	↓	↓
4	×	d

You can also write $4 \times d$ as $4 \cdot d$, $4(d)$, or $4d$.

MATH IDEA To **evaluate** an expression with a variable, replace the variable with a number. Then perform the operations to find the value of the expression for that number.

If there are 6 dominoes in each row, how many dominoes does Kirstin have altogether? What if there are 9 dominoes in each row?

▲ Ralf Laue stacked 529 dominoes on one supporting domino.

Find $4 \times d$ if $d = 6$.

4×6 ← Replace d with 6.

↓

24 So, with 6 dominoes in each row, Kirstin has 24 dominoes.

Find $4 \times d$ if $d = 9$.

4×9 ← Replace d with 9.

↓

36 So, with 9 dominoes in each row, Kirstin has 36 dominoes.

Write an expression. Then evaluate for the given value.

STEP 1

Ana bought some pencils and a pen. The pencils cost 12¢ each and the pen cost 50¢.

Let p = the number of pencils.

cost of pencils	**plus**	**cost of pen**
↓	↓	↓
$12p$	+	50

Expression: $12p + 50$

STEP 2

If she bought a total of 4 pencils and a pen, how much did she spend altogether?

Find $12p + 50$ if $p = 4$.

12×4	+	50
↓	↓	↓
48	+	50
	↓	
	98	

So, Ana spent 98¢ altogether.

- **What if** Ana bought a total of 3 pencils and a pen, how much did she spend altogether?

CALIFORNIA STANDARDS O━ **AF 1.2** Use a letter to represent an unknown number; write and evaluate simple algebraic expressions in one variable by substitution. **AF 1.0** Students use variables in simple expressions, compute the value of the expression for specific values of the variable, and plot and interpret the results. *also* **MR 2.0, MR 2.3, MR 2.4, MR 3.0, MR 3.2**

Check

1. **Explain** how you evaluate the expression $(8 \times n) - 5$ if $n = 3$.

Write an expression. Tell what each variable represents.

2. Naomi walked 2 miles on Monday and 3 miles today.

3. Jaime read the same number of books each week for 4 weeks.

Evaluate each expression if $n = 10, 11,$ and 12.

4. $4n$

5. $(n + 3) \times 4$

6. $2 \times n \times 5$

Practice and Problem Solving

Write an expression. Tell what each variable represents.

7. Lin placed 5 magazines on each shelf. There were several shelves.

8. Jason had 18 cars. He gave the same number of cars to each of 4 friends.

Copy and complete the table.

9.

n	$21 \times n$
2	■
3	■
4	■

10.

y	$y + 28$
12	■
14	■
20	■

11.

n	$9n - 6$
6	■
7	■
8	■

12.

y	$(y - 1) \times 2$
8	■
10	■
12	■

Algebra **Let $n = 6$. Write <, >, or = for each ●.**

13. $4 \times n$ ● $4 + n$

14. $2 \times n$ ● $10 + 6 + n$

15. $20 \times n$ ● $5 \times 4 \times n$

16. **What's the Question?** Don drove 3 hours at n miles per hour. The answer is $3 \times n$.

17. Find the value of y that makes the expression $y \times 7$ equal to 49.

18. **Write About It** Write one or two sentences that can be modeled by the expression $(4 \times n) + 5$ and one or two that can be modeled by $4 \times (n + 5)$. Evaluate the expressions if $n = 2$.

19. After toppling 15,000,000 dominoes, only 117,889 were left standing. How many dominoes were not standing?

Mixed Review and Test Prep

20. What is the value of the blue digit in 57,894,072? (p. 4)

21. $3.1 + 1.8 + 4.4$ (p. 50)

22. Write 66.066 in word form. (p. 20)

23. $456{,}039 + 28{,}403$ (p. 38)

24. **TEST PREP** Which digits make 52■,308 < 524,803 true? (p. 8)

    A 1, 2, 3, 4, or 5 **C** 3, 4, 5, 6, or 7

    B 0, 1, 2, 3, or 4 **D** 2, 3, 4, 5, or 6

Extra Practice page H37, Set A

LESSON 2

Problem Solving Strategy

Write an Equation

Find the value of the expression if $y = 12$.

1. $12 + y$
2. $64 - y$
3. $y + 36$
4. $4 + y$
5. $y \times 7$

PROBLEM The largest flying flag in the world is 7,410 square feet and weighs 180 pounds. The flag is 65 feet high. There are a total of 13 stripes. What is the height of each stripe?

Remember

In an *equation,* the values on both sides of the equal sign are the same.
Example:

$$10 + 5 = 15$$
$$\downarrow \quad = \quad \downarrow$$
$$15 = 15$$

- What are you asked to find?
- What information will you use?
- Is there information you will not use? If so, what?

- What strategy can you use to solve the problem?

You can *write and solve an equation* to find the height of each stripe.

- How can you use the strategy to solve the problem?

Write an equation, and then use mental math to solve the equation.

Let h = the height of each stripe.

height of flag	=	number of stripes	×	height of stripe
↓	↓	↓	↓	↓
65	=	13	×	h

$5 = h$ **Think:** 13 times what number equals 65?

So, each stripe is 5 feet high.

- How can you decide if your answer is correct?
- What other strategy could you use?

▲ **The largest flying United States flag, in Gastonia, North Carolina, takes at least 6 people to raise and lower it.**

CALIFORNIA STANDARDS MR 2.0 Students use strategies, skills, and concepts in finding solutions. **AF 1.2** Use a letter to represent an unknown number; write and evaluate simple algebraic expressions in one variable by substitution. *also* **AF 1.0, AF 1.1, MR 1.0, MR 1.1, MR 2.6, MR 3.1, MR 3.2**

Problem Solving Practice

Write and solve an equation for each problem. Explain what the variable represents.

1. **What if** the flag is 130 feet high? What is the height of each stripe? **What if** the flag is 143 feet high?
2. The 50 state flags are in rows at the park entrance. There are an equal number of flags in each row. If the total number of flags is 5 times the number of flags in one row, how many flags are in one row?

PROBLEM SOLVING STRATEGIES

- Draw a Diagram or Picture
- Make a Model or Act It Out
- Make an Organized List
- Find a Pattern
- Make a Table or Graph
- Predict and Test
- Work Backward
- Solve a Simpler Problem
- Write an Equation
- Use Logical Reasoning

USE DATA For 3–4, the table indicates the number of words each student spelled correctly on a spelling test.

SPELLING TEST			
Erin	**Jacob**	**Theresa**	**Tom**
19	18	15	16

3. Jacob received a 90 on his test. Which equation can you use to find the number of points each correct spelling word is worth?
   A $18 \times n = 90$ **C** $18 + n = 90$
   B $18 - n = 90$ **D** $90 - n = 18$
4. Which equation uses n to represent how many more words Erin spelled correctly than Theresa?
   F $n = 19 + 15$ **H** $15 = 19 + n$
   G $15 + n = 19$ **J** $19 = n - 15$

Mixed Strategy Practice

5. Goro arrived at Kennedy Space Center at 1:15 P.M. He spent 25 minutes in The Space Shop, 20 minutes in the Launch Status Center, and 35 minutes in the Rocket Garden before he left. What time did he leave the Kennedy Space Center?
6. Bob's car gets 20 miles to the gallon. He has about 4 gallons left in the tank. Does he have enough gas to drive 125 more miles? Explain.
7. **REASONING** Two numbers have a difference of 4 and a sum of 34. What are the two numbers?

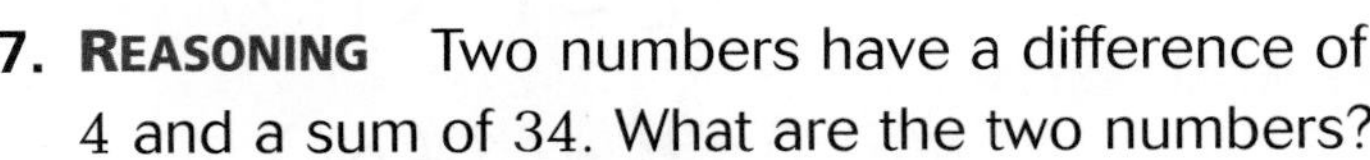

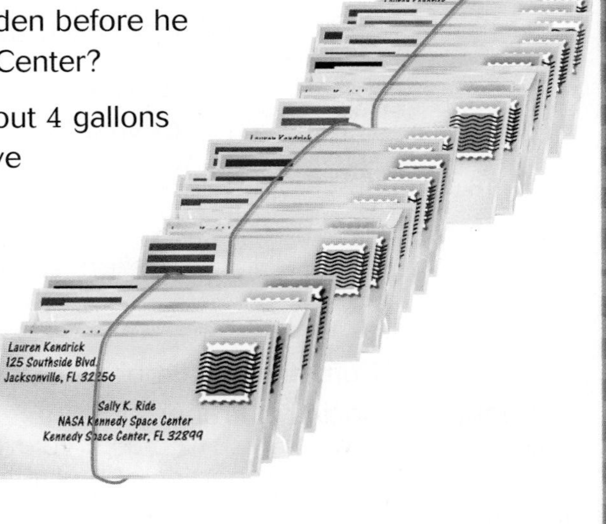

8. The astronaut received 23 fan letters on Monday, 54 on Tuesday, and more on Friday. If she got 127 letters in all, how many did she get on Friday? Write and solve an equation. Explain what the variable represents.
9. 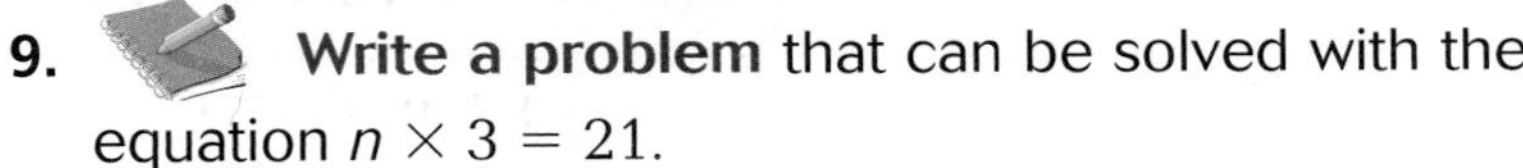 **Write a problem** that can be solved with the equation $n \times 3 = 21$.

LESSON 3

Use Multiplication Properties

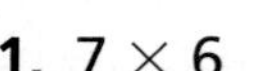

Quick Review

1. 7×6
2. 70×6
3. 700×6
4. $7{,}000 \times 6$
5. $70{,}000 \times 6$

VOCABULARY

Commutative Property
Associative Property

Learn

DIGITAL DILEMMA Keri has 8 shelves with the same number of DVDs on each shelf. Malik has 5 shelves with 8 DVDs on each shelf. They have the same number of DVDs. How many DVDs does Keri have on each shelf?

You can write an equation to model this problem.

Let d = the number of DVDs Keri has on each shelf.

Keri's DVDs			=	Malik's DVDs		
number of shelves	×	DVDs on each shelf		number of shelves	×	DVDs on each shelf
↓	↓	↓		↓	↓	↓
8	×	d	=	5	×	8

$8 \times d = 5 \times 8$ Think about the Order Property.
$8 \times 5 = 5 \times 8$ d must equal 5.
$d = 5$

So, Keri has 5 DVDs on each shelf.

The Order Property is also called the **Commutative Property**. The Grouping Property is also called the **Associative Property**.

Remember

You can use mental math and the properties of multiplication to solve problems.

Order Property
$4 \times 3 = 3 \times 4$

Grouping Property
$(8 \times 2) \times 6 = 8 \times (2 \times 6)$

Property of One
$7 \times 1 = 7$

Zero Property
$9 \times 0 = 0$

Examples

A $(8 \times 6) \times 5 = 8 \times (6 \times p)$ Associative Property
$(8 \times 6) \times 5 = 8 \times (6 \times 5)$
So, $p = 5$.

B $16 \times c = 16$ Property of One
$16 \times 1 = 16$
So, $c = 1$.

- How is the Commutative Property different from the Associative Property?

MATH IDEA The properties of multiplication can help you solve problems mentally.

CALIFORNIA STANDARDS O—n **AF 1.2** Use a letter to represent an unknown number; write and evaluate simple algebraic expressions in one variable by substitution. **AF 1.0** Students use variables in simple expressions, compute the value of the expression for specific values of the variable, and plot and interpret the results. *also* **AF 1.1, MR 2.0, MR 2.2, MR 2.4, MR 3.2, MR 3.3**

Check

1. **Tell** which property can help make it easier to find the value of n. $(16 \times 72) \times n = 0$

Solve the equation. Identify the property used.

2. $(7 \times y) \times 8 = 7 \times (3 \times 8)$

3. $n \times 3 = 3 \times 9$

4. $h \times 7 = 0$

Practice and Problem Solving

Solve the equation. Identify the property used.

5. $4 \times (8 \times d) = (4 \times 8) \times 7$

6. $9 \times y = 0$

7. $n \times 43 = 43 \times 3$

8. $(6 \times 7) \times 2 = 6 \times (n \times 2)$

9. $5 \times n = 5$

10. $1 \times k = 165$

11. $5 \times (n \times 13) = (5 \times 4) \times 13$

12. $112 \times 9 = n \times 112$

13. $n \times 1 = n$

Algebra **Identify the property or properties shown.**

14. $a \times b = b \times a$

15. $a \times 0 = 0$

16. $a \times 1 = a$

17. $a \times (b \times c) = (a \times b) \times c$

18. $(a \times b) \times c = b \times (a \times c)$

USE DATA **For 19–20, use the graph.**

19. Estimate the number of DVD players sold from 1997 through 1999.

20. In what year were the most DVD players sold?

21. Tell where to place parentheses to make the equation $24 \times 3 + 24 \times 7 = 240$ true.

22. Tamar took her pulse for 20 seconds. She counted 29 heartbeats. How many beats per minute is that?

23. A group of goats and ducks have a total of 99 heads and legs among them. There are twice as many ducks as there are goats. How many are there of each?

24. **Write About It** Give an example to show how the Associative Property of Multiplication can help you solve problems mentally.

Mixed Review and Test Prep

Write the decimal as a fraction. (p. 20)

25. 0.021

26. 0.704

27. Order 3.52, 3.25, and 2.35 from least to greatest. (p. 24)

28. $23 + 47 + 19 + 10 + 11$

29. **TEST PREP** Nancy is 6 years younger than her sister, who is 19. Which equation can you use to find how old Nancy is? (p. 72)

A $6 + n = 19$

B $19 + 6 = n$

C $6 + 19 = n$

D $19 + n = 6$

LESSON 4

The Distributive Property

Quick Review

1. 8 + 29
2. 239 – 15
3. (6 × 2) × 8
4. 8,500 ÷ 10
5. (9 × 20) + (9 × 4)

VOCABULARY

Distributive Property

Learn

PIECE BY PIECE The **Distributive Property** states that multiplying a sum by a number is the same as multiplying each addend in the sum by the number and then adding the products.

How can you use the Distributive Property to find the product 5×16?

STEP 1

On grid paper, outline a rectangle that is 5 units high and 16 units wide. Think of the area as the product.

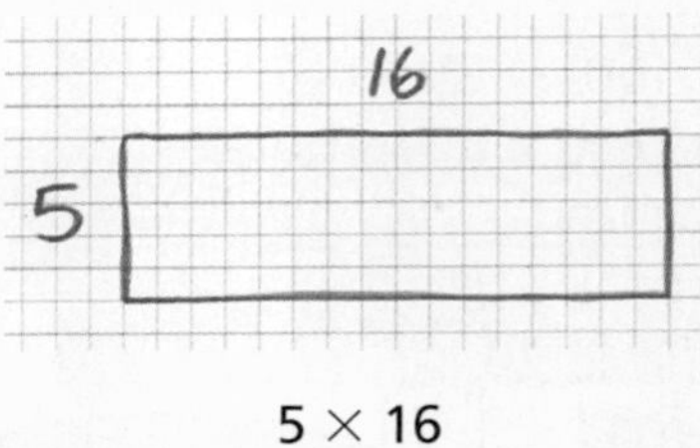

5 × 16

STEP 2

Count over 10 units from the left and draw a line to break apart the rectangle.

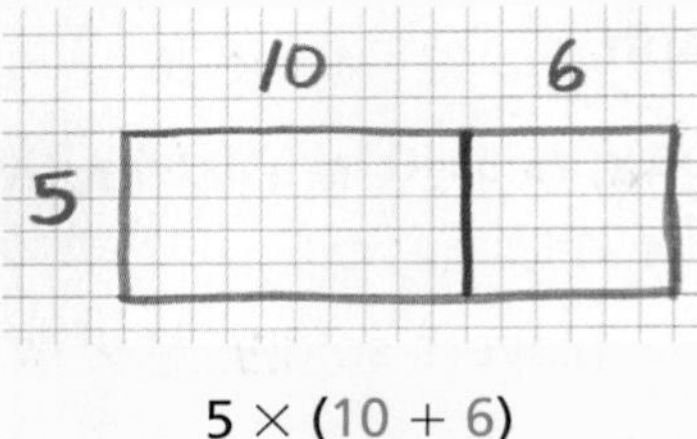

5 × (10 + 6)

STEP 3

Use the Distributive Property to show a sum of two products. Multiply what is in parentheses first. Add the products.

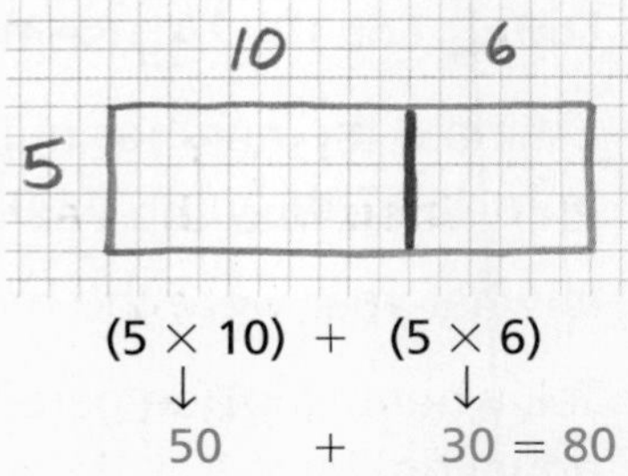

(5 × 10) + (5 × 6)
↓ ↓
50 + 30 = 80

So, $5 \times 16 = 80$.

- Use grid paper to find $20 \times 17 = n$.

You can also use the Distributive Property to find the value of expressions with variables.

Find the value of 6 × (*n* + 5) if *n* = 10.

$6 \times (n + 5) = (6 \times n) + (6 \times 5)$	Use the Distributive Property.
$= (6 \times 10) + (6 \times 5)$	Replace *n* with 10.
$= 60 + 30$	
$= 90$	

So, $6 \times (n + 5) = 90$ if $n = 10$.

- Use the Distributive Property to check if the equation $7 \times (n + 2) = 84$ is true if $n = 5$ and if $n = 10$.

TECHNOLOGY LINK

More Practice: Use E-Lab, *Using the Distributive Property*.

www.harcourtschool.com/elab2002

CALIFORNIA STANDARDS AF 1.3 Know and use the distributive property in equations and expressions with variables. **AF 1.0** Students use variables in simple expressions, compute the value of the expression for specific values of the variable, and plot and interpret the results. **AF 1.2** Use a letter to represent an unknown number; write and evaluate simple algebraic expressions in one variable by substitution. *also* **MR 1.2, MR 2.3, MR 2.4, MR 3.0, MR 3.2**

Check

1. **Explain** how you can use the Distributive Property to find the value of 7×82. Show your steps.

Use grid paper to find the product.

2. 10×18
3. 20×15
4. 23×6
5. 30×14

Use the Distributive Property to restate each expression. Find the product.

6. 7×26
7. 9×24
8. 26×3
9. 7×15

Practice and Problem Solving

Use grid paper to find the product.

10. 5×17
11. 9×16
12. 20×14
13. 33×6

Use the Distributive Property to restate each expression. Find the product.

14. 14×8
15. 8×34
16. 9×37
17. 6×35
18. 9×28
19. 10×32
20. 12×26
21. 50×12

Algebra **Restate the expression, using the Distributive Property. Then find the value of the expression.**

22. $9 \times (y + 8)$ if y is 10
23. $8 \times (4 + n)$ if n is 20
24. $3 \times (6 + n)$ if n is 40
25. $5 \times (d + 7)$ if d is 80
26. Ida has 4 tomato plants. For 3 weeks in a row, each plant produced 4 tomatoes. What is the total number of tomatoes her plants produced in 3 weeks?

27. **What's the Error?** Jill wrote $7 \times (30 + 7) = 210 + 7 = 217$. Identify and correct her error.
28. Use the Distributive Property to check if the equation $5 \times (n + 6) = 130$ is true if $n = 20$.
29. **Write About It** Explain the Distributive Property in your own words. Give an example.

Mixed Review and Test Prep

30. Find $(17 - n) + 27$ if $n = 4$. (p. 68)
31. Round to the nearest tenth. Then add. $7.39 + 2.46$ (p. 46)
32. $0.34 + 2.51$ (p. 50)
33. $90.8 - 56.9$ (p. 50)
34. **TEST PREP** Write the value of the blue digit in 2,805,704. (p. 4)
    - A 800
    - B 8,000
    - C 80,000
    - D 800,000

Extra Practice page H37, Set C

CHAPTER 6

Review/Test

CHECK VOCABULARY AND CONCEPTS

Choose the best term from the box.

Associative Property
Distributive Property
equation
evaluate
expression

1. The __?__ states that multiplying a sum by a number is the same as multiplying each addend in the sum by the number and then adding the products. (p. 92)
2. The values on both sides of an equal sign are the same in an __?__. (p. 88)
3. When you replace the variable with a number and perform the operations, you __?__ an expression. (p. 86)

CHECK SKILLS

Evaluate each expression. (pp. 86–87)

4. $n + 56$ if $n = 11$
5. $12 \times y$ if $y = 8$
6. $(5 \times n) \times 9$ if $n = 4$
7. $(90 - n) \times 2$ if $n = 45$
8. $n \times 10 \times 4$ if $n = 3$
9. $(0 \times s) \times 5$ if $s = 7$

Solve the equation. Identify the property. (pp. 90–91)

10. $0 = 19 \times n$
11. $(5 \times 4) \times 9 = 5 \times (n \times 9)$
12. $16 = m \times 16$
13. $6 \times f = 13 \times 6$
14. $(3 \times 2) \times 5 = n \times (2 \times 5)$
15. $105 \times 1 = n$

Use the Distributive Property to restate each expression. Find the product. (pp. 92–93)

16. 24×7
17. 9×44
18. 11×23
19. 6×32
20. 54×5

Restate the expression, using the Distributive Property. Then find the value of the expression. (pp. 92–93)

21. $6 \times (n + 6)$ if $n = 40$
22. $k \times (50 + 2)$ if $k = 11$
23. $8 \times (30 + n)$ if $n = 8$

CHECK PROBLEM SOLVING

Write and solve an equation. Explain what the variable represents. (pp. 88–89)

24. The librarian packed 24 books. He put the same number of books in each of 4 boxes. How many books did he pack in each box?
25. Ben walked the same number of miles each day. After 5 days, he had walked a total of 15 miles. How many miles did he walk each day?

Cumulative Review

TIP! **Decide on a plan.**
See item **8.**
Work backward from what you know to find what you don't know. Use the information about Hal to find out how many stamps Ethan has before you find how many stamps Joe has.

Also see problem **4,** p. H63.

For 1–10, choose the best answer.

For 1–3, use the bar graph.

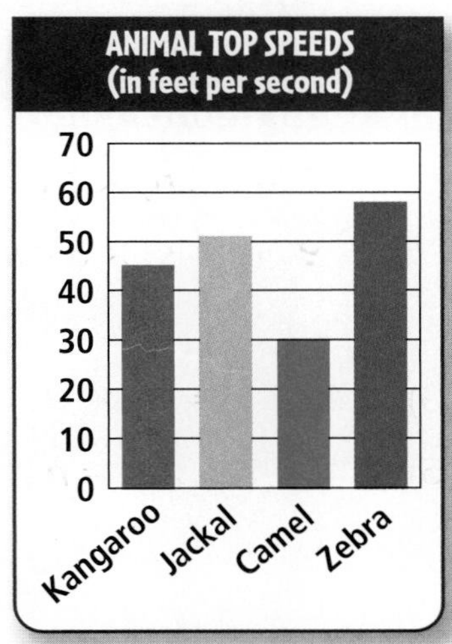

1. Which animal runs almost two times as fast as a camel?

   A Kangaroo **C** Jackal
   B Zebra **D** None of the above

2. If a kangaroo runs for 60 seconds, how many seconds will a camel have to run to go the same distance?

   F 60 seconds **H** 90 seconds
   G 75 seconds **J** 120 seconds

3. Which equation tells how many feet a jackal runs in 40 seconds?

   A $40 \times 51 = 2{,}040$ **C** $40 + 51 = 98$
   B $40 \times n = 51$ **D** $40 \times 51 = 232$

4. Which illustrates the Distributive Property with $5 \times (n + 4)$?

   F $(5 + n) \times 4$ **H** $(5 \times n) + (5 \times 4)$
   G $(5 + 4) \times n$ **J** $(5 + n) \times (5 + 4)$

5. Which multiplication property does $(6 \times 7) \times 2 = 6 \times (7 \times 2)$ illustrate?

   A Commutative Property
   B Zero Property
   C Associative Property
   D Distributive Property

6. Patrice has 14 stuffed animals and Nada has 11. Jill has twice as many as Patrice and Nada have combined. Which expression describes this?

   F $(14 + 11) + 2$ **H** $(14 + 11) \times 2$
   G $(14 - 11) \times 2$ **J** $(14 \times 2) + 11$

7. What is the value of the 5 in 254,896,103?

   A 5,000 **C** 5,000,000
   B 500,000 **D** 50,000,000

8. Ethan has three times as many stamps as Joe, but only half as many as Hal. If Hal has 150 stamps, how many stamps does Joe have?

   F 25 **H** 50
   G 30 **J** 75

9. What is the value of $10 \times n \times 3$ if $n = 2$?

   A 23 **C** 36
   B 30 **D** Not here

10. What is the value of $12 + 18 - n$ if $n = 15$?

    F 12 **H** 17
    G 15 **J** 20

CHAPTER 7 Analyze Data and Graphs

Dr. James Naismith invented basketball in 1891. The first official college basketball game took place in 1897, when Yale University defeated the University of Pennsylvania, 32–10. The graph shows the average points per game scored by college basketball's top scorers for 1995 through 1999. How much greater was Kurt Thomas's average than Alvin Young's? Can you tell from the graph how many total points Charles Jones scored during the 1997 season? Explain.

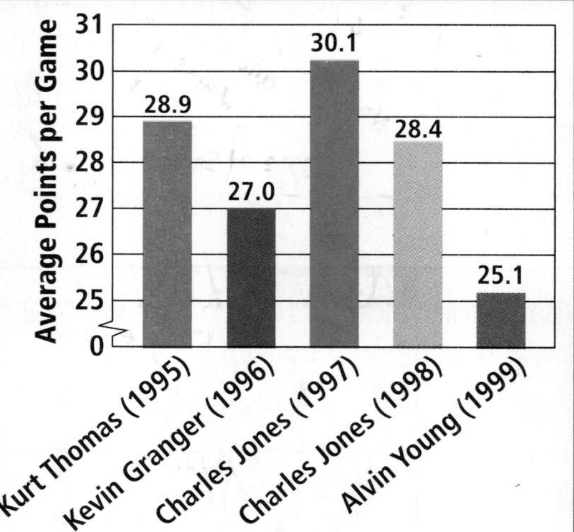

Use this page to help you review and remember important skills needed for Chapter 7.

VOCABULARY

Choose the best term from the box.

bar graph
circle graph
range

1. A graph that shows data as parts of a whole circle is a ___?___.
2. When you have several categories of data, it is easy to compare the data by using a ___?___.

FREQUENCY TABLES (See p. H9.)

For 3–6, use the frequency table.

TEST SCORES

Score	Frequency
70	12
80	15
90	17
100	12

3. How many students scored 90 on the test?
4. How many students scored either 90 or 100?
5. How many students took the test?
6. How many students scored less than 90?

READ GRAPHS (See p. H9.)

For 7–12, use the bar graph.

7. How many types of supplies are shown on the graph?
8. How many markers does the store have in stock?
9. How many items does the store have in stock altogether?
10. How many more sticker sets are there than rulers?
11. Which item does the store have the most of?
12. If markers sell for $2.00, how much will all the markers cost?

For 13–15, use the circle graph.

13. What flavor received the most votes?
14. How many people chose vanilla?
15. How many people voted?

LESSON 1

Collect and Organize Data

Quick Review

1. 9 + 14
2. 13 + 24
3. 40 + 18
4. 15 + 220
5. 240 + 160

VOCABULARY

cumulative frequency
outlier
range

Learn

SCHOOL STORE The tally table shows the results of sales of binders during the first five weeks of school. How can you find the total number of binders sold?

You can make a frequency table to help you organize the data from a tally table.

Activity

Copy the frequency table below.

STEP 1

Count the tally marks for each week. Place that sum next to the week number in the frequency column of the frequency table below.

BINDERS SOLD AT SCHOOL STORE

Weeks	Number of Binders
Week 1	𝍸 𝍸 𝍸 𝍸
Week 2	𝍸 𝍸 𝍸
Week 3	𝍸 𝍸 ||||
Week 4	𝍸 𝍸 ||
Week 5	𝍸 𝍸 |

STEP 2

The **cumulative frequency** shows a running total of the number of binders sold. In the cumulative frequency column, write the sum of the frequencies as each new line of data is entered.

BINDERS SOLD AT SCHOOL STORE

Weeks	Frequency (Number of Binders Sold)	Cumulative Frequency
Week 1	20	20
Week 2	15	20 + 15 = 35
Week 3	■	■
Week 4	■	■
Week 5	■	■

The last entry in the cumulative frequency column will tell you the total number of binders sold.

So, 72 binders were sold.

- How many binders were sold by Week 3?

CALIFORNIA STANDARDS SDAP 1.0 Students display, analyze, compare, and interpret different data sets, including data sets of different sizes. **SDAP 1.2** Organize and display single variable data in appropriate graphs and representations, and explain which types of graphs are appropriate for various data sets. *also* **MR 1.0, MR 2.0, MR 2.3, MR 2.4, MR 3.2**

Use a Line Plot

A kid's magazine took a poll of fifth graders to find out how many items they put in a backpack. You can use a line plot to record the data.

> **Remember**
> A line plot is a diagram that shows the frequency of data.

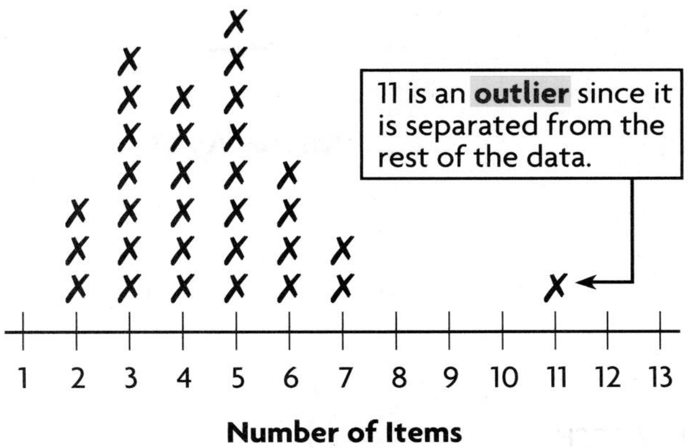

You can use the line plot to find the range of the number of items fitting in a backpack.

The **range** is the difference between the greatest and least numbers in a set of data.

greatest number		least number		range
↓		↓		↓
11	−	2	=	9

So, the range is 9 items.

- How many students said 5 items? Explain how you know.

MATH IDEA When you record data in a table or a line plot, it is easier to analyze.

Check

1. **Compare and contrast** a tally table and a frequency table.

For 2–5, use the frequency table.

2. What number of folder covers was most frequently purchased by students?
3. What number of folder covers was least frequently purchased by students?
4. How many students purchased less than 4 folder covers?
5. How many students purchased folder covers?

FOLDER COVERS PURCHASED

Number of Covers	Frequency (Number of Students)	Cumulative Frequency
1	31	31
2	22	53
3	16	69
4	15	84
5	12	96

LESSON CONTINUES 

Practice and Problem Solving

For 6–9, use the frequency table.

SOCCER LEAGUE SIGN-UPS		
Week	**Frequency**	**Cumulative Frequency**
1	7	7
2	13	20
3	5	25
4	14	39

6. How many people signed up in Week 1? in Week 3?
7. By the end of Week 2, how many people had signed up?
8. How many people signed up during the 4 weeks?
9. What is the range of the number of people signing up each week?

Find the range for each set of data.

10. 8, 12, 17, 15, 20, 9
11. 34, 26, 37, 31, 37, 22
12. 100, 110, 105, 103
13. 96, 74, 83, 81, 64
14. 156, 147, 145, 168, 155
15. 38, 45, 59, 32, 47, 51

For 16–19, use the line plot.

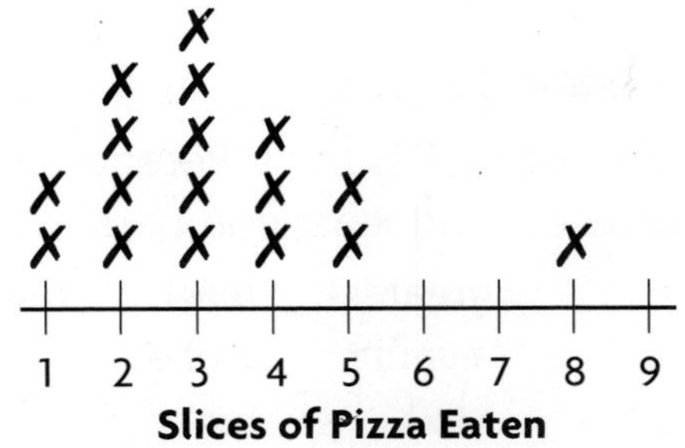

16. How many slices did the greatest number of students eat?
17. How many students ate 4 or more slices of pizza?
18. Which value is an outlier?
19. One slice of pizza costs $1.50. How much would the two students who each ate 5 pieces pay in all for pizza?

For 20–23, use the tally table.

SIGN-UP FOR FIELD EVENT TIMES	
Event Times	**Number of Students**
9:30 A.M.	𝍸 𝍸 𝍸 |
10:30 A.M.	𝍸 𝍸 |||
1:30 P.M.	𝍸 𝍸 𝍸 ||
2:30 P.M.	𝍸 𝍸 𝍸 |||

20. Make a cumulative frequency table for the data.
21. Which time was selected by the greatest number of students?
22. How many students signed up for Field Day events that would happen before noon?
23. If each student took part in only one event, how many students participated in Field Day?
24. **What's the Question?** In a survey of favorite colors, 11 students chose blue, 7 chose red, and the rest chose green. Twenty-four students took part in the survey. The answer is 6.

Take a survey of at least 10 people to find their favorite sandwich. Make a frequency table from your tallies. Use your table for 25–26.

25. List the favorite sandwiches in order from the most popular one to the least popular one.
26. How many more people chose the most popular sandwich than chose the least popular sandwich?

Mixed Review and Test Prep

27. Write an expression for this situation. Lisa read 4 fewer pages than Karen. Karen read n pages. (p. 68)

28. Use mental math to solve $n - 9 = 21$. (p. 74)

29. $\begin{array}{r} 4.32 \\ +3.57 \\ \hline \end{array}$ (p. 50)

30. $\begin{array}{r} 10.5 \\ -\ 1.46 \\ \hline \end{array}$ (p. 50)

31. Round 2,497,629 to the nearest million. (p. 32)

32. Round 6.058 to the nearest tenth. (p. 46)

33. $4.2 - 3.1$ (p. 50)

34. $1.02 - 0.09$ (p. 54)

For 35–36, choose the letter of the number written in standard form. (p. 18)

35. **TEST PREP** five and nine tenths

  A 5.09
  B 5.9
  C 9.5
  D 509

36. **TEST PREP** twenty-six hundredths

  F 0.026
  G 0.26
  H 1.26
  J 2.6

LINKUP to Reading

STRATEGY • USE GRAPHIC AIDS
Examining each part of a table will help you understand the data presented. The title tells you what data are represented. The labels on the columns and rows give you more detail about the data. As you read a problem about data in a table, match what you read with what you see.

For 1–6, use the frequency table.

1. What does the title tell you?
2. What does the frequency column tell you?
3. Where would you look in the table to find the total number of students represented by the data?
4. How many students read for 60 minutes during the week?
5. How many students read for 60 minutes or less during the week?
6. **Explain** how using a graphic aid, such as a table or graph, can help you analyze data.

MINUTES STUDENTS SPENT READING MAGAZINES (1 WEEK)

Minutes	Frequency	Cumulative Frequency
20	5	5
30	9	14
40	10	24
50	14	38
60	11	49

Extra Practice page H38, Set A

LESSON

2 Find the Mean

Quick Review

1. 90 ÷ 3
2. 180 ÷ 3
3. 56 ÷ 8
4. 75 ÷ 3
5. 88 ÷ 4

VOCABULARY

mean

Learn

"WE THE PEOPLE . . ." The table shows the ages of some of the youngest signers of the Declaration of Independence. What is the mean age of these signers?

SIGNERS OF DECLARATION OF INDEPENDENCE

Signer	Age
Edward Rutledge	26
Benjamin Rush	30
Elbridge Gerry	31
Thomas Jefferson	33

The mean, or average, is one number that is representative of the data. You can find the **mean** by adding all the data and then dividing the sum by the number of addends.

STEP 1

Add all the data.

```
  26
  30
  31
 +33
 120
```

STEP 2

Divide the sum by the number of addends.

```
   30
4)120
```

So, the mean age is 30 years old.

▲ Edward Rutledge

▲ Starting on August 2, 1776, 56 members of Congress signed the Declaration of Independence.

- **What if** the table included the oldest signer, Benjamin Franklin, who was 70 years old? How would that affect the mean?

Examples **Find the mean for each set of data.**

A 74, 91, 63, 92, 85

```
  74                      81
  91  number of  →  5)405
  63  addends             40
  92                      05
+ 85                       5
 405                       0
```

B 357, 562, 411, 584, 294, 372

```
  357                     430
  562  number of  →  6)2580
  411  addends            24
  584                      18
  294                      18
+ 372                      00
2,580                       0
                            0
```

▲ Benjamin Franklin

CALIFORNIA STANDARDS SDAP 1.1 Know the concepts of mean, median, and mode; compute and compare simple examples to show that they may differ. **NS 2.2** Demonstrate proficiency with division, including division with positive decimals and long division with multidigit divisors. *also* **NS 1.0, SDAP 1.0, MR 1.0, MR 2.0, MR 2.3, MR 2.4, MR 3.0**

Check

1. **Look** at this set of data: 5, 16, 17, 17, 17, 18. Do you think the mean will be closer to 5 or to 16? Explain.

Find the mean for each set of data.

2. 18, 22, 14
3. 58, 105, 172, 45
4. 33, 36, 35, 37, 39

Practice and Problem Solving

Find the mean for each set of data.

5. 9, 6, 5, 9, 11
6. $6, $7, $9, $10
7. 19, 25, 28, 32
8. 1,250; 980; 350
9. 110, 75, 135, 160
10. 147, 116, 148, 128, 116
11. 95, 84, 72, 73, 76
12. 124, 130, 100, 122, 124
13. 202, 213, 315, 285, 285
14. 100, 150, 200, 170, 250, 300
15. 1,200; 2,400; 1,320; 2,560

Use the given mean to find the missing number in each data set.

16. $1, $3, $5, ■; mean: $4
17. 5, 7, 10, 13, ■; mean: 10
18. 12, 15, 18, 10, ■; mean: 13
19. 20, 25, 30, ■, 32, 18; mean: 25

For 20–24, use the table.

HISTORY TEST SCORES				
Name	**Test 1**	**Test 2**	**Test 3**	**Test 4**
Ally	87	93	100	80
Andrew	95	89	90	78
Karen	81	86	82	absent
Lin	85	85	85	85

20. What is the average of Andrew's test scores?
21. How can you tell what Lin's test average is without calculating it?
22. Who had the highest test mean, or average, Ally, Andrew, or Lin?
23. **REASONING** What score does Karen need to get on Test 4 to raise her average to 85?
24. **Write a problem** using the information in the table. Then solve the problem.
25. Ally, Andrew, Karen, and Lin sit in the same row. Ally sits in front of Karen and behind Lin. Andrew sits in the front seat. Who sits right behind Andrew?

Mixed Review and Test Prep

26. Find the range for the scores 88, 79, 84, and 88. (p. 98)
27. $\begin{array}{r} 2{,}685{,}941 \\ +1{,}255{,}357 \\ \hline \end{array}$ (p. 38)
28. $\begin{array}{r} 3{,}105{,}568 \\ -1{,}387{,}555 \\ \hline \end{array}$ (p. 38)
29. Round 32.5876 to the nearest thousandth. (p. 46)
30. **TEST PREP** Which is the solution of the equation $13 + n = 29$? (p. 74)

   A $n = 13$
   B $n = 16$
   C $n = 29$
   D $n = 42$

Extra Practice page H38, Set B

LESSON 3

Find the Median and Mode

Quick Review

Give the number halfway between each pair of numbers.

1. 6 and 8
2. 12 and 14
3. 14 and 18
4. 32 and 36
5. 68 and 72

VOCABULARY

median **mode**

Learn

WHAT'S THE SCORE? The Eagles basketball team has played 9 games so far this season. The coach recorded their scores.

To analyze the data, the coach can find the median and mode of the scores.

The **median** is the middle number when the data are arranged in order.

The **mode** is the number or numbers that occur most often in a set of data. There may be one mode, more than one mode, or no mode.

EAGLES

GAME	POINTS	GAME	POINTS
1	42	6	43
2	50	7	40
3	45	8	42
4	42	9	45
5	44	10	

Activity

STEP 1

Put the basketball points in order from least to greatest. Find the median.

STEP 2

Make a line plot of the points.

- Draw a horizontal line.
- Label vertical tick marks in units of 1 from 40–50.
- Plot the data.

STEP 3

Use the line plot to find the mode.

- Where are the data clustered on the line plot?
- What if the Eagles scored 45 points in their tenth game? Would that change the mode? Explain.

- What part of the basketball points are greater than the median? What part are less than the median?

TECHNOLOGY LINK

More Practice: Use E-Lab, *Finding the Median and Mode.*

www.harcourtschool.com/elab2002

When there are two middle numbers, the median is the mean of the two numbers.

The median for 3, 5, 7, 9 is between 5 and 7.

$(5 + 7) \div 2 = 12 \div 2 = 6$ The median is 6.

CALIFORNIA STANDARDS SDAP 1.1 Know the concepts of mean, median, and mode; compute and compare simple examples to show that they may differ. **SDAP 1.2** Organize and display single variable data in appropriate graphs and representations, and explain which types of graphs are appropriate for various data sets. *also* **SDAP 1.0, MR 1.1, MR 2.0, MR 2.3, MR 2.4, MR 3.2**

Check

1. **Explain** how the median and the mode differ from the mean of a set of data.

Find the median and the mode for each set of data.

2. 12, 14, 11, 11, 9, 16, 17, 19

3. 24, 32, 28, 45, 19, 24, 50, 32

Practice and Problem Solving

Find the median and the mode for each set of data.

4.

KATIE'S TEST SCORES							
Test	1	2	3	4	5	6	7
Score	94	93	95	87	94	78	85

5.

CARD COLLECTION					
Name	Tony	Kara	Ray	Sue	Jay
Number	240	200	200	265	285

Find the mean, median, mode, and range for each set of data.

6. 9, 6, 5, 9, 11

7. $4, $2, $5, $6, $3, $4

8. 124, 130, 100, 122, 124

9. 95, 84, 72, 73, 76

10. 202, 213, 315, 285, 285

11. 749, 751, 1,298, 202

For 12–14, use the line plot.

12. How many players were measured, and what is the range of their heights?

13. What is the most common height?

14. What is the median height for the basketball team?

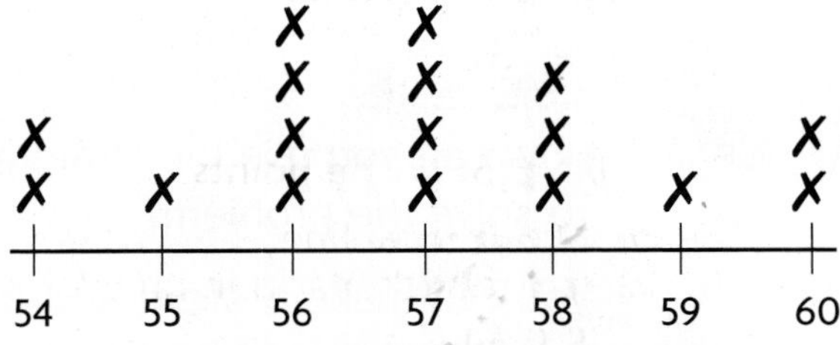

Height of Fifth-Grade Basketball Players (in.)

15. **What's the Error?** Julio said the median of the scores 90, 80, 82, 95, 78, 84, and 86 is 95. Explain Julio's error. What is the correct median of the scores?

16. **Write About It** Find and compare the mean, median, and mode of 27, 33, 33, 31, 34, 28.

Mixed Review and Test Prep

17. 345,000 + 9,004,356 (p. 38)

18. Round 34,576 to the nearest thousand. (p. 32)

19. Find the values of $15 + b$ if $b = 12$ and if $b = 20$. (p. 68)

20. Solve: $120 + t = 200$. (p. 74)

21. **TEST PREP** Which is greater than 5.698? (p. 24)

A 4.567
B 5.098
C 5.599
D 5.708

Extra Practice page H38, Set C

LESSON 4

Problem Solving Strategy

Make a Graph

Quick Review

Find the following for this data set: 14, 15, 18, 14, 14, 16, 18, 19, 16.

1. Mean
2. Mode
3. Median
4. Range
5. Least number

VOCABULARY

stem-and-leaf plot

PROBLEM A store recorded the ages of all the people who bought a pair of in-line skates last month. Which age group bought the most skates: people 10–19, people in their 20's, 30's, 40's, or 50's?

- What are you asked to find?
- What information will you use?

- What strategy can you use to solve the problem?

You can *make a graph* to organize the data.

Solve

- How can you use the strategy to solve the problem?

You can make a **stem-and-leaf plot** to help you see how data are clustered, or grouped.

Ages of People Who Bought In-Line Skates

26	18	41	51
12	34	23	19
31	44	45	14
34	16	37	23
13	27	12	22

List the tens digits of the data in order from least to greatest. These are the stems.

Beside each tens digit, record the ones digits of the data, in order from least to greatest. These are the leaves.

So, the most skate buyers were in the 10–19 age group.

The tens digit of each number is its stem.

Stem	Leaves
1	2 2 3 4 6 8 9
2	2 3 3 6 7
3	1 4 4 7
4	1 4 5
5	1

2|7 represents 27.

The ones digit of each number is its leaf.

Ages of In-Line Skate Buyers

- Look back. Does the answer make sense? Explain.

CALIFORNIA STANDARDS **MR 2.3** Use a variety of methods, such as words, numbers, symbols, charts, graphs, tables, diagrams, and models, to explain mathematical reasoning. **AF 1.1** Use information taken from a graph or equation to answer questions about a problem situation. *also* **SDAP 1.0, SDAP 1.2, MR 1.0, MR 1.1, MR 1.2, MR 2.0, MR 2.6**

Problem Solving Practice

Make a graph to solve.

1. **What if** three more people, ages 25, 28 and 32, bought skates? Which age group would have bought the most skates then?
2. Angela's bowling scores are 71, 87, 96, 73, 76, 95, 84, 95, and 97. Does she more often bowl in the 70's, 80's, or 90's?

PROBLEM SOLVING STRATEGIES

- Draw a Diagram or Picture
- Make a Model or Act It Out
- Make an Organized List
- Find a Pattern
- Make a Table or Graph
- Predict and Test
- Work Backward
- Solve a Simpler Problem
- Write an Equation
- Use Logical Reasoning

Mr. Brown recorded the class test scores. Make a stem-and-leaf plot, using the table at the right.

TEST SCORES										
84	81	85	75	81	87	70	90	82	93	86

3. Which is true about the test scores?
   - **A** Most of the test scores are in the 70's.
   - **B** The highest test score is 90.
   - **C** The range for the test scores is 13.
   - **D** The median test score is 84.
4. Jim and Tamara got the same score on the test. What was their score?
   - **F** 90
   - **G** 85
   - **H** 81
   - **J** 75

Mixed Strategy Practice

For 5–7, use the table.

AMERICAN LEAGUE HOME-RUN RECORDS 1989–1998			
Year	**Home Runs**	**Year**	**Home Runs**
1989	36	1994	40
1990	51	1995	50
1991	44	1996	52
1992	43	1997	56
1993	46	1998	56

5. Make a stem-and-leaf plot of the number of home runs. Which stem has the most leaves?
6. What are the range, median, and mode for the data set?
7. My home-run record has a 6 in the ones place. I hit more than 40 home runs and my record is *not* the mode. In which year was I the home run champion?
8. Brenda is older than June. Alice is younger than June. Cindy's age is between Alice's and June's. Write the names in order from oldest to youngest.
9. The Richardos family paid $14.00 for parking. Parking cost $5.00 for the first hour and $1.50 for each additional hour. How many hours were they parked?
10. Antonio has 6 coins that are dimes and quarters. He has a total of $1.05. What combination of coins does he have?

LESSON

5 Analyze Graphs

Quick Review

Give the next number in the pattern.

1. 16, 17, 18, ■
2. 0, 2, 4, 6, ■
3. 7, 10, 13, ■
4. 4, 8, 12, 16, ■
5. 80, 84, 88, ■

VOCABULARY

pictograph **bar graph**
line graph **circle graph**

Learn

GRAPHS GALORE! A **pictograph** displays countable data with symbols or pictures. Pictographs have a *key* to show how many each picture represents.

AEROBICS CLASS SIZE

3:00	
4:00	
5:00	
6:00	

Key: = 2 people

Look at the Aerobics Class Size pictograph. How many people are in the 3:00 and 4:00 aerobics classes altogether?

There are 8 whole symbols and 1 half symbol in the 3:00 and 4:00 classes combined.

$$(8 \times 2) + 1 = 17$$

So, there are 17 people altogether.

A **bar graph** displays countable data with horizontal or vertical bars. Bar graphs allow you to compare facts about groups of data.

Look at the exercise bar graph. Which type of exercise is more popular, jogging or weights?

The jogging bar is taller, so that exercise is more popular.

How many more students chose that exercise?

There were 14 students who chose it and 10 who chose weights.

$$14 - 10 = 4$$

So, 4 more students chose jogging.

CALIFORNIA STANDARDS SDAP 1.0 Students display, analyze, compare, and interpret different data sets, including data sets of different sizes. **AF 1.1** Use information taken from a graph or equation to answer questions about a problem situation. *also* **SDAP 1.2, SDAP 1.3, MR 1.0, MR 1.1, MR 2.0, MR 2.3, MR 3.2**

More Graphs

A **line graph** shows how data change over time.

Look at Brad's exercise line graph. How long do you predict Brad's average daily exercise time will be in Week 7?

Brad increases his exercise time by 5 minutes each week. In Week 6, he exercised an average of 45 minutes. So, in Week 7 he should increase his time to 50 minutes.

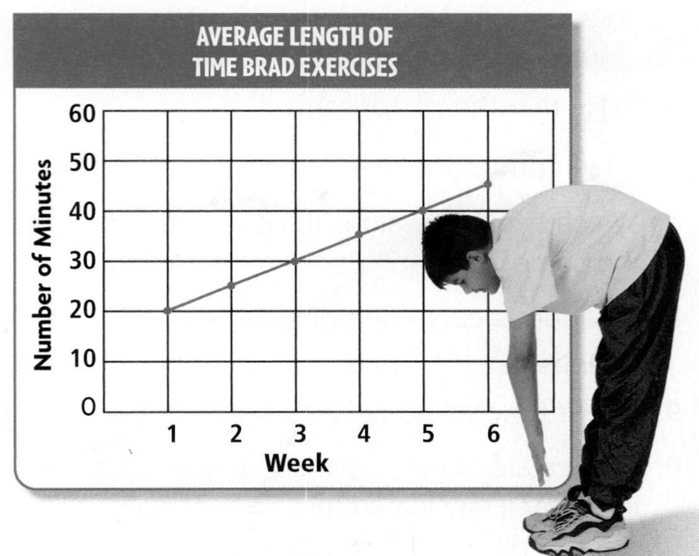

A **circle graph** shows how parts of the data are related to the whole and to each other.

Look at the circle graph of Susan's workout. On which activity does Susan spend the most time? How long does she spend on this activity?

Jogging fills half of the circle. So, Susan spends the most time jogging. Half of 60 minutes is 30 minutes. So, Susan jogs for 30 minutes.

MATH IDEA When you know how to analyze graphs, you can draw conclusions, answer questions, and make predictions about the data.

Check

1. **Explain** which type of graph would be the most appropriate to compare the high temperatures for a week.

For 2–5, use the graphs above.

2. What was Brad's average daily exercise time in Week 3?
3. In which week do you predict Brad will exercise an average of 1 hour?
4. How does the amount of time Susan jogs compare to the amount of time she stretches?
5. What fraction of her workout does Susan spend warming up?

LESSON CONTINUES 

Practice and Problem Solving

For 6–10, use the pictograph.

PROFESSIONS WITH MUCH WALKING	
Police Officer	
Mail Carrier	
TV Reporter	
Doctor	

Key: = 200 mi/yr

6. Which profession has the most walking?
7. Which two professions have about the same amount?
8. What does each footprint symbol represent?
9. About how far do doctors walk in a year?
10. Real estate agents walk about 600 miles in a year. How would this amount be displayed on the pictograph?

For 11–13, use the line graph.

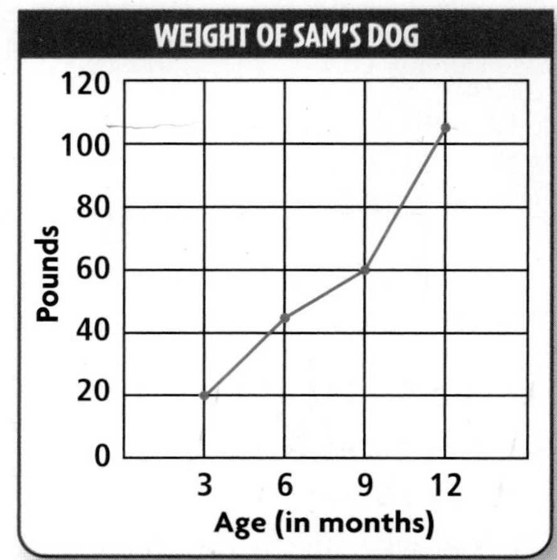

11. Between which ages did Sam's dog gain the most weight?
12. How much weight did Sam's dog gain from 3 months to 9 months?
13. **What if** Sam's dog stays at 105 pounds for the next 3 months? What will the line graph look like for these 3 months?
14. Sam spent $30 on dog food, $20 on a collar, $15 on toys, and $35 for a dog bed. What type of graph would best display how Sam spent $100?

For 15–16, use the circle graph.

15. About what part of all the pets in the United States are either cats, birds, or fish?
16. Do dogs or fish account for about one quarter of all the pets in the United States?

17. Rosa recorded her height each month for the year she was in fifth grade. What type of graph would best represent her growth during that year? Explain.

18. Of the 25 students in Mr. Reeve's class, 7 ride a bus to school, 8 ride in a carpool, and 10 walk to school. What type graph would best display this data?

Mixed Review and Test Prep

19. $\begin{array}{r} 1{,}369{,}501 \\ +3{,}695{,}224 \\ \hline \end{array}$ (p. 38)

20. $\begin{array}{r} 4{,}812{,}954 \\ -2{,}600{,}387 \\ \hline \end{array}$ (p. 38)

21. $11 + n = 57$ (p. 74)

22. $n - 16 = 38$ (p. 74)

23. TEST PREP Find the mean for 26, 35, 87, 13, 4, 25, 20. (p. 102)

A 83 C 30
B 38 D 25

24. $\begin{array}{r} 12.7945 \\ -\ 1.68 \\ \hline \end{array}$ (p. 50)

25. $\begin{array}{r} 0.69 \\ +1.577 \\ \hline \end{array}$ (p. 50)

26. $42 - n = 28$ (p. 74)

27. $n + 8 = 25$ (p. 74)

28. TEST PREP Round 36,582,367 to the nearest hundred thousand. (p. 32)

F 36,500,000 H 36,600,000
G 36,580,000 J 37,000,000

LINKUP to Science

The cheetah is the fastest land animal, reaching speeds of 70 miles per hour. The slowest land animal is the snail, moving at top speeds of 3 feet per minute.

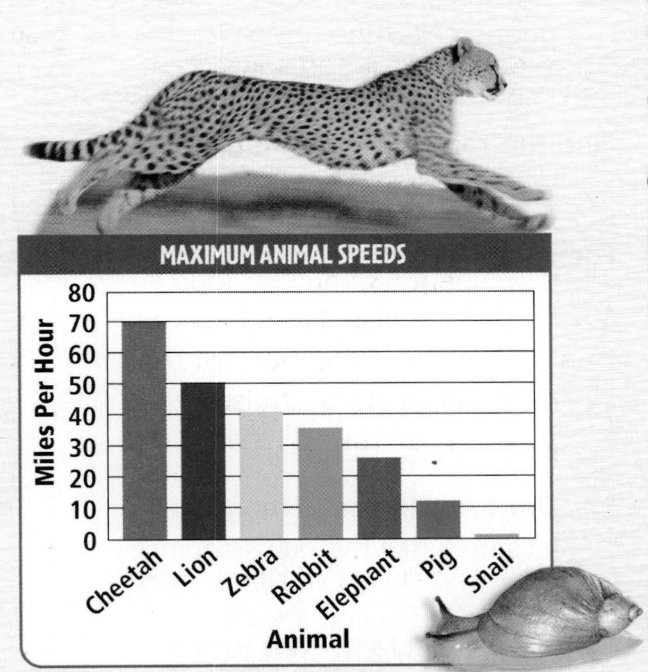

For 1–6, use the bar graph.

1. Which animal runs faster, a rabbit or a zebra?
2. About how fast can an elephant run?
3. A slow-moving truck is traveling at 37 miles per hour. Which animals can run faster than the truck is moving?
4. A greyhound runs faster than a rabbit but slower than a zebra. How fast can a greyhound run?
5. What's the Error? Lisa read the graph and said that a pig can run about 20 miles per hour. How fast can a pig run? Explain Lisa's error.
6. What's the Question? The answer is 40 miles per hour.

CHAPTER 7

Review/Test

CHECK VOCABULARY AND CONCEPTS

Choose the best term from the box.

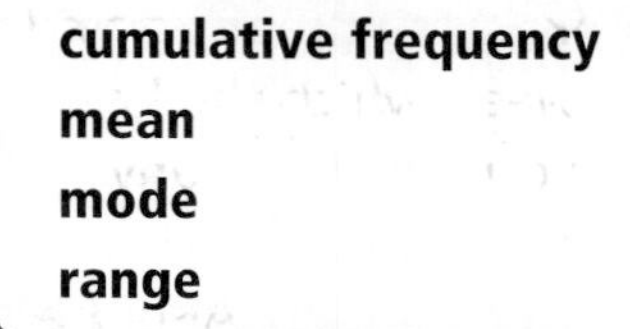

1. When you add all of the data and divide by the number of addends, you find the __?__ of the data. (p. 102)
2. To find the __?__ for a data set, subtract the least value from the greatest value. (p. 99)
3. The number or numbers that appear most often in a set of data are called the __?__ of the data. (p. 104)

CHECK SKILLS

For 4–6, use the tally table. (pp. 98–101)

CASEY'S PHONE CALLS	
Day	**Number of Calls**
Monday	𝍸 IIII
Tuesday	IIII
Wednesday	𝍸 𝍸
Thursday	𝍸 𝍸 IIII
Friday	𝍸 III

4. Make a cumulative frequency table for the data.
5. On which day did Casey make the most phone calls? the fewest?
6. How many phone calls did Casey make?

Find the range, mean, median, and mode for each. (pp. 98–105)

7. \$8, \$9, \$11, \$5, \$8, \$7
8. 15, 11, 20, 12, 12
9. 635, 499, 717, 601

For 10–12, use the circle graph. (pp. 108–111)

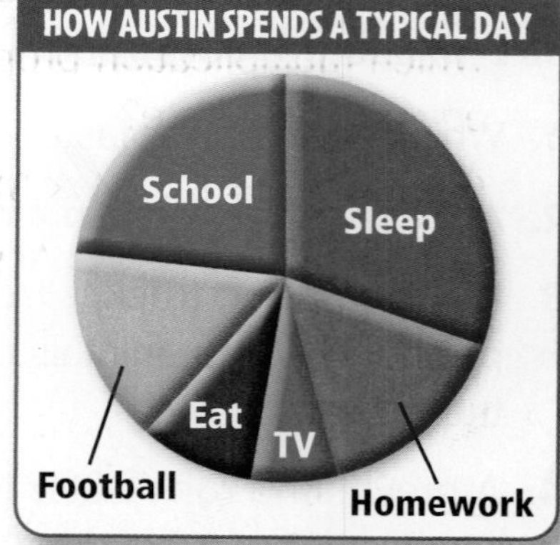

10. In which activity does Austin spend the most time?
11. Which two activities take up about the same part of Austin's day?
12. Does Austin spend more than or less than half the day sleeping?

CHECK PROBLEM SOLVING

Solve. Use the test score data. (pp. 106–107)

5th Grade Test Scores			
76	92	82	88
89	68	72	
70	94	89	

13. Make a stem-and-leaf plot for the data.
14. What are the range, median, and mode for the test scores?
15. Two more students score 67 and 94 on the test. How does this affect the range, median, and mode of the scores? Explain.

Cumulative Review

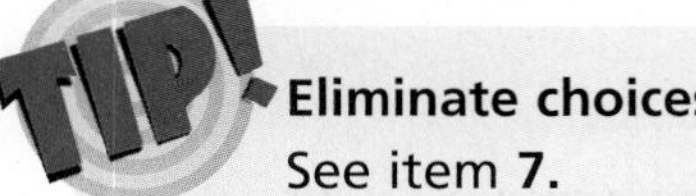

Eliminate choices.
See item **7.**

Look at the graph to find two days between which the line shows a sharp rise. You can eliminate any choice that does not have one of those days.

Also see problem **5,** p. H64.

For 1–8, choose the best answer.

1. The list shows the weights in pounds of several fifth graders.

88, 94, 91, 79, 84, 100, 92, 90, 87

What is the median weight?

A 84 pounds
B 89 pounds
C 90 pounds
D 91 pounds

2. What is the estimate for 6,493 + 8,804 when the numbers are rounded to the nearest thousand?

F 14,000
G 15,000
H 15,300
J 16,000

3. Which multiplication property does the equation illustrate?

$(8 \times 4) \times 3 = 8 \times (4 \times 3)$

A Zero
B Distributive
C Associative
D Commutative

4. A post office clerk recorded these weights in ounces for some packages.

5, 7, 12, 17, 9, 4

What is the mean weight of the packages?

F 13 oz
G 9 oz
H 6 oz
J Not here

5. Tamika kept track of the number of visits to the class website for several days. She recorded 15, 22, 29, 30, 37, and another number she was not able to read later. Which value would make the range of her data set 32?

A 54
B 47
C 32
D 22

For 6–8, use the line graph.

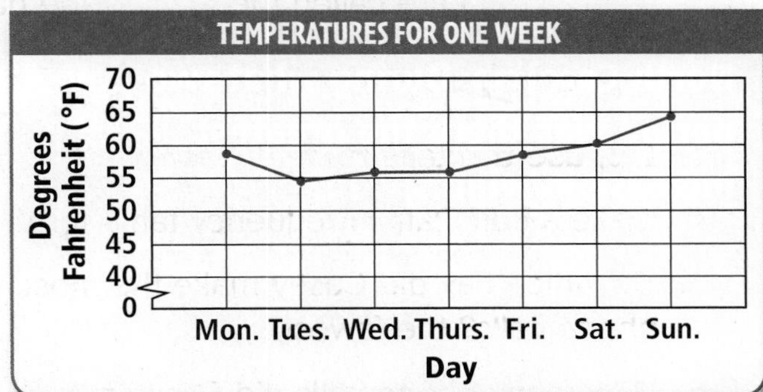

6. What was the temperature on Tuesday?

F 60°
G 58°
H 56°
J 54°

7. Between which two days did the temperature increase the most?

A Wednesday and Thursday
B Thursday and Friday
C Friday and Saturday
D Saturday and Sunday

8. Between which two days did the temperature remain unchanged?

F Monday and Tuesday
G Tuesday and Wednesday
H Wednesday and Thursday
J Thursday and Friday

CHAPTER 8 Make Graphs

A hurricane is a large revolving storm with winds that stay at 74 miles per hour or more. In 1999 Hurricane Floyd damaged many areas of North Carolina and the Bahamas. Floyd was a Category 4 storm while it had wind speeds of 131 to 155 miles per hour. The line graph shows Floyd's wind speeds during the time it affected North Carolina. During what two-day period was there the greatest increase in wind speed?

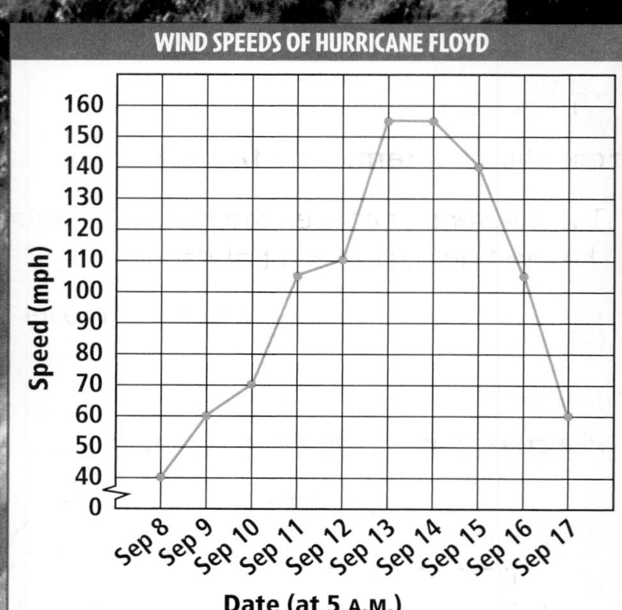

1999, Hurricane Floyd

Use this page to help you review and remember important skills needed for Chapter 8.

VOCABULARY

Choose the best term from the box.

range
line graph
bar graph

1. The difference between the greatest number and the least number in a set of data is the ___?___.
2. A graph that shows how data change over time is a ___?___.

SKIP-COUNT ON A NUMBER LINE (See p. H10.)

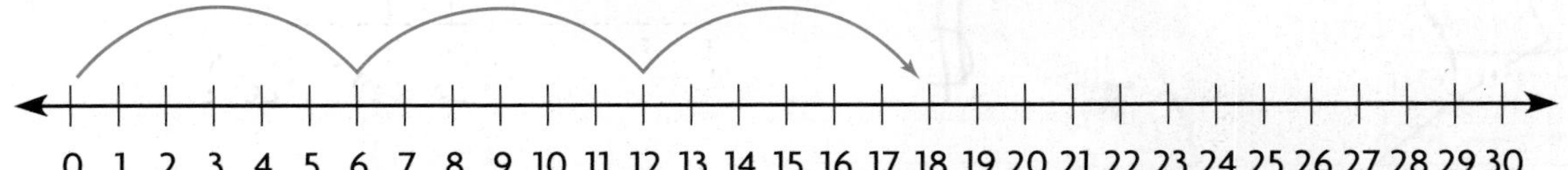

3. Use the number line above to complete the pattern.

   0, 6, 12, 18, ■, ■

For 4–7, use the number line below. List the numbers you count from zero.

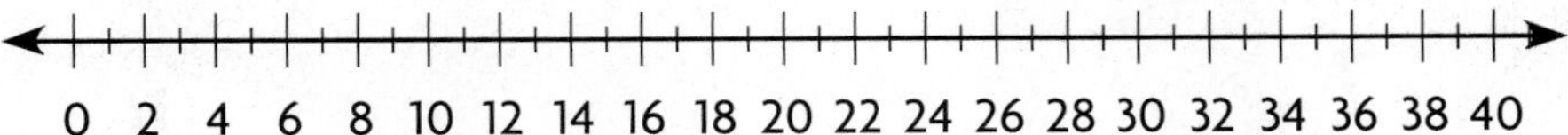

4. Count by fives.
5. Count by twos.
6. Count by tens.
7. Count by fours.

ORDERED PAIRS (See p. H10.)

For 8–17, use the ordered pair to name the point on the grid.

8. (1,3)
9. (2,6)
10. (5,4)
11. (4,1)
12. (6,8)
13. (0,5)
14. (7,5)
15. (2,9)
16. (9,10)
17. (8,0)

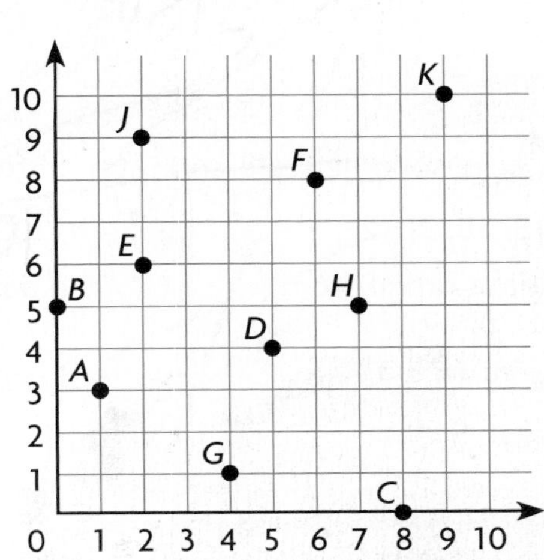

LESSON 1

Choose a Reasonable Scale

Quick Review

Use the pattern to find the next number.

1. 0, 5, 10, ■
2. 0, 10, 20, ■
3. 0, 15, 30, ■
4. 0, 20, 40, ■
5. 0, 25, 50, ■

VOCABULARY

scale interval

Learn

SEAL SIZES For her report, Helen made a graph of the lengths of 5 different types of seals. Her data are shown in the table.

First, Helen decided what scale and interval she would use. The **scale** is the series of numbers starting at 0 placed at fixed distances on a graph. The **interval** is the difference between one number and the next on the scale.

SEALS	
Type	**Average Length**
Ribbon seal	58 in.
Harbor seal	65 in.
Ringed seal	67 in.
Hooded seal	95 in.
Bearded seal	85 in.

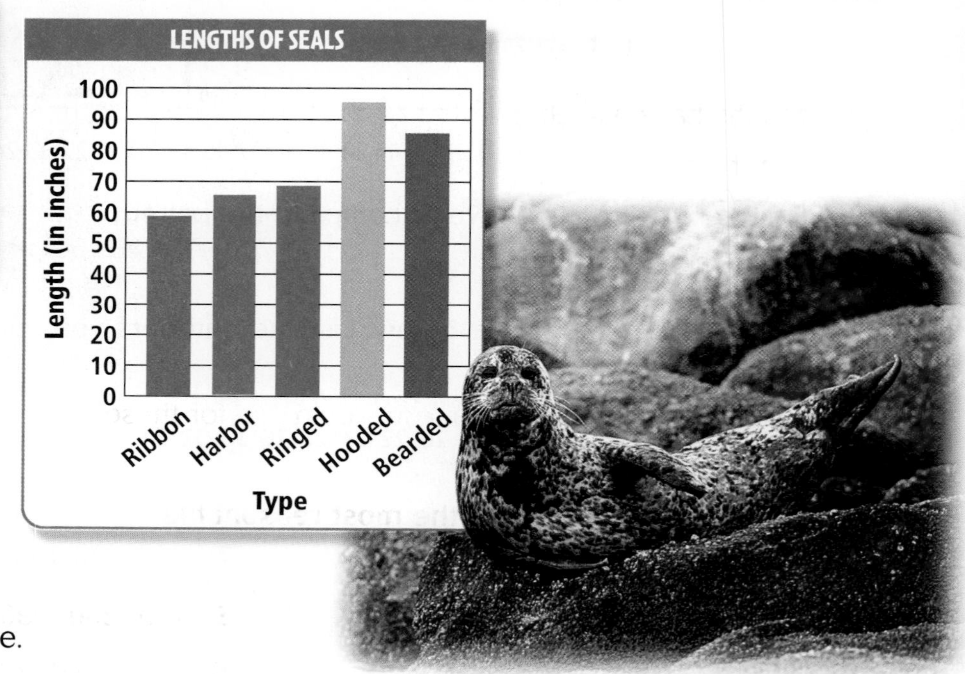

▲ Harbor seal

The data range from 58 to 95 inches. So, Helen used a scale from 0–100 because the greatest number in the data was 95. She used an interval of 10 because most of the data had different digits in the tens place.

Examples

Choose a reasonable scale and interval for the data.

A 12, 15, 18, 30, 22, 16

The data range from 12 to 30.

Scale: 0–35

Interval: 5

B 50, 100, 200, 225, 120, 60

The data range from 50 to 225.

Scale: 0–250

Interval: 50

CALIFORNIA STANDARDS **SDAP 1.0** Students display, analyze, compare, and interpret different data sets, including data sets of different sizes. **MR 1.1** Analyze problems by identifying relationships, distinguishing relevant from irrelevant information, sequencing and prioritizing information, and observing patterns. *also* **SDAP 1.2, MR 2.3, MR 3.1, MR 3.2**

Horizontal or Vertical

The scale on a bar graph can run horizontally or vertically.

The fifth graders at Westhill School voted for their favorite sea mammal. Will's and Anthony's bar graphs show the results in different ways.

Anthony's Bar Graph

Will's Bar Graph

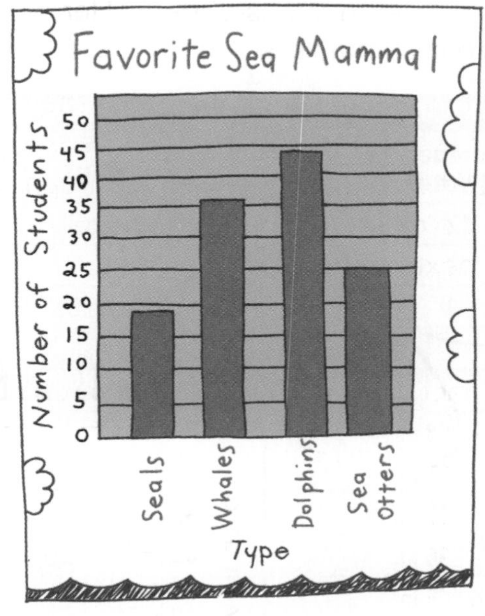

Both graphs have a scale of 0 to 50.

Each graph shows that dolphins are the most popular sea mammal.

- How are the graphs different?

Check

1. **Tell** what scale and interval you would use for these data: 16, 20, 24, 28, 35, 42, 50.

Choose 10, 25, 50, or 100 as the most reasonable interval for the data.

2. 10, 62, 18, 21, 31
3. 105, 200, 990, 800, 620
4. 45, 100, 95, 50, 150
5. 20, 31, 40, 78, 85

Choose the more reasonable scale for the data.

6.

GAME SCORES	
Game	**Score**
1	32
2	15
3	28
4	45

7.

FAVORITE PLACE	
Place	**Votes**
Zoo	9
Beach	5
Park	12
Museum	7

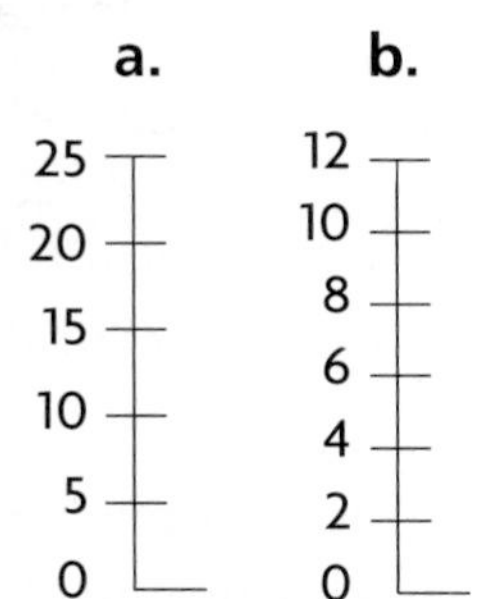

LESSON CONTINUES

Practice and Problem Solving

TECHNOLOGY LINK
To learn more about graphs watch the Harcourt Math Newsroom Video *Roller Coaster.*

Choose 1, 5, 10, or 25 as the most reasonable interval for the data.

8. 2, 5, 6, 3, 1, 4, 7

9. 6, 3, 3, 5, 4

10. 75, 25, 50, 100, 110

11. 10, 35, 40, 20, 30

12. 27, 24, 50, 74, 101

13. 10, 5, 20, 15, 17, 25

Choose the more reasonable scale for the data.

14.

FAVORITE ZOO ANIMALS	
Animal	Number of Votes
Lion	12
Bear	19
Elephant	8
Zebra	5

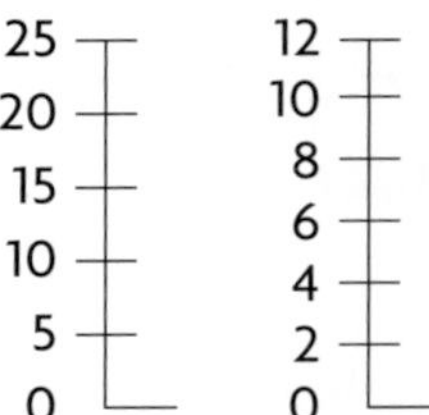

15.

FAVORITE SPORTS	
Sport	Number of Votes
Baseball	20
Soccer	28
Basketball	16
Softball	25
Other	5

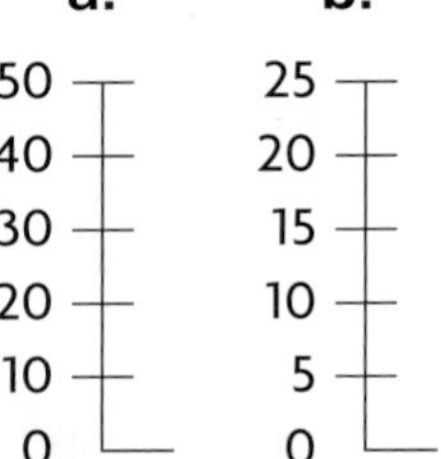

16.

PETS AT HOME	
Pet	Number of Students
Dog	9
Cat	12
Bird	8
Fish	5

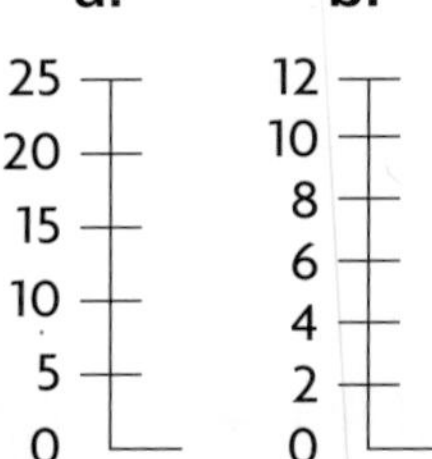

USE DATA For 17–21, use the table.

LENGTHS OF FISH	
Name	Length
Salmon	60 in.
Cod	80 in.
Haddock	40 in.
Sole	25 in.
Halibut	160 in.

17. What is a reasonable scale for the data?

18. What is a reasonable interval for the data?

19. **REASONING** What would be the difference in the heights of two vertical graphs of the same data if one used an interval of 50 and the other used an interval of 10?

20. Make a bar graph for the data. Then compare your graph to that of a classmate. Did you use the same scale and interval?

21. **Write a problem** that can be answered from the data in the table. Solve the problem.

22. During regular seasons, Michael Jordan's lowest average points per game was 22.7. His highest was 37.1. What scale and interval would you use to graph his 13 yearly averages?

23. The U.S. won the women's soccer World Cup in 1999. A record crowd of 90,185 for any U.S. women's sport saw the event. What scale and interval would you use to graph crowds at women's sporting events of this type?

Mixed Review and Test Prep

24. Use mental math to solve the equation. $72 - n = 50$ (p. 74)

25. Order from greatest to least. 4.2805, 4.2058, 4.2508 (p. 24)

26. $\begin{array}{r} 560{,}071 \\ +276{,}188 \\ \hline \end{array}$ (p. 38)

27. **TEST PREP** What is the value of the blue digit? 3,479,045 (p. 4)

A 700 B 7,000 C 70,000 D 700,000

28. $5.3 + 0.8 + 3.7 + 1.4$ (p. 50)

29. Find the mean. 64.5, 72, 60.5, 75 (p. 102)

30. $\begin{array}{r} 87{,}023 \\ -\ 9{,}584 \\ \hline \end{array}$ (p. 38)

31. **TEST PREP** Which is the median of the data 390, 500, 495, 395, 450, 410, 475? (p. 104)

F 395 G 410 H 445 J 450

Thinker's Corner

REASONING The graph represents the top speeds of three roller coasters in the United States.

1. Use the clues to determine the scale and interval for the graph.
   - The scale starts at 0.
   - There is a difference of 10 miles per hour between the top speeds of the Scorpion and the Montu.
2. What is the top speed of the Scorpion roller coaster?
3. What is the top speed of the Python roller coaster?
4. Would an interval of 1 make sense for this graph? Explain your answer.

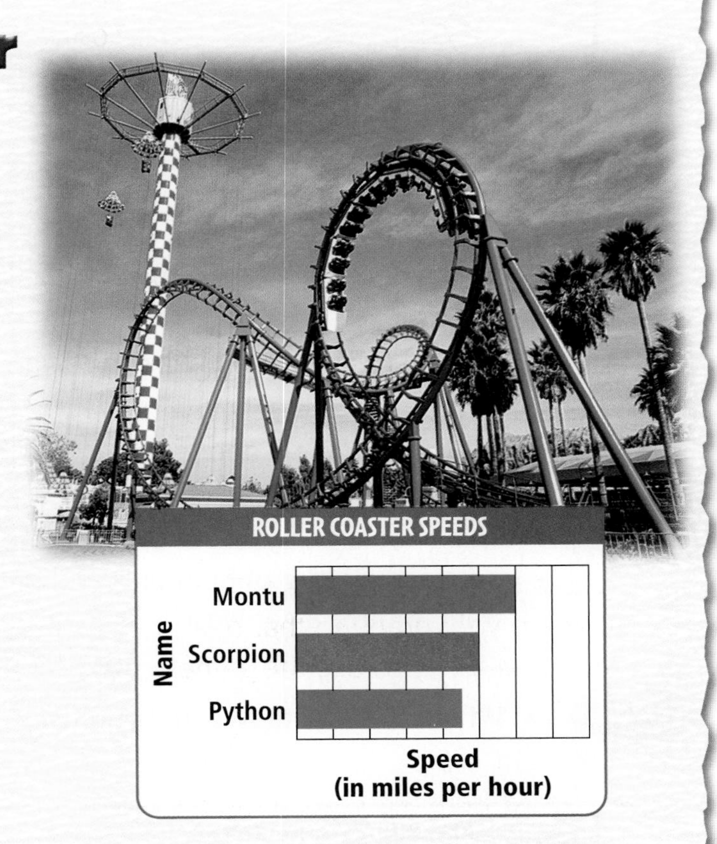

LESSON 2

Problem Solving Strategy

Make a Graph

Quick Review

Find the range.

1. 8, 6, 15, 12, 9, 11

2. 24, 11, 21, 24, 13, 18

3. 58, 46, 55, 54, 51

4. 81, 78, 84, 80, 82

5. 110, 105, 111, 97, 100

PROBLEM Jan surveyed the students in the Science Club to find out what type of fund-raiser they want to have. She organized the data in the table below. Show another display Jan can use to compare the data.

FUND-RAISER CHOICES			
	Bake Sale	**Car Wash**	**Dinner**
Girls	4	6	5
Boys	4	8	3

- What are you asked to do?
- What information will you use?
- Is there information you will not use? If so, what?

- What strategy can you use to solve the problem?

You can *make a graph* to display the data.

- Which graph can you make?

You can make a double-bar graph to compare the girls' responses to the boys' responses. A double-bar graph compares two sets of data in the same graph.

This double-bar graph shows that the greatest number of both girls and boys want to have a car wash.

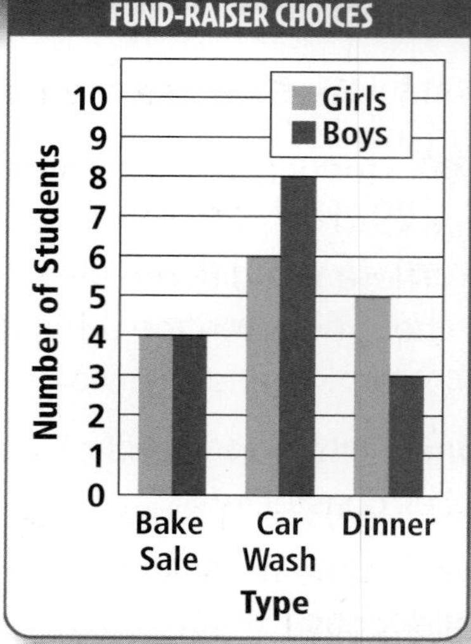

- What other strategy could you use?

CALIFORNIA STANDARDS SDAP 1.0 Students display, analyze, compare, and interpret different data sets, including data sets of different sizes. **SDAP 1.2** Organize and display single-variable data in appropriate graphs and representations and explain which types of graphs are appropriate for various data sets. **MR 2.3** Use a variety of methods, such as words, numbers, symbols, charts, graphs, tables, diagrams, and models, to explain mathematical reasoning. *also* **SDAP 1.3, MR 1.1, MR 2.4, MR 3.2**

Problem Solving Practice

1. **What if** 2 more girls and 4 more boys voted for the car wash? Would the scale or interval change?

USE DATA For 2–4, use the table. Ms. Cook, the librarian, asked students what type of books they wanted her to buy. She recorded the data in the table below.

NEW BOOKS FOR THE LIBRARY			
	Fiction	Biography	Science
Boys	20	14	25
Girls	31	10	28

2. Make a double-bar graph to compare the boys' choices to the girls' choices.

3. What is the most reasonable scale and interval for the graph?
   - **A** scale: 0 to 60; interval: 10
   - **B** scale: 0 to 35; interval: 5
   - **C** scale: 10 to 100; interval 10
   - **D** scale: 0 to 100; interval 5

4. Which of the following is *not* true for the library book data?
   - **F** Fiction books were the most popular choice of the girls.
   - **G** More boys than girls chose biographies.
   - **H** More boys than girls voted for science books.
   - **J** More girls than boys voted.

PROBLEM SOLVING STRATEGIES

- Draw a Diagram or Picture
- Make a Model or Act It Out
- Make an Organized List
- Find a Pattern
- Make a Table or Graph
- Predict and Test
- Work Backward
- Solve a Simpler Problem
- Write an Equation
- Use Logical Reasoning

Mixed Strategy Practice

USE DATA For 5–6, use the table.

5. Both students and parents voted for their choice of food. Make a double-bar graph showing their votes. What was the favorite choice of students? of parents?

6. How many people voted altogether?

7. The Tigers sold 1,018 tickets to the first game and 650 tickets to the second game. Their other games had sales of 835; 1,186; and 946 tickets. What is the mean and the range of the data?

FALL FESTIVAL FOOD CHOICES		
Food	Parents	Students
Burgers	60	72
Hot Dogs	48	65
Chicken	78	75
Ribs	50	42

8. **REASONING** Jason forgot his locker number, but he remembered that the sum of its digits is 11 and that the digits are all odd numbers. The locker numbers are from 1 to 120. What is Jason's locker number?

LESSON 3

Graph Ordered Pairs

Quick Review

Find the median.

1. 8, 12, 8, 9, 7, 11

2. 37, 35, 40, 41, 38

3. 63, 60, 60, 71, 68

4. 53, 58, 53, 54, 58, 60

5. 24, 18, 26, 24, 29

VOCABULARY

ordered pair

Learn

SIGHTSEEING Reading a map of a city is like finding a point on a coordinate grid. A grid is formed by horizontal lines and vertical lines.

Each point on the coordinate grid can be located by using an **ordered pair** of numbers.

- The first number tells how far to move horizontally.
- The second number tells how far to move vertically.

The horizontal and vertical axes intersect at the point (0,0). To get to point *A*, start at (0,0). Move 4 units to the right and 2 units up. The ordered pair (4,2) gives the location of point *A*.

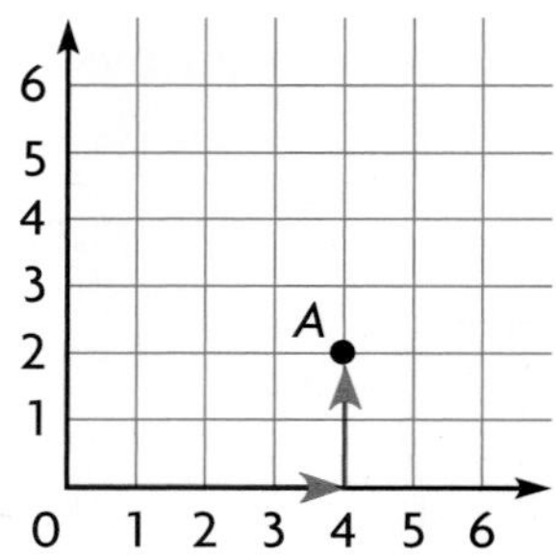

Graph the ordered pair (5,3).

STEP 1
Start at (0,0).

STEP 2
Move 5 units to the right.

STEP 3
Move 3 units up.

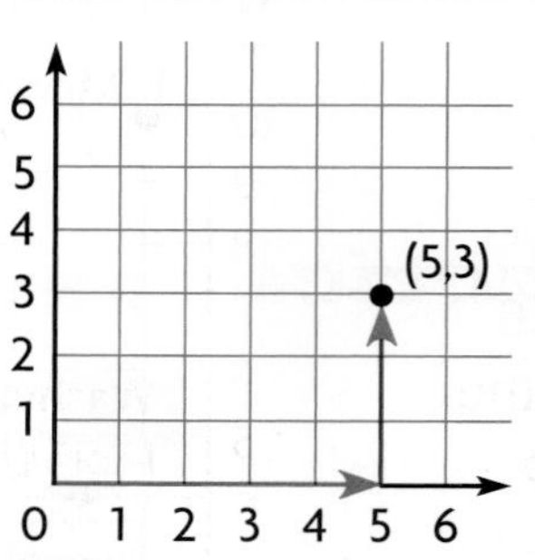

L'Enfant's design for Washington, D.C. is based on a grid pattern of streets. ▼

MATH IDEA A point on a grid can be graphed or identified by using an ordered pair of numbers.

Check

1. **Explain** why you think the numbers used to graph a point on a grid are called an ordered pair.

Graph and label the following points on a coordinate grid.

2. *R* (3,1) **3.** *S* (1,3) **4.** *T* (0,5) **5.** *U* (2,7)

CALIFORNIA STANDARDS **SDAP 1.4** Identify ordered pairs of data from a graph and interpret the meaning of the data in terms of the situation depicted by the graph. **SDAP 1.5** Know how to write ordered pairs correctly. **MR 2.3** Use a variety of methods, such as words, numbers, symbols, charts, graphs, tables, diagrams, and models, to explain mathematical reasoning. *also* **MR 2.0, MR 2.4, MR 3.0, MR 3.2**

Practice and Problem Solving

Graph and label the following points on a coordinate grid.

6. *L* (1,6) **7.** *M* (2,1) **8.** *N* (3,4) **9.** *P* (3,8)

10. *Q* (8,6) **11.** *R* (5,7) **12.** *S* (0,3) **13.** *T* (4,6)

14. *U* (6,3) **15.** *V* (6,0) **16.** *W* (8,2) **17.** *Z* (7,5)

Name the ordered pair for each point.

18. *A* **19.** *B*

20. *C* **21.** *D*

22. *E* **23.** *F*

24. *G* **25.** *H*

26. *I* **27.** *J*

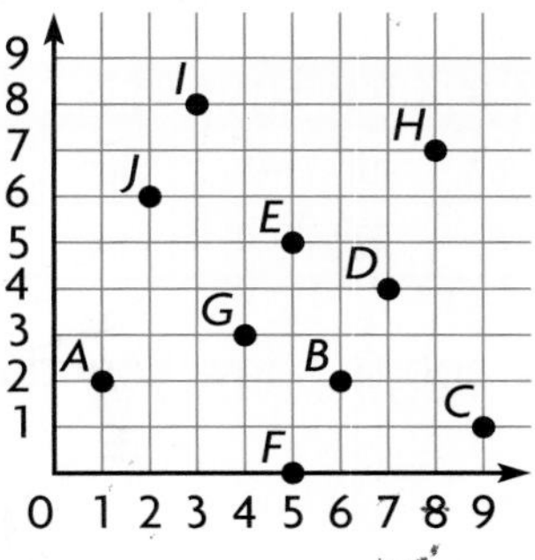

Use the map grid for 28–32.

28. What ordered pair gives the location of the Capitol?

29. The Air and Space Museum is 6 blocks east and 1 block south of the Washington Monument. What ordered pair gives the location of the Air and Space Museum?

30. Write directions on how to walk from the White House to the FBI Building.

31. **What's the Error?** Myra walked 6 blocks east and 1 block north from the Washington Monument to get to the White House. What did Myra do wrong?

32. **Write a problem** using the information in the graph.

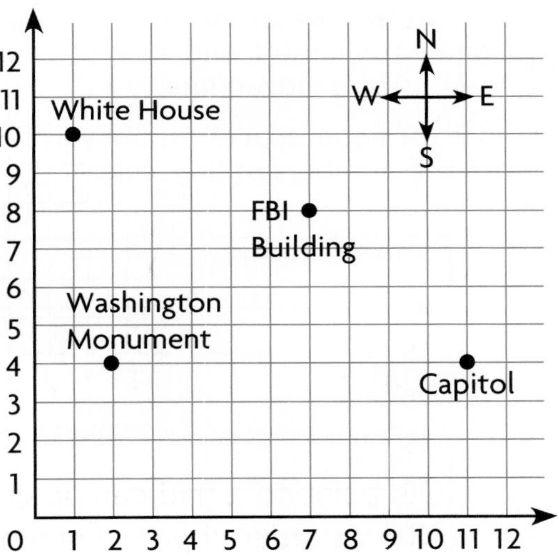

Mixed Review and Test Prep

Order from least to greatest. (p. 24)

33. 0.471, 0.74, 0.56, 0.052

34. 3.57, 3.5, 3.601, 3.498

35. 5.61 + 3.85 (p. 50)

36. 9.05 − 4.126 (p. 50)

37. **TEST PREP** In which number does 3 have the greatest value? (p. 4)

A 8,710,543 **C** 436,125
B 1,910,328 **D** 345,602

Extra Practice page H39, Set B

LESSON 4

Make Line Graphs

Learn

BAKED APPLE! The table below shows normal monthly temperatures for New York City, nicknamed the Big Apple.

NEW YORK CITY MONTHLY NORMAL TEMPERATURE (°F)						
Month	Jun	Jul	Aug	Sep	Oct	Nov
Temperature	72	77	76	68	58	48

You can make a line graph to show how data change over time.

STEP 1

Decide on a scale for the data. The temperatures range from 48° to 77°. Scale: 0–80

Choose an interval that would work well with the data. Interval: 5

STEP 2

Scales on graphs always start at zero. Since there are no data between 0°F and 48°F, show a *break in the scale* using a zigzag line.

Then mark off equal spaces on the vertical axis and write the numbers of the scale.

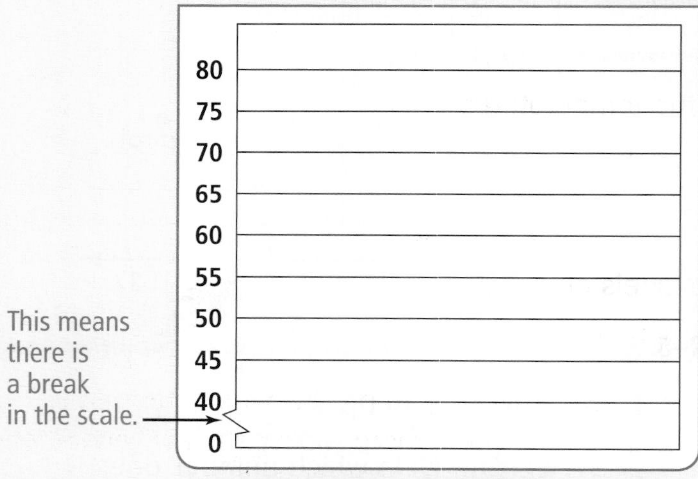

▲ Statue of Liberty in July

Quick Review

What scale would you use to graph the data?

1. 24, 10, 15, 8, 19
2. 2, 5, 7, 8, 11
3. 50, 48, 51, 54, 51
4. 75, 78, 70, 66, 71
5. 95, 80, 101, 87, 99

CALIFORNIA STANDARDS SDAP 1.0 Students display, analyze, compare, and interpret different data sets, including data sets of different sizes. **SDAP 1.2** Organize and display single-variable data in appropriate graphs and representations and explain which types of graphs are appropriate for various data sets. **MR 2.3** Use a variety of methods, such as words, numbers, symbols, charts, graphs, tables, diagrams, and models, to explain mathematical reasoning. *also* **MR 1.0, MR 1.1, MR 2.4, MR 3.2**

STEP 3

Mark off equal spaces on the horizontal axis and write the months along the bottom of the graph.

Label the vertical and horizontal axes.

Write a title for the graph.

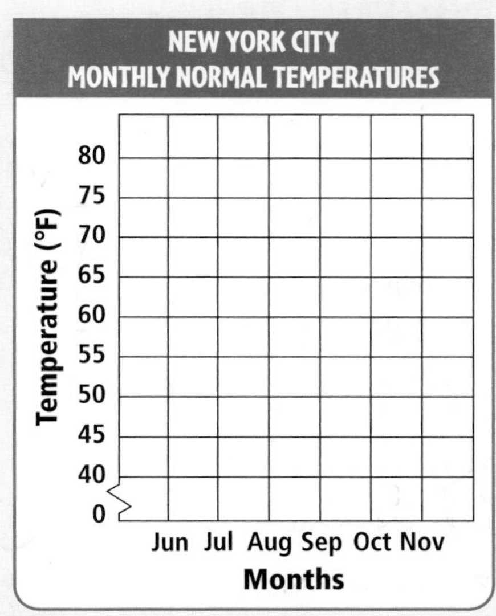

STEP 4

Write ordered pairs for the data.

(Jun,72), (Jul,77), (Aug,76), (Sep,68), (Oct,58), (Nov,48)

Plot the points. Place a point at 72°F above June. Do the same for the other months and temperatures.

Connect the points with straight lines.

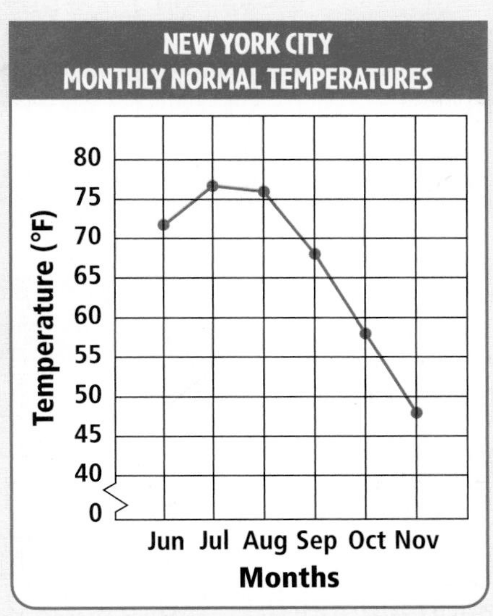

▲ Central Park in October

- What does the graph tell you about how the temperatures in New York City change over time?

Check

1. **Explain** the purpose of the title and the labels on a graph.

USE DATA Use the line graph above for 2–5.

2. Which three months have about the same normal temperature?

3. What do you predict would happen to the line if there were data for December?

4. Does the normal temperature in New York City increase or decrease from August to November? How do you know from the graph?

5. **What if** you used an interval of 1 on the graph? How would the placement of the points change?

LESSON CONTINUES 

Practice and Problem Solving

USE DATA For 6–8, use the data in the table.

6. How would you label and title a graph of this data?
7. What would be an appropriate scale and interval for the wages?
8. What years would you show on the bottom of the graph?

FEDERAL MINIMUM WAGE					
Year	1960	1970	1980	1990	2000
Amount	\$1.15	\$2.30	\$3.25	\$3.80	\$5.15

Make a line graph for each set of data. Describe the data shown on your graph.

9.

INCHES OF RAINFALL				
Month	Jun	Jul	Aug	Sep
Inches	3	5	6	4

10.

MONTHLY CAR SALES				
Month	Jan	Feb	Mar	Apr
Cars Sold	25	18	23	32

11.

SKATING RINK ATTENDANCE					
Week	1	2	3	4	5
People	150	155	165	180	185

12.

WEEKLY FOOD COST					
Week	1	2	3	4	5
Amount	\$80	\$100	\$95	\$90	\$105

USE DATA For 13–19, use the graph.

13. What does the ordered pair (Jul,3.5) tell you?
14. Does the median monthly precipitation in New York City increase or decrease from July to October? How can you tell from the graph?
15. **What if** the data included the month of December with a median precipitation amount of 2.2 inches? How would the graph change? How would the line change on the graph?
16. What is the difference in the median precipitation amounts for July and October?
17. Describe the precipitation in New York City from June through November.
18. **Write About It** Tell what the zigzag line on the left of the graph represents.
19. **Write a problem** using the data in the graph.

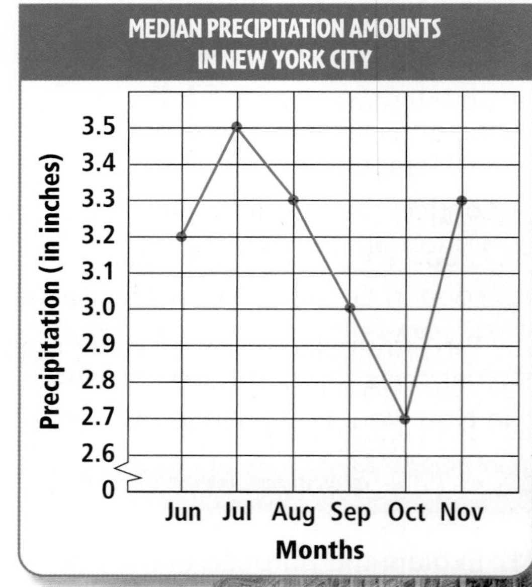

20. **REASONING** **What if** a line graph showed the number of fifth-grade students from 1980 to 2000. Do you think the line on the graph would go up or down? Explain.

21. The temperature on Friday at 12 noon was 23°F. A cold front moved through and by midnight, the temperature had dropped 25 degrees. What was the temperature at midnight?

Mixed Review and Test Prep

22. Order 5.02, 5.22, 5.022, and 5.002 from least to greatest. (p. 24)

23. Find the mean, median, and mode for 17, 14, 10, 17, 18, 22, and 14. (p. 104)

24. Marla practices her flute 20 minutes longer than usual for a total time of 65 minutes. How long does she usually practice? Write an equation and solve. (p. 74)

25. $\begin{array}{r} 49{,}506 \\ +\ 7{,}506 \\ \hline \end{array}$ (p. 38)

26. $\begin{array}{r} 35{,}221 \\ -\ 7{,}261 \\ \hline \end{array}$ (p. 38)

27. Evaluate $25 + n$ for $n = 36$. (p. 68)

28. Solve the equation $x - 5 = 23$. (p. 74)

29. **TEST PREP** Round 3.4506 to the nearest hundredth. (p. 46)

    A 3.45 **C** 3.456
    B 3.451 **D** 3.5

30. **TEST PREP** Round 8,745,123 to the nearest ten thousand. (p. 32)

    F 8,700,000 **H** 8,750,000
    G 8,745,000 **J** 9,000,000

LINKUP to Reading

STRATEGY • DRAW CONCLUSIONS This line graph shows the bald eagle population recorded in Georgia over a 15-year period.

Use the information shown on the graph to draw conclusions.

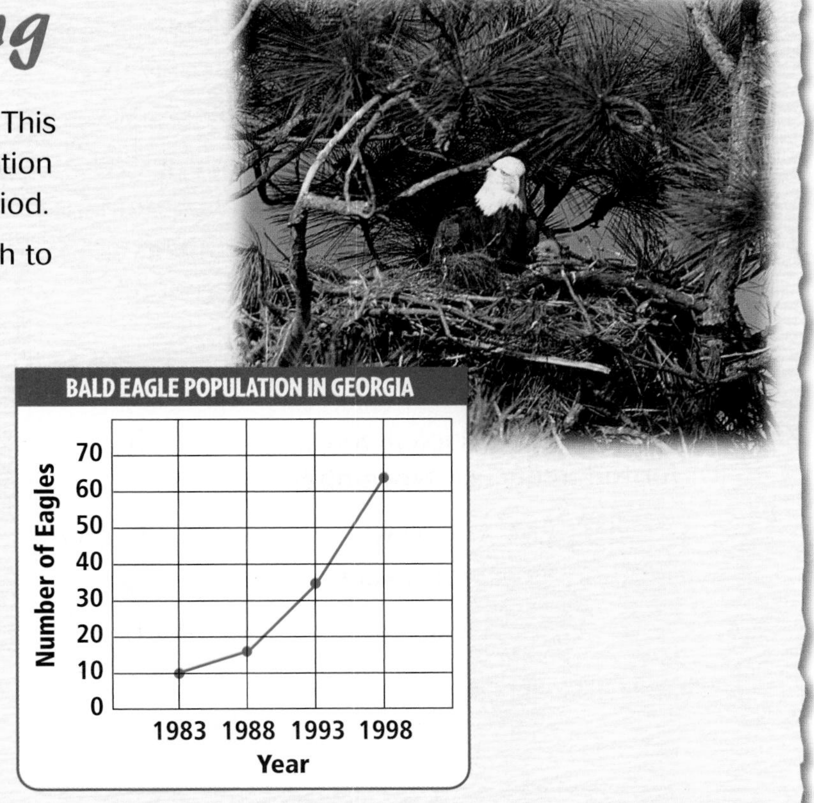

1. What happened to the bald eagle population in Georgia from 1983 to 1998?

2. During which 5-year period did the bald eagle population increase the most?

3. Based on the data, what do you predict the bald eagle population in Georgia will be in 2003? How did you make your prediction?

Extra Practice page H39, Set C

LESSON 5

HANDS ON
Histograms

Quick Review

What interval would you use to graph the data?

1. 3, 8, 4, 2, 5
2. 25, 52, 75, 99
3. 10, 16, 35, 42, 26
4. 20, 15, 30, 50, 41
5. 4, 10, 7, 18, 21

VOCABULARY

histogram

Explore

A **histogram** is a bar graph that shows the number of times data occur within intervals.

The data show the number of minutes some students spent on homework each night.

MINUTES SPENT ON HOMEWORK						
15	20	25	35	10	10	45
30	20	15	40	25	5	10
20	25	20	30	45	35	

Follow the steps to make a histogram of the data.

STEP 1

- Find the range of the data. — Range: 45 − 5 = 40
- Decide on the scale. — The scale could be 0 to 50 minutes.
- Select an interval to divide the data equally. — Use 5 intervals of 10 minutes.

The intervals are used to show the number of times data occur within them.

STEP 2

- Make a frequency table with the intervals.
- Tally the data for each interval.
- Record the frequencies.

Number of Minutes Frequency	Tally	
0–10 minutes	IIII	4
11–20 minutes	~~IIII~~ I	6
21–30 minutes	~~IIII~~	5
31–40 minutes	III	3
41–50 minutes	II	2

STEP 3

- Use the frequency table to make the histogram.
- The intervals are along one axis. The scale for the frequencies is along the other axis.

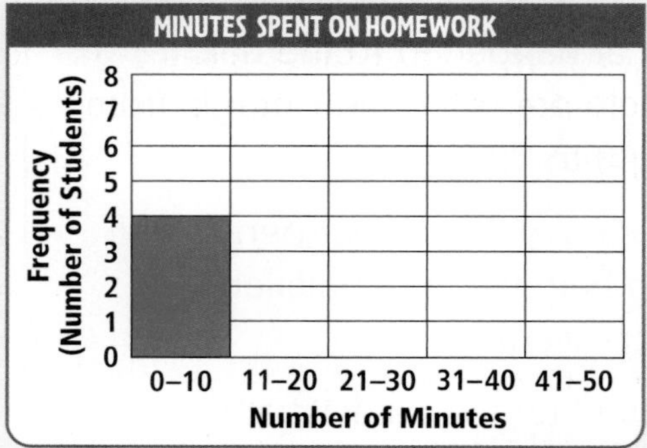

CALIFORNIA STANDARDS SDAP 1.2 Organize and display single-variable data in appropriate graphs and representations and explain which types of graphs are appropriate for various data sets. **MR 2.3** Use a variety of methods, such as words, numbers, symbols, charts, graphs, tables, diagrams, and models, to explain mathematical reasoning. *also* **SDAP 1.0, MR 2.4, MR 3.2, MR 3.3**

Try it

a. Copy the histogram on page 128.

b. Graph the frequency for each interval. The bars should be connected side-by-side.

Connect

Look at the bar graph and histogram at the right.

- In which graph can you compare the number of people listening to each type of music?
- What is the same about both graphs? What is different?

Practice

Decide which graph would better represent the data below, a bar graph or histogram. Then make each graph.

1.

Heights	Number of Students
48–51 inches	2
52–55 inches	5
56–59 inches	12
60–63 inches	7

2.

Name of Trail	Time to Complete
Nature	30 minutes
Challenge	90 minutes
Stroller	25 minutes
Scenic	35 minutes

3. Give five intervals that you could use to make a histogram for this data.

MINUTES SPENT DOING CHORES							
20	25	30	15	40	55	60	42
90	80	45	30	25	20	40	65

4. **REASONING** How do you think the number of students in each interval would change in Table 1 if you made six height intervals?

Mixed Review and Test Prep

5. $6.16 + 0.9$ (p. 50)

6. $13.9 - 3.0$ (p. 50)

7. Find the value of $x - 8$ for $x = 32$. (p. 68)

8. $107{,}211 + 47{,}998$ (p. 38)

9. **TEST PREP** Which is less than 3.178? (p. 24)

A 3.278 **C** 3.187

B 3.198 **D** 3.173

LESSON 6

Choose the Appropriate Graph

Quick Review

Find the range.

1. 24, 18, 22, 31, 27
2. 56, 57, 60, 59, 55
3. 40, 44, 50, 56, 60
4. 19, 30, 36, 24, 27
5. 72, 85, 97, 81, 90

Learn

BEST CHOICE To choose the best graph for a set of data, determine the type of data and how to analyze them. Using the correct graph makes it easier to interpret data and solve problems.

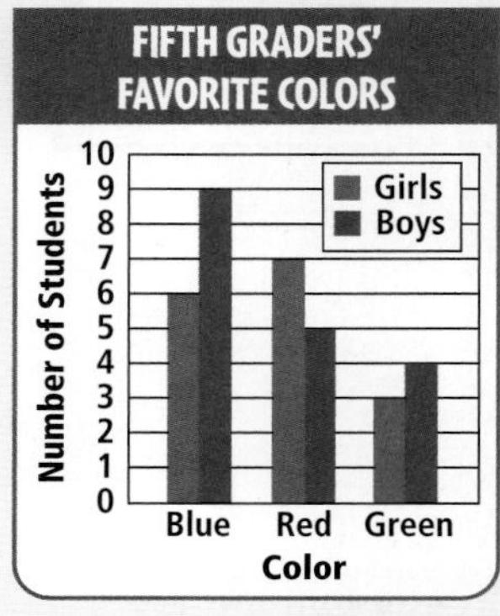

Use a bar graph to compare data by category.

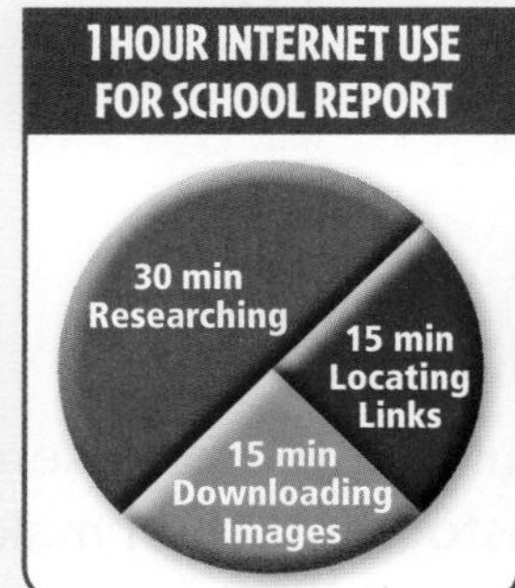

Use a circle graph to compare parts of a group to the whole group.

Stem	Leaves
7	1 3 6
8	4 7
9	5 5 6 7

Angela's Bowling Scores

Use a stem-and-leaf plot to organize data by place value.

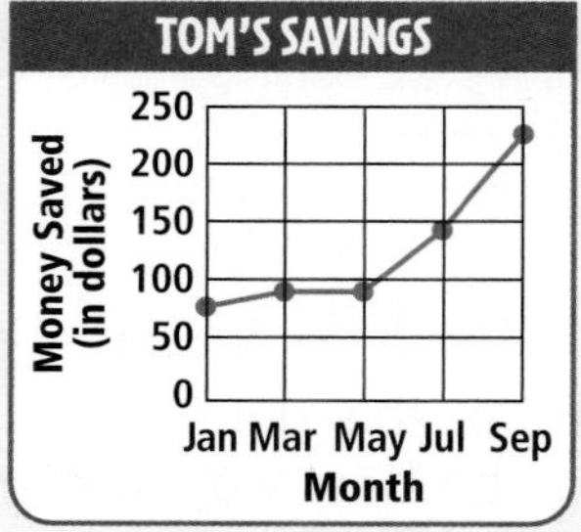

Use a line graph to show how data change over time.

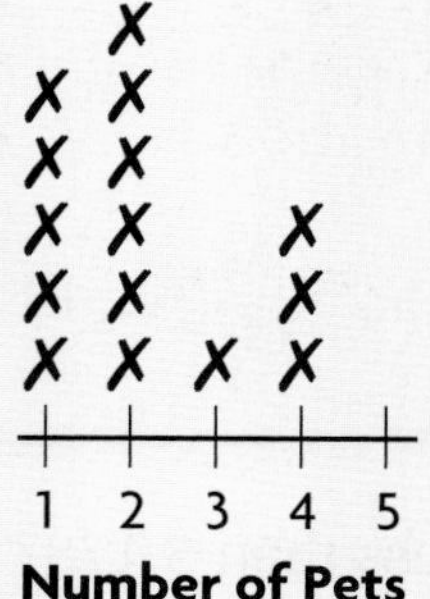

Use a line plot to keep count of data as they are collected or to show frequencies of repeated amounts.

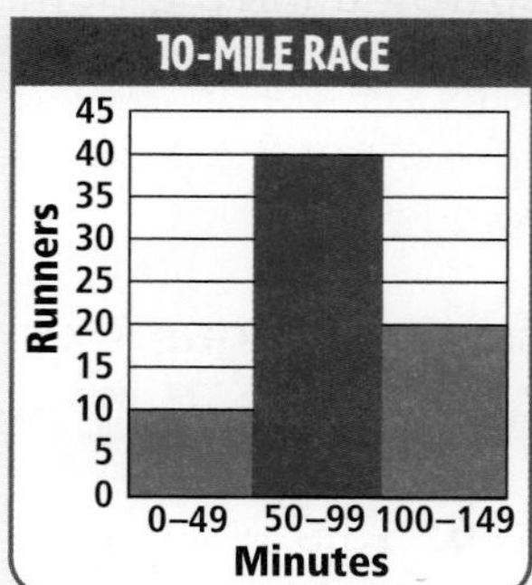

Use a histogram to show the number of times data occur within certain intervals.

CALIFORNIA STANDARDS SDAP 1.2 Organize and display single-variable data in appropriate graphs and representations (e.g., histogram, circle graphs) and explain which types of graphs are appropriate for various data sets. **MR 2.3** Use a variety of methods, such as words, numbers, symbols, charts, graphs, tables, diagrams, and models, to explain mathematical reasoning. *also* **SDAP 1.0, MR 2.4, MR 3.2, MR 3.3**

What is the Best Graph for the Data?

Fernando, Tamara, and Joshua are giving a report on weather. They each used different ways to display the data in the table. Who chose the best graph or plot?

MEAN TEMPERATURE ON APRIL 11	
City	**Temperature**
Chicago, IL	48°F
Los Angeles, CA	60°F
Miami, FL	75°F
Washington, DC	57°F

Fernando's Line Plot

FERNANDO'S LINE PLOT

One X above 75°F means there was one city with that temperature.

48°F 57°F 60°F 75°F

Temperature

TEMPERATURE ON APRIL 11

You do not need to count the temperature frequencies. You need to list the cities. So, a line plot is *not* the best way to display the data.

Tamara's Bar Graph

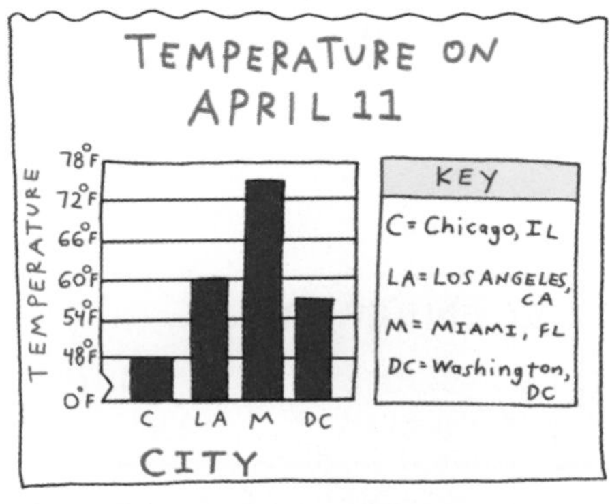

The table compares the temperatures in different cities on the same day. So, a bar graph is an appropriate way to display the data.

Joshua's Line Graph

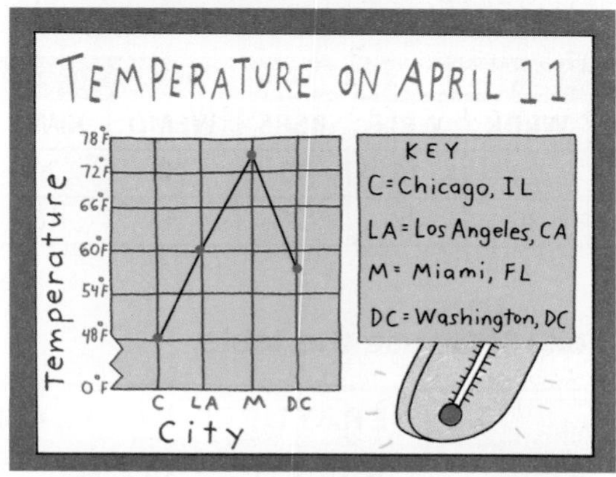

The temperatures are for different cities, not over periods of time. So, a line graph is *not* the best way to display the data.

- Why is Tamara's graph the best way to display the data?

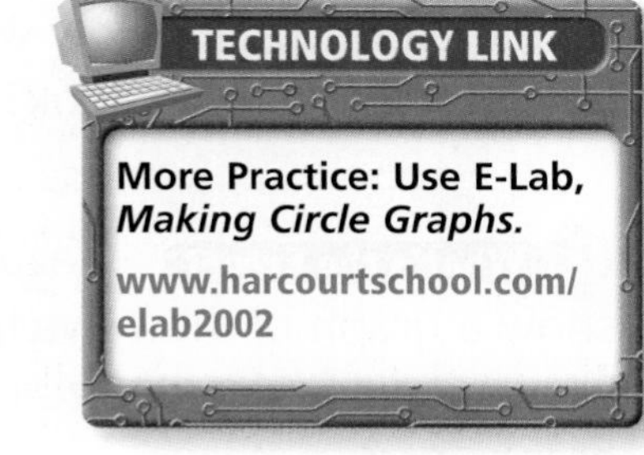

TECHNOLOGY LINK
More Practice: Use E-Lab, *Making Circle Graphs.*
www.harcourtschool.com/elab2002

Check

1. **Describe** the kind of temperature data that would be appropriate for a line graph.

For 2–5, choose the best type of graph or plot for the data. Explain your choice.

2. scores for a math test
3. number of students in four schools
4. yearbook sales over a four-week period
5. minutes spent on each activity in a one-hour class

Practice and Problem Solving

For 6–9, choose the best type of graph or plot for the data. Explain your choice.

6. favorite ice-cream flavors of fifth and sixth graders
7. money a business earned from January to June
8. survey of the number of CDs your friends own
9. number of houses built each decade in the United States

Draw the graph or plot that best displays each set of data.

10.

NUMBER OF FAMILY MEMBERS					
Number of Members	2	3	4	5	6
Frequency	9	7	4	2	1

11.

DISTANCE HEATHER TRAVELED ON A TRIP						
Day	1	2	3	4	5	6
Miles	250	100	0	50	100	250

12.

FAVORITE RADIO STATIONS					
Station	**WRDK**	**WRLF**	**KSRS**	**WEMD**	**KWEB**
Boys	24	16	12	28	20
Girls	36	22	20	4	18

USE DATA **For 13–16, use the table.**

TEMPERATURES IN ATLANTA, GA, ON DECEMBER 31						
Time	6:30 A.M.	9:30 A.M.	12:30 P.M.	3:30 P.M.	6:30 P.M.	9:30 P.M.
Temperature	35°F	45°F	58°F	62°F	55°F	48°F

13. What type of graph would you choose to display the data? Explain.
14. Would it be reasonable to use an interval of 1°? Explain.
15. When did the least change in temperature occur? How would the line between those hours look?
16. **What if** another row of data is added to the table to show the temperatures on December 30? Could you display the new data on the same graph? Explain.
17. **What's the Error?** Kazuo wants to show a graph of the ways he spends his allowance. He decided to make a bar graph.
18. You recorded the high and low temperatures for your city during one week. What type of graph will show this data?
19. Reina surveyed students to find their favorite type of birthday party. There were 5 choices. What type of graph will show the choices of boys and girls?
20. **Write a problem** using the data about favorite radio stations.

Mixed Review and Test Prep

USE DATA For 21–24, use the histogram.

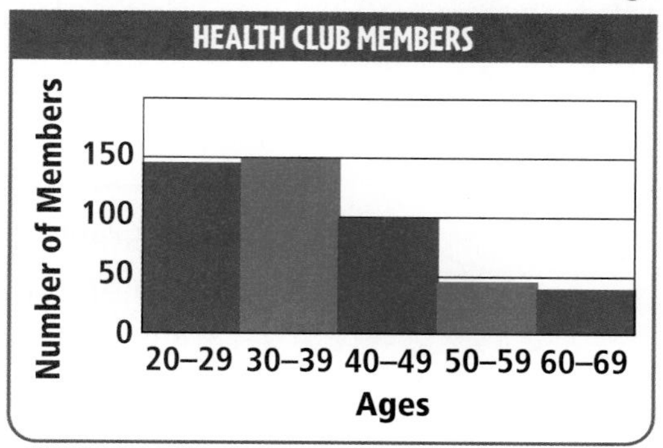

21. Which age range has the greatest number of health club members? (p. 128)
22. About how many members are younger than 40? (p. 128)
23. Which age range has the least number of members? (p. 128)
24. Which two age ranges have almost the same number of members? (p. 128)
25. Find the range, mean, median, and mode for 8, 3, 15, 12, 17. (p. 104)
26. Find the range, mean, median, and mode for 46, 14, 5, 22, 92, 54, 5. (p. 104)
27. Write 19.0876 in expanded form. (p. 20)
28. Write two million, fifty-five thousand, sixty-eight in standard form. (p. 4)
29. **TEST PREP** John is twice as old as Enrico. The sum of their ages is 21. How old is John?

    A 3 **C** 14
    B 7 **D** 18

30. **TEST PREP** Lisa bought a book for $2 and 3 pens. She spent $3.50. How much did each pen cost? (p. 86)

    F $0.50 **H** $1.50
    G $1.00 **J** $5.50

LINKUP to Science

Many of the waterfalls in Yosemite National Park were formed by ancient glaciers. Glaciers shape the land by erosion and by moving and depositing rock debris.

The graph shows the heights of several waterfalls in Yosemite National Park.

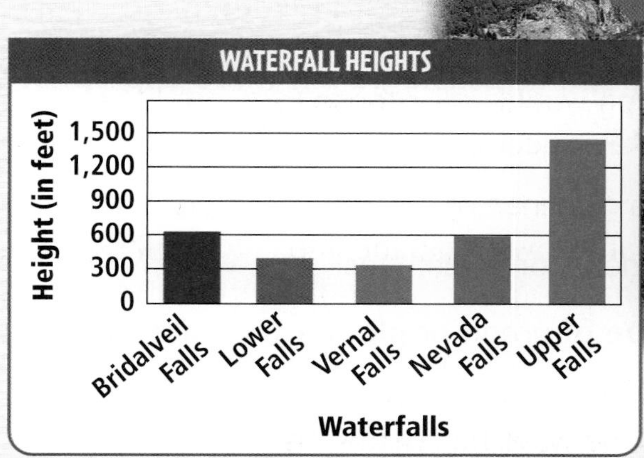

USE DATA For 1–4, use the graph.

1. What are the scale and interval for the graph?
2. How much higher is Upper Falls than Lower Falls?
3. How would you change the graph to include Sentinel Falls at a height of 2,000 feet?
4. **What's the Question?** The answer is Vernal Falls.

Review/Test

CHECK VOCABULARY AND CONCEPTS

Choose the best term from the box.

histogram
line graph
scale

1. The series of numbers placed at fixed distances on a graph is the __?__. (p. 116)
2. A bar graph that shows the number of times data occur within certain intervals is a __?__. (p. 128)

CHECK SKILLS

Choose 1, 5, 20, or 50 as the most reasonable interval for the data. (pp. 116–119)

3. 35, 15, 20, 45, 60, 55
4. 75, 48, 100, 125, 150, 200
5. 85, 100, 61, 20, 45, 80
6. 3, 13, 10, 7, 5, 12

Name a reasonable scale for the data. (pp. 116–119)

7. Josie found information on the price of in-line skates. The prices were $30, $45, $48, and $53.
8. The average miles per gallon for several new cars are 26, 17, 32, 12, and 28.

Graph and label the following points on a coordinate grid. (pp. 122–123)

9. *A* (1,2)
10. *B* (2,4)
11. *C* (3,6)
12. *D* (4,5)
13. *E* (0,3)

Tell whether a bar graph or a histogram is more appropriate. (pp. 128–129)

14. ages of 75 ice-skating competitors
15. frequency of five girls' names
16. scores of 50 students on a math test

Choose the best type of graph or plot for the data. Explain your choice. (pp. 130–133)

17. money earned working one summer
18. number of skyscrapers in four cities

CHECK PROBLEM SOLVING

For 19–20, use the table. (pp. 120–121)

19. Construct the best type of graph to display the data.
20. Which category had the biggest difference between the girls' and boys' choices?

FAVORITE TYPE OF MOVIE

	Girls	Boys
Comedy	10	16
Mystery	15	20
Science Fiction	3	25
Animated	15	15

Cumulative Review

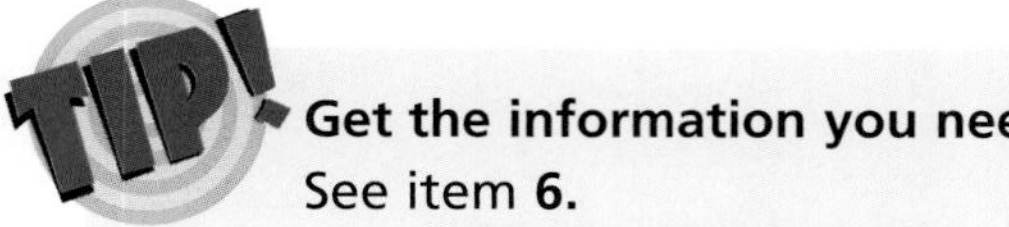

Get the information you need.
See item **6.**

A circle graph represents a whole. The whole is 24 hours, so find the fraction words that represent 8 out of 24 hours.

Also see problem **3,** p. H63.

For 1–9, choose the best answer.

For 1–3, use the table.

LARGEST SNAILS	
Snail	**Length in Inches**
California sea hare	30
Apple snail	20
Trumpet snail	15
African land snail	10

1. Which graph or table is most appropriate to display the snail data?

 A bar graph **C** line graph
 B frequency table **D** circle graph

2. Which is a reasonable interval for graphing the snail data?

 F 1 **H** 15
 G 5 **J** 20

3. What is the range of the snail lengths?

 A 10 **C** 20
 B 15 **D** 30

4. After 6 people quit, there were 20 people on the debate team. Which equation can be used to find n, the number of people that were on the team before some quit?

 F $20 + n = 6$ **H** $6 + n = 20$
 G $20 - 6 = n$ **J** $n - 6 = 20$

5. $\begin{array}{r} 980{,}368 \\ -654{,}325 \\ \hline \end{array}$

 A 336,043 **C** 326,043
 B 326,049 **D** Not here

6. Tim made a circle graph to show how he spent his day yesterday. He worked for 8 hours. What part of his graph should be for work?

 F one-half **H** one-fourth
 G one-third **J** three-fourths

7. What ordered pair gives the location of a point three spaces to the right of (0,0) and four spaces up?

 A (0,3) **C** (4,0)
 B (4,3) **D** (3,4)

8. Which data set would best fit on a graph with a scale 0, 2, 4, 6, 8, 10, 12, 14?

 F 5, 11, 10, 8, 4 **H** 9, 3, 6, 12, 15
 G 2, 10, 20, 4, 8 **J** 8, 4, 16, 12, 1

9. This table shows the heights of some trees in a park.

TREE HEIGHT	
Kind of Tree	**Height (in feet)**
Pine	67
Oak	56
Elm	75
Maple	66
Cedar	51

 Which scale would be best to use when making a bar graph for the heights of these trees?

 A 0 to 45, increasing by 5
 B 0 to 80, increasing by 10
 C 0 to 200, increasing by 50
 D 0 to 400, increasing by 100

UNIT 2

MATH DETECTIVE

Build Means

REASONING Apply your number sense and what you know about means to crack each case. Use the clues to unravel each mystery. Find the missing numbers in each set of numbers.

Case 1

Clue 1: Four different numbers have a mean of 80.

Clue 2: The least is 70 and the greatest is 90.

Clue 3: Each of the numbers is a multiple of 5.

The missing numbers are ■ and ■.

Case 2

Clue 1: Four different numbers have a mean of 25.

Clue 2: Each of the numbers is a multiple of 10.

Clue 3: The difference between the least and greatest numbers is 30.

The missing numbers are ■, ■, ■, and ■.

Case 3

Clue 1: The mean of four consecutive odd numbers is an even number.

Clue 2: Each of the numbers is between 70 and 80.

Clue 3: Two of the numbers round to 70. The others round to 80.

The missing numbers are ■, ■, ■, and ■.

Think It Over!

- **Write About It** Explain how you solved each case.
- **STRETCH YOUR THINKING** Each of three different whole numbers is greater than 4 and less than 10. Each of another three whole numbers is greater than 14 and less than 20. What are the least possible and greatest possible means of the set of six different numbers?

UNIT 2

Challenge

Relationships in Graphs

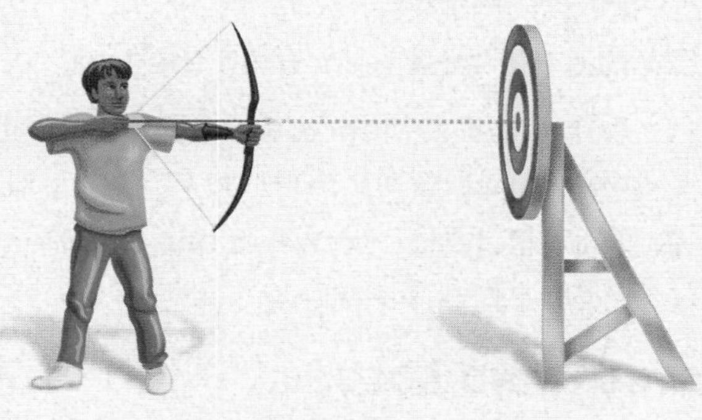

Graphs can show relationships between two different variables, such as distance and time.

If you shoot an arrow toward a target, you can use a graph to show the relationship between the distance of the arrow from you and the time the arrow is moving to the target.

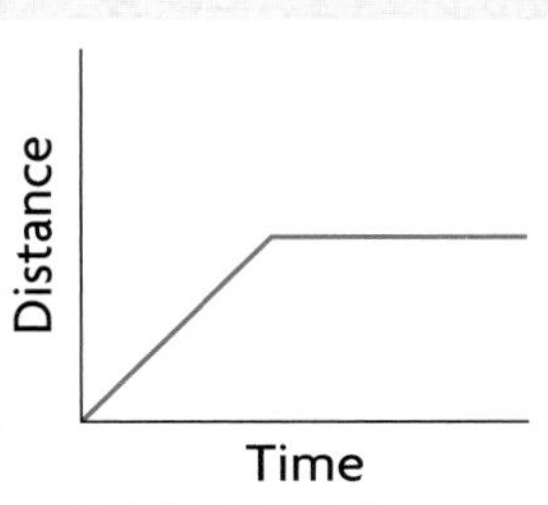

This graph shows the relationship between the distance of the arrow from you and time.

The graph begins with a sloped line, which shows that the distance of the arrow from you is increasing after it is shot.

The flat part of the graph begins when the arrow hits the target and stops. After this time, the distance does not change.

Talk About It

- How would the appearance of the graph change if the target were only half as far away?

Try It

For 1–4, tell which graph illustrates the described situation. Explain.

a.

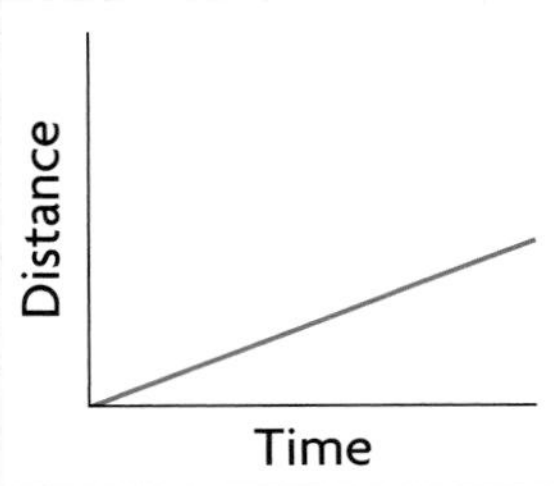

b.

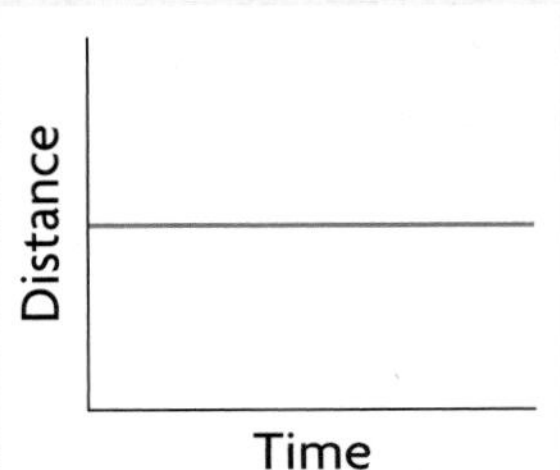

c.

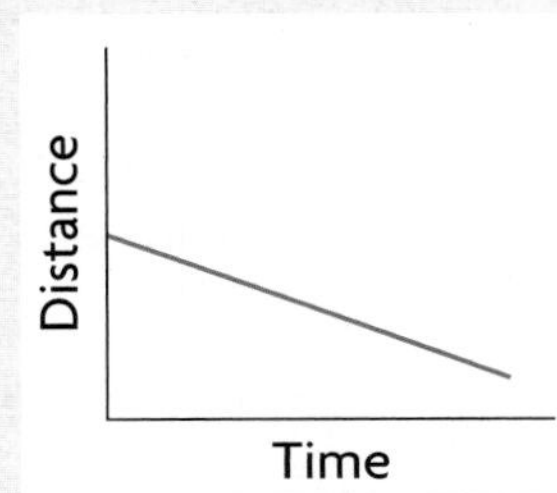

1. a rocket's distance from the launchpad during liftoff

2. a carousel rider's distance from the center of the carousel

3. a hot-air balloon's distance from the ground when descending

4. a golf ball's distance from the golfer immediately after it is hit

UNIT 2

Study Guide and Review

VOCABULARY

Choose the best term from the box.

equation evaluate interval mean median

1. A number sentence that uses the equal sign to show that two amounts are equal is an __?__. (p. 72)
2. The difference between one number and the next on a scale is the __?__. (p. 116)

STUDY AND SOLVE

Chapter 5

Write an expression with a variable.

Chris had 12 eggs. He cooked some of them for breakfast. Let b = number of eggs cooked.

12 eggs	**minus**	**some eggs**
↓	↓	↓
12	−	b

Expression: $12 - b$

Write an expression with a variable. (pp. 68–71)

3. Susan had collected 20 rocks. Then she found some more.
4. Michael had 20 pencils. He bought 7 more and gave some to Pauline.

Solve addition and subtraction equations.

Solve the equation.

$17 + m = 30$ **Think:** 17 plus what number equals 30?
$m = 13$

Solve the equation. (pp. 74–77)

5. $52 = n + 6$
6. $n + 9 = 13$
7. $n - 14 = 39$
8. $48 = 60 - n$

Chapter 6

Evaluate algebraic expressions.

$8 \times d$ Evaluate the expression if $d = 4$.
↓
8×4 Replace d with 4.
↓
32

Evaluate each expression. (pp. 86–87)

9. $15 \times n$ if $n = 3$
10. $37 - n$ if $n = 15$
11. $(7 - n) \times 2$ if $n = 1$
12. $(4 \times n) - 3$ if $n = 3$

Solve equations by using multiplication properties.

$8 \times n = 5 \times 8$ Commutative Property
$8 \times 5 = 5 \times 8$

Solve the equation. Identify the property used. (pp. 90–93)

13. $1 \times n = 4$
14. $37 \times 0 = n$

Chapter 7

Interpret data by finding the mean, median, mode, and range.

Data: 4, 6, 7, 8, 8, 9
mean: Add data; then divide by number of addends.
$4 + 6 + 7 + 8 + 8 + 9 = 42$
$42 \div 6 = 7$
median: Arrange the data in order then find the middle number. 7.5
mode: Find the number that occurs most often. 8
range: Subtract the least number from the greatest.
$9 - 4 = 5$

Find the mean, median, mode, and range for each set of data. (pp. 98–105)

15. 24, 10, 4, 10

16. 11, 12, 31, 28

17. 44, 36, 38, 38, 39

18. 84, 82, 84, 89, 92, 95, 90

19. 113, 115, 126, 101, 99, 98, 118

Chapter 8

Choose appropriate scales and intervals.

Choose a reasonable scale and interval for this data set: 12, 15, 18, 30, 22, 16.

The data range from 12 to 30.

Scale: 0 to 30

Interval: 5

Name a reasonable scale and interval for the data. (pp. 116–119)

20. 74, 100, 55, 125, 27

21. 20, 10, 5, 15, 30, 25

22. 50, 10, 40, 20, 30, 60

23. 120, 80, 60, 100, 40

Display data in tables and graphs.

SAN FRANCISCO FIVE-DAY FORECAST					
April	1	2	3	4	5
Temperature	60	62	51	55	57

You can make a line graph to show how data change over time.

Use the data in the table. (pp. 124–127)

24. How would you label the graph?

25. What dates would you show along the bottom of the graph?

26. Make a line graph to determine when the greatest change in temperature is expected to occur.

PROBLEM SOLVING PRACTICE

Solve. (pp. 80–81, 88–89)

27. Mr. MacDonald is walking around the park for exercise. He walks 2 miles along each side. The perimeter of the park forms a five-sided figure. How far does he walk?

28. Ken scored a 96 on his test and had 24 correct answers. Write an equation to find the mean number of points each correct answer was worth, and solve.

California Connections

YOSEMITE IN THE SUMMER

More than 3 million people visit Yosemite National Park each year. Visitors come in both summer and winter to view the spectacular scenery and to participate in many outdoor activities.

Most visitors come to Yosemite in the summer, when the entire park is open. ▼

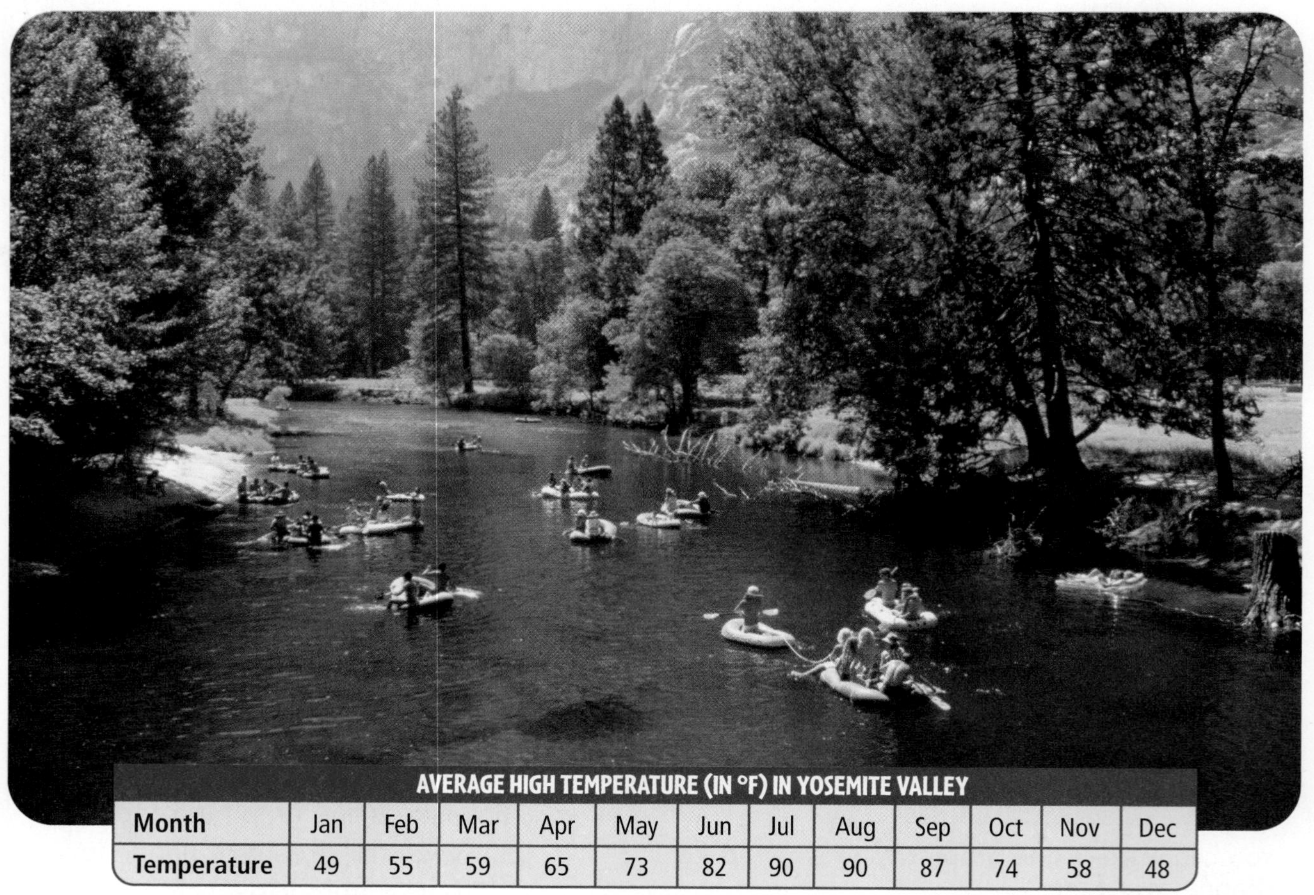

AVERAGE HIGH TEMPERATURE (IN °F) IN YOSEMITE VALLEY

Month	Jan	Feb	Mar	Apr	May	Jun	Jul	Aug	Sep	Oct	Nov	Dec
Temperature	49	55	59	65	73	82	90	90	87	74	58	48

1. Find the median, mode, and range for the temperature data.
2. Find the mean temperature for the first four months of the year. Explain how you found the mean.
3. What kind of graph would be most appropriate for displaying the temperature data? Explain.
4. What would be a reasonable scale for the data? a reasonable interval for the data?
5. Make a line graph of the data. Then, compare your graph to that of a classmate. How are your graphs alike? How are they different?
6. **Write a problem** that can be answered by looking at your graph. Solve the problem.

YOSEMITE IN THE WINTER

Many people prefer to visit Yosemite in the winter, when it is less crowded and when they can enjoy winter activities, such as downhill and cross-country skiing and snowboarding.

Yosemite has over 500 kilometers of ski trails and roads. ▼

For 1–2, write an expression with a variable. Explain what the variable represents.

1. Badger Pass ski area has 1 cable tow and some chairlifts.
2. Some people pay $30 each to rent a snowboard.

For 3–4, write and solve an equation. Explain what the variable represents.

3. Suppose you spend $44 to buy an adult's and a child's ski lift ticket at Badger Pass. The adult's ticket costs $28. How much does the child's ticket cost?
4. The most popular cross-country ski trail is a 32-kilometer track between Badger Pass ski area and Glacier Point. If you ski for 20 kilometers and then stop to rest, how many kilometers do you have left?

▲ Badger Pass ski area is California's oldest ski area.

CHAPTER 9 Multiply Whole Numbers

Reflecting telescopes use mirrors to help us see distant objects in our solar system and beyond. Each telescope has an opening that allows light to be gathered by the mirrors. The larger the opening, the more powerful the telescope, and the farther you can see. The largest reflecting telescopes are the twin Keck Telescopes. Their mirrors are about 2 times the size of the Hale Telescope's mirror, which is shown in the graph. How big are the mirrors of the Keck I and Keck II?

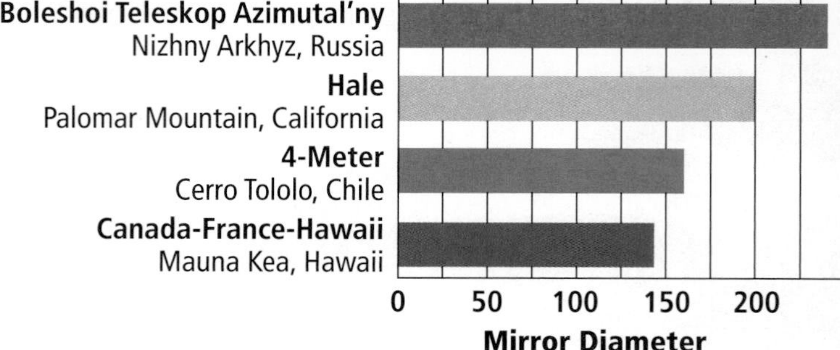

The twin Keck Telescopes stand eight stories tall on the 13,800-foot Mauna Kea summit and weigh 300 tons each.

Check What You Know

Use this page to help you review and remember important skills needed for Chapter 9.

MULTIPLY BY TENS AND HUNDREDS (See p. H11.)

Find the product.

1. 1×100 **2.** 2×10 **3.** 5×40

4. 5×400 **5.** 8×50 **6.** 8×500

Algebra Evaluate the expression.

7. $3 \times n$ if $n = 10$ **8.** $6 \times n$ if $n = 10$ **9.** $8 \times n$ if $n = 200$

10. $4 \times n$ if $n = 100$ **11.** $80 \times n$ if $n = 10$ **12.** $500 \times n$ if $n = 40$

MULTIPLY 2 DIGITS (See p. H14.)

Find the product.

13. 12×6 **14.** 19×9 **15.** 14×12 **16.** 17×4

17. 50×9 **18.** 24×15 **19.** 28×3 **20.** 20×6

21. 32×7 **22.** 64×2 **23.** 54×23 **24.** 94×7

25. 29×24 **26.** 34×27 **27.** 72×31 **28.** 93×45

MULTIPLY 3 DIGITS BY 1 DIGIT (See p. H11.)

Find the product.

29. 130×6 **30.** 134×4 **31.** 104×5 **32.** 260×9

33. 250×2 **34.** 111×3 **35.** 561×9 **36.** 903×8

37. 115×8 **38.** 123×7 **39.** 214×3 **40.** 776×4

LESSON 1

Estimation: Patterns in Multiples

Quick Review

Round to the nearest 10.

1. 93 **2.** 16

3. 139 **4.** 147 **5.** 998

VOCABULARY

multiple

Learn

ORBITING NUMBERS A **multiple** of a whole number is the product of a given whole number and another whole number. For example, multiples of 10 include 10, 20, 30, and 40 since $1 \times 10 = 10$, $2 \times 10 = 20$, $3 \times 10 = 30$, and $4 \times 10 = 40$.

Use multiples of 10 to help you estimate mentally.

Each hour, the Hubble Space Telescope transmits enough data to fill about 83 computer floppy disks. About how many computer floppy disks are filled with data in a day?

Estimate. 24×83
↓ ↓
20×80 Round each factor.

$2 \times 8 = 16$ **Think:** Use the basic fact 2×8.
$2 \times 80 = 160$ Look for patterns in multiples of 10.
$20 \times 80 = 1{,}600$

So, about 1,600 computer floppy disks are filled.

- Look for a pattern in the factors and products. What do you notice?

▲ **Each day, the Hubble Space Telescope collects enough data to fill an encyclopedia. Since 1990, over 2.5 terabytes of data (1 terabyte = 1 million megabytes) have been collected.**

Examples

A Estimate. 8×522

8×500 **Think:** $8 \times 5 = 40$
$8 \times 500 = 4{,}000$
↑ ↑
2 zeros 2 zeros

So, 8×522 is about 4,000.

B Estimate. 73×862

70×900 **Think:** $7 \times 9 = 63$
$70 \times 900 = 63{,}000$
↑ ↑ ↑
1 zero + 2 zeros 3 zeros

So, 73×862 is about 63,000.

MATH IDEA You can use multiples of 10 to estimate products mentally. Count the number of zeros in the rounded factors. Place that number of zeros at the end of the product of the basic fact.

CALIFORNIA STANDARDS NS 1.1 Estimate, round, and manipulate very large and very small numbers. **MR 1.1** Analyze problems by identifying relationships, distinguishing relevant from irrelevant information, sequencing and prioritizing information, and observing patterns. *also* **MR 2.0, MR 2.2, MR 2.4, MR 3.0, MR 3.2**

Check

1. **REASONING** **Explain** why 400×5 has more zeros in the product than 300×5.

TECHNOLOGY LINK
To learn more about multiplication, watch the Harcourt Math Newsroom Video *Hubble Images*.

Estimate each product.

2. $4 \times 1{,}453$
3. 67×824
4. $32 \times 1{,}839$
5. $6 \times 3{,}609$
6. 42×690
7. $99 \times 2{,}092$

Practice and Problem Solving

Estimate each product.

8. 34×123
9. 5×960
10. 82×53
11. 69×681
12. $58 \times 2{,}234$
13. 2×490
14. $7 \times 2{,}589$
15. 42×524
16. 301×12
17. 291×8
18. 341×86
19. 915×52
20. $5{,}451 \times 83$
21. 290×82
22. 723×3
23. 963×42
24. 389×49
25. $6{,}715 \times 39$
26. The product 20×600 has 3 zeros. How many zeros does the product $20 \times 600{,}000$ have?
27. How can you tell that $3{,}000{,}000 \times 6{,}000 = 30{,}000 \times 600{,}000$ without multiplying?
28. **What's the Question?** Juan bought 52 bicycles for his store. Each bicycle cost \$117. The estimate is \$5,000.
29. The product 20×50 has 3 zeros. How many zeros does the product 20×500 have?
30. Gail received \$10 as a gift. She bought a magazine that cost \$2 and a marker set. She has \$3.45 left. How much did she spend on the marker set?
31. **Write About It** Write the steps you would follow to find $8 \times 5 \times 9$ using the multiplication properties and multiples of 10.

Mixed Review and Test Prep

Evaluate each expression.

32. $8 \times n$ if $n = 3$ (p. 86)
33. $(6 \times 4) + n$ if $n = 9$ (p. 86)
34. Identify the addition property you use to solve $20 + (3 + 2) = (20 + n) + 2$. (p. 78)
35. Write four and six hundred fifty-nine ten-thousandths in standard form. (p. 20)
36. **TEST PREP** Which number is between 10 million and 100 million? (p. 4)
   - **A** 999,999
   - **B** 9,999,999
   - **C** 99,999,999
   - **D** 999,999,999

Extra Practice page H40, Set A

LESSON 2

Multiply by 1-Digit Numbers

Quick Review

1. 29 × 3
2. 39 × 3
3. 41 × 5
4. 49 × 5
5. 92 × 6

Learn

HEAVYWEIGHTS An African elephant, the heaviest land mammal, can weigh as much as 14,632 pounds. A humpback whale weighs as much as 4 African elephants. About how much does a humpback whale weigh?

Find $4 \times 14{,}632$.

$4 \times 15{,}000 = 60{,}000$ Estimate. Round 14,632 to the nearest thousand.

MATH IDEA To multiply a greater number by a 1-digit number, multiply the same way you multiply a 2-digit number. Just repeat the same steps for all the digits.

▲ The heaviest known 90-foot female blue whale weighed 418,000 pounds. That's about 29 African elephants!

STEP 1	STEP 2	STEP 3	STEP 4
Multiply the ones and tens. 4 × 2 = 8 ones 4 × 3 = 12 tens Regroup.	Multiply the hundreds. 4 × 6 = 24 hundreds Add the 1 regrouped hundred. Regroup.	Multiply the thousands. 4 × 4 = 16 thousands Add the 2 regrouped thousands. Regroup.	Multiply the ten thousands. 4 × 1 = 4 ten thousands Add the 1 regrouped ten thousand.
1 14,632 × 4 28	2 1 14,632 × 4 528	1 2 1 14,632 × 4 8,528	1 2 1 14,632 × 4 58,528

So, a humpback whale weighs about 58,528 pounds. This is close to the estimate of 60,000, so the answer is reasonable.

Example **Use the Distributive Property.**

$5 \times 3{,}214 = n$

$(5 \times 3{,}000) + (5 \times 200) + (5 \times 10) + (5 \times 4) = n$

$15{,}000 + 1{,}000 + 50 + 20 = n$

$16{,}070 = n$

So, $n = 16{,}070$.

CALIFORNIA STANDARDS **NS 1.0** Students compute with very large and very small numbers, positive integers, decimals, and fractions and understand the relationship between decimals, fractions, and percents. They understand the relative magnitudes of numbers. **MR 2.1** Use estimation to verify the reasonableness of calculated results. **MR 2.2** Apply strategies and results from simpler problems to more complex problems. *also* **NS 1.1, AF 1.3, MR 1.2, MR 2.4, MR 2.6, MR 3.1, MR 3.2**

Check

1. **Tell** how you know that 115,275 is a reasonable product for $38,425 \times 3$.

Find the product. Estimate to check.

2. $\begin{array}{r} 2,231 \\ \times \quad 3 \\ \hline \end{array}$

3. $\begin{array}{r} 348 \\ \times \quad 5 \\ \hline \end{array}$

4. $\begin{array}{r} 52,113 \\ \times \quad 2 \\ \hline \end{array}$

5. $\begin{array}{r} 608,123 \\ \times \quad 7 \\ \hline \end{array}$

6. $\begin{array}{r} 2,391,224 \\ \times \quad 4 \\ \hline \end{array}$

7. $36,825 \times 9$

8. $5 \times 309,224$

9. $4,649,321 \times 4$

Practice and Problem Solving

Find the product. Estimate to check.

10. $\begin{array}{r} 24,567 \\ \times \quad 8 \\ \hline \end{array}$

11. $\begin{array}{r} 88,112 \\ \times \quad 7 \\ \hline \end{array}$

12. $\begin{array}{r} 240,531 \\ \times \quad 9 \\ \hline \end{array}$

13. $\begin{array}{r} 6,417,263 \\ \times \quad 4 \\ \hline \end{array}$

14. $\begin{array}{r} 5,882,917 \\ \times \quad 6 \\ \hline \end{array}$

15. $8 \times 102,337$

16. $826,243 \times 3$

17. $7,952,344 \times 5$

Algebra **Use the Distributive Property to solve for *n*.**

18. $5,200,450 \times 6 = n$

19. $292,453 \times 9 = n$

USE DATA **For 20–21, use the table.**

ANIMAL WEIGHTS	
Animal	**Average Weight (in pounds)**
Hippopotamus	5,512
White rhinoceros	7,937
Whale shark	46,297
Blue whale	286,600

20. To ship animals to new locations, a zoo needs to determine how much the animals weigh. How much would 8 hippopotamuses weigh altogether?

21. **MENTAL MATH** Suppose the shipping weight limit is 50,000 pounds. Can a zoo ship 6 white rhinoceroses altogether? Explain.

22. **Write About It** Explain the steps you follow to multiply by a 1-digit number.

Mixed Review and Test Prep

23. $\begin{array}{r} 52,954 \\ -48,393 \\ \hline \end{array}$ (p. 38)

24. $\begin{array}{r} 62,752 \\ +\ 3,694 \\ \hline \end{array}$ (p. 38)

25. Write 2.057 in word form. (p. 20)

26. Find the next number in the pattern 20, 16, 12, 8, ■. Describe the pattern. (p. 36)

27. **TEST PREP** Identify the addition property you use to solve $n + 0 = 24$. (p. 78)

A Zero Property
B Associative Property
C Property of One
D Commutative Property

Extra Practice page H40, Set B

LESSON 3

Multiply by 2-Digit Numbers

Quick Review

1. 22 × 4
2. 40 × 6
3. 11 × 5
4. 600 × 70
5. 32 × 30

Learn

PEDAL POWER Eugene is starting a cycling training program. He plans to ride a total of 315 minutes each week for the next 26 weeks. How many minutes does he plan to ride altogether?

Find 26 × 315.

30 × 300 = 9,000 Estimate. Round each factor.

STEP 1

Multiply by the ones.

```
   3
  315
×  26
 1890 ← 6 × 315
```

STEP 2

Multiply by the tens.

```
   1
   3
  315
×  26
 1890
 6300 ← 20 × 315
```

STEP 3

Add the partial products.

```
    1
    3
   315
 ×  26
  1890 ← partial products
+ 6300 ← partial products
 8,190
```

So, Eugene will ride 8,190 minutes altogether. This is close to the estimate of 9,000, so the answer is reasonable.

Examples

A

```
    7
   29
  ×18
  232 ←  8 × 29
 +290 ← 10 × 29
  522
```

B

```
      1
    704
  ×  23
   2112 ←  3 × 704
 +14080 ← 20 × 704
 16,192
```

C

```
    2 2 1
    1 1
    2,342
  ×    64
     9368 ←  4 × 2,342
 +140520 ← 60 × 2,342
  149,888
```

- In Example B, what happens to the regrouped digit, 1, when there is a zero in the factor?

MATH IDEA To multiply by a 2-digit number, first multiply by the ones and then the tens. Add the partial products.

▲ **The fastest average speed in the Tour de France was 24.547 mph, by Miguel Indurain of Spain in 1992.**

CALIFORNIA STANDARDS NS 1.0 Students compute with very large and very small numbers, positive integers, decimals, and fractions and understand the relationship between decimals, fractions, and percents. They understand the relative magnitudes of numbers. **MR 2.1** Use estimation to verify the reasonableness of calculated results. *also* **NS 1.1, MR 1.0, MR 2.0, MR 2.3, MR 3.0, MR 3.2, MR 3.3**

Check

1. **REASONING** **Explain** why the second partial product is always greater than the first partial product when you multiply two 2-digit numbers.

TECHNOLOGY LINK

More Practice: Use Mighty Math Calculating Crew, *Nick Knack*, Level M.

Find the product. Estimate to check.

2. 91×43
3. 348×15
4. 24×16
5. $1,924 \times 21$
6. 710×13
7. 801×61
8. $4,506 \times 94$

Practice and Problem Solving

Find the product. Estimate to check.

9. 78×53
10. 159×26
11. 500×43
12. $2,391 \times 62$
13. $8,705 \times 98$
14. 16×52
15. 119×54
16. $2,610 \times 63$
17. $3,059 \times 72$

Evaluate each expression for $n = 10, 20, \text{and } 30$.

18. $11n$
19. $23 \times n$
20. $n \times 200$
21. $32n$
22. $14 \times n$

USE DATA For 23–26, copy and complete the table.

WINGBEATS IN FLIGHT

	Animal	Beats per min	Beats per hr
23.	Hummingbird	5,400	■
24.	Bat	1,200	■
25.	Butterfly	640	■
26.	Stork	180	■

For 27–28, use the table.

27. **REASONING** A swift beats its wings 2 times as fast as a stork. On average, how many times does a swift beat its wings in three hours?

28. **Write a problem** that can be solved using multiplication with the data about wingbeats.

Mixed Review and Test Prep

Write an equivalent decimal. (p. 22)

29. 0.023
30. 8.020
31. Evaluate $(365 \times n) - 12$ if $n = 8$. (p. 86)
32. $1,369,521 + 3,987,564$ (p. 38)
33. **TEST PREP** I drove 220 miles at 55 mph. How many hours did I drive? Which equation models this problem? (p. 86)

 A $55 + h = 220$
 B $55 \times h = 220$
 C $h \times 4 = 55$
 D $55 - h = 220$

Extra Practice page H40, Set C

LESSON 4

Multiply Greater Numbers

Quick Review

1. 7 × 3 × 3
2. 6 × 6 × 10
3. 5 × 3 × 100
4. 60 × 2 × 2
5. 200 × 2 × 3

Learn

STARDUST Every day 121 tons of cosmic dust—debris from outer space—enter Earth's atmosphere. If it could be shoveled onto one heap, it would be as big as a two-story house. How many tons of cosmic dust enter Earth's atmosphere every year?

Find 365 × 121.

400 × 100 = 40,000 Estimate.

◀ In 1986 the Vega probes were hit and damaged by 12,000 dust specks from Halley's Comet.

STEP 1

Multiply by the ones.

```
   365
  ×121
   365 ← 1 × 365
```

STEP 2

Multiply by the tens.

```
   11
   365
  ×121
   365
  7300 ← 20 × 365
```

STEP 3

Multiply by the hundreds.

```
    11
    365
   ×121
    365
   7300
  36500 ← 100 × 365
```

STEP 4

Add the partial products.

```
    11
    365
   ×121
    365
   7300
 +36500
 44,165
```

So, 44,165 tons of cosmic dust enter Earth's atmosphere every year. This is close to the estimate of 40,000, so the answer is reasonable.

Examples

A

```
     23,483
  ×      12
     46 966 ← 2 × 23,483
  +234 830 ← 10 × 23,483
    281,796
```

B

```
      530,721
  ×        36
    3 184 326 ← 6 × 530,721
  +15 921 630 ← 30 × 530,721
   19,105,956
```

C

```
       8,246,974
  ×          340   These zeros can
       0 000 000 ← be omitted.
     329 878 960 ← 40 × 8,246,974
  +2 474 092 200 ← 300 × 8,246,974
   2,803,971,160
```

- In Example C, what multiplication property was used to multiply the ones?

CALIFORNIA STANDARDS **NS 1.1** Estimate, round, and manipulate very large and very small numbers. **MR 2.1** Use estimation to verify the reasonableness of calculated results. **MR 2.2** Apply strategies and results from simpler problems to more complex problems. *also* **NS 1.0, MR 2.0, MR 2.3, MR 2.4, MR 3.0, MR 3.1, MR 3.2, MR 3.3**

Check

1. **Describe** how you multiply a 3-digit number by a 3-digit number. How do you multiply a 4-digit number by a 4-digit number?

Find the product. Estimate to check.

2. 289×124
3. $70{,}152 \times 71$
4. 400×789
5. $105{,}874 \times 36$
6. 923×202
7. 482×195
8. $10{,}248 \times 32$
9. $295{,}804 \times 55$

Practice and Problem Solving

Find the product. Estimate to check.

10. 481×142
11. $28{,}407 \times 572$
12. 403×830
13. $103{,}982 \times 395$
14. $908{,}056 \times 60$
15. $973{,}204 \times 31$
16. $28 \times 69{,}434$
17. $21{,}002 \times 93$
18. $2{,}318 \times 47$
19. $9{,}289 \times 106$
20. $3{,}008 \times 23$
21. $361{,}204 \times 837$
22. $4{,}271 \times 1{,}252$
23. $6{,}482{,}763 \times 89$
24. $6{,}800 \times 5{,}056$
25. $5{,}472{,}315 \times 958$
26. **NUMBER SENSE** Which estimate, 140,000,000 or 240,000,000, is a closer estimate for the product of 793,321 and 260? Explain.
27. Mr. Calvin sells sheets of stamps at the post office. There are 100 stamps on each sheet. Each sheet costs $33. Last week he sold a total of 192 sheets. How much money did he collect for the sheets?
28. Each day last week, people from 15 schools went to Kennedy Space Center. From each school, 210 students and 6 teachers attended. How many students and teachers attended during the 7 days?
29. 

Why is it a good idea to estimate the product when finding the exact product?
30. Which circle is larger?

Mixed Review and Test Prep

31. Find the mean for the data set 1,162; 34; 875; 933. (p. 102)
32. Identify the multiplication property you use to solve $3 \times 4 = n \times 3$. (p. 90)
33. $20{,}638 \times 5$ (p. 146)
34. $5{,}693 \times 23$ (p. 148)
35. **TEST PREP** Find the value of n.
$31 - (n + 3) = 19$ (p. 74)
A 5 **B** 7 **C** 9 **D** 11

Extra Practice page H40, Set D

LESSON 5

Problem Solving Skill

Evaluate Answers for Reasonableness

Quick Review

1. 500 × 30
2. 600 × 400
3. 40 × 50
4. 70 × 8,000
5. 6,000,000 × 50

PAPER PRODUCTS An important part of solving any problem is to check whether or not your answer is reasonable. You can use estimation to check the reasonableness of an answer.

One cord of hardwood weighs about 2 tons. Each cord can make an average of 4,384,000 commemorative-sized postage stamps. Tyler says 36 cords will make 157,824,000 stamps. Elena says 36 cords will make 39,456,000 stamps. Whose answer is reasonable?

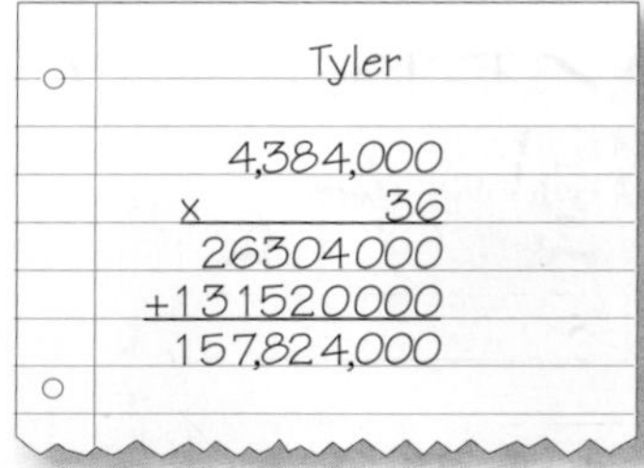

Elena

4,384,000
x 36
26304000
+13152000
39,456,000

Estimate 4,384,000 × 36.

4,000,000 × 40 = 160,000,000 Estimate. Round each factor.

Compare Tyler's and Elena's answers to the estimate. Tyler's answer of 157,824,000 is close to the estimate of 160,000,000. Elena's answer of 39,456,000 is not close to the estimate of 160,000,000. So, Tyler's answer is reasonable.

▲ **More than 5 new trees are planted each year for every man, woman, and child in the United States, and millions more regrow naturally from seeds and sprouts.**

Talk About It

- Why is it a good idea to determine if an answer is reasonable?
- What error do you think Elena made?

CALIFORNIA STANDARDS MR 2.1 Use estimation to verify the reasonableness of calculated results. **MR 3.1** Evaluate the reasonableness of the solution in the context of the original situation. *also* **NS 1.0, NS 1.1, O━NS 2.1, MR 1.0, MR 1.1, MR 1.2, MR 2.0, MR 2.4, MR 2.6, MR 3.0**

Problem Solving Practice

1. In 1996, Sweden had more phones per person than any other country. There were about 68 phones for each 100 inhabitants. There were about 8,863,000 people in Sweden. Peggy says there were about 6,026,840 phones in Sweden. Is her answer reasonable? Explain.

2. **REASONING** **What if** you round both factors in a multiplication problem up to a greater number? How does your estimate compare to the exact answer? How do you know?

Choose the most reasonable answer without solving.

3. One cord of hardwood makes 942 hardcover books. Each book has 100 pages. How many pages does one cord make altogether?

   A 942 pages
   B 9,420 pages
   C 94,200 pages
   D 942,000 pages

4. A newspaper company prints 818,231 newspapers daily. About how many newspapers does it print in a year?

   F 5,727,617 newspapers
   G 42,548,012 newspapers
   H 298,654,315 newspapers
   J 2,986,543,150 newspapers

Mixed Applications

USE DATA **For 5–6, use the graph.**

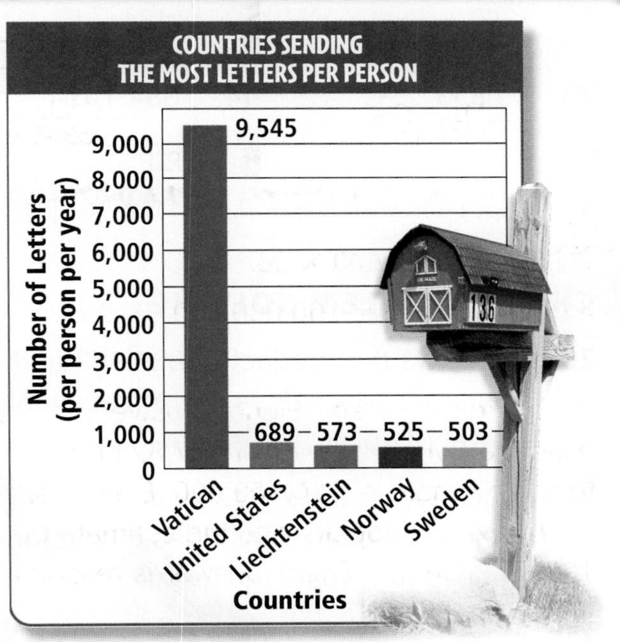

5. How many more letters are mailed in the Vatican than in the U.S. per person each year?

6. **What's the Error?** There are about 4,404,740 people in Norway. Lee says they mail about 2,292,667,170 letters each year. Describe and correct his error.

   Lee
   4,404,740
   × 525
   2202370
   88094800
   +2202370000
   2,292,667,170

7. The first place to issue postage stamps was Great Britain in 1840. How many years ago did Great Britain first issue postage stamps?

8. Luis worked for 18 hours. He earned $5.00 each hour. He bought a helmet for $15.95 and a skateboard for twice that amount. How much money did Luis have left?

9. **Write About It** How does estimation help you determine if your answer is reasonable?

Review/Test

CHECK VOCABULARY AND CONCEPTS

Choose the best term from the box.

estimate
factor
multiple

1. The product of a given number and another whole number is a _?_. (p. 144)
2. You can use an _?_ to determine if an exact answer is reasonable. (p. 152)

CHECK SKILLS

Estimate the product. (pp. 144–145)

3. $6 \times 1,239$
4. $84 \times 1,849$
5. $57 \times 6,236$

Find the product. Estimate to check. (pp. 146–151)

6. $542,290 \times 3$
7. $195,283 \times 6$
8. $9,312,648 \times 7$
9. 43×27
10. 682×484
11. $2,167 \times 51$
12. $4,260 \times 83$
13. $9,853 \times 37$
14. $384,101 \times 500$
15. $8,760 \times 45$
16. $4,756,533 \times 620$
17. $69,833 \times 9$
18. $911,265 \times 88$
19. $786,685 \times 67$

CHECK PROBLEM SOLVING

20. Myra says the product of 826,547 and 269 is 222,341,143. Is this a reasonable answer? Explain. (pp. 152–153)
21. Bert says the product of 6,704 and 43 is 46,928. Is this a reasonable answer? Explain. (pp. 152–153)
22. A factory produces 342 vans every week. How many vans does the factory produce in a year? (pp. 148–149)
23. The product of 60 and 900 has 3 zeros. How many zeros does the product of 60 and 9,000,000 have? (pp. 144–145)
24. On average, a museum has 182,469 visitors per year. How many visitors does the museum have in 3 years? in 8 years? (pp. 146–147)
25. For the past 15 years, Tanya has jogged 10 miles per day. How many miles has she jogged altogether? Use 365 days for a year. (pp. 148–149)

Cumulative Review

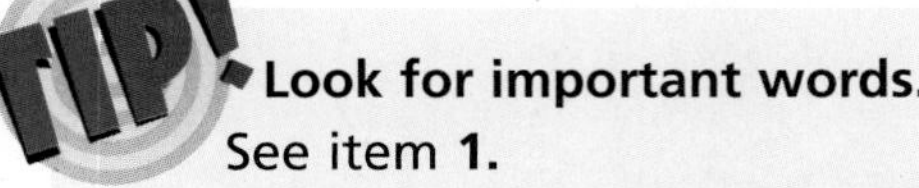

Look for important words.
See item **1**.

Estimate is an important word. It tells you to round the factors to estimate rather than find the exact answer.

Also see problem **2**, p. H63.

For 1–10, choose the best answer.

1. What is a good estimate for 625×29?

A 18,000
B 1,850
C 1,800
D 1,200

2. What is the value of n for $1{,}500 \times n = 60{,}000$?

F 400
G 45
H 40
J Not here

3. Which numbers are in order from *least* to *greatest*?

A 5,387; 5,373; 5,874; 5,840
B 5,373; 5,387; 5,840; 5,874
C 5,840; 5,874; 5,387; 5,373
D 5,373; 5,387; 5,874; 5,840

4. Shana bought 16 packs of beads. Each pack contained 25 beads. How many beads did Shana buy?

F 40
G 150
H 175
J 400

5. $3{,}000 \times 7$

A 210
B 2,100
C 21,000
D Not here

6. There are 5,280 feet in one mile and 12 inches in one foot. Which is the best estimate for the number of inches in one mile?

F 6,000
G 60,000
H 600,000
J 6,000,000

7. Denzel made a 13-mile bicycle trip on 4 Saturdays in a row. Which is a reasonable total distance for all of his trips?

A 0.52 mile
B 5.2 miles
C 52 miles
D 520 miles

8. A muffin recipe makes 18 muffins. Julian tripled the recipe. What is a reasonable number of muffins Julian could have made?

F 20
G 36
H 40
J 60

For 9–10, use the table.

Planet	Mean Distance from the Sun
Mercury	36 million miles
Mars	141 million miles
Jupiter	480 million miles
Saturn	900 million miles

9. How many miles farther from the sun is Jupiter than Mercury?

A 454 million
B 444 million
C 344 million
D 44.4 million

10. Which does **not** describe the distance from the sun to Mars?

F 14,100,000 miles
G $100 \times 1{,}410{,}000$ miles
H $141 \times 1{,}000{,}000$ miles
J about 4 times as far as Mercury

CHAPTER 10

Multiply Decimals

The bee hummingbird measures 60 millimeters long.

There are about 10 million different kinds of animals in the world. The range of animal weights can vary greatly. The table lists some of the lightest and heaviest animals. A bee hummingbird and a golf ball are about the same size, but the weight of a golf ball is about 28 times as great as that of a bee hummingbird. About how much does a golf ball weigh?

ANIMAL WEIGHTS		
Record	**Animal**	**Weight (kilograms)**
Lightest bird	Bee hummingbird	0.0016
Lightest land mammal	Bumblebee bat	0.002
Heaviest insect	Goliath beetle	0.099
Heaviest bird	Ostrich	148.5
Heaviest bony fish	Ocean sunfish	1,980
Heaviest land mammal	African elephant	7,000

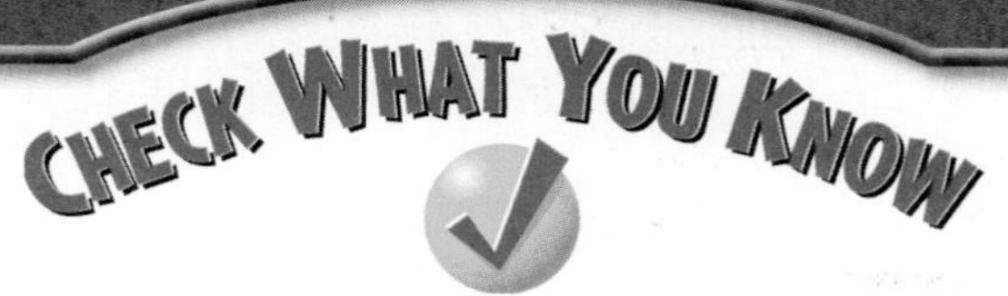

Use this page to help you review and remember important skills needed for Chapter 10.

VOCABULARY

Choose the best term from the box.

equivalent decimals
hundredth
ten
tenth
thousandth

1. One part of 10 equal parts is a ___?___.
2. One part of 100 equal parts is a ___?___.
3. Decimals that name the same number or amount are called ___?___.

REPEATED ADDITION OF DECIMALS (See p. H12.)

Add.

4. $0.1 + 0.1 + 0.1 + 0.1$
5. $0.6 + 0.6 + 0.6$
6. $0.3 + 0.3 + 0.3 + 0.3$
7. $0.7 + 0.7 + 0.7 + 0.7$
8. $0.2 + 0.2 + 0.2$
9. $0.5 + 0.5$
10. $0.9 + 0.9$
11. $1.0 + 1.0 + 1.0 + 1.0$
12. $0.8 + 0.8 + 0.8$
13. $1.1 + 1.1 + 1.1$
14. $0.9 + 0.9 + 0.9$
15. $1.2 + 1.2$
16. $\begin{array}{r} 0.2 \\ +0.2 \\ \hline \end{array}$
17. $\begin{array}{r} 0.3 \\ 0.3 \\ +0.3 \\ \hline \end{array}$
18. $\begin{array}{r} 0.5 \\ 0.5 \\ 0.5 \\ +0.5 \\ \hline \end{array}$

MULTIPLY MONEY (See p. H12.)

Multiply.

19. $\$0.10 \times 2$
20. $\$0.05 \times 20$
21. $\$0.25 \times 4$
22. $\$0.25 \times 3$
23. $\$0.10 \times 10$
24. $\$0.50 \times 5$
25. $\begin{array}{r} \$0.01 \\ \times \quad 9 \\ \hline \end{array}$
26. $\begin{array}{r} \$0.05 \\ \times \quad 5 \\ \hline \end{array}$
27. $\begin{array}{r} \$0.25 \\ \times \quad 8 \\ \hline \end{array}$
28. $\begin{array}{r} \$0.25 \\ \times \quad 2 \\ \hline \end{array}$
29. $\begin{array}{r} \$0.10 \\ \times \quad 8 \\ \hline \end{array}$
30. $\begin{array}{r} \$0.05 \\ \times \quad 9 \\ \hline \end{array}$
31. $\begin{array}{r} \$0.01 \\ \times \quad 12 \\ \hline \end{array}$
32. $\begin{array}{r} \$0.25 \\ \times \quad 6 \\ \hline \end{array}$
33. $\begin{array}{r} \$0.50 \\ \times \quad 6 \\ \hline \end{array}$

LESSON 1

HANDS ON
Multiply Decimals and Whole Numbers

Quick Review

1. 50¢ × 9
2. $60 × 8
3. $12 × 5
4. 75¢ × 4
5. $50 × 8

MATERIALS
decimal models
markers or colored pencils

Explore

The red kangaroo, the world's largest marsupial, uses its tail for balance when jumping. Its tail is about 0.53 times as long as its body. Its body is about 2 meters long. How long is its tail?

Make a model to show how to multiply 2 by 0.53.

What is 2×0.53?

STEP 1

Use hundredths models. Shade 0.53, or 53 hundredths, two times. Use a different color each time.

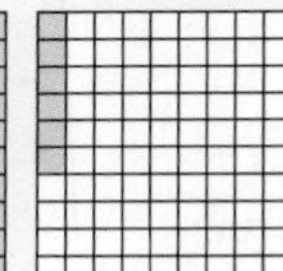

STEP 2

Count the number of shaded hundredths. There are 106 shaded hundredths. This is 1 whole and 6 hundredths.

So, $2 \times 0.53 = 1.06$.

So, the red kangaroo's tail is about 1.06 meters long.

- How is multiplying 2×0.53 similar to multiplying 2×53?
- Is the product of 3 and 0.53 greater than or less than 3? Why?

The model shows 1 × 0.12. How do you show 4 × 0.12?

Try It

Make a model to find the product.

a. 4×0.12
b. 3×0.03
c. 5×0.5
d. 3×0.3

CALIFORNIA STANDARDS **NS 2.1** Add, subtract, multiply, and divide with decimals; add with negative numbers; subtract positive integers from negative integers; and verify the reasonableness of the results. *also* **NS 1.0, NS 2.0, MR 2.3, MR 2.4, MR 3.2**

Connect

You can write a multiplication sentence for your model.

What is 3×0.6?

Model

Use tenths models. Shade 0.6, or 6 tenths, three times.

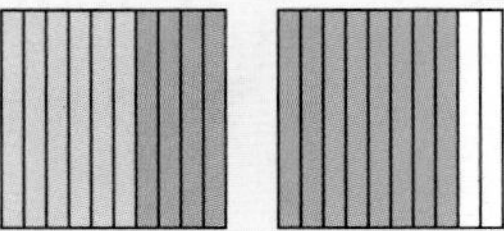

Use a different color each time.

Count the number of shaded tenths. There are 18 shaded tenths, or 1 whole and 8 tenths.

Record

Record.

$$\begin{array}{r} 0.6 \\ \times \ \ 3 \\ \hline 1.8 \end{array}$$

Use the model to place the decimal point.

So, $3 \times 0.6 = 1.8$.

TECHNOLOGY LINK

More Practice: Use E-Lab, ***Multiplying Decimals and Whole Numbers.***

www.harcourtschool.com/elab2002

Practice

Make a model to find each product.

1. 3×0.8
2. 6×0.4
3. 5×0.6
4. 2×0.82
5. 7×0.03
6. 2×0.12
7. 4×0.19
8. 9×0.18

Joy is shopping at the pet store. For 9–13, use the table to find the total cost.

TROPICAL FISH AND SUPPLIES	
Item	**Price**
Danio tetra	\$0.79
Zebra	\$0.50
Molly	\$0.88
Algae eater	\$0.94
Food flakes	\$0.99

9. 3 danio tetras, 2 zebras
10. 2 mollies, 3 zebras
11. 4 zebras, 2 mollies, 1 food flakes
12. 5 zebras, 3 mollies, 2 algae eaters
13. 8 danio tetras, 6 zebras, 1 molly
14. **Write About It** Explain how to draw a model to find 3×0.9.

Mixed Review and Test Prep

15. Aponi recorded temperatures of 78°, 82°, 65°, 66°, and 83°. What is the range of the data? (p. 99)
16. What is the value of the expression $9 \times (5 + n)$ if n is 6? (p. 68)
17. Round 82,394,021 to the nearest million. (p. 32)
18. Write an equivalent decimal for 8.02. (p. 22)
19. **TEST PREP** Which digits make this statement true? 67■,891 > 677,819 (p. 8)

   A 5, 6, and 7 **C** 7, 8, and 9
   B 6, 7, and 8 **D** 8, 9, and 0

LESSON 2

Algebra: Patterns in Decimal Factors and Products

Quick Review

1. 10 × 12
2. 100 × 12
3. 1,000 × 12
4. 10,000 × 12
5. 100,000 × 12

Learn

MEASURE UP The smallest frog in the Northern Hemisphere, found in Cuba, grows to about 0.98 centimeters in body length. The photos show the frog at its actual size and enlarged 5 times. If you could enlarge the image 1,000 times for a billboard, what would be the length of the image of the frog?

What is 0.98 × 1,000?

▲ Actual size of frog

▲ Frog enlarged 5 times

Look for a pattern.

0.98 × 1 = 0.98
0.98 × 10 = 9.80 ← The decimal point moves 1 place to the right.
0.98 × 100 = 98.00 ← The decimal point moves 2 places to the right.
0.98 × 1,000 = 980.00 ← The decimal point moves 3 places to the right.

So, the image of the frog would be 980 centimeters long.

- **What if** you enlarge the frog photo 100 times? What will be the length of the image of the frog?

MATH IDEA When you find the products of decimals and 10, 100, and 1,000, the decimal point moves one more place to the right each time.

Examples

A	B	C
\$3.25 × 1 = \$3.25	0.478 × 1 = 0.478	0.0009 × 1 = 0.0009
\$3.25 × 10 = \$32.50	0.478 × 10 = 4.78	0.0009 × 10 = 0.009
\$3.25 × 100 = \$325.00	0.478 × 100 = 47.8	0.0009 × 100 = 0.09
\$3.25 × 1,000 = \$3,250.00	0.478 × 1,000 = 478	0.0009 × 1,000 = 0.9

- **REASONING** In Example A, how can you use the pattern to place the decimal point in the product?

CALIFORNIA STANDARDS MR 1.1 Analyze problems by identifying relationships, distinguishing relevant from irrelevant information, sequencing and prioritizing information, and observing patterns. **NS 1.0** Students compute with very large and very small numbers, positive integers, decimals, and fractions and understand the relationship between decimals, fractions, and percents. They understand the relative magnitudes of numbers. *also* **NS 2.0, O━NS 2.1, O━AF 1.2, MR 1.1, MR 2.0, MR 2.3, MR 3.2, MR 3.3**

Check

1. **Look** at the pattern of the decimal points in the products in Examples A–C. Write a rule to mentally place the decimal point in your answer.

Use mental math to complete.

2. $1 \times 0.3 = 0.3$
   $10 \times 0.3 = 3$
   $100 \times 0.3 = ■$
   $1{,}000 \times 0.3 = ■$

3. $1 \times 2.845 = 2.845$
   $10 \times 2.845 = 28.45$
   $100 \times 2.845 = ■$
   $1{,}000 \times 2.845 = 2{,}845$

4. $1 \times 0.3459 = 0.3459$
   $10 \times 0.3459 = ■$
   $100 \times 0.3459 = 34.59$
   $1{,}000 \times 0.3459 = 345.9$

Practice and Problem Solving

Use mental math to complete.

5. $1 \times 0.005 = 0.005$
   $10 \times 0.005 = ■$
   $100 \times 0.005 = 0.5$
   $1{,}000 \times 0.005 = 5$

6. $1 \times 0.3459 = 0.3459$
   $10 \times 0.3459 = 3.459$
   $100 \times 0.3459 = ■$
   $1{,}000 \times 0.3459 = ■$

7. $1 \times 0.45 = 0.45$
   $10 \times 0.45 = ■$
   $100 \times 0.45 = ■$
   $1{,}000 \times 0.45 = ■$

Multiply each number by 10, by 100, and by 1,000.

8. 0.9
9. 0.51
10. 0.007
11. 0.015
12. 0.2178
13. $0.25
14. 45.69
15. 1.0608

Algebra Find the value of n.

16. $10 \times n = 0.09$
17. $n \times 0.08 = 0.8$
18. $100 \times n = 4.5$
19. $1.5 \times n = 15$
20. $n \times 100 = 1.9$
21. $n \times 1{,}000 = 3.1$
22. A half dollar is 0.50 of a dollar. What is the value of 100 half dollars? of 1,000 half dollars?
23. If a snail moves 8 inches a minute, how many minutes would it take it to move 12 feet?
24. **Write About It** Explain why multiplying a decimal by 10, by 100, and by 1,000 is easy to compute mentally.
25. **REASONING** How does the position of the decimal point change when you multiply 6.7 by 10,000?

Mixed Review and Test Prep

26. $1{,}456 \times 8$ (p. 146)
27. 345×23 (p. 148)

Write <, >, or = for each ●. (p. 24)

28. 0.89 ● 0.98
29. 3.9 ● 3.09
30. **TEST PREP** Find the value of n.
    $90 = n + 32 + 28$ (p. 74)
    A $n = 20$
    B $n = 30$
    C $n = 28$
    D $n = 32$

LESSON 3

Model Decimal Multiplication

Quick Review

1. $0.4 + 0.4$
2. $0.6 + 0.6$
3. $0.7 + 0.7$
4. $0.4 + 0.4 + 0.4$
5. $0.1 + 0.1 + 0.1$

Learn

FAST FOOD The bee hummingbird weighs about 0.2 dekagram. It needs to eat half its body weight in food every day to stay alive. About how much food does a bee hummingbird need to eat every day?

Find 0.2×0.5. ← **Think:** One half can be written as 0.5.

STEP 1

Divide a square into 10 equal columns. Shade 5 of the columns to show 0.5.

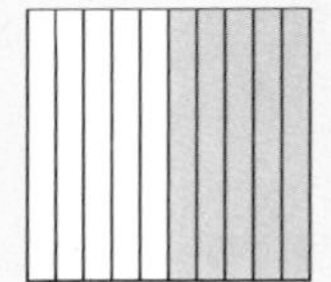

STEP 2

Divide the square into 10 equal rows to make 100 equal parts. Shade 2 of the rows to show 0.2.

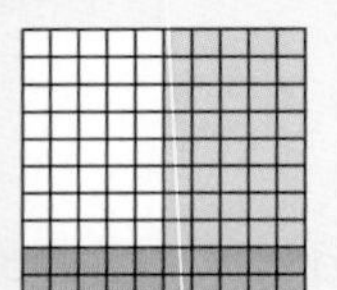

STEP 3

The area in which the shading overlaps shows the product, or 0.2×0.5.

So, $0.2 \times 0.5 = 0.10$.

Since hummingbirds can rotate their wings in a figure-eight pattern, they can fly backward, and even hover like a helicopter! ▼

So, a bee hummingbird eats about 0.10, or 0.1 dekagram, of food every day.

- **REASONING** What relationship do you see between the product and the size of the decimal factors less than 1?

Check

1. **Tell** whether the product 0.2×0.4 is greater than or less than 1. Explain your reasoning.

Complete the multiplication sentence for each model.

2. 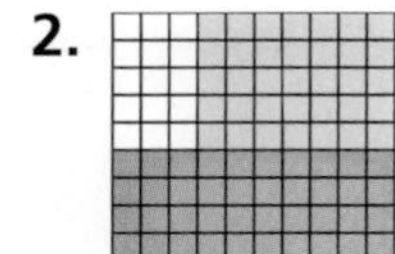

$0.5 \times 0.7 = n$

3. 

$y \times 0.8 = 0.24$

4. 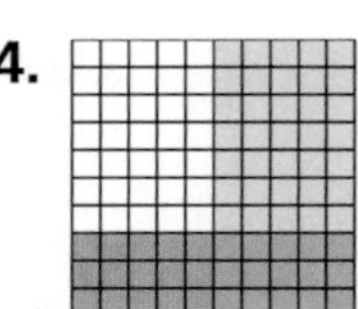

$0.3 \times 0.5 = p$

5. 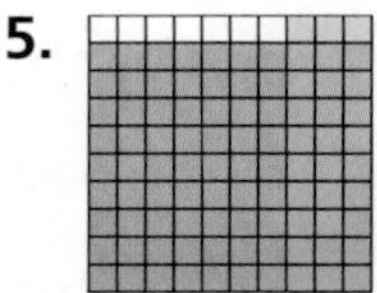

$0.9 \times n = 0.27$

CALIFORNIA STANDARDS **O━n NS 2.1** Add, subtract, multiply, and divide with decimals; add with negative numbers; subtract positive integers from negative integers; and verify the reasonableness of the results. **MR 2.3** Use a variety of methods, such as words, numbers, symbols, charts, graphs, tables, diagrams, and models, to explain mathematical reasoning. *also* **NS 1.0, NS 2.0, MR 2.0, MR 2.4, MR 3.0, MR 3.2**

Make a model to find the product.

6. 0.1×0.5 7. 0.4×0.7 8. 0.3×0.3 9. 0.8×0.4

Practice and Problem Solving

Complete the multiplication sentence for each model.

10. 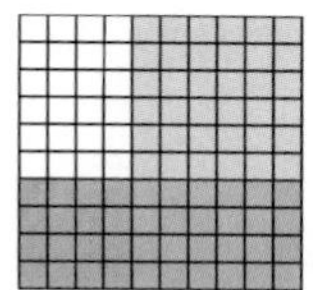
$0.4 \times 0.6 = n$

11. 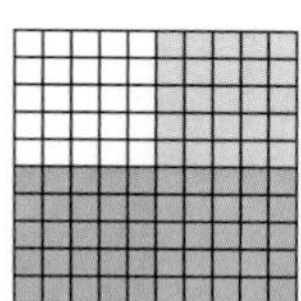
$0.5 \times y = 0.25$

12. 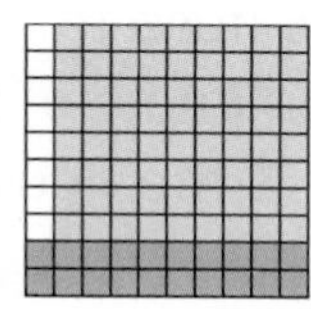
$p \times 0.9 = 0.18$

13. 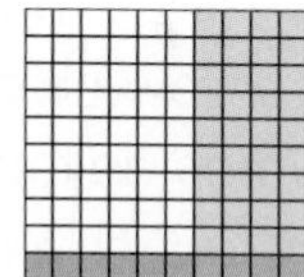
$0.1 \times 0.4 = n$

Make a model to find the product.

14. 0.8×0.8 15. 0.1×0.9 16. 0.6×0.2 17. 0.8×0.3

Find the product.

18. 0.7×0.8 19. 0.2×0.8 20. 0.9×0.6 21. 0.6×0.7

22. 0.5×0.6 23. 0.8×0.9 24. 0.5×0.4 25. 0.3×0.6

Algebra Find the value of n.

26. $n \times 0.2 = 0.14$ 27. $0.6 \times n = 0.48$ 28. $0.9 \times n = 0.63$ 29. $n \times 0.7 = 0.35$

30. The largest frog is the Goliath frog. With its legs stretched out, it is about 0.8 meter long. When its legs are not stretched out, its body is about 0.5 times that length. How long is its body?

31. When you multiply a decimal less than one by a whole number, is the product greater than or less than the whole-number factor? Draw a model to explain your answer.

32. What's the Error? Marco said $0.1 \times 0.6 = 0.6$. Describe his error. Draw a model.

33. A grasshopper can leap about 0.7 meter. A flea can jump 0.5 times as far. About how far can a flea jump?

Mixed Review and Test Prep

Round to the nearest million. (p. 32)

34. 4,099,999 35. 16,399,999

Order from least to greatest. (p. 10)

36. 87,314; 87,413; 81,341

37. 109,721; 190,271; 109,271

38. TEST PREP Which shows three and two tenths written as a decimal and a fraction? (p. 18)

A 3.2; $3\frac{2}{10}$ C 3.12; $3\frac{12}{100}$

B 2.3; $2\frac{3}{10}$ D 3.02; $3\frac{2}{100}$

LESSON 4

Place the Decimal Point

Quick Review

Round to the nearest whole number.

1. 3.4 **2.** 99.2 **3.** 0.38
4. 0.68 **5.** 5.3

Learn

TIP TO TIP From the tip of its nose to the end of its tail, a pygmy shrew measures 6.1 centimeters long. A house mouse is 2.7 times as long. How long is the house mouse?

You can use estimation to help you place the decimal point in a decimal product and to determine if your answer is reasonable.

Find 2.7×6.1.

STEP 1

Estimate the product. Round each factor.

$$\begin{array}{ccc} 2.7 & \times & 6.1 \\ \downarrow & & \downarrow \\ 3 & \times & 6 = 18 \end{array}$$

STEP 2

Multiply as with whole numbers.

$$\begin{array}{r} 6.1 \\ \times 2.7 \\ \hline 427 \\ +1220 \\ \hline 1647 \end{array}$$

STEP 3

Use the estimate to place the decimal point in the product.

$$\begin{array}{r} 6.1 \\ \times 2.7 \\ \hline 427 \\ +1220 \\ \hline 16.47 \end{array}$$

Since the estimate is 18, place the decimal point so there is a two-digit whole number in the product.

So, the house mouse is 16.47 centimeters long.

The pygmy shrew, one of the world's smallest mammals, could sleep in a spoon! And it weighs about as much as a table-tennis ball! ▼

Examples

A Find 12×0.48.

$10 \times 0.5 = 5 \leftarrow$ Estimate.

$$\begin{array}{r} 0.48 \\ \times \quad 12 \\ \hline 96 \\ +480 \\ \hline 5.76 \end{array}$$

Since the estimate is 5, place the decimal point so there is a one-digit whole number in the product.

B Find $0.75 \times \$1.25$.

$0.8 \times \$1 = \$0.8$, or $\$0.80 \leftarrow$ Estimate.

$$\begin{array}{r} \$1.25 \\ \times \; 0.75 \\ \hline 625 \\ +8750 \\ \hline \$0.9375 \end{array}$$

Since the estimate is \$0.80, place the decimal point so there is less than 1 dollar in the product.

CALIFORNIA STANDARDS **NS 2.1** Add, subtract, multiply, and divide with decimals; add with negative numbers; subtract positive integers from negative integers; and verify the reasonableness of the results. **NS 1.1** Estimate, round, and manipulate very large and very small numbers. *also* **NS 1.0, NS 2.0, SDAP 1.0, MR 1.0, MR 1.1, MR 2.1, MR 3.2**

Count Decimal Places

You can also place the decimal point by finding the total number of decimal places in the factors. Then count that many places from the right in the product.

Find 0.7×0.2.

STEP 1

Multiply as with whole numbers.

$$\begin{array}{r} 0.7 \\ \times 0.2 \\ \hline 14 \end{array}$$

STEP 2

Find the total number of decimal places in the factors. Place the decimal point that number of places from the right in the product.

$$\begin{array}{rl} 0.7 & \leftarrow \text{1 decimal place in the factor} \\ \times 0.2 & \leftarrow \text{1 decimal place in the factor} \\ \hline 0.14 & \leftarrow 1 + 1\text{, or 2 decimal places in the product} \end{array}$$

So, 0.7×0.2 is 0.14.

Examples

C

$$\begin{array}{rl} 23 & \leftarrow \text{0 decimal places in the factor} \\ \times 0.04 & \leftarrow \text{2 decimal places in the factor} \\ \hline 0.92 & \leftarrow 0 + 2\text{, or 2 decimal places in the product} \end{array}$$

D

$$\begin{array}{rl} 7.52 & \leftarrow \text{2 decimal places in the factor} \\ \times 0.23 & \leftarrow \text{2 decimal places in the factor} \\ \hline 2256 & \\ +15040 & \\ \hline 1.7296 & \leftarrow 2 + 2\text{, or 4 decimal places in the product} \end{array}$$

MATH IDEA You can use estimation or the total number of decimal places in the factors to determine where to place the decimal point in the product.

▼ House mouse

Check

1. **Tell** how you can check that the answer to Example D is reasonable.

Choose the better estimate. Write *a* or *b*.

2. 34×0.8 **a.** 24 **b.** 2.4

3. 4.2×3.9 **a.** 16 **b.** 1.6

Copy each exercise. Place the decimal point in the product.

4. $\begin{array}{r} 29 \\ \times 0.7 \\ \hline 203 \end{array}$

5. $\begin{array}{r} 2.98 \\ \times\ 0.7 \\ \hline 2086 \end{array}$

6. $\begin{array}{r} 1.8 \\ \times 0.2 \\ \hline 36 \end{array}$

7. $\begin{array}{r} 0.37 \\ \times 0.64 \\ \hline 2368 \end{array}$

Find the product. Estimate to check.

8. 9×1.7
9. 0.2×12
10. 95×0.64
11. 1.25×0.5
12. 0.9×0.4
13. 0.37×0.6
14. 0.211×18

LESSON CONTINUES ▶

Practice and Problem Solving

Choose the better estimate. Write *a* or *b*.

15. 22×0.6 **a.** 12 **b.** 1.2

16. 2.3×4.8 **a.** 10 **b.** 1.0

17. 0.82×6 **a.** 0.48 **b.** 4.8

18. 42×0.5 **a.** 20 **b.** 2.0

Copy each exercise. Place the decimal point in the product.

19. $\begin{array}{r} 3.4 \\ \times\ 5 \\ \hline 170 \end{array}$

20. $\begin{array}{r} 0.58 \\ \times\ 2 \\ \hline 116 \end{array}$

21. $\begin{array}{r} 5.48 \\ \times 0.726 \\ \hline 397848 \end{array}$

22. $\begin{array}{r} 2.32 \\ \times 4.68 \\ \hline 108576 \end{array}$

Find the product. Estimate to check.

23. 0.3×14 **24.** $0.5 \times 1{,}206$ **25.** 6.8×4.5 **26.** 7.25×3.8

27. 5.1×2.7 **28.** 12.92×7.2 **29.** 19×0.21 **30.** 38.8×4.62

31. 0.7×0.8 **32.** 6.9×3.1 **33.** 0.12×0.9 **34.** 0.325×82

35. 0.68×0.4 **36.** 2.36×0.9 **37.** 0.3×0.918 **38.** 9.2×0.07

39. 0.25×0.75 **40.** 0.8×0.201 **41.** 2.7×5.6 **42.** 43.3×6.2

43. 24.37×0.8 **44.** 0.8481×3.2 **45.** 9.748×0.42 **46.** 436.3×0.181

47. Which product will have 6 decimal places?

a. 0.12×0.0456 **b.** 1.2×0.0456 **c.** 12×0.0456

48. Which product will have 5 decimal places?

a. 0.283×1.078 **b.** 2.83×1.078 **c.** 28.3×1.078

49. **Write About It** How can you use an estimate to help you place the decimal point? Give an example.

USE DATA **For 50–53, use the graph.**

50. How many more birds than cats have lived in the White House?

51. The number of goats that have lived in the White House is 0.25 times as many as which other type of animal?

52. Of which type of animal have the least number lived in the White House?

53. How many more birds than cows have lived in the White House?

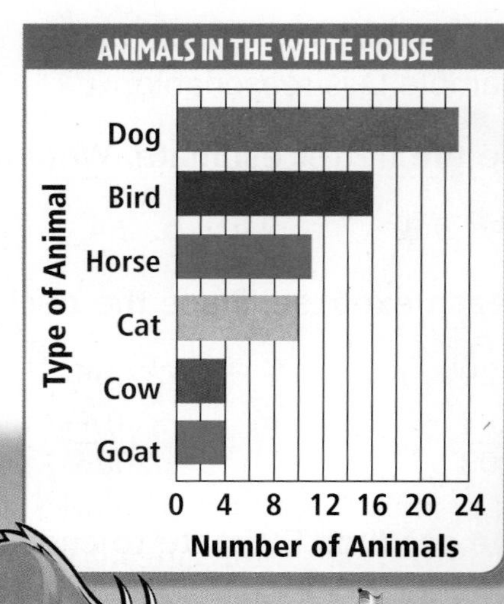

54. Four students shared 6 boxes of markers equally. Each of the boxes had the same number of markers. Each student received 6 markers. How many markers were in each box?

55. A flea can jump about 130 times as high as its own height. A flea is about 0.16 centimeter tall. About how high can a flea jump?

Mixed Review and Test Prep

56. $7{,}981 + 6{,}909 + 2{,}574$ (p. 36)

57. Solve $87 + n = 147$. (p. 74)

58. **TEST PREP** Write eight thousandths as a decimal. (p. 20)

A 0.80 C 0.008
B 0.08 D 0.0008

59. Solve $17 \times n = 0$. Identify the property used. (p. 90)

60. **TEST PREP** In which number does the digit 4 have the least value? (p. 20)

F 7.0492 H 7.4902
G 7.2049 J 7.9204

LINKUP to Reading

STRATEGY • SUMMARIZE To *summarize,* you restate information in a shortened form. You include only the most important information.

Gwen runs her computer 8 hours each day, 5 days each week. How much does the electricity cost for 4 weeks?

Summary	Solve
A. Electricity costs $0.01 per hour.	
B. The computer runs 8 hours a day.	8 × $0.01 = $0.08
C. The computer runs 5 days a week.	5 × $0.08 = $0.40
D. How much does it cost for 4 weeks?	4 × $0.40 = $1.60

ELECTRICITY COST (based on a 1999 average electric rate)

Appliances	Cost per hour
Computer	$0.01
Lamp, 100 watt	$0.01
Refrigerator/freezer	$0.06
Hair dryer	$0.15
Television	$0.02
Water heater	$0.45

So, it will cost Gwen $1.60.

USE DATA For 1–2, use the table.

1. Mr. Ramos' water heater is on a timer. It runs a total of 3 hours each day. How much does it cost to run his water heater each week?

2. A fluorescent lamp costs $0.003 per hour to run. How much more does it cost to run the 100 watt lamp for 10 hours than the fluorescent lamp for 10 hours?

Extra Practice page H41, Set C

LESSON

5 Zeros in the Product

Quick Review

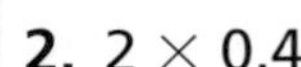

1. 1×0.4
2. 2×0.4
3. 10×0.4
4. 20×0.4
5. 125×0.5

Learn

Sometimes when you multiply with decimals, there are zeros in the product.

SUPER ANT A *Formica japonica* worker ant weighs 0.004 gram. It can walk while holding in its mouth an object weighing 5 times as much as its own body. How many grams can a worker ant carry?

Find 5×0.004.

$$\begin{array}{r} 0.004 \\ \times \quad 5 \\ \hline 0.020 \end{array}$$

0.004 ← Since there are 3 decimal places in the factors, you will need 3 decimal places in the product.

0.020 ↑ Write a zero at the left in the product to place the decimal point.

So, a worker ant can carry 0.020, or 0.02 gram.

Examples

A Find 0.003 × $18.

$$\begin{array}{r} \$18 \\ \times 0.003 \\ \hline \$0.054 \end{array}$$

Since 3 decimal places are needed in the product, write a zero for this place.

B Find 0.09 × 0.07.

$$\begin{array}{r} 0.07 \\ \times 0.09 \\ \hline 0.0063 \end{array}$$

Since 4 decimal places are needed in the product, write zeros in these places.

C Find 0.002 × 9.27.

$$\begin{array}{r} 9.27 \\ \times 0.002 \\ \hline 0.01854 \end{array}$$

Since 5 decimal places are needed in the product, write a zero in this place.

- In Example A, what is the product to the nearest cent?

MATH IDEA You may need to insert zeros at the left in the product to keep the same number of decimal places in the product as in the factors.

CALIFORNIA STANDARDS **NS 2.1** Add, subtract, multiply, and divide with decimals; add with negative numbers; subtract positive integers from negative integers; and verify the reasonableness of the results. **NS 1.0** Students compute with very large and very small numbers, positive integers, decimals, and fractions and understand the relationship between decimals, fractions, and percents. They understand the relative magnitudes of numbers. *also* **NS 1.1, NS 2.0, MR 2.0, MR 2.4, MR 3.2, MR 3.3**

Check

1. **Tell** how you know when to insert a zero in a product to place a decimal point.

Find the product.

2. 8×0.003
3. 0.04×0.4
4. 0.018×9
5. 6×0.0006

Practice and Problem Solving

Find the product.

6. 9×0.007
7. 0.02×2
8. 0.016×0.5
9. 0.8×0.03
10. 0.006×9
11. 0.08×0.02
12. 0.004×24
13. 0.12×0.09
14. 54.07×0.04
15. 0.007×8
16. 0.032×17
17. 0.014×0.06

Find the product. Round to the nearest cent.

18. $0.89 × 0.08
19. $0.95 × 0.05
20. $3.09 × 0.05
21. $5.05 × 0.06

Write <, >, or = for each ●.

22. 0.008×9 ● 0.009×8
23. 3×0.025 ● 3.01×0.02
24. **What's the Question?** Eileen bought 2 gallons of milk at $3.02 per gallon and one loaf of bread at $0.99. The answer is $7.03.
25. **Write a problem** including this information: a dozen eggs costs $0.78, a loaf of bread costs $1.03, a pound of bananas costs $0.63.
26. You can draw a line 35 miles long with a standard pencil. That's 0.9 of the distance from Baltimore to Washington, D.C. How long a line can you draw with 3 pencils?
27. The price of an ant farm is $28.00. The computer multiplies the price by 1.07 to find the total cost including tax. What is the total cost of the ant farm?

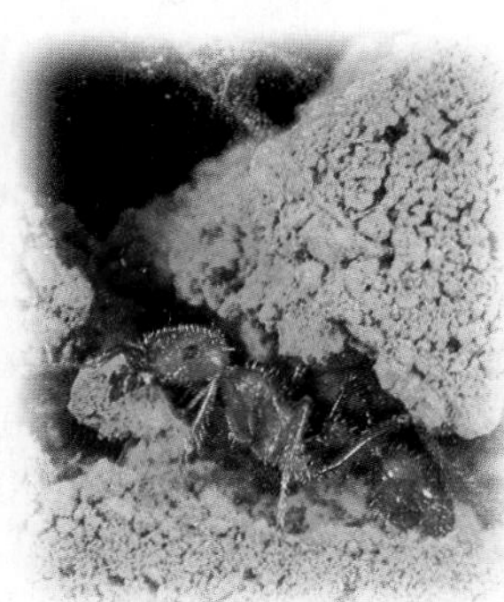

28. **Write About It** Tell the steps you would use to find 0.05×0.06.

Mixed Review and Test Prep

29. 9,126,543 − 3,972,645 (p. 38)
30. 19,282,443 + 723,451 (p. 38)
31. 456 − 309 (p. 36)
32. 13 × $1.25 (p. 164)
33. **TEST PREP** Round 2.2178 to the place of the blue digit. (p. 46)

   A 2 **B** 2.2 **C** 2.22 **D** 2.218

Extra Practice page H41, Set D

LESSON 6

Problem Solving Skill

Make Decisions

Quick Review

Write $<$, $>$, or $=$ for each ●.

1. $36.52 ● $36.25
2. 49¢ ● 51¢
3. $7.81 ● $7.84
4. $0.98 ● $0.89
5. 4 × $1.50 ● 6 × $1.25

ARTIST'S CHOICE Suppose you are planning to take art lessons. You can sign up for either painting or sculpture. To help you make a decision on which to take, there are several things that you need to think about before you decide.

THINGS TO CONSIDER	PAINTING	SCULPTURE
Time	Wednesday, 4:00 P.M. to 6:00 P.M.	Saturday, 10 A.M. to 11:30 A.M.
Number of lessons	6	9
Cost per lesson	$12.75	$9.50
Cost of general supplies	$10	$25
Cost per project	$2.35 per canvas panel	$1.45 per 5-lb bag of clay

Use the information in the table to help you answer the following questions.

A. For each choice, how much will it cost for lessons and general supplies? Which costs less?

B. What is the difference in cost for general supplies for each choice?

C. How many total hours will each choice last? What is the difference in the amount of time?

D. You will complete a total of 4 painting projects or 6 sculpture projects. How much will you spend on project supplies for painting? for sculpture? In which project would supplies cost less?

Talk About It

- If you had to take the one that costs less, would you take painting or sculpture? Explain the reasons for your decision.
- Make a list of other things you may consider before deciding which to take.

MATH IDEA To help you make a decision, compare facts and data.

CALIFORNIA STANDARDS MR 2.4 Express the solution clearly and logically by using the appropriate mathematical notation and terms and clear language; support solutions with evidence in both verbal and symbolic work. **MR 3.1** Evaluate the reasonableness of the solution in the context of the original situation. *also* **NS 1.0, NS 2.0, 🔑NS 2.1, MR 1.0, MR 2.0, MR 2.3, MR 3.3**

Problem Solving Practice

USE DATA **For 1–3, use the information in the table.**

1. You have a coupon from Store A for $2.00 off if you purchase $10.00 worth of art supplies. You plan to buy 12 sheets of charcoal paper and 10 pastels. Would you spend more at Store A or Store B? How much more?

WHICH STORE HAS THE BEST BUY?			
Items	**Store A**	**Store B**	**Store C**
Charcoal	2 for $0.75	$0.33 each	4 for $1.40
Kneaded eraser	$0.47 each	3 for $1.50	$0.51 each
Charcoal paper	3 for $1.89	$0.57 each	2 for $1.36
Pastels	$0.98 each	2 for $1.80	$0.92 each

You are buying art supplies for a drawing class. You need 4 sticks of charcoal, 3 kneaded erasers, and 2 pastels. You do not have any store coupons.

2. You need to buy all the supplies from the same store. At which store would you spend the least?

   A Store A
   B Store B
   C Store C
   D The cost is the same at each store.

3. **What if** you could go to more than one store? What is the least amount you could spend?

   F $4.53 **H** $4.77
   G $4.62 **J** $4.87

Mixed Applications

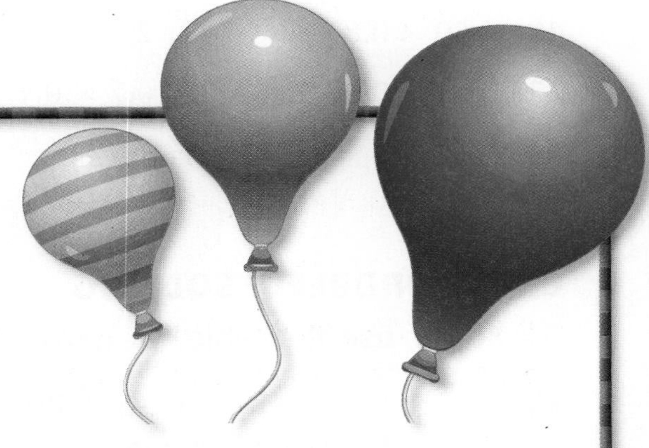

4. On an average day, 450 cars pass through the toll plaza. Val says about 164,250 cars pass through the toll plaza in a year. Chuck says about 42,750 pass through the toll plaza in a year. Whose answer is reasonable? Explain.

5. **Write a problem** in which you need to make a decision that can be answered with information in the table above.

6. Use the diagram above. Al's balloon is not next to Maggie's. Clem's balloon is larger than Maggie's. Who has the striped balloon?

CHAPTER 10

Review/Test

CHECK VOCABULARY AND CONCEPTS

Choose the best term from the box.

equal
one tenth
one hundredth
one thousandth

1. 0.1 represents __?__. (p. 158)
2. The number of decimal places in the product is __?__ to the total number of decimal places in the factors. (p. 165)
3. 0.01 represents __?__. (p. 158)

Make a model to find the product. (pp. 162–163)

4. 0.6×5
5. 0.51×3
6. 2×0.25
7. 0.82×4

CHECK SKILLS

Multiply each number by 10, by 100, and by 1,000. (pp. 160–161)

8. 0.3
9. 0.64
10. 0.002
11. 0.0225

Copy each exercise. Place the decimal point in each product. (pp. 164–167)

12. $0.5 \times 0.5 = 25$
13. $1.6 \times 3.34 = 5344$
14. $0.152 \times 0.78 = 11856$
15. $0.28 \times 0.7 = 196$
16. $0.9 \times 2.186 = 19674$
17. $25.4 \times 0.92 = 23368$

Find the product. Estimate to check. (pp. 164–169)

18. 6.50×0.8
19. 2.85×22
20. 4.15×0.6
21. 9.8×2.3
22. 0.62×0.07
23. 0.7×0.1
24. 0.4×0.5
25. 0.76×65
26. $\begin{array}{r} 0.96 \\ \times \quad 5 \\ \hline \end{array}$
27. $\begin{array}{r} 7.62 \\ \times 0.08 \\ \hline \end{array}$
28. $\begin{array}{r} 0.9 \\ \times 0.09 \\ \hline \end{array}$
29. $\begin{array}{r} 3.96 \\ \times \quad 28 \\ \hline \end{array}$

CHECK PROBLEM SOLVING

For 30–31, use the table. (pp. 170–171)

SERVICE	DELIVERY SCHEDULE	RATE (in 2000)
First-class Mail	varies	\$0.33 for the first oz, 0.22 for each additional oz
Priority Mail	2 days	\$3.20
Express Mail	next day	\$11.75

30. Janet needs to mail a 1-oz letter. Today is Monday. What is the least amount she could spend and guarantee it will arrive by Wednesday?
31. Which service is the fastest? How much does it cost?
32. Jorge earns \$24.65 a week. How much does he earn in 12 weeks? (pp. 164–167)
33. What is the value of 100 quarters? of 1,000 quarters? (pp. 160–161)

Cumulative Review

Check your work.
See item 6.

Ask yourself if your answer seems reasonable. Then be sure you followed the rounding instructions in the problem.

Also see problem 7, p. H65.

For 1–10, choose the best answer.

1. $0.6 \times 10{,}000 =$

   A 600 **C** 60,000
   B 6,000 **D** Not here

2. Which multiplication property is illustrated in $1.25 \times 0 = 0$?

   F Associative Property
   G Zero Property
   H Commutative Property
   J Distributive Property

3. How is 1.29 written in word form?

   A one hundred twenty-nine
   B one hundred twenty-nine tenths
   C one and twenty-nine tenths
   D one and twenty-nine hundredths

4. Which fraction has the same value as the 8 in 4.582?

   F $\frac{8}{10}$ **H** $\frac{8}{1{,}000}$
   G $\frac{8}{100}$ **J** $\frac{1}{8}$

5. Tasha put 75 cents in her bank each week for one year. How much did she put in the bank?

   A \$39.00 **C** \$3,900
   B \$390.00 **D** \$3,939

6. Tim runs 9 mph and it takes him 2.91 hours to finish a race. To the nearest tenth of a mile, how far does he run?

   F 2.6 mi **H** 26.2 mi
   G 26 mi **J** 261.9 mi

7. How many decimal places are in the answer?

   $$0.003 \times 0.06$$

   A 2 **C** 5
   B 4 **D** Not here

For 8–10, use the table.

INSECT SPEEDS	
Insect	**Speed in Feet Per Second**
Honeybee	10.5
Hornet	19.6
Dragonfly	26.1
Horsefly	13.2

8. Which shows the insect speeds in order from slowest to fastest?

   F 10.5, 13.2, 19.6, 26.1
   G 10.5, 13.2, 26.1, 19.6
   H 26.1, 13.2, 19.6, 10.5
   J 26.1, 19.6, 13.2, 10.5

9. How many feet can a dragonfly fly in 6 seconds?

   A 156.6 ft **C** 79.2 ft
   B 117.6 ft **D** 63 ft

10. If a honeybee and a horsefly fly for 21 seconds in the same direction, how far apart will they be?

   F 297.7 ft **H** 56.7 ft
   G 57.7 ft **J** 36 ft

UNIT 3

MATH DETECTIVE

Product Production

REASONING To solve each case, replace each triangle with a digit from 0 to 9. Use each digit no more than once for each case. Think carefully before selecting each digit. Remember, a good detective will test several possibilities before saying the case is closed!

Case 1

Produce the greatest possible product.

▲▲▲ × ▲▲

Case 2

Produce the least possible product.

▲▲▲ × ▲▲

Case 3

Produce the greatest possible product divisible by 5.

▲▲▲ × ▲▲

Case 4

Produce the least product that rounds to 15.

▲.▲▲× ▲

Case 5

Produce a product between 2 and 3.

0.▲▲▲ × ▲.▲

Case 6

Produce the greatest possible product.

0.▲▲▲ × 0.▲▲

Think It Over!

- **Write About It** Explain how you solved each case.
- **Write a Problem** Make up a case of your own about a product. Ask a friend to solve it.

UNIT 3

Challenge

Lattice Multiplication

Lattice multiplication is another way to find products.

Multiply. 295 × 731

STEP 1

Draw a 3 × 3 lattice like the one shown. Write 295 across the top. Write 731 down the right side. Write an X above 731 as shown. Draw a diagonal line through each of the boxes.

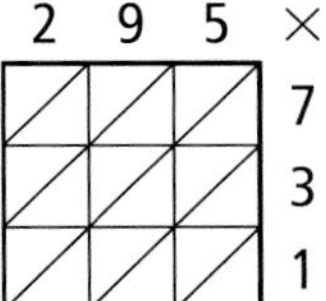

STEP 2

For each box, multiply the number at the top of that column by the number at the right of that row. Write the digits of the product above and below the diagonal. If the product has only one digit, write a zero above the diagonal.

2	9	5	×
1/4	6/3	3/5	7
0/6	2/7	1/5	3
0/2	0/9	0/5	1

STEP 3

Add along the diagonals. Start with the lower right corner. If needed, regroup to the next diagonal. Write the sums along the bottom and left side of the lattice.

	2 1	9 1	5 1	×
2	1/4	6/3	3/5	1 7
1	0/6	2/7	1/5	3
5	0/2	0/9	0/5	1
	6	4	5	

STEP 4

Read the product by moving down the left side and across the bottom.

	2	9	5	×
2	1/4	6/3	3/5	7
1	0/6	2/7	1/5	3
5	0/2	0/9	0/5	1
	6	4	5	

So, 295 × 731 = 215,645.

Try It

Use lattice multiplication to find the product.

1. 124 × 562 **2.** 552 × 601 **3.** 759 × 106 **4.** 324 × 651 **5.** 944 × 293

6. 783 × 504 **7.** 342 × 175 **8.** 427 × 899 **9.** 135 × 975 **10.** 246 × 806

Study Guide and Review

VOCABULARY

Choose the best term from the box.

1. A number that is the product of a given whole number and another whole number is a(n) __?__. (p. 144)
2. To determine if an exact answer is reasonable, you can use a(n) __?__. (p. 152)

estimate
factor
multiple
decimal
product

STUDY AND SOLVE

Chapter 9

Estimate products.

43 × 184
↓ ↓
40 × 200

4 × 2 = 8 → basic fact
4 × 20 = 80
40 × 20 = 800
40 × 200 = 8,000 So, 43 × 184 is about 8,000.

Estimate the product. (pp. 144–145)

3. 32 × 145
4. 7 × 861
5. 56 × 724
6. 5,387 × 31
7. 864 × 43
8. 7,613 × 68

Multiply by 1-digit and 2-digit numbers.

```
   642
 × 23
  1 926   Multiply by the ones.
+12 840   Multiply by the tens.
 14,766   Add the partial products.
```

So, 642 × 23 is 14,766.

Find the product. (pp. 146–147, 148–149)

9. 12,354 × 4
10. 248,021 × 5
11. 617,324 × 6
12. 626 × 38
13. 597 × 54
14. 2,847 × 18
15. 59,702 × 8
16. 10,287 × 9
17. 24,984 × 32
18. 58,701 × 12

Multiply greater numbers.

```
   375
 ×149
  3 375   Multiply by the ones.
 15 000   Multiply by the tens.
+37 500   Multiply by the hundreds.
 55,875   Add the partial products.
```

So, 375 × 149 is 55,875.

Find the product. (pp. 150–151)

19. 539 × 843
20. 792 × 154
21. 4,703 × 238
22. 6,178 × 297
23. 5,087 × 964
24. 4,532 × 102
25. 5,908 × 309

Chapter 10

Use mental math and patterns to multiply decimals.

Multiply 0.32 by 10, 100, and 1,000.

$0.32 \times 10 = 3.2$
$0.32 \times 100 = 32$
$0.32 \times 1{,}000 = 320$

Multiply decimals.

35.48×0.72 Multiply as with whole numbers. Find the total number of decimal places in the factors. Place the decimal point that number of places from the right in the product.

$$\begin{array}{r} 35.48 \\ \times\ 0.72 \\ \hline 7096 \\ +\ 24\,8360 \\ \hline 25.5456 \end{array}$$

35.48 ←2 decimal places in the factor
× 0.72 ←2 decimal places in the factor
25.5456 ←2 + 2, or 4 decimal places in the product

Multiply each number by 10, 100, and 1,000. (pp. 160–161)

26. 0.6
27. 0.82
28. 0.003
29. 0.213
30. 3.145
31. 0.014

Find the product. (pp. 162–169)

32. 0.6×0.8
33. 0.4×0.7
34. 0.08×8
35. 7×0.09
36. $\$0.76 \times 4$
37. $\$0.94 \times 0.5$
38. $\begin{array}{r} 0.05 \\ \times\ \ 4 \\ \hline \end{array}$
39. $\begin{array}{r} 1.50 \\ \times\ 2.5 \\ \hline \end{array}$
40. $\begin{array}{r} 0.007 \\ \times\ \ 2 \\ \hline \end{array}$
41. $\begin{array}{r} 1.006 \\ \times\ \ 8 \\ \hline \end{array}$
42. $\begin{array}{r} 2.506 \\ \times\ 0.37 \\ \hline \end{array}$
43. $\begin{array}{r} 84.11 \\ \times\ 0.12 \\ \hline \end{array}$
44. $\begin{array}{r} \$25.68 \\ \times\ \ 0.5 \\ \hline \end{array}$
45. $\begin{array}{r} \$3.99 \\ \times\ \ 6 \\ \hline \end{array}$
46. $\begin{array}{r} 47 \\ \times\ 0.05 \\ \hline \end{array}$
47. $\begin{array}{r} 6.03 \\ \times\ 0.34 \\ \hline \end{array}$
48. $\begin{array}{r} 20.09 \\ \times\ \ 17 \\ \hline \end{array}$
49. $\begin{array}{r} 9.8 \\ \times 0.4 \\ \hline \end{array}$

PROBLEM SOLVING PRACTICE

Solve. (pp. 152–153; 170–171)

50. Ted works for a magazine that prints 712,245 copies each week. He calculates that 37,036,740 copies are printed each year. Is this reasonable? Explain.

51. Jiffy Market sells 12 cans of juice for $4.10. C&G Store sells each can of juice for $0.35. Which store has the better price for 12 cans of juice?

California Connections

THE LOWEST POINT: DEATH VALLEY NATIONAL PARK

Death Valley National Park has more than 3,300,000 acres of desert scenery and wildlife. Badwater, located in the valley at 282 feet below sea level, is the lowest point in the western hemisphere.

Death Valley National Park is the largest national park outside of Alaska. ▼

1. In Death Valley National Park there are 14 square miles of sand dunes and about 15 times as many square miles of salt flats. About how many square miles of salt flats are there?

2. The actual valley in Death Valley National Park averages about 10 miles east to west and about 10 times as many miles in length. About how long is the valley?

3. At the Furnace Creek Visitor's Center, a slide show about the region is shown every half hour. If the first slide show begins at 8:30 A.M. and the last show begins at 5:30 P.M., how many times is the slide show presented in a day?

4. There are 23 species of plants that are known to grow only in the Death Valley region. In all, there are about 45 times this number of different species of plants. About how many different kinds of plants are found in the region?

5. June is the driest month in Death Valley. The average precipitation in June is 0.04 inches. In January the average precipitation is 6 times as great, in February it is 10 times as great, and in May it is 2 times as great. What is the average precipitation in January? in February? in May?

THE HIGHEST POINT: MT. WHITNEY

Mount Whitney, in the Sierra Nevada mountain range, is the tallest mountain in the continental United States.

1. At an average speed of 35 miles per hour, it takes 3 hours to drive from Death Valley to Mt. Whitney. What is the driving distance between Death Valley and Mt. Whitney?

2. Suppose the straight-line distance between Death Valley and Mt. Whitney on a map is 10.3 cm. If each centimeter on the map represents 8 miles, what is the actual straight-line distance?

3. The peak of Mt. Whitney can be reached by a 10.7-mile trail from Whitney Portal. How long is a round trip to the peak and back to Whitney Portal on this trail?

4. The elevation at Whitney Portal is 8,365 feet. The highest elevation on the trail is 14,494 feet. What is the difference between these two elevations?

5. To hike on the Mt. Whitney Trail you need a wilderness permit. The number of daily permits is limited to 50 for overnight hikers and 3 times as many for day hikers. In all, how many daily permits can be issued?

6. It costs $15.00 per person to apply for a permit to hike on the Mt. Whitney Trail. How much does it cost a group of 5 hikers to apply for a permit?

Mt. Whitney is the most frequently climbed peak in the Sierra Nevada. ▼

Divide by 1-Digit Divisors

Since history was first recorded, beads have been used for trade, adornment, and decoration. Beads can be made from stone, wood, bone, seeds, shells, metal, plastic, glass, or similar materials. Depending on the size and shape of each bead, the number of beads you need varies. The table below will help you figure out how many beads you would need to string a bracelet approximately 7 inches long. For each bead size, about how many beads are needed for 1 inch?

ABOUT HOW MANY BEADS PER 7-INCH BRACELET?

Bead Size	Number of Beads
2 mm	91
3 mm	56
4 mm	42
6 mm	28
8 mm	21

A Cherokee Indian wearing a beaded garment.

CHECK WHAT YOU KNOW

Use this page to help you review and remember important skills needed for Chapter 11.

VOCABULARY

Choose the best term from the box.

dividend
divisor
evaluate
factor
inverse
quotient
remainder

1. Multiplication and division are __?__ operations.
2. The number that divides the dividend is called the __?__.
3. When you replace the variable with a number and perform the operations, you __?__ an expression.
4. The number that is being divided in a division problem is called the __?__.
5. In the problem on the right, the __?__ is 7 and the __?__ is 3.

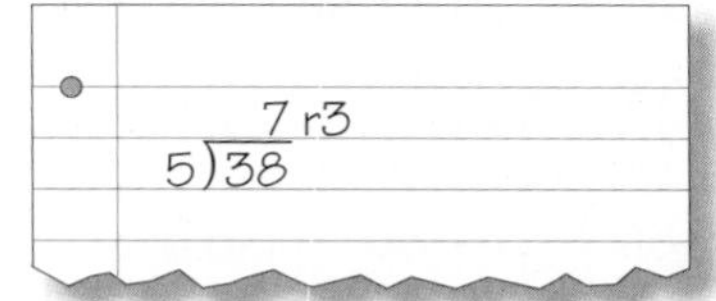

DIVIDE 2 DIGITS BY 1 DIGIT (See p. H13.)

Divide.

6. $3\overline{)85}$ **7.** $7\overline{)93}$ **8.** $2\overline{)61}$ **9.** $6\overline{)92}$ **10.** $6\overline{)29}$

11. $4\overline{)83}$ **12.** $8\overline{)54}$ **13.** $3\overline{)81}$ **14.** $9\overline{)95}$ **15.** $8\overline{)36}$

16. $7\overline{)84}$ **17.** $2\overline{)58}$ **18.** $6\overline{)44}$ **19.** $8\overline{)97}$ **20.** $5\overline{)52}$

CHECK DIVISION (See p. H13.)

Check the division by using multiplication.

21. $\begin{array}{r}23\\2\overline{)46}\end{array}$ **22.** $\begin{array}{r}11\text{ r5}\\8\overline{)93}\end{array}$ **23.** $\begin{array}{r}124\text{ r6}\\7\overline{)874}\end{array}$ **24.** $\begin{array}{r}47\text{ r1}\\4\overline{)189}\end{array}$

25. $\begin{array}{r}267\\3\overline{)801}\end{array}$ **26.** $\begin{array}{r}31\text{ r7}\\8\overline{)255}\end{array}$ **27.** $\begin{array}{r}10\text{ r2}\\3\overline{)32}\end{array}$ **28.** $\begin{array}{r}62\\6\overline{)372}\end{array}$

29. $\begin{array}{r}10\text{ r3}\\6\overline{)63}\end{array}$ **30.** $\begin{array}{r}13\\7\overline{)91}\end{array}$ **31.** $\begin{array}{r}188\text{ r2}\\4\overline{)754}\end{array}$ **32.** $\begin{array}{r}17\\5\overline{)85}\end{array}$

LESSON 1

Estimate Quotients

Quick Review

Write the closest basic division fact.

1. $28 \div 7$
2. $48 \div 8$
3. $20 \div 4$
4. $36 \div 6$
5. $24 \div 3$

VOCABULARY

compatible numbers

Learn

CLOSE ENCOUNTERS Musa Manarov, one of the top ten astronauts by total space flight time, spent 541 days in space. About how many total weeks was he in space?

Estimate. $541 \div 7$ $\quad 7\overline{)541}$

Compatible numbers are numbers that are easy to compute mentally. Compatible numbers for 7 are divisible by 7, such as 14, 21, 28, 35, 42, 49, and 56.

Use compatible numbers to estimate $541 \div 7$.

$541 \div 7 \rightarrow 490 \div 7 = 70$ or $541 \div 7 \rightarrow 560 \div 7 = 80$

Think: 541 is between 490 and 560. So, use these compatible numbers as dividends.

So, Manarov spent between 70 and 80 weeks in space.

MATH IDEA Use basic facts and compatible numbers to estimate quotients.

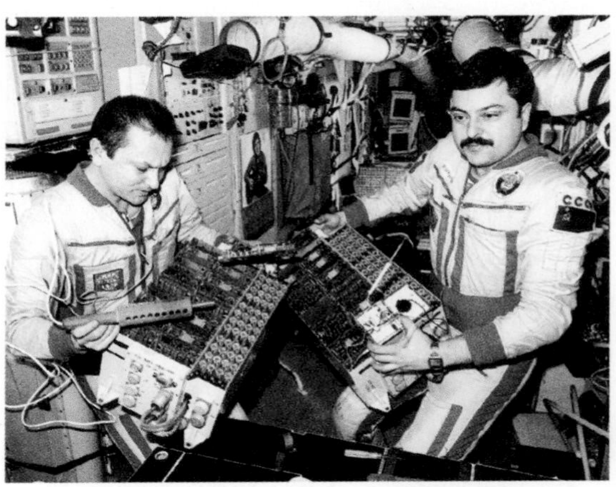

▲ Titov and Manarov's *Soyuz TM-4* mission lasted 365 days, 22 hours, 39 minutes. This was the first flight duration to exceed one year in space.

Examples

A Estimate $4,126 \div 8$.

Think: 40 is the closest compatible number to 41.

$4,126 \div 8 \rightarrow 4,000 \div 8 = 500$

So, $4,126 \div 8$ is about 500.

B Estimate $23,078 \div 4$.

Think: 24 is the closest compatible number to 23.

$23,078 \div 4 \rightarrow 24,000 \div 4 = 6,000$

So, $23,078 \div 4$ is about 6,000.

C Estimate $4,620,028 \div 7$.

Think: 49 is the closest compatible number to 46.

$4,620,028 \div 7 \rightarrow 4,900,000 \div 7 = 700,000$

So, $4,620,028 \div 7$ is about 700,000.

- **What if** you use two sets of compatible numbers, $180 \div 6$ and $240 \div 6$, to estimate $239 \div 6$? Which estimate will be closer to the exact answer? Explain.

CALIFORNIA STANDARDS NS 1.1 Estimate, round, and manipulate very large and very small numbers. 🔑 NS 2.2 Demonstrate proficiency with division, including division with positive decimals and long division with multidigit divisors. *also* NS 1.0, MR 1.1, MR 2.2, MR 2.4, MR 3.1, MR 3.3

Check

1. **Explain** how using basic facts and compatible numbers helps you estimate quotients. Give an example.

Estimate the quotient. Tell what compatible numbers you used.

2. $5\overline{)263}$
3. $7\overline{)1,502}$
4. $8\overline{)1,389}$
5. $9\overline{)61,409}$
6. $11,583 \div 2$
7. $451,133 \div 9$
8. $337,028 \div 4$
9. $1,123,506 \div 9$

Practice and Problem Solving

Estimate the quotient. Tell what compatible numbers you used.

10. $8\overline{)316}$
11. $9\overline{)5,536}$
12. $4\overline{)3,306}$
13. $7\overline{)34,072}$
14. $372 \div 6$
15. $125,703 \div 6$
16. $78,305 \div 7$
17. $2,500,823 \div 4$

Estimate the quotient, using two sets of compatible numbers.

18. $8,219 \div 9$
19. $25,316 \div 7$
20. $50,068 \div 4$
21. $172,388 \div 5$
22. $6,358 \div 8$
23. $71,269 \div 9$
24. $1,369,168 \div 5$
25. $800,625 \div 9$

USE DATA **For 26–28, use the table.**

TIME TO REVOLVE AROUND THE SUN (based on Earth days)

Planet	Number of Days
Mars	687
Mercury	88
Pluto	90,950

▲ Mars

26. About how many Earth weeks does it take Mars to complete one revolution around the sun?
27. Estimate how many Earth weeks it takes Pluto to complete one revolution around the sun.
28. About how many times will Mercury revolve around the sun while the Earth revolves around the sun once?
29. **REASONING** When I am divided by 4, the quotient has two digits and the remainder is 0. When I am divided by 6, the quotient has one digit and the remainder is 0. My ones digit is 8. What number am I?

Mixed Review and Test Prep

30. $71,053 \times 28$ (p. 150)
31. $192,550 \times 47$ (p. 150)
32. Use the Distributive Property to restate 20×16. Then find the product. (p. 92)
33. Write 505.05 in word form. (p. 18)
34. **TEST PREP** Which number is between 0.5 and 0.711? (p. 24)
    - **A** 0.071
    - **B** 0.157
    - **C** 0.571
    - **D** 0.750

Extra Practice page H42, Set A

LESSON 2

Divide 3-Digit Dividends

Quick Review

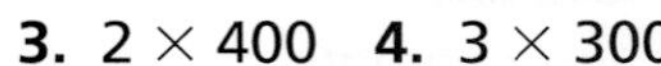

1. 4×70
2. 8×90
3. 2×400
4. 3×300
5. 6×90

Learn

WEIGHT LOSS The moon's weaker gravity causes things to weigh one sixth as much on the moon as they weigh on Earth. Suppose some of the rocks brought back from the moon weigh 102 pounds on Earth. How much did they weigh on the moon?

◀ Between 1969 and 1972, Apollo missions brought back 382 kg (843 lb) of lunar rocks and soil from the Moon's surface.

One Way

Use estimation to place the first digit in the quotient.
Find $102 \div 6$.

STEP 1

Estimate.

Think: $6\overline{)60}$ = 10 or $6\overline{)120}$ = 20

$6\overline{)102}$ ■ So, place the first digit in the tens place.

STEP 2

Divide the 10 tens.

```
   1
6)102
 - 6
   4
```

Divide. $6\overline{)10}$
Multiply. 6×1
Subtract. $10 - 6$
Compare. $4 < 6$

STEP 3

Bring down the 2 ones.
Divide the 42 ones.

```
   17
6)102
 - 6↓
   42
 - 42
    0
```

Divide. $6\overline{)42}$
Multiply. 6×7
Subtract. $42 - 42$
Compare. $0 < 6$

So, the rocks weighed 17 pounds on the moon.

Another Way

Use place value to place the first digit.
Find $893 \div 9$.

STEP 1

Look at the hundreds.

$9\overline{)893}$ $8 < 9$, so look at the tens.

$9\overline{)893}$ ■ $89 > 9$, so use 89 tens. Place the first digit in the tens place.

STEP 2

Divide the 89 tens.

```
   9
9)893
 -81
   8
```

Divide.
Multiply.
Subtract.
Compare.

STEP 3

Bring down the 3 ones. Divide the 83 ones.

```
   99 r2
9)893
 -81↓
   83
  -81
    2
```

Divide.
Multiply.
Subtract.
Compare.

CHECK ✓

Multiply the quotient by the divisor. Then add the remainder.

```
   99
 ×  9
  891
 +  2
  893
```

CALIFORNIA STANDARDS **O━ NS 2.2** Demonstrate proficiency with division, including division with positive decimals and long division with multidigit divisors. **NS 1.0** Students compute with very large and very small numbers, positive integers, decimals, and fractions and understand the relationship between decimals, fractions, and percents. They understand the relative magnitudes of numbers. *also* **NS 1.1, MR 1.0, MR 2.0, MR 2.2, MR 3.0, MR 3.2**

Check

1. **Explain** how you can tell without dividing whether the quotient of $539 \div 6$ will have 2 digits or 3.

Name the position of the first digit of the quotient.

2. $9\overline{)117}$ **3.** $5\overline{)213}$ **4.** $2\overline{)472}$ **5.** $6\overline{)917}$ **6.** $9\overline{)526}$

Practice and Problem Solving

Name the position of the first digit of the quotient.

7. $8\overline{)504}$ **8.** $2\overline{)724}$ **9.** $4\overline{)183}$ **10.** $7\overline{)857}$ **11.** $6\overline{)176}$

Divide.

12. $2\overline{)408}$ **13.** $7\overline{)489}$ **14.** $5\overline{)516}$ **15.** $4\overline{)844}$ **16.** $8\overline{)246}$

17. $8\overline{)113}$ **18.** $9\overline{)912}$ **19.** $3\overline{)607}$ **20.** $2\overline{)526}$ **21.** $4\overline{)223}$

22. $348 \div 8$ **23.** $905 \div 5$ **24.** $124 \div 6$ **25.** $414 \div 5$ **26.** $807 \div 4$

Algebra **Find the value of *n*.**

27. $357 \div 3 = n$ **28.** $n \div 5 = 35$ r4 **29.** $n \div 7 = 47$ r3 **30.** $269 \div n = 134$ r1

USE DATA **For 31–32, use the table.**

SPACE FLIGHT FOOD (based on a standard 7-day shuttle menu)		
Food Type	**Number of Servings**	**Servings Per Crew Member**
Rehydratable Beverage	203	■
Fresh Food	175	■
Rehydratable Food	238	■

31. If there are 7 crew members, how many servings of each food type does each crew member receive? Copy and complete the table.

32. What is the range of the number of servings of each type of food?

33. The weight allowed for food is limited to 3.8 pounds per crew member per day. If there are 7 crew members on a 6-day mission, how much does their food supply weigh?

34. **Write a problem** that has a divisor of 6 and a quotient with a remainder. Use the table above.

Mixed Review and Test Prep

35. $0.8 + 6.8$ (p. 50)

36. $6.16 - 3.47$ (p. 50)

37. Find the next number in the pattern: 24, 36, 48, 60, ... Describe the pattern. (p. 18)

38. Find $4 \times n$ if $n = 16.5$. (p. 164)

39. **TEST PREP** Without multiplying, estimate which product is least. (p. 160)

A 0.9×10
B 0.9×1
C 0.9×0.1
D 0.9×0.01

Extra Practice page H42, Set B

LESSON 3

Zeros in Division

Quick Review

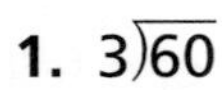
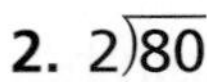

1. $3\overline{)60}$
2. $2\overline{)80}$
3. $4\overline{)80}$
4. $9\overline{)90}$
5. $10\overline{)100}$

Learn

LOONY TUNES Mr. Knowles owns 4 music stores. He ordered 432 kazoos. He wants each store to have the same number of kazoos. How many will each store receive?

Divide. $432 \div 4$ $\quad 4\overline{)432}$

$\downarrow \quad \downarrow$

Estimate. $440 \div 4 = 110$ ← Use compatible numbers.

STEP 1

Divide the 4 hundreds.

Think: 4 hundreds can be divided by 4. So, place the first digit in the hundreds place in the quotient.

```
    1
 4)432   Divide.
  -4     Multiply.
   0     Subtract.
         Compare. 0 < 4
```

STEP 2

Bring down the tens.
Divide the 3 tens.

Think: Since $4 > 3$, write 0 in the quotient.

```
    10
 4)432   Divide.
  -4↓    Multiply.
    3    Subtract.
   -0    Compare. 3 < 4
    3
```

STEP 3

Bring down the ones.
Divide the 32 ones.

Think: $4 \times n = 32$

```
    108
 4)432   Divide.
  -4     Multiply.
    3    Subtract.
   -0↓   Compare. 0 < 4
    32
   -32
     0
```

Since 108 is close to the estimate of 110, the answer is reasonable. So, each store will receive 108 kazoos.

- What happens to your answer if you forget to write the zero in the quotient?

MATH IDEA Place a zero in the quotient when the divisor is greater than the number to be divided at that step.

Check

1. **Tell** how you decide when to write a zero in the quotient. Give an example.

CALIFORNIA STANDARDS O━ NS 2.2 Demonstrate proficiency with division, including division with positive decimals and long division with multidigit divisors. **MR 2.1** Use estimation to verify the reasonableness of calculated results. *also* **NS 1.0, NS 1.1, MR 2.3, MR 2.4, MR 3.0, MR 3.2, MR 3.3**

Divide. Estimate to check.

2. $5\overline{)250}$ 3. $7\overline{)73}$ 4. $4\overline{)801}$

TECHNOLOGY LINK

More Practice: Use **Mighty Math Calculating Crew,** ***Captain Nick Knack,*** **Level Q.**

Practice and Problem Solving

Divide. Estimate to check.

5. $9\overline{)87}$ 6. $7\overline{)440}$ 7. $4\overline{)82}$

8. $3\overline{)309}$ 9. $5\overline{)54}$ 10. $6\overline{)635}$ 11. $4\overline{)816}$ 12. $3\overline{)908}$

13. $6\overline{)612}$ 14. $8\overline{)255}$ 15. $5\overline{)525}$ 16. $6\overline{)624}$ 17. $7\overline{)745}$

18. $61 \div 2$ 19. $528 \div 9$ 20. $744 \div 8$ 21. $606 \div 3$ 22. $401 \div 5$

23. $122 \div 4$ 24. $504 \div 5$ 25. $824 \div 8$ 26. $631 \div 6$ 27. $801 \div 2$

Algebra Find the value of *n*.

28. $305 \div 5 = n$ 29. $n \div 8 = 110$ r7 30. $n \div 7 = 101$ r3 31. $209 \div n = 104$ r1

Algebra Complete each table and write a division rule.

32.

Input, *n*	49	107	239	564
Output	7	15 r2	■	■

33.

Input, *n*	83	■	404	412
Output	20 r3	24 r3	101	■

34. Felipe's family is planning to see a "Kazoo Band" show at a festival 306 miles away. It takes them 6 hours to drive there. On average, how many miles do they drive each hour?

36. **REASONING** Are there any zeros in the quotient $884 \div 4$? How can you tell without dividing?

37. **What's the Error?** Jackie says $6 \div 6$ equals 1, so $606 \div 6$ equals 11. Find her error and write the correct division sentence.

35. At a "Kazoo Band" show, everyone in the audience is given a kazoo for the grand finale. If 1,634 people attend the first show and 2,582 people attend the second show, how many more kazoos will be given out in the second show?

Mixed Review and Test Prep

38. $100 - 47$ (p. 36) 39. $500 - 176$ (p. 36)

40. $2{,}000 - 786$ (p. 36)

41. Write 246.975 in expanded form. (p. 20)

42. **TEST PREP** Choose an equivalent decimal for 0.04. (p. 22)

A 4.0 **B** 0.4 **C** 0.040 **D** 0.004

Extra Practice page H42, Set C

LESSON 4

Divide Greater Numbers

Quick Review

1. 23 − 18
2. 90 − 72
3. 44 − 36
4. 62 − 56
5. 39 − 32

Learn

DRAMATIC PARTS The costume department has a box filled with 12,575 five-mm beads. The members plan to make necklaces for costumes for the school play. What if the beads were divided equally among 9 members? How many beads would each member be able to use? Would there be any beads left over?

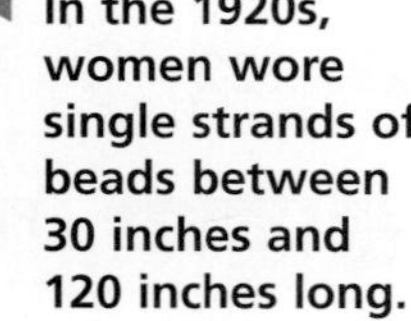

◀ In the 1920s, women wore single strands of beads between 30 inches and 120 inches long.

Divide. $12{,}575 \div 9$ $\quad 9\overline{)12{,}575}$

Estimate. $9{,}000 \div 9 = 1{,}000$ ← Use compatible numbers.

STEP 1

The first digit will be in the thousands place. Divide the 12 thousands.

$$\begin{array}{r} 1 \\ 9\overline{)12{,}575} \\ -9 \\ \hline 3 \end{array}$$

Divide.
Multiply.
Subtract.
Compare.
$3 < 9$

STEP 2

Bring down the 5 hundreds. Divide the 35 hundreds.

$$\begin{array}{r} 13 \\ 9\overline{)12{,}575} \\ -9\downarrow \\ \hline 35 \\ -27 \\ \hline 8 \end{array}$$

Divide.
Multiply.
Subtract.
Compare.
$8 < 9$

STEP 3

Bring down the 7 tens. Divide the 87 tens.

$$\begin{array}{r} 139 \\ 9\overline{)12{,}575} \\ -9 \\ \hline 35 \\ -27\downarrow \\ \hline 87 \\ -81 \\ \hline 6 \end{array}$$

Divide.
Multiply.
Subtract.
Compare.
$6 < 9$

STEP 4

Bring down the 5 ones. Divide the 65 ones.

$$\begin{array}{r} 1{,}397 \text{ r2} \\ 9\overline{)12{,}575} \\ -9 \\ \hline 35 \\ -27 \\ \hline 87 \\ -81\downarrow \\ \hline 65 \\ -63 \\ \hline 2 \end{array}$$

Divide.
Multiply.
Subtract.
Compare.
$2 < 9$

Since there is a remainder of 2, the beads cannot be divided equally. So, each member of the costume department could use 1,397 beads to make necklaces, and 2 beads would be left over.

- How can you tell if your answer is reasonable?

MATH IDEA For greater numbers, divide the same way you would divide a 3-digit number by a 1-digit number. Just repeat the same steps for all the digits.

CALIFORNIA STANDARDS O━━n NS 2.2 Demonstrate proficiency with division, including division with positive decimals and long division with multidigit divisors. **NS 1.0** Students compute with very large and very small numbers, positive integers, decimals, and fractions and understand the relationship between decimals, fractions, and percents. They understand the relative magnitudes of numbers. *also* **NS 1.1, MR 2.0, MR 2.1, MR 3.0, MR 3.1, MR 3.2**

Check

1. **REASONING** **Explain** how you can tell before you find 34,682 ÷ 8 that the first digit in the quotient is in the thousands place.

Divide.

2. $3\overline{)2,624}$
3. 18,152 ÷ 4
4. $2\overline{)283,462}$
5. 19,145 ÷ 5

Practice and Problem Solving

Divide.

6. $6\overline{)4,632}$
7. $3\overline{)3,016}$
8. $6\overline{)16,459}$
9. $8\overline{)3,348}$
10. $8\overline{)34,762}$
11. $2\overline{)83,761}$
12. $4\overline{)711,286}$
13. $7\overline{)603,194}$
14. 624,294 ÷ 3
15. 55,489 ÷ 7
16. 4,932,700 ÷ 4
17. 4,384,623 ÷ 5
18. 9,088 ÷ 3
19. 816,422 ÷ 6
20. 49,092 ÷ 9
21. 7,074,608 ÷ 6

USE DATA **For 22–23, use the circle graph.**

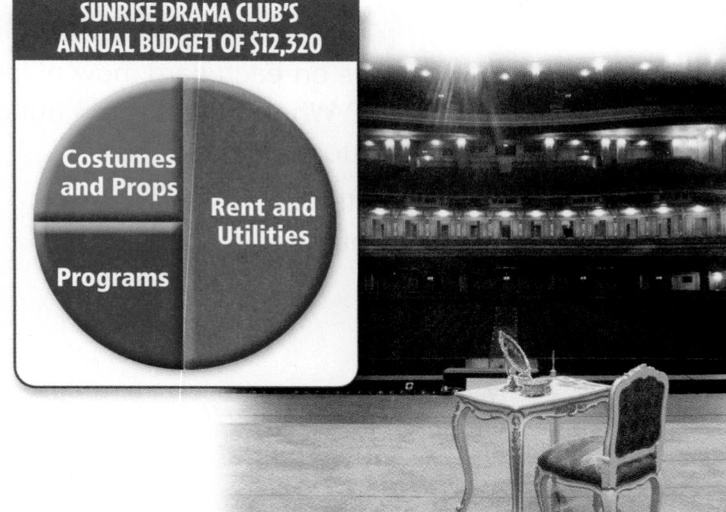

22. The Booster Club donates money for costumes and props each year. If the Booster Club matches the costumes and props budget dollar for dollar, how much money will the Drama Club have for costumes and props?

23. The club members plan to print 49,100 programs. If they print an equal number for each of 4 plays, how many will they print for 1 play? How much money is budgeted per play for printing programs?

24. **What's the Question?** An embossing machine stamps 20 sheets of paper in a minute. The answer is 300 sheets.

25. **Write a problem** that has a 4-digit number divided by a 1-digit number with a 4-digit quotient.

Mixed Review and Test Prep

26. Write the value of the blue digit in 7,543,229. (p. 4)

27. Evaluate $26 + n$ if $n = 20$. (p. 68)

28. Find the value of n. $4 \times 5 \times n = 40$ (p. 86)

29. Find 24,548 × 4. (p. 146)

30. **TEST PREP** Find 0.4 × 0.8. (p. 162)

A 3.2
B 1.2
C 0.32
D 0.12

LESSON **5**

Algebra: Expressions and Equations

Quick Review

Evaluate each expression if $n = 5$.

1. $25 + n$ **2.** $38 - n$
3. $40 \times n$ **4.** $7n$
5. $6 \times (5 + n)$

Learn

BUTTON UP Renée has 576 buttons in her button box. She would like to display her collection on individual cards. If she puts an equal number of buttons on each card, how many cards will she need?

Write and evaluate an expression to find out how many cards she will need.

Let c = the number of buttons on each card.

number of buttons	**divided by**	**number of buttons on each card**
↓	↓	↓
576	÷	c

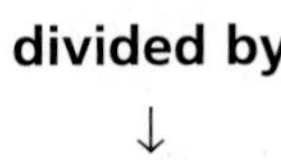

If there are 4 buttons on each card, how many cards will she need? **What if** there are 6 buttons on each card? 8 buttons on each card?

To find the value of the expression, replace the variable with a number.

Remember
An *expression* is a mathematical phrase that combines numbers, operation signs, and sometimes variables. A *variable* is a letter or symbol that stands for a number.

$9n$ **$n \div 6$** **$n + 7$**

Examples

Evaluate the expression for each value of c.

A Find $576 \div c$ if $c = 4$.

$576 \div 4$ Replace c with 4.

$$\begin{array}{r} 144 \\ 4\overline{)576} \\ -4 \\ \hline 17 \\ -16 \\ \hline 16 \\ -16 \\ \hline 0 \end{array}$$

$576 \div 4 = 144$

So, Renée will need 144 cards to have 4 buttons on each card.

B Find $576 \div c$ if $c = 6$.

$576 \div 6$ Replace c with 6.

$$\begin{array}{r} 96 \\ 6\overline{)576} \\ -54 \\ \hline 36 \\ -36 \\ \hline 0 \end{array}$$

$576 \div 6 = 96$

So, Renée will need 96 cards to have 6 buttons on each card.

C Find $576 \div c$ if $c = 8$.

$576 \div 8$ Replace c with 8.

$$\begin{array}{r} 72 \\ 8\overline{)576} \\ -56 \\ \hline 16 \\ -16 \\ \hline 0 \end{array}$$

$576 \div 8 = 72$

So, Renée will need 72 cards to have 8 buttons on each card.

- In the examples above, why are there three different values for the expression $576 \div c$?

CALIFORNIA STANDARDS **AF 1.2** Use a letter to represent an unknown number; write and evaluate simple algebraic expressions in one variable by substitution. **AF 1.0** Students use variables in simple expressions, compute the value of the expression for specific values of the variable, and plot and interpret the results. *also* **NS2.2, MR 1.0, MR 1.1, MR 2.3, MR 2.4, MR 3.1**

Write and Solve Equations

Antoni has collected 84 autographs. He filled 14 pages in his new autograph album. Each page holds an equal number of autographs. How many autographs fit on a page?

▲ **Experts believe that William Shakespeare's autograph would sell for $5,000,000.**

Write an equation with a variable to model this problem.

Let a = the number of autographs per page.

number of autographs	**divided by**	**number of autographs per page**	**equals**	**number of pages**
↓	↓	↓	↓	↓
84	÷	a	=	14

When you solve an equation, you find the value of the variable that makes the equation true.

Examples Which of the values 3, 4, or 6 is the solution of $84 \div a = 14$?

Substitute each of the values for a.

D $84 \div a = 14$

$84 \div 3 \stackrel{?}{=} 14$ Replace a with 3.
$28 \neq 14$

The two sides are not the same, so $a = 3$ is not the solution.

E $84 \div a = 14$

$84 \div 4 \stackrel{?}{=} 14$ Replace a with 4.
$21 \neq 14$

The two sides are not the same, so $a = 4$ is not the solution.

F $84 \div a = 14$

$84 \div 6 \stackrel{?}{=} 14$ Replace a with 6.
$14 = 14$ √

The two sides are the same, so $a = 6$ is the solution.

So, each page of Antoni's autograph album holds 6 autographs.

You can also solve an equation by using mental math.

Examples

G Solve $72 \div n = 12$.

$72 \div n = 12$
$n = 6$

Think: 72 divided by what number equals 12?

Check: $72 \div 6 = 12$ Replace n with 6.
$12 = 12$ √

So, $n = 6$.

H Solve $n \div 8 = 11$.

$n \div 8 = 11$
$n = 88$

Think: What number divided by 8 equals 11?

Check: $88 \div 8 = 11$ Replace n with 88.
$11 = 11$ √

So, $n = 11$.

- **REASONING** How could you use inverse operations to solve Examples G and H?

LESSON CONTINUES ▶

Check

1. **Tell** how you can find whether $d = 6$, $d = 5$, or $d = 4$ is the solution of $78 \div d = 13$.

Evaluate the expression $n \div 8$ for each value of n.

2. $n = 96$
3. $n = 856$
4. $n = 112$
5. $n = 488$

Determine which value is a solution for the given equation.

6. $72 \div n = 9$; $n = 7, 8,$ or 9
7. $147 \div y = 21$; $y = 6, 7,$ or 8
8. $n \times 4 = 92$; $n = 23$ or 24
9. $m \div 7 = 16$; $m = 112$ or 119

Practice and Problem Solving

Evaluate the expression $1{,}080 \div n$ for each value of n.

10. $n = 2$
11. $n = 3$
12. $n = 5$
13. $n = 9$

Evaluate the expression for each value of n.

14. $2{,}700 \div n$; $n = 3, 4, 5$
15. $n \times 13$; $n = 14, 15, 16$
16. $168 \div n$; $n = 6, 7, 8$
17. $n \div 3$; $n = 237, 243$

Determine which value is the solution for the given equation.

18. $27 \div n = 3$; $n = 7, 8,$ or 9
19. $248 \div n = 62$; $n = 4, 5,$ or 6
20. $p \div 6 = 31$; $p = 184$ or 186
21. $n \div 9 = 23$; $n = 207$ or 208
22. $60 \div n = 12$; $n = 3, 4,$ or 5
23. $n \div 5 = 28$; $n = 135$ or 140
24. $n \div 8 = 31$; $n = 248$ or 256
25. $189 \div n = 63$; $n = 2, 3,$ or 4

Solve the equation. Then check the solution.

26. $24 \div n = 12$
27. $45 \div m = 5$
28. $n \div 9 = 12$
29. $s \div 6 = 3$
30. $81 \div n = 9$
31. $n \div 8 = 4$
32. $n \div 3 = 12$
33. $55 \div n = 5$
34. **REASONING** Find the value of each symbol by using the given equations. A symbol always has the same value.
    ♥ $\div 4 = 10$ $\qquad$ $5 \times$ ♦ $=$ ♥ $\qquad$ (♥ $+ 8) \div 3 =$ ♣ $\qquad$ ♠ $\div 2 =$ ♣

USE DATA **For 35–37, use the table.**

35. How many kids were surveyed?
36. Collecting cards is how many more times as popular as collecting stamps?
37. What type of graph would you use to show this data? Explain.
38. **Write About It** Describe how you evaluate $p \div 11$ if $p = 132$.

KIDS' FAVORITE COLLECTIBLES

Collectible	Number of Kids
Cards	36
Coins	20
Dolls	24
Model cars	14
Stamps	6

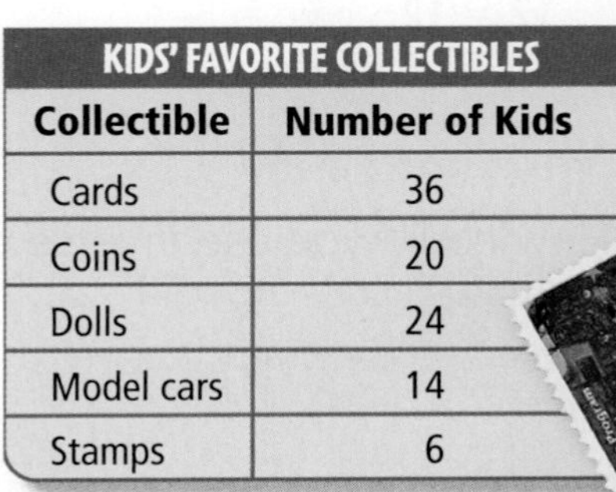

39. Debbie divided her collection of 120 shells equally into n boxes. How many could she put in each box? Write an expression to represent the number of shells in each box.

40. Silvia divided her stickers into 9 subjects. Each subject had 7 stickers. Write and solve an equation that represents the total number of stickers, s, she has.

Mixed Review and Test Prep

41. 44×37 (p. 148)

42. 1.8×56 (p. 164)

43. Compare. $30 \times 4,000$ ● 300×400 (p. 144)

44. Elena has a $20 bill. She buys 5 cookies for $2.00 each, a cup of tea for $1.50, and a sandwich for $2.35. How much change should she get? (p. 50)

45. **TEST PREP** Round 2,698,734 to the nearest ten thousand. (p. 32)
   - **A** 2,698,700
   - **B** 2,699,000
   - **C** 2,700,000
   - **D** 3,000,000

46. **TEST PREP** Which is an equivalent decimal for 3.20? (p. 22)
   - **F** 0.0320
   - **G** 0.32
   - **H** 3.200
   - **J** 32.0

LINKUP to Reading

STRATEGY • ANALYZE INFORMATION When you *analyze information*, look for the important details. Then think about how to use the details to solve the problem.

By 1999, Gemma Dickmann of Holland had collected over 4,275 pencil sharpeners. She had bronze, plastic, wood, stone, and rubber sharpeners. She had about 2,200 plastic sharpeners. She had twice as many plastic sharpeners as stone sharpeners. About how many stone sharpeners did Gemma have?

- List the important details. about 2,200 plastic sharpeners; twice as many plastic sharpeners as stone sharpeners
- Tell what you need to find. the number of stone sharpeners she had
- Look for clues. *Twice as many* tells you to multiply or divide. Since you know the number of plastic sharpeners, divide.
- Solve. $2,200 \div 2 = 1,100$; So, Gemma had about 1,100 stone sharpeners.

Write the details you need to solve each problem. Then solve.

1. Rob and Jennifer both collect basketball cards. Rob has collected 177 cards. Rob has 3 times as many cards as Jennifer. How many cards does Jennifer have?

2. In the past year, Jack doubled his collection of origami paper shapes. He now has a total of 568 origami shapes. How many shapes did he have a year ago?

Extra Practice page H42, Set E

LESSON 6

Problem Solving Skill

Interpret the Remainder

Quick Review

1. 29 ÷ 8
2. 39 ÷ 9
3. 60 ÷ 7
4. 22 ÷ 5
5. 25 ÷ 6

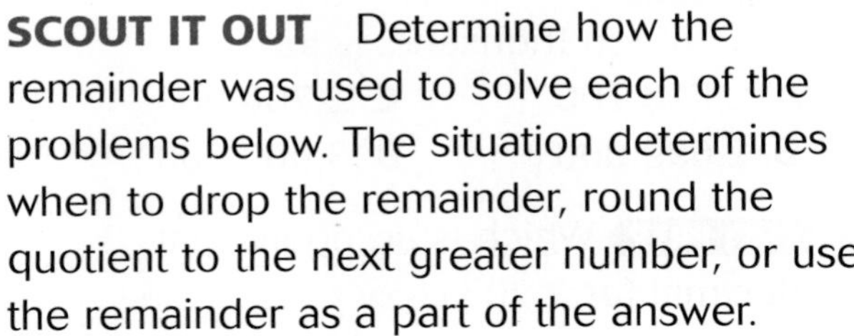

SCOUT IT OUT Determine how the remainder was used to solve each of the problems below. The situation determines when to drop the remainder, round the quotient to the next greater number, or use the remainder as a part of the answer.

MATH IDEA When you solve a division problem with a remainder, the way you interpret the remainder depends on the situation.

A

There are 1,853 Boy Scouts and parents signed up for the spring banquet. Each table will seat 6 people. How many tables will be needed?

```
    308 r5
 6)1,853
  -18
    05
   - 0
    53
   -48
     5
```

Since 308 tables will not be enough to seat everyone, 309 tables will be needed.

B

Erik made punch with 3 ounces of lemon juice, 32 ounces of orange juice, 32 ounces of lemonade, and 64 ounces of ginger ale. How many 6-ounce servings did the punch recipe make?

```
                21 r5
   3       → 6)131
  32           -12
  32            11
 +64           - 6
 131 ←           5
```

It made 21 six-ounce servings. The 5 ounces left over are not enough for another 6-ounce serving.

C

Mrs. Clarke, the troop leader, brought in a 145-inch rope for the Boy Scouts to demonstrate tying knots. She divided the rope into 4 pieces of equal length. How long was each piece of rope?

$36\frac{1}{4}$ ← Use the remainder as the numerator and the divisor as the denominator.

```
 4)145
  -12
    25
   -24
     1
```

So, each piece of rope was $36\frac{1}{4}$ inches long.

Talk About It

- Explain how the remainder was used to answer Problems A–C.
- Explain why a fractional remainder was appropriate for Problem C, but not for Problems A or B.

CALIFORNIA STANDARDS MR 3.1 Evaluate the reasonableness of the solution in the context of the original situation. **MR 2.0** Students use strategies, skills, and concepts in finding solutions. *also* **NS 1.0, NS 2.2, MR 2.2, MR 2.3, MR 2.6, MR 3.2, MR 3.3**

Problem Solving Practice

Solve and then explain how you interpreted the remainder.

1. Jake has 170 books to display on 5 shelves in the bookstore. If Jake puts an equal number of books on each shelf, what is the maximum number of books on each? How many books will be left over?

2. Girl Scout Troop 7 is learning how to arrange flowers. There are 18 girls in the troop. They have 124 roses. Will there be enough roses for each scout to have 7 roses in her arrangement?

USE DATA Use the table. At the concession stand, the popcorn machine pops 80 cups of popcorn at a time.

BAGS OF POPCORN	
Size	**Amount per serving**
Large	4 cups
Jumbo	6 cups

3. Marci calculated that she could fill 13 jumbo popcorn bags from one batch. Explain how she interpreted the remainder.
   - **A** There is no remainder.
   - **B** She dropped the remainder.
   - **C** She rounded the quotient.
   - **D** She used the remainder.

4. Marci sold 8 jumbo bags. Does she have enough popcorn left to fill 9 large popcorn bags? How many large bags can she fill?
   - **F** No; 8 bags
   - **G** No; 3 bags
   - **H** Yes; 10 bags
   - **J** Yes; 8 bags

Mixed Applications

5. A group of 213 students signed up for ski lessons. There are a maximum of 9 students per instructor. What is the least number of instructors needed?

6. Chiano bought a 50-page photo album. Each page holds 6 photos. She took 288 photos on her trip. Will the album hold all of her photos? Explain.

7. Mr. Jones is planning for buses for the school trip. A total of 105 students signed up. If each bus holds 42 students, will 2 buses be enough for the trip? Explain.

8. **PATTERNS** Charlie is designing a necklace. She plans to repeat the pattern 1 red bead, 1 blue bead, and 1 green bead until she strings a total of 60 beads. What color will the 31st bead be?

9. Brian's model train track is 4 times as long as Mario's track. Brian's track is 1,247 inches long. How long is Mario's track?

10. **Write a problem** in which you divide and you interpret the remainder to answer the question. Explain how the remainder helps answer the question.

CHAPTER 11

Review/Test

CHECK VOCABULARY AND CONCEPTS

Choose the best term from the box.

compatible numbers
place value
remainder
zero

1. To decide where to place the first digit in the quotient, you can estimate or use __?__. (p. 184)
2. Pairs of numbers close to the actual numbers that are easy to compute mentally are __?__. (p. 182)
3. The amount left over when you find a quotient is called the __?__. (p. 194)

CHECK SKILLS

Estimate the quotient. Tell what compatible numbers you used. (pp. 182–183)

4. $3\overline{)3{,}296}$
5. $5\overline{)13{,}158}$
6. $2\overline{)23{,}498}$
7. $4\overline{)29{,}746}$

Divide. (pp. 184–189)

8. $9\overline{)504}$
9. $7\overline{)4{,}308}$
10. $8\overline{)998}$
11. $6\overline{)1{,}241}$
12. $5\overline{)421}$
13. $5\overline{)120{,}250}$
14. $4\overline{)28{,}750}$
15. $9\overline{)1{,}785{,}889}$

Determine which value is the solution for the given equation. (pp. 190–193)

16. $360 \div n = 72$; $n = 4, 5,$ or 6
17. $m \div 8 = 43$; $m = 336$ or 344
18. $x \div 3 = 97$; $x = 291$ or 294
19. $504 \div n = 72$; $n = 7, 8,$ or 9
20. Solve $96 \div n = 8$ for n. Then check the solution. (pp. 190–193)
21. Solve $n \div 5 = 7$ for n. Then check the solution. (pp. 190–193)

CHECK PROBLEM SOLVING

Solve. Explain how you interpreted the remainder. (pp. 194–195)

22. A group of 123 students will take a swimming course. Each class can have 11 students. What is the least number of classes needed?
23. Mark has 300 miniature cars. He would like to display them in 7 cases with 42 spaces each. Will the 7 cases hold all of his collection? Explain.
24. The bakery sells muffins in trays of 6 each. If the bakery has 226 muffins, how many trays can be filled?
25. Stefan divided a 60-yard roll of fabric evenly into 8 pieces. How long was each piece?

Cumulative Review

Decide on a plan.
See item **5.**

First find how many pancakes 5 boxes will make. Write the number sentence that tells how many people will be served if each person eats 3 pancakes. Then solve the problem.

Also see problem **4,** p. H63.

For 1–9, choose the best answer.

For 1–3, use the table.

COUNTRIES WITH THE MOST MOVIE THEATERS	
Country	**Number of Theaters**
India	8,975
United States	23,662
China	4,364
Ukraine	14,960

1. If a film is shown in half of all the theaters in the United States and China, how many theaters show the film?

A 1,413 **C** 11,831
B 2,182 **D** 14,013

2. How many more theaters are in the United States than in the Ukraine?

F 38,662 **H** 11,302
G 18,702 **J** Not here

3. For which country will the number of theaters have a remainder when divided by 2?

A United States **C** India
B China **D** Ukraine

4. Marbles come in sets of 50. Cal wants to give an equal number of marbles to 6 friends. How many sets does he need so that there are no marbles left over?

F 3 **H** 5
G 4 **J** 6

5. A box of pancake mix contains enough mix to make 65 pancakes. If each person eats 3 pancakes, how many people will 5 boxes serve?

A 300 **C** 108
B 125 **D** 100

6. For a theater with 345 seats and 23 in each row, what does *n* represent in the equation $345 \div n = 23$?

F number of rows without seats
G number of rows in the theater
H number of seats in each row
J number of seats in the theater

7. $815 \div 8$

A 101 r7 **C** 100 r15
B 101 r5 **D** Not here

8. What is the best estimate for $534 \div 6$ when the dividend is rounded to a compatible number?

F 100 **H** 60
G 90 **J** 50

9. Which numbers are in order from *greatest* to *least*?

A 65,248, 65,482, 62,500, 65,208
B 65,482, 65,208, 65,248, 62,500
C 65,482, 65,248, 65,208, 62,500
D 65,482, 62,500, 65,248, 65,208

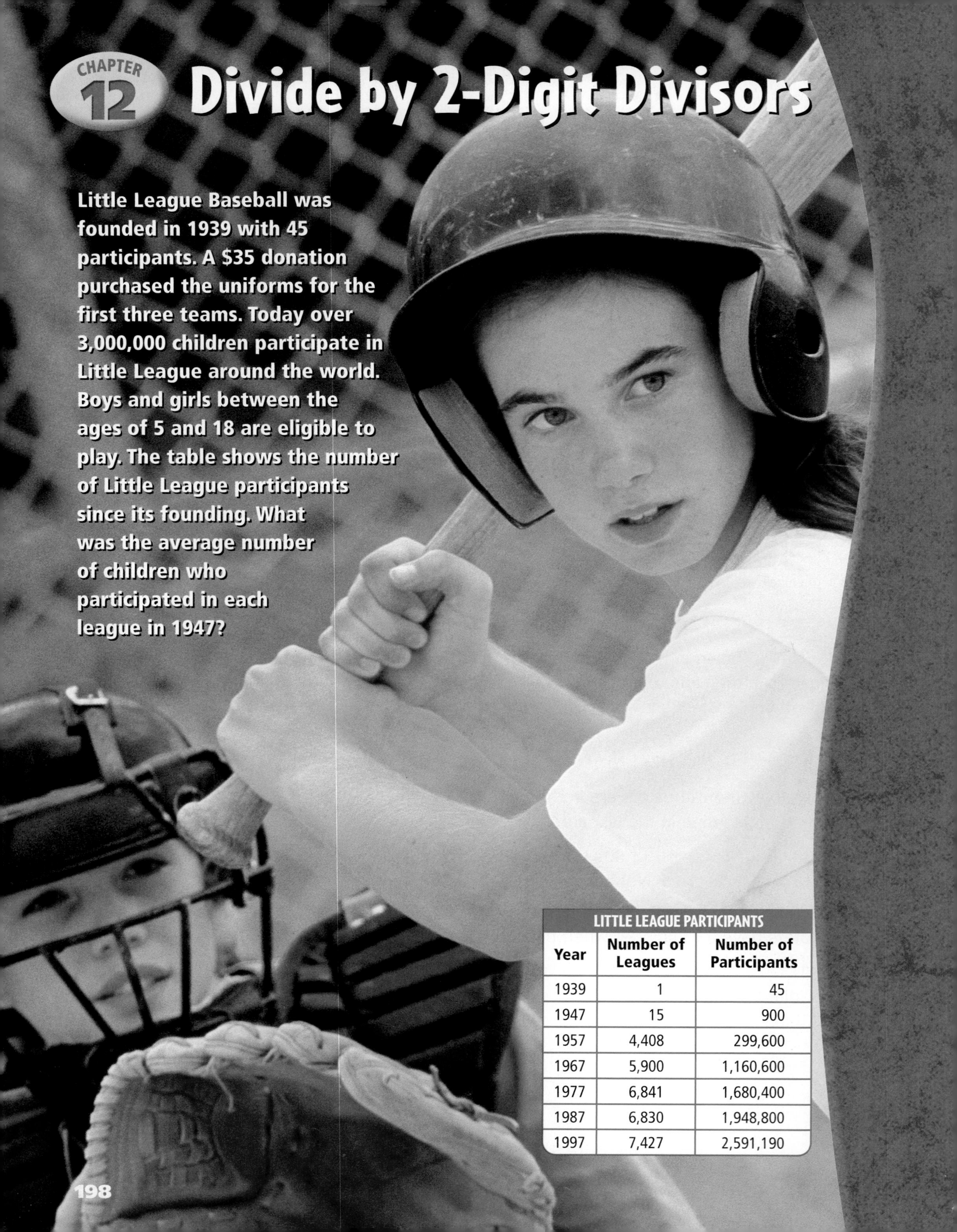

CHAPTER 12

Divide by 2-Digit Divisors

Little League Baseball was founded in 1939 with 45 participants. A $35 donation purchased the uniforms for the first three teams. Today over 3,000,000 children participate in Little League around the world. Boys and girls between the ages of 5 and 18 are eligible to play. The table shows the number of Little League participants since its founding. What was the average number of children who participated in each league in 1947?

LITTLE LEAGUE PARTICIPANTS

Year	Number of Leagues	Number of Participants
1939	1	45
1947	15	900
1957	4,408	299,600
1967	5,900	1,160,600
1977	6,841	1,680,400
1987	6,830	1,948,800
1997	7,427	2,591,190

CHECK WHAT YOU KNOW

Use this page to help you review and remember important skills needed for Chapter 12.

MULTIPLY BY 2-DIGIT NUMBERS (See p. H14.)

Find the product.

1. $\begin{array}{r} 14 \\ \times 10 \\ \hline \end{array}$ **2.** $\begin{array}{r} 93 \\ \times 47 \\ \hline \end{array}$ **3.** $\begin{array}{r} 24 \\ \times 16 \\ \hline \end{array}$ **4.** $\begin{array}{r} 79 \\ \times 45 \\ \hline \end{array}$

5. $\begin{array}{r} 295 \\ \times\ 24 \\ \hline \end{array}$ **6.** $\begin{array}{r} 150 \\ \times\ 74 \\ \hline \end{array}$ **7.** $\begin{array}{r} 949 \\ \times\ 16 \\ \hline \end{array}$ **8.** $\begin{array}{r} 845 \\ \times\ 67 \\ \hline \end{array}$

9. $\begin{array}{r} 5{,}200 \\ \times\ 45 \\ \hline \end{array}$ **10.** $\begin{array}{r} 7{,}034 \\ \times\ 86 \\ \hline \end{array}$ **11.** $\begin{array}{r} 2{,}611 \\ \times\ 69 \\ \hline \end{array}$ **12.** $\begin{array}{r} 8{,}864 \\ \times\ 14 \\ \hline \end{array}$

13. $\begin{array}{r} 49{,}205 \\ \times\ 83 \\ \hline \end{array}$ **14.** $\begin{array}{r} 32{,}295 \\ \times\ 77 \\ \hline \end{array}$ **15.** $\begin{array}{r} 21{,}093 \\ \times\ 25 \\ \hline \end{array}$ **16.** $\begin{array}{r} 48{,}003 \\ \times\ 39 \\ \hline \end{array}$

DIVIDE BY 10 (See p. H14.)

Divide.

17. $40 \div 10$ **18.** $400 \div 10$ **19.** $4{,}000 \div 10$ **20.** $330 \div 10$

21. $670 \div 10$ **22.** $6{,}700 \div 10$ **23.** $7{,}000 \div 10$ **24.** $700 \div 10$

25. $10\overline{)60}$ **26.** $10\overline{)600}$ **27.** $10\overline{)850}$ **28.** $10\overline{)8{,}550}$

DIVIDE BY 1-DIGIT NUMBERS (See p. H15.)

Divide.

29. $24 \div 6$ **30.** $45 \div 5$ **31.** $72 \div 9$ **32.** $80 \div 8$

33. $312 \div 2$ **34.** $450 \div 5$ **35.** $264 \div 8$ **36.** $896 \div 7$

37. $3\overline{)1{,}017}$ **38.** $9\overline{)4{,}104}$ **39.** $6\overline{)8{,}028}$ **40.** $4\overline{)9{,}684}$

LESSON 1

Algebra: Patterns in Division

Quick Review

1. (2 × 4) × 10
2. (3 × 6) × 20
3. (8 × 5) × 100
4. (6 × 7) × 100
5. (8 × 8) × 1,000

Learn

ON THE ROAD AGAIN Last summer, Emilie's family drove 6,000 miles round trip from Orlando to San Diego. On the trip, the car averaged 30 miles per gallon (mpg). How many gallons of gasoline did the car use?

Divide. $6{,}000 \div 30$ $\quad 30\overline{)6{,}000}$

Look for a pattern.

$6 \div 3 = 2$ ← **Think:** Use the basic fact.
$60 \div 30 = 2$
$600 \div 30 = 20$
$6{,}000 \div 30 = 200$

So, the car used 200 gallons of gasoline.

- What pattern do you see in $60 \div 30 = 2$, $600 \div 30 = 20$, and $6{,}000 \div 30 = 200$?

▲ The average American car got about 15 mpg in 1980 and about 23 mpg in 1996.

Examples

A

$8 \div 2 = 4$ ← basic fact
$80 \div 20 = 4$
$800 \div 20 = 40$
$8{,}000 \div 20 = 400$
$80{,}000 \div 20 = 4{,}000$

B

$12 \div 6 = 2$ ← basic fact
$120 \div 60 = 2$
$1{,}200 \div 60 = 20$
$12{,}000 \div 60 = 200$
$120{,}000 \div 60 = 2{,}000$

C

$30 \div 6 = 5$ ← basic fact
$300 \div 60 = 5$
$3{,}000 \div 60 = 50$
$30{,}000 \div 60 = 500$
$300{,}000 \div 60 = 5{,}000$
$3{,}000{,}000 \div 60 = 50{,}000$

MATH IDEA Use basic facts and patterns to divide by multiples of 10.

Check

1. **REASONING** **Compare** the basic facts and patterns in Examples B and C. How are they different?

Use mental math to complete. Write the basic fact you use.

2. $320 \div 40 = 8$
   $3{,}200 \div 40 = 80$
   $32{,}000 \div 40 = ■$

3. $420 \div 60 = 7$
   $4{,}200 \div 60 = 70$
   $42{,}000 \div 60 = ■$

4. $400 \div 50 = ■$
   $4{,}000 \div 50 = 80$
   $40{,}000 \div 50 = 800$

CALIFORNIA STANDARDS **MR 1.1** Analyze problems by identifying relationships, distinguishing relevant from irrelevant information, sequencing and prioritizing information, and observing patterns. **NS 2.2** Demonstrate proficiency with division, including division with positive decimals and long division with multidigit divisors. *also* **NS 1.0, MR 2.0, MR 2.3, MR 2.4**

Practice and Problem Solving

Use mental math to complete. Write the basic fact you use.

5. $90 \div 30 = ■$
$900 \div 30 = 30$
$9{,}000 \div 30 = 300$

6. $100 \div 50 = 2$
$1{,}000 \div 50 = 20$
$10{,}000 \div 50 = ■$

7. $450 \div 90 = 5$
$4{,}500 \div 90 = ■$
$45{,}000 \div 90 = ■$

Algebra **Use basic facts and patterns to solve for *n*.**

8. $90 \div 3 = n$ **9.** $160 \div 40 = n$ **10.** $280 \div 40 = n$ **11.** $360 \div 40 = n$

12. $540 \div 90 = n$ **13.** $3{,}000 \div 60 = n$ **14.** $7{,}200 \div 90 = n$ **15.** $36{,}000 \div 90 = n$

16. $n \div 80 = 30$ **17.** $1{,}000 \div n = 50$ **18.** $3{,}500 \div n = 70$ **19.** $18{,}000 \div n = 600$

20. $330{,}000 \div 10 = n$ **21.** $4{,}800{,}000 \div 80 = n$ **22.** $4{,}000{,}000 \div n = 50$

Compare. Use <, >, or = for the ●.

23. $400 \div 20$ ● $400 \div 200$ **24.** $5{,}600 \div 70$ ● $560 \div 7$ **25.** $80{,}000 \div 40$ ● $800 \div 4$

USE DATA **For 26–27, use the map.**

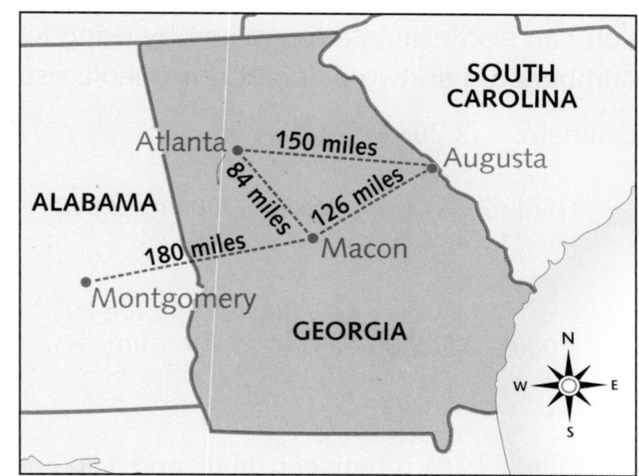

26. Mr. Fiori's truck route is round trip from Macon to Montgomery, 5 days a week. On his route, he drives an average of 60 miles per hour. How many hours does he drive each week on his route?

27. Jo Jo's Delivery Service's van averages 30 miles per gallon. How many gallons of gas does the van use on the delivery route from Macon to Atlanta to Augusta and back to Macon?

28. You will walk about 70,000 miles between childhood and old age. If you live a total of 70 years during this time span, how many miles do you walk, on average, a year?

29. **What's the Error?** Tina says the quotient of $200{,}000 \div 40$ has four zeros. Describe Tina's error and find the correct quotient.

Mixed Review and Test Prep

30. $596 \div 4$ (p. 184) **31.** $434 \div 7$ (p. 184)

32. Find the value of *n*: $12n = 72$. (p. 88)

33. Write sixty million, five hundred two thousand, four in standard form. (p. 4)

34. **TEST PREP** Kiyo bought 3 six-packs of canned juice and gave some of the cans to Sara. Which expression models this situation? (p. 86)

A $(3 \times 6) + n$
B $(3 \times 6) - n$
C $(n \times 3) + 6$
D $(n - 3) \times 6$

Extra Practice page H43, Set A

LESSON 2

Estimate Quotients

Quick Review

1. 36 ÷ 4
2. 81 ÷ 9
3. 56 ÷ 8
4. 48 ÷ 12
5. 30 ÷ 5

Learn

WILD BLUE YONDER In 1927 Charles Lindbergh was the first person to fly solo across the Atlantic Ocean. He flew the 3,610 miles from New York to Paris in about 34 hours. About how many miles did Lindbergh fly per hour?

MATH IDEA You can use compatible numbers and multiples of 10 to estimate quotients.

Estimate. 3,610 ÷ 34
↓ ↓
3,600 ÷ 30 = 120 ← 3,600 and 30 are compatible numbers, since 36 ÷ 3 = 12.

So, Lindbergh flew about 120 miles per hour.

You can also estimate a quotient by using two sets of compatible numbers to find two different reasonable estimates.

Estimate. 3,281 ÷ 52

Think: 3,281 is between 3,000 and 3,500. Use 3,000.

3,281 ÷ 52
↓ ↓
3,000 ÷ 50 = 60 ← 3,000 and 50 are compatible numbers, since 30 ÷ 5 = 6.

Think: 3,281 is between 3,000 and 3,500. Use 3,500.

3,281 ÷ 52
↓ ↓
3,500 ÷ 50 = 70 ← 3,500 and 50 are compatible numbers, since 35 ÷ 5 = 7.

So, two reasonable estimates for 3,281 ÷ 52 are 60 and 70.

- Would 65 be a good estimate for 3,281 ÷ 52? Explain.

Lindbergh's flight helped make air travel popular. Between 1926 and 1930, the number of people traveling by plane grew from about 6,000 to about 400,000. ▼

Check

1. **Show** two reasonable estimates for 2,652 ÷ 43.

CALIFORNIA STANDARDS NS 1.1 Estimate, round, and manipulate very large and very small numbers. *also* NS 1.0, 🔑NS 2.2, MR 2.0, MR 2.3, MR 2.5, MR 3.0

Write two pairs of compatible numbers for each. Give two possible estimates.

2. $524 \div 68$ **3.** $329 \div 26$ **4.** $171 \div 34$ **5.** $2{,}548 \div 65$

Practice and Problem Solving

Write two pairs of compatible numbers for each. Give two possible estimates.

6. $186 \div 62$ **7.** $1{,}275 \div 47$ **8.** $20{,}725 \div 49$

Estimate the quotient.

9. $42\overline{)508}$ **10.** $27\overline{)849}$ **11.** $64\overline{)532}$ **12.** $73\overline{)620}$ **13.** $86\overline{)743}$

14. $23\overline{)1{,}260}$ **15.** $47\overline{)3{,}524}$ **16.** $59\overline{)4{,}636}$ **17.** $77\overline{)8{,}199}$ **18.** $31\overline{)6{,}468}$

19. $81\overline{)2{,}417}$ **20.** $92\overline{)5{,}583}$ **21.** $34\overline{)27{,}925}$ **22.** $36\overline{)33{,}842}$ **23.** $53\overline{)48{,}574}$

Name the compatible numbers used to find the estimate.

24. $652 \div 18$ estimate: 30 **25.** $423 \div 21$ estimate: 20 **26.** $2{,}993 \div 75$ estimate: 40 **27.** $37{,}642 \div 38$ estimate: 900

USE DATA For 28–30, copy and complete the table. Estimate each flight speed.

HISTORIC FLIGHTS

	Year	Flight	Miles	Time (hours)	Speed (mph)
28.	1931	First Nonstop Transpacific Flight	4,458	42	■
29.	1932	First Woman's Transatlantic Solo	2,026	15	■
30.	1949	First Nonstop Round-the-World Flight	23,452	94	■

31. REASONING A single-engine airplane can safely carry 5 people weighing 857 pounds or less. In this situation, would it be better to estimate or to find an exact answer for the amount each of 5 people weighs? Explain.

32. **Write About It** How does using compatible numbers and basic facts help you estimate? Give an example.

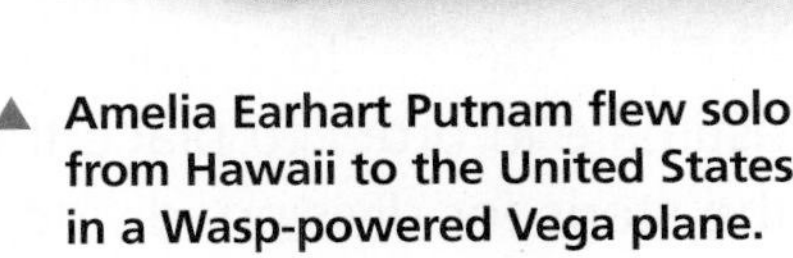

▲ Amelia Earhart Putnam flew solo from Hawaii to the United States in a Wasp-powered Vega plane.

Mixed Review and Test Prep

33. Order 36.9581, 36.5981, and 36.9851 from least to greatest. (p. 24)

34. Round 37,968,534 to the nearest hundred thousand. (p. 32)

35. $360 \div 90$ (p. 200) **36.** $720 \div 80$ (p. 200)

37. **TEST PREP** Which is NOT true about the data set 78, 94, 63, 94, 75, 66, 90? (p. 104)

A mean: 80
B median: 78
C mode: 94
D range: 30

Extra Practice page H43, Set B

LESSON 3

Divide by 2-Digit Divisors

Quick Review

1. 34 × 10
2. 25 × 9
3. 20 × 30
4. 50 × 12
5. 50 × 80

Learn

TAKE ME OUT TO THE BALL GAME Henry Louis "Hank" Aaron is American baseball's all-time champion home-run hitter. In his 23-year major-league career, Aaron hit 755 home runs. How many home runs did he average per year?

Divide. 755 ÷ 23 23)755

STEP 1

Estimate to place the first digit in the quotient.

Think:

$25\overline{)750}$ = 30 or $20\overline{)800}$ = 40

$23\overline{)755}$ So, place the first digit in the tens place.

STEP 2

Divide the 75 tens.

```
    3
23)755
  -69
    6
```

Divide. 23)75
Multiply. 23 × 3
Subtract. 75 − 69
Compare. 6 < 23

STEP 3

Bring down the 5 ones. Divide the 65 ones.

```
   32 r19
23)755
  -69↓
    65
   -46
    19
```

Divide. 23)65
Multiply. 23 × 2
Subtract. 65 − 46
Compare. 19 < 23

So, Hank Aaron averaged between 32 and 33 home runs per year.

- Explain how to use multiplication to check 755 ÷ 23 = 32 r19.

MATH IDEA To divide by a 2-digit divisor, first estimate in order to place the first digit. Then follow the steps for division: divide, multiply, subtract, and compare. Repeat the steps until the remainder is zero or less than the divisor.

Hank Aaron entered the record books on April 8, 1974, by breaking Babe Ruth's record of 714 home runs. Aaron's lifetime batting average was .305. ▶

CALIFORNIA STANDARDS O→ **NS 2.2** Demonstrate proficiency with division, including division with positive decimals and long division with multidigit divisors. **NS 1.1** Estimate, round, and manipulate very large and very small numbers. *also* **NS 1.0, MR 1.1, MR 2.1, MR 2.3**

Check by Multiplying

Examples

A

Divide.

```
    95 r15
16)1,535
  -1 44
     95
    -80
     15
```

Check ✓

```
   95
  ×16
  570
 +950
1,520
+  15
1,535✓
```

B

Divide.

```
  1,303 r22
24)31,294
  -24
    72
   -72
    09
   - 0
    94
   -72
    22
```

Check ✓

```
  1,303
×    24
  5 212
+26 060
 31,272
+    22
 31,294✓
```

- In Example B, what does the zero in the quotient indicate?

TECHNOLOGY LINK

More Practice: Use E-Lab, *Modeling Division with 2-Digit Divisors.*

www.harcourtschool.com/elab2002

Check

1. **Describe** how you can use an estimate to help you divide.

Name the position of the first digit of the quotient.

2. 36)209	**3.** 14)1,624	**4.** 53)2,369	**5.** 47)7,395
6. 89)8,969	**7.** 43)205	**8.** 38)7,268	**9.** 13)24,613

Divide. Check by multiplying.

10. 831 ÷ 38	**11.** 2,816 ÷ 56	**12.** 3,974 ÷ 23	**13.** 694 ÷ 81
14. 42)365	**15.** 38)4,031	**16.** 74)3,568	**17.** 54)34,078

Practice and Problem Solving

Name the position of the first digit of the quotient.

18. 47)310	**19.** 25)3,205	**20.** 63)1,824	**21.** 71)9,546
22. 42)42,785	**23.** 68)54,330	**24.** 21)458	**25.** 34)3,894

Divide. Check by multiplying.

26. 942 ÷ 49	**27.** 5,690 ÷ 78	**28.** 458 ÷ 45	**29.** 10,675 ÷ 11
30. 4,369 ÷ 74	**31.** 12,362 ÷ 35	**32.** 8,149 ÷ 41	**33.** 30,362 ÷ 62

LESSON CONTINUES

Divide.

34. $24\overline{)824}$ **35.** $50\overline{)545}$ **36.** $73\overline{)7{,}809}$ **37.** $29\overline{)6{,}932}$

38. $33\overline{)669}$ **39.** $43\overline{)759}$ **40.** $37\overline{)4{,}801}$ **41.** $58\overline{)9{,}286}$

42. $73\overline{)5{,}841}$ **43.** $29\overline{)31{,}803}$ **44.** $26\overline{)5{,}330}$ **45.** $37\overline{)3{,}363}$

46. $399 \div 26$ **47.** $2{,}904 \div 45$ **48.** $4{,}598 \div 61$ **49.** $35{,}718 \div 13$

Match each check with a division problem.

50. Check: $(66 \times 14) + 60 = 984$ **a.** $7{,}982 \div 69 = 115$ r47

51. Check: $(82 \times 108) + 71 = 8{,}927$ **b.** $984 \div 66 = 14$ r60

52. Check: $(98 \times 1{,}010) + 9 = 98{,}989$ **c.** $8{,}927 \div 82 = 108$ r71

53. Check: $(69 \times 115) + 47 = 7{,}982$ **d.** $98{,}989 \div 98 = 1{,}010$ r9

Algebra **For 54–57, find the missing digit that makes each equation true.**

54. $1{,}288 \div 23 = 5\blacksquare$ **55.** $3{,}580 \div (\blacksquare \times 10) = 89$ r20

56. $(456 - 18) \div 42 = 1\blacksquare$ r18 **57.** $452 \div 22 = 20$ r$\blacksquare$

58. **What's the Error?** In the student paper on the right, which is incorrect, the problem or the check? Describe and correct the error.

```
      15 r12      Check ✓
18 ) 282              15
     18             x 18
     102             120
   − 90             +150
      12             270
```

59. **REASONING** What is the greatest remainder you can have if you divide by 48? Explain.

60. **NUMBER SENSE** Two numbers have a sum of 64. Their quotient is 7. What are the two numbers?

61. An adult ticket to the baseball game costs \$9. A student ticket is \$6. Anna paid a total of \$153 for tickets for a company outing. She bought 11 adult tickets. How many student tickets did she buy?

USE DATA **For 62–64, use the table.**

62. Estimate how many games Cy Young won in an average season.

63. **What's the Question?** The answer is Warren Spahn won 10 fewer games.

64. **REASONING** Who averaged more wins per season, Alexander or Spahn?

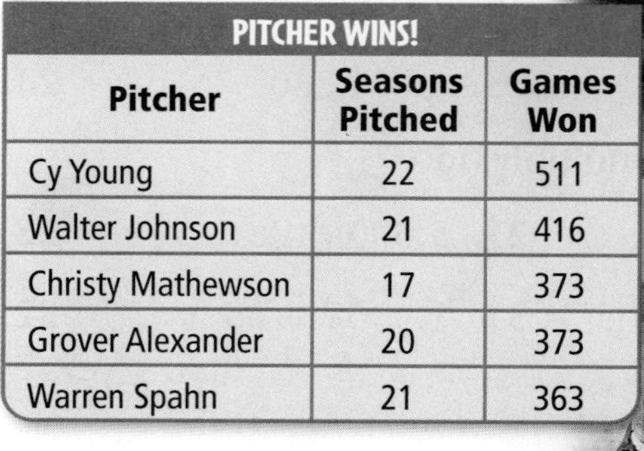

PITCHER WINS!

Pitcher	Seasons Pitched	Games Won
Cy Young	22	511
Walter Johnson	21	416
Christy Mathewson	17	373
Grover Alexander	20	373
Warren Spahn	21	363

Walter Johnson ▶

65. A professional baseball team has an average of 10 pitchers and plays an average of 160 games a year. If all the pitchers pitch an equal number of games, how many games does each pitcher pitch?

66. Suppose a pitcher is able to pitch for 20 seasons. What is the average number of games the pitcher would need to win each season to win a total of 300 games?

Mixed Review and Test Prep

For 67–70, identify the multiplication property shown. (p. 90)

67. $2 \times (3 \times 5) = (2 \times 3) \times 5$

68. $2 \times 7 = 7 \times 2$

69. $8 \times 1 = 8$

70. $2 \times 0 = 0$

71. $\begin{array}{r} 6.829 \\ +8.551 \\ \hline \end{array}$ (p. 50)

72. $\begin{array}{r} 19.05 \\ -\ 5.637 \\ \hline \end{array}$ (p. 54)

73. $265 \div 5$ (p. 184)

74. $299 \div 9$ (p. 184)

75. **TEST PREP** How many zeros will be in the quotient of $72{,}000 \div 90$? (p. 200)

A 1 zero
B 2 zeros
C 3 zeros
D 4 zeros

76. **TEST PREP** Pam had 7 bracelets and got 4 more for her birthday. Which expression models this situation? (p. 68)

F $7 - 4$
G $7 \div 4$
H $7 + 4$
J 7×4

Thinker's Corner

REMAINDER SCOREBOARD

Materials: paper, pencil, 5 blank number cubes

Getting Ready: Label 2 number cubes 0–5. Label 3 number cubes 4–9.

Let's Play!

- One player tosses all the cubes, using the results as digits to write a 3-digit by 2-digit division problem that has a high remainder. Then the player divides.
- The other players check the quotient and remainder. If they are correct, the player records the remainder as the score for that round. If they are incorrect, the remainder is not recorded.
- Players continue alternating turns, writing division problems, and adding the remainders (if correct) to their scores. The player with the highest score after ten rounds wins.

$90\overline{)346}$ = 3 r76

$(90 \times 3) + 76 = 346$

LESSON

Correcting Quotients

Quick Review

Compare.
Use <, >, or = for the ●.

1. 45 ● 49
2. 23 ● 21
3. 55 ● 56
4. 87 ● 78
5. 21 ● 19

Learn

DRAMATIC DIVISION The Civic Theater plans to have 238 seats in all on the main floor and in the balcony. Each row on the main floor will have 32 seats. The rest of the seats will be in the balcony. What is the maximum number of rows the theater can have on the main floor? How many balcony seats will the theater have?

Divide. $238 \div 32$ $\quad 32\overline{)238}$

Estimate. $240 \div 30 = 8$ or $210 \div 30 = 7$

STEP 1

Divide, using your first estimate.

$$\begin{array}{r} 8 \\ 32\overline{)238} \\ -256 \end{array}$$

Divide. $238 \div 32$
Multiply. 32×8

Think: Since $256 > 238$, this estimate is too high.

STEP 2

Adjust. Divide, using your second estimate.

$$\begin{array}{r} 7 \text{ r14} \\ 32\overline{)238} \\ -224 \\ \hline 14 \end{array}$$

Divide. $238 \div 32$
Multiply. 32×7
Subtract. $238 - 224$
Compare. $14 < 32$

So, the main floor can have 7 rows, and there would be 14 seats in the balcony.

Example **Divide.** $5{,}867 \div 65$ $\quad 65\overline{)5{,}867}$

Estimate. Use compatible numbers.

$5{,}867 \div 65$

$5{,}600 \div 70 = 80$
or
$6{,}300 \div 70 = 90$

Divide.

$$\begin{array}{r} 8 \\ 65\overline{)5{,}867} \\ -5\,20 \\ \hline 66 \end{array}$$

Think: Since $66 > 65$, the estimate is too low.

Adjust.

$$\begin{array}{r} 90 \text{ r17} \\ 65\overline{)5{,}867} \\ -5\,85 \\ \hline 17 \\ -\ 0 \\ \hline 17 \end{array}$$

- How do you adjust a quotient that is too high? too low?

MATH IDEA Sometimes when you divide, your estimate is too high or too low. Then you need to adjust the first digit of your quotient to complete the division.

Shakespeare's plays were first performed in London, England's Globe Theatre in 1599. It seated about 1,500 people, with a ground area and three tiers of seats. ▼

CALIFORNIA STANDARDS O—n **NS 2.2** Demonstrate proficiency with division, including division with positive decimals and long division with multidigit divisors. *also* **NS 1.0, NS 1.1, MR 2.0, MR 2.3, MR 3.0, MR 3.2, MR 3.3**

Check

1. **Explain** how you can tell if each digit in the quotient is large enough.

Write *too high, too low,* or *just right* for each estimate.

2. $25\overline{)83}$ estimate 4
3. $32\overline{)61}$ estimate 1
4. $78\overline{)6,778}$ estimate 90
5. $57\overline{)239}$ estimate 3
6. $95\overline{)54,362}$ estimate 500

Practice and Problem Solving

Write *too high, too low,* or *just right* for each estimate.

7. $42\overline{)321}$ estimate 7
8. $64\overline{)519}$ estimate 7
9. $88\overline{)3,265}$ estimate 20
10. $39\overline{)34,527}$ estimate 800
11. $54\overline{)34,563}$ estimate 700

Choose the better estimate to use for the quotient. Write *a* or *b*.

12. $19\overline{)84}$ **a.** 3 **b.** 4
13. $34\overline{)276}$ **a.** 8 **b.** 9
14. $24\overline{)158}$ **a.** 6 **b.** 7
15. $46\overline{)463}$ **a.** 9 **b.** 10
16. $59\overline{)283}$ **a.** 3 **b.** 4
17. $62\overline{)506}$ **a.** 8 **b.** 9
18. $78\overline{)314,674}$ **a.** 3,000 **b.** 4,000
19. $38\overline{)2,736,214}$ **a.** 60,000 **b.** 70,000

Divide.

20. $32\overline{)154}$
21. $25\overline{)278}$
22. $29\overline{)3,275}$
23. $28\overline{)3,582}$
24. $53\overline{)45,320}$
25. $47\overline{)842}$
26. $15\overline{)485}$
27. $26\overline{)5,206}$
28. $31,827 \div 56$
29. $89,345 \div 89$
30. $38,568 \div 47$
31. $749 \div 62$

32. A total of 131 students are planning to go to Asolo Theatre in Florida to see *The Rivals.* For every 14 students, one adult chaperone is required. How many chaperones are required?

33. **NUMBER SENSE** If the dividend stays the same, what happens to the quotient if you increase the divisor?

34. **Write About It** Explain how finding two estimates for a division problem helps you find a quotient.

Mixed Review and Test Prep

35. Shante earned test scores of 84, 76, 92, and 88. What is the mean of her test scores? (p. 102)

36. Round 216,025 to the nearest thousand. (p. 32)

37. $24,435 \times 34$ (p. 148)

38. $1,608 \div 8$ (p. 186)

39. **TEST PREP** Barnie has \$25. He buys a shirt for \$12.75 and a hat for \$8.39. How much change should he get back? (p. 50)

A \$3.86
B \$4.14
C \$12.25
D \$16.61

Extra Practice page H43, Set D

LESSON

5 Practice Division

Quick Review

1. 345 + 4
2. 781 + 11
3. 628 + 9
4. 6,760 + 38
5. 1,232 + 12

Learn

ALL YOU CAN EAT An adult giant panda in the wild eats about 10,220 pounds of bamboo shoots and leaves a year. On average, how many pounds of bamboo does a giant panda eat each month?

Divide.

10,220 ÷ 12   12)10,220

◀ **Bamboo makes up 99 percent of a giant panda's diet. A panda sometimes consumes as much as 80 pounds of fresh bamboo shoots per day.**

STEP 1

Estimate.

Think:

```
   800           900
12)9,600 or 12)10,800

   ■
12)10,220
```

So, place the first digit in the hundreds place.

STEP 2

Divide the 102 hundreds. Use your first estimate.

```
     8
12)10,220
  - 9 6
      6
```

STEP 3

Bring down the 2 tens. Divide the 62 tens.

```
     85
12)10,220
  - 9 6↓
      62
     -60
       2
```

STEP 4

Bring down the 0 ones. Divide the 20 ones.

```
     851 r8
12)10,220
  - 9 6
      62
     -60↓
       20
      -12
        8
```

Express the remainder as a fraction. So, a giant panda eats an average of $851\frac{8}{12}$, or $851\frac{2}{3}$, pounds of bamboo each month.

Examples

A

```
    24 r3 or 24 3/35
35)843
  -70
   143
  -140
     3
```

Check ✓

```
   24
  ×35
  120
 +720
  840
 +  3
  843✓
```

B

```
    61 r17 or 61 17/42
42)2,579
  -2 52
     59
    -42
     17
```

Check ✓

```
     61
    ×42
    122
 +2 440
  2,562
 +   17
  2,579✓
```

- How can you tell if each digit in the quotient is large enough?

CALIFORNIA STANDARDS **NS 2.2** Demonstrate proficiency with division, including division with positive decimals and long division with multidigit divisors. **NS 1.0** Students compute with very large and very small numbers, positive integers, decimals, and fractions and understand the relationship between decimals, fractions, and percents. They understand the relative magnitudes of numbers. *also* **NS 1.1, MR 1.1, MR 3.1, MR 3.2, MR 3.3**

Check

1. **REASONING** **Describe** how you would divide a 4-digit number by a 3-digit number.

TECHNOLOGY LINK
To learn more about division, watch the **Harcourt Math Newsroom Video** *Giant Pandas*.

Divide. Check by multiplying.

2. $851 \div 15$
3. $93 \div 14$
4. $1{,}242 \div 48$
5. $42\overline{)23{,}898}$
6. $61\overline{)36{,}182}$
7. $29\overline{)1{,}073}$

Practice and Problem Solving

Divide.

8. $402 \div 65$
9. $4{,}161 \div 19$
10. $4{,}591 \div 37$
11. $89 \div 23$
12. $71\overline{)7{,}171}$
13. $33\overline{)1{,}877}$
14. $51\overline{)4{,}095}$
15. $92\overline{)46{,}235}$
16. $91\overline{)5{,}396}$
17. $69\overline{)2{,}109}$
18. $48\overline{)24{,}916}$
19. $42\overline{)23{,}911}$
20. $25\overline{)424{,}867}$
21. $32\overline{)306{,}452}$
22. $41\overline{)8{,}269{,}079}$
23. $37\overline{)6{,}313{,}642}$

Algebra **For 24–29, find the missing digit that makes each equation true.**

24. $4■4 \div 23 = 18$
25. $1{,}065 \div ■1 = 15$
26. $7{,}■22 \div 99 = 78$
27. $12{,}084 \div 76 = 15■$
28. $(1{,}188 - 18) \div 90 = 1■$
29. $1{,}860 \div (■ \times 15) = 62$

30. About 31,000 people visited the San Diego Zoo's website the day the live "Panda Cam" was launched. On average, how many people visited the website each hour?

31. **NUMBER SENSE** What is the least 4-digit number that can be divided by 20 and have a remainder of 5?

32. Giant pandas are active about 5,110 hours a year. They generally rest 2–4 hours at a time. How many hours is the giant panda active each day?

Mixed Review and Test Prep

For 33–34, identify the addition property shown. (p. 78)

33. $10 + 7 = 7 + 10$
34. $(4 + 2) + 3 = 4 + (2 + 3)$
35. $24.18 - 16.09$ (p. 50)
36. $16.82 + 7.48$ (p. 50)

37. **TEST PREP** A magazine sells for \$2.75. If a store sells 62 copies, what is the total amount of the sales? (p. 158)

A \$17.05 **C** \$170.50
B \$71.50 **D** \$1,705.00

Extra Practice page H43, Set E

LESSON 6

Problem Solving Strategy
Predict and Test

Quick Review

1. 144 ÷ 12
2. 88 ÷ 44
3. 132 ÷ 11
4. 65 ÷ 2
5. 123 ÷ 11

PROBLEM A collector has 158 world's fair souvenirs. He wants to sell them in equal groups with no more than 25 in each group. After he divides them into equal groups, there are 20 souvenirs left over that he will sell individually. How many groups was he able to make? How many souvenirs are in each group?

- What are you asked to find?
- What information will you use?

- What strategy can you use to solve the problem?

Since you know the dividend is 158, you can use *predict and test* to find equal groups with no more than 25 souvenirs in each.

- How can you use the strategy to solve the problem?

Since there are 20 souvenirs left over, there must be more than 20 souvenirs in each group. So, start by predicting different divisors greater than 20. Then test each prediction to see if the remainder is 20.

Predict the number of souvenirs in each group.	Divide to find the number of groups.	Compare the remainder to 20.	Does the remainder equal 20?
21	158 ÷ 21 = 7 r11	11 < 20	No
22	158 ÷ 22 = 7 r4	4 < 20	No
23	158 ÷ 23 = 6 r20	20 = 20 √	Yes

So, the collector made 6 groups, each with 23 souvenirs.

- How can you decide if your answer is correct?

CALIFORNIA STANDARDS MR 2.6 Make precise calculations and check the validity of the results from the context of the problem. **MR 1.0** Students make decisions about how to approach problems. **MR 2.0** Students use strategies, skills, and concepts in finding solutions. *also* **NS 1.0, O━ NS 2.2, MR 1.1, MR 2.3, MR 2.4, MR 3.0, MR 3.2**

Problem Solving Practice

Predict and test to solve.

1. **What if** there were 18 souvenirs left over? How many groups would he be able to make? How many souvenirs would be in each group?
2. Tim spent $22.50 on souvenirs at the fair. He bought 2 souvenirs. One of them cost $4.50 more than the other. How much was each souvenir?

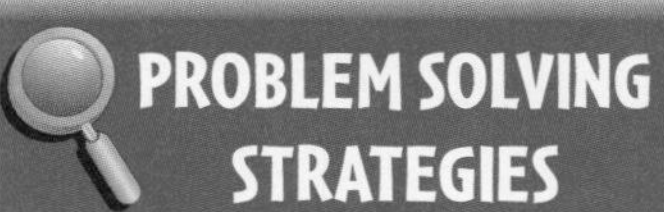

- Draw a Diagram or Picture
- Make a Model or Act It Out
- Make an Organized List
- Find a Pattern
- Make a Table or Graph
- ▶ Predict and Test
- Work Backward
- Solve a Simpler Problem
- Write an Equation
- Use Logical Reasoning

The fifth-grade students had 181 paintings to be displayed in equal groups on walls throughout the school. No more than 30 paintings would fit on each wall. After the paintings were placed, 20 were left over and were placed on the office wall.

3. How many groups were formed? How many paintings were in each group?

   A 6 groups of 30 paintings
   B 7 groups of 23 paintings
   C 8 groups of 21 paintings
   D 9 groups of 20 paintings

4. Which equation models this situation?

   F $181 \div 30 = 6$ r1
   G $181 \div 20 = 9$ r1
   H $181 \div 23 = 7$ r20
   J $181 \div 7 = 25$ r6

Mixed Strategy Practice

USE DATA **For 5–6, use the table.**

5. If a tour boat travels 19 miles per hour down the Colorado and another tour boat travels 25 miles per hour down the Rio Grande, which tour boat will travel the length of its river first?
6. Suppose there are rescue stations every 35 miles along the Yenisey. If the first station is located at the beginning of the river, how many stations are there in all?
7. Karen left her house and drove 35 miles north and then 15 miles west to the mall. When she left the mall, she drove 5 miles east and then 20 miles south to the library. How far north and west was Karen from her house?
8. **NUMBER SENSE** The sum of two numbers is 36. Their product is 320. What are the two numbers?

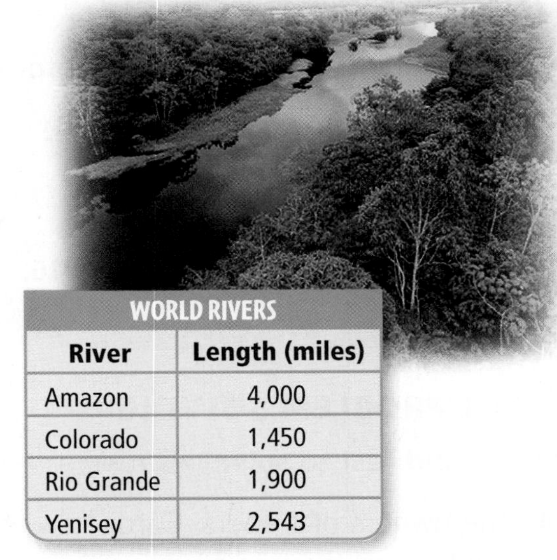

WORLD RIVERS

River	Length (miles)
Amazon	4,000
Colorado	1,450
Rio Grande	1,900
Yenisey	2,543

9. **Write a problem** using the table above. Explain the strategy you would use to solve the problem.

CHAPTER 12

Review/Test

CHECK VOCABULARY AND CONCEPTS

Choose the best term from the box.

estimate
multiples
multiplication
patterns

1. You can use basic facts and __?__ to divide by multiples of 10. (p. 200)
2. You can use basic facts and compatible numbers to __?__ quotients. (p. 202)
3. You can use __?__ to check division. (p. 204)

CHECK SKILLS

Use mental math to complete. Write the basic fact you use. (pp. 200–201)

4. $490 \div 70 = 7$
   $4{,}900 \div 70 = 70$
   $49{,}000 \div 70 = \blacksquare$
5. $640 \div 80 = 8$
   $6{,}400 \div 80 = \blacksquare$
   $64{,}000 \div 80 = 800$
6. $810 \div 90 = \blacksquare$
   $8{,}100 \div 90 = \blacksquare$
   $81{,}000 \div 90 = 900$

Write two pairs of compatible numbers for each. Give two possible estimates. (pp. 202–203)

7. $263 \div 43$
8. $1{,}456 \div 37$
9. $5{,}063 \div 81$

Write *too high, too low,* or *just right* for each estimate. (pp. 208–209)

10. $13\overline{)667}$ estimate 40
11. $26\overline{)3{,}521}$ estimate 90
12. $41\overline{)37{,}651}$ estimate 900
13. $75\overline{)49{,}036}$ estimate 700

Divide. (pp. 204–207 and pp. 210–211)

14. $17\overline{)64}$
15. $22\overline{)90}$
16. $36\overline{)184}$
17. $47\overline{)2{,}943}$
18. $34\overline{)3{,}294}$
19. $27\overline{)344}$
20. $36\overline{)645}$
21. $45\overline{)4{,}292}$
22. $23\overline{)66{,}804}$
23. $71\overline{)685{,}210}$

CHECK PROBLEM SOLVING

Predict and test to solve. (pp. 212–213)

24. The owners of Traders' Cards have 161 baseball cards. They want to sell them in equal groups, with no more than 30 cards in each group. After they divide the cards into equal groups, there are 17 cards left over that they will sell individually. How many groups are they able to make? How many cards are in each group?
25. **NUMBER SENSE** The sum of two numbers is 35 and their difference is 7. What are the two numbers?

Cumulative Review

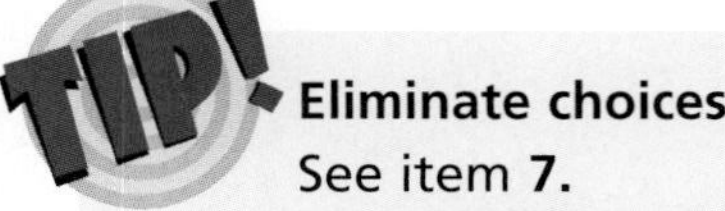

Eliminate choices.
See item **7.**

Think about what is a reasonable height for each story of a building. Estimate by finding compatible numbers for 386 ÷ 29.

Also see problem **5,** p. H64.

For 1–10, choose the best answer.

1. Sara has 312 stamps. She can display 48 stamps on each album page. How many stamps will be left over if she fills 6 pages?

A 12 **C** 24
B 18 **D** 36

2. The gas tank in Scott's car holds 24 gallons. He traveled 1,800 miles and used 3 tanks of gas. How many miles per gallon did his car average?

F between 23 and 24
G exactly 24
H between 24 and 25
J exactly 25

3. How is forty-two million, six hundred five thousand, one hundred twenty-six written in standard form?

A 420,605,026 **C** 42,605,126
B 42,650,126 **D** 42,605,120.6

4. Steve earns $18 each time he mows the lawn. If he earned $72 last month, how many times did he mow the lawn?

F 4 **H** 54
G 6 **J** 90

5. 60,000 ÷ 2

A 300,000 **C** 3,000
B 30,000 **D** Not here

6. Where should the first digit of the quotient be placed for 633 ÷ 72?

F thousands place **H** tens place
G hundreds place **J** ones place

For 7–9, use the table.

SOME UNITED STATES SKYSCRAPERS		
Building	**Number of Stories**	**Height (in feet)**
Park Row Building, NY	29	386
Empire State Building, NY	102	1,472
World Trade Center, NY	110	1,710
Metropolitan Life, NY	50	700

7. About how tall is each story in the Park Row Building?

A 13 feet **C** 23 feet
B 16 feet **D** 29 feet

8. The height of which building can be divided by the number of stories and give an answer with no remainder?

F Park Row **H** Metropolitan Life
G Empire State **J** World Trade Center

9. How long would it take an elevator moving 25 feet per second to get to the top of the Metropolitan Life Building?

A 2 seconds **C** 50 seconds
B 28 seconds **D** 200 seconds

10. Which of the following is true for the data set 12, 7, 4, 23, 18, 15, 9?

F The median is 12.
G The mode is 12.
H The range is 16.
J The mean is 15.

Divide Decimals by Whole Numbers

In some grocery stores, signs on the shelves tell the price of an item per pound or per ounce. Consumers can use this information to see which size of a product is the best buy. The table shows orange juice sold in different-size containers. Which one is the best buy? Explain.

ORANGE JUICE

Name	Cost	Container Size
Orange Delight	$4.00	64 ounces
Alba's Orange Juice	$1.00	8 ounces
Fresh Squeezed	$0.75	6 ounces
Sweet Orange	$2.50	16 ounces

Use this page to help you review and remember important skills needed for Chapter 13.

VOCABULARY

Choose the best term from the box.

1. In 396 ÷ 3 = 132, the number 3 is the __?__.
2. In 65 ÷ 5 = 13, the number 65 is the __?__.
3. 0.6 and 0.60 are examples of __?__.

equivalent decimals
divisor
dividend
quotient

DIVISION PATTERNS (See p. H15.)

Complete the pattern.

4. 350 ÷ 50 = 7
3,500 ÷ 50 = 70
35,000 ÷ 50 = *n*

5. 180 ÷ 30 = 6
1,800 ÷ 30 = *n*
18,000 ÷ 30 = 600

6. 360 ÷ 40 = *n*
3,600 ÷ 40 = 90
36,000 ÷ 40 = 900

7. 300 ÷ 60 = 5
3,000 ÷ 60 = 50
n ÷ 60 = 500

8. 320 ÷ 80 = 4
n ÷ 80 = 40
32,000 ÷ 80 = 400

9. 400 ÷ 50 = 8
4,000 ÷ *n* = 80
40,000 ÷ 50 = 800

Find the quotient.

10. $20\overline{)80}$ **11.** $10\overline{)60}$ **12.** $60\overline{)420}$ **13.** $80\overline{)160}$

14. $30\overline{)900}$ **15.** $40\overline{)800}$ **16.** $40\overline{)2{,}400}$ **17.** $20\overline{)1{,}000}$

DIVIDE WHOLE NUMBERS (See p. H13.)

Divide.

18. $7\overline{)84}$ **19.** $4\overline{)92}$ **20.** $5\overline{)485}$ **21.** $9\overline{)207}$

22. $6\overline{)288}$ **23.** $4\overline{)824}$ **24.** $3\overline{)1{,}620}$ **25.** $8\overline{)3{,}248}$

26. 144 ÷ 16 **27.** 832 ÷ 32 **28.** 3,825 ÷ 45 **29.** 1,365 ÷ 91

EQUIVALENT DECIMALS (See p. H16.)

Write two equivalent decimals for each.

30. 0.8 **31.** 0.25 **32.** 1.3 **33.** 2.60

LESSON 1

Algebra: Patterns in Decimal Division

Quick Review

1. 540 ÷ 9
2. 480 ÷ 8
3. 240 ÷ 4
4. 5,600 ÷ 8
5. 3,000 ÷ 6

Learn

POP! POP! POP! Roderic read that the average person in the United States eats about 16 gallons of popcorn each year. Roderic bought 4 cans of popcorn for $2. What was the cost for one can?

Find 2 ÷ 4.

Look for a pattern in these quotients.

200 ÷ 4 = 50 ← Pattern in the zeros
20 ÷ 4 = 5 ← Basic fact
2 ÷ 4 = 0.5 ← Decimal quotient

So, one can of popcorn costs $0.50.

Examples

A Find 2 ÷ 5.

200 ÷ 5 = 40
20 ÷ 5 = 4
2 ÷ 5 = 0.4

B Find 1.8 ÷ 6.

180 ÷ 6 = 30
18 ÷ 6 = 3
1.8 ÷ 6 = 0.3

C Find 3.2 ÷ 4.

320 ÷ 4 = 80
32 ÷ 4 = 8
3.2 ÷ 4 = 0.8

MATH IDEA When the divisor is greater than the dividend, the quotient is a decimal less than 1.

Remember
In any whole number, a decimal point is to the right of the ones place.

50 = 50.
↑

Check

1. **Explain** what happens to the position of the decimal point in the quotients in Examples A–C.

Copy and complete each pattern.

2. 100 ÷ 5 = ■
   10 ÷ 5 = ■
   1 ÷ 5 = ■

3. 400 ÷ 8 = ■
   40 ÷ 8 = ■
   4 ÷ 8 = ■

4. 500 ÷ 2 = ■
   50 ÷ 2 = ■
   5 ÷ 2 = ■

CALIFORNIA STANDARDS NS 2.0 Students perform calculations and solve problems involving addition, subtraction, and simple multiplication and division of fractions and decimals. **NS 2.1** Add, subtract, multiply, and divide with decimals; add with negative integers; subtract positive integers from negative integers; and verify the reasonableness of the results. *also* **NS 2.2, MR 1.1, MR 2.2, MR 2.3, MR 3.2**

Practice and Problem Solving

Copy and complete each pattern.

5. $300 \div 5 = ■$
$30 \div 5 = ■$
$3 \div 5 = ■$

6. $160 \div 4 = ■$
$16 \div 4 = ■$
$1.6 \div 4 = ■$

7. $1{,}500 \div 6 = ■$
$150 \div 6 = ■$
$15 \div 6 = ■$

8. $400 \div 5 = ■$
$40 \div 5 = ■$
$4 \div 5 = ■$

9. $2{,}000 \div 8 = ■$
$200 \div 8 = ■$
$20 \div 8 = ■$

10. $1{,}000 \div 8 = ■$
$100 \div 8 = ■$
$10 \div 8 = ■$

Algebra **Copy and complete each table. Use patterns and mental math.**

11.

n	$n \div 25$
10,000	■
1,000	■
100	■
10	■

12.

n	$n \div 40$
28,000	■
■	70
■	7
28	■

13.

n	$n \div 8$
2,000	■
200	■
■	2.5
2	■

Write the check for each division problem.

14. $10 \div 2 = 5$
$1 \div 2 = 0.5$

15. $5{,}000 \div 10 = 500$
$5 \div 10 = 0.5$

16. $400 \div 5 = 80$
$4 \div 5 = 0.8$

17. **REASONING** Continue the pattern until the quotient extends to the hundredths place:

$560 \div 7 = 80$
$56 \div 7 = 8$

18. Amy bought 5 bags of peanuts for \$3.00. Did she pay more or less than \$0.50 for each bag? Explain.

19. **Write About It** Explain how a pattern beginning with $300 \div 6 = 50$ leads to a decimal quotient of 0.5.

20. A bus traveled 3,000 miles over 6 days. If it covered the same route every day, how long was the route?

21. Find values for ♦ and ▼ that make the equation true. ♦ ÷ ▼ = 0.5

Mixed Review and Test Prep

22. $152{,}073 \times 6$ (p. 146)

23. $81{,}514 \times 28$ (p. 150)

24. Evaluate $24 + (n \div 60)$ if n is 180. (p. 86)

25. Which property is illustrated in $8 \times (3 + 9) = 24 + 72$? (p. 92)

26. **TEST PREP** Which is the quotient for $1{,}400 \div 7$? (p. 188)

A 2
B 20
C 200
D 2,000

Extra Practice page H44, Set A

LESSON 2

HANDS ON

Decimal Division

Quick Review

1. $9\overline{)27}$
2. $8\overline{)72}$
3. $8\overline{)64}$
4. $6\overline{)54}$
5. $7\overline{)56}$

MATERIALS
decimal models
markers
scissors

Explore

You can make a model to find the quotient of a decimal divided by a whole number.

Find $1.5 \div 3$. $3\overline{)1.5}$

STEP 1

Show 1.5 by shading decimal models.

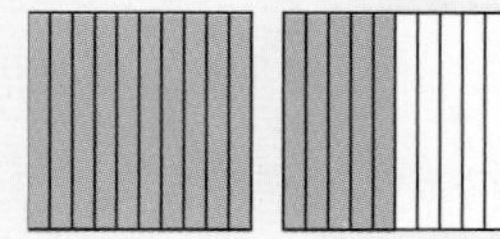

STEP 2

Cut each model to show the tenths.

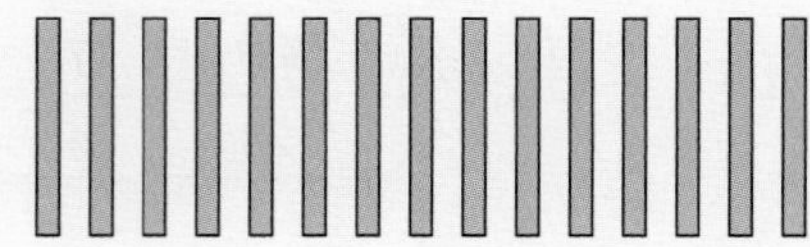

STEP 3

Divide the tenths into 3 groups of the same size. Each group has 5 tenths.

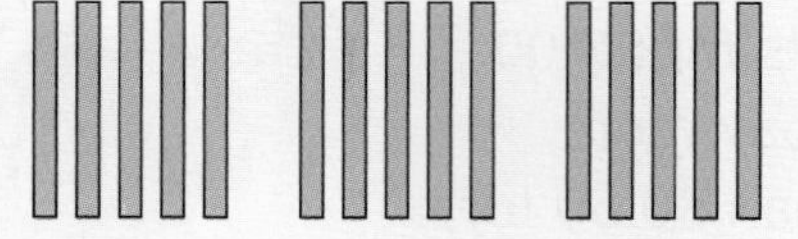

STEP 4

Record a division number sentence for your model.

$1.5 \div 3 = 0.5$

How could you show equal groups for $1.4 \div 2$?

So, $1.5 \div 3$ is 0.5.

MATH IDEA Using models can help you visualize division of decimals.

- Explain how you can use this pattern to show that your quotient is correct.

$150 \div 3 = 50$
$15 \div 3 = 5$
$1.5 \div 3 = \blacksquare$

Try It

Make a model to show the division. Record a number sentence for the model.

a. $1.4 \div 2$ **b.** $0.9 \div 3$
c. $2.5 \div 5$ **d.** $1.6 \div 4$

CALIFORNIA STANDARDS **NS 2.1** Add, subtract, multiply, and divide with decimals; add with negative integers; subtract positive integers from negative integers; and verify the reasonableness of the results. **MR 2.3** Use a variety of methods, such as words, numbers, symbols, charts, graphs, tables, diagrams, and models, to explain mathematical reasoning. *also* **NS 2.0, NS 2.2, MR 1.0, MR 1.1, MR 2.0, MR 2.4**

Connect

You can use a hundredths model to show $0.24 \div 6$.

STEP 1

Show 0.24 by shading a hundredths model.

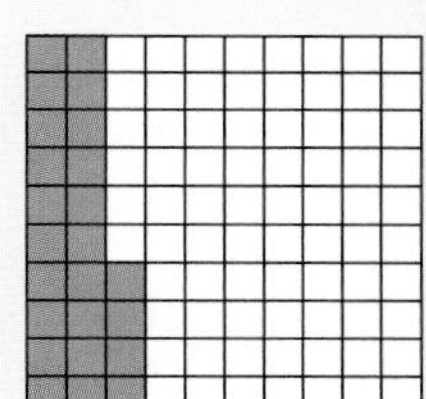

STEP 2

Cut the model to show the hundredths.

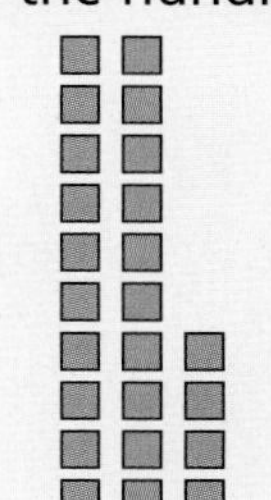

STEP 3

Divide the hundredths into 6 groups of the same size. Each group has 4 hundredths.

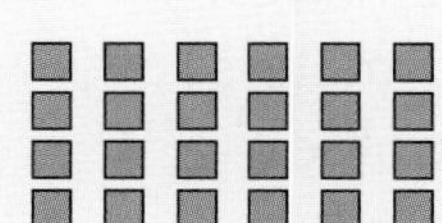

STEP 4

Record a division number sentence for your model.

$0.24 \div 6 = 0.04$

Practice

Make a model and find the quotient.

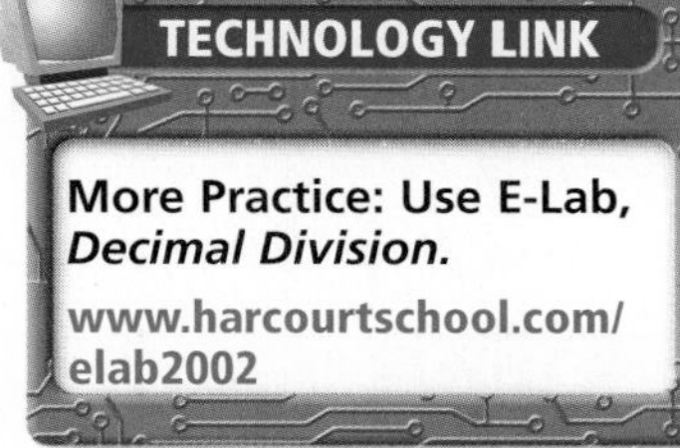

TECHNOLOGY LINK
More Practice: Use E-Lab, *Decimal Division.*
www.harcourtschool.com/elab2002

1. $0.08 \div 2$
2. $\$0.24 \div 2$
3. $0.15 \div 3$
4. $1.2 \div 4$
5. $\$0.12 \div 4$
6. $3.5 \div 5$
7. $2.4 \div 3$
8. $0.08 \div 8$
9. $\$0.39 \div 3$
10. $\$0.44 \div 2$
11. $0.18 \div 9$
12. $0.18 \div 6$

Use the model to complete the number sentence.

13. 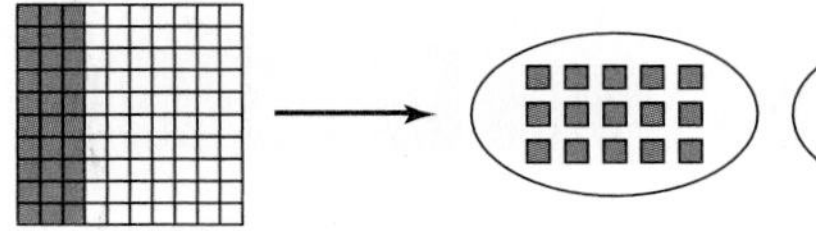

$0.30 \div 2 = n$

14. 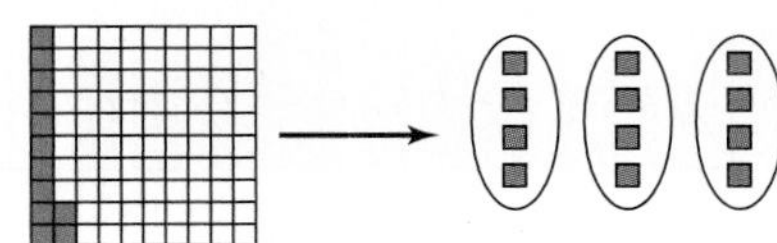

$0.12 \div 3 = n$

15. John is building a cabinet for his basketball trophies. For the trim, he bought 4 pieces of wood for a total of \$7.20. If the pieces each cost the same, how much was one piece?

16. **What's the Error?** Look at the model below for $0.16 \div 4$. Describe and correct the error.

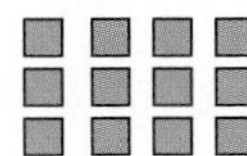

Mixed Review and Test Prep

17. $5{,}000 - 3{,}436$ (p. 36)

18. $4.846 + 1.404$ (p. 50)

19. Write four and one thousandth as a decimal. (p. 20)

20. Write an equivalent decimal for 0.90. (p. 22)

21. **TEST PREP** Solve for n. $58 = n + 39$ (p. 74)

A $n = 9$
B $n = 19$
C $n = 20$
D $n = 90$

LESSON 3

Divide Decimals by Whole Numbers

Quick Review

1. 246 ÷ 6
2. 520 ÷ 5
3. 1,260 ÷ 3
4. 564 ÷ 4
5. 4,806 ÷ 6

Learn

UP, UP AND AWAY Paulo and his friends are making 3 kites. They will share equally a narrow piece of cloth that is 4.2 meters long to make the tails for the kites. How long will each piece of cloth be?

Find $4.2 \div 3$.
Estimate. $3 \div 3 = 1$

◀ The largest kite ever flown had an area of 5,952 square feet. It was first flown in the Netherlands on August 8, 1991.

One Way You can divide by using a model.

STEP 1

Shade decimal models to show 4.2.

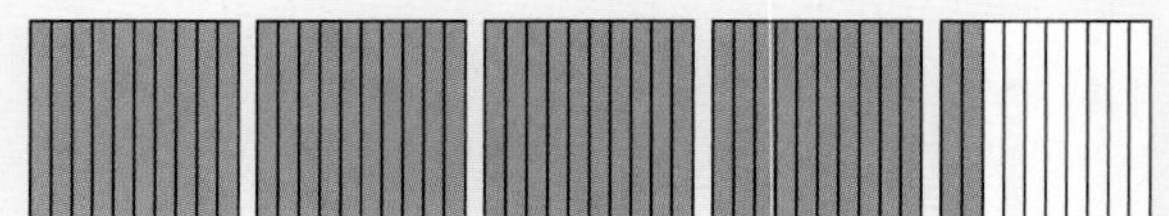

STEP 2

Divide the models into 3 groups of the same size.

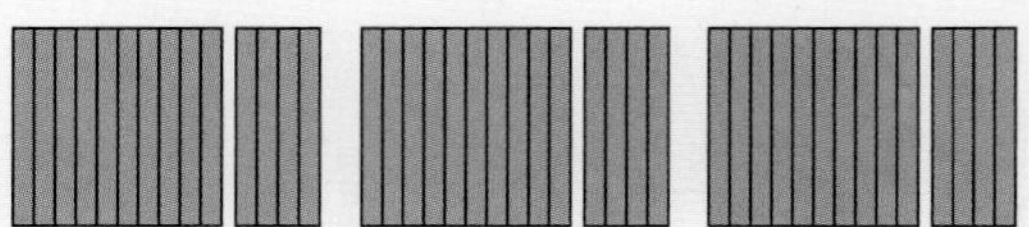

- Explain how you know how many wholes and how many tenths to put in each of the 3 equal groups.

Another Way You can divide by using paper and pencil.

STEP 1

Write the decimal point of the quotient above the decimal point of the dividend.

$$3\overline{)4.2}$$ (decimal point placed above the decimal point of 4.2)

STEP 2

Divide as you would divide whole numbers.

$$\begin{array}{r} 1.4 \\ 3\overline{)4.2} \\ -3 \\ \hline 1\,2 \\ -1\,2 \\ \hline 0 \end{array}$$

So, each piece of cloth will be 1.4 meters long. Since the answer is close to the estimate, the answer is reasonable.

CALIFORNIA STANDARDS **NS 2.1** Add, subtract, multiply, and divide with decimals; add with negative integers; subtract positive integers from negative integers; and verify the reasonableness of the results. **NS 2.2** Demonstrate proficiency with division, including division with positive decimals and long division with multidigit divisors. *also* **NS 2.0, MR 2.0, MR 2.1, MR 2.3, MR 3.0, MR 3.2**

Multiply to Check

You can multiply to check your answer.

Examples

A Find 2.49 ÷ 3.

```
  0.83
3)2.49
 -0
  24
  24
  09
  -9
   0
```

← The divisor is greater than the dividend, so place a zero in the ones place.

Check ✓

```
  0.83
×    3
  2.49
```

B Find \$18.45 ÷ 15.

```
    $1.23
15)$18.45
   -15
     34
    -30
     45
    -45
      0
```

← When you divide money, remember to include a dollar sign in the quotient.

Check ✓

```
  $1.23
×    15
    615
   123
 $18.45
```

C Find 423.5 ÷ 7.

```
  60.5
7)423.5
 -42
  03
  -0
   35
  -35
    0
```

← Since 7 > 3, place a zero in the ones place.

Check ✓

```
  60.5
×    7
 423.5
```

MATH IDEA When the dividend is a decimal and the divisor is a whole number, divide as with whole numbers. Line up the decimal point in your quotient with the decimal point in the dividend.

Check

1. **Tell** where you place the decimal point in the quotient when you divide a decimal by a whole number.

Draw a picture to show the quotient.

2. 1.26 ÷ 3

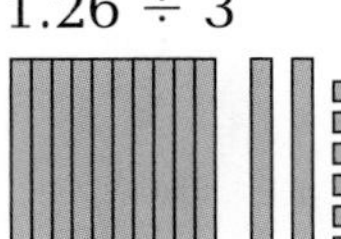

3. 2.38 ÷ 7

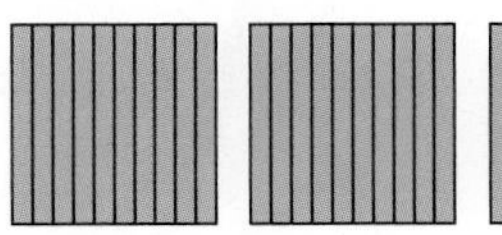

Copy the quotient and correctly place the decimal point.

4. $2\overline{)1.06}$ quotient: 053
5. $5\overline{)0.045}$ quotient: 0009
6. $7\overline{)49.0}$ quotient: 70
7. $4\overline{)32.60}$ quotient: 815

Find the quotient. Check by multiplying.

8. $6\overline{)4.8}$
9. $9\overline{)\$9.09}$
10. $4\overline{)12.04}$
11. $2\overline{)\$1.04}$
12. 35.7 ÷ 7
13. \$0.80 ÷ 5
14. 4.355 ÷ 5
15. 115.2 ÷ 24

LESSON CONTINUES

Practice and Problem Solving

Copy the quotient and correctly place the decimal point.

16. $\begin{array}{r} 09 \\ 7\overline{)6.3} \end{array}$	**17.** $\begin{array}{r} 202 \\ 4\overline{)8.08} \end{array}$	**18.** $\begin{array}{r} 075 \\ 8\overline{)6.00} \end{array}$	**19.** $\begin{array}{r} 027 \\ 15\overline{)4.05} \end{array}$
20. $\begin{array}{r} 0712 \\ 6\overline{)4.272} \end{array}$	**21.** $\begin{array}{r} 0122 \\ 3\overline{)0.366} \end{array}$	**22.** $\begin{array}{r} 341 \\ 8\overline{)27.28} \end{array}$	**23.** $\begin{array}{r} 1236 \\ 5\overline{)61.80} \end{array}$

Find the quotient. Check by multiplying.

24. $8\overline{)74.4}$	**25.** $3\overline{)27.9}$	**26.** $5\overline{)\$55.55}$	**27.** $2\overline{)19.64}$
28. $7\overline{)85.4}$	**29.** $70\overline{)854.0}$	**30.** $3\overline{)\$93.66}$	**31.** $5\overline{)6.15}$
32. $8\overline{)\$33.04}$	**33.** $4\overline{)21.56}$	**34.** $14\overline{)59.22}$	**35.** $19\overline{)\$234.84}$
36. $12\overline{)39.0}$	**37.** $10\overline{)212.5}$	**38.** $9\overline{)343.89}$	**39.** $11\overline{)570.9}$
40. $24.6 \div 6$	**41.** $\$2.46 \div 6$	**42.** $0.93 \div 3$	**43.** $8.05 \div 7$
44. $0.54 \div 6$	**45.** $5.04 \div 8$	**46.** $72.64 \div 8$	**47.** $48.6 \div 9$
48. $48.3 \div 7$	**49.** $18.5 \div 5$	**50.** $21.6 \div 4$	**51.** $\$4.24 \div 8$

USE DATA For 52–55, use the data in the list.

52. How much did Blair pay for 1 paintbrush?

53. How much did Bryan pay for 2 yards of ribbon?

54. Ian bought 6 silk flowers and 2 glue sticks. How much did he spend?

55. **Write a problem** that uses data in the list and is solved by dividing a decimal. Solve your problem.

56. Eric earned $17.25 raking leaves for 5 hours. How much did he earn in one hour?

57. **What's the Question?** The difference between the prices of two bikes is $16. The sum of their prices is $232. The answer is $124.

58. Tennis balls are 3 for $2.79 or $5.00 for a half dozen. Which is the better buy?

59. Three pounds of peanuts cost $6.99. Mark has $2.00. Can he buy 1 pound of peanuts? Explain.

Mixed Review and Test Prep

60. Divide 4,024,804 by 4. (p. 188)

61. $345 \div 2$ (p. 184)

62. $4 \times (25 + 2)$ (p. 90)

63. Write 213,675 in word form. (p. 4)

64. Compare 0.09898 and 0.9898. Use $<$. (p. 24)

65. **TEST PREP** Find *y*. $y + 42 = 80$ (p. 74)

A $y = 38$ **C** $y = 118$
B $y = 48$ **D** $y = 122$

66. Charlie recorded the daily high temperature for nine days: 79°F, 82°F, 74°F, 82°F, 78°F, 84°F, 76°F, 80°F, 82°F. What was the median temperature? (p. 104)

67. **TEST PREP** Janell bought 2 shirts for $11.98 each, a sweater for $18.98, and 3 pairs of socks for $2.05 a pair. How much did she spend in all? (p. 50)

F $49.00 **H** $61.07
G $49.09 **J** $140.05

Thinker's Corner

REASONING Mr. Renard had $120 to buy supplies for his kite-making classes. After he finished shopping, he had $7.82 left. Mr. Renard decided to record what he could remember about his purchases.

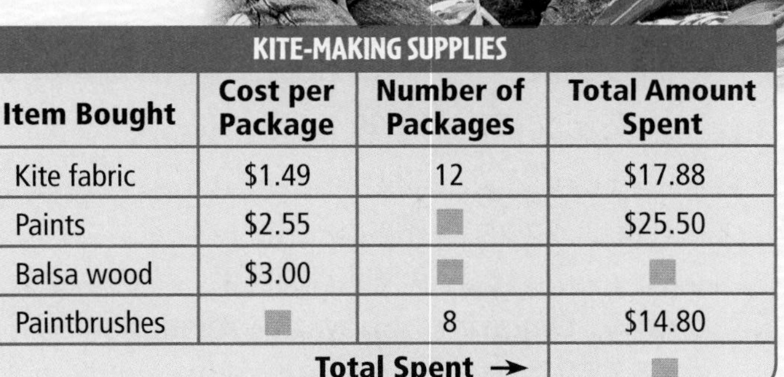

KITE-MAKING SUPPLIES

Item Bought	Cost per Package	Number of Packages	Total Amount Spent
Kite fabric	$1.49	12	$17.88
Paints	$2.55	■	$25.50
Balsa wood	$3.00	■	■
Paintbrushes	■	8	$14.80
		Total Spent →	■

For 1–2, choose the best answer. Then copy and complete the chart.

1. How can you find the total amount spent on all the supplies?

A Find $17.88 + $25.50 + $14.80. Then subtract the sum from $120.00.

B Subtract $7.82 from $120.00.

C Add $120.00 and $7.82.

D Add all the numbers in the table.

2. **Algebra** If *b* equals the cost per package for paintbrushes, which equation can you use to solve for *b*?

F $b = \$14.80 \times 8$ **H** $b \div \$14.80 = 8$
G $\$14.80 - b = 8$ **J** $\$14.80 \div 8 = b$

Extra Practice page H44, Set B

LESSON 4

Problem Solving Strategy

Compare Strategies

Quick Review

1. $(3 + 4) \times 2$
2. $10 + (25 \div 5)$
3. $(42 \div 6) + 4$
4. $84 \div (7 + 5)$
5. $4 \times (3 + 6)$

PROBLEM Erica and her father paid $7.50 for tickets to the Hot-Air Balloon Festival. An adult's ticket costs $3.00 more than a child's ticket. What was the cost of each ticket?

- What are you asked to find?
- What information will you use?

- What strategy can you use?

Often you can use more than one strategy to solve a problem. Use *work backward* or *draw a diagram.*

- How will you solve the problem?

Work Backward Show how to find the total, and use *work backward* and inverse operations to find the cost of the child's ticket.

(■	×	2)	+	\$3.00	=	\$7.50
↑		↑		↑		↑
price of child's ticket		total number of tickets		additional cost for adult ticket		total cost

■ × 2 = \$7.50 − \$3.00
■ × 2 = \$4.50
■ = \$4.50 ÷ 2
■ = \$2.25 ← the cost of a child's ticket

\$2.25
+ 3.00
\$5.25 ← the cost of an adult's ticket

Draw a Diagram Show the relationship of 2 tickets for \$7.50. Let t = the cost of the child's ticket.

adult	t	\$3.00
child	t	

← $2t + \$3.00 = \$7.50$
Subtract \$3.00 from the total cost to find the cost of 2 tickets.
Divide \$4.50 by 2 to find the cost of 1 child's ticket.

So, a child's ticket costs \$2.25 and an adult's ticket costs \$5.25.

- Look back at the problem. How can you decide if your answer makes sense for the problem? Explain.

CALIFORNIA STANDARDS **MR 1.0** Students make decisions about how to approach problems. **MR 2.0** Students use strategies, skills, and concepts in finding solutions. **MR 2.3** Use a variety of methods, such as words, numbers, symbols, charts, graphs, tables, diagrams, and models, to explain mathematical reasoning. **MR 2.6** Make precise calculations and check the validity of the results from the context of the problem. *also* **MR 1.2, MR 2.4, MR 3.1**

Problem Solving Practice

Work backward or draw a diagram to solve.

1. **What if** Erica and her father had paid $7.00 for tickets and the adult's ticket was $1.50 more than the child's ticket? How much would each ticket have cost?

2. After buying 3 postcards for $0.25 each and a book about hot-air balloons for $3.65, Erica had $5.60 left. How much did Erica have before she made her purchases?

PROBLEM SOLVING STRATEGIES

- ▶ Draw a Diagram or Picture
- Make a Model or Act It Out
- Make an Organized List
- Find a Pattern
- Make a Table or Graph
- Predict and Test
- ▶ Work Backward
- Solve a Simpler Problem
- Write an Equation
- Use Logical Reasoning

Brian came home from a baseball game with $7.26. At the game, he spent $7.50 for a ticket, $2.25 for lunch, and $3.99 for a souvenir. During the game, his brother gave him $1.00.

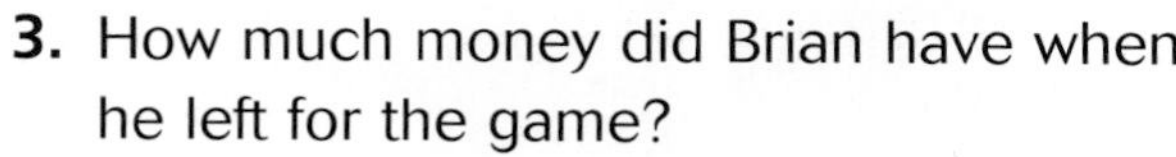

3. How much money did Brian have when he left for the game?

   A $19.00 **C** $21.00
   B $20.00 **D** $22.00

4. How much money would Brian have come home with if he had also spent $1.73 for a poster but not had lunch?

   F $11.24 **H** $7.78
   G $9.51 **J** $6.74

Mixed Strategy Practice

5. The pretzels booth at the fair is on the left end of the row. The popcorn booth is between pretzels and ice cream. The cotton candy booth is between ice cream and popcorn. The lemonade booth is between cotton candy and popcorn. List the booths in order from left to right.

6. Ali has been keeping track of her plant's growth since she bought it 2 weeks ago. It grew 2.5 centimeters the first week and twice as much the second week. If it is 12.0 centimeters tall now, how tall was it when Ali bought it?

7. At the carnival, there were 108 prizes at the ring-toss booth. Nine prizes were given away during every 15-minute period, starting at noon. At what time were all the prizes gone?

8. Ana needs $130 to buy a coat. She has $45. If she saves $5 each week, how many weeks will it take her to save enough money to buy the coat?

9. Two buses will transport 1,225 people to the stadium. If each bus holds 52 people, how many trips will each bus make?

LESSON 5

Divide to Change a Fraction to a Decimal

Quick Review

Name two equivalent decimals for each.

1. 0.5 **2.** 3.4
3. 2.70 **4.** 0.14
5. 0.700

Learn

HAVING FUN! The students at Lakeville School are helping plan their school carnival. Last Friday, $\frac{2}{5}$ of the students planned the game booths and $\frac{1}{4}$ of the students planned the food booths. Did more students plan the food or the games?

When fractions have different denominators, you can compare them by changing them to decimals.

MATH IDEA Change a fraction to an equivalent decimal by dividing the numerator by the denominator.

STEP 1

Write each fraction as a division problem.

$\frac{2}{5} \rightarrow 5\overline{)2}$

$\frac{1}{4} \rightarrow 4\overline{)1}$

STEP 2

Divide as with whole numbers. Place a decimal point and zeros as needed in the dividend. Place a decimal point and a zero in the quotient.

$$\begin{array}{r} 0.4 \\ 5\overline{)2.0} \\ -2\,0 \\ \hline 0 \end{array} \qquad \begin{array}{r} 0.25 \\ 4\overline{)1.00} \\ -8 \\ \hline 20 \\ -20 \\ \hline 0 \end{array}$$

Remember
$\frac{1}{4}$ can be read as one fourth, one out of four, or one divided by four.

Since $0.4 > 0.25$, $\frac{2}{5} > \frac{1}{4}$. So, more students planned the games.

Examples

A

$$\frac{1}{5} \rightarrow \begin{array}{r} 0.2 \\ 5\overline{)1.0} \\ -1\,0 \\ \hline 0 \end{array}$$

B

$$\frac{3}{4} \rightarrow \begin{array}{r} 0.75 \\ 4\overline{)3.00} \\ -2\,8 \\ \hline 20 \\ -20 \\ \hline 0 \end{array}$$

C

$$\frac{7}{8} \rightarrow \begin{array}{r} 0.875 \\ 8\overline{)7.000} \\ -6\,4 \\ \hline 60 \\ -56 \\ \hline 40 \\ -40 \\ \hline 0 \end{array}$$

CALIFORNIA STANDARDS O→ **NS 1.2** Interpret percents as a part of a hundred; find decimal and percent equivalents for common fractions and explain why they represent the same value; compute a given percent of a whole number. O→ **NS 2.2** Demonstrate proficiency with division, including division with positive decimals and long division with multidigit divisors. *also* **NS 2.0,** O→ **NS 2.1, MR 2.3, MR 2.4, MR 3.2, MR 3.3**

Check

1. **Explain** how you know that $\frac{1}{4}$ and 0.25 are equivalent.

Write as a decimal.

2. $\frac{2}{10}$ 3. $\frac{3}{5}$ 4. $\frac{3}{8}$ 5. $\frac{4}{8}$ 6. $\frac{6}{10}$ 7. $\frac{3}{20}$

Practice and Problem Solving

Write as a decimal.

8. $\frac{7}{10}$ 9. $\frac{23}{25}$ 10. $\frac{7}{16}$ 11. $\frac{2}{5}$ 12. $\frac{2}{4}$ 13. $\frac{2}{8}$

14. $\frac{3}{10}$ 15. $\frac{17}{20}$ 16. $\frac{5}{8}$ 17. $\frac{1}{8}$ 18. $\frac{4}{5}$ 19. $\frac{6}{8}$

USE DATA **For 20–22, use the circle graph.**

20. Use a decimal to name the part of the money Jake's family spent on souvenirs.

21. Use a decimal to name the part of the money Jake's family spent on rides and games.

22. **What if** Jake's family had bought snacks for $15 and saved the extra money? Use a decimal to name the part of their money they would have saved.

23. Jenny ran $\frac{6}{25}$ mile. Greg ran 0.21 mile. Who ran farther? Explain how you know.

24. Write a fraction that has a decimal equivalent that is greater than 0.5 and less than 0.75.

Mixed Review and Test Prep

25. Order 6.010, 6.101, and 6.001 from least to greatest. (p. 24)

26. From home, Josie rode 4 blocks east, 2 blocks north, 3 blocks west, and 2 blocks south. How far from home was she then? Draw a diagram to solve. (p. 226)

27. $756{,}213 \times 52$ (p. 148)

28. $421{,}559 \times 74$ (p. 148)

29. **TEST PREP** Which is the quotient for $630{,}662 \div 82$? (p. 204)

A 82 **C** 7,691
B 760 **D** 70,196

Review/Test

CHECK CONCEPTS

Copy and complete each pattern. (pp. 218–219)

1. $210 \div 7 = \blacksquare$
$21 \div 7 = \blacksquare$
$2.1 \div 7 = \blacksquare$

2. $400 \div 5 = \blacksquare$
$40 \div 5 = \blacksquare$
$4 \div 5 = \blacksquare$

3. $240 \div 6 = \blacksquare$
$24 \div 6 = \blacksquare$
$2.4 \div 6 = \blacksquare$

CHECK SKILLS

Find the quotient. Check by multiplying. (pp. 223–224)

4. $8\overline{)7.2}$ **5.** $7\overline{)5.6}$ **6.** $9\overline{)0.81}$

7. $2\overline{)3.66}$ **8.** $3.95 \div 5$ **9.** $109.2 \div 6$

10. $218.4 \div 8$ **11.** $5{,}716.8 \div 9$ **12.** $33.68 \div 8$

13. $28.44 \div 9$ **14.** $81.84 \div 6$ **15.** $58.75 \div 5$

Write as a decimal. (pp. 228–229)

16. $\frac{4}{16}$ **17.** $\frac{6}{12}$ **18.** $\frac{5}{25}$

19. $\frac{3}{16}$ **20.** $\frac{6}{24}$ **21.** $\frac{7}{16}$

CHECK PROBLEM SOLVING

Solve.

22. Sally came home from the fair with $3.86. She spent $5.50 for a ticket, $4.35 for food, and $7.29 for rides. How much money did Sally have before she went to the fair? (pp. 226–227)

23. The total rainfall for the week was 2.14 inches. For 4 days it did not rain at all. On 2 days it rained 0.5 inch each day. How much rain fell on the last day? (pp. 226–227)

24. At the fruit market, cherries cost $5.28 for 3 pounds. How much does one pound cost? (pp. 222–225)

25. Jim earned $22.00 for mowing lawns for 4 hours. How much did he earn in one hour? (pp. 222–225)

Cumulative Review

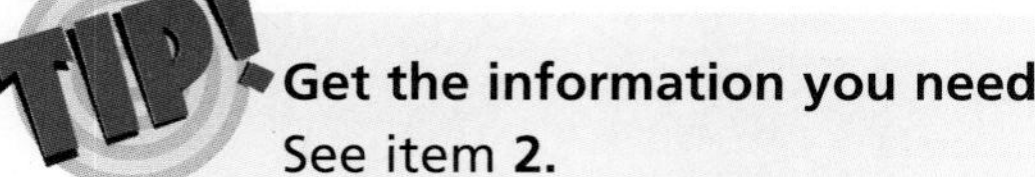

Get the information you need.
See item **2.**

Remember how to change a fraction to a decimal. One way is to find an equivalent fraction with a denominator of 10 or 100.

Also see problem **3,** p. H63.

For 1–11, choose the best answer.

1. How is 0.125 written in expanded form?

A $100 + 20 + 5$
B $1 + 2 + 5$
C $0.1 + 0.2 + 0.5$
D $0.1 + 0.02 + 0.005$

---

2. How is $\frac{3}{5}$ written as a decimal?

F 0.35 **H** 0.6
G 0.53 **J** 3.5

---

3. Which number is equivalent to $\frac{1}{5}$?

A 0.05 **C** 0.20
B 0.15 **D** 2

---

4. Jack bought 5 small jars of paint for a total of $\$4.45$. How much did each jar cost?

F $\$0.94$ **H** $\$0.85$
G $\$0.89$ **J** $\$0.80$

---

5. $0.2 \div 4$

A 5 **C** 0.5
B 2 **D** 0.05

---

6. $0.56 \div 7$

F 8.0 **H** 0.08
G 0.8 **J** 0.008

---

7. One carat equals 200 milligrams. The Blue Hope diamond weighs 44.5 carats. How many milligrams does the Blue Hope diamond weigh?

A $1,700$ mg **C** $17,000$ mg
B $8,900$ mg **D** $89,000$ mg

For 8–9, use the circle graph.

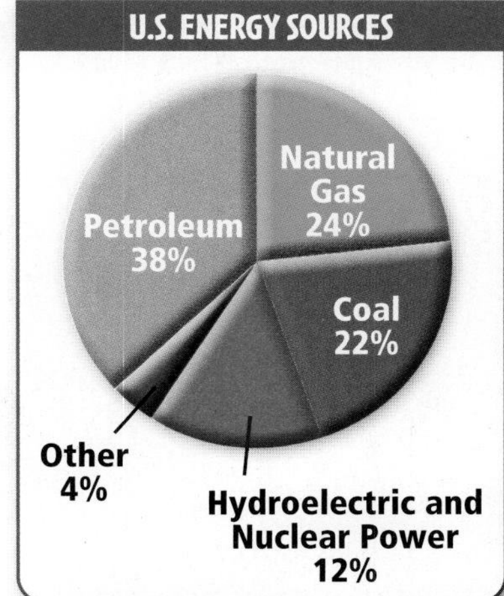

8. Hydroelectric and nuclear energy are used half as much as what other energy source?

A coal **C** petroleum
B natural gas **D** other sources

9. What percentage of energy comes from petroleum and coal together?

F 16% **H** 38%
G 22% **J** 60%

---

10. $8\overline{)4.8}$

A 0.06 **C** 6
B 0.6 **D** Not here

---

11. During a 24-hour period, the average person will speak 4,800 words. How many words per hour is that?

F 200 **H** $20,000$
G $2,000$ **J** Not here

CHAPTER 14

Divide Decimals by Decimals

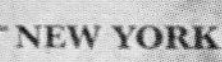

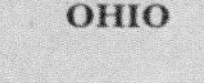

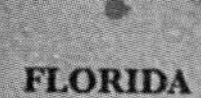

There are about 24 billion dollars worth of coins in circulation in the United States. Four friends have decided to collect the U.S. quarters that are being made for each state. They are going to trade their savings in for quarters so they can start their collections. How many quarters will each person have?

SAVINGS	
Name	**Amount of Money**
Carlo	$31.75
Kaylee	$23.25
Lauren	$19.00
David	$46.50

Each year from 1999–2008, quarters for five states will be released. They will be minted in the order in which the states ratified the Constitution or joined the Union.

Check What You Know

Use this page to help you review and remember important skills needed for Chapter 14.

Related Facts (See p. H16.)

For each multiplication fact, write a related division fact.

1. $3 \times 4 = 12$	**2.** $2 \times 5 = 10$	**3.** $8 \times 11 = 88$	**4.** $9 \times 7 = 63$
5. $6 \times 3 = 18$	**6.** $9 \times 4 = 36$	**7.** $12 \times 6 = 72$	**8.** $8 \times 9 = 72$
9. $4 \times 7 = 28$	**10.** $5 \times 12 = 60$	**11.** $9 \times 10 = 90$	**12.** $5 \times 5 = 25$

Multiply by 10 and 100 (See p. H17.)

Find the products.

13. 7×10 7×100	**14.** 5.8×10 5.8×100	**15.** 7.23×10 7.23×100	**16.** 3.04×10 3.04×100
17. 19×10 19×100	**18.** 2.94×10 2.94×100	**19.** 6.4×10 6.4×100	**20.** 0.4×10 0.4×100
21. 22×10 22×100	**22.** 0.7×10 0.7×100	**23.** 0.62×10 0.62×100	**24.** 0.09×10 0.09×100

Divide Decimals by Whole Numbers (See p. H17.)

Copy the problem. Place the decimal point correctly in the quotient.

25. $\begin{array}{r} 342 \\ 3\overline{)10.26} \end{array}$	**26.** $\begin{array}{r} 163 \\ 5\overline{)81.5} \end{array}$	**27.** $\begin{array}{r} 83 \\ 23\overline{)190.9} \end{array}$	**28.** $\begin{array}{r} 061 \\ 4\overline{)2.44} \end{array}$
29. $\begin{array}{r} 56 \\ 4\overline{)22.4} \end{array}$	**30.** $\begin{array}{r} 09 \\ 9\overline{)8.1} \end{array}$	**31.** $\begin{array}{r} 84 \\ 34\overline{)285.6} \end{array}$	**32.** $\begin{array}{r} 45 \\ 62\overline{)279.0} \end{array}$

Find the quotient.

33. $6\overline{)4.38}$	**34.** $8\overline{)20.64}$	**35.** $7\overline{)6.58}$	**36.** $4\overline{)9.2}$
37. $26.75 \div 5$	**38.** $3.16 \div 4$	**39.** $1.08 \div 9$	**40.** $315.20 \div 5$

LESSON 1

Algebra: Patterns in Decimal Division

Quick Review

Find the product. Then name a related division fact.

1. 9×7
2. 8×9
3. 9×4
4. 4×7
5. 10×7

Learn

FOLLOW THAT DECIMAL! First, look at the patterns in the multiplication equations. Then, look at the patterns in the related division equations.

MULTIPLICATION EQUATIONS Factor × Factor = Product		RELATED DIVISION EQUATIONS Dividend ÷ Divisor = Quotient
$12 \times 4 = 48$	→	$48 \div 12 = 4$
$1.2 \times 4 = 4.8$	→	$4.8 \div 1.2 = 4$
$0.12 \times 4 = 0.48$	→	$0.48 \div 0.12 = 4$

- In the multiplication equations, what happens in the product as the decimal point moves left in one factor?
- What do you notice about the quotients in the division equations?

MATH IDEA You can use multiplication and patterns to help you find the quotient in decimal division.

Remember
You can use patterns to place the decimal point in the product.

$1 \times 10 = 10$
$0.1 \times 10 = 1.0$
$0.01 \times 10 = 0.10$
$0.001 \times 10 = 0.010$

Examples

A Find $0.35 \div 0.05$.

$5 \times 7 = 35 \rightarrow 35 \div 5 = 7$
$0.5 \times 7 = 3.5 \rightarrow 3.5 \div 0.5 = 7$
$0.05 \times 7 = 0.35 \rightarrow 0.35 \div 0.05 = 7$

So, $0.35 \div 0.05 = 7$.

B Find $0.56 \div 0.07$.

$7 \times 8 = 56 \rightarrow 56 \div 7 = 8$
$0.7 \times 8 = 5.6 \rightarrow 5.6 \div 0.7 = 8$
$0.07 \times 8 = 0.56 \rightarrow 0.56 \div 0.07 = 8$

So, $0.56 \div 0.07 = 8$.

Check

1. **Tell** what multiplication equations you could use to help you find $0.48 \div 1.2$.

Copy and complete each multiplication pattern. Then write the related division pattern.

2. $6 \times 9 = 54$
$0.6 \times 9 = ■$
$0.06 \times 9 = ■$

3. $15 \times 5 = 75$
$1.5 \times 5 = ■$
$0.15 \times 5 = ■$

4. $16 \times 4 = 64$
$1.6 \times 4 = ■$
$0.16 \times 4 = ■$

CALIFORNIA STANDARDS MR 1.1 Analyze problems by identifying relationships, distinguishing relevant from irrelevant information, sequencing and prioritizing information, and observing patterns. O-π NS 2.2 Demonstrate proficiency with division, including division with positive decimals and long division with multidigit divisors. *also* NS 1.0, NS 2.0, O-π NS 2.1, O-π AF 1.2, MR 2.0, MR 2.2, MR 2.4, MR 3.0

Practice and Problem Solving

Copy and complete each multiplication pattern. Then write the related division pattern.

5. $7 \times 3 = 21$
$0.7 \times 3 = $ ■
$0.07 \times 3 = $ ■

6. $81 \times 7 = 567$
$8.1 \times 7 = $ ■
$0.81 \times 7 = $ ■

7. $26 \times 4 = 104$
$2.6 \times 4 = $ ■
$0.26 \times 4 = $ ■

8. $5 \times 2 = 10$
$0.5 \times 2 = $ ■
$0.05 \times 2 = $ ■

9. $35 \times 5 = 175$
$3.5 \times 5 = $ ■
$0.35 \times 5 = $ ■

10. $57 \times 8 = 456$
$5.7 \times 8 = $ ■
$0.57 \times 8 = $ ■

Complete each division pattern.

11. $50 \div 25 = 2$
$5.0 \div 2.5 = $ ■
$0.50 \div 0.25 = $ ■

12. $75 \div 15 = 5$
$7.5 \div 1.5 = $ ■
$0.75 \div 0.15 = $ ■

13. $144 \div 12 = 12$
$14.4 \div 1.2 = $ ■
$1.44 \div 0.12 = $ ■

14. $360 \div 4 = 90$
$36 \div 0.4 = $ ■
$3.6 \div 0.04 = $ ■

15. $72 \div 9 = 8$
$7.2 \div$ ■ $= 8$
$0.72 \div$ ■ $= 8$

16. $120 \div 40 = 3$
$12 \div$ ■ $= 3$
$1.2 \div$ ■ $= 3$

Algebra **Use basic facts and patterns to solve for *n*.**

17. $0.09 \div 0.03 = n$

18. $0.16 \div 0.04 = n$

19. $3.0 \div 0.05 = n$

20. $1.6 \div n = 2$

21. $n \div 0.02 = 7$

22. $0.81 \div n = 9$

USE DATA **For 23–25, use the table.**

23. A garden snail traveled 0.9 mile. How long did that take?

24. A giant tortoise traveled 1.7 miles. How long did that take?

25. Which would travel a mile faster, a garden snail or a sloth? Explain.

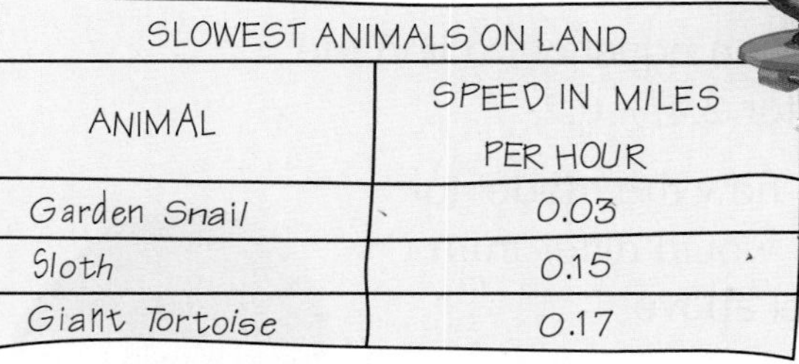

SLOWEST ANIMALS ON LAND	
ANIMAL	SPEED IN MILES PER HOUR
Garden Snail	0.03
Sloth	0.15
Giant Tortoise	0.17

26. **What's the Error?** Sheryl saved \$7.50 in nickels. She says she has 1,500 nickels. Describe and correct her error.

27. **NUMBER SENSE** Which is greater?
a. $0.12 \div 0.4$ or $0.12 \div 0.04$?
b. $0.4 \div 0.2$ or $0.004 \div 0.02$?

Mixed Review and Test Prep

28. $65{,}984 \div 44$ (p. 204)

29. $294{,}861 \div 98$ (p. 204)

30. Solve for *n*. $n \times 5 = 45$ (p. 90)

31. Find the mode for this set of data. 3.2, 4.6, 2.3, 6.4, 3.2, 4.6, 5 (p. 104)

32. **TEST PREP** Which expression describes this problem? Steve made 20 necklaces. Each necklace had 25 beads. Then he gave 1 necklace to Sara. (p. 86)

A $(20 \times 25) + 25$
B $(20 \times 25) + 1$
C $(20 \times 25) - 25$
D $(20 \times 25) - 1$

Extra Practice page H45, Set A

LESSON 2

HANDS ON
Divide with Decimals

Quick Review

Tell how many hundredths are in:

1. 0.10 **2.** 0.04
3. 0.38 **4.** 3.56
5. 2.45

MATERIALS
decimal models, markers, scissors

Explore

Make a model to find decimal quotients.

Divide 2.4 by 0.2. Write: $2.4 \div 0.2$, or $0.2\overline{)2.4}$

STEP 1

Show 2.4 by shading three decimal models.

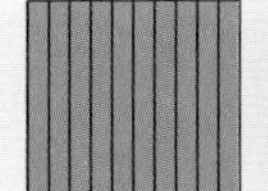

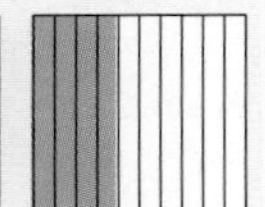

STEP 2

Cut the model into groups of 0.2.

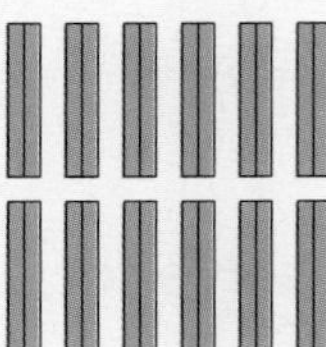

STEP 3

Record a division equation for the model.

Think: How many groups of 0.2 are in 2.4?

There are 12 groups of 0.2 in 2.4.

$2.4 \div 0.2 = 12$

So, $2.4 \div 0.2 = 12$.

- Explain how you can use a related multiplication equation to find the quotient for $2.4 \div 0.2$.
- Describe how the model for $2.4 \div 0.6$ would differ from the model above.

How can you cut 1.2 into groups of 0.3?

Try It

Make a model to find the quotient. Record a division equation for each model.

a. $1.2 \div 0.3$
b. $4.8 \div 1.2$
c. $3.9 \div 0.3$
d. $2.4 \div 0.4$

CALIFORNIA STANDARDS O━ NS 2.1 Add, subtract, multiply, and divide with decimals; add with negative numbers; subtract positive integers from negative integers; and verify the reasonableness of the results. O━ NS 2.2 Demonstrate proficiency with division, including division with positive decimals and long division with multidigit divisors. *also* **NS 1.0, NS 2.0, MR 1.0, MR 2.0, MR 2.3, MR 3.0, MR 3.2**

Connect

Use a hundredths model to show $0.75 \div 0.25$.

STEP 1

Show 0.75 by shading a hundredths model.

STEP 2

Cut the model into groups of 0.25.

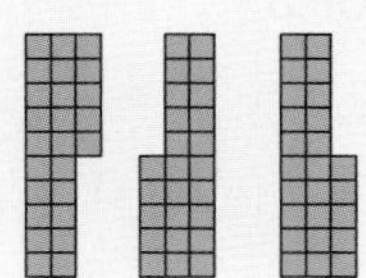

STEP 3

Record a division equation for your model.

There are 3 groups of 0.25 in 0.75.

$0.75 \div 0.25 = 3$

Practice

TECHNOLOGY LINK

More Practice: Use E-Lab, *Exploring Division of Decimals.*

www.harcourtschool.com/elab2002

Make a model to find the quotient. Record a division equation for each model.

1. $2.1 \div 0.7$
2. $0.24 \div 0.12$
3. $0.64 \div 0.08$
4. $3.5 \div 0.5$
5. $3.6 \div 0.6$
6. $0.09 \div 0.03$
7. $0.42 \div 0.07$
8. $2.8 \div 0.4$
9. $3.6 \div 1.2$

Use the model. Copy and complete the equation.

10. 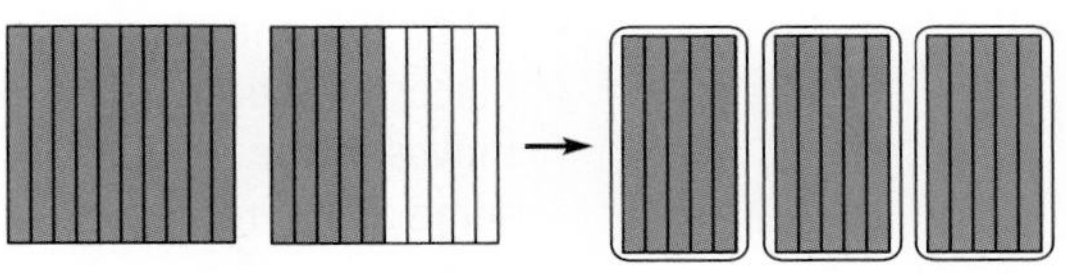

$1.5 \div 0.5 = ■$

11. 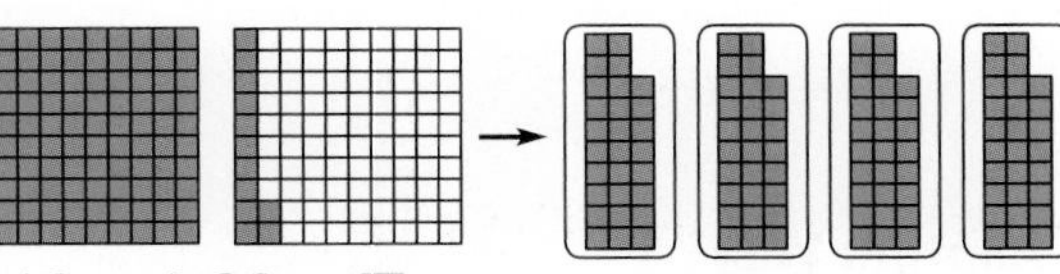

$1.12 \div 0.28 = ■$

12. 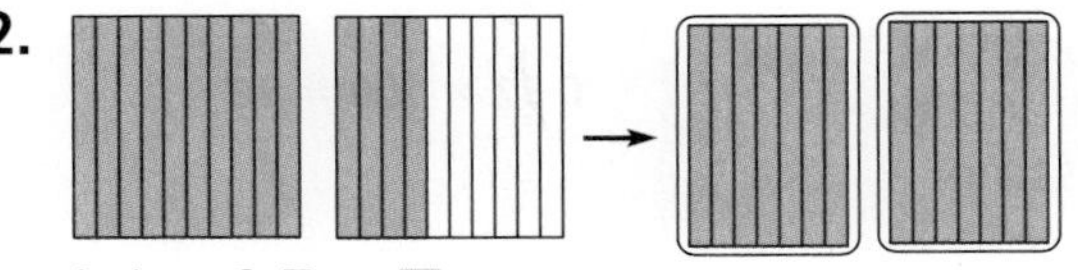

$1.4 \div 0.7 = ■$

13. 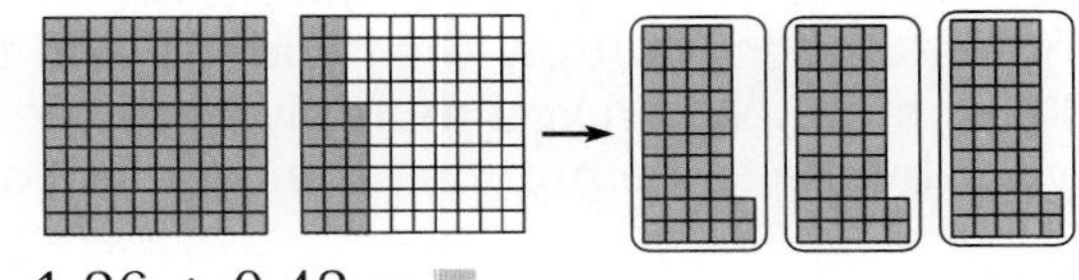

$1.26 \div 0.42 = ■$

14. **What's the Error?** Lin drew these models to divide 1.6 by 0.08. Describe and correct Lin's error.

15. **Write About It** Explain how you could use dimes to model $2.4 \div 0.8$. What coins could you use to model $2.4 \div 0.08$?

Mixed Review and Test Prep

16. $4.79 - 2.84$ (p. 50)

17. 6.14×2.5 (p. 164)

18. Order 2.11, 2.01, 2.1, and 2.001 from least to greatest. (p. 24)

19. $5.75 \div 5$ (p. 222)

20. **TEST PREP** Solve for n. $44 = 55 - n$ (p. 74)

A $n = 10$

B $n = 11$

C $n = 22$

D $n = 33$

LESSON 3

Decimal Division

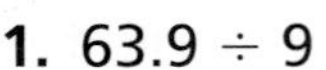
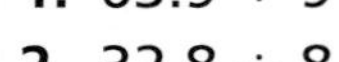
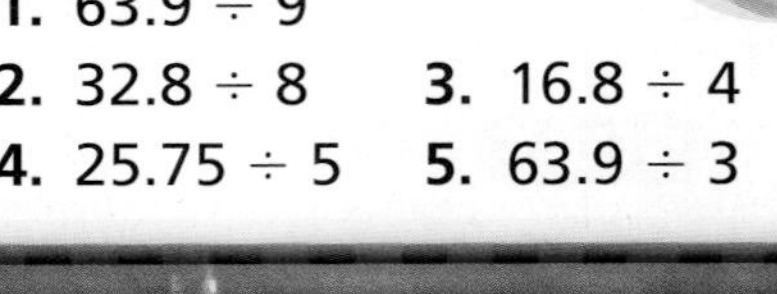

Quick Review

1. 63.9 ÷ 9
2. 32.8 ÷ 8
3. 16.8 ÷ 4
4. 25.75 ÷ 5
5. 63.9 ÷ 3

Learn

PIZZA SPICE Oregano, "the pizza herb," was not a favorite seasoning in the United States until 1945, when returning American soldiers introduced the Italian pizza. If 0.7 ounce of oregano costs $1.40, how much does 1 ounce cost?

Find 1.4 ÷ 0.7. **Think:** 1.40 and 1.4 are equivalent decimals.

▲ **In the U.S., the popularity of oregano increased by 1,500% in the decade after World War II.**

One Way

Use a model to find the quotient.

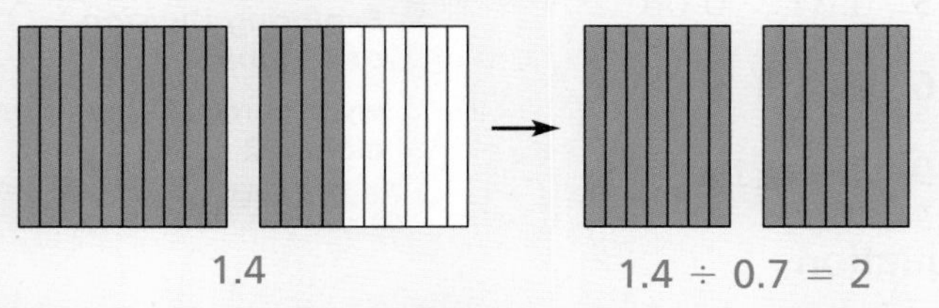

Remember

When you multiply by 10 or 100, the number of places the decimal point moves to the right is the same as the number of zeros.

0.32 × 10 = 3.2

0.32 × 100 = 32

Another Way

To divide a decimal by a decimal, change the divisor to a whole number. Multiply the dividend *and* the divisor by 10, 100, or more. When you multiply both the dividend and the divisor by the same number, the quotient stays the same.

STEP 1

0.7 has one decimal place. So, to change it to a whole number, multiply it by 10.

$0.7 \times 10 = 7$

Multiply the dividend **and** the divisor by 10.

$0.7\overline{)1.4}$

STEP 2

In the quotient, place the decimal point directly above the decimal point in the dividend.

$7.\overline{)14.}$

Divide as with whole numbers.

$$\begin{array}{r} 2. \\ 7.\overline{)14.} \\ -14 \\ \hline 0 \end{array}$$

So, 1 ounce of oregano costs $2.

CALIFORNIA STANDARDS **O━ NS 2.1** Add, subtract, multiply, and divide with decimals; add with negative numbers; subtract positive integers from negative integers; and verify the reasonableness of the results. **O━ NS 2.2** Demonstrate proficiency with division, including division with positive decimals and long division with multidigit divisors. *also* **NS 1.0, NS 1.1, NS 2.0, MR 1.0, MR 2.0, MR 2.3, MR 2.4, MR 3.3**

Whole Numbers and Decimals

MATH IDEA To divide a number by a decimal, change the divisor to a whole number by multiplying both the divisor and the dividend by 10, 100, or 1,000.

Since many of Lee's favorite recipes call for vanilla extract, she bought an 8-ounce bottle of it. One teaspoon of vanilla extract is about 0.16 fluid ounce. How many teaspoons of vanilla extract are in an 8-ounce bottle?

Find $8 \div 0.16$.

STEP 1

0.16 has two decimal places. To change the divisor to a whole number, multiply the divisor and the dividend by 100.

$0.16\overline{)8.00}$ ← Place the decimal point in the dividend. You can write zeros to the right of the decimal point without changing the value.

STEP 2

$$\begin{array}{r} 50. \\ 16.\overline{)800.} \\ -80 \\ \hline 0 \end{array}$$

Place the decimal point in the quotient.

Divide as with whole numbers.

So, there are about 50 teaspoons in an 8-ounce bottle.

Examples

A Find $6 \div 0.025$.

$0.025\overline{)6.000}$ ← Place the decimal point in the dividend. Make the divisor a whole number.

$$\begin{array}{r} 240. \\ 25.\overline{)6000.} \\ -50\downarrow \\ \hline 100 \\ -100 \\ \hline 0 \end{array}$$

B Find $\$6.64 \div \$0.08$.

$\$0.08\overline{)\$6.64}$ ← Make the divisor a whole number by multiplying the divisor and the dividend by 100.

$$\begin{array}{r} 83. \\ 8.\overline{)664.} \\ -64\downarrow \\ \hline 024 \\ -\ 24 \\ \hline 0 \end{array}$$

C Find $2.08 \div 0.4$.

$0.4\overline{)2.08}$ ← Make the divisor a whole number by multiplying the divisor and the dividend by 10.

$$\begin{array}{r} 5.2 \\ 4.\overline{)20.8} \\ -20\downarrow \\ \hline 0\,8 \\ -\ \ 8 \\ \hline 0 \end{array}$$

Check

1. **Explain** how you know whether the decimal point in the divisor should be moved one place or two places.

Copy the problem. Place the decimal point in the quotient. Draw arrows to help you.

2. $\begin{array}{r} 13 \\ 0.2\overline{)2.6} \end{array}$ 3. $\begin{array}{r} 12 \\ 0.16\overline{)1.92} \end{array}$ 4. $\begin{array}{r} 23 \\ 0.5\overline{)1.15} \end{array}$ 5. $\begin{array}{r} 4 \\ 0.28\overline{)1.12} \end{array}$

LESSON CONTINUES 

Practice and Problem Solving

Copy the problem. Place the decimal point in the quotient. Draw arrows to help you.

6. $0.2\overline{)2.12}$ quotient 106	**7.** $0.7\overline{)8.05}$ quotient 115	**8.** $0.25\overline{)0.225}$ quotient 09	**9.** $3.4\overline{)46.512}$ quotient 1368
10. $0.7\overline{)5.32}$ quotient 76	**11.** $0.08\overline{)0.544}$ quotient 68	**12.** $0.7\overline{)5.32}$ quotient 76	**13.** $0.25\overline{)0.325}$ quotient 13

Divide.

14. $0.7\overline{)3.22}$	**15.** $0.4\overline{)6}$	**16.** $1.6\overline{)18.112}$	**17.** $0.07\overline{)0.84}$
18. $\$0.52\overline{)\$6.24}$	**19.** $\$0.07\overline{)\$1.33}$	**20.** $\$0.05\overline{)\$1.75}$	**21.** $\$0.35\overline{)\$5.25}$
22. $0.7\overline{)5.32}$	**23.** $0.2\overline{)2.12}$	**24.** $1.7\overline{)5.44}$	**25.** $0.25\overline{)12}$
26. $71.4 \div 2.1$	**27.** $42 \div 1.2$	**28.** $5.25 \div 0.7$	**29.** $\$0.42 \div \$0.14$
30. $25 \div 0.5$	**31.** $0.34 \div 3.4$	**32.** $0.64 \div 0.4$	**33.** $\$9.00 \div \$1.50$

PATTERNS **Divide. Then describe a pattern in the quotients.**

34. a. $6.6 \div 1.2$ **b.** $6.6 \div 0.12$ **c.** $6.6 \div 0.012$

35. a. $7.5 \div 2.5$ **b.** $0.75 \div 2.5$ **c.** $0.075 \div 2.5$

USE DATA **For 36–38, use the table.**

36. How much does 1 ounce of parsley flakes cost?

37. **What's the Question?** The answer is 1 ounce costs $3.65.

38. Compare and order the costs of the spices per ounce from least expensive to most expensive.

39. REASONING In the expression $6.02 \div 0.2$, if you multiply the dividend by 100 and the divisor by 10, will the quotient remain the same? Explain.

40. **What's the Error?** Lamar divided 7.25 by 0.25 and got a quotient of 0.29. Describe and correct his error.

41. REASONING If you divide 7.56 by 7.6, will the quotient be more than or less than 1? Explain.

THE SPICE RACK

Spice	Amount	Cost
Oregano	1.37 ounces	$2.74
Basil	0.6 ounce	$2.19
Parsley flakes	0.25 ounce	$1.19
Bay leaves	0.12 ounce	$1.98
Chili powder	2.5 ounces	$1.45

42. You have $21.75 and want to buy 15 posters. Each poster costs $1.50. Do you have enough money? Explain.

43. The coin-operated dryer costs $1.75 per load. How many quarters are needed to dry 2 loads of laundry?

Mixed Review and Test Prep

44. 903×8 (p. 146)

45. $507{,}648 \times 3$ (p. 146)

46. $9\overline{)0.081}$ (p. 222)

47. 85.639×2.5 (p. 164)

48. Evaluate $n + (7 \times 14)$ if $n = 33$. (p. 68)

49. Tyler slept 9 hours, worked 8 hours, exercised 1 hour, read 2 hours, and went out with friends for 4 hours today. What type of graph would be best to display how Tyler spent his day? Explain. (p.130)

50. **TEST PREP** What is the value of the blue digit in 5.681? (p. 18)

A 8 ones
B 8 tenths
C 8 hundredths
D 8 thousandths

USE DATA **For 51–52, use the table.**

CARLO'S BOWLING SCORES

Game	1	2	3	4	5
Score	85	97	82	73	76

51. What is Carlo's mean bowling score? (p. 102)

52. What are the range, median, and mode for his scores? (p. 104)

53. **TEST PREP** Lara paid $12.58 for a turkey that cost $2 per pound. How much did the turkey weigh? (p. 90)

A 6.29 pounds
B 10.58 pounds
C 14.58 pounds
D 25.16 pounds

Thinker's Corner

1	2	3
4	5	6
7	8	9

BREAKING THE CODE How good are you at decoding secret symbols? For these puzzles, the key to the code is at the right. Numbers are represented by the shapes of the lines they are within. The code for 3 is └.

Example: Find ┐.□ ÷ ⊔.□.

┐ = 7, □ = 5 → ┐.□ = 7.5 ⊔ = 2, □ = 5 → ⊔.□ = 2.5

$7.5 \div 2.5 = 3$

So, ┐.□ ÷ ⊔.□ = 3.

Decode the problem. Then solve. Write your answer as a number.

1. ┘⊐.⊐ ÷ ┘.⊔
2. ┘┌.⊔ ÷ ⊔.⊐
3. □.└ × ⊓
4. ⊐□.⊏ − ┘⊓.⊐
5. □.┘⊔ ÷ ┘.⊏
6. └┘.⊏⊔ ÷ ┌.└
7. Find $12 \div 1.5$. Then write the problem and quotient in code.

LESSON 4

Problem Solving Skill

Choose the Operation

Quick Review

1. 23.06 + 4.5
2. 39.05 − 6.1
3. 5.25 × 4
4. 55.5 ÷ 0.5
5. (2.5 × 10) − 7

"WEIGH OUT" IN SPACE Beth used data she found on a class trip to the Space Museum to make this table. The table shows how much a 1-pound weight on Earth would weigh on other planets.

ONE EARTH POUND IN SPACE	
Planet	**Weight (in pounds)**
Mercury	0.38
Venus	0.91
Earth	1
Mars	0.38
Jupiter	2.53
Saturn	1.07
Uranus	0.90
Neptune	1.10
Pluto	0.07

MATH IDEA The way numbers in a problem are related can help you choose the operation needed to solve the problem.

Add	To join groups.
Subtract	To take away or compare groups.
Multiply	To join equal-sized groups.
Divide	To separate into equal-sized groups. To find out how many in each group.

Decide how the numbers in the problem are related. Name the operation needed to solve the problem. Then solve.

A If a cocker spaniel weighs 35 pounds on Earth, how much would it weigh on Jupiter?

B If a gorilla weighs 33.95 pounds on Pluto, how much would it weigh on Earth?

C How much more would a kitten that weighs 1 pound on Earth weigh on Jupiter?

D How much would 2 hamsters that each weigh 1 pound on Earth weigh altogether on Mercury?

Talk About It

- Explain how you decided which operation you used to solve each problem.
- **What if** the gorilla in Problem B weighed 15.75 pounds on Pluto? How much would it weigh on Earth?

CALIFORNIA STANDARDS **MR 1.0** Students make decisions about how to approach problems. **MR 2.0** Students use strategies, skills, and concepts in finding solutions. **MR 3.2** Note the method of deriving the solution and demonstrate a conceptual understanding of the derivation by solving similar problems. *also* **NS 1.0, NS 2.0, O━NS 2.1, O━NS 2.2**

Problem Solving Practice

Solve. Name the operation or operations you used.
For 1–2, use the table on page 242.

1. If a blue whale weighs 286,600 pounds on Earth, how much would it weigh on Mars?
2. If a tiger weighs 727.1 pounds on Neptune, how much would it weigh on Earth?
3. Suppose a polar bear weighs 1,323 pounds and a walrus weighs 1,503 pounds. How much more does the walrus weigh than the polar bear?
4. Suppose 3 giant pandas weigh a total of 1,056 pounds. What is the average weight of the pandas?

In 1999, the record was set for the most animals on a space mission. On board the shuttle *Columbia* were 1,500 crickets, 18 mice, 152 rats, 135 snails, 4 oyster toadfish, and 229 swordtail fish.

Laika, the first living creature in space, was launched aboard *Sputnik 2* in 1957.

5. What operation would you use to find how many creatures were on the mission in all?
   A multiplication
   B division
   C subtraction
   D addition
6. How many more crickets were there than all other animals combined?
   F 538
   G 872
   H 962
   J 1,500

Mixed Applications

USE DATA For 7–8, use the bar graph.

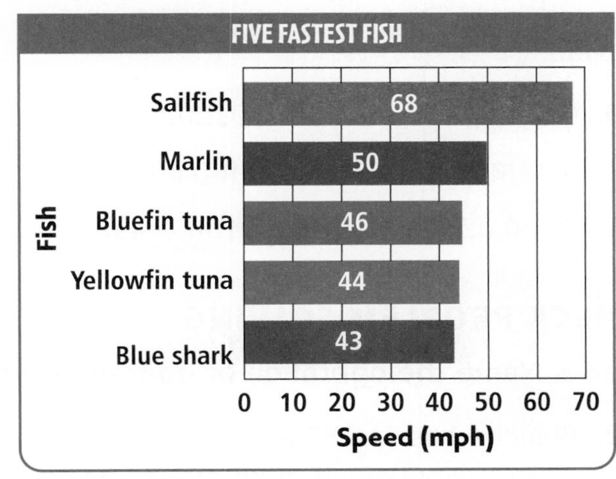

7. What is the difference in speed between the fastest and slowest fish on the graph?
8. How long would it take a marlin to travel 1,000 miles?
9. The sixth-fastest fish is the wahoo. It swims 66 kilometers per hour. How many miles per hour does it swim? (1 kilometer equals 0.62 mile)
10. Marian made a phone call that cost $0.99 for the first 10 minutes and $0.25 for each additional minute. She was on the phone for 20 minutes. What was the cost of the phone call?
11. **Write a problem** using the information in the graph above. Exchange problems with a classmate and solve.

CHAPTER 14

Review/Test

CHECK VOCABULARY AND CONCEPTS

Choose the best term from the box.

decimal point
multiplication
quotient
whole number

1. To divide a number by a decimal, you change the divisor to a __?__. (p. 238)
2. When you multiply the decimal divisor and the decimal dividend by the same multiple of 10, the __?__ is the same as in the whole-number division. (p. 238)
3. To help you find the quotient in decimal division, use related __?__ equations. (p. 234)

CHECK SKILLS

Complete each division pattern. (pp. 234–235)

4. $204 \div 12 = 17$
   $20.4 \div 1.2 = \blacksquare$
   $2.04 \div 0.12 = \blacksquare$
5. $216 \div 3 = 72$
   $21.6 \div 0.3 = \blacksquare$
   $2.16 \div \blacksquare = 72$
6. $420 \div 70 = 6$
   $42.0 \div \blacksquare = 6$
   $4.2 \div \blacksquare = 6$

Use a basic fact and patterns to solve for *n*. (pp. 234–235)

7. $5.6 \div 0.8 = n$
8. $1.8 \div 0.6 = n$
9. $0.90 \div n = 30$

Divide. (pp. 238–241)

10. $0.5\overline{)4.5}$
11. $\$0.28\overline{)\$1.40}$
12. $6\overline{)20.4}$
13. $9.9\overline{)16.83}$
14. $0.32\overline{)1.824}$
15. $0.9\overline{)0.405}$
16. $23.5\overline{)5.405}$
17. $0.03\overline{)1.566}$
18. $8 \div 0.025$
19. $0.138 \div 4.6$
20. $\$4.75 \div \$0.25$
21. $54 \div 4.5$

CHECK PROBLEM SOLVING

Solve. Name the operation or operations you used. (pp. 242–243)

22. Blanche has $16.87 and Carl has $19.54. How much more money does Carl have than Blanche?
23. Alice saves $4.25 every week. If she saves for 15 weeks, how much money will she have?
24. Jenna paid $3.54 for apples. The apples cost $0.59 per pound. How many pounds did she buy?
25. Arturo worked for 5 hours at $5.75 per hour. Christine worked 3 hours at $6.50 per hour. Who made more money? How much more?

Cumulative Review

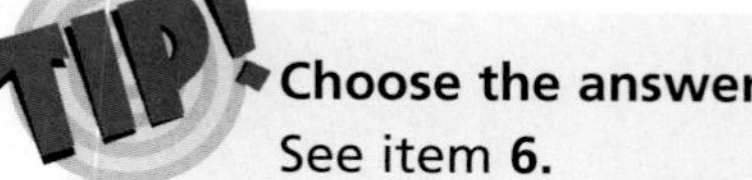

Choose the answer.
See item **6.**

If your answer doesn't match one of the choices, check your computation and the placement of the decimal point. If your computation is correct, mark the letter for Not here.

Also see problem **6,** p. H64.

For 1–10, choose the best answer.

1. $2.4 \div 0.6$

A 0.04 **C** 4
B 0.4 **D** Not here

---

2. A spring toy moves down the stairs at about 0.9 mile per hour. At that rate, how long would it take to travel 0.45 mile?

F 5 hours **H** 0.5 hour
G 4.05 hours **J** 0.05 hour

---

3. Sam paid $2.58 for pears that cost $0.43 per pound. How many pounds did he buy?

A 6 **C** 5
B 5.5 **D** 4

---

4. In the last 5 games, the Rockets scored 58, 62, 58, 67, and 70 points. What is the median score?

F 63 **H** 58
G 62 **J** 12

---

5. How is 12.605 written in expanded form?

A 12 + 6 + 5
B 10 + 2 + 0.6 + 0.005
C 10 + 2 + 0.6 + 0.05
D 12 + 0.6 + 0.5

6. $0.45 \div 0.03$

F 0.015 **H** 1.5
G 0.15 **J** Not here

---

7. Angie gets paid $2.25 for each customer she delivers newspapers to during the week. How could Angie find out how much she would earn for delivering papers to 130 customers?

A Multiply $2.25 and 130
B Add $2.25 and 130
C Divide 130 by $2.25
D Subtract 130 from $2.25

For 8–10, use the table.

COUNTRIES WITH THE MOST VCRs	
Country	**Number of VCRs**
China	40,000,000
United States	86,825,000
Germany	26,328,000
Japan	34,309,000

8. Which operation should be used to find how many more VCRs there are in the United States than in Germany?

F Addition **H** Multiplication
G Subtraction **J** Division

9. How many VCRs are there in China and Japan altogether?

A 74,309,000 **C** 7,430,900
B 56,910,000 **D** 5,691,000

10. What is the order of countries from *least* to *greatest* number of VCRs?

F U.S., China, Germany, Japan
G China, Germany, U.S., Japan
H Japan, Germany, China, U.S.
J Germany, Japan, China, U.S.

UNIT 4

MATH DETECTIVE

Missing Operations

Reasoning Two students from your school have invented symbols for a new mathematical language. They are seeking your help in solving some mathematical problems. Be a detective and decode what each symbol represents. When you figure out what each symbol represents, use this information to find the solution to each problem. Good luck!

CASE 1

Circulation means A ⊙ B ♒ C.

12 ⊙ 4 ♒ 3

54 ⊙ 9 ♒ 6

72 ⊙ 8 ♒ 9

1. 121 ⊙ *B* ♒ 11 *B* = ____
2. *A* ⊙ 7 ♒ 12 *A* = ____
3. 207 ⊙ 9 ♒ *C* *C* = ____

CASE 2

Starvision means A ✷ B ♒ C.

28 ✷ 3 ♒ 1

69 ✷ 5 ♒ 4

74 ✷ 8 ♒ 2

1. *A* ✷ 6 ♒ 5 *A* = ____
2. 87 ✷ *B* ♒ 6 *B* = ____
3. 117 ✷ 9 ♒ *C* *C* = ____

Think It Over!

- Explain what each symbol represents.
- **Write About It** How did you determine the rule for each operation?
- **Write a problem** using the new symbols. Give your solution.

UNIT 4

Challenge

Terminating and Repeating Decimals

To find the decimal equivalent of a fraction, you divide the numerator by the denominator.

Find the decimal equivalent of $\frac{3}{4}$.

$$\frac{3}{4} = \begin{array}{r} 0.75 \\ 4\overline{)3.00} \\ -2\,8 \\ \hline 20 \\ -20 \\ \hline 0 \end{array}$$

When you divide and the remainder is zero, the quotient can be written as a **terminating decimal.**

$$\frac{3}{4} = 0.75$$

Find the decimal equivalent of $\frac{5}{6}$.

$$\frac{5}{6} = \begin{array}{r} 0.8333 \\ 6\overline{)5.0000} \\ -4\,8 \\ \hline 20 \\ -18 \\ \hline 20 \\ -18 \\ \hline 20 \\ -18 \\ \hline 2 \end{array}$$

Sometimes when you divide, you continue to get the same remainder. The quotient is called a **repeating decimal.**

To show that the digit or digits repeat endlessly, you can write the decimal as $0.8333\ldots$

Or you can draw a bar over the digit or digits that repeat.

$$\frac{5}{6} = 0.8\overline{3}$$

Every fraction can be renamed as either a repeating or a terminating decimal.

Examples

A $\frac{1}{2} = 0.5$ terminating	B $\frac{2}{3} = 0.\overline{6}$ repeating	C $\frac{5}{8} = 0.625$ terminating	D $\frac{6}{11} = 0.\overline{54}$ repeating

- What is the remainder in a terminating decimal?

Try It

Rename each fraction as a terminating or a repeating decimal. Write *terminating* or *repeating*.

1. $\frac{1}{9}$ **2.** $\frac{1}{4}$ **3.** $\frac{4}{9}$ **4.** $\frac{11}{25}$ **5.** $\frac{4}{15}$

UNIT 4

Study Guide and Review

VOCABULARY

Choose the best term from the box.

1. Numbers that are easy to compute mentally are __?__. (p. 182)
2. You can use basic facts and __?__ to divide by multiples of 10. (p. 200)
3. To divide a number by a decimal, you change the divisor to a __?__ . (p. 238)

compatible numbers
patterns
remainder
quotient
whole number

STUDY AND SOLVE

Chapter 11

Divide by 1-digit divisors.

$652 \div 8$

$\downarrow \quad \downarrow$

$640 \div 8 = 80$ Estimate.

$$\begin{array}{r} 81 \text{ r4} \\ 8\overline{)652} \\ -64 \\ \hline 12 \\ -8 \\ \hline 4 \end{array}$$

Divide. (pp. 184–189)

4. $4\overline{)628}$
5. $3\overline{)250}$
6. $6\overline{)687}$
7. $7\overline{)219}$
8. $6\overline{)3,348}$
9. $8\overline{)2,867}$
10. $5\overline{)68,515}$
11. $8\overline{)436,408}$
12. $2\overline{)409,007}$
13. $9\overline{)1,389,394}$

Solve equations that use division.

Solve. $72 \div n = 8$

Think: 72 divided by what number equals 8?

$72 \div 9 = 8$

$n = 9$

Solve the equation. Then check the solution. (pp. 190–193)

14. $36 \div n = 12$
15. $n \div 5 = 30$
16. $n \div 3 = 12$
17. $22 \div n = 11$

Chapter 12

Divide by 2-digit divisors.

$625 \div 22$

$\downarrow \quad \downarrow$

$600 \div 20 = 30$ Estimate.

$$\begin{array}{r} 28 \text{ r9} \\ 22\overline{)625} \\ -44 \\ \hline 185 \\ -176 \\ \hline 9 \end{array}$$

Divide the 62 tens.

Divide the 185 ones.

Divide. (pp. 202–211)

18. $15\overline{)249}$
19. $26\overline{)13,761}$
20. $37\overline{)5,082}$
21. $41\overline{)22,506}$
22. $52\overline{)63,124}$
23. $88\overline{)91,669}$
24. $30\overline{)608,432}$
25. $63\overline{)4,863,405}$

Chapter 13

Divide decimals by whole numbers.

$$\begin{array}{r} 64.1 \\ 5\overline{)320.5} \\ -30 \\ \hline 20 \\ -20 \\ \hline 05 \\ -5 \\ \hline 0 \end{array}$$

Place the decimal point for the quotient, above the decimal point in the dividend.

Divide 32 tens by 5.

Divide 20 ones by 5.

Divide 5 tenths by 5.

Divide. (pp. 220–225)

26. $73.6 \div 8$ **27.** $42.6 \div 5$

28. $18.6 \div 4$ **29.** $53.2 \div 7$

30. $6\overline{)36.072}$ **31.** $16\overline{)134.4}$

32. $9\overline{)1{,}148.4}$ **33.** $17\overline{)209.1}$

Write fractions as decimals.

$$\frac{3}{4} \rightarrow \begin{array}{r} 0.75 \\ 4\overline{)3.00} \\ -2\,8 \\ \hline 20 \\ -20 \\ \hline 0 \end{array}$$

Divide the numerator by the denominator.

Place zeros in the dividend.
Place the decimal point. Divide.

Write as a decimal. (pp. 228–229)

34. $\frac{13}{20}$ **35.** $\frac{1}{16}$

36. $\frac{5}{8}$ **37.** $\frac{1}{20}$

38. $\frac{4}{5}$ **39.** $\frac{3}{8}$

Chapter 14

Divide by decimals.

$0.2\overline{)1.6}$
$0.2 \times 10 = 2$
$1.6 \times 10 = 16$

Multiply the divisor by a multiple of 10 to change it to a whole number. Multiply the dividend by the same number.

$$\begin{array}{r} 8. \\ 2.\overline{)16.} \\ -16 \\ \hline 0 \end{array}$$

Place the decimal point. Divide.

Divide. (pp. 236–241)

40. $0.03\overline{)0.042}$ **41.** $0.8\overline{)0.32}$

42. $0.12\overline{)0.18}$ **43.** $0.75\overline{)9.3}$

44. $0.24\overline{)144}$ **45.** $0.56\overline{)224}$

46. $0.34\overline{)272}$ **47.** $\$0.55\overline{)\$3.30}$

PROBLEM SOLVING PRACTICE

Solve. (pp. 226–227, 242–243)

48. Henry spent \$6.50 on a movie ticket, \$2.50 on popcorn, and \$1.25 for a drink. He had \$3.85 left. How much money did he have at the start?

49. Ana has 7.5 feet of rope that she would like to divide into 0.5 foot-long pieces. How many pieces of rope will she have? What operation(s) did you use to solve?

California Connections

THE GOLDEN GATE BRIDGE

When the Golden Gate Bridge was opened in 1937, it was the longest suspension bridge in the world. Today it is one of the most visited sites in California and is a symbol of San Francisco.

1. The length of one main cable suspending the bridge is 91,800 inches. How long is the cable in feet? in yards?
2. The height of each tower is 746 feet above the water and 500 feet above the roadway. What is the height in yards of each tower, above the water? above the roadway?
3. There are a total of 24 sidewalk lights around the two towers. On each tower there are 12 decorative lights above the roadway, 12 lights below the roadway, and a beacon with 2 lights. In all, how many lights are on and around each tower? Explain.
4. There are two foghorns on the bridge to help guide vessels when fog rolls in. The horns operate about 15 hours per month in March. However, during the foggy season from July through October, they sound about 150 hours per month. About how many hours per day do the horns operate in March? in September?

At night, the Golden Gate Bridge is illuminated by 234 lights.▼

CROSSING THE GOLDEN GATE BRIDGE

The Golden Gate Bridge has six lanes for vehicle traffic, and it has east and west sidewalks for walkers and cyclists.

On weekdays, when cyclists and walkers share the east sidewalk, it can become very crowded.▼

1. About 42 million vehicles cross the Golden Gate Bridge in a year. About how many vehicles cross the bridge each month? each day?

2. On a typical weekday, about 13,600 people cross the bridge as part of carpools. If each carpool has 4 people in it, about how many cars are needed for the carpools?

3. The total length of the bridge including the approaches is 8,981 feet, or 1.7 miles. If you walk at an average speed of 264 feet per minute, how many minutes will it take you to cross the bridge? Explain.

4. All vehicles traveling south across the bridge pay a $3 toll. In a recent year, $58,123,992 in tolls was collected. What was the average monthly amount collected in tolls that year?

5. **REASONING** The speed limit on the bridge is 45 miles per hour. If you drive at the speed limit, will it take you more than or less than 5 minutes to cross the 1.7-mile bridge? Explain.

CHAPTER 15 Number Theory

Scientists use the Richter scale to measure the strength of earthquakes, which vary enormously. An increase of 1 on the Richter scale indicates a tenfold (ten times as great) increase in earthquake strength. For example, an earthquake registering 3.5 has vibrations which are 10 times as strong as one registering 2.5 and 100 times as strong as one registering 1.5. How many times as strong was the earthquake in Kobe, Japan, as the one in Los Angeles, California?

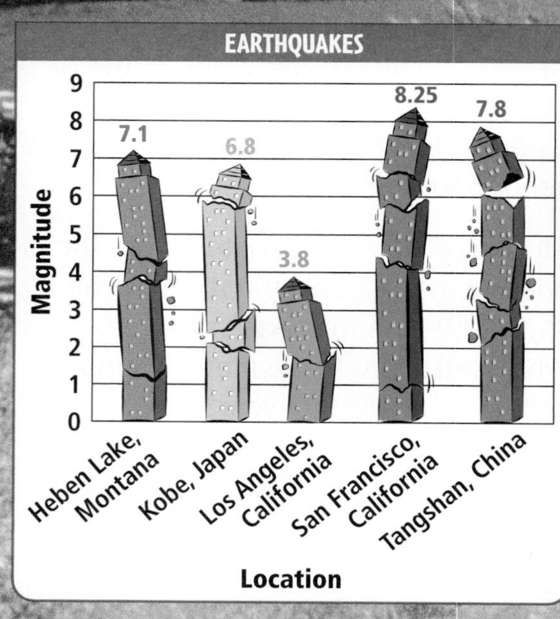

Quake Lake was formed as a result of an earthquake in Heben Lake, Montana, in 1959. You can see how the roadway disappears into the new lake.

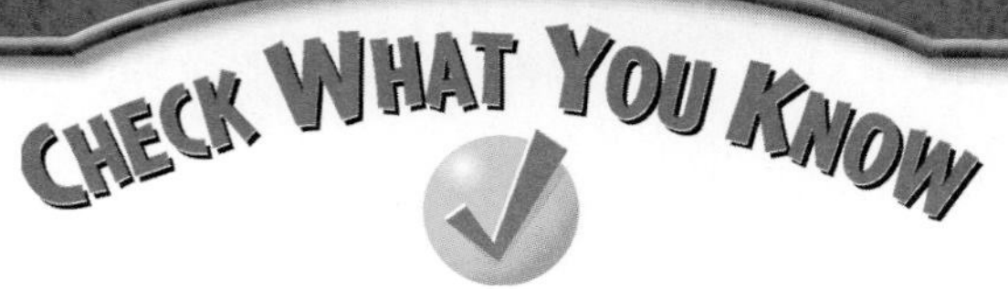

Use this page to help you review and remember important skills needed for Chapter 15.

VOCABULARY

Choose the best term from the box.

factor
Commutative Property
multiple
Property of One
Venn diagram

1. $2 \times 2 \times 3 = 2 \times 3 \times 2$ illustrates the __?__ of Multiplication.
2. $5 \times 1 = 5$ illustrates the __?__ for multiplication.
3. In $6 \times 8 = 48$, the number 6 is a __?__.
4. A __?__ is shown at the right.

WHOLE NUMBERS LESS THAN 25

÷ by 2 ÷ by 3

1 4 20 10 14 22 2 8 16 6 18 24 12 9 3 21 15 17 5 13 7 19 23 11

MULTIPLICATION AND DIVISION FACTS (See p. H18.)

Find the product.

5. 3×5	6. 5×12	7. 7×2	8. 9×1
9. 2×0	10. 4×8	11. 6×7	12. 8×9
13. 11×6	14. 9×9	15. 8×5	16. 4×4

Find the quotient.

17. $16 \div 2$	18. $21 \div 7$	19. $36 \div 6$	20. $44 \div 11$
21. $108 \div 9$	22. $144 \div 12$	23. $80 \div 10$	24. $64 \div 8$
25. $48 \div 8$	26. $72 \div 9$	27. $6 \div 6$	28. $60 \div 5$

FACTORS (See p. H18.)

Write all the factors for each number.

29. 9	30. 12	31. 18	32. 21
33. 25	34. 14	35. 10	36. 8
37. 36	38. 32	39. 20	40. 24

LESSON 1

Divisibility

Learn

BRANCH OUT Duncan and Leila are planning this year's Arbor Day celebration. They have 1,764 seedlings that they want to divide evenly among the 9 schools in the county. Will they have any seedlings left over?

If 1,764 is divisible by 9, there will not be any seedlings left over. A whole number is **divisible** by another whole number when the quotient is a whole number and there is a remainder of zero.

Some numbers have a divisibility rule. A number is divisible by 9 if the sum of the digits is divisible by 9.

$1{,}764 \rightarrow 1 + 7 + 6 + 4 = 18$ **Think:** 18 is divisible by 9, so 1,764 is divisible by 9.

So, there will not be any seedlings left over.

Look at the divisibility rules in the table.

A number is divisible by		Divisible	Not Divisible
2	if the last digit is an even number.	324	753
3	if the sum of the digits is divisible by 3.	612	7,540
4	if the last two digits form a number divisible by 4.	6,028	861
5	if the last digit is 0 or 5.	5,875	654
6	if the number is divisible by 2 and 3.	3,132	982
9	if the sum of the digits is divisible by 9.	9,423	6,032
10	if the last digit is 0.	450	7,365

MATH IDEA Knowing the rules of divisibility will help you determine if one number is divisible by another number.

Quick Review

Find the quotient. Tell if the quotient is *even* or *odd*.

1. 54 ÷ 9
2. 48 ÷ 4
3. 93 ÷ 3
4. 180 ÷ 9
5. 153 ÷ 3

VOCABULARY

divisible

Check

1. **Explain** why a number that is divisible by 10 is also divisible by 2 and by 5.

Tell if each number is divisible by 2, 3, 4, 5, 6, 9, or 10.

2. 10
3. 18
4. 30
5. 56
6. 72

TECHNOLOGY LINK

More Practice: Use E-Lab, *Divisibility.*

www.harcourtschool.com/elab2002

CALIFORNIA STANDARDS NS 2.2 Demonstrate proficiency with division, including division with positive decimals and long division with multidigit divisors. *also* **MR 2.0, MR 2.3, MR 3.3**

Practice and Problem Solving

Tell if each number is divisible by 2, 3, 4, 5, 6, 9, or 10.

7. 24 **8.** 45 **9.** 108 **10.** 130 **11.** 185

12. 308 **13.** 297 **14.** 519 **15.** 728 **16.** 2,604

17. 3,750 **18.** 3,756 **19.** 605 **20.** 1,896 **21.** 12,035

Leap years have 366 days. Every year whose number is divisible by 4 is a leap year, except years that end in two zeros. These years are leap years only when they are divisible by 400. Write *yes* or *no* to tell whether the year was a leap year.

22. 1985 **23.** 2000 **24.** 1776 **25.** 1492 **26.** 1700

Is the statement *true* or *false*? Explain.

27. All numbers that are divisible by 9 are also divisible by 3.

28. All numbers that are divisible by 3 are also divisible by 9.

29. All numbers that are divisible by 3 and 4 are also divisible by 6.

30. All even numbers are divisible by 4.

USE DATA For 31, use the table.

31. The printer wants to package the "How to Care for a Tree" brochures into stacks with none left over. Can he put them in stacks of 6? stacks of 10?

32. A total of 118 children received "What Tree Am I?" books at the celebration. Josh said the books came in packages of 9 and there were no books left over. Could he be correct? Explain.

ARBOR DAY CELEBRATION MATERIALS

Item	Number
Tree Seedlings	1,764
Brochures	504
Shovels	9

33. **NUMBER SENSE** I am a number between 50 and 70. I am divisible by 2, 3, and 9. What number am I?

34. **Write About It** What is the least number that is divisible by 2, 3, 5, and 10? Explain.

Mixed Review and Test Prep

35. 6.03 ÷ 0.9 (p. 238) **36.** 0.12 ÷ 0.03 (p. 238)

37. Write $\frac{3}{4}$ as a decimal. (p. 228)

38. Write four million, two hundred sixty thousand, four hundred twelve in standard form. (p. 4)

39. **TEST PREP** In which number is 6 in the greatest place-value position? (p. 4)

A 4,346,470 **C** 387,650
B 396,324 **D** 360,034

LESSON 2

HANDS ON

Multiples and Least Common Multiples

Quick Review

Count by

1. fives from 5 to 30.
2. threes from 3 to 18.
3. fours from 4 to 24.
4. sixes from 6 to 36.
5. eights from 8 to 40.

VOCABULARY

common multiple
least common multiple (LCM)

MATERIALS

red and yellow counters

Explore

A multiple of a whole number is the product of two or more nonzero whole numbers.

When a number is a multiple of 2 or more numbers in a set, it is a **common multiple**.

The least number that is a common multiple of two or more nonzero whole numbers is the **least common multiple**, or **LCM**.

Make a model to find the least common multiple of 3 and 5.

STEP 1

Place 3 red counters in a row. Place 5 yellow counters in a row directly below.

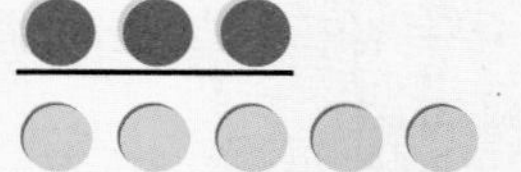

STEP 2

Continue placing groups of 3 red counters and groups of 5 yellow counters until both rows have the same number of counters. At that point, the number of counters in each row is the least common multiple, or LCM, of 3 and 5.

There are 15 counters in each row. So, the least common multiple of 3 and 5 is 15.

Try It

Use counters to find the least common multiple for each set of numbers.

a. 2, 7 **b.** 4, 5 **c.** 4, 8

d. 3, 4 **e.** 2, 3, 6 **f.** 2, 3, 9

What should you do next to find the LCM of 2 and 7?

CALIFORNIA STANDARDS MR 2.4 Express the solution clearly and logically by using the appropriate mathematical notation and terms and clear language; support solutions with evidence in both verbal and symbolic work. *also* **MR 2.0, MR 2.3**

Connect

You can also find the LCM of two or more numbers by making a list or by using a number line.

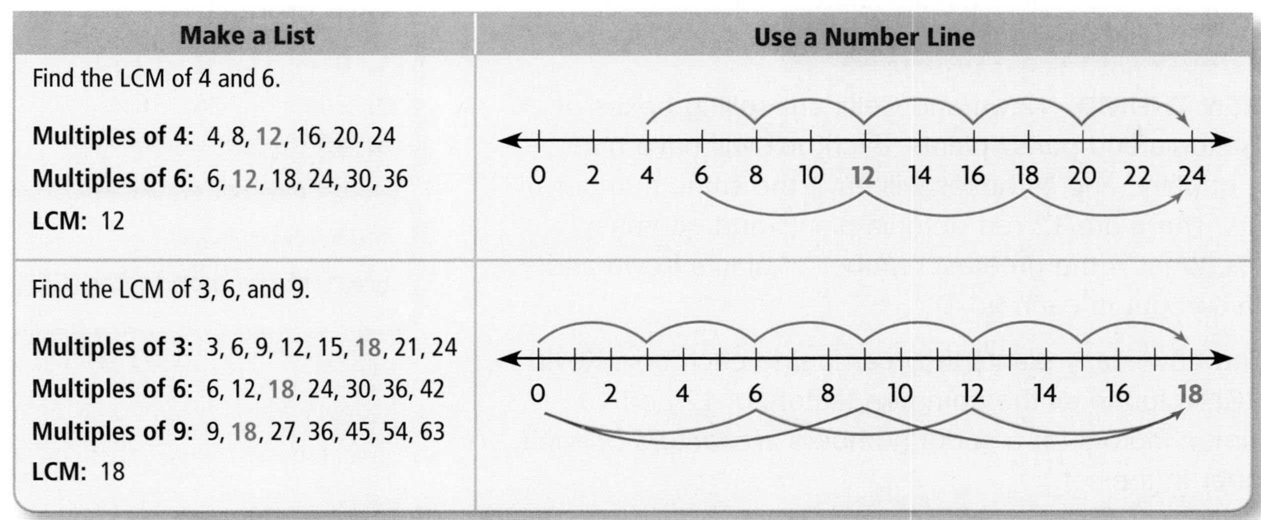

Make a List	Use a Number Line
Find the LCM of 4 and 6. **Multiples of 4:** 4, 8, 12, 16, 20, 24 **Multiples of 6:** 6, 12, 18, 24, 30, 36 **LCM:** 12	0 2 4 6 8 10 12 14 16 18 20 22 24
Find the LCM of 3, 6, and 9. **Multiples of 3:** 3, 6, 9, 12, 15, 18, 21, 24 **Multiples of 6:** 6, 12, 18, 24, 30, 36, 42 **Multiples of 9:** 9, 18, 27, 36, 45, 54, 63 **LCM:** 18	0 2 4 6 8 10 12 14 16 18

Practice

TECHNOLOGY LINK
More Practice: Use E-Lab, *Multiples and Least Common Multiples.*
www.harcourtschool.com/elab2002

List the first six multiples of each number.

1. 4 **2.** 5 **3.** 6 **4.** 8 **5.** 9

Write the least common multiple for each set of numbers.

6. 2, 3 **7.** 3, 7 **8.** 6, 9 **9.** 3, 5, 12 **10.** 2, 8, 10

USE DATA For 11–12, use the table.

PACKS OF BEADS

Type of Bead	Number per Pack
Glass	3
Wooden	5
Ceramic	8
Silver	6

11. What are the least numbers of packs of glass beads and ceramic beads you have to buy to have the same number of each type of bead?

12. What are the least numbers of packs of glass beads, wooden beads, and silver beads you have to buy to have the same number of each type of bead?

Mixed Review and Test Prep

13. 23.45×3.5 (p. 162) **14.** $3.85 \div 0.7$ (p. 238)

15. Compare. Write <, >, or =.
67.031 ● 67.103 (p. 24)

16. Solve for n. $35 = n - 21$ (p. 74)

17. TEST PREP After buying 3 markers for \$1.25 each, Sarah had \$3.45 left. How much did Sarah have before she made her purchases? (p. 226)

A \$3.75 **B** \$4.70 **C** \$7.20 **D** \$10.00

LESSON

3 Greatest Common Factor

Learn

GREEN THUMB Kevin and Celia are selling boxes of red petunia and pansy plants. Each box will have one type of plant, and all boxes will have the same number of plants. There are 12 red petunia plants and 18 pansy plants. What is the greatest number of plants Kevin and Celia can put in each box?

To find how many plants they can put in each box, Kevin and Celia found all the common factors of 12 and 18. Common factors for a set of numbers are factors of each number in the set.

Quick Review

Name two multiplication facts for each product.

1. 6 **2.** 12
3. 24 **4.** 36
5. 45

VOCABULARY

greatest common factor (GCF)

Kevin listed the factors.

Kevin

Factors of 12: 1, 2, 3, 4, 6, 12

Factors of 18: 1, 2, 3, 6, 9, 18

Common factors: 1, 2, 3, 6

Celia made a Venn diagram.

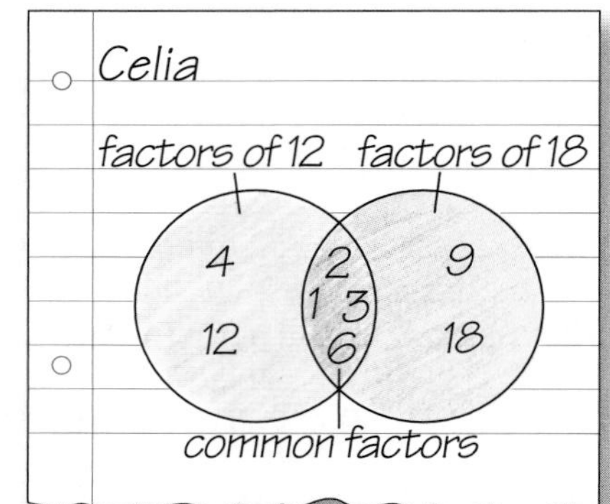

The greatest number of plants can be found by using the greatest common factor. The **greatest common factor**, or **GCF**, is the greatest number that is a factor of each of two or more numbers. The greatest common factor of 12 and 18 is 6.

So, the greatest number of plants they can put in each box is 6 plants.

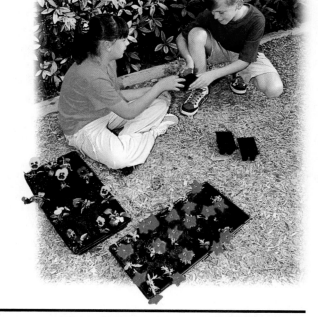

CALIFORNIA STANDARDS **MR 2.4** Express the solution clearly and logically by using the appropriate mathematical notation and terms and clear language; support solutions with evidence in both verbal and symbolic work. *also* **MR 2.0, MR 2.3, MR 3.3**

GCF of Three Numbers

What if there are 15 purple petunia plants to put into boxes? Now what is the greatest number of plants Kevin and Celia can put in each box?

Factors of 12: 1, 2, 3, 4, 6, 12
Factors of 18: 1, 2, 3, 6, 9, 18
Factors of 15: 1, 3, 5, 15

The GCF of 12, 18, and 15 is 3. So, they can put 3 plants in each box.

- **REASONING** Explain how you can tell if a number is a factor of another number.

MATH IDEA You can find the greatest common factor of two or more numbers by listing the factors and finding the greatest factor shared by all of the numbers.

Examples

A Find the GCF of 10 and 25.

10: 1, 2, 5, 10
25: 1, 5, 25

GCF: 5

B Find the GCF of 36 and 81.

36: 1, 2, 3, 4, 6, 9, 12, 18, 36
81: 1, 3, 9, 27, 81

GCF: 9

C Find the GCF of 8, 24, and 30.

8: 1, 2, 4, 8
24: 1, 2, 3, 4, 6, 8, 12, 24
30: 1, 2, 3, 5, 6, 10, 15, 30

GCF: 2

- Explain the difference between the least common multiple and the greatest common factor.

Check

1. **Explain** how divisibility rules can help you find common factors.

List the factors for each number.

2. 32 **3.** 55 **4.** 21 **5.** 17 **6.** 100

7. 37 **8.** 42 **9.** 18 **10.** 63 **11.** 77

Write the common factors for each set of numbers.

12. 6, 14 **13.** 8, 16 **14.** 30, 45 **15.** 18, 22 **16.** 5, 9, 18

17. 5, 25 **18.** 4, 24 **19.** 36, 54 **20.** 13, 19 **21.** 8, 12, 18

Write the greatest common factor for each set of numbers.

22. 4, 12 **23.** 5, 25 **24.** 3, 12, 18

LESSON CONTINUES 

▶ Practice and Problem Solving

List the factors for each number.

25. 16 **26.** 20 **27.** 23 **28.** 28 **29.** 80

30. 15 **31.** 51 **32.** 31 **33.** 49 **34.** 50

Write the common factors for each set of numbers.

35. 2, 14 **36.** 11, 15 **37.** 30, 50 **38.** 13, 26 **39.** 5, 20, 45

40. 16, 20 **41.** 12, 14 **42.** 36, 45 **43.** 34, 51 **44.** 8, 16, 20

Write the greatest common factor for each set of numbers.

45. 6, 15 **46.** 12, 21 **47.** 7, 35 **48.** 18, 45 **49.** 6, 8, 12

50. 9, 15 **51.** 12, 36 **52.** 16, 24 **53.** 42, 49 **54.** 6, 50, 60

55. 14, 21 **56.** 12, 28 **57.** 8, 32 **58.** 21, 56 **59.** 4, 8, 16

USE DATA For 60–61, use the table. Mr. Torres collected marbles. He is placing the marbles in bags to give to his grandchildren. Each bag will have only one kind of marble. All bags of the same color will have the same number of marbles.

MR. TORRES'S MARBLE COLLECTION

Type	Swirl	Helmet Patch	Patch	Slag	Sunburst
Number	18	20	12	32	30

60. Mr. Torres is placing swirl and sunburst marbles in green bags. What is the greatest number of marbles he can put in a bag?

61. Mr. Torres is placing helmet patch, patch, and slag marbles in red bags. What is the greatest number of marbles he can put in a bag?

62. **REASONING** I am thinking of two numbers. Each number is between 20 and 30. The greatest common factor of the numbers is 4. What are the numbers?

63. **What's the Question?** Eva is making gift baskets. She has 36 barrettes, 12 friendship bracelets, and 18 ribbons. All baskets must have the same number of each item. Eva made 6 gift baskets.

64. A total of 90 students were divided into 15 groups. In each group, there were 3 boys. How many girls were there altogether?

65. **REASONING** What is the greatest factor any number can have? What is the least factor any number can have?

66. **Write About It** Explain how to find the greatest common factor of three numbers.

67. Marcy has $\$10.50$ to spend on seeds from a garden catalog. Zinnia seeds cost $\$1.95$ per package. Does she have enough money to buy 6 packages of zinnia seeds? If not, how much more does she need?

68. Four friends plan to hike a 102-mile section of the Appalachian Trail in 9 days. If they walk 10 miles each day, will they walk the length of the entire 102-mile section? If not, how many more miles will they have to walk?

Mixed Review and Test Prep

69. $92{,}001 \times 9$ (p. 146)

70. $704{,}056 \times 30$ (p. 148)

71. $0.9\overline{)320.58}$ (p. 238)

72. $11\overline{)75.24}$ (p. 222)

73. If a number is divisible by 1 and 4, what other number is it also divisible by? (p. 254)

74. Order 3.652, 3.562, and 3.256 from least to greatest. (p. 24)

75. **TEST PREP** Evaluate $(n - 6) \times 2$ if $n = 14$. (p. 86)

A 2 **B** 10 **C** 14 **D** 16

76. What is the least common multiple of 9 and 7? (p. 256)

77. Write six billion, five hundred six thousand, twenty-eight in standard form. (p. 4)

78. Write an equivalent decimal for 6.03. (p. 22)

79. **TEST PREP** There are some rows with 25 seats in each row. Which expression models this situation? (p. 86)

F $25 + n$ **H** $n \times 25$
G $n \div 25$ **J** $25 \div n$

LINKUP to History

Euclid, a Greek mathematician who lived more than 2,000 years ago, described another method for finding the greatest common factor in his book *Elements*. The method is now known as the **Euclidean algorithm**. He used division to find the GCF. Follow the steps to find the GCF for 220 and 60.

Divide the greater number by the lesser number.

$60\overline{)220}$ = 3; -180; remainder 40

STEP 2

Divide the divisor, 60, by the remainder, 40.

$40\overline{)60}$ = 1; -40; remainder 20

STEP 3

Continue dividing the divisors by the remainders until the remainder is 0.

$20\overline{)40}$ = 2; -20; remainder 0

STEP 4

The divisor in the last division problem is the GCF.

Since $40 \div 20$ has a remainder of 0, the GCF of 220 and 60 is 20.

Use the Euclidean algorithm to find the GCF for each pair of numbers.

1. $16, 60$
2. $190, 36$
3. $27, 40$
4. $45, 330$
5. $90, 105$

LESSON 4

Problem Solving Skill

Identify Relationships

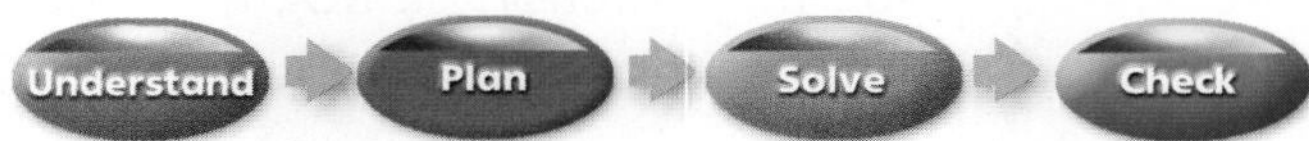

Quick Review

Find the least common multiple.

1. 3, 6
2. 7, 4
3. 12, 4
4. 10, 15
5. 4, 13

THEY'RE RELATED! When you test numbers for divisibility or find the least common multiple (LCM) or greatest common factor (GCF), you are *identifying relationships* between those numbers.

Lynn and Lynda used this table to help them find relationships between two numbers and their LCM and GCF.

First Number, *a*	Second Number, *b*	$a \times b$	GCF	LCM	GCF × LCM
2	3	6	1	6	6
4	6	24	2	12	24
5	10	50	5	10	50
7	3	21	1	21	21
9	3	27	3	9	27
60	3	180	3	60	180

Talk About It

- Look at 2 and 3 in the first row of the table. When 1 is the GCF of two numbers, what is the LCM?
- Look at 5 and 10. If the GCF of two numbers is one of the numbers, what is the LCM?
- Look at 9 and 3. If one number is divisible by the other number, what are the GCF and LCM?
- What is the relationship between the product of two numbers and the product of their GCF and LCM?

Examples

A The GCF of 3 and another number is 1. The LCM is 15. What is the other number?

Think: When 1 is the GCF of two numbers, the LCM is the product of the numbers, $a \times b$.

$3 \times b = 15$ ← 3 times what number equals 15?

$b = 5$

So, the other number is 5.

B The GCF of 4 and 20 is 4. What is the LCM of 4 and 20?

Think: When the GCF of two numbers is one of the numbers, the LCM is the other number.

So, the LCM is 20.

CALIFORNIA STANDARDS MR 1.1 Analyze problems by identifying relationships, distinguishing relevant from irrelevant information, sequencing and prioritizing information, and observing patterns. *also* **SDAP 1.1, MR 1.0, MR 2.0, MR 3.0, MR 3.2, MR 3.3**

Problem Solving Practice

Use the relationships between the given numbers to find the missing number.

1. The GCF of 7 and another number is 1. The LCM is 28. What is the other number?

2. The GCF of 3 and 21 is 3. What is the LCM of 3 and 21?

3. The LCM of 25 and 75 is 75. What is the GCF of 25 and 75?

4. The LCM of 12 and 36 is 36. What is the GCF of 12 and 36?

5. Which statement describes a relationship between 14 and 17?

   A The GCF is 2.
   B The LCM is 238.
   C Both numbers are multiples of 2.
   D Both numbers are even.

6. Which statement does *not* describe a relationship between 8 and 14?

   F The GCF is 2.
   G The LCM is 112.
   H GCF $\times$ LCM $= 112$
   J Both numbers are even.

Mixed Applications

7. Carlos divided 60 by a number and got 2.5 for an answer. Is the number greater than or less than 70? Explain.

8. What is the number of pieces of string you can get by using 16 cuts to cut apart a straight piece of string?

9. How many angles are in 3 triangles, 4 squares, and 1 circle?

10. What is the average of all the numbers from 1 to 50 that are divisible by 4?

USE DATA **For 11–12, use the table.**

11. Find the range, mean, median, and mode for the data.

12. **What if** the membership of each club doubles? What are the range, mean, median, and mode for the new set of data?

13. Dave packed oranges into boxes, 36 oranges per box. He then packed the boxes into crates, 16 boxes per crate. When he finished, he had 4 crates, 11 boxes, and 23 loose oranges. How many oranges did he have when he started?

14. **Write a problem** about a relationship between a pair of numbers. Use the GCF or LCM.

MOTHERS OF TWINS AND TRIPLETS CLUBS

Club	Number of Members
1	27
2	35
3	32
4	25
5	27
6	34

LESSON 5

HANDS ON

Prime and Composite Numbers

Quick Review

List the factors for each number.

1. 4 **2.** 8 **3.** 12
4. 36 **5.** 15

VOCABULARY
composite number
prime number

MATERIALS
square tiles

Explore

You can make arrays to find factors.

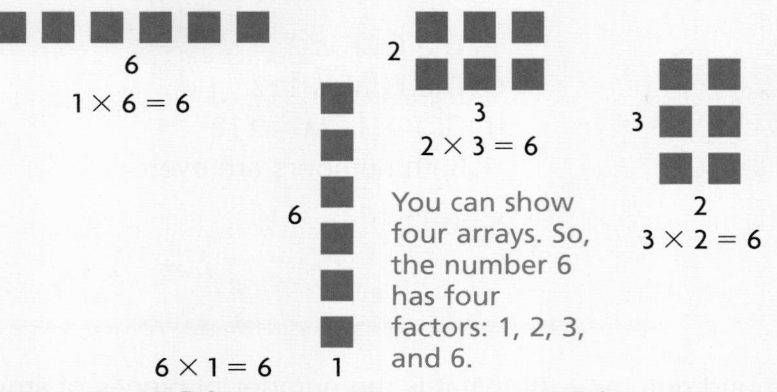

If a number has more than two factors, it is a **composite number**. So, 6 is a composite number.

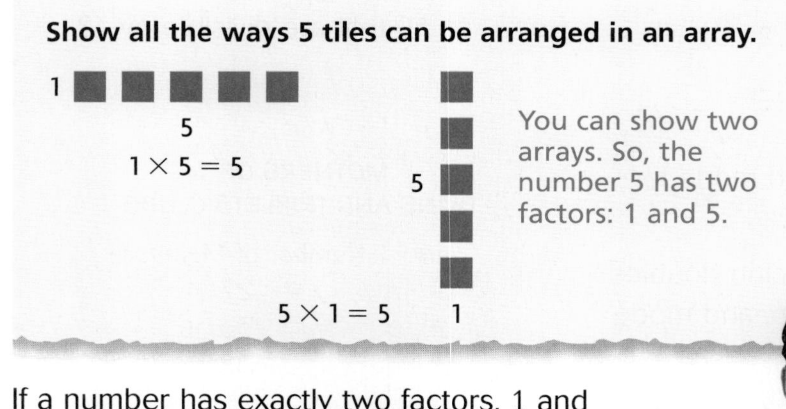

Show all the ways 5 tiles can be arranged in an array.
1
5
1 × 5 = 5
5
5 × 1 = 5
1
You can show two arrays. So, the number 5 has two factors: 1 and 5.

If a number has exactly two factors, 1 and the number itself, it is a **prime number**. So, 5 is a prime number.

How many arrays can you show for the number 15?

Try It

Use tiles to show all the arrays for each number. Write *prime* or *composite* for each number.

a. 15 **b.** 9 **c.** 13 **d.** 7

CALIFORNIA STANDARDS O━ NS 1.4 Determine the prime factors of all numbers through 50 and write the numbers as the product of their prime factors by using exponents to show multiples of a factor. *also* MR 2.3, MR 3.3

Connect

Use tiles to show all the arrays you can make for the numbers 2–12. Copy and complete the table below.

Number	Rows × Columns for Arrays	Factors	Prime or Composite?
2	1 × 2 2 × 1	1, 2	prime
3	1 × 3 3 × 1	1, 3	prime
4	1 × 4 ■ × ■ ■ × ■	■, ■, ■	?

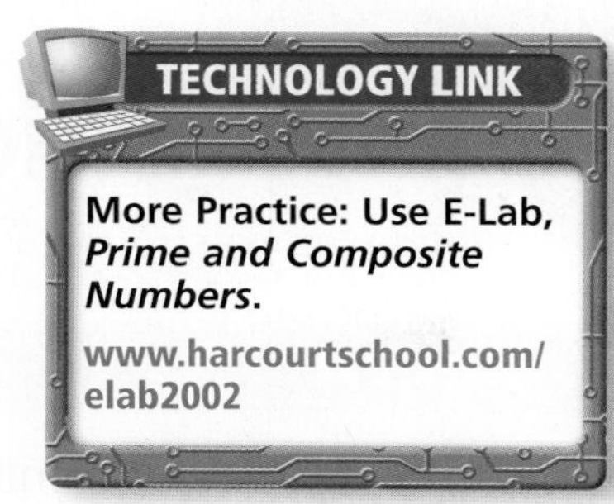

TECHNOLOGY LINK
More Practice: Use E-Lab, *Prime and Composite Numbers.*
www.harcourtschool.com/elab2002

MATH IDEA Every whole number greater than 1 is either prime or composite. The number 1 is neither prime nor composite.

- **REASONING** Explain why 2 is the only even prime number.

Practice

Use tiles to show all the arrays for each number. Write *prime* or *composite* for each number.

1. 3 **2.** 10 **3.** 11 **4.** 12 **5.** 14

Write *prime* or *composite* for each number.

6. 7 **7.** 14 **8.** 21 **9.** 25 **10.** 23

11. 41 **12.** 32 **13.** 37 **14.** 49 **15.** 19

16. List all the prime numbers from 1 to 20.

17. List all the composite factors of 72.

18. There are 110 lockers in Kelly's gym. Her locker number has three digits and is a prime number. What are the four possible numbers for her locker?

Mixed Review and Test Prep

For 19–20, identify the addition property illustrated in the equation. (p. 78)

19. $(8 + 2) + 3 = 8 + (2 + 3)$

20. $6{,}423 + 578 = 578 + 6{,}423$

21. $537{,}892 - 295{,}221$ (p. 38)

22. Write $3{,}000{,}000 + 6{,}000 + 700 + 2$ in standard form. (p. 4)

23. TEST PREP Which number is 612 NOT divisible by? (p. 254)

A 2 **B** 3 **C** 4 **D** 10

LESSON

6 Introduction to Exponents

Quick Review

1. 10×1
2. 10×10
3. $10 \times 10 \times 10$
4. $10 \times 10 \times 10 \times 10$
5. $10 \times 10 \times 10 \times 10 \times 10$

VOCABULARY

exponent
base
square number

Learn

OFF THE SCALE! Scientists measure the strength of earthquakes with the Richter scale. Each whole number on the scale represents ground vibration 10 times as great as that represented by the next lower whole number.

In an earthquake of magnitude 6, for example, the ground vibrates 10×10, or 100, times as much as in a quake of magnitude 4.

You can use an exponent to show repeated multiplication. An **exponent** shows how many times a number, called the **base**, is used as a factor.

exponent ↓

$$100 = 10 \times 10 = 10^2$$

↓ ↓ 2 factors; ↑ base

The exponent shows how many times the base is a factor.

Read as "the second power of ten," or as "ten squared."

A **square number** is the product of a number and itself. A square number can be expressed with the exponent 2.

- Explain how 10^2 relates to the area of a square with sides that are 10 units long.

▲ **The strongest earthquake in U.S. history, measuring 8.4 on the Richter scale, was near Prince William Sound in Alaska on March 27, 1964.**

Examples

A Write 1,000 in exponent form.
$1{,}000 = 10 \times 10 \times 10 = 10^3$
Read: the third power of ten, or ten cubed

B Find the value of 10^4.
Read: the fourth power of ten
$10^4 = 10 \times 10 \times 10 \times 10 = 10{,}000$

MATH IDEA Expressing numbers as powers of ten makes it easier to work with large numbers.

CALIFORNIA STANDARDS **NS 1.3** Understand and compute positive integer powers of nonnegative integers; compute examples as repeated multiplication. *also* **AF 1.2, MR 1.1, MR 2.3, MR 2.4, MR 3.3**

Check

1. **Explain** how the exponent for a power of ten is related to the number of zeros in its standard form.

Write in exponent form.

2. 100
3. 100,000
4. 10,000
5. 1,000,000
6. 10,000,000

Find the value.

7. 10^2
8. 10^4
9. 10^1
10. 10^5
11. 10^3

Practice and Problem Solving

Write in exponent form.

12. 1,000
13. 100,000,000
14. 1,000,000,000
15. 10

Find the value.

16. 10^6
17. 10^7
18. 10^8
19. 10^9

Algebra Find the value of *n*.

20. $10 \times 10 \times 10 = 10^n$
21. $10^4 = n$
22. $10 \times 10 \times n = 10^3$
23. $10 \times 10 \times 10 \times 10 = 10^n$
24. $10^5 = n$
25. $10 \times n = 10^2$

Algebra Compare. Write $<$, $>$, or $=$ for each ●.

26. 1,000 ● 10^2
27. 10^4 ● 10,000
28. $10 \times 10 \times 10$ ● 10^3
29. 10^5 ● 10^7
30. 1,000 ● 10^3
31. 10^4 ● 100,000
32. 10×10 ● 10^1
33. 10^4 ● 10^3

34. Elianne earned $10 each week for 10 weeks of baby-sitting. Ricardo earned $10 each week for 10 weeks of mowing lawns. How much money did they earn altogether?

35. After Kelly paid $26.50 for books, $3.55 for paper, and $4.36 for pens, she had $5.98. How much money did she have to start?

36. An earthquake caused $750 million in damage. Was the number of dollars in damage greater than or less than 10^8?

37. **Write About It** How would you explain to a classmate that 10^5 is the same as 100,000?

Mixed Review and Test Prep

Write *prime* or *composite* for each. (p. 264)

38. 97
39. 99
40. 91

41. Evaluate $16 + n - 5$ if $n = 20$. (p. 68)

42. **TEST PREP** Which number is divisible by 4? (p. 254)

A 367
B 673
C 706
D 736

Extra Practice page H46, Set C

LESSON 7

Evaluate Expressions with Exponents

Quick Review

1. $2 \times 2 \times 2$
2. 10^3
3. 10^1
4. $5 \times 5 \times 5$
5. $4 \times 4 \times 4$

Learn

ONE, TWO, FOUR, EIGHT . . . An amoeba is a one-celled organism that reproduces by splitting. After 6 splittings in the laboratory, there were 2^6 amoebas. How many amoebas were there?

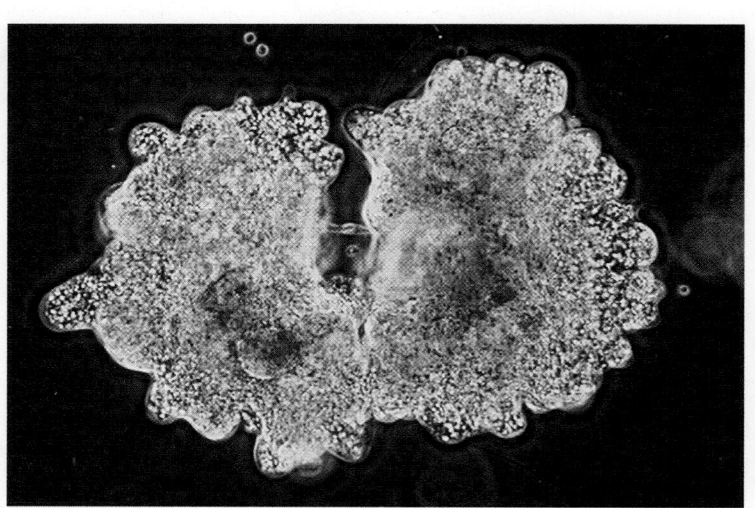

▲ **Some freshwater amoebas travel between 0.5 and 3 microns per second. A micron is 0.001 millimeter. So, the fastest time for an amoeba 1-millimeter dash is about 5 minutes.**

Exponents can be used with bases other than 10.

$$\overset{\text{exponent}}{\downarrow}$$
$$2^6 = \underbrace{2 \times 2 \times 2 \times 2 \times 2 \times 2}_{\text{6 factors}} = 64$$
$$\underset{\text{base}}{\uparrow}$$

Read 2^6 as "the sixth power of two."

So, there were 64 amoebas.

MATH IDEA You can use multiplication to find the value of a number with an exponent.

Examples

A Find the value of 7^3.

$7^3 = 7 \times 7 \times 7$
$= (7 \times 7) \times 7$
$= 49 \times 7$
$= 343$

Think: Rewrite 7^3 as a multiplication problem.

So, $7^3 = 343$.

B Find the value of 3^4.

$3^4 = 3 \times 3 \times 3 \times 3$
$= (3 \times 3) \times (3 \times 3)$
$= 9 \times 9$
$= 81$

Think: Rewrite 3^4 as a multiplication problem.

So, $3^4 = 81$.

- How does knowing that $5^2 = 25$ help you find 5^3?

CALIFORNIA STANDARDS O━ NS 1.3 Understand and compute positive integer powers of nonnegative integers; compute examples as repeated multiplication. *also* **MR 1.1, MR 2.2, MR 2.4, MR 3.0**

Powers of 2

You can fold a sheet of paper to illustrate powers of 2.

Activity

MATERIALS: paper

Fold a sheet of paper in half and record the number of layers. Continue folding the paper in half until it has been folded five times. Copy and complete the table to show the results.

Number of Folds	Number of Layers (Equal Factors)	Number of Times 2 is a Factor	Exponent Form
0	1	0	2^0
1	2	1	2^1
2	4, or 2×2	2	2^2
3	8, or $2 \times 2 \times 2$	■	■
4	■	■	■
5	■	■	■

- How many layers would you have if you folded the paper six times? Write the number in exponent form and in standard form.
- Explain how the splitting of the amoebas described on page 268 is related to the folding of the paper in this activity.

NOTE: The first power of any number equals that number.

$2^1 = 2 \qquad 9^1 = 9 \qquad 10^1 = 10$

The zero power of any number, except zero, is 1.

$2^0 = 1 \qquad 9^0 = 1 \qquad 10^0 = 1$

Check

1. **Explain** how you would rewrite 5^4 as a multiplication problem. What is the value of 5^4?

Write the equal factors.

2. 6^3
3. 8^3
4. 3^4
5. 9^2
6. 10^5

Write each expression by using an exponent.

7. $9 \times 9 \times 9$
8. $6 \times 6 \times 6 \times 6$
9. $4 \times 4 \times 4 \times 4 \times 4 \times 4$

Find the value.

10. 4^3
11. 5^2
12. 4^4
13. 7^2
14. 2^4
15. 2^7
16. 1^5
17. 3^3
18. 5^3

LESSON CONTINUES

Practice and Problem Solving

Write the equal factors.

19. 8^2 **20.** 6^4 **21.** 11^2 **22.** 15^3 **23.** 20^4

24. 2^7 **25.** 5^3 **26.** 6^2 **27.** 12^4 **28.** 25^5

Write each expression by using an exponent.

29. $7 \times 7 \times 7 \times 7 \times 7 \times 7$

30. $3 \times 3 \times 3 \times 3 \times 3 \times 3 \times 3$

31. $2 \times 2 \times 2 \times 2 \times 2 \times 2 \times 2 \times 2$

32. $15 \times 15 \times 15 \times 15 \times 15$

Find the value.

33. 3^2 **34.** 8^2 **35.** 1^4 **36.** 4^1 **37.** 4^2

38. 11^2 **39.** 1^8 **40.** 3^4 **41.** 6^3 **42.** 10^2

43. 10^4 **44.** 0^5 **45.** 7^3 **46.** 14^2 **47.** 5^4

48. 9^3 **49.** 1^{15} **50.** 8^4 **51.** 10^3 **52.** 13^3

Algebra **Find the value of *n*.**

53. $3^n = 81$ **54.** $n^5 = 32$ **55.** $10^n = 1{,}000$ **56.** $n^3 = 125$

57. $n^2 = 121$ **58.** $8^n = 64$ **59.** $10^n = 100{,}000$ **60.** $n^4 = 1$

For 61–64, match the word expression with its value.

61. four cubed

62. four squared

63. the fifth power of four

64. the sixth power of four

a. $4 \times 4 \times 4 \times 4 \times 4 = 4^5 = 1{,}024$

b. $4 \times 4 = 4^2 = 16$

c. $4 \times 4 \times 4 = 4^3 = 64$

d. $4 \times 4 \times 4 \times 4 \times 4 \times 4 = 4^6 = 4{,}096$

65. A whole number squared is 49. What is the number?

66. A number cubed is 27. What is the number?

67. Jacob has 18.75 meters of fencing that he needs to cut into 3 equal sections. How long will each section be?

68. Kay has 11,268 ants to divide equally among 12 ant farms. How many ants will live in each farm?

69. **REASONING** Maria thinks that $5^4 = 625$. How can she use division to check her work?

70. **What's the Error?** Describe the error in $9^4 = 36$. Write the correct answer.

71. Marshall is counting the number of people at the assembly. So far he has counted 8 rows, with 8 people in each row, in 8 different sections. How many people has he counted so far?

72. A bacterium cell splits into 2 every hour. How many cells will there be in 4 hours? in 7 hours?

Mixed Review and Test Prep

Write as a decimal. (p. 228)

73. $\frac{6}{8}$ **74.** $\frac{7}{14}$ **75.** $\frac{8}{10}$ **76.** $\frac{3}{8}$

77. TEST PREP Which number is divisible by 9? (p. 254)

A 5,684 C 38,654
B 22,691 D 1,096,353

78. TEST PREP What is the GCF of 12 and 15? (p. 258)

F 3 G 6 H 60 J 180

79. Write an equivalent decimal for 0.010. (p. 22)

USE DATA For 80–81, use the stem-and-leaf plot of points scored.

80. What is the range of points scored? (p. 98)

81. What is the median for the points scored? (p. 104)

Stem	Leaves
5	8
6	8 9
7	2 2 8
8	0 4 5 6 8

Thinker's Corner

PATTERNS WITH EXPONENTS You can make models of squares and cubes to help visualize numbers with exponents of 2 and 3. Look at the figures at the right. You can write the area of the square as 4^2 and the volume of the cube as 4^3.

Copy and complete each table.

1.

Model	■	■	■	■	■	■
Equal Factors	1×1	2×2	3×3	■	■	■
Exponent Form	1^2	2^2	■	■	■	■
Standard Form	1	4	■	■	■	■

2.

Model						
Equal Factors	$1 \times 1 \times 1$	$2 \times 2 \times 2$	$3 \times 3 \times 3$	■	■	■
Exponent Form	1^3	2^3	■	■	■	■
Standard Form	1	8	■	■	■	■

3. What is the exponent form for the area of a square with 50 tiles on a side?

4. How many small cubes do you need to make a large cube that is 25 cubes along each edge?

Extra Practice page H46, Set D

LESSON 8

Exponents and Prime Factors

Quick Review

Write *prime* or *composite* for each.

1. 6
2. 5
3. 8
4. 12
5. 13

VOCABULARY

prime factorization

factor tree

Learn

PRIME TIME All composite numbers can be written as the product of prime factors. This is called the **prime factorization** of the number. To find the prime factorization of a number, you can use a diagram called a **factor tree**.

To use a factor tree to find the prime factorization of 36, first choose any two factors of 36. Continue factoring until only prime factors are left. Since there is only one prime factorization of a number, you can start with any pair of factors of the number.

David and Luisa each made a factor tree.

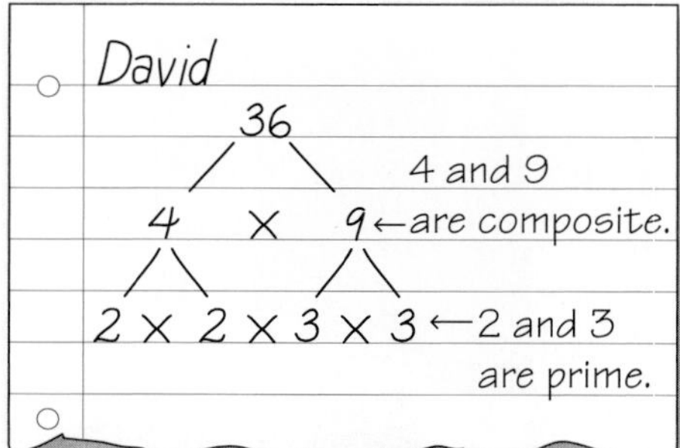

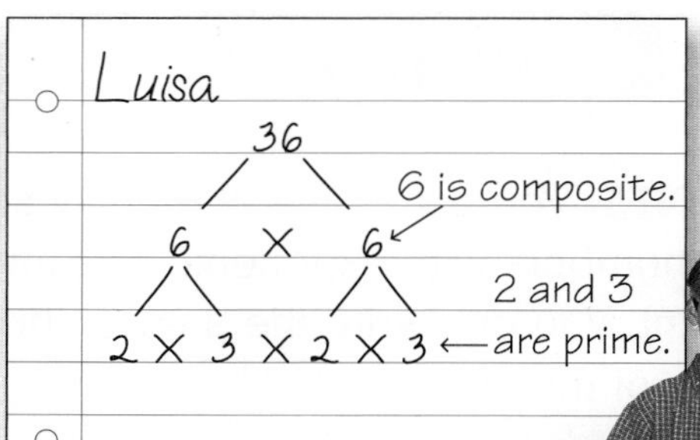

Because of the Commutative Property of Multiplication, $2 \times 2 \times 3 \times 3 = 2 \times 3 \times 2 \times 3$. Order the prime factors from least to greatest. So, the prime factorization of 36 is $2 \times 2 \times 3 \times 3$.

Examples

A What is the prime factorization of 40?

```
    40
    / \
   8 × 5
  / \
 4 × 2
 / \
2 × 2
```

So, $40 = 2 \times 2 \times 2 \times 5$.

B What is the prime factorization of 27?

```
  27
  / \
 9 × 3
/ \
3 × 3
```

So, $27 = 3 \times 3 \times 3$.

C What is the prime factorization of 630?

```
      630
      / \
  63  ×  10
  / \    / \
 9 × 7 × 2 × 5
/ \
3 × 3
```

So, $630 = 2 \times 3 \times 3 \times 5 \times 7$.

CALIFORNIA STANDARDS **NS 1.3** Understand and compute positive integer powers of nonnegative integers; compute examples as repeated multiplication. **NS 1.4** Determine the prime factors of all numbers through 50 and write the numbers as the product of their prime factors by using exponents to show multiples of a factor. *also* **AF 1.2, MR 2.0, MR 2.3, MR 2.4, MR 3.0**

Exponents in Prime Factorizations

Sometimes you can use exponents to write the prime factorization of a number.

When a prime factor is repeated in a factorization, use the prime as the base and use the number of times it is multiplied as the exponent.

$36 = \underbrace{2 \times 2} \times \underbrace{3 \times 3}$ ← 2 and 3 are each factors **two** times.

$36 = \quad 2^2 \quad \times \quad 3^2$ ← So, use 2 for the exponents.

Examples D, E, and F show how you can write the prime factorization by using exponents for the numbers in Examples A–C on page 272.

Examples

D $40 = \underbrace{2 \times 2 \times 2} \times 5$

$40 = \quad 2^3 \times 5$

E $27 = \underbrace{3 \times 3 \times 3}$

$27 = \quad 3^3$

F $630 = 2 \times \underbrace{3 \times 3} \times 5 \times 7$

$630 = 2 \times \quad 3^2 \quad \times 5 \times 7$

- **Explain** why exponents are not needed to write the prime factorization of 15.

Check

1. **Explain** why you would not use the number 1 as a factor in a factor tree.

Copy and complete.

2. $5 \times 4 = 5 \times$ ■ $\times$ ■
3. $6 \times 9 =$ ■ $\times 3 \times 3 \times 3$
4. $4 \times 4 =$ ■ $\times$ ■ $\times$ ■ $\times 2$
5. $2 \times 6 = 2 \times$ ■ $\times$ ■
6. $35 \times 2 =$ ■ $\times$ ■ $\times 2$
7. $98 = 2 \times$ ■ $\times$ ■

Rewrite the prime factorization by using exponents.

8. $5 \times 5 \times 5$
9. $2 \times 2 \times 2 \times 3$
10. $2 \times 2 \times 5 \times 5$
11. $3 \times 7 \times 7$
12. $2 \times 3 \times 5 \times 5$
13. $7 \times 7 \times 11 \times 11$

Find the prime factorization of the number. Use exponents when possible.

14. 4
15. 21
16. 24
17. 19
18. 65
19. 10
20. 16
21. 28
22. 100
23. 155

LESSON CONTINUES

Practice and Problem Solving

Copy and complete.

24. $3 \times 9 = 3 \times ■ \times ■$ **25.** $5 \times 6 = 5 \times ■ \times ■$ **26.** $8 \times 7 = ■ \times ■ \times ■ \times 7$

27. $50 = 2 \times ■ \times ■$ **28.** $6 \times 7 = ■ \times ■ \times 7$ **29.** $39 = ■ \times ■$

Rewrite the prime factorization by using exponents.

30. $11 \times 11 \times 11$ **31.** $5 \times 2 \times 2 \times 5$ **32.** $2 \times 3 \times 13 \times 13$

33. $11 \times 5 \times 5$ **34.** $2 \times 5 \times 2 \times 11$ **35.** $2 \times 3 \times 11 \times 19 \times 19$

Find the prime factorization of the number. Use exponents when possible.

36. 18 **37.** 46 **38.** 22 **39.** 35 **40.** 243

41. 38 **42.** 44 **43.** 26 **44.** 196 **45.** 225

46. 125 **47.** 294 **48.** 135 **49.** 224 **50.** 686

Algebra **Complete the prime factorization. Find the value of the variable.**

51. $2 \times 2 \times 2 = 2^n$ **52.** $2^2 \times m = 28$ **53.** $3^s \times 5 = 45$

54. $3 \times 3 \times r = 3^3$ **55.** $5^n \times 11 = 275$ **56.** $5^2 \times s = 1{,}225$

57. NUMBER SENSE What is the greatest square number that is a factor of 72 and a whole number?

58. Find the prime factorization of all the composite numbers from 4 to 50. Use exponents when possible.

59. NUMBER SENSE The prime factors of a number are the first three prime numbers. No factor is repeated. What is the number?

60. REASONING Jen says the prime factorization of 24 is $2 \times 2 \times 2 \times 3$. Ralph says it is $2 \times 3 \times 2 \times 2$. Who is right? Explain.

61. Tamika spent \$4.75, \$14.65, and \$19.95 at the mall. How much did she spend in all?

62. Bill has to wait 98 days for his vacation. How many weeks does Bill have to wait?

63. Ohio has a population of about 11 million. Is the population greater than or less than 10^7?

64. Is 10^4 greater than, less than, or equal to $2^4 \times 5^4$? Explain.

65. **What's the Error?** Tyler made this factor tree for 72. Describe and correct his error.

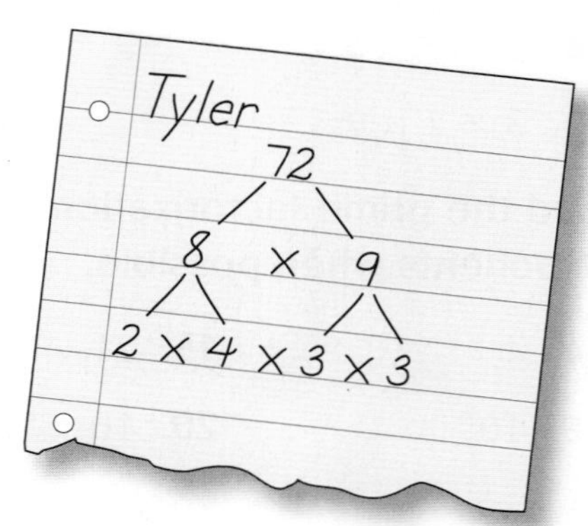

66. Josh is shipping 12 cases of seltzer water. Each case holds 24 bottles. Each bottle weighs 8 ounces. How many ounces do the bottles weigh altogether?

67. After Kelly paid \$26.50 for books, \$3.55 for paper, and \$4.36 for pens, she had \$5.98. How much money did she have to start?

Mixed Review and Test Prep

USE DATA **For 68–70, use the pictograph.**

FLOWERS IN SUE'S GARDEN

Roses	✿✿✿✿✿✿
Tulips	✿✿✿½✿
Carnations	✿✿✿✿✿✿✿✿✿½✿
Daisies	✿✿✿✿✿✿✿✿✿✿✿

Each ✿ = 10 flowers.

68. How many flowers are in Sue's garden altogether? (p. 108)

69. **TEST PREP** Sue wants to divide the roses equally among bouquets and do the same with the daisies. What is the greatest possible number of bouquets of roses and daisies without any roses or daisies left over? (p. 258)

A 5 bouquets **C** 10 bouquets
B 8 bouquets **D** 16 bouquets

70. **TEST PREP** Sue sells her carnations for \$4.50 per box. Each box has 5 flowers. How much will she get for all the carnations? (p. 152)

F \$8.55 **H** \$58.50
G \$19.00 **J** \$85.50

71. $2{,}000 \div 5 =$ ■
$200 \div 5 =$ ■
$20 \div 5 =$ ■
$2 \div 5 =$ ■
(p. 200)

72. $2{,}400 \div 8 =$ ■
$240 \div 8 =$ ■
$24 \div 8 =$ ■
$2.4 \div 8 =$ ■
(p. 200)

73. What is the least common multiple and the greatest common factor of 24 and 36? (p. 258)

74. List all the numbers from 1 to 10 by which 48 is divisible. (p. 254)

Thinker's Corner

REASONING You can use prime factorizations to find the greatest common factor of two numbers.

Find the GCF of 18 and 30.

STEP 1

Factor each number into primes.

18 = 2 × 9, 9 = 3 × 3
30 = 10 × 3, 10 = 5 × 2

STEP 2

Find the common prime factors.

18 = 2 × 3 × 3
30 = 2 × 3 × 5

Common factors: 2, 3

STEP 3

Multiply the common factors.

2 × 3 = 6

So, the GCF of 18 and 30 is 6.

Use prime factorization to find the GCF of each pair of numbers.

1. 8, 28 **2.** 32, 60 **3.** 12, 48 **4.** 16, 18 **5.** 75, 125

Review/Test

CHECK VOCABULARY AND CONCEPTS

Choose the best term from the box.

base
common multiples
composite number
exponent
prime factorization
prime number

1. The __?__ of a number is the number written as the product of prime numbers. (p. 272)
2. Multiples of one number that are also multiples of another number are called __?__. (p. 256)
3. For 10^2, 10 is the __?__ and 2 is the __?__. (p. 266)
4. A number that has more than two factors is a __?__. (p. 264)

CHECK SKILLS

Tell if each number is divisible by 2, 3, 4, 5, 6, 9, or 10. (pp. 254–255)

5. 56
6. 90
7. 264
8. 6,460

Write the least common multiple for each set of numbers. (pp. 256–257)

9. 5, 30
10. 8, 36
11. 3, 6, 20
12. 12, 4, 10

Write the greatest common factor for each set of numbers. (pp. 258–261)

13. 6, 12
14. 9, 15
15. 18, 32
16. 3, 5, 15

Write *prime* or *composite* for each number. (pp. 264–265)

17. 13
18. 21
19. 101
20. 75
21. 2

Find the value. (pp. 266–271)

22. 10^2
23. 2^7
24. 10^3
25. 6^5
26. 3^6

Find the prime factorization of the number. Use exponents when possible. (pp. 272–275)

27. 33
28. 30
29. 63
30. 42
31. 225

CHECK PROBLEM SOLVING

32. Baseball cards come in packs of 20; race car cards, packs of 40; and football cards, packs of 50. What is the least number of packs of each that will give you the same number of each kind of card? (pp. 258–261)
33. The GCF of 11 and another number is 1. The LCM is 99. What is the other number? (pp. 262–263)

Cumulative Review

Eliminate choices.
See item **3.**

First, eliminate choices with any factors that are not prime numbers. Then eliminate any choices that do not equal 42.

Also see problem **5**, p. H64.

For 1–12, choose the best answer.

1. What is the least common multiple of 6 and 8?

A 2 **C** 24
B 16 **D** 48

2. How is six and five hundred eight thousandths written in standard form?

F 0.658 **H** 6.508
G 0.6508 **J** 6,508

3. Which is the prime factorization of 42?

A 6×7 **C** $2^2 \times 7$
B $2 \times 3 \times 7$ **D** 2×3^2

4. What is the value of 2^5?

F 10 **H** 25
G 16 **J** Not here

5. On average, a human's heart beats 100,000 times a day. To the nearest thousand, how many times does it beat per hour?

A 2,000 **C** 4,000
B 3,000 **D** 5,000

6. Which number is equivalent to $\frac{1}{20}$?

F 0.5 **H** 0.06
G 0.12 **J** 0.05

7. Which is another way to write $10 \times 10 \times 10 \times 10$?

A 10×4 **C** 10^4
B 4^{10} **D** 1,000

For 8–9, use the table.

Item	Number Per Pack	Price Per Pack
Hot dogs	10	$2.75
Buns	8	$1.50

8. What are the least numbers of packs of hot dogs and buns you need to buy to have as many hot dogs as buns?

F 4 and 5 **H** 7 and 9
G 5 and 6 **J** 8 and 10

9. How much will 5 packs of hot dogs and 6 packs of buns cost together?

A $4.75 **C** $13.75
B $9.00 **D** $22.75

10. $0.3\overline{)1.23}$

F 4.1 **H** 0.041
G 0.41 **J** Not here

11. What is the value of 3^4?

A 9 **C** 27
B 12 **D** 81

12. Which number is *not* prime?

F 13 **H** 23
G 21 **J** 29

CHAPTER 16

Fraction Concepts

SLICES OF BREAD IN A LOAF

Types of Bread

White

Wheat

Rye

Raisin

Sourdough

Each [bread symbol] = 3 slices

The first loaf of bread was probably baked about 12,000 years ago. It wasn't until 1928 that Otto Frederick Rohwedder invented a machine that sliced bread and put it in a bag. Today we can buy many different kinds of sliced bread. Look at the pictograph. What fraction of each loaf of bread would you use to make 4 sandwiches?

CHECK WHAT YOU KNOW

Use this page to help you review and remember important skills needed for Chapter 16.

VOCABULARY

Choose the best term from the box.

denominator
equivalent fraction
factor
numerator

1. $\frac{1}{2}$ is an __?__ for the decimal 0.5.
2. In the fraction $\frac{3}{4}$, the 3 is the __?__ and the 4 is the __?__.
3. A number multiplied by another number to find the product is a __?__.

UNDERSTAND FRACTIONS (See p. H19.)

Write the fraction modeled.

4. 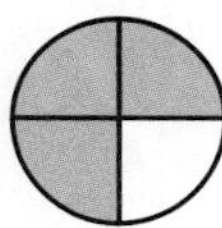

5. 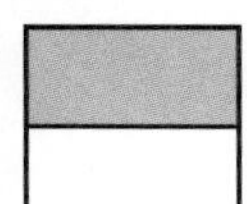

6. 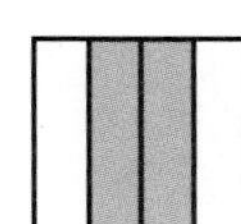

7. 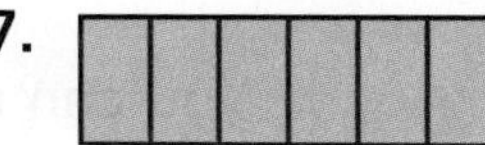

Write in words.

8. $\frac{2}{3}$

9. $\frac{4}{5}$

10. $\frac{1}{6}$

11. $\frac{3}{8}$

COMPARE FRACTIONS (See p. H19.)

Compare the fractions. Write <, >, or = for each ●.

12. 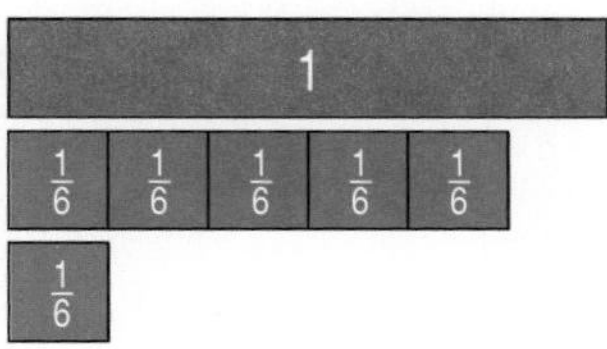

$\frac{5}{6}$ ● $\frac{1}{6}$

13. 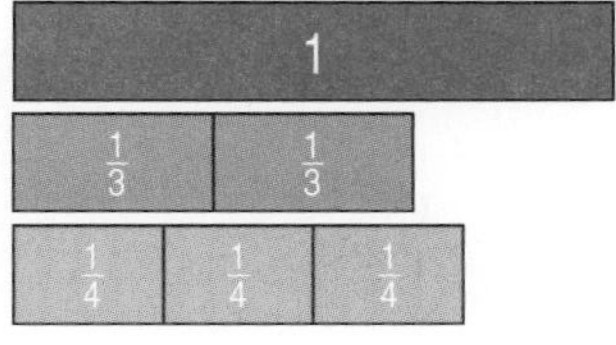

$\frac{2}{3}$ ● $\frac{3}{4}$

14. 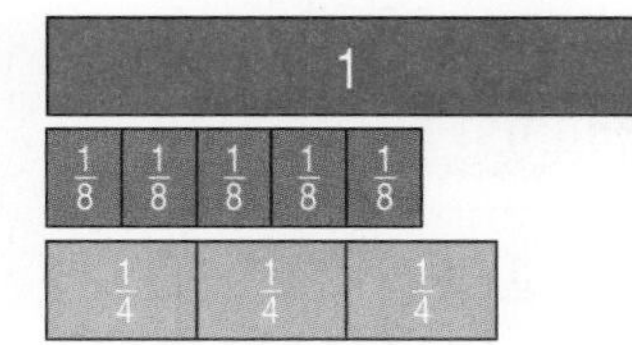

$\frac{5}{8}$ ● $\frac{3}{4}$

15. 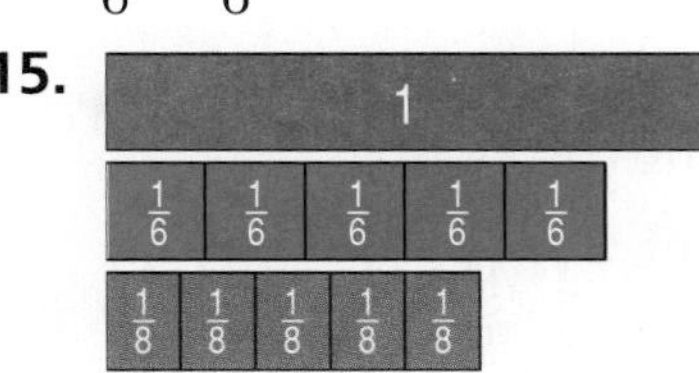

$\frac{5}{6}$ ● $\frac{5}{8}$

16. 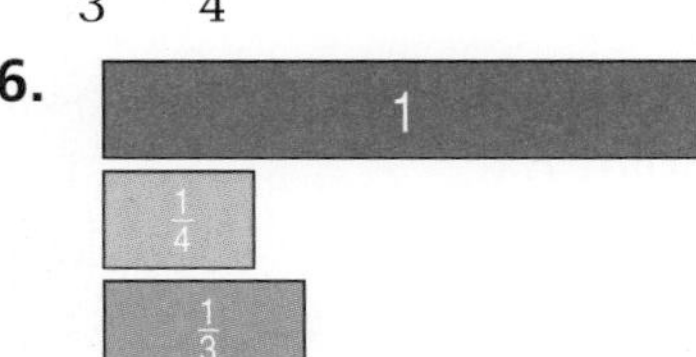

$\frac{1}{4}$ ● $\frac{1}{3}$

17. 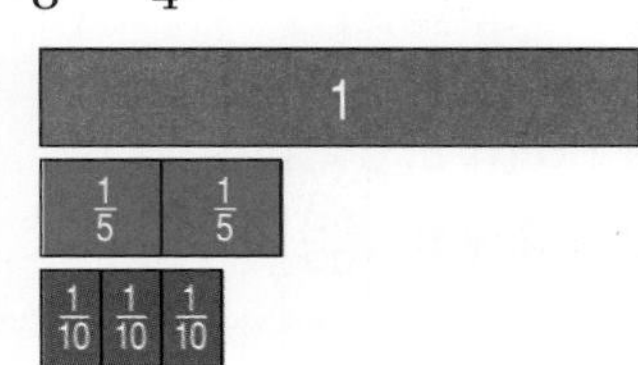

$\frac{2}{5}$ ● $\frac{3}{10}$

LESSON 1

Relate Decimals to Fractions

Quick Review

What fraction is shaded? What fraction is *not* shaded?

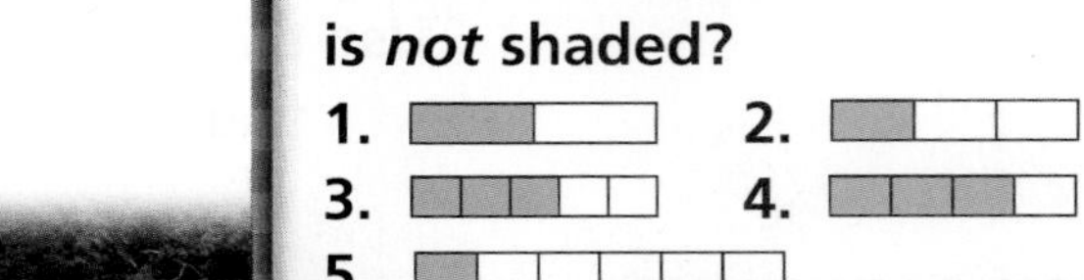

Learn

KEEP ON TRACK! During track practice, Melinda ran 0.75 of a mile and Kaspar ran $\frac{3}{4}$ of a mile. You can use a number line to show the distances they ran.

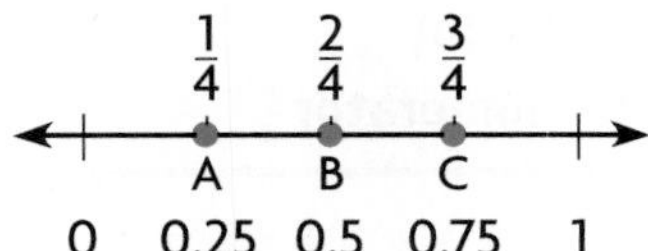

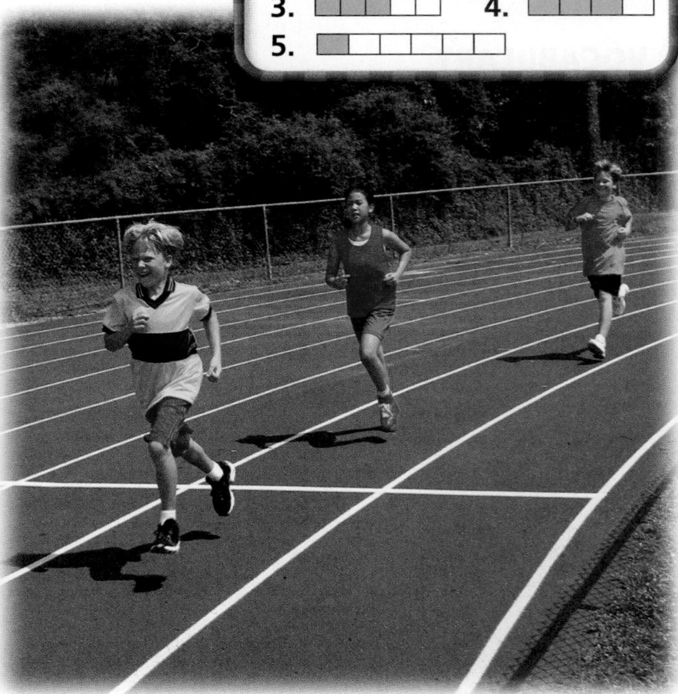

Point *C* represents both the distance Melinda ran and the distance Kaspar ran. 0.75 and $\frac{3}{4}$ represent the same number. So, $0.75 = \frac{3}{4}$.

MATH IDEA You can use a number line to represent a decimal or a fraction.

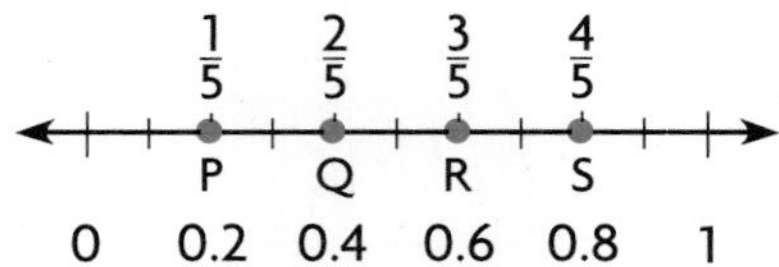

- Point *S* can be named $\frac{4}{5}$ or 0.8. Describe two ways you can name Point *Q*.

Remember

There are three ways to think of a fraction. $\frac{3}{4}$ is three fourths, three out of four, or three divided by four.

Examples

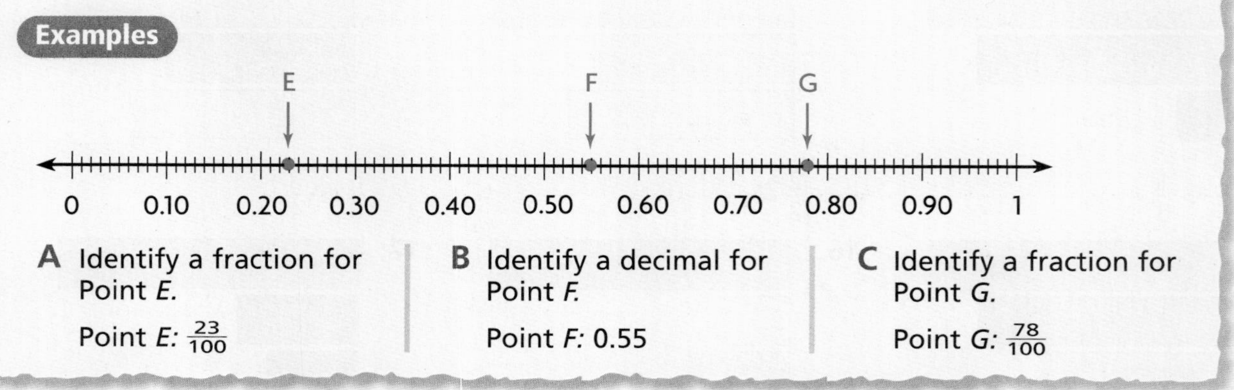

A Identify a fraction for Point *E*.

Point *E*: $\frac{23}{100}$

B Identify a decimal for Point *F*.

Point *F*: 0.55

C Identify a fraction for Point *G*.

Point *G*: $\frac{78}{100}$

- Explain where you would place a point on the line above for 0.45 and where you would place a point for 0.045.

CALIFORNIA STANDARDS **O━ NS 1.5** Identify and represent on a number line decimals, fractions, mixed numbers, and positive and negative integers. **MR 2.3** Use a variety of methods, such as words, numbers, symbols, charts, graphs, tables, diagrams, and models, to explain mathematical reasoning. *also* **NS 1.0, O━ NS 1.2, MR 2.0, MR 2.4, MR 3.0, MR 3.1, MR 3.2**

Place Value and Division

Here are other ways you can change a decimal to a fraction or a fraction to a decimal.

Decimal to Fraction

You can use place value to change a decimal to a fraction.

0.24 Identify the place value of the last digit. The 4 is in the hundredths place.

$\frac{24}{100}$ Use that place value for the denominator.

So, $0.24 = \frac{24}{100}$.

Fraction to Decimal

You can use division to change a fraction to a decimal.

$\frac{2}{5}$ Divide the numerator by the denominator.

$$\begin{array}{r} 0.4 \\ 5\overline{)2.0} \\ -2\,0 \\ \hline 0 \end{array}$$

Place the decimal point. Since 5 does not divide into 2, place a zero. Then divide as with whole numbers.

So, $\frac{2}{5} = 0.4$.

Examples

A Write 0.375 as a fraction.

0.375 The 5 is in the thousandths place.

So, $0.375 = \frac{375}{1,000}$.

B Write $\frac{3}{4}$ as a decimal.

$$\begin{array}{r} 0.75 \\ 4\overline{)3.00} \\ -2\,8 \\ \hline 20 \\ -20 \\ \hline 0 \end{array}$$

So, $\frac{3}{4} = 0.75$.

C Write $\frac{5}{8}$ as a decimal.

$$\begin{array}{r} 0.625 \\ 8\overline{)5.000} \\ -4\,8 \\ \hline 20 \\ -16 \\ \hline 40 \\ -40 \\ \hline 0 \end{array}$$

So, $\frac{5}{8} = 0.625$.

Check

1. **Describe** two ways you can change a decimal to a fraction and two ways you can change a fraction to a decimal.

Identify a decimal and a fraction for the point.

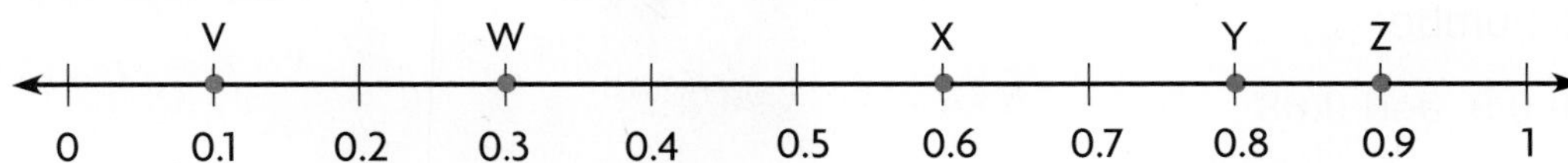

2. Point X 3. Point W 4. Point Z 5. Point V 6. Point Y

Write a fraction for each decimal.

7. 0.3 8. 0.63 9. 0.10 10. 0.425 11. 0.55

LESSON CONTINUES

Write a decimal for each fraction.

12. $\frac{1}{100}$ **13.** $\frac{3}{10}$ **14.** $\frac{1}{5}$ **15.** $\frac{2}{4}$ **16.** $\frac{3}{8}$

Practice and Problem Solving

Write the fraction and decimal for each point.

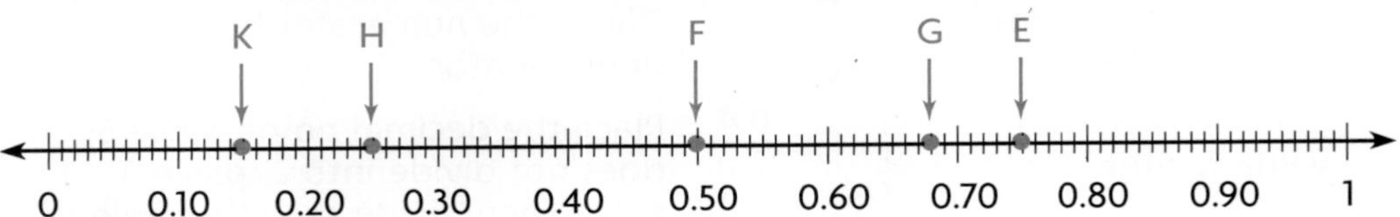

17. Point E **18.** Point F **19.** Point G **20.** Point H **21.** Point K

Write a fraction for each decimal.

22. 0.8 **23.** 0.37 **24.** 0.90 **25.** 0.125 **26.** 0.33

Write a decimal for each fraction.

27. $\frac{8}{100}$ **28.** $\frac{7}{10}$ **29.** $\frac{1}{4}$ **30.** $\frac{4}{5}$ **31.** $\frac{7}{8}$

Complete the table to show equivalent decimals and fractions.

		32.	33.	34.	35.	36.	37.	38.	39.
Decimal	0.56	0.4	■	■	0.295	■	0.8	■	0.73
Fraction	$\frac{56}{100}$	■	$\frac{1}{2}$	$\frac{3}{5}$	■	$\frac{3}{4}$	■	$\frac{35}{1,000}$	■

For 40–43, draw a number line and label points for the numbers described.

40. the number that is halfway between 0 and 1

41. the number that is halfway between 0.2 and the number that is twice its value

42. the number $\frac{7}{10}$ and the number that is 0.3 less than this number

43. the numbers 0.8, 0.9, and 0.88

44. **What's the Error?** Harvey says $\frac{1}{4}$ is equivalent to 0.4 because they both have a 4. Explain and correct the error.

45. Kelly walks her dog 0.8 of a mile every day. Write as a fraction the number of miles Kelly walks her dog.

46. Mariko hoped to sell a used book for $0.80 but accepted $\frac{3}{4}$ of a dollar for it. What is the difference in the two amounts?

47. **Write a problem** about someone who walks less than 1 mile. Express as both a fraction and a decimal.

Mixed Review and Test Prep

Find the mean for each set of data. (p. 102)

48. 23, 53, 46, 75, 48
49. 0.8, 0.4, 0.2, 0.6
50. 16, 37, 68, 84, 95
51. 53, 68, 35, 60

Find the difference. (p. 38)

52. $\begin{array}{r} 462{,}459 \\ -231{,}157 \\ \hline \end{array}$

53. $\begin{array}{r} 564{,}316 \\ -168{,}945 \\ \hline \end{array}$

Copy and complete the pattern. (p. 200)

54. $21{,}000 \div 7 = \blacksquare$
$2{,}100 \div 7 = \blacksquare$
$210 \div 7 = \blacksquare$
$21 \div 7 = \blacksquare$

55. $18{,}000 \div 600 = \blacksquare$
$18{,}000 \div 60 = \blacksquare$
$18{,}000 \div 6 = \blacksquare$
$18{,}000 \div 0.6 = \blacksquare$

Find the difference. (p. 50)

56. $3.75 - 1.29$
57. $23.35 - 1.07$

Find the quotient. (p. 204)

58. $2{,}772 \div 11$
59. $4{,}875 \div 15$
60. $7{,}196 \div 14$
61. $19{,}506 \div 6$

Solve for *n*. (p. 74)

62. TEST PREP $n - 12 = 30$

A $n = 18$
B $n = 22$
C $n = 32$
D $n = 42$

63. TEST PREP $(3 \times 10) - 2 = n$

F $n = 24$
G $n = 28$
H $n = 30$
J $n = 32$

Thinker's Corner

REASONING Think about the number of numbers that you can represent on a number line.

1. Are there numbers between 0 and 0.1? Name some if you can.
2. Are there numbers between 0 and 0.01? Name some if you can.
3. What are some numbers between 0.08 and 0.09? between 0.09 and 0.10?
4. How many numbers can be represented on a number line? Explain.

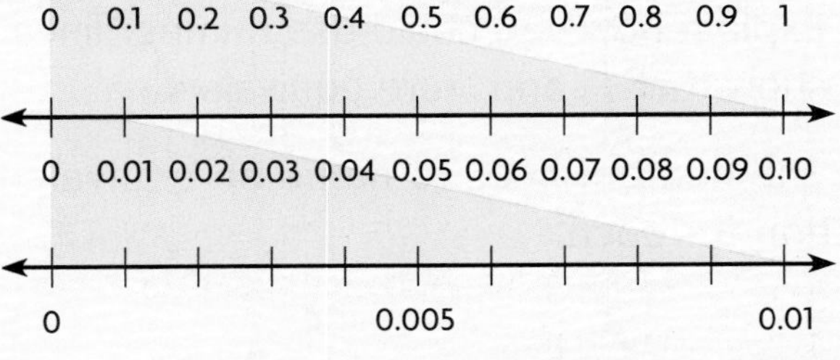

Extra Practice page H47, Set A

LESSON 2

Equivalent Fractions

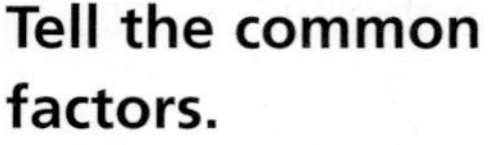

Quick Review

Tell the common factors.

1. 3 and 6 **2.** 4 and 8
3. 9 and 12 **4.** 12 and 15
5. 16 and 24

VOCABULARY

equivalent fractions

Learn

SUB DIVISION! Christina bought two sub sandwiches. She cut one into halves and the other into fourths. She noticed that one half of a sub and two fourths of a sub are the same amount.

On the number lines, the fractions $\frac{1}{2}$ and $\frac{2}{4}$ name the same amount because they are the same distance from 0.

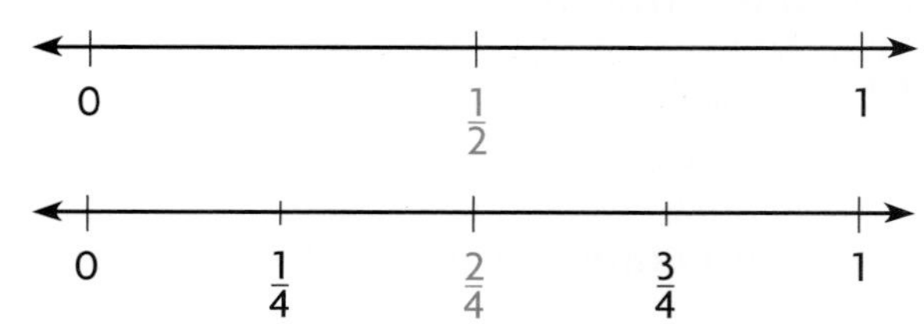

Fractions that name the same amount are called **equivalent fractions**. So, $\frac{1}{2} = \frac{2}{4}$.

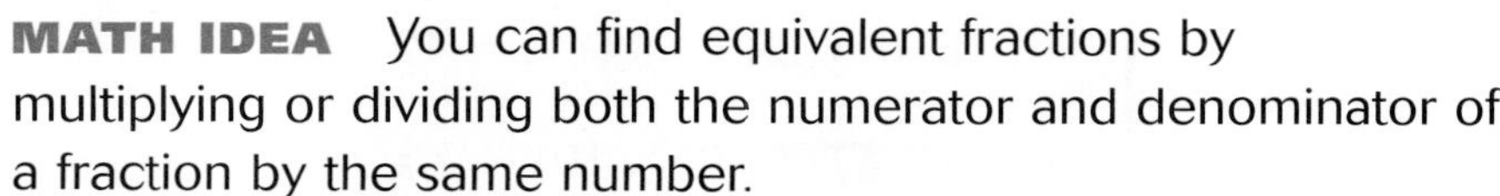

MATH IDEA You can find equivalent fractions by multiplying or dividing both the numerator and denominator of a fraction by the same number.

Multiply the numerator and the denominator by the same number.

$$\frac{4}{6} = \frac{4 \times 2}{6 \times 2} = \frac{8}{12}$$

Divide the numerator and the denominator by the same common factor.

$$\frac{6}{18} = \frac{6 \div 3}{18 \div 3} = \frac{2}{6}$$

Examples

A Find a fraction equivalent to $\frac{5}{8}$.

$$\frac{5}{8} = \frac{5 \times 3}{8 \times 3} = \frac{15}{24}$$

B Find a fraction equivalent to $\frac{8}{12}$.

$$\frac{8}{12} = \frac{8 \div 4}{12 \div 4} = \frac{2}{3}$$

Check

1. Explain how you could use number lines to show that $\frac{8}{12}$ and $\frac{2}{3}$ are equivalent.

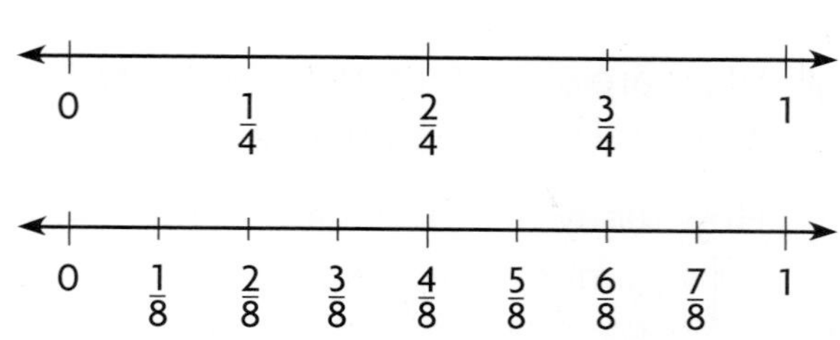

Use the number lines to name an equivalent fraction for each.

2. $\frac{3}{4}$ **3.** $\frac{2}{8}$ **4.** $\frac{2}{4}$ **5.** $\frac{6}{8}$

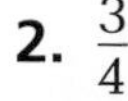

CALIFORNIA STANDARDS **NS 1.5** Identify and represent on a number line decimals, fractions, mixed numbers, and positive and negative integers. *also* **NS 1.0, MR 2.0, MR 2.3, MR 2.4, MR 3.0, MR 3.2**

Practice and Problem Solving

Use the number lines to name an equivalent fraction for each.

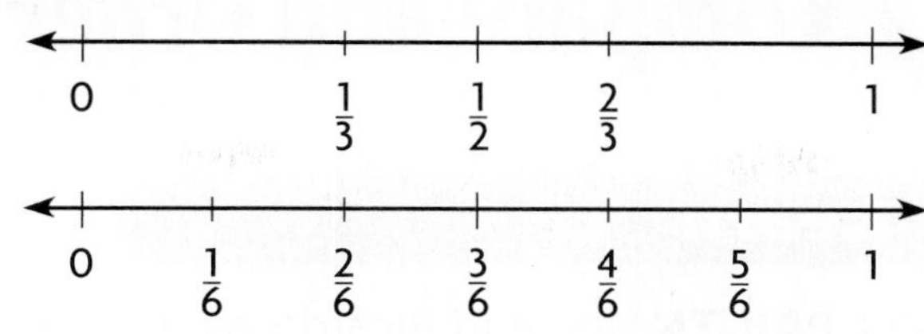

6. $\frac{1}{3}$ **7.** $\frac{2}{3}$ **8.** $\frac{3}{6}$ **9.** $\frac{2}{6}$

Write an equivalent fraction. Use multiplication or division.

10. $\frac{2}{6}$ **11.** $\frac{2}{12}$ **12.** $\frac{1}{3}$ **13.** $\frac{4}{8}$ **14.** $\frac{5}{25}$ **15.** $\frac{9}{12}$

16. $\frac{2}{9}$ **17.** $\frac{5}{6}$ **18.** $\frac{3}{15}$ **19.** $\frac{7}{21}$ **20.** $\frac{2}{7}$ **21.** $\frac{4}{11}$

Which fraction is *not* equivalent to the given fraction? Write *a*, *b*, or *c*.

22. $\frac{5}{15}$ **a.** $\frac{2}{3}$ **b.** $\frac{1}{3}$ **c.** $\frac{4}{12}$

23. $\frac{1}{2}$ **a.** $\frac{4}{8}$ **b.** $\frac{4}{7}$ **c.** $\frac{7}{14}$

24. $\frac{8}{12}$ **a.** $\frac{2}{3}$ **b.** $\frac{10}{18}$ **c.** $\frac{4}{6}$

25. $\frac{3}{4}$ **a.** $\frac{6}{8}$ **b.** $\frac{12}{16}$ **c.** $\frac{6}{12}$

Copy and complete to make a true statement.

26. $\frac{2}{7} = \frac{\blacksquare}{14}$ **27.** $\frac{7}{8} = \frac{\blacksquare}{24}$ **28.** $\frac{9}{12} = \frac{3}{\blacksquare}$ **29.** $\frac{3}{4} = \frac{6}{\blacksquare}$

30. $\frac{2}{\blacksquare} = \frac{6}{9}$ **31.** $\frac{10}{25} = \frac{\blacksquare}{5}$ **32.** $\frac{\blacksquare}{12} = \frac{1}{3}$ **33.** $\frac{1}{6} = \frac{3}{\blacksquare}$

USE DATA For 34, use the table at the right.

34. Sayre has $7.50 to spend at the bakery. She wants to buy 2 loaves of wheat bread. Does she have enough money to also buy one each of the other weekly specials? Explain.

35. Mr. Florez had 10 cans of juice. He gave 3 cans to Greg and 5 cans to Meg. Write 2 equivalent fractions to describe the fraction of the cans that are left.

36. Describe the pattern. Then write the next two numbers in the pattern.

$\frac{2}{5}, \frac{4}{10}, \frac{6}{15}, \frac{8}{20}, \blacksquare, \blacksquare, \ldots$

37. **What's the Error?** Rita used $\frac{2}{5}$ of her pet-food supply last week. She said that was the same as $\frac{4}{25}$. Describe and correct her error.

Mixed Review and Test Prep

38. 23×30 (p. 150)

39. $29{,}541 \times 96$ (p. 150)

40. $0.5\overline{)17}$ (p. 238)

41. $0.9\overline{)10.8}$ (p. 238)

42. **TEST PREP** Which is equivalent to 0.05? (p. 22)

A 5.0
B 0.5
C 0.050
D 0.005

LESSON 3

Compare and Order Fractions

Quick Review

Find the LCM for each set of numbers.

1. 3, 4 **2.** 2, 4

3. 2, 3, 6 **4.** 2, 3, 5

5. 6, 8

VOCABULARY

compare **contrast**

Learn

WET PAINT! Joe and Ricardo are painting the walls of their clubhouse. By noon Joe had painted $\frac{5}{6}$ of his wall and Ricardo had painted $\frac{3}{4}$ of his wall. Who had painted more of his wall?

One Way

You can use fraction bars to compare.

$\frac{1}{6}$	$\frac{1}{6}$	$\frac{1}{6}$	$\frac{1}{6}$	$\frac{1}{6}$
$\frac{1}{4}$	$\frac{1}{4}$	$\frac{1}{4}$		

$\frac{5}{6} > \frac{3}{4}$

So, Joe had painted more of his wall.

Remember

To compare fractions with like denominators, compare the numerators.

Since $3 > 1$, $\frac{3}{4} > \frac{1}{4}$.

Another Way

You can rename fractions with unlike denominators, such as $\frac{5}{6}$ and $\frac{3}{4}$, so they have like denominators for easy comparison.

STEP 1

Find the least common multiple, or LCM, of the denominators.

6: 6, 12, 18, 24
4: 4, 8, 12, 16

So, the LCM is 12.

STEP 2

Rename as equivalent fractions with denominators of 12.

$\frac{5 \times 2}{6 \times 2} = \frac{10}{12}$

$\frac{3 \times 3}{4 \times 3} = \frac{9}{12}$

STEP 3

Compare the numerators of the new fractions.

Since $10 > 9$, $\frac{10}{12} > \frac{9}{12}$.

So, $\frac{5}{6} > \frac{3}{4}$.

- Which pair of fractions is easier to compare, $\frac{2}{3}$ and $\frac{5}{9}$ or $\frac{2}{3}$ and $\frac{5}{7}$? Explain.

CALIFORNIA STANDARDS **NS 1.0** Students compute with very large and very small numbers, positive integers, decimals, and fractions and understand the relationship between decimals, fractions, and percents. They understand the relative magnitudes of numbers. *also* **MR 2.0, MR 2.3, MR 3.0, MR 3.2, MR 3.3**

Order Fractions

By noon Mr. Banak had painted $\frac{5}{8}$ of his wall. Order the fractions $\frac{5}{6}$, $\frac{3}{4}$, and $\frac{5}{8}$ from least to greatest to find who painted the least amount.

When you have three or more fractions to order, rename the fractions so they have like denominators. Then put them in order.

STEP 1

Find the LCM of 6, 4, and 8.

6: 6, 12, 18, 24, 30, 36
4: 4, 8, 12, 16, 20, 24, 28, 32
8: 8, 16, 24, 32, 40, 48

The LCM is 24.

STEP 2

Rename as equivalent fractions with denominators of 24.

$\frac{5 \times 4}{6 \times 4} = \frac{20}{24}$

$\frac{3 \times 6}{4 \times 6} = \frac{18}{24}$

$\frac{5 \times 3}{8 \times 3} = \frac{15}{24}$

STEP 3

Compare the numerators. Put them in order from least to greatest.

Since $15 < 18 < 20$,

$\frac{15}{24} < \frac{18}{24} < \frac{20}{24}$.

So, the order is $\frac{5}{8}$, $\frac{3}{4}$, and $\frac{5}{6}$.

So, Mr. Banak painted the least amount.

- Order the fractions above from greatest to least.

MATH IDEA To compare and order fractions, first rename the fractions so they have like denominators. Then compare the numerators.

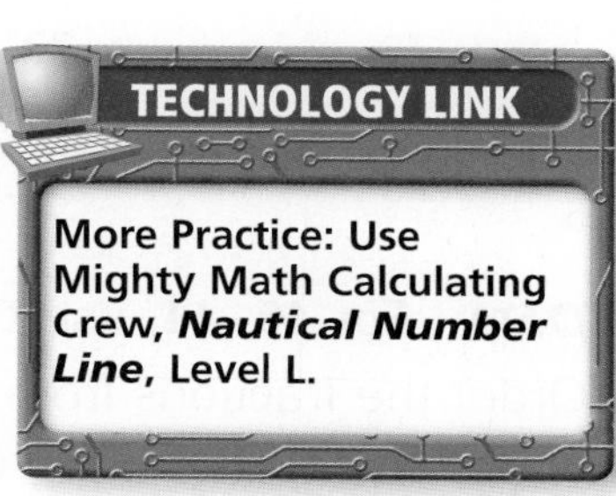

More Practice: Use Mighty Math Calculating Crew, ***Nautical Number Line***, Level L.

Check

1. **Describe** a situation when the LCM of two numbers is one of the numbers.

Compare the fractions. Write <, >, or = for each ●.

2. 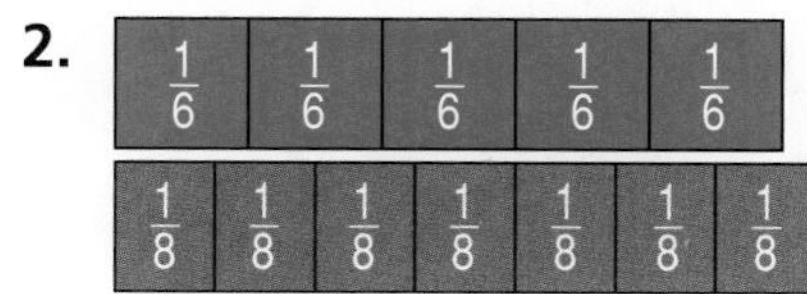

$\frac{5}{6}$ ● $\frac{7}{8}$

3. 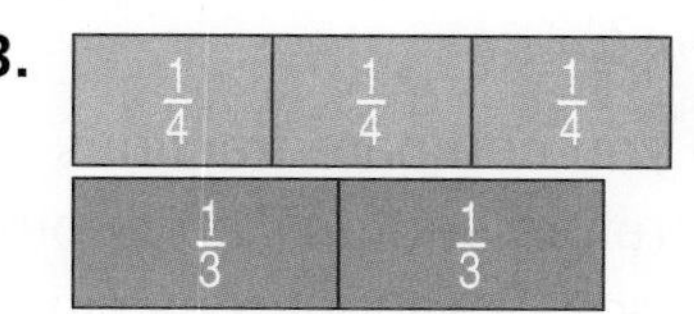

$\frac{3}{4}$ ● $\frac{2}{3}$

Rename, using the LCM, and compare. Write <, >, or = for each ●.

4. $\frac{2}{6}$ ● $\frac{1}{9}$
5. $\frac{1}{5}$ ● $\frac{1}{4}$
6. $\frac{3}{10}$ ● $\frac{2}{5}$
7. $\frac{3}{8}$ ● $\frac{1}{6}$

Rename the fractions, using the LCM as the denominator.

8. $\frac{1}{8}, \frac{3}{4}, \frac{1}{2}$
9. $\frac{7}{12}, \frac{3}{4}, \frac{7}{8}$
10. $\frac{2}{3}, \frac{2}{5}, \frac{4}{5}$
11. $\frac{5}{6}, \frac{7}{9}, \frac{2}{2}$

LESSON CONTINUES

Practice and Problem Solving

Rename, using the LCM, and compare. Write <, >, or = for each ●.

12. $\frac{3}{10}$ ● $\frac{1}{4}$
13. $\frac{2}{3}$ ● $\frac{3}{4}$
14. $\frac{4}{9}$ ● $\frac{2}{6}$
15. $\frac{1}{3}$ ● $\frac{3}{8}$

16. $\frac{4}{6}$ ● $\frac{8}{12}$
17. $\frac{7}{8}$ ● $\frac{3}{4}$
18. $\frac{2}{7}$ ● $\frac{5}{14}$
19. $\frac{1}{4}$ ● $\frac{2}{8}$

Rename the fractions, using the LCM as the denominator.

20. $\frac{1}{3}, \frac{2}{9}, \frac{1}{6}$
21. $\frac{1}{2}, \frac{3}{4}, \frac{5}{8}$
22. $\frac{1}{3}, \frac{1}{5}, \frac{3}{5}$
23. $\frac{2}{3}, \frac{5}{6}, \frac{2}{2}$

Write in order from least to greatest.

24. $\frac{1}{10}, \frac{3}{5}, \frac{1}{2}$
25. $\frac{2}{3}, \frac{3}{4}, \frac{7}{12}$
26. $\frac{9}{14}, \frac{2}{4}, \frac{5}{7}$
27. $\frac{4}{5}, \frac{7}{10}, \frac{1}{2}$

Write in order from greatest to least.

28. $\frac{2}{3}, \frac{5}{9}, \frac{1}{2}$
29. $\frac{1}{6}, \frac{1}{12}, \frac{1}{24}$
30. $\frac{6}{10}, \frac{2}{5}, \frac{10}{20}$
31. $\frac{2}{5}, \frac{3}{5}, \frac{1}{2}$

Rename either the fraction or the decimal. Then compare. Write <, >, or = for each ●.

32. $\frac{1}{2}$ ● 0.75
33. 0.25 ● $\frac{2}{8}$
34. $\frac{3}{5}$ ● 0.52
35. 0.65 ● $\frac{3}{4}$

36. $\frac{1}{5}$ ● 0.23
37. 0.75 ● $\frac{7}{10}$
38. $\frac{1}{4}$ ● 0.50
39. 0.35 ● $\frac{3}{10}$

USE DATA For 40–41, use the table.

40. Order the fractions from greatest to least. On which day were the most rooms painted? On which day were the fewest rooms painted?

41. There are 20 classrooms in all, and 9 still need to be painted. What fraction of the total is that?

CLASSROOMS PAINTED

Day	Fraction of Total
Monday	$\frac{1}{5}$
Tuesday	$\frac{1}{4}$
Wednesday	$\frac{1}{10}$

42. How can you compare fractions with *like* numerators and *unlike* denominators, such as $\frac{2}{3}$ and $\frac{2}{5}$, without renaming or using fraction strips?

43. **Write About It** In science class, Molly recorded her data in fractions. Kim recorded her data in decimals. Now they must put all the numbers in order, least to greatest. What should they do?

Mixed Review and Test Prep

Write *equivalent* or *not equivalent* to describe each pair of decimals. (p. 22)

44. 1.034 and 1.0340

45. 0.1230 and 0.023

Write an expression for each situation. (p. 68)

46. Chris ate two apples every day for ten days.

47. Maria shared fifteen cookies equally among five friends.

48. State which addition property is illustrated by this equation: $(11 + 4) + 5 = 11 + (4 + 5)$ (p. 8)

Write as a decimal. (p. 280)

49. $\frac{3}{5}$ **50.** $\frac{5}{8}$ **51.** $\frac{3}{4}$

USE DATA For 52–53, use the circle graph. (p. 108)

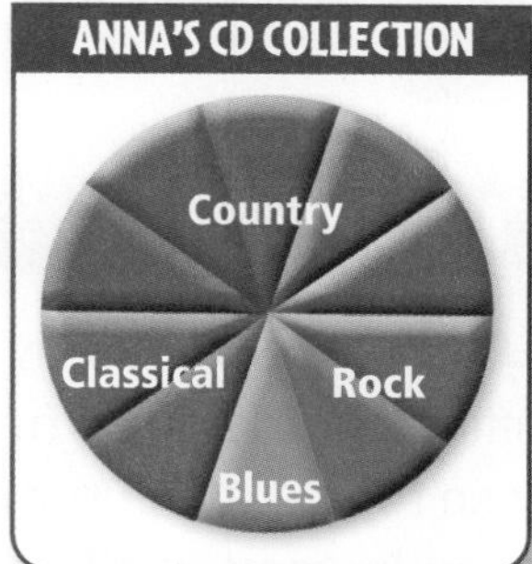

52. TEST PREP Which fraction represents the part that is country music?

A $\frac{1}{5}$ B $\frac{5}{10}$ C $\frac{5}{8}$ D $\frac{5}{5}$

53. TEST PREP Of which kind of music does Anna have the fewest CDs?

F Country H Classical
G Rock J Blues

LINKUP to Reading

STRATEGY • COMPARE AND CONTRAST You can **compare** to see how things are alike. You can also **contrast** to see how they are different. Read the following problem.

A recipe for hot chocolate calls for $\frac{1}{2}$ cup of cocoa, $\frac{1}{8}$ cup of sugar, and $\frac{3}{4}$ cup of milk. List the ingredients in order from greatest amount to least amount.

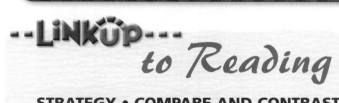

Compare	Contrast
All the ingredients are measured in cups.	The amounts are not equal.
All the amounts are fractions less than 1.	The fractions have different denominators.

Order the fractions. Think: $\frac{3}{4} > \frac{1}{2} > \frac{1}{8}$
So, the ingredients in order are milk, cocoa, and sugar.

Compare and contrast the data to solve.

1. Latoya used $\frac{1}{2}$ cup of milk for icing, $\frac{7}{8}$ cup for a cake, and $\frac{3}{4}$ cup for cocoa. For which item did she use the most milk?

2. A recipe for apple pie calls for $\frac{5}{8}$ tsp of nutmeg. Kara has $\frac{1}{2}$ tsp of nutmeg left. Does she have enough for the apple pie? Explain.

LESSON 4

Simplest Form

Learn

MUFFIN MAN! Andy increased a recipe for muffins and decided he needs $\frac{4}{8}$ cup of butter. He doesn't have a $\frac{1}{8}$-cup measure. What fraction can he use instead of $\frac{4}{8}$?

A fraction is in **simplest form** when the numerator and denominator have 1 as their only common factor.

Quick Review

Find the greatest common factor, or GCF, for each set of numbers.

1. 6, 9
2. 12, 24
3. 3, 9, 15
4. 12, 16, 20
5. 6, 12, 18, 24

VOCABULARY

simplest form

One Way

You can divide by common factors to write $\frac{4}{8}$ in simplest form.

Divide both the numerator and the denominator by a common factor of 4 and 8.

Try 2. $\frac{4 \div 2}{8 \div 2} = \frac{2}{4}$ ← not in simplest form

Try 2 again. $\frac{2 \div 2}{4 \div 2} = \frac{1}{2}$ The numerator and denominator have 1 as their only common factor.

So, $\frac{4}{8}$ in simplest form is $\frac{1}{2}$. Andy needs $\frac{1}{2}$ cup of butter.

▲ Measuring cups used for cooking are often in sets of 1, $\frac{3}{4}$, $\frac{2}{3}$, $\frac{1}{2}$, $\frac{1}{3}$, and $\frac{1}{4}$ cup.

Examples

A Write $\frac{6}{12}$ in simplest form.

$\frac{6 \div 2}{12 \div 2} = \frac{3}{6}$

$\frac{3 \div 3}{6 \div 3} = \frac{1}{2}$

So, $\frac{6}{12}$ in simplest form is $\frac{1}{2}$.

B Write $\frac{15}{18}$ in simplest form.

$\frac{15 \div 3}{18 \div 3} = \frac{5}{6}$

So, $\frac{15}{18}$ in simplest form is $\frac{5}{6}$.

C Write $\frac{30}{45}$ in simplest form.

$\frac{30 \div 3}{45 \div 3} = \frac{10}{15}$

$\frac{10 \div 5}{15 \div 5} = \frac{2}{3}$

So, $\frac{30}{45}$ in simplest form is $\frac{2}{3}$.

- How does finding simplest form compare to finding equivalent fractions? How does it contrast?

CALIFORNIA STANDARDS O━n**NS 2.3** Solve simple problems, including ones arising in concrete situations, involving the addition and subtraction of fractions and mixed numbers (like and unlike denominators of 20 or less), and express answers in the simplest form. *also* **SDAP 1.0, SDAP 1.2, MR 2.0, MR 2.3, MR 2.4, MR 3.0, MR 3.2**

Use the GCF

Another Way

You can use the greatest common factor, or GCF, to help you write a fraction in simplest form.

Write $\frac{12}{16}$ in simplest form.

STEP 1

List the factors of 12 and 16. Find the GCF.

Factors of 12: 1, 2, 3, 4, 6, 12
Factors of 16: 1, 2, 4, 8, 16
The GCF is 4.

STEP 2

Divide the numerator and the denominator of $\frac{12}{16}$ by the GCF.

$$\frac{12 \div 4}{16 \div 4} = \frac{3}{4}$$

So, $\frac{12}{16}$ in simplest form is $\frac{3}{4}$.

Examples

A Write $\frac{9}{16}$ in simplest form.

$\frac{9}{16}$ 1 is the only common factor for 9 and 16.

So, $\frac{9}{16}$ is in simplest form.

B Write $\frac{21}{42}$ in simplest form.

$\frac{21 \div 21}{42 \div 21} = \frac{1}{2}$

So, $\frac{21}{42}$ in simplest form is $\frac{1}{2}$.

C Write $\frac{18}{45}$ in simplest form.

$\frac{18 \div 9}{45 \div 9} = \frac{2}{5}$

So, $\frac{18}{45}$ in simplest form is $\frac{2}{5}$.

MATH IDEA You can divide using common factors until there is no common factor but 1, or you can divide with the GCF one time to write a fraction in simplest form.

Check

1. **Explain** how you would find the simplest form of $\frac{16}{24}$ by using both the common factor method and the GCF method.

Tell whether the fraction is in simplest form. Write *yes* or *no*.

2. $\frac{2}{6}$ 3. $\frac{1}{10}$ 4. $\frac{5}{12}$ 5. $\frac{6}{18}$ 6. $\frac{8}{20}$ 7. $\frac{25}{32}$

Name the GCF of the numerator and the denominator.

8. $\frac{2}{6}$ 9. $\frac{8}{24}$ 10. $\frac{12}{15}$ 11. $\frac{15}{45}$ 12. $\frac{18}{30}$ 13. $\frac{50}{100}$

Write each fraction in simplest form.

14. $\frac{4}{10}$ 15. $\frac{8}{14}$ 16. $\frac{8}{20}$ 17. $\frac{12}{36}$ 18. $\frac{8}{8}$ 19. $\frac{24}{32}$

LESSON CONTINUES

Practice and Problem Solving

Tell whether the fraction is in simplest form. Write *yes* or *no*.

20. $\frac{3}{8}$ **21.** $\frac{4}{10}$ **22.** $\frac{10}{32}$ **23.** $\frac{7}{15}$ **24.** $\frac{20}{45}$ **25.** $\frac{48}{50}$

Name the GCF of the numerator and the denominator.

26. $\frac{8}{22}$ **27.** $\frac{9}{30}$ **28.** $\frac{8}{12}$ **29.** $\frac{21}{33}$ **30.** $\frac{9}{54}$ **31.** $\frac{36}{60}$

Write each fraction in simplest form.

32. $\frac{6}{16}$ **33.** $\frac{14}{49}$ **34.** $\frac{40}{75}$ **35.** $\frac{24}{26}$ **36.** $\frac{15}{45}$ **37.** $\frac{8}{12}$

38. $\frac{27}{36}$ **39.** $\frac{4}{4}$ **40.** $\frac{8}{72}$ **41.** $\frac{48}{54}$ **42.** $\frac{30}{25}$ **43.** $\frac{32}{60}$

Complete.

44. $\frac{4}{16} = \frac{1}{■}$ **45.** $\frac{21}{24} = \frac{■}{8}$ **46.** $\frac{■}{36} = \frac{1}{2}$

47. $\frac{15}{18} = \frac{■}{6}$ **48.** $\frac{■}{20} = \frac{3}{5}$ **49.** $\frac{7}{■} = \frac{1}{8}$

USE DATA Use the graph for 50–53. The graph shows the number of boxes of cookies four students have sold. Each student set a goal of selling 20 boxes of each kind of cookie.

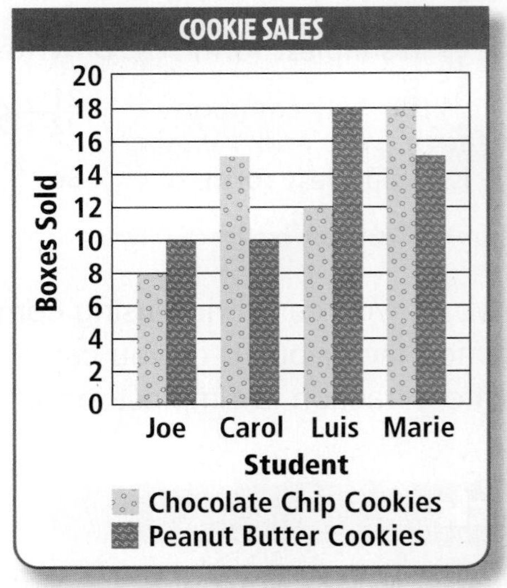

50. What fraction of Carol's goal for the sale of chocolate chip cookies has she met? Write the fraction in simplest form.

51. What fraction of Luis's goal for the sale of peanut butter cookies has he *not* met? Write the fraction in simplest form.

52. How many more boxes of cookies does Marie need to sell to meet her goal for both kinds of cookies?

53. The cookies sell for $2.50 a box. How much money have the students collected so far?

54. REASONING Colette looked at fractions in simplest form and said that any fraction that had both an odd and an even number was in simplest form. Do you agree with Colette?

55. Kyle earns $12.00 per week for baby-sitting. He saves $8.00 per week. Write the fraction, in simplest form, for the part he saves.

56. Hannah used $\frac{3}{4}$ cup of sugar, $\frac{2}{3}$ cup of cornmeal, and $\frac{1}{2}$ cup of flour for muffins. Of which ingredient did she use the greatest amount?

57. Write About It Describe the way you would find the simplest form of $\frac{32}{40}$.

Mixed Review and Test Prep

Write <, >, or = for each ●. (p. 8)

58. 5,650 ● 5,650

59. 451,300 ● 452,030

Write an equation for the problem. Use the variable n and state what n represents. (p. 72)

60. Walter has 12 pencils. There are 5 red, 3 green, and some blue. How many pencils are blue?

Find the product. (p. 150)

61. $\begin{array}{r} 12{,}118 \\ \times \quad 117 \\ \hline \end{array}$

62. $\begin{array}{r} 26{,}463 \\ \times \quad 213 \\ \hline \end{array}$

Complete each pattern. (p. 200)

63. $117 \div 13 = 9$
$117 \div 1.3 = ■$
$117 \div 0.13 = ■$

64. $32 \div 4 = 8$
$32 \div 0.4 = ■$
$32 \div 0.04 = ■$

Solve each equation. (p. 90)

65. **TEST PREP** $38 \times n = 152$

A $n = 4$ **C** $n = 190$
B $n = 114$ **D** $n = 5{,}776$

66. **TEST PREP** $24 \div n = 4$

F $n = 6$ **H** $n = 28$
G $n = 20$ **J** $n = 96$

Thinker's Corner

DATA AND GRAPHING You can show factors graphically and then use the graph to find the greatest common factor of two numbers.

MATERIALS: grid paper, ruler

Make a line graph. Show the numbers 1–20 on the horizontal axis and the factors 1–20 on the vertical axis. Use points to show all the factors of each number. The graph at the right shows the factors for 1–10. The factors for 4 are 1, 2, and 4.

1. Study your graph. Look up the vertical lines for 6 and 9. What factors have points in both lines? What is the greatest common factor?
2. Use your graph to find the greatest common factor for 8 and 20.
3. Explain how to use your graph to help you write $\frac{12}{18}$ in simplest form.

Extra Practice page H47, Set D

LESSON 5

Understand Mixed Numbers

Quick Review

1. $(3 \times 9) + 2$
2. $(2 \times 8) + 1$
3. $14 \div 3$
4. $15 \div 2$
5. $24 \div 5$

VOCABULARY

mixed number

Learn

BAGEL MANIA A baker packages bagels in boxes that each hold one dozen. Stephanie bought one full box of bagels and 5 more bagels, or $1\frac{5}{12}$ dozen bagels.

A **mixed number** is made up of a whole number and a fraction.

Read: one and five twelfths **Write:** $1\frac{5}{12}$

Use models and number lines to show mixed numbers.

three and five sixths

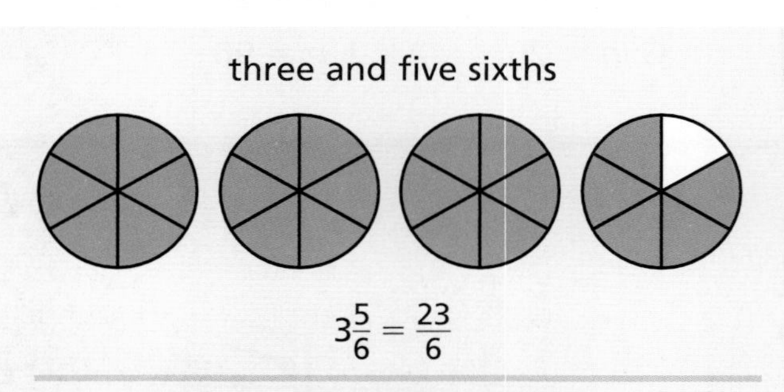

$3\frac{5}{6} = \frac{23}{6}$

two and one fourth

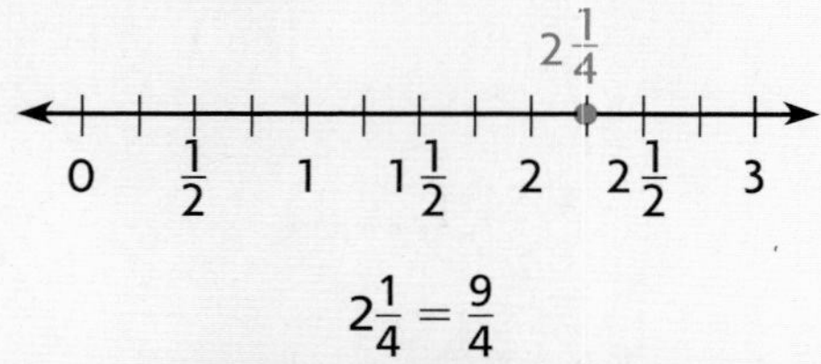

$2\frac{1}{4} = \frac{9}{4}$

MATH IDEA A fraction that is greater than 1 can be renamed as a mixed number. A mixed number can be renamed as a fraction.

Examples

A Rename $2\frac{3}{4}$ as a fraction.

$2\frac{3}{4} = \frac{2 \times 4}{1 \times 4} + \frac{3}{4}$

$= \frac{8}{4} + \frac{3}{4}$

$= \frac{11}{4}$

Write a fraction for the whole number by using the denominator, 4. Find the sum of the fractions.

B Rename $\frac{11}{4}$ as a mixed number.

$$\begin{array}{r} 2 \\ 4\overline{)11} \\ -8 \\ \hline 3 \end{array} \rightarrow 2\frac{3}{4}$$

Divide the numerator by the denominator. Write the remainder as a fraction.

A fraction greater than 1 is sometimes called an *improper fraction*.

CALIFORNIA STANDARDS O⚿NS 1.5 Identify and represent on a number line decimals, fractions, mixed numbers, and positive and negative integers. *also* **MR 2.0, MR 2.3, MR 2.4, MR 3.0, MR 3.2**

Check

1. **Explain** how you can tell whether a fraction is greater than or less than 1.

Write a mixed number and a fraction for each.

2. four and three fourths

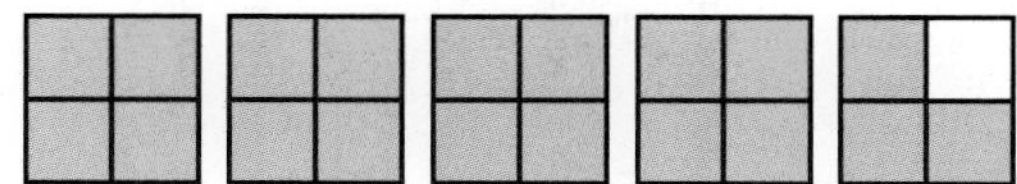

3. one and five eighths

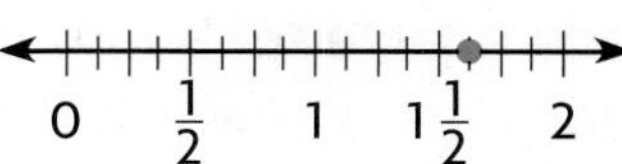

Write each mixed number as a fraction. Write each fraction as a mixed number.

4. $4\frac{1}{4}$ 5. $3\frac{2}{5}$ 6. $2\frac{3}{7}$ 7. $\frac{9}{8}$ 8. $\frac{31}{6}$ 9. $\frac{7}{3}$

Practice and Problem Solving

Write a mixed number and a fraction for each.

10. three and two thirds

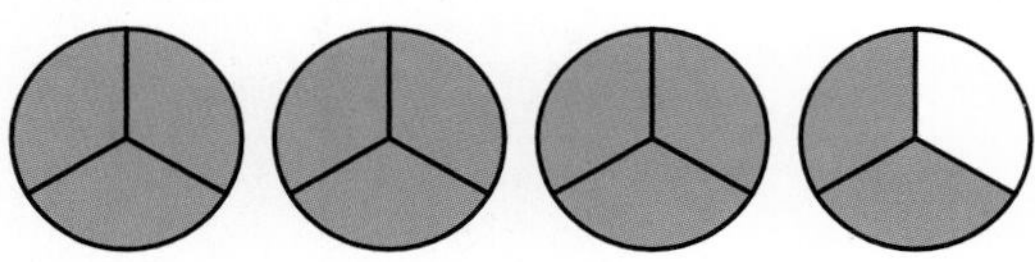

11. two and one fifth

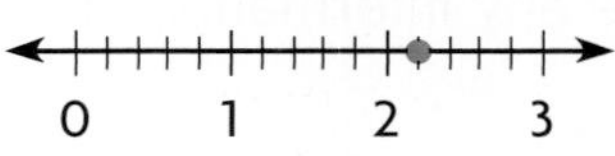

Write each mixed number as a fraction.

12. $4\frac{1}{3}$ 13. $6\frac{3}{5}$ 14. $2\frac{7}{11}$ 15. $9\frac{2}{9}$ 16. $1\frac{11}{12}$ 17. $5\frac{3}{4}$

Write each fraction as a mixed number.

18. $\frac{5}{2}$ 19. $\frac{10}{3}$ 20. $\frac{11}{6}$ 21. $\frac{15}{4}$ 22. $\frac{11}{8}$ 23. $\frac{11}{5}$

24. Dan used $2\frac{1}{2}$ cups of flour to make bread. Write as a fraction the number of cups he used.

25. Who wrote a fraction less than 1?

Chris	Ashley	Danny	Jimet
$\frac{13}{8}$	$\frac{16}{8}$	$\frac{8}{8}$	$\frac{7}{8}$

26. **REASONING** Explain how to write 2.25 as a mixed number.

Mixed Review and Test Prep

27. $4.61 - 2.49$ (p. 50)

28. $3.16 - 2.78$ (p. 50)

Solve each equation for *n*. (p. 88)

29. $3n = 180$

30. $42 = 6n$

31. **TEST PREP** Write an expression using the variable *n* to show that Jan's sister is 3 years younger than Jan. (p. 68)

A $3n$
B $n - 3$
C $n + 3$
D $n + n$

LESSON 6

Problem Solving Strategy
Make a Model

Quick Review

Compare. Write <, >, or = for each ●.

1. $\frac{4}{4}$ ● $\frac{3}{4}$
2. $\frac{2}{4}$ ● $\frac{1}{2}$
3. $\frac{1}{2}$ ● $\frac{1}{3}$
4. $\frac{2}{3}$ ● $\frac{3}{4}$
5. $\frac{1}{4}$ ● $\frac{2}{9}$

PROBLEM Jill made a table to record the number of miles she walked each day. On which day did Jill walk the greatest distance? On which day did she walk the least distance?

Day	Mon	Tue	Wed	Thu
Miles Walked	$2\frac{1}{2}$	$3\frac{1}{4}$	$2\frac{3}{4}$	$2\frac{3}{8}$

- What are you asked to find?
- What information will you use?
- Is there any information you will not use?

- What strategy can you use to solve the problem?

You can *make a model* with fraction bars.

- How can you use the strategy to solve the problem?

You can use fraction bars to help you compare the numbers.

First, look at the whole numbers: $2\frac{1}{2}$, $3\frac{1}{4}$, $2\frac{3}{4}$, $2\frac{3}{8}$. Since 3 is the greatest whole number, $3\frac{1}{4}$ miles is the greatest distance.

Now use fraction bars to compare the fractional parts of $2\frac{1}{2}$, $2\frac{3}{4}$, and $2\frac{3}{8}$. Since $\frac{3}{8}$ is the shortest, $2\frac{3}{8}$ miles is the least distance.

$\frac{1}{2}$ | $\frac{1}{2}$

$\frac{3}{4}$ | $\frac{1}{4}$ | $\frac{1}{4}$ | $\frac{1}{4}$

$\frac{3}{8}$ | $\frac{1}{8}$ | $\frac{1}{8}$ | $\frac{1}{8}$

So, Jill walked the greatest distance on Tuesday and the least distance on Thursday.

- What other strategy can you use?

CALIFORNIA STANDARDS MR 2.0 Students use strategies, skills, and concepts in finding solutions. **MR 2.3** Use a variety of methods, such as words, numbers, symbols, charts, graphs, tables, diagrams, and models, to explain mathematical reasoning. *also* **NS 1.0, MR 1.0, MR 1.1, MR 2.4, MR 3.2**

Problem Solving Practice

Make a model to solve.

1. **What if** Jill had walked only $2\frac{1}{4}$ miles on Tuesday? How would this change the answers?
2. The five planets closest to the sun are Earth, Venus, Mars, Mercury, and Jupiter. Earth is between Venus and Mars. Mercury is between Venus and the sun. Which of these planets is next to Jupiter?

PROBLEM SOLVING STRATEGIES

- Draw a Diagram or Picture
- Make a Model or Act It Out
- Make an Organized List
- Find a Pattern
- Make a Table or Graph
- Predict and Test
- Work Backward
- Solve a Simpler Problem
- Write an Equation
- Use Logical Reasoning

USE DATA **Use the table for 3 and 4.**

MONTHLY RAINFALL				
Month	Jan	Feb	Mar	Apr
Rainfall (in inches)	$3\frac{3}{8}$	$2\frac{2}{4}$	$3\frac{1}{2}$	$5\frac{7}{8}$

3. How do you know that April had the most rainfall?
   - A The greatest rainfall is always in April.
   - B April has the greatest whole-number part.
   - C April has the greatest fractional part.
   - D April is the last month listed.
4. What is the order of the months from greatest rainfall to least rainfall?
   - F Apr, Mar, Feb, Jan
   - G Jan, Mar, Feb, Apr
   - H Apr, Mar, Jan, Feb
   - J Apr, Jan, Mar, Feb

Mixed Strategy Practice

5. **GEOMETRY** Alex is enclosing his pentagon-shaped garden with fencing. If each side is 8 meters long, how much fencing does he need?
6. If 2 people can be seated on each side of a square table, how many people can be seated at 12 square tables that are pushed together end to end to form a rectangle?
7. Aiko had $20.00 to buy candles. She returned 2 candles for which she had paid $4.75 each. Then she bought 3 candles for $3.50 each and 1 candle for $5.00. How much money did Aiko have then?
8. In Ted's class, students were asked to name their favorite sport. Football was the response of $\frac{1}{8}$ of them. If 3 students said football, how many students are in Ted's class?

Review/Test

CHECK VOCABULARY AND CONCEPTS

Choose the best term from the box.

equivalent fractions
fraction
mixed number
simplest form

1. Fractions that name the same amount are called _?_. (p. 284)
2. A fraction is in _?_ when the greatest common factor of the numerator and denominator is 1. (p. 290)
3. A number, like $2\frac{1}{2}$, that has a whole-number part and a fraction part is a(n) _?_. (p. 294)

CHECK SKILLS

Write a fraction for each decimal. Write a decimal for each fraction. (pp. 280–283)

4. 0.4
5. $\frac{3}{5}$
6. 0.625
7. $\frac{3}{4}$
8. 0.65

Write an equivalent fraction. (pp. 284–285)

9. $\frac{1}{7}$
10. $\frac{3}{9}$
11. $\frac{6}{42}$
12. $\frac{2}{9}$
13. $\frac{21}{49}$

Rename the fractions, using the LCM, and compare. Write <, >, or = for each ●. (pp. 286–289)

14. $\frac{2}{4}$ ● $\frac{5}{8}$
15. $\frac{3}{4}$ ● $\frac{6}{12}$
16. $\frac{7}{28}$ ● $\frac{4}{7}$
17. $\frac{7}{8}$ ● $\frac{14}{16}$

Write in order from greatest to least. (pp. 286–289)

18. $\frac{1}{6}, \frac{15}{18}, \frac{2}{3}$
19. $\frac{3}{6}, \frac{1}{3}, \frac{8}{12}$
20. $\frac{1}{4}, \frac{1}{2}, \frac{1}{3}$
21. $\frac{2}{5}, \frac{1}{2}, \frac{3}{4}$

Write each fraction in simplest form. (pp. 290–293)

22. $\frac{4}{16}$
23. $\frac{8}{12}$
24. $\frac{20}{45}$
25. $\frac{16}{56}$
26. $\frac{63}{81}$

Write each fraction as a mixed number. Write each mixed number as a fraction. (pp. 294–295)

27. $\frac{9}{4}$
28. $\frac{17}{8}$
29. $\frac{13}{4}$
30. $2\frac{1}{5}$
31. $3\frac{3}{8}$

CHECK PROBLEM SOLVING

Make a model to solve. (pp. 296–297)

32. Dave's practice swims were $1\frac{3}{8}$ miles, $1\frac{2}{6}$ miles, and $1\frac{3}{4}$ miles. Which was the greatest distance?
33. Rosa bought $1\frac{2}{5}$ lb of chocolate, $2\frac{1}{2}$ lb of fudge, and $1\frac{5}{8}$ lb of caramels. Which was the least amount?

Cumulative Review

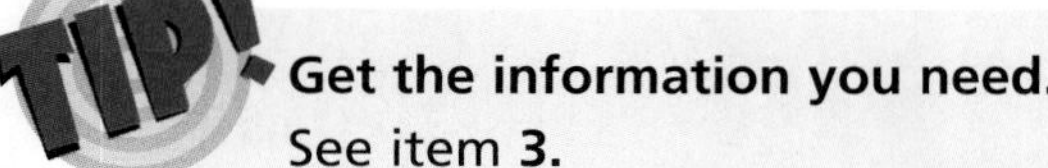

Get the information you need.
See item **3**.
Think about where each of the fractions would be on a number line. Then compare the fractions to determine the symbol needed.

Also see problem **3**, p. H63.

For 1–12, choose the best answer.

1. Which number is equivalent to $\frac{4}{25}$?

   A 1.6 **C** 0.016
   B 0.16 **D** 0.0106

2. How is $5\frac{2}{3}$ written as a fraction?

   F $\frac{17}{3}$ **H** $\frac{15}{3}$
   G $\frac{17}{5}$ **J** $\frac{10}{3}$

3. Which symbol makes this a true number sentence? $\frac{3}{5}$ ● $\frac{5}{10}$

   A $>$ **C** $=$
   B $<$ **D** $+$

4. What is the value of 4^3?

   F 12 **H** 32
   G 16 **J** 64

5. Which of these fractions is in simplest form?

   A $\frac{2}{4}$ **C** $\frac{7}{15}$
   B $\frac{9}{12}$ **D** $\frac{5}{20}$

6. Which fractions are in order from *greatest* to *least?*

   F $\frac{1}{2}, \frac{3}{4}, \frac{2}{3}$ **H** $\frac{3}{4}, \frac{1}{2}, \frac{2}{3}$
   G $\frac{3}{4}, \frac{2}{3}, \frac{1}{2}$ **J** $\frac{2}{3}, \frac{1}{2}, \frac{3}{4}$

7. 3.92×4.1

   A 16 **C** 16.072
   B 16.7 **D** Not here

8. Which fraction is equivalent to $\frac{3}{9}$?

   F $\frac{1}{9}$ **H** $\frac{6}{12}$
   G $\frac{12}{36}$ **J** $\frac{4}{21}$

9. In a recent survey of favorite snacks $\frac{2}{5}$ of the people surveyed chose chips. Which number is equivalent to that amount?

   A 0.2 **C** 0.4
   B 0.25 **D** 0.5

10. In the same survey, 5 out of 100 of the people surveyed chose nuts. What fraction names that amount?

    F $\frac{1}{20}$ **H** $\frac{1}{100}$
    G $\frac{1}{25}$ **J** $\frac{1}{50}$

11. How is $\frac{25}{4}$ written as a mixed number?

    A $6\frac{3}{4}$ **C** 6.4
    B $6\frac{1}{2}$ **D** $6\frac{1}{4}$

12. If a car travels at an average of 54 miles per hour, how far will it go in 9 hours at that rate?

    F 6 miles **H** 486 miles
    G 456 miles **J** Not here

Ratio

HOCKEY GAME RESULTS

Team	Goals	Team	Goals
Badgers	2	Pelicans	0
Ducks	4	Eagles	2
Bears	5	Penguins	1
Sharks	4	Otters	3

Hockey players use a special stick to drive a puck forward at speeds that can exceed 100 miles per hour. The table lists the results of four ice hockey games. What is the ratio of the number of goals the Sharks made to the number of goals the Otters made?

Use this page to help you review and remember important skills needed for Chapter 17.

VOCABULARY

Choose the best term from the box.

compare
parts
simplify
whole

1. You are able to __?__ a fraction when the numerator and denominator have a common factor.
2. The numerator of a fraction names the number of equal __?__ of the whole being considered.
3. The total number of equal parts in the __?__ is shown in the denominator of a fraction.

WRITE FRACTIONS (See p. H20.)

Write a fraction for the following.

4. 

What fraction of the triangles are red?

5. 

What fraction of the circles are blue?

6. 

What fraction of the squares are *not* red?

7. 

What fraction of the circles are red?

8. 

What fraction of the squares are red?

9. 

What fraction of the triangles are either red or blue?

Write each fraction in simplest form.

10. $\frac{5}{15}$
11. $\frac{4}{20}$
12. $\frac{14}{63}$
13. $\frac{6}{36}$
14. $\frac{10}{14}$
15. $\frac{8}{12}$
16. $\frac{7}{7}$
17. $\frac{25}{45}$
18. $\frac{11}{44}$
19. $\frac{6}{18}$

Write an equivalent fraction for each.

20. $\frac{3}{6}$
21. $\frac{2}{5}$
22. $\frac{50}{100}$
23. $\frac{6}{16}$
24. $\frac{8}{15}$

LESSON

HANDS ON

Understand Ratios

Quick Review

Write a fraction for the shaded part.

1. 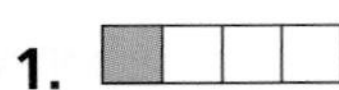

2. 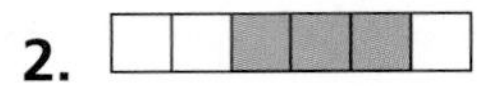

3. 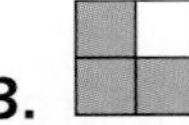

4. 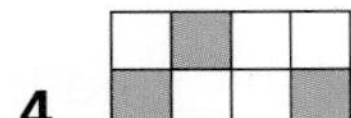 **5.** 

VOCABULARY
ratio

MATERIALS
two-color counters

Explore

A **ratio** is a comparison of two quantities. For example, when you compare the number of red counters to the number of yellow counters, the ratio is 2 to 3.

In a group of 10 students, 4 are boys and 6 are girls. Find the ratio of the number of boys to the number of girls.

STEP 1

Use the yellow side to represent all 10 students in the group. Each counter represents one student.

STEP 2

Turn the red side up to represent the number of students who are boys.

The ratio of boys to girls is the same as the ratio of red counters to yellow counters, 4 to 6.

Try It

Use two-color counters to represent 4 baseballs and 3 basketballs. Write each ratio.

a. baseballs to basketballs

b. basketballs out of all balls

c. baseballs to all balls

How many counters do you turn red side up to represent the ratio 4 baseballs to 3 basketballs?

CALIFORNIA STANDARDS **MR 2.0** Students use strategies, skills, and concepts in finding solutions. **MR 1.1** Analyze problems by identifying relationships, distinguishing relevant from irrelevant information, sequencing and prioritizing information, and observing patterns. *also* **NS 1.0, MR 3.0, MR 3.2, MR 3.3**

Connect

You can use a ratio to compare two numbers in *three ways:* whole to part, part to whole, and part to part. Sumi has 6 yellow counters and 3 red counters. Here are three examples of ratios that she can form.

Compare	Ratio	Type of Ratio
all counters to yellow counters	9 to 6	whole to part
red counters to all counters	3 to 9	part to whole
red counters to yellow counters	3 to 6	part to part

- Which type of ratio has the same meaning as a fraction?

TECHNOLOGY LINK

More Practice: Use E-Lab, *Understanding Ratios.*
www.harcourtschool.com/elab2002

Practice

Use two-color counters. Write each ratio and name the type of ratio.

1. Tennis is played by 3 out of 7 students.

2. There were 2 rainy and 5 sunny days.

3. There were 5 dogs and 6 cats in the room.

Write each ratio.

4. 

wheels to skate

5. 

baseballs to players

6. 

tires to bicycles

7. What is the ratio of vowels to consonants in the word PARALLELOGRAM? What kind of ratio is this?

8. What is the ratio of vowels to the number of letters in the alphabet? What kind of ratio is this?

9. **What's the Error?** Jordan has 3 goldfish, 2 turtles, and a snail. He says that the ratio of turtles to the total number of animals is 2 to 5. Find and correct Jordan's error.

10. **GEOMETRY** What is the ratio of number of sides of a triangle to number of sides of an octagon?

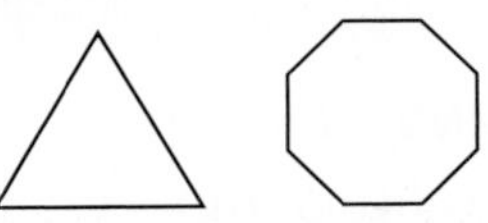

Mixed Review and Test Prep

11. Write $\frac{8}{12}$ in simplest form. (p. 290)

12. Write the LCM of 6 and 8. (p. 256)

13. Find the mean, median, and mode.
43, 64, 51, 64, 78 (p. 104)

14. $62.16 \div 7$ (p. 222)

15. **TEST PREP** Solve for n.
$8 + (7 + 9) = (n + 7) + 9$ (p. 78)

A $n = 8$ **C** $n = 6$
B $n = 7$ **D** $n = 4$

LESSON 2

Express Ratios

Quick Review

Write each fraction in simplest form.

1. $\frac{4}{8}$ **2.** $\frac{3}{9}$ **3.** $\frac{6}{24}$ **4.** $\frac{10}{2}$ **5.** $\frac{6}{8}$

Learn

READING RATIOS At the library, 5 out of 8 books Loren checked out are about dogs. The other 3 books are about cats. The ratio of dog books to the total number of books he checked out is part to whole.

Write: 5 to 8 5:8 $\frac{5}{8}$

Read: "five to eight"

Find other ratios by considering the number of cat books Loren checked out.

A cat books:all books
3 to 8, 3:8, $\frac{3}{8}$
part to whole

B all books:cat books
8 to 3, 8:3, $\frac{8}{3}$
whole to part

C cat books:dog books
3 to 5, 3:5, $\frac{3}{5}$
part to part

MATH IDEA You can use ratios to express the relationships between sets. The relationships of part to whole, whole to part, and part to part can be written in three ways.

Check

1. **REASONING** **Explain** why the order in which you write a ratio is important.

Write each ratio in three ways. Then name the type of ratio.

2. circles to squares
3. all figures to red
4. circles to all figures
5. blue to red
6. all figures to squares
7. green to all figures

CALIFORNIA STANDARDS NS 1.0 Students compute with very large and very small numbers, positive integers, decimals, and fractions and understand the relationship between decimals, fractions, and percents. They understand the relative magnitudes of numbers. *also* **MR 2.3, MR 3.0, MR 3.2, MR 3.3**

Practice and Problem Solving

Write each ratio in three ways. Then name the type of ratio. Use the circles, squares, and stars on page 304.

8. squares to stars

9. red to green

10. stars to all figures

Write *a* or *b* to show which fraction represents each ratio.

11. 5 to 2
a. $\frac{2}{5}$ **b.** $\frac{5}{2}$

12. 5:9
a. $\frac{9}{5}$ **b.** $\frac{5}{9}$

13. 8 to 4
a. $\frac{8}{4}$ **b.** $\frac{4}{8}$

14. 1:20
a. $\frac{20}{1}$ **b.** $\frac{1}{20}$

USE DATA **For 15–17, use the circle graph. Write each ratio in three ways.**

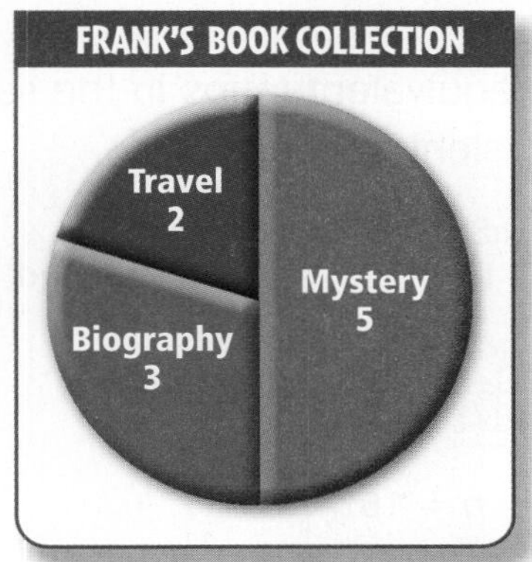

15. What is the ratio of mystery books to biography books?

16. What is the ratio of all books to biography books?

17. What is the ratio of travel books to all books?

USE DATA **For 18–22, use this information: Out of every 60 people in the world, 12 live in China, 10 live in India, and 3 live in the United States. For 18–21, match each ratio with its comparison.**

a. 60:3 **b.** 10:60 **c.** 12:10 **d.** 3:12

18. India:world

19. U.S.:China

20. world:U.S.

21. China:India

22. **REASONING** There are about 6 billion people in the world. About how many people live in India?

23. **Algebra** Find the value of n.
$\frac{5}{7} = \frac{(5 \times n)}{56}$

24. In one bag, there were 4 red, 6 brown, 3 yellow, 6 green, and 2 blue candies. What does the ratio 5:21 represent? What does the ratio 19:2 represent?

25. **Write About It** Two out of every 4 dogs adopted last month were Dalmatians. Explain why you can compare the number of Dalmatians to the total number with the ratio $\frac{1}{2}$.

Mixed Review and Test Prep

26. What is the greatest common factor of 24 and 60? (p. 258)

27. $90{,}211 \times 34$ (p. 150)

28. $358{,}976 \times 25$ (p. 150)

29. Evaluate $4n^2 - (3 \times n)$ if $n = 5$. (p. 268)

30. **TEST PREP** Which number is between 2.011 and 2.022? (p. 20)

A 2.150
B 2.033
C 2.015
D 2.010

Extra Practice page H48, Set A

LESSON

3 Equivalent Ratios

Quick Review

Write an equivalent fraction.

1. $\frac{2}{3}$ **2.** $\frac{3}{8}$ **3.** $\frac{14}{10}$ **4.** $\frac{10}{15}$ **5.** $\frac{7}{12}$

VOCABULARY

equivalent ratios

Learn

POWER PLAY Last year Mike's ice-hockey team won 8 out of 10 games. This year's team will play 20 games. The players hope to have the same ratio of games won to games played as last year. How many games must they win?

Find equivalent ratios in the same way you find equivalent fractions.

$\frac{8}{10} = \frac{n}{20}$ ← games won ← games played

$\frac{8 \times 2}{10 \times 2} = \frac{16}{20}$ **Think:** $10 \times 2 = 20$

$n = 16$

So, Mike's team must win 16 games this year.

Remember

Equivalent fractions name the same amount.

Examples:

$\frac{2}{3} = \frac{4}{6} = \frac{6}{9} = \frac{8}{12}$

MATH IDEA To find **equivalent ratios**, divide or multiply the numerator and the denominator of a ratio by the same number, using a fraction equal to 1.

Examples

A $\frac{3 \times 3}{5 \times 3} = \frac{9}{15}$

$\frac{3}{5}$ and $\frac{9}{15}$ are equivalent ratios.

B $\frac{20 \div 4}{16 \div 4} = \frac{5}{4}$

$\frac{20}{16}$ and $\frac{5}{4}$ are equivalent ratios.

Check

1. Show how you would decide if $\frac{4}{10}$ and $\frac{2}{5}$ are equivalent ratios.

CALIFORNIA STANDARDS NS 1.0 Students compute with very large and very small numbers, positive integers, decimals, and fractions and understand the relationship between decimals, fractions, and percents. They understand the relative magnitudes of numbers. **AF 1.2** Use a letter to represent an unknown number; write and evaluate simple algebraic expressions in one variable by substitution. *also* **MR 1.0, MR 2.0, MR 3.0, MR 3.2**

Tell whether the ratios are equivalent. Write *yes* or *no*.

2. $\frac{2}{4}$ and $\frac{6}{12}$

3. 2:5 and 5:10

4. 4 to 12 and 1 to 3

Write three ratios that are equivalent to the given ratio.

5. 3:1

6. $\frac{2}{5}$

7. 1 to 4

8. 3:5

Practice and Problem Solving

Tell whether the ratios are equivalent. Write *yes* or *no*.

9. $\frac{3}{4}$ and $\frac{12}{20}$

10. 5:10 and 1:2

11. 1 to 4 and 25 to 100

12. 3:9 and 9:27

13. 4 to 9 and 18 to 8

14. $\frac{5}{8}$ and $\frac{20}{32}$

Write three ratios that are equivalent to the given ratio.

15. 10 to 1

16. $\frac{1}{7}$

17. 9 to 3

18. 26:36

19. $\frac{1}{5}$

20. 50:100

21. $\frac{2}{2}$

22. 100 to 1

Copy and complete the ratio table.

23.

Number of Apples	3	■	■	■
Cups of Apple Cider	1	3	9	27

24.

Number of Canoes	1	4	7	10
Number of People	3	■	■	■

Algebra **Find *n* to make the ratios equivalent.**

25. $\frac{1}{n} = \frac{2}{6}$

26. $\frac{n}{4} = \frac{6}{12}$

27. $\frac{10}{n} = \frac{20}{4}$

28. $\frac{3}{5} = \frac{n}{25}$

29. During the ice-hockey game, 3 of the 20 shots were goals. What is the ratio of goals made to all the shots taken?

30. **NUMBER SENSE** Find the missing numbers in the pattern of equivalent ratios: $\frac{2}{■}, \frac{4}{10}, \frac{8}{■}, \frac{■}{40}$. Describe the pattern.

For 31–32, use the ratio of 4 oranges to 1 glass of juice.

31. Brice is making orange juice for some friends. If he wants to make 3 glasses of juice, how many oranges will he need?

32. Write three ratios equivalent to the ratio of oranges to glasses of orange juice.

Mixed Review and Test Prep

33. Attendance at the last three games was 325, 410, and 465. What was the average attendance for those three games? (p. 102)

34. Write $\frac{3}{4}$ as a decimal. (p. 280)

Write each ratio in two other ways. (p. 304)

35. 5 games to 7 games

36. 9 green to 4 red marbles

37. **TEST PREP** Which is the GCF of 18 and 45? (p. 258)

A 3 **B** 5 **C** 6 **D** 9

Extra Practice page H48, Set B

LESSON 4

Scale Drawings

Quick Review

Find n to make the ratios equivalent.

1. $\frac{7}{14} = \frac{n}{28}$
2. $\frac{8}{n} = \frac{16}{32}$
3. $\frac{5}{25} = \frac{n}{75}$
4. $\frac{3}{100} = \frac{30}{n}$
5. $\frac{36}{42} = \frac{n}{7}$

VOCABULARY

scale drawing **map scale**

Learn

GO THE DISTANCE! A **scale drawing** shows the correct relationship between distances or sizes, even though the drawing is larger or smaller than the actual object. Some maps are scale drawings. A ratio that compares the distance on the map to the actual distance is a **map scale**.

You can use the scale to compute the distance from Sacramento to Stockton.

Read the map scale.	The map shows a scale of 1 cm = 16 km, or $\frac{1 \text{ cm}}{16 \text{ km}}$.
Use a ruler to measure the distance from Sacramento to Stockton on the map.	The distance is 4.7 cm on the map.
Use equivalent ratios to find the actual distance.	$\frac{1}{16} = \frac{4.7}{n}$ $\frac{1 \times 4.7}{16 \times 4.7} = \frac{4.7}{75.2}$ $n = 75.2$

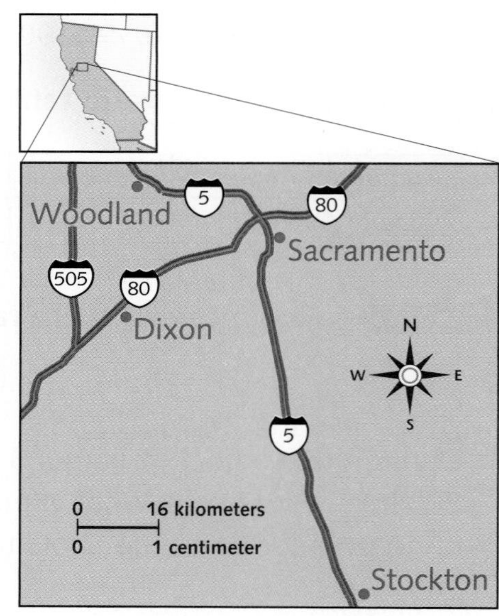

So, the distance is 75.2 kilometers.

MATH IDEA You can use scale drawings and map scales to determine actual sizes and distances.

Check

1. **Use** the map and map scale to find the distance from Sacramento to Woodland.

Copy and complete the map scale ratio table.

2.

Map Distance (inches)	1	■	2.6	■	7.8
Actual Distance (miles)	100	150	■	680	■

TECHNOLOGY LINK

To learn more about scale models, watch the **Harcourt Math Newsroom Video** ***3D Ocean Mapping.***

CALIFORNIA STANDARDS **NS 1.0** Students compute with very large and very small numbers, positive integers, decimals, and fractions and understand the relationship between decimals, fractions, and percents. They understand the relative magnitudes of numbers. **AF 1.2** Use a letter to represent an unknown number; write and evaluate simple algebraic expressions in one variable by substitution. *also* **MR 2.3, MR 3.0, MR 3.2**

Practice and Problem Solving

Copy and complete the ratio table.

3.

Scale Length (in.)	1	■	3.5	■	■
Actual Length (ft)	10	30	■	50	60

4.

Scale Distance (cm)	1	3	■	8	■
Actual Distance (m)	7.5	■	37.5	■	75

5.

Scale Length (in.)	1	2	4	■	■
Actual Length (ft)	12	■	■	60	72

6.

Scale Distance (cm)	1	2.6	■	7.6	■
Actual Distance (m)	75	■	375	■	600

7.

Scale Length (in.)	1	■	4	4.4	5
Actual Length (ft)	200	500	■	■	■

8.

Scale Distance (cm)	3	6	■	12	■
Actual Distance (m)	150	■	450	■	750

For 9–11, use the map and map scale to find the distance.

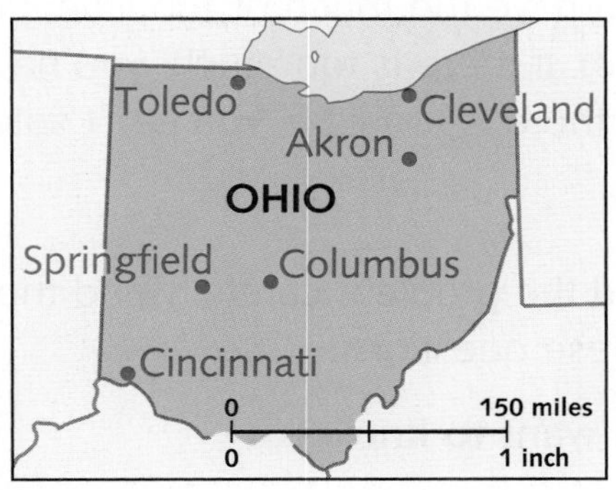

9. Cincinnati to Columbus
10. Toledo to Columbus
11. Columbus to Akron
12. **Write a problem** about driving between two cities that are less than 100 miles apart. Use the map of Ohio.
13. A rectangular garden is 48 ft long and 42 ft wide. Fencing for the garden costs $12 for each foot. How much will it cost to fence in the garden?
14. Choose a scale for this drawing. Then use it to find the actual length and width of the cabin.

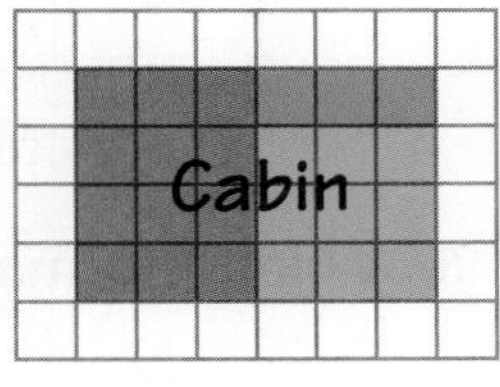

15. Maura has a dollhouse with a scale of 1 inch to 1.5 ft. The dimensions of the living room are 10 in. by 12 in. What would be the dimensions of this room in an actual house?
16. **REASONING** Joy planned a scale drawing with a scale of 1 in. = 4 ft. She found that her drawing would not fit on her paper. What part of the ratio should she change to make the drawing fit? Explain.

Mixed Review and Test Prep

17. List the prime numbers between 30 and 40. (p. 264)
18. Write three ratios equivalent to 3:2. (p. 306)
19. Solve for n. $17 = 4n - 3$ (p. 88)
20. $23.574 - 13.221$ (p. 50)
21. **TEST PREP** Which of the following is NOT equivalent to 1.5? (p. 22)
    - **A** $1\frac{1}{2}$
    - **B** 1.50
    - **C** $1\frac{5}{10}$
    - **D** 1.05

LESSON 5

Problem Solving Skill

Too Much/Too Little Information

Quick Review

Is the ratio equivalent to $\frac{3}{5}$?

1. $\frac{2}{3}$ **2.** $\frac{6}{10}$ **3.** $\frac{9}{15}$ **4.** $\frac{1}{2}$ **5.** $\frac{15}{20}$

CAN YOU SOLVE IT? Yoko found these people-to-pet ratios on the Internet. For her report she wants to know the ratio of pet dogs in the United States to pet cats.

Sometimes you have too much or too little information to solve a problem. If there is too much, you have to decide what to use. If there is too little, you can't solve the problem.

To decide, read the problem carefully and then ask yourself these questions:

1. **What do I want to know?**
   pet dogs:pet cats
2. **What do I know?**
   people:pet dogs = 30:6
   people:pet cats = 30:7
3. **What information is unnecessary?**
   people to bird ratios
4. **What necessary information is missing?**
   none

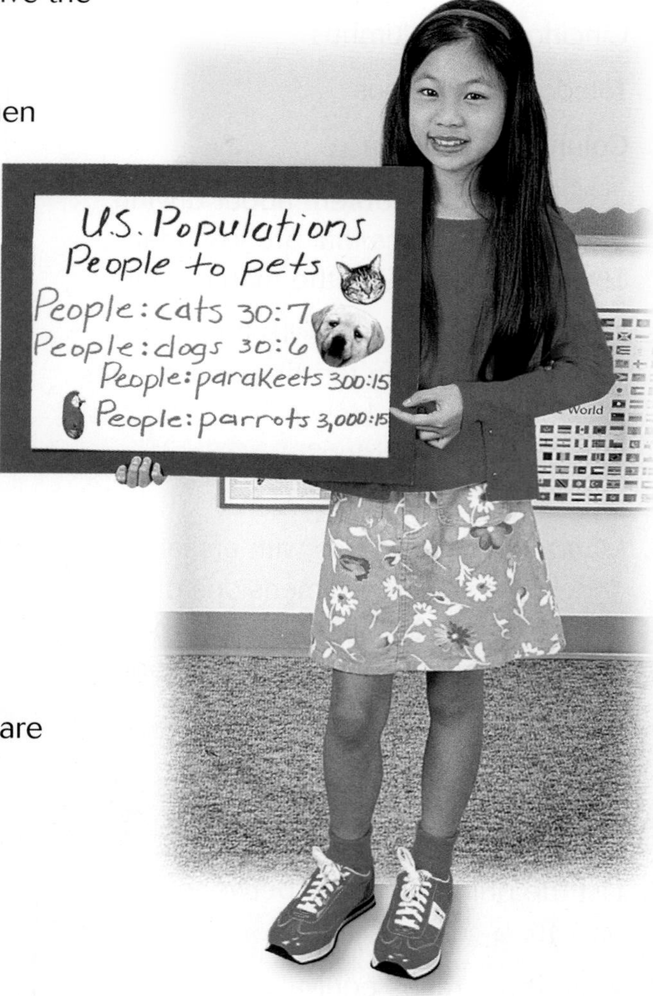

U.S. Populations
People to pets
People: cats 30:7
People: dogs 30:6
People: parakeets 300:15
People: parrots 3,000:15

Since the number of people compared to dogs and cats is the same, you can compare dogs to cats.

So, the ratio of pet dogs to pet cats is 6:7.

Talk About It

- How can you decide what information is necessary to solve a problem?

CALIFORNIA STANDARDS MR 1.0 Students make decisions about how to approach problems. **MR 1.1** Analyze problems by identifying relationships, distinguishing relevant from irrevelant information, sequencing and prioritizing information, and observing patterns. **MR 2.0** Students use strategies, skills, and concepts in finding solutions. *also* **MR 2.3, MR 2.4, MR 2.6, MR 3.0, MR 3.2**

Problem Solving Practice

USE DATA For 1–4, use the table on page 310. Write whether each problem has *too much* or *too little* information. Then solve if possible, or describe the additional information needed.

1. How many people are there in the U.S. for every one pet dog?
2. How many people are there in the U.S. for every one Siamese cat?
3. Which is the more popular pet in the U.S, a parrot or a goldfish?
4. How many pet parakeets are there for every 3,000 people in the U.S.?

Tomás needs to walk the dog, do the dishes, and finish his homework. It will take him 15 minutes to walk the dog, 30 minutes to do the dishes, and 1 hour 10 minutes to finish his homework. Does he have enough time to complete all three tasks?

5. What other information do you need to solve the problem?
   - **A** the name of Tomás's dog
   - **B** how much time Tomás has
   - **C** how many dishes are dirty
   - **D** what homework he needs to do
6. What is the least amount of time Tomás needs to complete all three tasks?
   - **F** 2 hours
   - **G** 1 hour 55 minutes
   - **H** 1 hour 45 minutes
   - **J** 1 hour 35 minutes

Mixed Applications

USE DATA For 7–8, use the table.

7. What is the weight difference between the largest and smallest dogs?
8. Which dog was bigger, the mastiff or the Great Dane?
9. The first year of a dog's life equals 15 "human years." The second year equals 10 human years. Every year thereafter equals 3 human years. Use this formula to find a 6-year-old dog's age in human years.

WORLD DOG RECORDS

Record	Breed	Measurement
Largest	Old English Mastiff	343 pounds
Tallest	Great Dane	3.5 feet
Smallest	Yorkshire Terrier	4 ounces

10. **What's the Question?** An African gray parrot named Prudle was found in 1958. From then until 1977, it learned nearly 1,000 words. The answer is about 50.
11. **Write a problem** that has too little information to be solved. Then write one that includes information that is not needed to solve the problem.

Review/Test

CHECK VOCABULARY AND CONCEPTS

Choose the best term from the box.

equivalent ratios
map scale
ratio
scale drawing

1. You can use a __?__ to compare two quantities. (p. 302)
2. A ratio that compares the distance on a map to the actual distance is called the __?__. (p. 308)
3. Ratios that describe the same comparison of quantities are called __?__. (p. 306)

CHECK SKILLS

For 4–6, use the letters in the word MULTIPLICATION to write each ratio. (pp. 302–303)

4. consonants to total letters
5. consonants to vowels
6. total letters to vowels

Write *a* or *b* to show which fraction represents each ratio. (pp. 304–305)

7. 5 to 1
   a. $\frac{1}{5}$ b. $\frac{5}{1}$
8. 9 to 5
   a. $\frac{9}{5}$ b. $\frac{5}{9}$
9. 12:36
   a. $\frac{1}{3}$ b. $\frac{3}{1}$
10. 25:100
   a. $\frac{4}{1}$ b. $\frac{1}{4}$
11. 1 to 50
   a. $\frac{1}{50}$ b. $\frac{50}{1}$

Write three ratios that are equivalent to the given ratio. (pp. 306–307)

12. 6 to 8
13. 3:7
14. 9:5

Tell whether the ratios are equivalent. Write *yes* or *no*. (pp. 306–307)

15. $\frac{3}{21}$ and $\frac{1}{7}$
16. 3:7 and 6:21

Copy and complete the map scale ratio table. (pp. 308–309)

17.

Scale Length (in.)	1	4	■	8
Actual Length (ft)	50	■	300	■

18.

Scale Length (cm)	1	■	4	■
Actual Length (m)	2.5	5	■	15

CHECK PROBLEM SOLVING

Write whether each problem has *too much* or *too little* information. Then solve if possible, or describe the additional information needed. (pp. 310–311)

19. At Jerry's Market a gallon of chocolate milk costs $2.15, a half gallon costs $1.20, and a gallon of skim milk costs $1.75. How much will 2 gallons of chocolate milk cost?

20. The music store is having a CD sale. If you buy two CDs, you get one free. How much will three CDs cost?

Cumulative Review

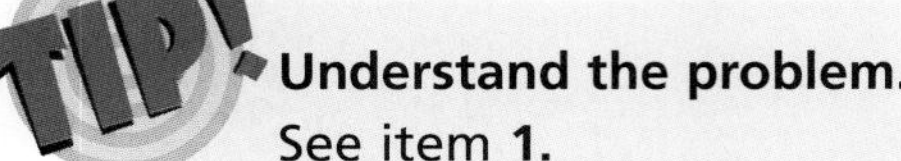

Understand the problem.
See item **1.**

To find a part-to-part ratio, you need to count the number of objects that are in each part. Be sure to express your ratio with the number of hexagons first.

Also see problem **1,** p. H62.

For 1–8, choose the best answer.

For 1–2, use the figures.

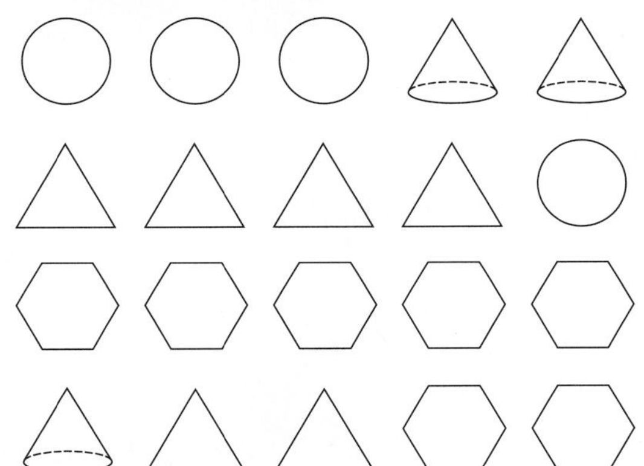

1. What is the ratio of the number of hexagons to the number of triangles?

A 7:6 **C** 6:7
B 7:3 **D** 3:7

2. What is the ratio of the number of solid figures to the number of polygons?

F 13:3 **H** 3:13
G 7:13 **J** 3:17

---

3. What is the greatest common factor of 24 and 60?

A 4 **C** 8
B 6 **D** 12

4. Jack has a map with a scale where 1 inch represents 10 miles. How many inches on this map represent 45 miles?

F 4 inches **H** 45 inches
G 4.5 inches **J** Not here

---

5. When Tim shared all of his cookies, each of his friends got 3. What else do you need to know to find out how many cookies Tim had?

A the cost of the cookies
B the number of leftover cookies
C the kind of cookies
D the number of friends

---

6. Last year Cara's soccer team won 7 out of 11 games. What is being compared in the ratio 7:4?

F the number of wins to the number of games
G the number of wins to the number of losses
H the number of losses to the number of games
J the number of losses to the number of wins

---

7. Which ratio is equivalent to 55:80?

A $\frac{80}{55}$ **C** $\frac{11}{16}$
B $\frac{16}{11}$ **D** $\frac{5}{8}$

---

8. Out of 500 fifth-graders, 260 are girls. Which ratio compares the number of fifth-graders to the number of boys?

F 500 to 260 **H** 500 to 240
G 260 to 500 **J** 240 to 500

CHAPTER 18 Percent

The United States government owns 632.7 million acres of land. The National Park System is part of that land. The largest park is Wrangell-St. Elias National Park and Preserve in Alaska. It covers 13,200,000 acres. The smallest park is the Thaddeus Kosciuszko National Memorial in Pennsylvania. It is only 0.02 acre. Look at the circle graph. Of the acres that are owned by the government, how many acres are in the National Park System?

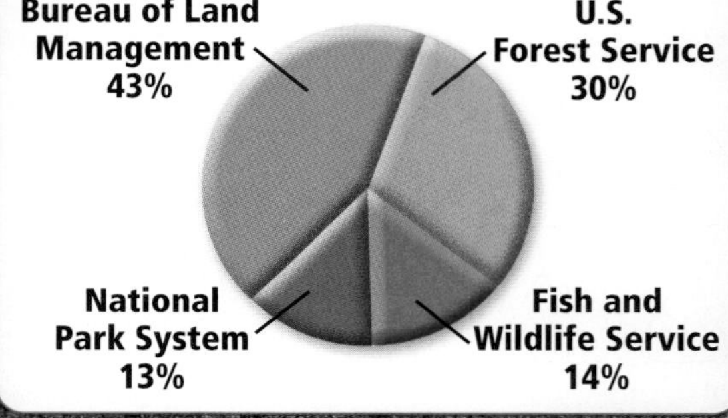

Denali National Park, Alaska

Use this page to help you review and remember important skills needed for Chapter 18.

VOCABULARY

Choose the best term from the box.

decimal
hundredth
ratio
simplest form
whole number

1. The comparison of two quantities is a __?__.
2. A decimal or fraction that names one part of 100 equal parts is one __?__.
3. A fraction is in __?__ when 1 is the greatest common factor of the numerator and the denominator.
4. A number that uses place value and a decimal point to show values less than 1, such as tenths and hundredths, is a __?__.

UNDERSTAND HUNDREDTHS (See p. H20.)

Write the standard form and word form for the decimal model.

5. 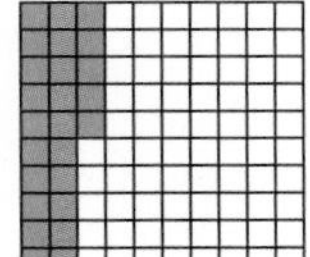
6. 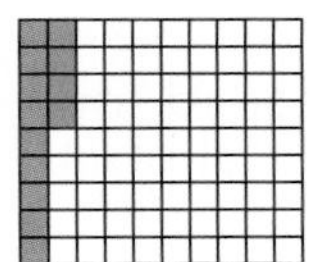
7. 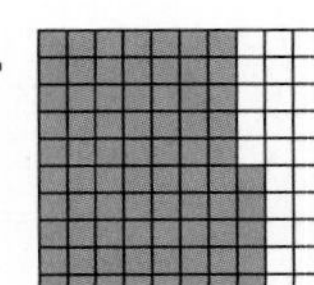
8. 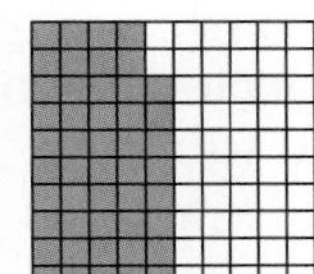

Write as a decimal and a fraction.

9. twenty-nine hundredths
10. sixteen hundredths
11. two and five hundredths
12. fifty hundredths

RELATE FRACTIONS AND DECIMALS (See p. H21.)

Write a decimal and a fraction for each model.

13.  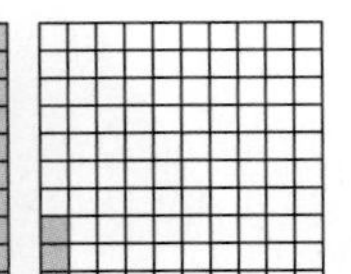 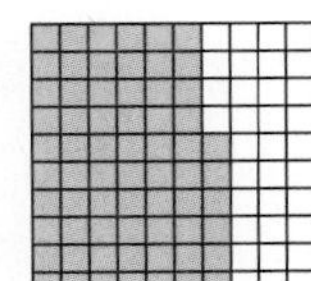
14. (see model)
15. 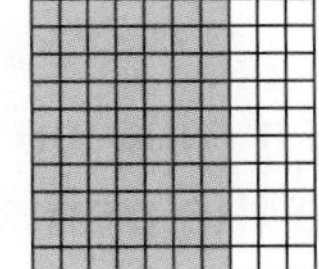

Write as a decimal.

16. $\frac{54}{100}$
17. $\frac{36}{100}$
18. $\frac{3}{4}$
19. $\frac{19}{1,000}$
20. $\frac{4}{5}$

Write as a fraction.

21. 0.25
22. 0.9
23. 0.5
24. 0.63
25. 1.2

LESSON 1

HANDS ON
Understand Percent

Explore

Did you know that fifty percent of the Earth's species live in rain forests? **Percent** means "per hundred." A percent is a ratio of a number to 100. The symbol for percent is %. 1% means "1 out of 100." So, "50% of the Earth's species" means that 50 out of every 100, or $\frac{50}{100}$, species on Earth live in rain forests.

Use a grid with 100 squares to model percents.

Out of every 100 known bird species on Earth, 30 live in rain forests. What percent of known bird species live in rain forests?

STEP 1

Let each grid square represent 1 bird species. To show 30 bird species out of 100, shade 30 squares.

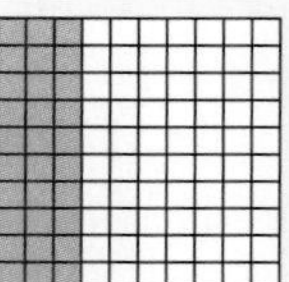

STEP 2

Write the ratio of shaded squares to the total squares. Then write the percent.

$$\frac{\text{shaded squares}}{\text{total squares}} = \frac{30}{100} = 30\%$$

So, 30% of all known bird species live in rain forests.

- The unshaded squares represent the bird species that do not live in rain forests. What percent of known bird species do not live in rain forests?

What ratio can I write to show 42 ducks out of 100 birds?

Try It

Model each ratio on grid paper. Then write the percent.

a. 42 ducks out of 100 birds

b. 50 lions out of 100 cats

c. 18 puppies out of 100 pets

Quick Review

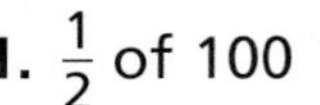
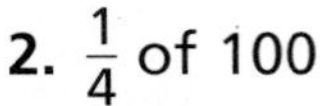

1. $\frac{1}{2}$ of 100
2. $\frac{1}{4}$ of 100
3. $\frac{1}{5}$ of 100
4. $\frac{1}{10}$ of 100
5. $\frac{3}{4}$ of 100

VOCABULARY

percent

MATERIALS

10 × 10 grid paper, colored pencils

CALIFORNIA STANDARDS **NS 1.2** Interpret percents as a part of a hundred; find decimal and percent equivalents for common fractions and explain why they represent the same value; compute a given percent of a whole number. **NS 1.0** Students compute with very large and very small numbers, positive integers, decimals, and fractions and understand the relationship between decimals, fractions, and percents. They understand the relative magnitudes of numbers. *also* **MR 2.0, MR 2.3**

Connect

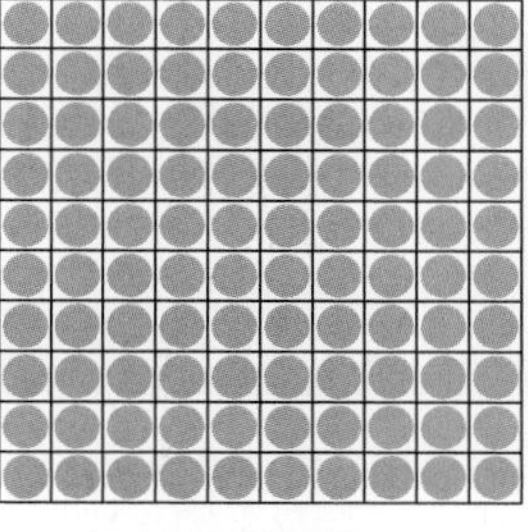

$100\% = \frac{100}{100} = 1$

Remember, a ratio is a comparison of two quantities. A percent is a special ratio because it always compares a part to 100. Percent is often used with money.

One dollar has 100 parts, or cents. Look at the 10×10 grid. There is 1 cent, or penny, in each of 100 squares, or a total of $1.00.

1% of $1.00 is 1 penny, or $0.01.

Talk About It

- What is 50% of $1.00? 10% of $1.00? 76% of $1.00?
- What part of a dollar do you have if you have 0% of it? 100% of it?
- **REASONING** What does 200% of a number mean?

TECHNOLOGY LINK

More Practice: Use E-Lab, *Understanding Percent.*

www.harcourtschool.com/elab2002

Practice

Model each ratio on grid paper. Then write the percent.

1. 31 tigers out of 100 animals
2. 100 red balls out of 100 balls
3. 4 dimes out of $1.00
4. 5 blue pens out of 100 pens

Write a percent to describe the shaded part.

5. 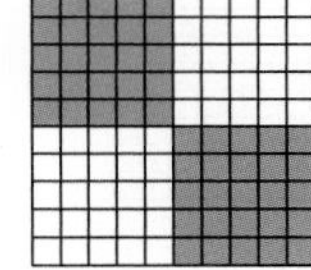
6. 
7. 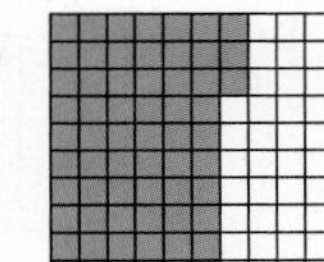
8. 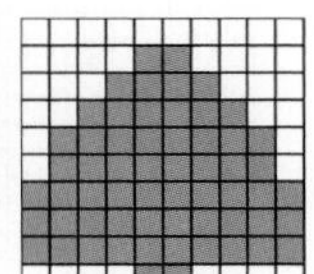

Choose the more reasonable percent. Write *a* or *b*.

9. "*Almost everyone* passed the test," Mrs. Philips said with pride.
   a. 95% passed. **b.** 15% passed.
10. "*Very few* children like spicy foods," said the chef.
   a. 40% like them. **b.** 8% like them.
11. **REASONING** A few months ago, Van had 100 days to wait until his birthday. His wait is 98% over. Today is Monday. What day is Van's birthday?
12. **Write About It** The *cent* in *percent* means "100." Write a list of words that contain *cent*, and explain how their definitions relate to 100.

Mixed Review and Test Prep

13. What is the greatest common factor of 12 and 15? (p. 258)
14. Write two ratios that are equivalent to 7:8. (p. 306)
15. 0.5×11 (p. 164)
16. $11.34 \div 7$ (p. 222)
17. **TEST PREP** Write $\frac{23}{1,000}$ as a decimal. (p. 228)
   A 23 **B** 2.3 **C** 0.23 **D** 0.023

LESSON 2

Relate Decimals and Percents

Quick Review

Write each decimal in word form.

1. 0.43 **2.** 2.04

3. 0.7 **4.** 0.18

5. 14.45

Learn

MONEY, MONEY, MONEY! A quarter of a dollar can be written as the decimal $0.25.

MATH IDEA You can write a decimal as a percent.

What percent of a dollar is $0.25?

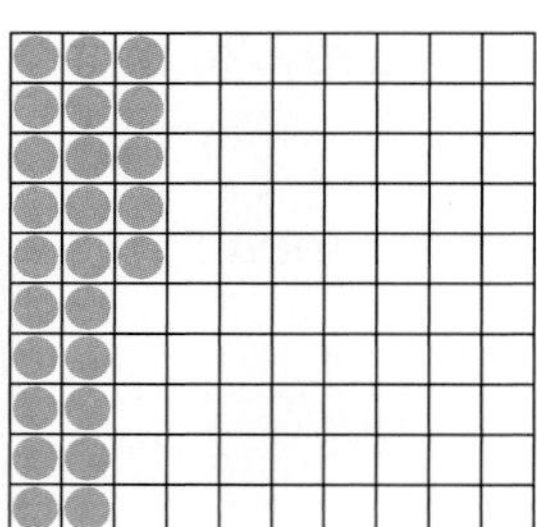

Write: 0.25

Read: twenty-five hundredths

Ratio: 25 out of 100, or $\frac{25}{100}$

Percent: 25%

So, $0.25 is 25% of a dollar.

Examples

A Write 0.07 as a percent.

0.07 = 7 out of 100

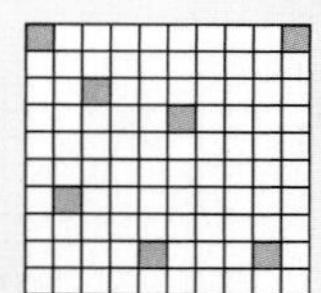

So, 0.07 is 7%.

B Write 90% as a decimal.

90% = 90 out of 100

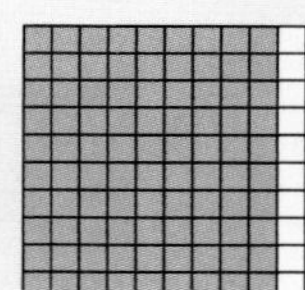

So, 90% is 0.90, or 0.9.

C Write 125% as a decimal.

125% = 125 out of 100

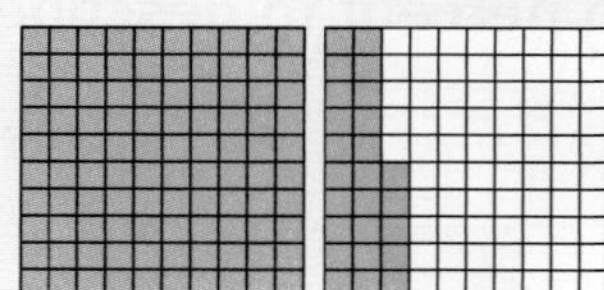

So, 125% is 1.25.

Check

1. **REASONING** **Tell** why you can write the decimal in Example B as 0.90 and as 0.9 but you can't write 90% as 9%.

For 2–4, write a decimal and a percent to describe each shaded part of the model.

2. purple squares
3. yellow and blue squares
4. all the shaded squares

The Federal Reserve estimates $395 million could be saved annually by replacing bills with coins like the Golden Dollar. ▶

CALIFORNIA STANDARDS **O—NS 1.2** Interpret percents as a part of a hundred; find decimal and percent equivalents for common fractions and explain why they represent the same value; compute a given percent of a whole number. **NS 1.0** Students compute with very large and very small numbers, positive integers, decimals, and fractions and understand the relationship between decimals, fractions, and percents. They understand the relative magnitudes of numbers. *also* **MR 2.0, MR 2.2, MR 2.3, MR 2.4**

Practice and Problem Solving

USE DATA For 5–7, use the circle graph. Write a decimal and a percent to describe each.

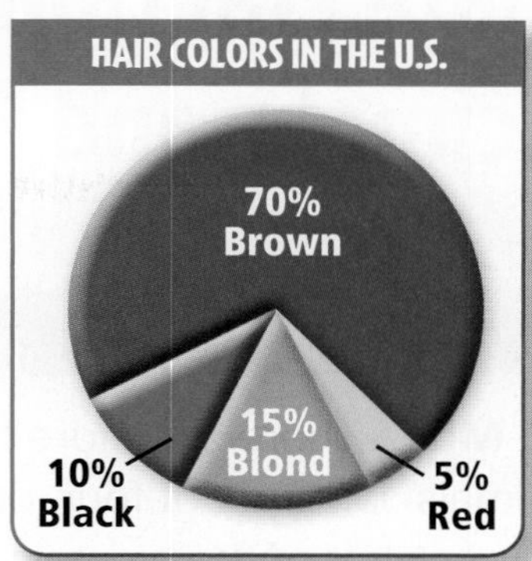

5. What part of the population has either brown or black hair?
6. What part of the population does not have red hair?
7. Fifteen hundredths of the population has the same color hair as I do. What color is my hair?

Write the number as a decimal and a percent.

8. eighty-one hundredths
9. twelve hundredths
10. thirty-seven hundredths
11. four hundredths

Write each decimal as a percent.

12. 0.78
13. 0.99
14. 0.06
15. 0.2
16. 1.2

Write each percent as a decimal.

17. 35%
18. 100%
19. 2%
20. 60%
21. 104%

REASONING Write the greater number of each pair.

22. 0.5, 5%
23. 0.14, 140%
24. 0.08, 80%
25. 1.9, 19%
26. 2, 2%

27. Out of 100 students in the spelling bee, 23 girls and 28 boys misspelled *aardvark*. What percent of the students spelled the word correctly?
28. **REASONING** Jeb read that 10% of people are left-handed. If he surveys 200 people, about how many left-handers should he expect to find?
29. In early 2000, a poll found 75 out of every 100 Americans opposed eliminating the $1 bill and using a $1 coin. What percent opposed eliminating the $1 bill?
30. **Write About It** What pattern do you see in the placement of the decimal point when you write a decimal as a percent? a percent as a decimal?

Mixed Review and Test Prep

31. Write the ratio 13:4 as a fraction. (p. 304)
32. Find the value of 4^3. (p. 268)
33. $22.44 \div 4$ (p. 222)
34. $3{,}005 \times 18$ (p. 148)
35. **TEST PREP** Solve for n.
$5 \times (n + 3) = 50$ (p. 92)
 A $n = 5$
 B $n = 7$
 C $n = 10$
 D $n = 50$

Extra Practice page H49, Set A

LESSON 3

Relate Fractions, Decimals, and Percents

Quick Review

Write the fraction in simplest form.

1. $\frac{4}{10}$
2. $\frac{32}{36}$
3. $\frac{88}{99}$
4. $\frac{45}{55}$
5. $\frac{14}{21}$

Learn

DON'T BUG ME! There are more beetles on Earth than any other animals. One out of four, or $\frac{1}{4}$, of all animals are beetles. What percent of the animals on Earth are beetles?

You can write a ratio in fraction form as a percent.

TECHNOLOGY LINK

More Practice: Use Mighty Math Number Heroes, *Fraction Fireworks*, Level Y.

One Way

Write an equivalent fraction with a denominator of 100. Then write the fraction as a percent.

$$\frac{1}{4} = \frac{1 \times 25}{4 \times 25} = \frac{25}{100}$$

$$\frac{25}{100} = 25\%$$

Another Way

Divide the numerator by the denominator. Then write the decimal as a percent.

$$\frac{1}{4} \rightarrow 4\overline{)1.00} = 0.25$$

$$0.25 = 25\%$$

So, 25% of all animals on Earth are beetles.

Examples **Write each fraction as a percent.**

A $\frac{7}{20} = \frac{7 \times 5}{20 \times 5} = \frac{35}{100}$

$\frac{35}{100} = 35\%$

B $\frac{4}{5} \rightarrow 5\overline{)4.00} = 0.80$

$0.80 = 80\%$

C $\frac{3}{40} = 40\overline{)3.000} = 0.075$

$0.075 = 7.5\%$

- **REASONING** Which way would you use to write $\frac{1}{8}$ as a percent? Explain your choice.

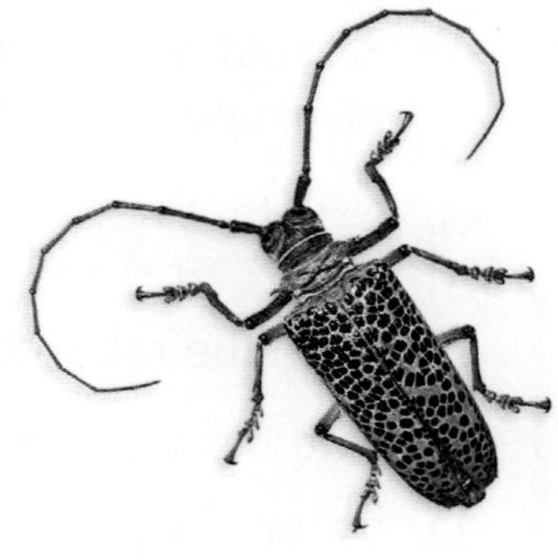

CALIFORNIA STANDARDS O-n **NS 1.2** Interpret percents as a part of a hundred; find decimal and percent equivalents for common fractions and explain why they represent the same value; compute a given percent of a whole number. **NS 1.0** Students compute with very large and very small numbers, positive integers, decimals, and fractions and understand the relationship between decimals, fractions, and percents. They understand the relative magnitudes of numbers. *also* **SDAP 1.0, MR 1.0, MR 2.0, MR 2.3, MR 2.4**

Percents as Fractions

Scientists estimate that ants make up 10% of the Earth's biomass, the amount of living matter. What fraction of the Earth's biomass are ants?

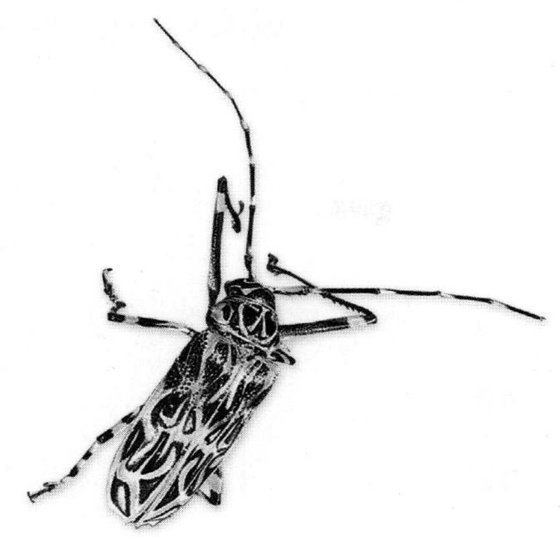

You can write a percent as a fraction.

STEP 1

Write the percent as a ratio in fraction form. Use the percent as the numerator and 100 as the denominator.

$$10\% = \frac{10}{100}$$

STEP 2

Write the fraction in simplest form.

$$\frac{10}{100} = \frac{10 \div 10}{100 \div 10} = \frac{1}{10}$$

So, ants make up $\frac{1}{10}$ of the Earth's biomass.

Examples **Write each percent as a fraction.**

D $8\% = \frac{8}{100}$

$$\frac{8}{100} = \frac{8 \div 4}{100 \div 4} = \frac{2}{25}$$

E $17\% = \frac{17}{100}$

F $155\% = \frac{155}{100}$

$$\frac{155}{100} = \frac{155 \div 5}{100 \div 5} = \frac{31}{20}$$

MATH IDEA To show comparisons, you can write a ratio as a fraction, as a decimal, or as a percent.

Ratio	Fraction	Decimal	Percent
25 to 100	$\frac{25}{100}$, or $\frac{1}{4}$	0.25	25%

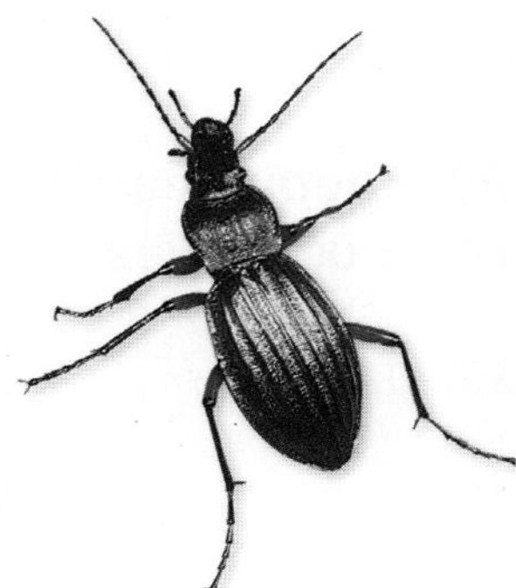

Check

1. **Show** how a fraction, a decimal, and a percent are related. Write $\frac{12}{25}$ as a decimal. Then rewrite your answer as a percent.

Copy and complete the tables. Write each fraction in simplest form.

	Fraction	Decimal	Percent
2.	$\frac{8}{25}$	■	■
4.	■	0.80	80%

	Fraction	Decimal	Percent
3.	■	0.04	■
5.	$\frac{147}{100}$	■	147%

LESSON CONTINUES

Practice and Problem Solving

Copy and complete the tables. Write each fraction in simplest form.

	Fraction	Decimal	Percent
6.	■	0.35	35%
8.	■	■	20%
10.	$\frac{1}{20}$	■	■
12.	■	0.60	60%
14.	$\frac{4}{25}$	■	16%
16.	$\frac{1}{40}$	■	2.5%

	Fraction	Decimal	Percent
7.	$\frac{29}{100}$	■	■
9.	$\frac{3}{50}$	■	■
11.	■	0.11	■
13.	■	2.0	■
15.	$\frac{117}{100}$	■	117%
17.	$\frac{3}{8}$	■	■

Express the shaded part of each model as a decimal, a percent, and a fraction in simplest form.

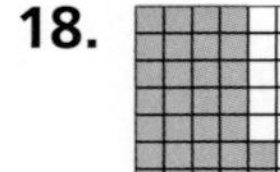

18. 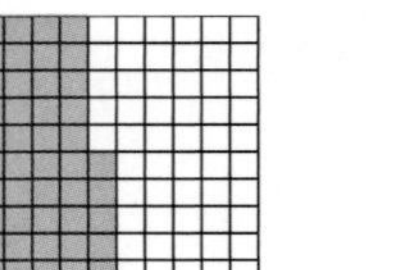

19. 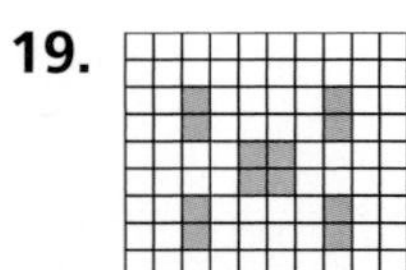

20. 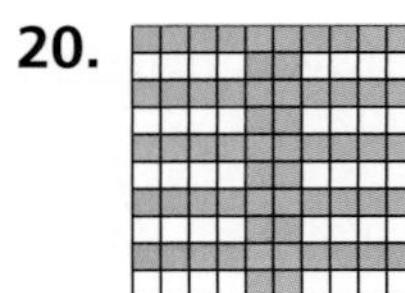

21.

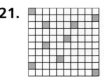

Compare. Write <, >, or = for each ●.

22. 30% ● 0.03

23. $\frac{4}{5}$ ● 75%

24. 50% ● 0.60

25. 105% ● $\frac{5}{5}$

26. 80% ● 0.8

27. $\frac{4}{25}$ ● 20%

28. 0.01 ● 0.1%

29. $\frac{14}{7}$ ● 210%

Tell whether each fraction or decimal is greater than 100% or between 1% and 100%. Write *greater* or *between*.

30. 0.64

31. $\frac{24}{50}$

32. 2.50

33. $\frac{300}{100}$

34. $\frac{1}{5}$

35. 0.72

36. 3.0

37. $\frac{24}{50}$

38. $\frac{1}{4}$

39. $\frac{125}{100}$

40. NUMBER SENSE Ted scored 85% on his spelling test. Kenya spelled 4 out of 5 words correctly. Rosa scored $\frac{16}{20}$ on the test. Who got the highest test score?

41. At the concert, 100 students performed. Out of that group, 35% played the flute. How many students played a different instrument?

42. NUMBER SENSE Order $\frac{9}{4}$, 35%, 0.3, and $2\frac{1}{2}$ from greatest to least.

43. NUMBER SENSE Order 0.42, $4\frac{1}{4}$, 420%, and 4.02 from least to greatest.

44. **Write About It** How is writing a percent as a fraction like writing a decimal as a fraction?

45. The Andes Mountains are over twice as long as North America is wide. The Andes Mountains are about 7,240 kilometers long. How wide is North America at its widest?

46. Bill's car uses 0.875 of a tank of gas each week. His car has a 20-gallon tank. Gas costs $1.40 per gallon. How much does he spend each week for gas?

Mixed Review and Test Prep

Write each ratio. Then write *part to whole, whole to part,* or *part to part* to describe the ratio. (p. 302)

47. Out of 15 students, 6 wear glasses.

48. There are 3 silver cars to 4 white cars parked on the street.

Write an equivalent fraction. (p. 280)

49. 0.45

50. 1.03

51. **TEST PREP** Find $3,068 \times 400$. (p. 150)

A 12,272
B 122,720
C 1,224,200
D 1,227,200

52. **TEST PREP** What is the range for the data set $80.00, $82.50, $75.00, $76.50? (p. 98)

F $5
G $7.50
H $78.50
J $79.50

STRATEGY • USE GRAPHIC AIDS
Sometimes the information you need to solve a problem is provided in a graphic aid, such as a graph, diagram, or table. To use a graphic aid, look at each part of it and study the relationships that exist among the facts.

MAIN COMPONENTS OF THE EARTH'S CRUST

47% Oxygen
28% Silicon
8% Aluminum
17% Other Elements

The title tells you the subject of the graph.

The whole circle represents the whole crust.

Each section of the circle represents one part of the crust. The label for each section names the element that makes up the part of the crust. The size of the section tells you how many parts of the whole it makes up.

Use the graphic aid to solve each problem.

1. Which element composes about $\frac{1}{4}$ of the Earth's crust?

2. What fraction of the Earth's crust is composed of aluminum?

3. What percent of the Earth's crust is composed of oxygen or silicon?

4. There is about 2.25 times as much oxygen in the Earth's crust as in the air we breathe. About what percent of the air we breathe is composed of oxygen?

Extra Practice page H49, Set B

LESSON 4

Find a Percent of a Number

Quick Review

Write each percent as a decimal.

1. 70% **2.** 55%
3. 4% **4.** 99%
5. 100%

Learn

ZZZZZ . . . The students in Andrew's science class hope 5,000 people visit their *Sleeptime* home page by the end of the school year. According to the latest tracking report, they have reached 30% of their goal. How many people have visited the home page? What is 30% of 5,000?

Make a model to find 30% of 5,000.

Activity

MATERIALS: index cards

STEP 1

Let each card represent 10% of the number of visitors. Put down 10 cards to represent 100%, or 5,000. Each 10% represents 500 since $10 \times 500 = 5{,}000$.

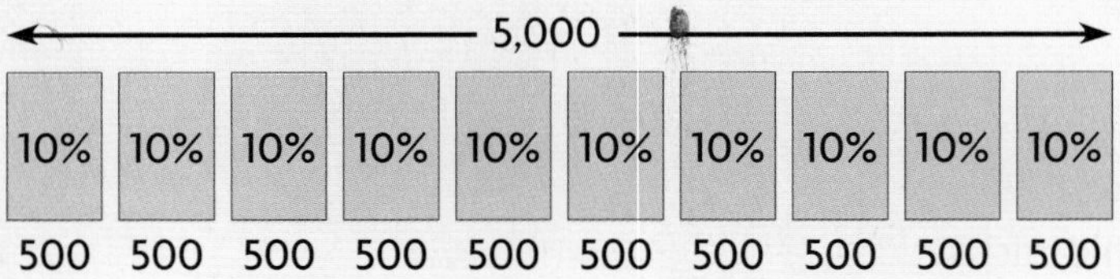

STEP 2

Now separate 3 cards to show $3 \times 10\%$, or 30%.

10% 10% 10%   10% 10% 10% 10% 10% 10% 10%

500 500 500

Since each card represents 500, the 3 cards that make up 30% represent 3×500, or 1,500.

So far, 1,500 people have visited the *Sleeptime* home page.

- How many cards show 50% of 5,000? 80% of 5,000?

CALIFORNIA STANDARDS **O—NS 1.2** Interpret percents as a part of a hundred; find decimal and percent equivalents for common fractions and explain why they represent the same value; compute a given percent of a whole number. **NS 1.0** Students compute with very large and very small numbers, positive integers, decimals, and fractions and understand the relationship between decimals, fractions, and percents. They understand the relative magnitudes of numbers. *also* **MR 2.0, MR 2.3**

Change the Percent and Multiply

You can find a percent of a number by changing the percent to a decimal and multiplying.

When asleep, the average person dreams about 25% of the time. If you sleep for 9 hours, how long do you dream?

Find 25% of 9.

STEP 1

Change the percent to a decimal.

25% = 0.25

STEP 2

Multiply the number by the decimal.

0.25 × 9 = 2.25

25% of 9 equals 2.25.

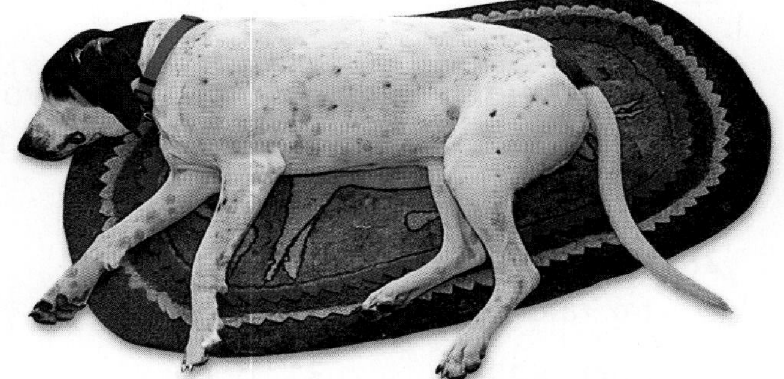

▲ **Since everyone dreams for about a quarter of his or her nightly sleep, the total world dream time per night is one million years!**

So, you dream about 2.25, or $2\frac{1}{4}$, hours on average.

Example

About 12% of people snore. Out of 600 people, how many of them snore?

Find 12% of 600.

12% = 0.12 ← Change 12% to a decimal.

0.12 × 600 = 72 ← Multiply the number by the decimal.

12% of 600 equals 72.

So, about 72 people snore out of 600 people.

MATH IDEA You can find a percent of a number by changing the percent to a decimal and multiplying.

Check

1. **Tell** how to find a percent of a number.

Find the percent of the number.

2. 5% of 80
3. 25% of 64
4. 15% of 120
5. 50% of 92
6. 40% of 60
7. 75% of 120
8. 35% of 39
9. 150% of 400

LESSON CONTINUES 

Practice and Problem Solving

Find the percent of the number.

10. 30% of 130 **11.** 15% of 40 **12.** 8% of 44 **13.** 35% of 160

14. 90% of 64 **15.** 100% of 15 **16.** 23% of 175 **17.** 200% of 190

18. 65% of 100 **19.** 70% of 210 **20.** 40% of 20 **21.** 15% of 60

22. 2% of 37 **23.** 60% of 60 **24.** 85% of 42 **25.** 150% of 14

You can find the sales tax for any item you buy by finding a percent of the price. Find the sales tax for each price to the nearest cent.

26. price: $25.00
tax rate: 5%

27. price: $8.50
tax rate: 4%

28. price: $0.99
tax rate: 7%

29. price: $198.23
tax rate: 9%

30. price: $32.00
tax rate: 7.5%

31. price: $1.79
tax rate: 6%

32. price: $45.00
tax rate: 8.25%

33. price: $79.80
tax rate: 8.5%

USE DATA **For 34–36, use the graph.**

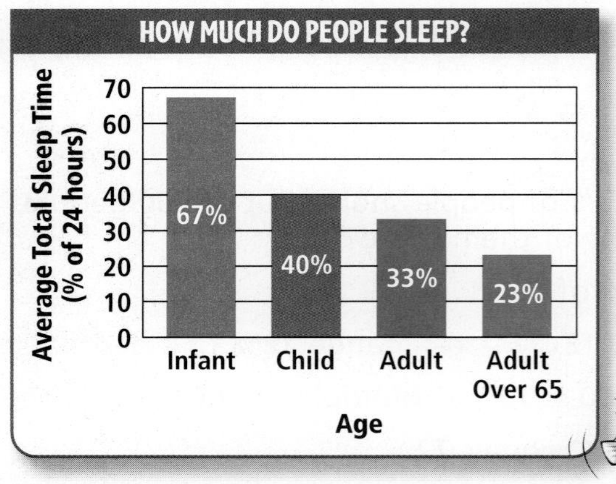

34. On average, how many hours a day does a child sleep?

35. On average, how many more hours a day does an infant sleep than an adult over 65?

36. On average, how many hours does an adult sleep in a year?

37. **What's the Question?** Karen scored 85% on the test. The test had 20 questions. The answer is 3 questions.

38. **REASONING** Eduardo showed Jim 50% of his card collection. If he showed him 25 cards, how many cards does Eduardo have in all?

39. **NUMBER SENSE** Tom says that when you find any percent of a number, the answer is always less than the number. Do you agree? Explain.

40. On average, a cat sleeps 12.1 hours a day, a dog sleeps 10.6 hours a day, and a guppy sleeps 7.0 hours a day. Order the animals from least amount of sleep to greatest amount of sleep.

41. Huang is stringing 8-mm beads for a garland. There are about 3.25 beads per inch. How many beads will he need to make a garland 48 inches long?

42. Anya won 40% of her 20 games. Linda won 60% of her 15 games. Who won more games?

Mixed Review and Test Prep

43. Identify the property used.
$4 \times (20 + 3) = (4 \times 20) + (4 \times 3)$ (p. 92)

44. Find the value of *m*. $m + 25 = 76$ (p. 74)

45. Round 2,912,433 to the nearest hundred thousand. (p. 32)

46. **TEST PREP** Which number is NOT evenly divisible by 4? (p. 254)

A 119,022 **C** 137,032
B 119,444 **D** 331,728

47. $2{,}567 \times 84$ (p. 148)

48. Write $\frac{3}{8}$ as a decimal. (p. 280)

49. Round 0.2374 to the nearest hundredth. (p. 46)

50. **TEST PREP** Rita's school has a student to teacher ratio of 20:1. If there are 300 students in the school, how many teachers are there? (p. 306)

F 5 **G** 10 **H** 15 **J** 20

Thinker's Corner

SALE PRICE Now that you know how to find a percent of a number, you can find the sale price of any item. Follow these steps to find the sale price of a sleeping bag.

STEP 1
Change the percent to a decimal.
30% = 0.30

STEP 2
Multiply the regular price by the decimal.
$159.75 × 0.30 = $47.925
$47.93 ← Round to the nearest cent.

STEP 3
Subtract the discount from the regular price.
$159.75 − $47.93 = $111.82
So, the sale price is $111.82.

Find the sale price to the nearest cent.

1. Regular price: $15.00

20% OFF

2. Regular price: $17.50

3. Regular price: $24.89

Extra Practice page H49, Set C

LESSON 5

Mental Math: Percent of a Number

Quick Review

Write each fraction as a percent.

1. $\frac{1}{2}$ **2.** $\frac{1}{4}$ **3.** $\frac{1}{10}$

4. $\frac{3}{4}$ **5.** $\frac{1}{1}$

Learn

GO FIGURE! At Bow-Wow Pet Salon, Lucia was charged $30 to have her Airedale groomed. If she wants to leave a 20% tip, how much should she leave for the tip?

TIPPING GUIDE	
Job Title	**Tip**
Chauffeur	20% of charge
Waiter/Waitress	15%–20% of charge
Hair Stylist	10%–15% of charge
Cabin Steward	5% of cruise fare

You can use mental math to find a percent of a number.

Find 20% of 30. **Think:** 20% = 10% + 10%

$$20\% \times 30 = (10\% \text{ of } 30) + (10\% \text{ of } 30)$$
$$\downarrow \qquad\qquad \downarrow$$
$$(0.1 \times 30) + (0.1 \times 30)$$
$$\downarrow \qquad\qquad \downarrow$$
$$3 + 3 = 6$$

So, Lucia should leave a tip of $6.00.

Examples

A Find 150% of 60.

Think: 150% = 100% + 50%

$$150\% \times 60 = (100\% \text{ of } 60) + (50\% \text{ of } 60)$$
$$\downarrow \qquad\qquad \downarrow$$
$$(1 \times 60) + (0.5 \times 60)$$
$$\downarrow \qquad\qquad \downarrow$$
$$60 + 30 = 90$$

So, 150% of 60 = 90.

B Find 40% of 800.

Think: 40% = 4 × 10%

$$40\% \times 800 = 4 \times (10\% \text{ of } 800)$$
$$\downarrow$$
$$4 \times (0.1 \times 800)$$
$$\downarrow$$
$$4 \times 80 = 320$$

So, 40% of 800 = 320.

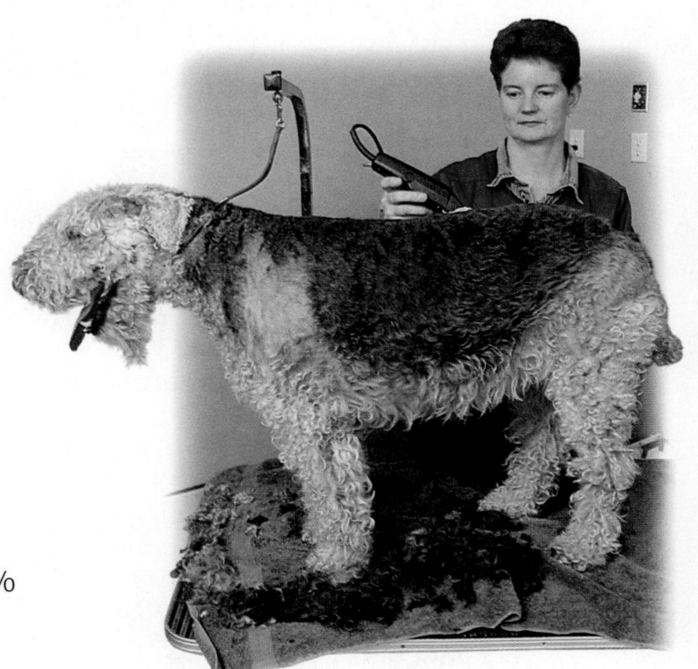

- **REASONING** How would you use multiples of 10% to find a 15% tip for a meal that costs $40?

CALIFORNIA STANDARDS **NS 1.2** Interpret percents as a part of a hundred; find decimal and percent equivalents for common fractions and explain why they represent the same value; compute a given percent of a whole number. **NS 1.0** Students compute with very large and very small numbers, positive integers, decimals, and fractions and understand the relationship between decimals, fractions, and percents. They understand the relative magnitudes of numbers. *also* **MR 1.0, MR 1.2, MR 2.0, MR 2.3**

Check

1. **Explain** how you can use mental math to find 50% of a number.

Use mental math to find the percent of each number.

2. 20% of 200
3. 75% of 300
4. 15% of 50
5. 10% of 1,000
6. 20% of 100
7. 50% of 1,000
8. 75% of 800
9. 10% of 10,000

Practice and Problem Solving

Use mental math to find the percent of each number.

10. 25% of 200
11. 10% of 25
12. 30% of 50
13. 60% of 2,000
14. 35% of 60
15. 85% of 200
16. 200% of 124
17. 150% of 500
18. 45% of 280
19. 40% of 25
20. 60% of 140
21. 175% of 400
22. 45% of 250
23. 20% of 330
24. 70% of 400
25. 25% of 600
26. 70% of 90
27. 60% of 300
28. 35% of 50
29. 130% of 2,000
30. Of 3,128 licensed cats, 23% of them are named Tiger. About how many are named Tiger?

31. **NUMBER SENSE** Write the percent that is nearly equivalent to each fraction.

    a. $\frac{8}{15}$ **b.** $\frac{19}{25}$ **c.** $\frac{23}{100}$ **d.** $\frac{16}{17}$

32. Lisa gave away 14 dolls, or 50% of her collection. How many did she have to start with?
33. **REASONING** The Scouts exceeded their \$5,000 goal in the fund-raiser by 10%. What percent of their goal did they meet? How much money did they raise?
34. **Write About It** If you know 10% of a number, how can you find 20% of the number and 5% of the number?

Mixed Review and Test Prep

35. $67.052 - 52.008$ (p. 50)
36. $363.905 - 89.54$ (p. 50)
37. Write 0.165 as a percent. (p. 318)
38. Find the value of n. $(6 - 3) \times n = 3$ (p. 90)
39. **TEST PREP** Shannon bought 2 yards of cloth for \$3.50 per yard and 4 yards of ribbon for \$0.90 per yard. Find the total cost. (p. 164)

    A \$7.10 **C** \$10.60
    B \$7.90 **D** \$10.90

LESSON 6

Problem Solving Strategy

Make a Graph

Quick Review

1. 100 × 33.45
2. 10 × 0.78
3. 100 × 3.59
4. 10 × 0.06
5. 10 × 1.05

PROBLEM Sonya surveyed 200 students in her school to find out their favorite type of music. She wants to display her data to show how the votes for each type of music relate to the total number of votes. What is the best way to display her data?

- What are you asked to find?
- What information will you use?

- What strategy can you use to solve the problem?

You can *make a graph* to display the data, showing the percent of students who prefer each type of music.

- What graph would be the best to make?

Since Sonya wants to show how the parts relate to the whole, a circle graph is best. Divide a circle into ten equal parts. Each part represents 10% of the 200 students. Find the percent of total votes for each type of music.

Pop Rock	Jazz, Classical, Rap	Country
$\frac{60}{200} = 30\%$	$\frac{20}{200} = 10\%$	$\frac{80}{200} = 40\%$

Shade the parts to represent each percent. Label and title the graph.

WHAT TYPE OF MUSIC IS YOUR FAVORITE?

Music	Student votes
Pop Rock	60
Jazz	20
Classical	20
Rap	20
Country	80

WHAT TYPE OF MUSIC IS YOUR FAVORITE?

- Why is this the best way to display the data?
- What is the sum of the percents that represents the whole, or 200 students?

CALIFORNIA STANDARDS SDAP 1.2 Organize and display single-variable data in appropriate graphs and representations and explain which types of graphs are appropriate for various data sets. **NS 1.2** Interpret percents as a part of a hundred; find decimal and percent equivalents for common fractions and explain why they represent the same value; compute a given percent of a whole number. *also* **NS 1.0, SDAP 1.0, SDAP 1.3, MR 2.3, MR 3.0, MR 3.1, MR 3.2**

Problem Solving Practice

Make a graph to solve.

1. **What if** 80 of the students liked pop rock, 40 liked country, 20 liked jazz, 40 liked classical, and 20 liked rap? What percent of the vote would pop rock receive? What percent of the vote would all the other types of music receive?

2. Out of $100 for the class trip, Kim spent $\frac{1}{10}$ for food, 20% for museum tickets, $\frac{3}{10}$ for souvenirs, and 40% for the bus ride. Which section is the largest? How much money does it represent?

PROBLEM SOLVING STRATEGIES

- Draw a Diagram or Picture
- Make a Model or Act It Out
- Make an Organized List
- Find a Pattern
- ▶ Make a Table or Graph
- Predict and Test
- Work Backward
- Solve a Simpler Problem
- Write an Equation
- Use Logical Reasoning

USE DATA Use the circle graph. Each year people in the United States throw away over 180 million tons of trash.

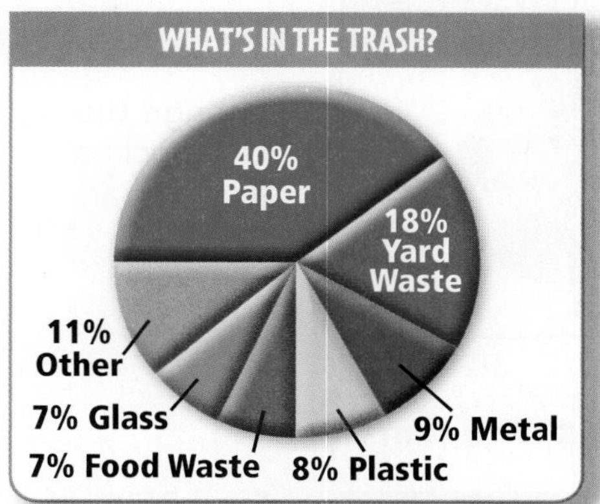

3. What fills $\frac{2}{25}$ of the trash?
   A paper **C** yard waste
   B plastic **D** other

4. How much paper goes in the trash in an average month?
   F 1 million tons **H** 6 million tons
   G 3 million tons **J** 72 million tons

Mixed Strategy Practice

USE DATA For 5–8, use the table to make a circle graph. There are 50 musicians in the school orchestra.

WHO PLAYS WHAT INSTRUMENT IN THE SCHOOL ORCHESTRA?

Section	Percent
Strings	60%
Percussion	10%
Brass	10%
Woodwind	20%

5. How many musicians play woodwind instruments?

6. How many more musicians play stringed instruments than brass instruments?

7. How many fewer musicians play woodwind instruments than stringed instruments?

8. **What if** there were 100 musicians in the orchestra. How many would play stringed instruments?

9. There are 56 teachers at Sunridge Elementary. Of the teachers, $\frac{1}{4}$ have no pets. Some teachers have one pet, and twice as many have two or more. How many have two or more?

10. **Write About It** Out of 50 students surveyed on favorite national parks, 30 chose Yellowstone, 10 chose Everglades, and 10 Death Valley. Explain how you would display this data.

LESSON 7

Compare Data Sets

Quick Review

Find 20% of each number.

1. 100 **2.** 75

3. 200 **4.** 10

5. 20

Learn

PET PICKS Molly conducted two surveys to find out the favorite pets of the fifth graders. Molly made two circle graphs to compare her data. In which survey did dogs get more votes?

Survey 1: 20 Students

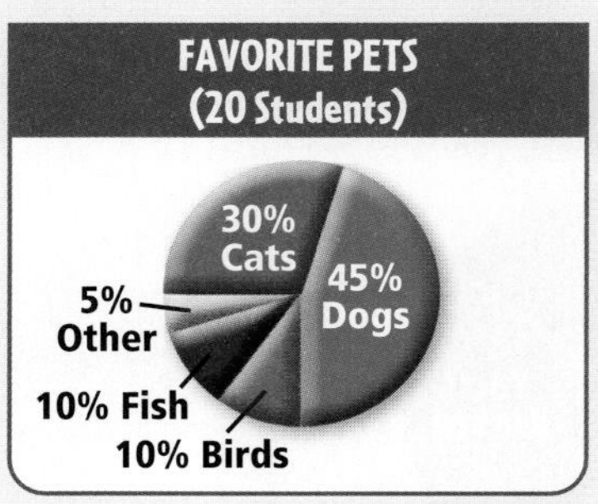

Find 45% of 20.

STEP 1

Change the percent to a decimal.

45% = 0.45

STEP 2

Multiply the total number of students surveyed by the decimal.

0.45 × 20 = 9

9 students voted for dogs.

Survey 2: 40 Students

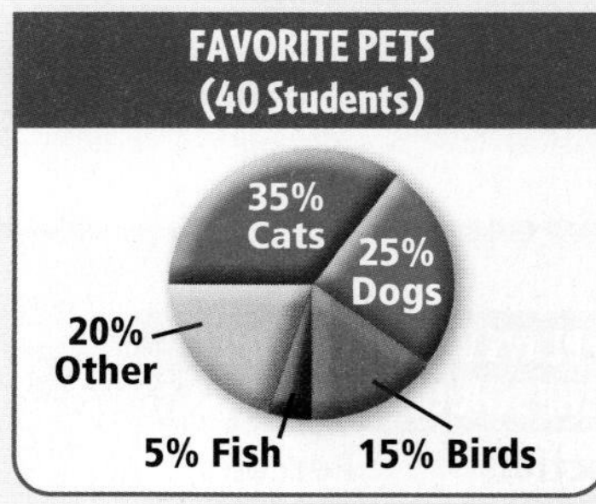

Find 25% of 40.

STEP 1

Change the percent to a decimal.

25% = 0.25

STEP 2

Multiply the total number of students surveyed by the decimal.

0.25 × 40 = 10

10 students voted for dogs.

Since $10 > 9$, dogs received more votes in Survey 2.

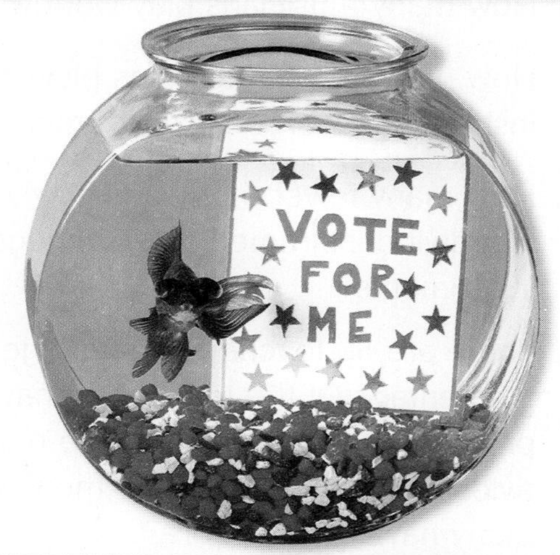

Check

1. REASONING **Compare** the two circle graphs. It appears that dogs received more votes in Survey 1 than Survey 2. Explain why.

CALIFORNIA STANDARDS SDAP 1.0 Students display, analyze, compare, and interpret different data sets, including data sets of different sizes. **SDAP 1.3** Use fractions and percentages to compare data sets of different sizes. *also* **NS 1.0, NS 1.2, AF 1.1, SDAP 1.2, MR 2.0, MR 2.3, MR 3.1**

USE DATA **For 2–3, use the circle graphs on page 332.**

2. In which survey did cats receive more votes? How many more votes?

3. Which pet received the same number of votes in both surveys?

Practice and Problem Solving

USE DATA **For 4–7, use the circle graphs. Both groups held the same fund-raisers.**

$300 RAISED BY SOCCER TEAM

40% Car Wash
25% Bake Sale
35% Walkathon

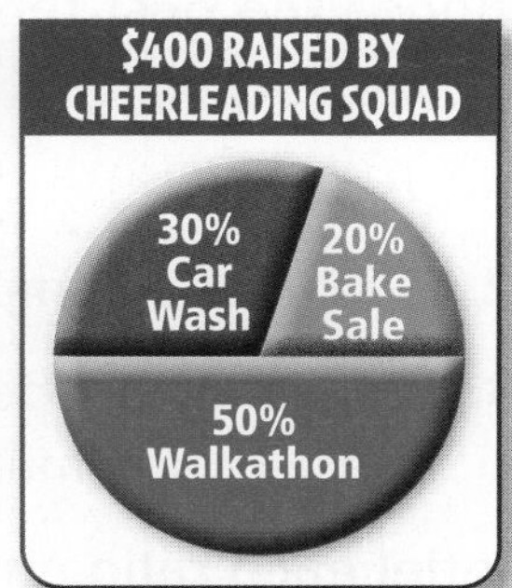

4. Which group raised more money from its bake sale?

5. Which fund-raiser made the most money?

6. Which fund-raiser made the same amount of money for both groups?

7. Altogether, how much money was raised from washing cars and walking?

USE DATA **For 8–10, use the table.**

WHAT IS YOUR DOG'S NAME?					
	Jake	Lady	Max	Sasha	Sam
Charlotte, NC	16%	30%	29%	8%	17%
United States	14%	13%	30%	18%	25%

8. Of the 1,615 dogs registered in Charlotte, how many are named Max?

9. List the names from the United States data in order from the least percent to the greatest percent.

10. The latest trend is to name pets after people. What are the top five names for your classmates' dogs? Take a survey and compare your data with the data above. Does your class follow the latest trend?

11. **What's the Error?** 50% of the 100 sixth graders walk to school. 50% of the 120 fifth graders walk to school. Bob says the same number of fifth and sixth graders walk to school. Find and correct Bob's error.

12. **Write a problem** that compares two sets of data. Collect your own data by taking two surveys.

Mixed Review and Test Prep

13. What is 25% of 200? (p. 324)

14. What is 50% of 84? (p. 324)

15. Use mental math to find 15% of 8,500. (p. 328)

16. List the factors of 36. (p. 258)

17. **TEST PREP** Find $9.1 + 0.053 + 12$. (p. 50)

 A 11.153
 B 21.053
 C 21.153
 D 22.053

Extra Practice page H49, Set E

CHAPTER 18

Review/Test

CHECK VOCABULARY AND CONCEPTS

decimal
percent
product
ratio

Choose the best term from the box.

1. You can write a _?_ as a fraction, a decimal, or a percent. (p. 321)
2. A ratio of a number to 100 is a _?_. (p. 316)
3. You can find a percent of a number by changing a percent to a _?_ and multiplying. (p. 325)

Model each ratio on grid paper. Then write the percent. (pp. 316–317)

4. 27 tiles out of 100 tiles
5. 8 bats out of 100 bats
6. 49 balls out of 100 balls

CHECK SKILLS

Write each decimal as a percent. (pp. 318–319)

7. 0.35
8. 0.75
9. 0.49
10. 1.23

Copy and complete the tables. Write each fraction in simplest form. (pp. 318–323)

11.

Fraction	Decimal	Percent
■	0.20	■

12.

Fraction	Decimal	Percent
■	■	43%

Find the percent of the number. (pp. 324–329)

13. 6% of 50
14. 200% of 75
15. 15% of 345
16. 99% of 99

CHECK PROBLEM SOLVING

For 17–18, make a circle graph to solve. 120 students were surveyed. (pp. 330–331)

FAVORITE BEVERAGES				
Beverage	Water	Milk	Soda	Juice
Student votes	12	36	48	24

17. What percent of the students liked soda the best?
18. What percent of the students liked milk the best?

Of 80 fifth graders, 65% buy lunch and 35% bring lunch. Of 70 sixth graders, 80% buy lunch and 20% bring lunch. (pp. 332–333)

19. Which grade has more students who buy lunch?
20. How many fifth graders bring lunch?

Cumulative Review

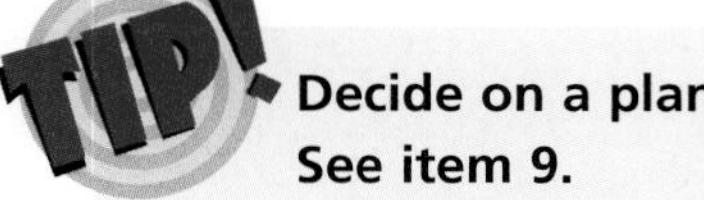

Decide on a plan.
See item 9.

Write the percent as a decimal. Then write the number sentence that will solve the problem. Find the answer.

Also see problem **4**, p. H63.

For 1–11, choose the best answer.

1. What is 98,671 rounded to the nearest hundred?

   A 98,000 C 98,670
   B 98,600 D 98,700

2. How is twenty-five and six thousand thirty-eight ten-thousandths written in standard form?

   F 25.6038 H 2.5638
   G 25.638 J Not here

For 3–5, use the circle graph.

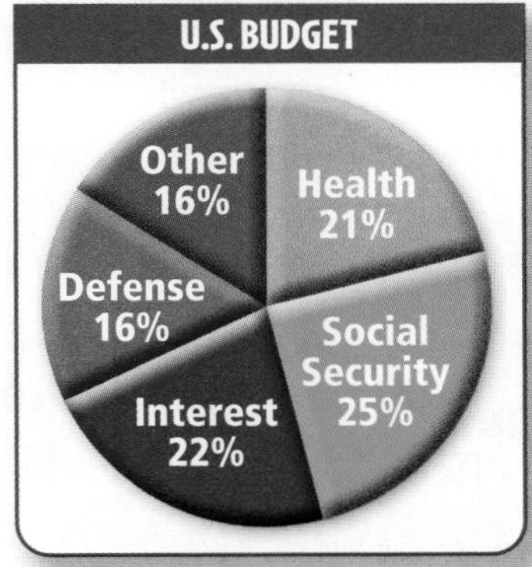

3. What percent of the budget is spent on health and social security?

   A 3% C 25%
   B 21% D 46%

4. What percent of the budget is *not* spent on defense?

   F 84% H 70%
   G 75% J 67%

5. Out of every dollar, how much is spent on interest?

   A \$2.20 C \$0.22
   B \$1.60 D \$0.16

6. 269×82

   F 21,058 H 23,058
   G 21,378 J Not here

7. Last week 30 out of 100 students at Wilson School brought their lunches. What is another way to express this?

   A $\frac{3}{10}$ C 3.0
   B $\frac{30}{70}$ D 70%

8. About 70% of the students at Jamie's school ride a bus to school. Which of the following is **not** equivalent to 70%?

   F 0.7 H $\frac{70}{100}$
   G 0.07 J $\frac{7}{10}$

9. If 90% of the population is right-handed, how many people in a group of 200 should be right-handed?

   A 180 C 20
   B 90 D 18

10. Which ratio is **not** equivalent to 6:18?

   F 2:6 H 1:3
   G 2:9 J 12:36

11. A magazine subscription regularly costs \$24.50 for a year. Darius got it for 20% off. How much did he pay?

   A \$4.50 C \$19.60
   B \$4.90 D \$20.60

UNIT 5

MATH DETECTIVE

Common Thread

REASONING You can use Venn diagrams to sort information. Venn diagrams show relationships among groups of items called sets. Look at the Venn diagram.

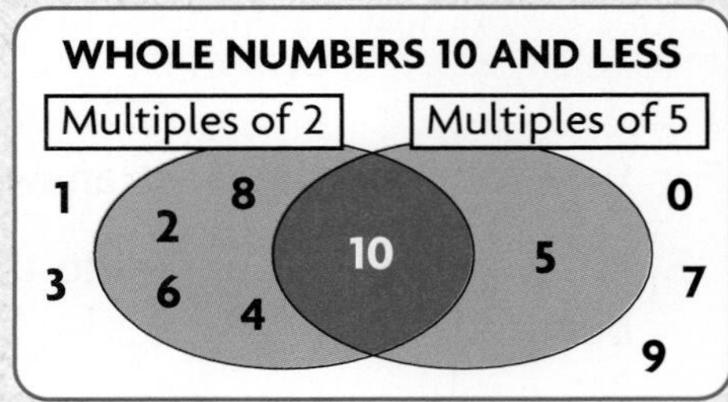

- The numbers 2, 4, 6, 8, and 10 belong to the set of numbers that are multiples of 2.
- The numbers 5 and 10 belong to the set of numbers that are multiples of 5.
- The number 10 is in the intersection of the two sets, so it belongs to both sets. It is a common multiple of 2 and 5.
- The numbers 0, 1, 3, 7, and 9 are outside the circles. They are not multiples of 2 or 5.

Case 1

Copy and complete the diagram at the right. Sort the numbers from 0 to 20.

- What numbers are in the intersection of the sets of multiples of 2 and multiples of 3?
- What numbers are outside the circles?

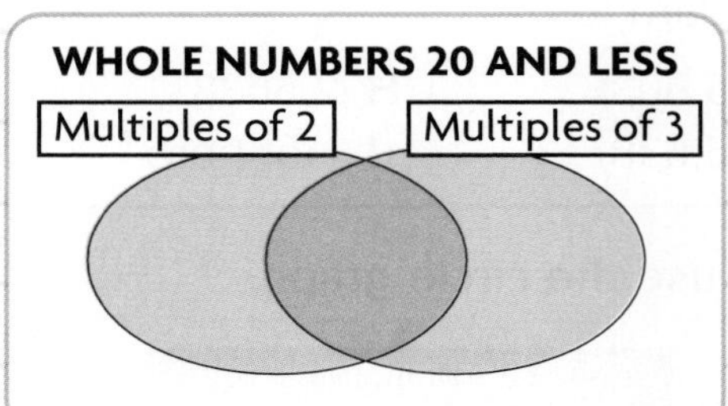

Case 2

Find all the relationships among the sets of numbers in the diagram at the right. Write labels to describe each part of the diagram.

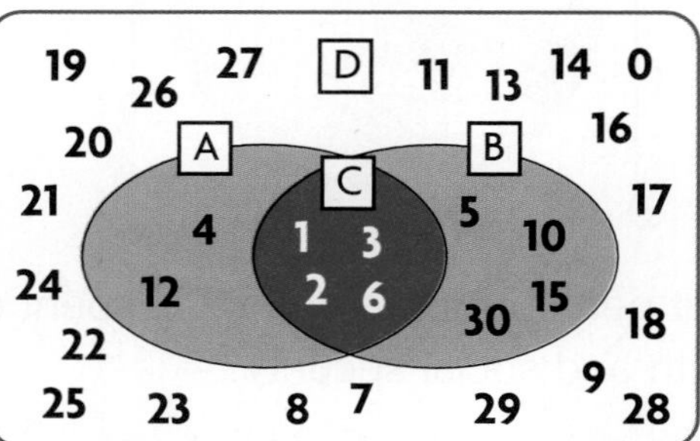

Think It Over!

- **Write a problem** Make a Venn diagram of your own. Ask a classmate to write labels to describe each part of the diagram.

UNIT 5

Challenge

Explore Patterns with Fractions

A **pattern** of fractions follows a rule. When you know the rule for the pattern, you can figure out the next fraction in the sequence.

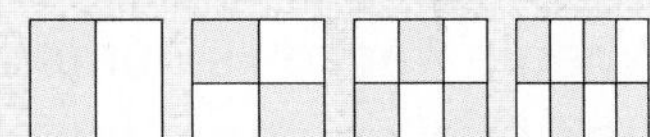

Find the next fraction in this pattern. $\frac{1}{2}, \frac{2}{4}, \frac{3}{6}, \frac{4}{8}, \ldots$

STEP 1

Look for a pattern to find a rule.

Think: What do I do to $\frac{1}{2}$ to get $\frac{2}{4}$?

$\frac{1+1}{2+2} \begin{matrix}\leftarrow\text{add } 1\\ \leftarrow\text{add } 2\end{matrix} = \frac{2}{4}$

STEP 2

Test the rule for all the terms.

Rule: Add 1 to the numerator. Add 2 to the denominator.

$\frac{2+1}{4+2} = \frac{3}{6}, \frac{3+1}{6+2} = \frac{4}{8}$

STEP 3

Use the rule to find the next term.

$\frac{4}{8} \rightarrow \frac{4+1}{8+2} = \frac{5}{10}$

So, the next fraction in the pattern is $\frac{5}{10}$.

Examples

A $\frac{1}{3}, \frac{2}{6}, \frac{3}{9}, \frac{4}{12}, \ldots$

Rule: Add 1 to the numerator. Add 3 to the denominator.

So, the next fraction in the pattern is $\frac{4+1}{12+3} = \frac{5}{15}$.

B $\frac{3}{6}, \frac{6}{12}, \frac{12}{24}, \frac{24}{48}, \ldots$

Rule: Multiply the numerator and the denominator by 2.

So, the next fraction in the pattern is $\frac{24 \times 2}{48 \times 2} = \frac{48}{96}$.

Talk About It

- What would you do if your rule in Step 1 did not work with the other fractions?

Try It

Find the next fraction in the pattern. Write a rule.

1. $\frac{3}{4}, \frac{9}{12}, \frac{27}{36}, \frac{81}{108}, \ldots$
2. $\frac{6}{7}, \frac{7}{8}, \frac{8}{9}, \frac{9}{10}, \ldots$
3. $\frac{1}{2}, \frac{3}{5}, \frac{5}{8}, \frac{7}{11}, \ldots$

Create a pattern for the rule. Start with any fraction.

4. **Rule:** Multiply the numerator by 2. Add 1 to the denominator.
5. **Rule:** Add 2 to the numerator. Add 1 to the denominator.
6. **Rule:** Add 2 to the numerator and the denominator.

UNIT 5

Study Guide and Review

VOCABULARY

Choose the best term from the box.

common multiples
composite number
least common multiples
simplest form
ratio
percent

1. When a fraction's numerator and denominator have 1 as their only common factor, the fraction is in __?__. (p. 290)
2. A comparison of two quantities is a __?__. (p. 302)
3. The ratio of a number to 100 is a __?__. (p. 316)

STUDY AND SOLVE

Chapter 15

Find the least common multiple of two numbers.

Find the LCM of 6 and 9.
Multiples of 6: 6, 12, 18, 24, 30, 36, . . .
Multiples of 9: 9, 18, 27, 36, 45, 54, . . .
So, the LCM is 18.

Write the LCM of each set of numbers. (pp. 256–257)

4. 2 and 14
5. 6 and 32
6. 3 and 18
7. 3, 5, and 20
8. 5 and 6
9. 2, 7, and 14

Write and evaluate exponents.

Exponent ↓
Base → $3^5 = 3 \times 3 \times 3 \times 3 \times 3 = 243$

Find the value. (pp. 266–271)

10. 10^3 11. 8^3 12. 2^2 13. 1^4
14. 4^3 15. 3^3 16. 5^2 17. 2^6

Chapter 16

Write fractions as decimals and decimals as fractions.

To write $\frac{2}{5}$ as a decimal, divide 2 by 5.
$2 \div 5 = 0.4$
To write 0.125 as a fraction, use place value.
$0.125 = \frac{125}{1{,}000}$ ← 125 thousandths

Write a fraction for each decimal. (pp. 280–283)

18. 0.16 19. 0.5 20. 0.012

Write a decimal for each fraction. (pp. 280–283)

21. $\frac{3}{4}$ 22. $\frac{1}{5}$ 23. $\frac{5}{8}$

Write equivalent fractions.

Multiply the numerator and the denominator by the same number.
$\frac{4}{6} = \frac{4 \times 2}{6 \times 2} = \frac{8}{12}$

Divide the numerator and the denominator by a common factor.
$\frac{8}{12} = \frac{8 \div 4}{12 \div 4} = \frac{2}{3}$

Write an equivalent fraction. (pp. 284–285)

24. $\frac{9}{21}$ 25. $\frac{10}{12}$
26. $\frac{2}{5}$ 27. $\frac{4}{7}$
28. $\frac{11}{12}$ 29. $\frac{16}{20}$

Chapter 17

Write ratios that are equivalent to a given ratio.

To find equivalent ratios, multiply or divide the numerator and the denominator by the same number.

$\frac{1 \times 2}{5 \times 2} = \frac{2}{10}$

So, $\frac{1}{5}$ and $\frac{2}{10}$, are equivalent ratios.

$\frac{12 \div 3}{15 \div 3} = \frac{4}{5}$

So, $\frac{12}{15}$ and $\frac{4}{5}$ are equivalent ratios.

Write three ratios that are equivalent to the given ratio. (pp. 304–307)

30. $\frac{2}{5}$
31. 10 to 20
32. 3 to 8
33. $\frac{6}{2}$
34. 1:4
35. 7:3

Chapter 18

Write a number as a decimal, percent, and fraction.

twenty-one hundredths

decimal: 0.21
percent: 21%
fraction: $\frac{21}{100}$

Write each number as a decimal, percent, and fraction. (pp. 318–323)

36. sixteen hundredths
37. ninety-one hundredths
38. four tenths
39. two

Use a decimal to find the percent of a number.

Find 20% of 70.

0.20 × 70 = 14 — Change the percent to a decimal and multiply.

So, 20% of 70 = 14.

Find the percent of each number. (pp. 324–329)

40. 15% of 30
41. 6% of 25
42. 45% of 115
43. 19% of 25
44. 18% of 300
45. 77% of 430
46. 7.5% of 20
47. 300% of 12

PROBLEM SOLVING PRACTICE

Solve. (pp. 262–263, 296–297)

48. The GCF of 8 and another number is 1. The LCM is 24. What is the other number?

49. Shawn rode his bike $\frac{5}{8}$ mi on Monday, $1\frac{2}{3}$ mi on Tuesday, $\frac{3}{2}$ mi on Wednesday, $2\frac{1}{9}$ mi on Thursday, and $\frac{7}{16}$ mi on Friday. On which day did he ride the least distance? the greatest distance?

California Connections

AGRICULTURE

For more than fifty years, California has been the number one food and agricultural producer in the United States.

California produces 350 different crops and agricultural products. ▼

1. California exports about 20% of the food and agricultural products it grows. What percent does it NOT export? Write each percent as a decimal and as a fraction in simplest form.

2. Twenty-one out of every twenty-five pounds of strawberries that are grown in the United States are grown in California. What percent of the strawberries are grown in California?

3. In California, 1 in every 10 jobs is related to agriculture. Write an equivalent ratio to compare the number of jobs related to agriculture to the total number of jobs.

4. California produces nearly 100% of the olives grown in the United States. Explain what this statement means.

USE DATA For 5–6, use the circle graph.

5. Of every 500 pounds of milk produced in California, how much is used to make butter and nonfat dry milk?

6. What is the most common use of the milk produced in California? Explain how you know.

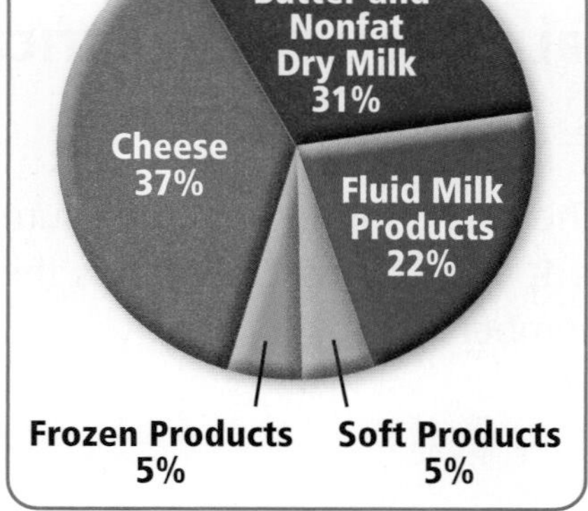

FLOWERS

More than 400 California growers market cut flowers and foliage. The favorable California climate allows for year-round production of an enormous variety of flowers.

1. Flowers are harvested by hand, cooled, and boxed to ship by truck or by air. About $\frac{2}{3}$ of the flowers are shipped by truck. What fraction of the flowers are shipped by air?
2. About 18% of all cut roses sold in the United States are grown in California. Write 18% as a decimal and as a fraction in simplest form.
3. About $\frac{3}{5}$ of all cut flowers grown in the United States are grown in California.
   a. Write $\frac{3}{5}$ as a decimal and as a percent.
   b. What fraction of the cut flowers grown in the United States are NOT grown in California?
   c. Are more cut flowers grown in California than in all the other states combined? Explain.

Cut flowers are grown in open fields and in covered greenhouses.▼

CHAPTER 19

Add and Subtract Fractions

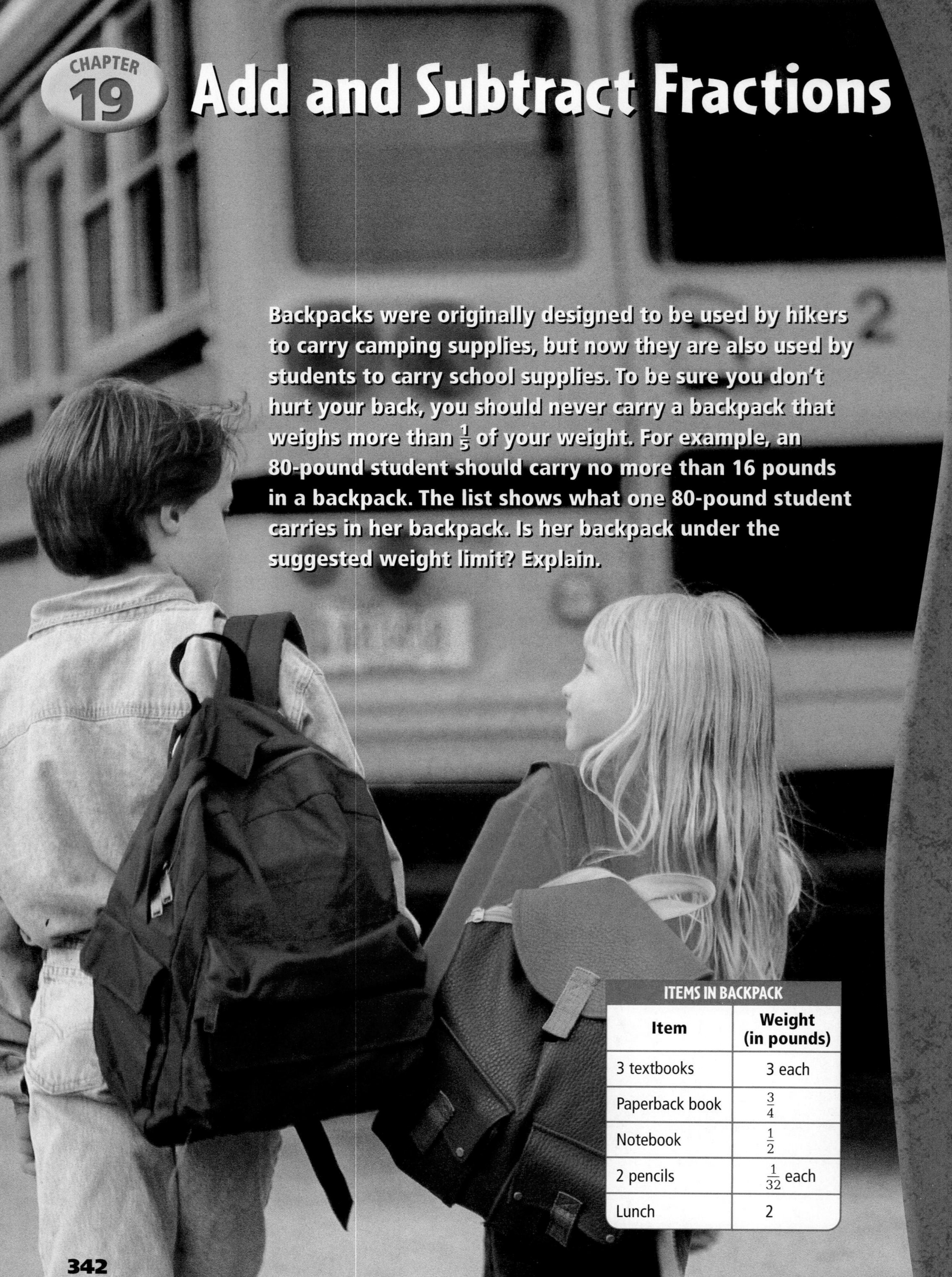

Backpacks were originally designed to be used by hikers to carry camping supplies, but now they are also used by students to carry school supplies. To be sure you don't hurt your back, you should never carry a backpack that weighs more than $\frac{1}{5}$ of your weight. For example, an 80-pound student should carry no more than 16 pounds in a backpack. The list shows what one 80-pound student carries in her backpack. Is her backpack under the suggested weight limit? Explain.

ITEMS IN BACKPACK

Item	Weight (in pounds)
3 textbooks	3 each
Paperback book	$\frac{3}{4}$
Notebook	$\frac{1}{2}$
2 pencils	$\frac{1}{32}$ each
Lunch	2

Use this page to help you review and remember important skills needed for Chapter 19.

VOCABULARY

Choose the best term from the box.

common multiples
fraction
least common multiple
whole number

1. A number that names part of a whole or part of a group is called a __?__.
2. Multiples of one number that are also multiples of another number are called __?__.
3. The least number that is a common multiple of two or more numbers is called the __?__.

SKIP COUNT FRACTIONS (See p. H21.)

Complete the pattern and write the rule.

4. $2\frac{1}{2}$, 3, $3\frac{1}{2}$, ■, ■, 5, $5\frac{1}{2}$, ■
5. $6\frac{1}{3}$, $6\frac{2}{3}$, ■, ■, ■, 8
6. $3\frac{3}{4}$, $3\frac{7}{8}$, 4, $4\frac{1}{8}$, ■, ■, $4\frac{1}{2}$, ■
7. 1, $1\frac{1}{4}$, ■, ■, 2, ■, $2\frac{1}{2}$
8. $\frac{1}{3}$, $1\frac{2}{3}$, ■, ■, $5\frac{2}{3}$, ■, $8\frac{1}{3}$
9. $\frac{3}{7}$, ■, ■, $2\frac{1}{7}$, $2\frac{5}{7}$, ■, $3\frac{6}{7}$

FRACTIONS ON A RULER (See p. H22.)

Write a fraction or a mixed number to name each point shown.

10.

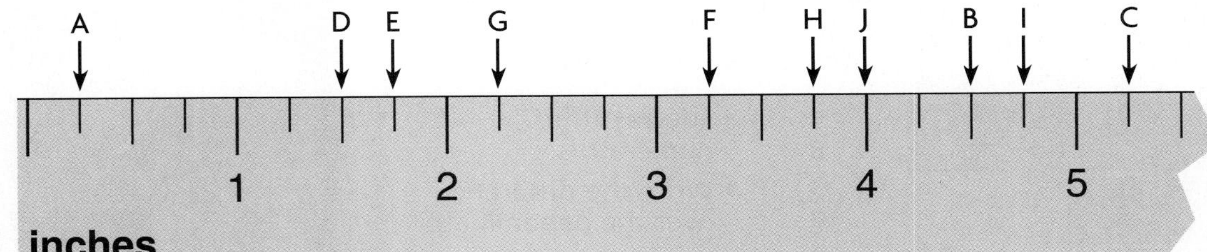

11.

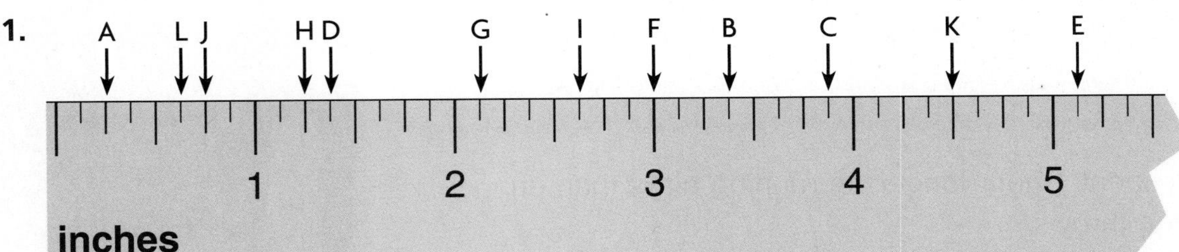

LESSON 1

Add and Subtract Like Fractions

Quick Review

Write each fraction in simplest form.

1. $\frac{2}{8}$
2. $\frac{3}{6}$
3. $\frac{3}{9}$
4. $\frac{8}{12}$
5. $\frac{12}{24}$

Learn

WHOOSH! Damon is on the basketball team. The team practices free throws for $\frac{3}{8}$ hour and runs plays for $\frac{5}{8}$ hour. How long does the team practice last?

Add. $\frac{3}{8} + \frac{5}{8}$

One Way

Use a model.

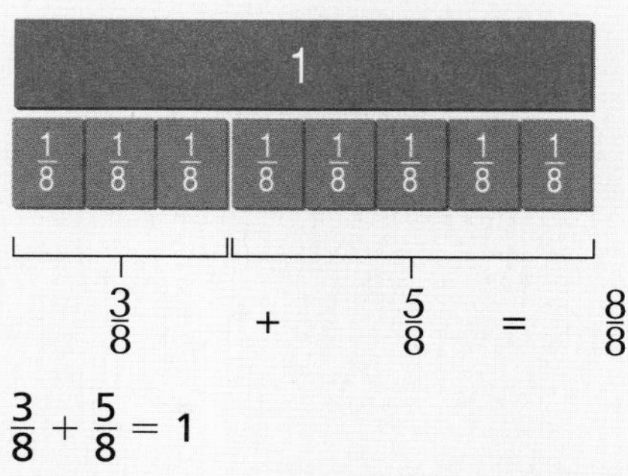

$\frac{3}{8} + \frac{5}{8} = 1$

Another Way

Use paper and pencil.

$$\frac{3}{8} + \frac{5}{8} = \frac{8}{8} = 1$$

- Add the numerators.
- Write the sum over the denominator.
- Write the sum in simplest form.

$\frac{3}{8} + \frac{5}{8} = 1$

So, the team practice lasts 1 hour.

How much more time did Damon spend practicing running plays than shooting free throws?

Subtract. $\frac{5}{8} - \frac{3}{8}$

One Way

Use a model.

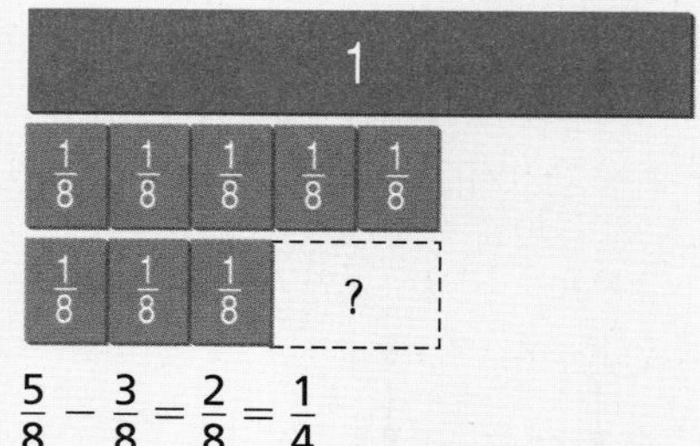

$\frac{5}{8} - \frac{3}{8} = \frac{2}{8} = \frac{1}{4}$

Another Way

Use paper and pencil.

$$\frac{5}{8} - \frac{3}{8} = \frac{2}{8} = \frac{1}{4}$$

- Subtract the numerators.
- Write the difference over the denominator.
- Write the difference in simplest form.

So, Damon spent $\frac{1}{4}$ hour longer on running plays than on shooting free throws.

On May 10, 1997, Karl Malone set a record for the most free throws made in a play-off game. He made 18 free throws without missing one! ▼

CALIFORNIA STANDARDS **NS 2.0** Students perform calculations and solve problems involving addition, subtraction, and simple multiplication and division of fractions and decimals. **NS 2.3** Solve simple problems, including ones arising in concrete situations, involving the addition and subtraction of fractions and mixed numbers, and express answers in the simplest form. *also* **NS 1.0, MR 1.0, MR 2.0, MR 2.3, MR 3.0, MR 3.3**

Check

1. **Explain** what is true about the numerators of two like fractions with a sum of 1.

Find the sum or difference. Write it in simplest form.

2. $\frac{2}{3} + \frac{1}{3}$
3. $\frac{1}{4} + \frac{2}{4}$
4. $\frac{5}{10} + \frac{2}{10}$
5. $\frac{2}{3} - \frac{1}{3}$
6. $\frac{7}{8} - \frac{1}{8}$
7. $\frac{5}{6} - \frac{2}{6}$

Practice and Problem Solving

Find the sum or difference. Write it in simplest form.

8. $\frac{1}{6} + \frac{3}{6}$
9. $\frac{3}{10} - \frac{1}{10}$
10. $\frac{5}{7} + \frac{2}{7}$
11. $\frac{4}{5} - \frac{1}{5}$
12. $\frac{7}{12} + \frac{1}{12}$
13. $\frac{4}{9} - \frac{1}{9}$
14. $\frac{2}{8} + \frac{3}{8}$
15. $\frac{5}{12} - \frac{2}{12}$
16. $\frac{1}{10} + \frac{4}{10}$

Algebra **Find the value of *n*.**

17. $n + \frac{4}{9} = \frac{7}{9}$
18. $\frac{9}{10} - n = \frac{7}{10}$
19. $1 - n = \frac{8}{11}$

20. **MEASUREMENT** Karla is using $\frac{3}{8}$ yard of ribbon for one craft project and $\frac{5}{8}$ yard for another. How much ribbon is she using altogether?

21. Suppose you ordered $\frac{4}{8}$ of a pizza with pepperoni, $\frac{3}{8}$ with mushrooms, $\frac{5}{8}$ with vegetables, and $\frac{4}{8}$ plain. How many whole pizzas would you get?

22. **REASONING** Pat made $\frac{3}{5}$ of her free throws in the basketball game. What part of her free throws did she not make?

23. **Write About It** When you add or subtract two like fractions, when is the answer a fraction with a different denominator?

Mixed Review and Test Prep

24. Order 0.101, 0.011, and 0.111 from greatest to least. (p. 24)

25. $32{,}734 + 12{,}196$ (p. 36)

26. $6{,}213 - 4{,}789$ (p. 36)

27. $7 \times 213{,}645$ (p. 146)

28. **TEST PREP** Choose the mean for the following set of data. 123, 64, 76, 149 (p. 102)

A 99
B 101
C 102
D 103

Extra Practice page H50, Set A

LESSON 2

HANDS ON
Add Unlike Fractions

Quick Review

Write an equivalent fraction.

1. $\frac{1}{2}$ **2.** $\frac{3}{9}$

3. $\frac{5}{20}$ **4.** $\frac{2}{5}$ **5.** $\frac{6}{24}$

MATERIALS
fraction bars

Explore

Steve is making 5-bean salad. He mixed the spices first by adding $\frac{1}{2}$ teaspoon of salt and $\frac{1}{4}$ teaspoon of pepper. How many teaspoons of spices has he added altogether?

Use fraction bars to add fractions with unlike denominators.

Add. $\frac{1}{2} + \frac{1}{4}$

STEP 1

Place one $\frac{1}{2}$-bar and one $\frac{1}{4}$-bar under a 1-whole bar.

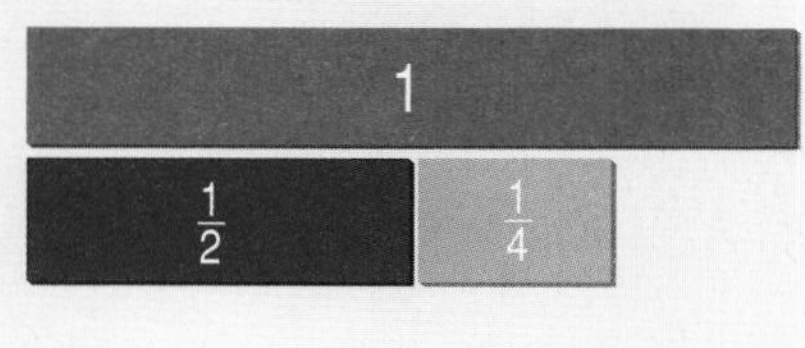

STEP 2

Find like fraction bars that fit exactly under the sum $\frac{1}{2} + \frac{1}{4}$.

1

$\frac{1}{2}$ $\frac{1}{4}$

$\frac{1}{4}$ $\frac{1}{4}$ $\frac{1}{4}$

$\frac{1}{2} + \frac{1}{4} = \frac{3}{4}$

So, Steve added $\frac{3}{4}$ teaspoon of spices.

- How many fourths did you use to fit exactly under the sum $\frac{1}{2} + \frac{1}{4}$?
- What is the least common multiple of 2 and 4?

Which like fraction bars would you use to find the sum $\frac{1}{8} + \frac{1}{4}$?

Try It

Find the sum.

a. $\frac{1}{8} + \frac{1}{4}$ **b.** $\frac{2}{3} + \frac{1}{6}$ **c.** $\frac{3}{4} + \frac{1}{2}$

CALIFORNIA STANDARDS NS 2.0 Students perform calculations and solve problems involving addition, subtraction, and simple multiplication and division of fractions and decimals. **NS 2.3** Solve simple problems, including ones arising in concrete situations, involving the addition and subtraction of fractions and mixed numbers, and express answers in the simplest form. *also* **NS 1.0, MR 2.0, MR 2.3, MR 2.4, MR 3.0, MR 3.2**

Connect

When you add fractions with unlike denominators, you need to find equivalent fractions with like denominators.

TECHNOLOGY LINK
More Practice:
Use E-Lab, ***Adding Unlike Fractions.***
www.harcourtschool.com/elab2002

Add. $\frac{1}{3} + \frac{1}{2}$

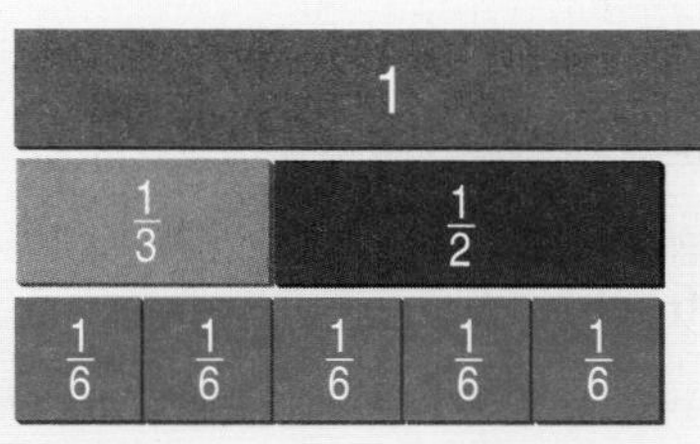

$\frac{1}{3} + \frac{1}{2}$

$\downarrow \quad \downarrow$

$\frac{2}{6} + \frac{3}{6}$

Think: $\frac{1}{3}$ is equivalent to $\frac{2}{6}$ and $\frac{1}{2}$ is equivalent to $\frac{3}{6}$.

So, $\frac{1}{3} + \frac{1}{2} = \frac{5}{6}$.

- Why do you need to find equivalent fractions before adding unlike fractions?

Practice

Use fraction bars to find the sum.

1. $\frac{1}{2}$ | $\frac{1}{6}$ | $\frac{1}{6}$

2. $\frac{1}{10}$ | $\frac{1}{10}$ | $\frac{1}{2}$

Use fraction bars to find the sum.

3. $\frac{1}{4} + \frac{1}{6}$ **4.** $\frac{1}{5} + \frac{3}{10}$ **5.** $\frac{5}{6} + \frac{1}{12}$ **6.** $\frac{3}{10} + \frac{1}{2}$

7. $\frac{2}{5} + \frac{1}{2}$ **8.** $\frac{4}{10} + \frac{1}{2}$ **9.** $\frac{5}{6} + \frac{1}{4}$ **10.** $\frac{1}{8} + \frac{3}{4}$

11. **REASONING** John has to mix $\frac{1}{2}$ cup of flour and $\frac{1}{4}$ cup of sugar. He has a container that holds $\frac{7}{8}$ cup. Can John mix the flour and sugar in the container? Explain.

12. **Write About It** Explain how to add fractions with unlike denominators.

Mixed Review and Test Prep

13. Order 31.019, 30.091, and 31.119 from least to greatest. (p. 24)

14. $\begin{array}{r} 108{,}097 \\ +112{,}336 \\ \hline \end{array}$ (p. 36)

15. $\begin{array}{r} 51{,}002 \\ -39{,}659 \\ \hline \end{array}$ (p. 36)

16. Find the value of n in $59 - 13 = n$. (p. 74)

17. **TEST PREP** Which fraction is between $\frac{1}{8}$ and $\frac{1}{2}$? (p. 286)

A $\frac{1}{10}$ **C** $\frac{2}{3}$

B $\frac{1}{4}$ **D** $\frac{3}{4}$

LESSON 3

HANDS ON
Subtract Unlike Fractions

Quick Review

Write an equivalent fraction.

1. $\frac{3}{12}$ 2. $\frac{1}{6}$ 3. $\frac{5}{10}$
4. $\frac{3}{4}$ 5. $\frac{3}{24}$

MATERIALS
fraction bars

Explore

At a crafts class, Liz is cutting colored string to make a friendship bracelet. A piece of blue string is $\frac{2}{3}$ yard long. Liz cuts $\frac{1}{2}$ yard from it. How much string is left?

Use fraction bars to subtract fractions with unlike denominators.

Subtract. $\frac{2}{3} - \frac{1}{2}$

STEP 1

Place two $\frac{1}{3}$-bars under a 1-whole bar. Then place one $\frac{1}{2}$-bar under the $\frac{1}{3}$-bars.

Compare the bars.

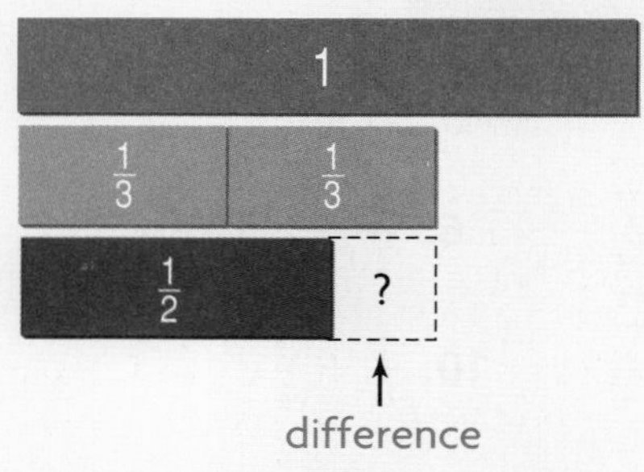

STEP 2

Find like fraction bars that fit exactly under the difference $\frac{2}{3} - \frac{1}{2}$.

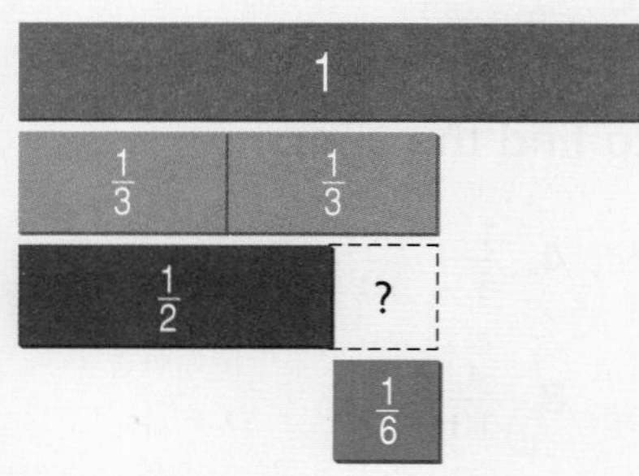

$\frac{2}{3} - \frac{1}{2} = \frac{1}{6}$

So, Liz had $\frac{1}{6}$ yard of ribbon left.

- How many sixths did you use to fit exactly under the difference $\frac{2}{3} - \frac{1}{2}$?
- What fraction bars would you use to subtract $\frac{1}{4}$ from $\frac{5}{8}$? Draw a picture to explain.

Which like fraction bars would you use to find the difference $\frac{3}{8} - \frac{1}{4}$?

Try It

Find the difference.

a. $\frac{3}{8} - \frac{1}{4}$ **b.** $\frac{5}{6} - \frac{1}{3}$

CALIFORNIA STANDARDS NS 2.0 Students perform calculations and solve problems involving addition, subtraction, and simple multiplication and division of fractions and decimals. **NS 2.3** Solve simple problems, including ones arising in concrete situations, involving the addition and subtraction of fractions and mixed numbers, and express answers in the simplest form. *also* **NS 1.0, MR 2.0, MR 2.3, MR 2.4, MR 3.2, MR 3.3**

Connect

When you subtract fractions with unlike denominators, you need to find equivalent fractions with like denominators.

Subtract. $\frac{7}{8} - \frac{1}{2}$

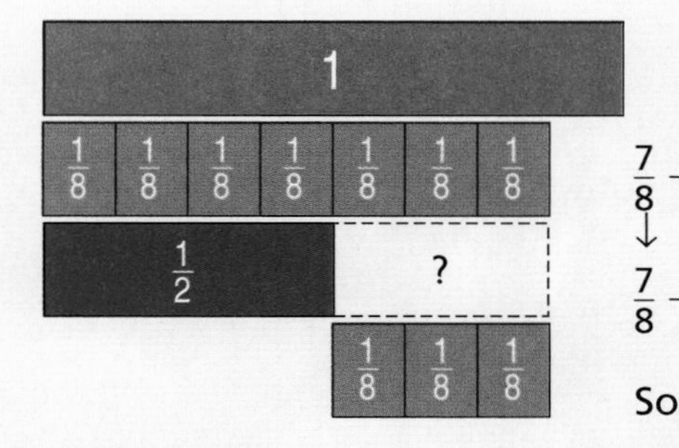

$\frac{7}{8} - \frac{1}{2}$

$\downarrow \quad \downarrow$

$\frac{7}{8} - \frac{4}{8}$

Think: $\frac{1}{2}$ is equivalent to $\frac{4}{8}$.

So, $\frac{7}{8} - \frac{1}{2} = \frac{3}{8}$.

- How is subtracting unlike fractions different from subtracting like fractions?

TECHNOLOGY LINK

More Practice:
Use E-Lab, *Subtracting Unlike Fractions.*

www.harcourtschool.com/elab2002

Practice

Use fraction bars to find the difference.

1. 1/4 1/4 1/4; 1/8 ?

2.

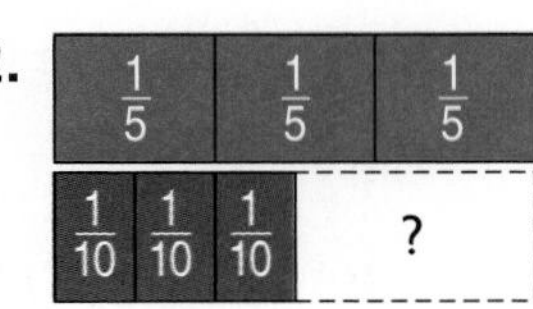

3.

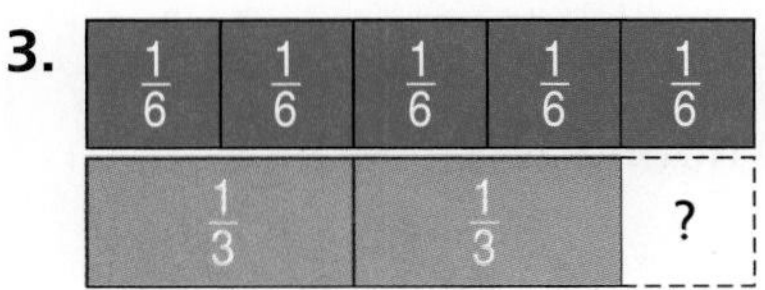

4. $\frac{4}{6} - \frac{1}{4}$ **5.** $\frac{5}{6} - \frac{1}{12}$ **6.** $\frac{7}{10} - \frac{1}{2}$ **7.** $\frac{6}{8} - \frac{1}{4}$

8. $\frac{2}{3} - \frac{1}{12}$ **9.** $\frac{1}{2} - \frac{1}{5}$ **10.** $\frac{11}{12} - \frac{3}{4}$ **11.** $\frac{4}{5} - \frac{1}{10}$

12. **What's the Question?** Carl worked for $\frac{5}{6}$ hour. He tilled his garden for $\frac{1}{2}$ hour, planted seeds for $\frac{1}{6}$ hour, and watered for the rest of the time. The answer is $\frac{1}{6}$ hour.

13. **What's the Error?** Ramon says $\frac{2}{3} - \frac{1}{2} = \frac{1}{1} = 1$. Describe his error. Write the correct answer.

Mixed Review and Test Prep

14. Write the factors of 30. (p. 264)

15. $90{,}116$ (p. 36) $-83{,}909$

16. $79{,}210$ (p. 36) $+92{,}885$

17. Evaluate $81 + n$ if n is 19. (p. 68)

18. **TEST PREP** Which 8 has the least value? (p. 20)

A 8.026 **C** 7.280

B 7.826 **D** 7.068

LESSON 4

Estimate Sums and Differences

Quick Review

Write an equivalent fraction.

1. $\frac{2}{4}$ **2.** $\frac{3}{12}$ **3.** $\frac{2}{5}$ **4.** $\frac{75}{100}$ **5.** $\frac{21}{49}$

Learn

HAPPY TRAILS On a weekend camping trip, Ken, Eric, and their dad went for a walk on the Appalachian Trail. The first hour, they walked $\frac{3}{8}$ mile. The second hour, they walked $\frac{4}{5}$ mile. About how many miles did the boys and their dad walk?

MATH IDEA Rounding fractions to benchmarks such as 0, $\frac{1}{2}$, or 1 on a number line can help you estimate sums and differences.

Estimate. $\frac{3}{8} + \frac{4}{5}$

STEP 1

The fraction $\frac{3}{8}$ is close to $\frac{1}{2}$. Round to $\frac{1}{2}$.

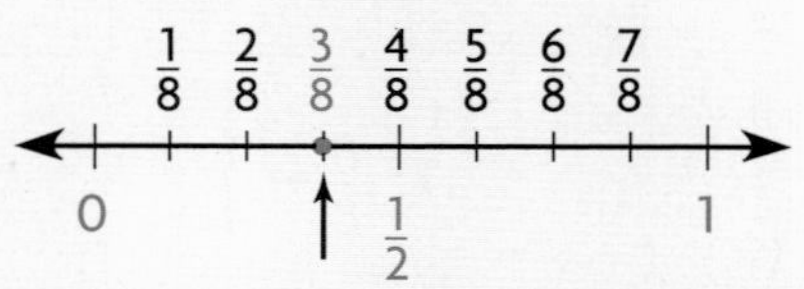

STEP 2

The fraction $\frac{4}{5}$ is close to 1. Round to 1.

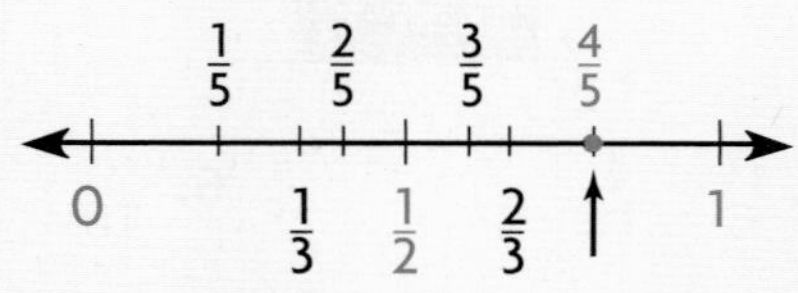

STEP 3

Add the rounded fractions.

$$\begin{array}{r} \frac{3}{8} \rightarrow \frac{1}{2} \\ +\frac{4}{5} \rightarrow 1 \\ \hline 1\frac{1}{2} \end{array}$$

So, Ken, Eric, and their dad walked about $1\frac{1}{2}$ miles.

Examples

A Estimate. $\frac{2}{3} + \frac{4}{5}$

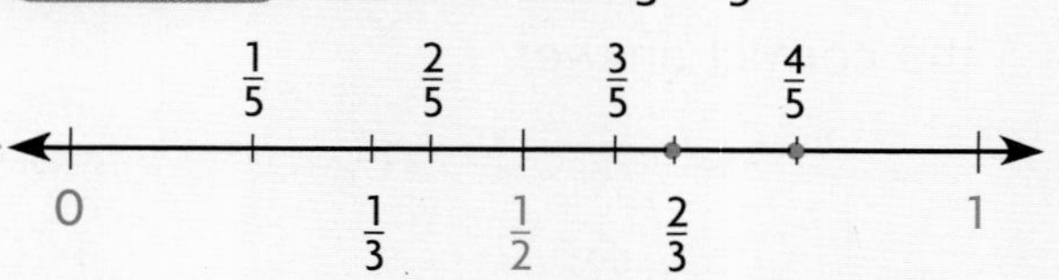

$\frac{2}{3} \rightarrow \frac{1}{2}$ $\frac{2}{3}$ is between $\frac{1}{2}$ and 1, but closer to $\frac{1}{2}$.

$+\frac{4}{5} \rightarrow 1$ $\frac{4}{5}$ is between $\frac{1}{2}$ and 1, but closer to 1.

$1\frac{1}{2}$ The sum is greater than 1, but less than 2.

B Estimate. $\frac{3}{8} - \frac{1}{6}$

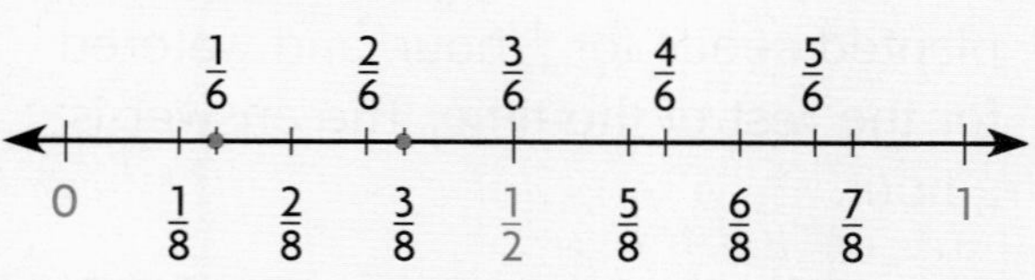

$\frac{3}{8} \rightarrow \frac{1}{2}$ $\frac{3}{8}$ is between 0 and $\frac{1}{2}$, but closer to $\frac{1}{2}$.

$-\frac{1}{6} \rightarrow 0$ $\frac{1}{6}$ is between 0 and $\frac{1}{2}$, but closer to 0.

$\frac{1}{2}$ The difference is greater than 0, but less than $\frac{1}{2}$.

CALIFORNIA STANDARDS NS 1.1 Estimate, round, and manipulate very large (e.g., millions) and very small (e.g., thousandths) numbers. **MR 2.3** Use a variety of methods, such as words, numbers, symbols, charts, graphs, tables, diagrams, and models, to explain mathematical reasoning. *also* **NS 1.0, NS 2.0, MR 2.0, MR 2.4, MR 3.0, MR 3.2, MR 3.3**

Check

1. **Compare** the numerators and denominators of fractions that round to 1, such as $\frac{11}{12}$, $\frac{8}{9}$, $\frac{7}{8}$, and $\frac{8}{10}$. What do you notice?

Write whether the fraction is closest to 0, $\frac{1}{2}$, or 1. Use the number lines.

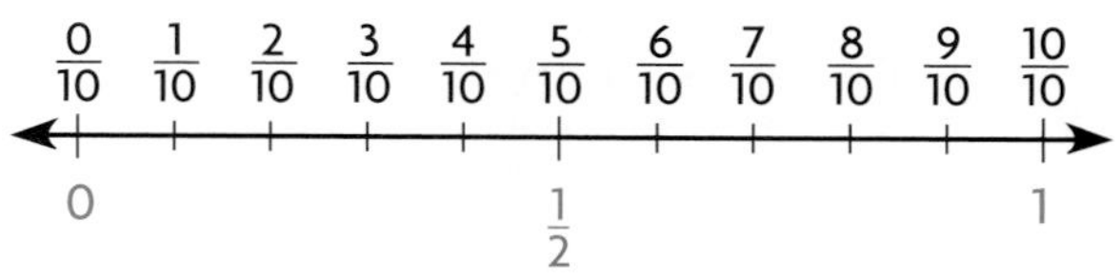

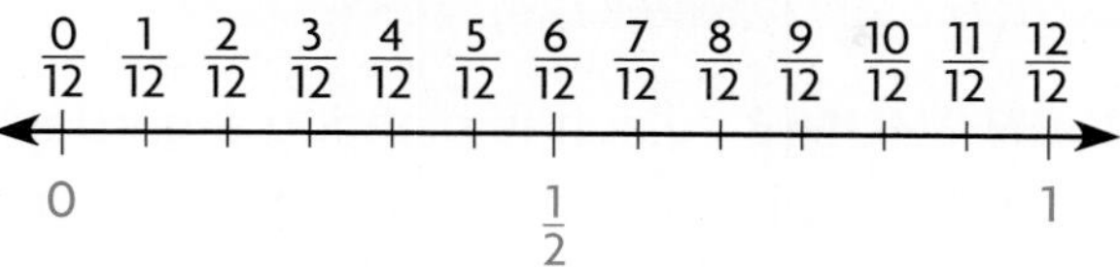

2. $\frac{7}{10}$
3. $\frac{1}{12}$
4. $\frac{3}{10}$
5. $\frac{5}{12}$
6. $\frac{9}{10}$
7. $\frac{2}{10}$

Practice and Problem Solving

Write whether the fraction is closest to 0, $\frac{1}{2}$, or 1.

8. $\frac{4}{7}$
9. $\frac{2}{9}$
10. $\frac{6}{7}$
11. $\frac{8}{9}$
12. $\frac{3}{7}$
13. $\frac{5}{6}$

Estimate each sum or difference.

14. $\frac{1}{9} + \frac{5}{6}$
15. $\frac{2}{3} + \frac{5}{7}$
16. $\frac{3}{8} - \frac{1}{5}$
17. $\frac{9}{10} - \frac{3}{8}$
18. $\frac{7}{12} + \frac{1}{7}$
19. $\frac{8}{10} - \frac{2}{3}$
20. $\frac{6}{7} + \frac{3}{5}$
21. $\frac{7}{9} - \frac{2}{6}$
22. $\frac{7}{8} + \frac{2}{3}$
23. $\frac{10}{12} - \frac{1}{10}$

Estimate to compare. Write $<$ or $>$ for each ●.

24. $\frac{6}{8} + \frac{2}{3}$ ● 1
25. $\frac{1}{2}$ ● $\frac{3}{5} - \frac{1}{4}$
26. $\frac{2}{5} + \frac{1}{7}$ ● 1

27. **REASONING** The number ♣ is three times the number ♥. What is the fraction $\frac{♥}{♣}$ in simplest form?

28. **Write About It** Explain how you can estimate the sum of $\frac{4}{5}$ and $\frac{1}{10}$.

29. Yoko added $\frac{3}{4}$ cup cashews, $\frac{1}{2}$ cup pecans, and $\frac{1}{3}$ cup peanuts to the trail mix. How many cups of nuts were added to the trail mix?

Mixed Review and Test Prep

Find the value of *n*. (p. 90)

30. $4 \times (6 \times n) = (4 \times 6) \times 8$
31. $(7 \times 3) \times n = 105$
32. Find the mean, median, and mode for the data set 90, 94, 89, 90, 92. (p. 104)
33. $96.54 \div 6$ (p. 222)
34. **TEST PREP** Choose the LCM and GCF of 4 and 50. (pp. 256 and 258)
    - **A** LCM: 150; GCF: 2
    - **B** LCM: 100; GCF: 4
    - **C** LCM: 100; GCF: 2
    - **D** LCM: 200; GCF: 4

Extra Practice page H50, Set B

LESSON 5

Use Least Common Denominators

Quick Review

Write the least common multiple for each pair.

1. 8 and 4
2. 6 and 4
3. 3 and 8
4. 7 and 4
5. 3 and 5

VOCABULARY

least common denominator (LCD)

Learn

GREEN THUMB Zach is planting a garden in his backyard. He planted daisies in $\frac{2}{3}$ of the garden and marigolds in $\frac{1}{4}$ of the garden. How much of the garden did Zach use?

To add unlike fractions, you need to rename them as like fractions. The least common multiple of two or more denominators is the **least common denominator**, or **LCD**.

Remember

The *least common multiple*, or *LCM*, is the least number that is a common multiple of two or more numbers.

Add. $\frac{2}{3} + \frac{1}{4}$

Estimate. $\frac{2}{3}$ is a little more than $\frac{1}{2}$ and $\frac{1}{4}$ is halfway between 0 and $\frac{1}{2}$, so the sum is close to 1.

One Way

Use a model.

Model the problem with fraction bars.

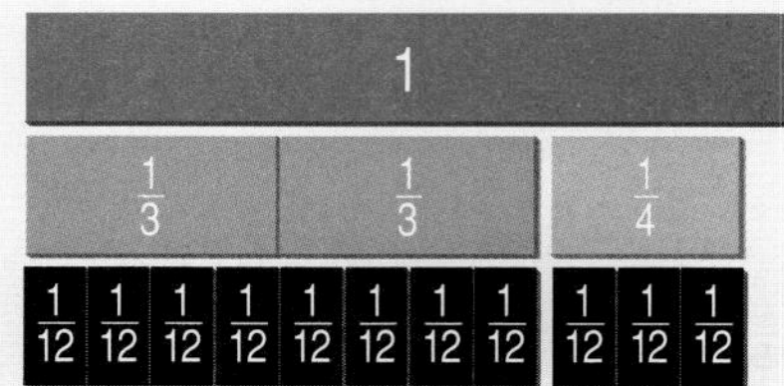

$$\frac{2}{3} + \frac{1}{4}$$
$$\downarrow \quad \downarrow$$
$$\frac{8}{12} + \frac{3}{12}$$

12 is the least common denominator (LCD).

So, $\frac{2}{3} + \frac{1}{4} = \frac{11}{12}$.

Another Way

Use multiples.

Find the least common multiple of the two denominators.

Multiples of 3: 3, 6, 9, 12, 15, 18
Multiples of 4: 4, 8, 12, 16, 20

The LCM is 12. So, the LCD of $\frac{2}{3}$ and $\frac{1}{4}$ is 12.

Write equivalent fractions.

$$\frac{2}{3} = \frac{2 \times 4}{3 \times 4} = \frac{8}{12}$$

$$+\frac{1}{4} = \frac{1 \times 3}{4 \times 3} = +\frac{3}{12}$$

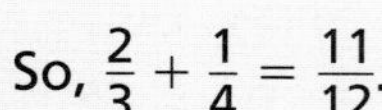

$$\frac{11}{12}$$

So, $\frac{2}{3} + \frac{1}{4} = \frac{11}{12}$.

So, Zach used $\frac{11}{12}$ of the garden. The answer is close to the estimate, so the answer is reasonable.

- What would the least common denominator be for $1 - \frac{2}{5}$? Explain.

MATH IDEA The least common multiple of the denominators is the least common denominator (LCD) for the fractions.

CALIFORNIA STANDARDS **NS 2.0** Students perform calculations and solve problems involving addition, subtraction, and simple multiplication and division of fractions and decimals. **O⊸ NS 2.3** Solve simple problems, including ones arising in concrete situations, involving the addition and subtraction of fractions and mixed numbers, and express answers in the simplest form. *also* **NS 1.0, MR 1.0, MR 2.1, MR 2.3, MR 2.4, MR 3.2**

Check

1. **Show** how you can use multiples to find the LCD for $\frac{2}{3} + \frac{3}{5}$.

Name the LCD. Then add to find the sum.

2. $\frac{1}{5} + \frac{1}{10}$
3. $\frac{1}{4} + \frac{1}{3}$
4. $\frac{1}{3} + \frac{2}{9}$
5. $\frac{1}{6} + \frac{2}{4}$
6. $\frac{1}{5} + \frac{5}{10}$

Name the LCD. Then subtract to find the difference.

7. $\frac{2}{3} - \frac{1}{2}$
8. $\frac{3}{4} - \frac{3}{8}$
9. $1 - \frac{1}{4}$
10. $\frac{11}{12} - \frac{1}{6}$
11. $\frac{7}{10} - \frac{1}{5}$

Practice and Problem Solving

Name the LCD. Then add or subtract.

12. $\frac{1}{8} + \frac{3}{4}$
13. $\frac{1}{2} + \frac{2}{5}$
14. $\frac{5}{6} - \frac{3}{4}$
15. $1 - \frac{3}{5}$
16. $\frac{7}{8} + \frac{1}{4}$

Find the sum or difference.

17. $\frac{2}{3} + \frac{1}{6}$
18. $\frac{2}{3} - \frac{1}{9}$
19. $\frac{7}{8} - \frac{1}{2}$
20. $\frac{1}{4} + \frac{2}{3}$
21. $\frac{1}{4} + \frac{2}{8}$
22. $\frac{4}{5} + \frac{2}{10}$
23. $\frac{3}{12} - \frac{1}{6}$
24. $\frac{9}{10} - \frac{1}{2}$
25. $\frac{1}{4} + \frac{1}{12}$
26. $\frac{1}{4} - \frac{3}{16}$

Algebra **Find the value of *n*.**

27. $\frac{2}{10} + n = 1$
28. $n + \frac{1}{12} = \frac{7}{12}$
29. $\frac{5}{8} - n = \frac{3}{8}$
30. $\frac{9}{10} - n = \frac{1}{5}$

USE DATA **For 31–32, use the table.**

GARDEN PLANT GROWTH (in inches)		
	First Week	Second Week
Tulips	$\frac{7}{8}$	$\frac{3}{4}$
Daffodils	$\frac{1}{4}$	$\frac{1}{2}$

31. Which plants grew the most during the first week?
32. **Write a problem** that uses subtraction, using the data from the second week.
33. Dan wrote a report. He found $\frac{1}{8}$ of the information for the report on the Internet and $\frac{1}{2}$ at the library. How much of his information did he get from sources other than the Internet and the library?

Mixed Review and Test Prep

Use mental math to find each sum. (p. 36)

34. $125 + 80 + 20 + 75$
35. $35{,}200 + 4{,}800 + 10{,}000$ (p. 36)
36. Round 3,950,964 to the nearest million. (p. 32)
37. Find the value of n in $102 - n = 57$. (p. 74)
38. **TEST PREP** Which number is between 0.285 and 0.29? (p.18)

    A 0.128 **C** 0.289

    B 0.278 **D** 0.291

LESSON

6 Add and Subtract Unlike Fractions

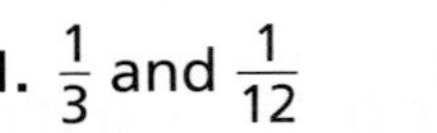

Quick Review

Write the LCD.

1. $\frac{1}{3}$ and $\frac{1}{12}$ **2.** $\frac{2}{3}$ and $\frac{1}{9}$

3. $\frac{2}{5}$ and $\frac{1}{7}$ **4.** $\frac{2}{8}$ and $\frac{7}{6}$

5. $\frac{2}{3}$ and $\frac{1}{6}$

Learn

PLAY THE PART Susan and Keith are making costumes for their play, *Mystery on the High Seas.* For each costume they need $\frac{1}{2}$ yard of white fabric and $\frac{1}{6}$ yard of blue fabric. How much fabric is needed for each costume?

Add. $\frac{1}{2} + \frac{1}{6}$

Estimate. $\frac{1}{6}$ is a little more than 0, so the sum is close to $\frac{1}{2}$.

STEP 1

The LCM of 2 and 6 is 6. So, the LCD of $\frac{1}{2}$ and $\frac{1}{6}$ is 6. Use the LCD to write like fractions.

$$\frac{1}{6} = \frac{1}{6}$$
$$+\frac{1 \times 3}{2 \times 3} = +\frac{3}{6}$$

STEP 2

Add the fractions. Write the answer in simplest form.

$$\frac{1}{6} = \frac{1}{6}$$
$$+\frac{1 \times 3}{2 \times 3} = +\frac{3}{6}$$
$$\frac{4}{6} = \frac{2}{3}$$

So, $\frac{2}{3}$ yard of fabric is needed for each costume. The exact answer is close to the estimate, so the answer is reasonable.

- Why didn't $\frac{1}{6}$ change?

TECHNOLOGY LINK

More Practice: ***Mighty Math Number Heroes, Fraction Fireworks*** **Levels T and U.**

CALIFORNIA STANDARDS NS 2.0 Students perform calculations and solve problems involving addition, subtraction, and simple multiplication and division of fractions and decimals. **NS 2.3** Solve simple problems, including ones arising in concrete situations, involving the addition and subtraction of fractions and mixed numbers, and express answers in the simplest form. *also* **NS 1.0, NS 1.1, MR 1.0, MR 2.0, MR 2.1, MR 2.4, MR 3.3**

Subtract

Subtract. $\frac{3}{4} - \frac{1}{6}$

Estimate. $\frac{3}{4}$ is halfway between $\frac{1}{2}$ and 1, and $\frac{1}{6}$ is a little more than 0, so the difference is close to $\frac{1}{2}$.

STEP 1

The LCM of 4 and 6 is 12. So, the LCD of $\frac{3}{4}$ and $\frac{1}{6}$ is 12. Use the LCD to change the fractions to like fractions.

$$\frac{3}{4} = \frac{3 \times 3}{4 \times 3} = \frac{9}{12}$$
$$-\frac{1}{6} = \frac{1 \times 2}{6 \times 2} = -\frac{2}{12}$$

STEP 2

Subtract the fractions.
Write the answer in simplest form.

$$\frac{3}{4} = \frac{3 \times 3}{4 \times 3} = \frac{9}{12}$$
$$-\frac{1}{6} = \frac{1 \times 2}{6 \times 2} = -\frac{2}{12}$$
$$\frac{7}{12} \leftarrow \text{simplest form}$$

Examples

A Find $\frac{3}{4} + \frac{5}{8}$.

$$\frac{3 \times 2}{4 \times 2} = \frac{6}{8}$$
$$+\frac{5}{8} = +\frac{5}{8}$$
$$\frac{11}{8}, \text{ or } 1\frac{3}{8}$$

B Find $1 - \frac{2}{5}$.

$$\frac{1 \times 5}{1 \times 5} = \frac{5}{5}$$
$$-\frac{2}{5} = -\frac{2}{5}$$
$$\frac{3}{5}$$

MATH IDEA To add or subtract unlike fractions, you need to find the least common denominator (LCD) and then add or subtract the numerators.

Check

1. **Tell** what the least common denominator is.

Find the LCD. Then add or subtract.

2. $\frac{6}{12} - \frac{1}{3}$
3. $\frac{1}{5} + \frac{1}{2}$
4. $\frac{5}{7} - \frac{1}{2}$
5. $\frac{8}{9} - \frac{2}{3}$
6. $\frac{1}{2} + \frac{1}{3}$

Find the sum or difference. Write the answer in simplest form.

7. $\frac{2}{3} + \frac{1}{6}$
8. $\frac{2}{3} - \frac{1}{9}$
9. $\frac{7}{8} - \frac{1}{2}$
10. $\frac{1}{4} + \frac{2}{3}$
11. $\frac{1}{4} + \frac{2}{8}$
12. $\frac{7}{12} - \frac{1}{3}$
13. $\frac{3}{8} + \frac{3}{4}$
14. $1 - \frac{5}{6}$
15. $\frac{3}{9} + \frac{1}{3}$
16. $1 - \frac{7}{10}$

LESSON CONTINUES

Practice and Problem Solving

Find the LCD. Then add or subtract.

17. $\frac{1}{6} + \frac{1}{3}$

18. $\frac{7}{10} + \frac{1}{2}$

19. $\frac{5}{9} - \frac{1}{6}$

20. $\frac{2}{3} + \frac{1}{4}$

21. $\frac{2}{3} - \frac{1}{4}$

Find the sum or difference. Write the answer in simplest form.

22. $\frac{1}{8} + \frac{3}{4}$ **23.** $\frac{1}{2} + \frac{2}{5}$ **24.** $\frac{5}{6} - \frac{3}{4}$ **25.** $1 - \frac{3}{5}$ **26.** $\frac{2}{3} + \frac{1}{4}$

27. $1 - \frac{5}{7}$ **28.** $\frac{4}{5} + \frac{3}{10}$ **29.** $\frac{7}{8} - \frac{1}{2}$ **30.** $\frac{3}{4} + \frac{1}{12}$ **31.** $\frac{4}{5} + \frac{1}{10}$

32. $\frac{7}{14} - \frac{1}{7}$ **33.** $\frac{2}{5} + \frac{6}{10}$ **34.** $1 - \frac{8}{9}$ **35.** $\frac{4}{18} + \frac{2}{6}$ **36.** $\frac{6}{8} - \frac{1}{4}$

37. $\frac{1}{3} + \frac{5}{12}$ **38.** $\frac{2}{3} - \frac{5}{9}$ **39.** $\frac{1}{3} + \frac{2}{5}$ **40.** $\frac{5}{6} - \frac{1}{8}$ **41.** $\frac{4}{9} + \frac{1}{3}$

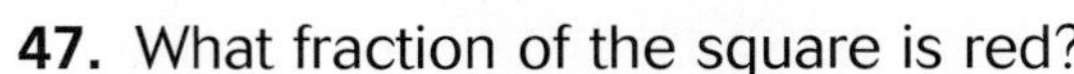

Algebra Compare. Write < or > for each ●.

42. $\frac{1}{2} + \frac{2}{3}$ ● $\frac{1}{6} + \frac{1}{3}$ **43.** $\frac{2}{4} - \frac{2}{8}$ ● $\frac{1}{2} - \frac{1}{8}$ **44.** $\frac{7}{12} - \frac{1}{2}$ ● $\frac{4}{6} + \frac{1}{12}$

45. **REASONING** John and Jack both made a prop for the play from the same board. John said he used $\frac{3}{8}$ of the board, and Jack said he used $\frac{3}{4}$ of the board. Is this possible? Explain.

46. **Write About It** Explain how to subtract fractions with unlike denominators.

For 47–50, use the drawing.

47. What fraction of the square is red?

48. What fraction of the square is green or orange?

49. What subtraction equation can you write to find the fraction of the square that is yellow? Solve the equation.

50. What fraction of the square is white, blue, or yellow?

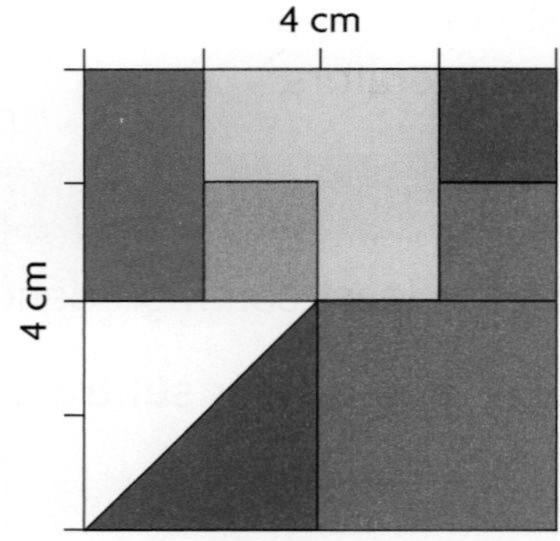

51. Tom, Bill, and Ann painted scenery. There were 8 equal-sized panels to paint. Tom painted $\frac{1}{4}$ of the panels, Bill painted $\frac{2}{8}$, and Ann painted $\frac{1}{2}$. How many panels of scenery did each person paint?

52. **MEASUREMENT** Eve rode her bike $\frac{1}{2}$ mile to school. After school she went to a friend's house $\frac{1}{4}$ mile away and then rode $\frac{3}{4}$ mile home. How far did Eve ride her bike in all?

53. **REASONING** Ed says that when you add two like fractions, the sum of the numerators will always be less than the denominator. Is this true? Explain.

54. Mitchell paid $94 for ten tickets to the concert. Adult tickets cost $15 each. Student tickets cost $8 each. How many of each kind of ticket did he buy?

Mixed Review and Test Prep

Find the mean, median, mode, and range for each set of data. (p. 102)

55. 75, 29, 83, 45, 53

56. 2, 4, 8, 14

57. Solve. Explain how you interpreted the remainder. (p. 190)
Ken has 32 books. Each shelf holds 5 books. How many shelves will he use?

58. **TEST PREP** Which is the value of 8^4? (p. 266)

A 64
B 512
C 4,096
D 32,768

Copy and complete each pattern. (p. 200)

59. $6{,}000 \div 4 = ■$
$600 \div 4 = ■$
$60 \div 4 = ■$
$6 \div 4 = ■$

60. $18{,}000 \div 3 = ■$
$1{,}800 \div 3 = ■$
$180 \div 3 = ■$
$18 \div 3 = ■$

61. **TEST PREP** What number is 10,000 more than 2,692,450? (p. 4)

F 2,690,450
G 2,693,450
H 2,702,450
J 2,720,450

Thinker's Corner

SOLVE THE RIDDLE! Write the sum or difference in simplest form. Then find the letter that matches each answer below. The letters will spell the answer to the riddle.

1. A $\frac{1}{4} + \frac{3}{8}$
2. E $\frac{2}{5} + \frac{4}{10}$
3. H $\frac{5}{6} - \frac{3}{8}$
4. U $\frac{2}{3} - \frac{3}{5}$
5. E $\frac{4}{6} + \frac{1}{4}$
6. N $\frac{8}{10} - \frac{2}{5}$
7. Q $\frac{2}{3} + \frac{2}{9}$
8. K $\frac{3}{4} - \frac{1}{3}$
9. R $\frac{5}{6} - \frac{1}{4}$
10. S $\frac{5}{12} + \frac{1}{3}$
11. C $\frac{7}{10} - \frac{1}{2}$
12. I $\frac{3}{6} + \frac{4}{9}$
13. M $\frac{5}{6} - \frac{5}{12}$
14. T $\frac{3}{4} - \frac{2}{5}$
15. B $\frac{3}{4} + \frac{2}{16}$

Riddle: Why did the football coach go to the bank?

To get $\frac{11}{24}$ ___ $\frac{17}{18}$ ___ $\frac{3}{4}$ ___ $\frac{8}{9}$ ___ $\frac{1}{15}$ ___ $\frac{5}{8}$ ___ $\frac{7}{12}$ ___ $\frac{7}{20}$ ___ $\frac{4}{5}$ ___ $\frac{7}{12}$ ___ $\frac{7}{8}$ ___ $\frac{5}{8}$ ___ $\frac{1}{5}$ ___ $\frac{5}{12}$ ___.

LESSON 7

Problem Solving Strategy

Work Backward

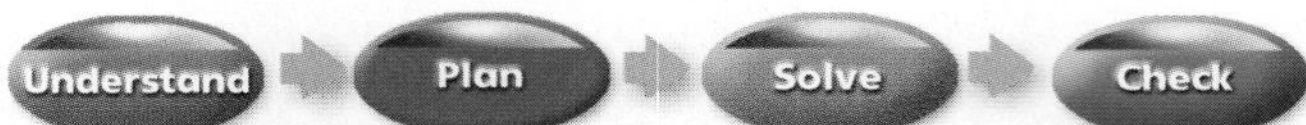

1. 52 + 19 + 21
2. 133 − 30 − 14
3. 609 + 91 + 170
4. $\frac{3}{4} - \frac{1}{6} - \frac{1}{12}$
5. $\frac{2}{3} + \frac{1}{6} + \frac{1}{9}$

PROBLEM Jessica built a tower out of building blocks for a craft fair. She worked on her tower for three weeks. After three weeks, the tower was $\frac{9}{10}$ meter tall. This was $\frac{1}{5}$ meter taller than it was after two weeks. The height at two weeks was $\frac{1}{2}$ meter taller than it was after the first week. How tall was Jessica's tower after the first week?

- What are you asked to find?
- What information will you use?

- What strategy can you use to solve this problem?

You can *work backward* to find out how tall Jessica's tower was after the first week.

- How can you work backward to solve?

The growth can be shown by this expression:

$n + \frac{1}{2} + \frac{1}{5} = \frac{9}{10}$

To work backward, start with $\frac{9}{10}$ and subtract.

$\frac{9}{10} - \frac{1}{5} - \frac{1}{2} = n$

$\frac{9}{10} - \frac{2}{10} - \frac{5}{10} = n$ Find a common denominator.

$\frac{7}{10} - \frac{5}{10} = n$ Subtract.

$\frac{2}{10} = n$, or $\frac{1}{5} = n$

So, Jessica's tower was $\frac{1}{5}$ meter tall after one week.

- How can you determine if your answer is reasonable?

CALIFORNIA STANDARDS MR 2.6 Make precise calculations and check the validity of the result from the context of the problem. **NS 2.3** Solve simple problems, including ones arising in concrete situations, involving the addition and subtraction of fractions and mixed numbers, and express answers in the simplest form. *also* **NS 1.0, NS 2.0, AF 1.0, MR 1.0, MR 2.0, MR 3.0, MR 3.1, MR 3.2**

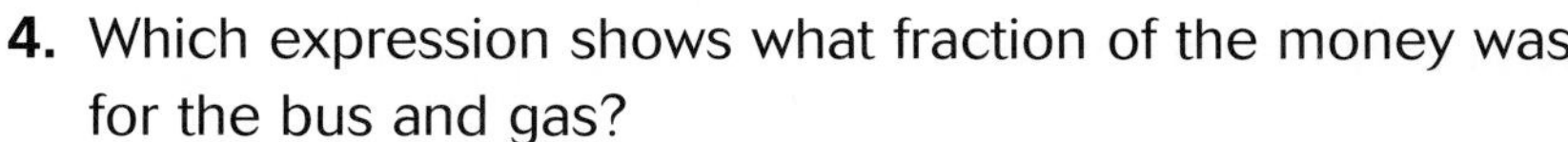

Problem Solving Practice

PROBLEM SOLVING STRATEGIES

- Draw a Diagram or Picture
- Make a Model or Act It Out
- Make an Organized List
- Find a Pattern
- Make a Table or Graph
- Predict and Test
- Work Backward
- Solve a Simpler Problem
- Write an Equation
- Use Logical Reasoning

Work backward to solve.

1. **What if** Jessica's tower was $\frac{11}{12}$ yard after 3 weeks? This was $\frac{1}{6}$ yard taller than it was after 2 weeks. The height at 2 weeks was $\frac{1}{4}$ yard taller than it was after the first week. How tall was the tower after the first week?
2. On Friday, Michael's plant was $\frac{5}{6}$ foot tall. It had grown $\frac{1}{3}$ foot from Wednesday to Friday. It had grown $\frac{3}{8}$ foot from Monday to Wednesday. How tall was Michael's plant on Monday?
3. Lisa came home from the toy store with $2.26. At the store she spent $7.38 on blocks, $4.17 on a book, and $8.55 on a science kit. She used a gift certificate worth $3.00 to help pay the bill. How much money did Lisa have before she went to the toy store?

USE DATA Mrs. Balog's class has raised money to go on a field trip. The circle graph shows their expenses for the trip as fractions of their earnings.

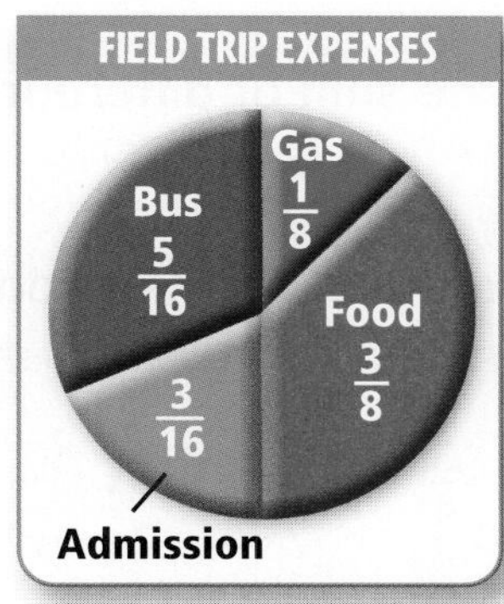

4. Which expression shows what fraction of the money was for the bus and gas?

   A $\frac{3}{16} - \frac{1}{8}$ **B** $\frac{5}{16} + \frac{1}{8}$ **C** $\frac{3}{16} + \frac{3}{8}$ **D** $\frac{5}{16} - \frac{1}{8}$

5. What fraction of the money was not for the bus and gas?

   F $\frac{5}{16}$ **G** $\frac{3}{8}$ **H** $\frac{7}{16}$ **J** $\frac{9}{16}$

Mixed Strategy Practice

6. Charity gave 12 baseball cards to Rhonda and 18 to James. Then she traded 5 of her cards for 3 of Andy's cards. Charity now has 48 cards. How many cards did she have to start with?
7. A rectangular fence has a perimeter of 20 feet. If the width is 2 feet, what is the length?
8. Carlotta bought 9 packages of lemonade for $1.10 each and 2 packages of cups for $1.09 each. She sold 23 cups of lemonade every hour for 4 hours at $0.40 per cup. How much more money did Carlotta earn than she spent on supplies?
9. Sherwood earned $35 last week. He earned twice as much for baby-sitting as he did for mowing lawns. He earned twice as much for mowing lawns as he did for washing cars. How much did Sherwood earn for each job?

Review/Test

CHECK VOCABULARY AND CONCEPTS

denominators
numerators
least common denominator (LCD)

Choose the best term from the box.

1. The least common multiple of two or more denominators is the ___?___. (p. 352)
2. When you add or subtract fractions with unlike denominators, write equivalent fractions with like ___?___. (p. 347)

CHECK SKILLS

Estimate each sum or difference. (pp. 350–351)

3. $\frac{5}{6} - \frac{3}{8}$
4. $\frac{3}{9} + \frac{5}{8}$
5. $\frac{7}{8} - \frac{2}{3}$
6. $\frac{1}{3} + \frac{1}{5}$

Find the sum or difference in simplest form. (pp. 344–345)

7. $\frac{2}{8} + \frac{4}{8}$
8. $\frac{7}{9} - \frac{4}{9}$
9. $\frac{4}{12} + \frac{2}{12}$
10. $\frac{5}{8} - \frac{1}{8}$

Name the LCD. Then add or subtract. Write the answer in simplest form. (pp. 352–353)

11. $\frac{3}{4} + \frac{2}{3}$
12. $\frac{3}{4} - \frac{4}{6}$
13. $\frac{1}{2} + \frac{4}{5}$
14. $\frac{2}{5} - \frac{1}{3}$
15. $\frac{1}{7} + \frac{1}{2}$
16. $\frac{4}{6} - \frac{1}{4}$
17. $1 - \frac{3}{8}$
18. $\frac{4}{5} - \frac{1}{4}$
19. $\frac{1}{4} + \frac{2}{3}$
20. $\frac{5}{8} - \frac{1}{3}$
21. $\frac{4}{5} + \frac{1}{3}$
22. $\frac{6}{7} - \frac{1}{3}$

CHECK PROBLEM SOLVING (pp. 358–359)

23. On Friday, Kayla's bean plant was $\frac{7}{8}$ in. tall. It had grown $\frac{3}{8}$ in. from Wednesday to Friday. It had grown $\frac{1}{4}$ in. from Monday to Wednesday. How tall was the plant on Monday?
24. From 9 A.M. to 4 P.M., $\frac{1}{3}$ foot of snow fell on the mountain. At 8 P.M., the snow was $\frac{5}{6}$ foot deep, which was $\frac{1}{4}$ foot deeper than it was at 4 P.M. How much snow was on the ground at 9 A.M.?
25. Michele gave away 21 stickers. She gave half as many stickers to her brother as she gave to her teacher. She gave twice as many to her best friend as she gave to her teacher. How many stickers did Michele give to each person?

Cumulative Review

Check your work.
See item **7**.

Check to be sure you found a common denominator for all the fractions. Ask yourself if your answer makes sense. Remember to name your answer in simplest form.

Also see problem **7**, p. H65.

For 1–12, choose the best answer.

1. What is the least common multiple of 3, 4, and 8?

A 12 **C** 24
B 16 **D** 32

2. Which fraction is the farthest from 1 on a number line?

F $\frac{15}{16}$ **H** $\frac{9}{10}$
G $\frac{4}{7}$ **J** $\frac{8}{9}$

3. Which is the best estimate for this sum?

$$\frac{5}{6} + \frac{7}{8} + \frac{2}{6}$$

A $3\frac{1}{2}$ **C** 2
B 3 **D** $1\frac{1}{2}$

4. What is the value of the 2 in 6.5217?

F 0.02 **H** 2
G 0.2 **J** 20

5. What is the value of n for $\frac{7}{10} - n = \frac{3}{10}$?

A $\frac{13}{10}$ **C** $\frac{3}{10}$
B $\frac{1}{2}$ **D** $\frac{4}{10}$

6. 5.63×3.5

F 1.95 **H** 19.705
G 1.9705 **J** Not here

7. Mark spent $\frac{1}{8}$ of his allowance at the movies, $\frac{3}{8}$ on food, and $\frac{1}{4}$ for baseball cards. How much of his allowance does he have left?

F $\frac{5}{8}$ **H** $\frac{1}{4}$
G $\frac{3}{8}$ **J** $\frac{1}{8}$

8. $\frac{10}{12} - \frac{3}{4}$

A $1\frac{7}{12}$ **C** $\frac{7}{12}$
B $1\frac{1}{2}$ **D** $\frac{1}{12}$

9. Which of these is closest to $\frac{1}{10}$?

A 0 **C** $\frac{3}{4}$
B $\frac{1}{2}$ **D** 1

10. Shania spent $45 for new shoes. She also spent $16 for school supplies. Then she had $13 left. How much money did she have when she started?

F $19 **H** $61
G $58 **J** $74

11. Katherine had a box of pencils. She gave 18 pencils to Danny. Then she gave some pencils to Miguel. After that, she had 12 pencils left. If she started with 40 pencils, how many did she give to Miguel?

A 4 **C** 22
B 10 **D** 30

12.
$$\begin{array}{r} 436{,}475 \\ -239{,}836 \\ \hline \end{array}$$

F 196,639 **H** 203,441
G 197,649 **J** Not here

Add and Subtract Mixed Numbers

Hiking and many outdoor activities require a greater amount of energy than most people think. Trail mix is a good high-energy snack. On a vigorous hike, many people will munch on trail mix and sip water to keep their energy levels fairly constant. In the 1970's, the term *trail mix* was first used to describe a snack of nuts, seeds, and dried fruit. Today, there are many different recipes for trail mix. How many cups of trail mix does the recipe below make?

Trail Mix

$2\frac{1}{2}$ cups wheat cereal

$1\frac{1}{3}$ cups granola

1 cup mini-pretzels

$1\frac{1}{4}$ cups peanuts

$2\frac{1}{4}$ cups raisins

$2\frac{1}{3}$ cups sunflower seeds

Directions: Mix everything. Store in resealable plastic bags.

CHECK WHAT YOU KNOW

Use this page to help you review and remember important skills needed for Chapter 20.

VOCABULARY

Choose the best term from the box.

simplest form
fraction
mixed number

1. A number with a whole-number part and a fraction part is a __?__.
2. When the numerator and denominator of a fraction have no common factors other than 1, the fraction is in __?__.

UNDERSTAND MIXED NUMBERS (See p. H22.)

Write a mixed number for each picture.

3. 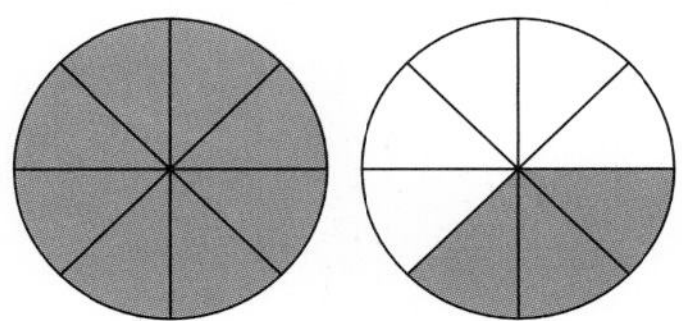

4. 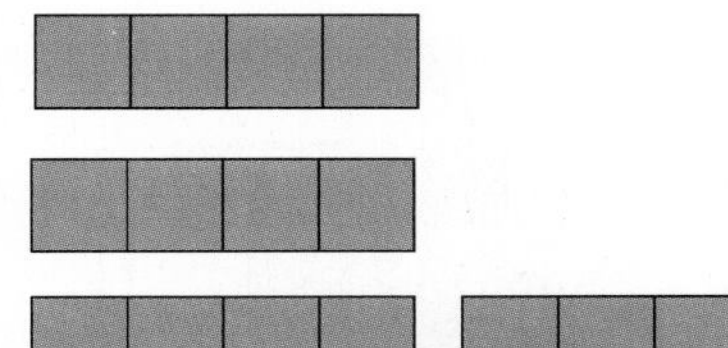

5. 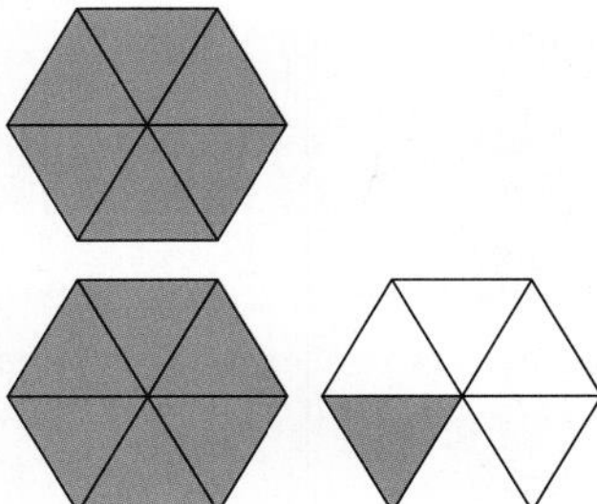

Rename each fraction as a mixed number. You may wish to draw a picture.

6. $\frac{4}{3}$ 7. $\frac{5}{2}$ 8. $\frac{11}{5}$ 9. $\frac{13}{4}$ 10. $\frac{22}{5}$

Rename each mixed number as a fraction. You may wish to draw a picture.

11. $1\frac{1}{4}$ 12. $2\frac{2}{3}$ 13. $4\frac{1}{6}$ 14. $3\frac{7}{8}$ 15. $5\frac{1}{2}$

ADD AND SUBTRACT FRACTIONS (See p. H23.)

Find the sum or difference. Write the answer in simplest form.

16. $\frac{3}{4} + \frac{1}{4}$ 17. $\frac{3}{5} + \frac{4}{5}$ 18. $\frac{5}{7} - \frac{1}{2}$ 19. $\frac{8}{9} - \frac{2}{3}$ 20. $\frac{1}{2} + \frac{1}{3}$

21. $\frac{2}{3} + \frac{5}{6}$ 22. $\frac{2}{3} - \frac{1}{9}$ 23. $\frac{7}{8} - \frac{1}{2}$ 24. $\frac{1}{4} + \frac{2}{8}$ 25. $\frac{3}{4} + \frac{2}{3}$

LESSON 1

Add Mixed Numbers

Quick Review

Find the least common denominator (LCD) for each pair of fractions.

1. $\frac{1}{2}$ and $\frac{1}{4}$ **2.** $\frac{1}{6}$ and $\frac{2}{3}$

3. $\frac{3}{4}$ and $\frac{3}{5}$ **4.** $\frac{7}{9}$ and $\frac{5}{6}$

5. $\frac{8}{11}$ and $\frac{4}{7}$

Learn

READY, SET, GO! The Boy Scouts are having their annual Pinewood Derby. Ahmad is making a banner for the event. He needs $2\frac{3}{8}$ yards of red fabric and $1\frac{1}{4}$ yards of yellow fabric to make a Boy Scout banner. How many yards of fabric does Ahmad need for the banner?

Activity

You can use fraction bars to find $2\frac{3}{8} + 1\frac{1}{4}$.

MATERIALS: fraction bars

STEP 1

Model the problem.

STEP 2

Find like fraction bars for $\frac{3}{8}$ and $\frac{1}{4}$. Add the fractions and add the whole numbers.

$2\frac{3}{8} + 1\frac{1}{4} = 2\frac{3}{8} + 1\frac{2}{8} = 3\frac{5}{8}$

So, Ahmad needs $3\frac{5}{8}$ yards of fabric for the banner.

- Use fraction bars to find the sum.

a. $3\frac{1}{8} + 1\frac{3}{4}$

b. $1\frac{2}{5} + 1\frac{7}{10}$

c. $2\frac{1}{3} + 2\frac{1}{4}$

At the Pinewood Derby, Scouts make a small car out of a block of wood and plastic wheels. The cars race down an inclined track that is 32 feet long. ▶

CALIFORNIA STANDARDS **NS 2.3** Solve simple problems, including ones arising in concrete situations, involving the addition and subtraction of fractions and mixed numbers (like and unlike denominators of 20 or less), and express answers in the simplest form. *also* **NS 1.0, NS 2.0, MR 2.1, MR 2.3, MR 2.4, MR 3.2**

Least Common Denominator

You can use the least common denominator (LCD) to add mixed numbers.

Find $2\frac{1}{3} + 1\frac{5}{12}$.

Estimate to the nearest whole number. $2 + 1 = 3$

STEP 1

Find the LCD. Write equivalent fractions.

$$\begin{array}{rcr} 2\frac{1}{3} & = & 2\frac{4}{12} \\ +1\frac{5}{12} & = & +1\frac{5}{12} \\ \hline \end{array}$$

STEP 2

Add the fractions.

$$\begin{array}{rcr} 2\frac{1}{3} & = & 2\frac{4}{12} \\ +1\frac{5}{12} & = & +1\frac{5}{12} \\ \hline & & \frac{9}{12} \end{array}$$

STEP 3

Add the whole numbers. Write the answer in simplest form if needed.

$$\begin{array}{rcr} 2\frac{1}{3} & = & 2\frac{4}{12} \\ +1\frac{5}{12} & = & +1\frac{5}{12} \\ \hline & & 3\frac{9}{12} = 3\frac{3}{4} \end{array}$$

So, $2\frac{1}{3} + 1\frac{5}{12} = 3\frac{3}{4}$.

- How is adding mixed numbers different from adding fractions?

Examples

A

$$\begin{array}{rcr} 2\frac{3}{4} & = & 2\frac{6}{8} \\ +1\frac{3}{8} & = & +1\frac{3}{8} \\ \hline & & 3\frac{9}{8} = 3 + 1\frac{1}{8} = 4\frac{1}{8} \end{array}$$

B

$$\begin{array}{rcr} 5\frac{7}{9} & = & 5\frac{7}{9} \\ +4\frac{1}{3} & = & +4\frac{3}{9} \\ \hline & & 9\frac{10}{9}, \text{ or } 10\frac{1}{9} \end{array}$$

C

$$\begin{array}{rcr} 14\frac{3}{4} & = & 14\frac{9}{12} \\ +12\frac{1}{6} & = & +12\frac{2}{12} \\ \hline & & 26\frac{11}{12} \end{array}$$

MATH IDEA You can make a model or use the LCD to add mixed numbers.

Check

1. **Explain** why in Example B, $9\frac{10}{9}$ was renamed as $10\frac{1}{9}$.

Find the sum in simplest form. Estimate to check.

2. $1\frac{1}{4} + 2\frac{1}{2}$

3. $2\frac{5}{8} + 1\frac{1}{2}$

4. $5\frac{1}{3} + 2\frac{1}{6}$

5. $4\frac{5}{9} + 2\frac{2}{3}$

6. $8\frac{5}{6} + 3\frac{1}{3}$

7. $4\frac{5}{12} + 1\frac{1}{6}$

8. $1\frac{1}{5} + 3\frac{2}{5}$

9. $9\frac{3}{4} + 8\frac{1}{2}$

10. $9\frac{4}{5} + 2\frac{3}{10}$

LESSON CONTINUES

Practice and Problem Solving

Find the sum in simplest form. Estimate to check.

11. $4\frac{2}{3} + 2\frac{1}{3}$

12. $3\frac{1}{2} + 1\frac{1}{4}$

13. $5\frac{7}{9} + 3\frac{1}{3}$

14. $5\frac{2}{5} + 1\frac{3}{10}$

15. $4\frac{7}{12} + 1\frac{2}{3}$

16. $2\frac{3}{5} + 2\frac{3}{10}$

17. $3\frac{2}{3} + 4\frac{1}{12}$

18. $5\frac{1}{6} + 1\frac{11}{12}$

19. $2\frac{1}{8} + 3\frac{1}{2}$

20. $1\frac{1}{5} + 1\frac{3}{4}$

21. $7\frac{3}{4} + 4\frac{7}{12}$

22. $4\frac{5}{8} + 2\frac{1}{4}$

23. $5\frac{5}{8} + 2\frac{1}{4}$

24. $4\frac{3}{4} + 3\frac{5}{12}$

25. $6\frac{1}{3} + 3\frac{1}{4}$

26. $5\frac{11}{12} + 2\frac{1}{4}$

27. $4\frac{3}{4} + 2\frac{3}{8}$

28. $7\frac{1}{2} + 1\frac{1}{16}$

Algebra Find the value of *n*.

29. $4\frac{4}{5} + n = 4\frac{4}{5}$

30. $n + n = 3\frac{1}{2}$

31. $n + 2\frac{3}{5} = 4$

32. $3\frac{1}{6} + n = 11\frac{5}{6}$

USE DATA For 33–35, use the table.

As part of the Citizenship in the World merit badge, the troop is studying two countries, Austria and Belgium. Mark wants to make the Austrian and Belgian flags.

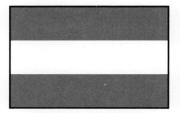

Austria

Belgium

33. How many yards of red and white fabric will he use altogether?

34. How many yards of fabric does Mark need to make both flags?

35. **Write a problem** using the information in the table, in which you add mixed numbers to solve.

FABRIC NEEDED FOR FLAGS

Color	Number of Yards
Red	$8\frac{3}{8}$
White	$2\frac{1}{8}$
Yellow	$3\frac{1}{2}$
Black	$3\frac{1}{2}$

36. **MEASUREMENT** This week, Amanda worked $2\frac{1}{2}$ hours on Monday, $1\frac{2}{3}$ hours on Tuesday, and $2\frac{1}{3}$ hours on Wednesday. How many hours did she work this week?

37. **REASONING** If you add two mixed numbers, will you always get a mixed number? Explain.

Mixed Review and Test Prep

For 38–39, order from greatest to least. (p. 24)

38. 3.157, 3.175, 3.1557, 3.1775

39. 1.168, 1.118, 1.181, 1.116

40. Write an expression to model this situation. Tell what the variable represents. (p. 68)

Jerry ran 10 miles, then he ran some more.

41. TEST PREP In which number does the digit 7 have the greatest value? (p. 20)

A 8.027 C 7.826
B 8.078 D 6.780

42. $3.489 \div 3 = n$ (p. 222)

43. $n \div 7 = 12$ (p. 190)

44. Michelle has 127 feet of yarn. She wants to divide the yarn into 4 equal pieces. How long will each piece be? Explain how you interpreted the remainder. (p. 194)

45. TEST PREP Which is $7\frac{4}{7}$ written as a fraction? (p. 294)

F $\frac{11}{7}$ G $\frac{18}{7}$ H $\frac{53}{7}$ J $\frac{49}{4}$

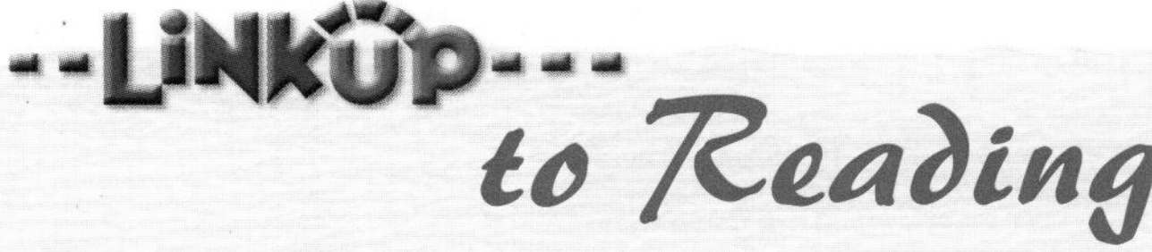

SOCCER PRACTICE SCHEDULE	
Sequence	**Time Spent**
1. stretching	■ hr
2. passing	$\frac{1}{3}$ hr
3. game	$\frac{2}{3}$ hr

STRATEGY • SEQUENCE Putting events in order, or in sequence, can help you solve a problem. Words like *first, next,* and *then* can help set the order. Compare the problem below with the sequence shown in the table.

Soccer practice starts at 3:00 P.M. Mary played a practice game for $\frac{2}{3}$ of an hour at the end of practice. She spent the first part of the practice stretching. Then she spent $\frac{1}{3}$ of an hour passing. Practice lasted $1\frac{1}{4}$ hours. What time did Mary finish stretching?

1. What must you find before you can find the time Mary finished stretching? Solve the problem.

2. Steve got to the mall at 11:30 A.M. and arrived back home at 4:00 P.M. After shopping at the mall, he walked 30 min to get to the baseball field. He stayed at the field for 1 hr. Next, he walked 20 min to get home. What time did Steve leave the mall?

Extra Practice page H51, Set A

LESSON 2

Subtract Mixed Numbers

Quick Review

1. $1 - \frac{1}{3}$ **2.** $\frac{5}{6} - \frac{2}{3}$

3. $\frac{2}{3} - \frac{2}{9}$ **4.** $\frac{3}{4} - \frac{2}{3}$

5. $1 - \frac{4}{5}$

Learn

ALL ABOARD! Jacob collects model trains. He is setting up the track for one of his models. He bought $3\frac{1}{2}$ feet of new track, but $1\frac{1}{4}$ feet of it was damaged in shipping. How many feet of new track for the model was not damaged?

Subtract. $3\frac{1}{2} - 1\frac{1}{4}$

First, get a rough estimate. $3 - 1 = 2$

One Way

Use a model.

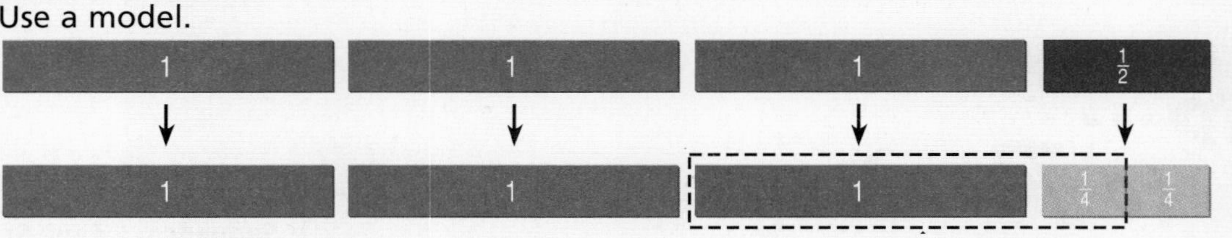

To subtract $1\frac{1}{4}$, replace the $\frac{1}{2}$ bar with $\frac{1}{4}$ bars.

Subtract $1\frac{1}{4}$. $3\frac{1}{2} - 1\frac{1}{4} = 3\frac{2}{4} - 1\frac{1}{4} = 2\frac{1}{4}$

Another Way

Use the LCD.

$$\begin{array}{rcr} 3\frac{1}{2} & = & 3\frac{2}{4} \\ -1\frac{1}{4} & = & -1\frac{1}{4} \\ \hline & & 2\frac{1}{4} \end{array}$$

Use the LCD to write the fractions with like denominators.

Subtract the fractions and the whole numbers.

Write the difference in simplest form.

The Toy Train Museum in Kenner, Louisiana, uses 275 feet of track.

So, $2\frac{1}{4}$ feet of new track was not damaged. Since $2\frac{1}{4}$ is close to the estimate of 2, $2\frac{1}{4}$ is reasonable.

- How is using the LCD similar to using fraction bars?

CALIFORNIA STANDARDS **NS 2.0** Students perform calculations and solve problems involving addition, subtraction, and simple multiplication and division of fractions and decimals. **NS 2.3** Solve simple problems, including ones arising in concrete situations, involving the addition and subtraction of fractions and mixed numbers, and express answers in the simplest form. *also* **NS 1.0, MR 2.0, MR 2.1, MR 2.3, MR 3.2**

Check

1. **Explain** how you can tell your answer is reasonable when subtracting mixed numbers.

Find the difference in simplest form. Estimate to check.

2. $5\frac{3}{4} - 2\frac{1}{8}$

3. $9\frac{1}{2} - 2\frac{2}{5}$

4. $5\frac{7}{9} - 3\frac{1}{9}$

5. $3\frac{2}{3} - 1\frac{5}{12}$

TECHNOLOGY LINK

To learn more about adding and subtracting mixed numbers, watch the Harcourt Math Newsroom Video *Model Trains.*

Practice and Problem Solving

Find the difference in simplest form. Estimate to check.

6. $7\frac{11}{12} - 4\frac{5}{6}$

7. $5\frac{5}{8} - 2\frac{1}{4}$

8. $6\frac{1}{3} - 3\frac{1}{4}$

9. $4\frac{7}{10} - 2\frac{3}{10}$

10. $8\frac{3}{5} - 3\frac{3}{10}$

11. $5\frac{7}{12} - 4\frac{1}{3}$

12. $6\frac{3}{4} - 2\frac{5}{16}$

13. $3\frac{8}{9} - 1\frac{5}{9}$

14. $7\frac{3}{5} - 2\frac{1}{4}$

Algebra Find the value of *n*.

15. $n - 1\frac{1}{2} = 3$

16. $3\frac{5}{n} - 1\frac{3}{8} = 2\frac{1}{4}$

17. $4\frac{3}{5} - n = 1\frac{2}{15}$

18. $8\frac{n}{6} - 3\frac{1}{6} = 5\frac{2}{3}$

USE DATA For 19–20, use the table.

19. How much longer is the S scale caboose than the HO scale caboose?

20. In which scale is the caboose $2\frac{7}{10}$ inch longer than a caboose in N scale?

MODEL TRAIN CABOOSE SIZES

Scale	O	S	HO	N	Z
Size (in.)	6	$4\frac{1}{2}$	$3\frac{1}{3}$	$1\frac{4}{5}$	$\frac{3}{4}$

21. **What's the Error?** Marty compared the length of four Z scale cabooses with the length of one HO scale caboose. He said they are the same. Describe his error and write the correct answer.

Mixed Review and Test Prep

22. $2\frac{1}{3} + 4\frac{1}{4}$ (p. 364)

23. $\frac{1}{2} - \frac{1}{3}$ (p. 354)

24. Write two fractions equivalent to $\frac{1}{4}$. (p. 284)

25. Write $\frac{4}{5}$ as a decimal. (p. 280)

26. **TEST PREP** Choose the least common multiple of 8 and 6. (p. 256)

A 8 **C** 16
B 14 **D** 24

Extra Practice page H51, Set B

LESSON 3

HANDS ON

Subtraction with Renaming

Quick Review

Write each mixed number as a fraction.

1. $1\frac{1}{2}$
2. $1\frac{5}{6}$
3. $3\frac{3}{4}$
4. $2\frac{7}{9}$
5. $3\frac{6}{7}$

MATERIALS
fraction bars

Explore

Sometimes you need to rename the whole number to subtract with mixed numbers.

Find $2 - 1\frac{3}{8}$.

STEP 1

Model 2 using two 1-whole bars.

1	1	2

STEP 2

To subtract $1\frac{3}{8}$, model 2 another way by renaming one of the whole bars with $\frac{1}{8}$ bars.

1	$\frac{1}{8}$	$\frac{1}{8}$	$\frac{1}{8}$	$\frac{1}{8}$	$\frac{1}{8}$	$\frac{1}{8}$	$\frac{1}{8}$	$\frac{1}{8}$	$1\frac{8}{8}$

STEP 3

Subtract $1\frac{3}{8}$. Write the answer in simplest form.

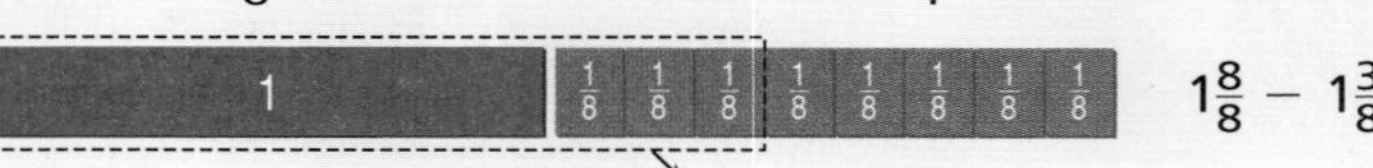

$1\frac{8}{8} - 1\frac{3}{8}$

So, $2 - 1\frac{3}{8} = 1\frac{8}{8} - 1\frac{3}{8} = \frac{5}{8}$.

- In Step 2, why did 2 have to be renamed as $1\frac{8}{8}$?

How can you rename one 1-whole bar to subtract $1\frac{4}{6}$?

Try It

Use fraction bars to find the difference.

a. $2\frac{1}{6} - 1\frac{4}{6}$

b. $6\frac{1}{4} - 2\frac{3}{4}$

c. $4\frac{3}{8} - 3\frac{7}{8}$

d. $7\frac{1}{3} - 2\frac{2}{3}$

CALIFORNIA STANDARDS **NS 2.0** Students perform calculations and solve problems involving addition, subtraction, and simple multiplication and division of fractions and decimals. **NS 2.3** Solve simple problems, including ones arising in concrete situations, involving the addition and subtraction of fractions and mixed numbers (like and unlike denominators of 20 or less), and express answers in the simplest form. *also* **NS 1.0, MR 2.0, MR 2.4, MR 3.0, MR 3.2**

Connect

Sometimes you need to rename when the denominators are *unlike*.

Find $2\frac{1}{3} - 1\frac{5}{6}$.

STEP 1

Model $2\frac{1}{3}$, using two 1-whole bars and one $\frac{1}{3}$ bar.

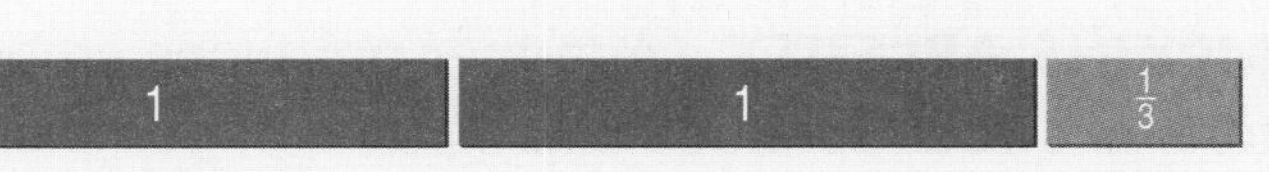

STEP 2

To subtract, think of the LCD for $\frac{5}{6}$ and $\frac{1}{3}$. Rename $\frac{1}{3}$ as $\frac{2}{6}$.

STEP 3

Rename one 1-whole bar as $\frac{6}{6}$.

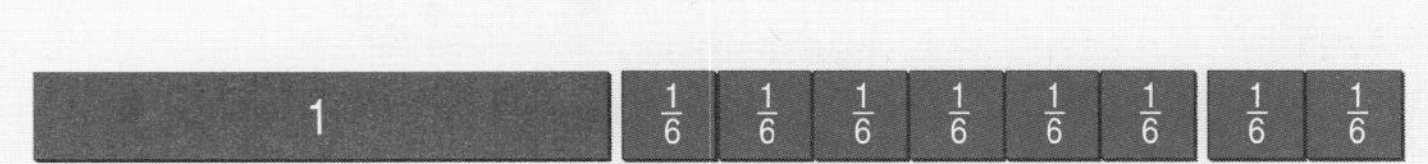

STEP 4

Subtract $1\frac{5}{6}$. Write the answer in simplest form.

So, $2\frac{1}{3} - 1\frac{5}{6} = 1\frac{8}{6} - 1\frac{5}{6} = \frac{3}{6} = \frac{1}{2}$.

MATH IDEA When subtracting mixed numbers, you can rename the whole-number part as a fraction part.

Practice

Use fraction bars to find the difference.

1. $5\frac{2}{9} - 4\frac{1}{3}$ **2.** $6\frac{5}{12} - 5\frac{3}{4}$ **3.** $3\frac{1}{4} - 1\frac{3}{4}$ **4.** $4\frac{2}{5} - 2\frac{1}{2}$

5. $3\frac{1}{8} - 2\frac{7}{8}$ **6.** $4\frac{2}{3} - 2\frac{5}{6}$ **7.** $8\frac{4}{10} - 2\frac{3}{5}$ **8.** $3\frac{1}{2} - 2\frac{2}{3}$

9. **What's the Error?** Larry had $4\frac{1}{2}$ quarts of apple cider. He used $3\frac{4}{5}$ quarts for punch. Larry says that he has $1\frac{3}{10}$ quarts left. Describe his error and write the correct amount.

10. **Write About It** Explain how to find $2\frac{1}{3} - 1\frac{2}{3}$.

Mixed Review and Test Prep

11. $\frac{5}{8} - \frac{1}{2}$ (p. 354) **12.** $\frac{1}{4} + \frac{7}{12}$ (p. 354)

13. Is 27 a prime number or a composite number? (p. 264)

14. Evaluate 4^2. (p. 268)

15. **TEST PREP** Which is the least common denominator of $\frac{1}{3}$ and $\frac{1}{4}$? (p. 352)

A 3 **C** 12
B 4 **D** 15

LESSON 4

Practice with Mixed Numbers

Quick Review

Write each as a mixed number in simplest form.

1. $\frac{7}{6}$
2. $\frac{12}{8}$
3. $\frac{10}{3}$
4. $\frac{17}{5}$
5. $\frac{26}{6}$

Learn

PUMPKINS APLENTY A farmer produces crops of $4\frac{3}{4}$ tons of sweet potatoes and $5\frac{1}{3}$ tons of pumpkins. How many tons does the farmer produce in all?

Add. $4\frac{3}{4} + 5\frac{1}{3}$ Estimate. $5 + 5 = 10$

STEP 1

Use the LCD to change the fractions to like fractions.

$$\begin{array}{r} 4\frac{3}{4} = 4\frac{9}{12} \\ +5\frac{1}{3} = +5\frac{4}{12} \\ \hline \end{array}$$

STEP 2

Add the fractions.

$$\begin{array}{r} 4\frac{3}{4} = 4\frac{9}{12} \\ +5\frac{1}{3} = +5\frac{4}{12} \\ \hline \frac{13}{12} \end{array}$$

STEP 3

Add the whole numbers. Write the answer in simplest form.

$$\begin{array}{r} 4\frac{3}{4} = 4\frac{9}{12} \\ +5\frac{1}{3} = +5\frac{4}{12} \\ \hline 9\frac{13}{12}, \text{ or } 10\frac{1}{12} \end{array}$$

So, $10\frac{1}{12}$ tons of sweet potatoes and pumpkins were produced. Since $10\frac{1}{12}$ is close to the estimate of 10, the answer is reasonable.

Subtract. $5\frac{1}{3} - 1\frac{3}{4}$ Estimate. $5 - 2 = 3$

STEP 1

Use the LCD to change the fractions to like fractions.

$$\begin{array}{r} 5\frac{1}{3} = 5\frac{4}{12} \\ -1\frac{3}{4} = -1\frac{9}{12} \\ \hline \end{array}$$

STEP 2

Rename the mixed numbers. Subtract the fractions.

$$\begin{array}{r} 5\frac{1}{3} = 5\frac{4}{12} = 4\frac{16}{12} \\ -1\frac{3}{4} = -1\frac{9}{12} = -1\frac{9}{12} \\ \hline \frac{7}{12} \end{array}$$

STEP 3

Subtract the whole numbers. Write the answer in simplest form.

$$\begin{array}{r} 5\frac{1}{3} = 5\frac{4}{12} = 4\frac{16}{12} \\ -1\frac{3}{4} = -1\frac{9}{12} = -1\frac{9}{12} \\ \hline 3\frac{7}{12} \end{array}$$

So, $5\frac{1}{3} - 1\frac{3}{4} = 3\frac{7}{12}$. Since $3\frac{7}{12}$ is close to the estimate of 3, the answer is reasonable.

The world produces 44,000 tons of pumpkins in a day. ▶

CALIFORNIA STANDARDS **NS 2.3** Solve simple problems, including ones arising in concrete situations, involving the addition and subtraction of fractions and mixed numbers (like and unlike denominators of 20 or less), and express answers in the simplest form. **MR 2.1** Use estimation to verify the reasonableness of calculated results. *also* **NS 1.0, MR 3.0, MR 3.2, MR 3.3**

Check

1. **Explain** how you can check your answer when adding or subtracting fractions.

Add or subtract. Write the answer in simplest form. Estimate to check.

2. $2\frac{1}{4} + 4\frac{7}{16}$

3. $7\frac{1}{8} - 2\frac{1}{2}$

4. $1\frac{2}{3} + 1\frac{7}{9}$

5. $5\frac{9}{10} - 4\frac{4}{5}$

Practice and Problem Solving

Add or subtract. Write the answer in simplest form. Estimate to check.

6. $6\frac{1}{2} + 7\frac{3}{5}$

7. $4\frac{5}{8} + 2\frac{14}{16}$

8. $6\frac{1}{4} - 3\frac{10}{12}$

9. $5\frac{2}{3} - 1\frac{3}{4}$

10. $4\frac{3}{8} - 3\frac{7}{8}$

11. $4\frac{1}{3} + 2\frac{3}{4}$

12. $9\frac{3}{8} - 3\frac{12}{16}$

13. $2\frac{2}{7} + 1\frac{1}{2} + 1\frac{3}{7}$

14. $6\frac{4}{15} - 2\frac{1}{5}$

15. $3\frac{3}{20} + 5\frac{3}{10}$

16. $8\frac{1}{18} - 2\frac{4}{9}$

17. $1\frac{2}{5} + 4\frac{2}{3} + 1\frac{1}{3}$

Algebra **Find the value of *n*.**

18. $4\frac{3}{8} + n = 6$

19. $n + 2\frac{1}{4} = 5$

20. $3\frac{1}{6} - 1\frac{2}{n} = 1\frac{1}{2}$

21. $n\frac{1}{3} + 1\frac{3}{4} = 4\frac{1}{12}$

22. Mrs. Kelly is making matching outfits. She needs $4\frac{1}{2}$ yards for her dress and 3 yards for her daughter's dress. If she buys 8 yards of fabric, will she have enough to make a matching doll dress that requires $\frac{1}{2}$ yard of fabric? Explain.

23. A $9 \times 1\frac{1}{2}$ inch round cake pan holds $5\frac{3}{4}$ cups of batter. Latasha's cake recipe makes 10 cups of batter. Will she need to use 2 or 3 pans to bake her cake? Explain.

24. The bakery used $6\frac{3}{4}$ cups of pumpkin for its pumpkin bread and $10\frac{1}{3}$ cups for pumpkin pies. How many cups of pumpkin did the bakery use altogether?

Mixed Review and Test Prep

25. $\frac{1}{3} - \frac{1}{4}$ (p. 354)

26. $\frac{3}{4} + \frac{3}{4}$ (p. 344)

27. Find the greatest common factor of 12 and 20. (p. 258)

28. Evaluate 2^3. (p. 268)

29. **TEST PREP** Which number is a prime number? (p. 264)

A 21 **B** 31 **C** 52 **D** 81

Extra Practice page H51, Set C

LESSON 5

Problem Solving Skill

Multistep Problems

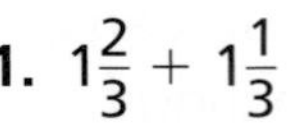

Quick Review

1. $1\frac{2}{3} + 1\frac{1}{3}$
2. $2\frac{1}{2} - 1\frac{1}{4}$
3. $3\frac{2}{5} + 2\frac{7}{10}$
4. $2\frac{3}{4} + 4\frac{5}{6}$
5. $5\frac{5}{8} - 3\frac{1}{4}$

STEPPIN' OUT Kaitlin is making costumes for the dance recital. She made a table to keep track of the amount of fabric she buys and uses. Kaitlin forgot to record some of the information in the table. How many yards of fabric does she have left?

FABRIC FOR COSTUMES

Color	Purchased	Used	Remaining
blue	$12\frac{7}{12}$ yd	$3\frac{5}{6}$ yd	■
gold	$11\frac{3}{4}$ yd	■	$9\frac{5}{12}$ yd
black	■	$5\frac{11}{12}$ yd	■
TOTAL	40 yd	■	■

STEP 1

Find how much gold fabric was used.

$11\frac{3}{4}$ gold fabric purchased

$-\ 9\frac{5}{12}$ gold fabric remaining

$2\frac{1}{3}$ yd gold fabric used

STEP 2

Add to find out how much total fabric was used.

$3\frac{5}{6}$ blue fabric used

$2\frac{1}{3}$ gold fabric used

$+5\frac{11}{12}$ black fabric used

$12\frac{1}{12}$ yd total fabric used

STEP 3

Subtract the total fabric used from the total purchased.

40

$-12\frac{1}{12}$

$27\frac{11}{12}$ yd

So, she has $27\frac{11}{12}$ yd left.

A *multistep problem* requires more than one step to solve.

Talk About It

- Is there another way this problem could have been solved? Explain.
- How could you find how many yards of black fabric Kaitlin has left?

CALIFORNIA STANDARDS **NS 2.3** Solve simple problems, including ones arising in concrete situations, involving the addition and subtraction of fractions and mixed numbers (like and unlike denominators of 20 or less), and express answers in the simplest form. **MR 1.1** Analyze problems by identifying relationships, distinguishing relevant from irrelevant information, sequencing and prioritizing information, and observing patterns. **MR 1.2** Determine when and how to break a problem into simpler parts. *also* **NS 1.0, NS 2.0, MR 2.0, MR 2.3, MR 3.0**

Problem Solving Practice

1. **Explain** how you know when a problem requires more than one step to solve.

2. **What if** Kaitlin had used $4\frac{1}{6}$ yards of gold fabric and $4\frac{3}{4}$ yards of blue fabric? How many total yards of fabric would she have left?

USE DATA For 3–4, use the table. At Snyder's Dance Supply Store, Vince records how many dozens of each item are sold every day. Some entries are missing from Vince's record book for today's sales. Choose the best answer. Then copy and complete the table.

SALES RECORDS

Item	Start	Sold	Remaining
Tights	■	$10\frac{2}{3}$	■
Sweat Bands	$8\frac{2}{3}$	■	$2\frac{5}{6}$
Leotards	$5\frac{5}{6}$	$2\frac{1}{2}$	■
TOTAL	28	■	■

3. What is the total number of dance items sold today? Remember, one dozen is the same as 12 items.

   A 2,228 **B** 228 **C** 128 **D** 28

4. What is the total number of dance items that Snyder's Dance Supply Store has left?

   F 1,118 **G** 208 **H** 180 **J** 108

Mixed Applications

USE DATA For 5–6, use the graph.

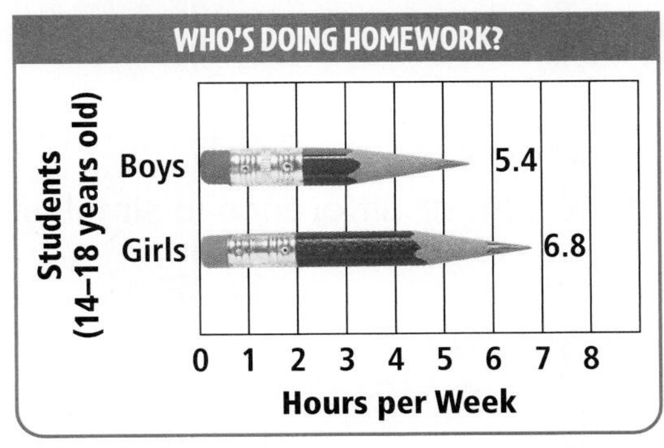

5. On average, how many more hours a week do girls spend doing homework than boys?

6. Find the average number of hours boys and girls spend doing homework each week.

7. Trina made a four-by-four grid. She colored the corner squares blue and any square that shares a side with a blue square green. The remaining squares she colored yellow. How many squares were colored green? yellow?

8. At the concession stand, customers can choose between a hot dog and a hamburger. The concession stand offers soda, lemonade, and water. How many different combinations of a sandwich and a drink are there?

9. **What's the Question?** Ellen has a goal of running 12 miles this week. She ran $2\frac{1}{3}$ miles Monday, $3\frac{1}{2}$ miles Tuesday, and $4\frac{1}{4}$ miles Thursday. The answer is $1\frac{11}{12}$ miles.

Review/Test

CHECK VOCABULARY AND CONCEPTS

Choose the best term from the box.

least common denominator (LCD)
mixed numbers
renaming
simplest form

1. When you add or subtract mixed numbers, make sure your answer is in __?__. (p. 365)
2. To add or subtract mixed numbers, first find the __?__ to make like fractions. (p. 365)
3. When you change $9\frac{1}{4}$ to $8\frac{5}{4}$, you are __?__ the mixed number. (p. 370)

CHECK SKILLS

Find the sum in simplest form. (pp. 364–369)

4. $3\frac{1}{4} + 6\frac{1}{8}$
5. $3\frac{5}{12} + 1\frac{1}{6}$
6. $5\frac{7}{8} + 3\frac{1}{4}$
7. $2\frac{1}{3} + 5\frac{8}{9}$
8. $6\frac{1}{8} + 1\frac{1}{4}$

Find the difference in simplest form. (pp. 368–369)

9. $4\frac{3}{4} - 1\frac{1}{7}$
10. $5\frac{17}{18} - 4\frac{8}{9}$
11. $10\frac{11}{12} - 5\frac{8}{9}$
12. $6\frac{1}{2} - 3\frac{2}{3}$
13. $9\frac{1}{6} - 3\frac{1}{2}$

Find the sum or difference in simplest form. (pp. 364–373)

14. $4\frac{1}{4} + 1\frac{1}{2}$
15. $5\frac{2}{5} + 3\frac{1}{2}$
16. $4\frac{3}{4} - 1\frac{1}{3}$
17. $8\frac{1}{8} - 4\frac{1}{4}$

CHECK PROBLEM SOLVING (pp. 374–375)

18. Mary volunteered $3\frac{2}{3}$ hours on Monday, 5 hours on Tuesday, and $2\frac{1}{12}$ hours on Wednesday. How many more hours must she work to meet her weekly goal of 15 hours?

19. On April 3, Jan's plant was $3\frac{1}{4}$ in. tall and Bob's was $3\frac{3}{8}$ in. tall. On May 3, Jan's had grown to $8\frac{7}{8}$ in. and Bob's measured $8\frac{5}{8}$ in. Whose plant grew more during the month? How much more?

20. Grandma baked 2 apple pies and 1 peach pie. The family ate one apple pie and $\frac{1}{3}$ of the second apple pie. They ate $\frac{3}{8}$ of the peach pie. What part of 3 pies is left?

Cumulative Review

Decide on a plan.
See item **9**.
You know you need to use the table to get the numbers you need. Since the question asks about 2 loaves of 2 different kinds of bread, you will need more than one step to solve the problem.

Also see problem **4**, p. H63.

For 1–10, choose the best answer.

1. Which fraction is equivalent to $3\frac{6}{7}$?

A $\frac{23}{7}$ **C** $\frac{25}{7}$

B $\frac{24}{7}$ **D** $\frac{27}{7}$

2. Mr. Bennett sold $55\frac{1}{2}$ bushels of apples, $20\frac{3}{8}$ bushels of pears, and $35\frac{1}{4}$ bushels of peaches at his fruit stand. How many bushels of fruit did he sell in all?

F $110\frac{1}{8}$ **H** $111\frac{1}{8}$

G $110\frac{1}{2}$ **J** 112

3. $2\frac{3}{8} + \frac{3}{4}$

A $3\frac{1}{4}$ **C** $3\frac{1}{2}$

B $3\frac{1}{8}$ **D** $3\frac{5}{8}$

4. $2\frac{5}{7} - \frac{6}{7}$

F $\frac{6}{7}$ **H** 2

G $1\frac{6}{7}$ **J** $2\frac{1}{7}$

5. $0.561 - 0.43$

A 0.131 **C** 0.522

B 0.518 **D** Not here

6. Which illustrates the Associative Property of Multiplication?

F $6 \times n = n \times 6$

G $(12 \times n + 4) \times 1 = 12 \times n + 4$

H $5 \times (3 + n) = (5 \times 3) + (5 \times n)$

J $5 \times (6 \times n) = (5 \times 6) \times n$

For 7–9, use the table.

FLOUR AND SUGAR FOR ONE LOAF		
Type of Bread	**Flour Needed**	**Sugar Needed**
banana	$1\frac{1}{2}$ cups	1 cup
pumpkin	2 cups	$\frac{3}{4}$ cup
wheat	$2\frac{1}{4}$ cups	$\frac{1}{2}$ cup

7. How many cups of flour are needed to make 1 loaf of banana bread and 1 loaf of wheat bread?

A $3\frac{1}{4}$ **C** $3\frac{3}{4}$

B $3\frac{1}{2}$ **D** $4\frac{1}{4}$

8. How many loaves of pumpkin bread can be made with 11 cups of flour?

F 5 **H** 7

G 6 **J** 8

9. How much more sugar is in 2 loaves of pumpkin bread than in 2 loaves of wheat bread?

A $\frac{1}{4}$ cup **C** $\frac{3}{4}$ cup

B $\frac{1}{2}$ cup **D** 1 cup

10.
$$\begin{array}{r} \$22.95 \\ -\quad 4.76 \\ \hline \end{array}$$

F \$12.21 **H** \$18.29

G \$18.19 **J** Not here

CHAPTER 21 Multiply Fractions

Community and school gardens are becoming more and more popular. In Ohio, for example, more than 500 volunteer leaders helped begin and maintain 212 community gardens. Gardeners may choose to grow flowers, herbs, or vegetables and are encouraged to share their harvest with a neighborhood family. Look at the garden diagram. If $\frac{1}{4}$ of the garden is planted with tomatoes, how many square feet are planted with tomatoes?

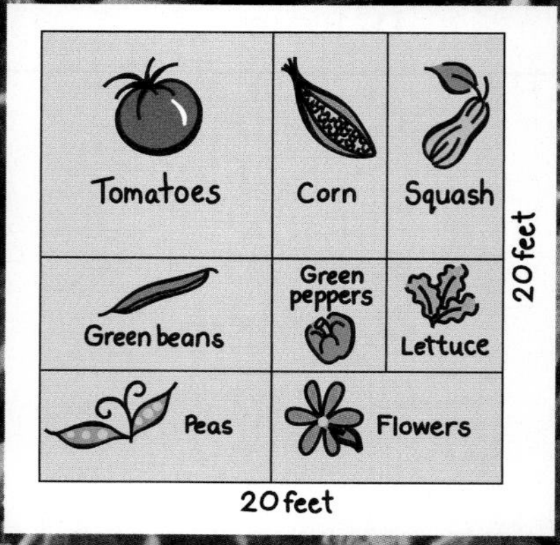

Tomatoes
Corn
Squash
Green beans
Green peppers
Lettuce
Peas
Flowers
20 feet
20 feet

Use this page to help you review and remember important skills needed for Chapter 21.

VOCABULARY

Choose the best term from the box.

fraction
mixed number
simplest form
whole number

1. A _?_ is made up of a whole number and a fraction.
2. A fraction is in _?_ when the greatest common factor of the numerator and denominator is 1.
3. You can use a _?_ to name equal parts of a whole or a group.

FRACTIONS OF A WHOLE OR A GROUP (See p. H23.)

Write a fraction to represent the shaded part.

4. 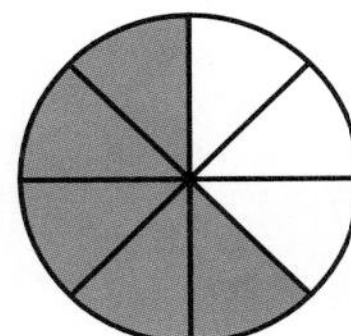

5. 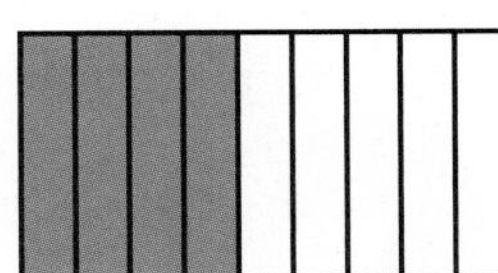

6.     

RENAME MIXED NUMBERS (See p. H22.)

Write the mixed number as a fraction.

7. $1\frac{4}{5}$ **8.** $2\frac{1}{3}$ **9.** $3\frac{2}{5}$ **10.** $5\frac{1}{2}$ **11.** $4\frac{2}{3}$

12. $6\frac{1}{4}$ **13.** $1\frac{9}{10}$ **14.** $8\frac{5}{6}$ **15.** $3\frac{7}{8}$ **16.** $2\frac{5}{9}$

MODEL DECIMAL MULTIPLICATION (See p. H24.)

Write the number sentence each model represents.

17. 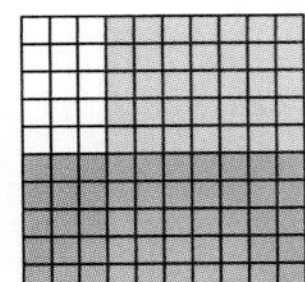

18. 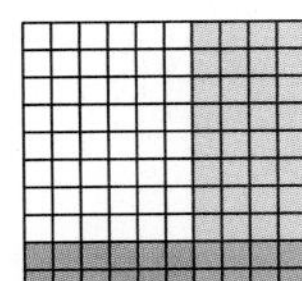

19. 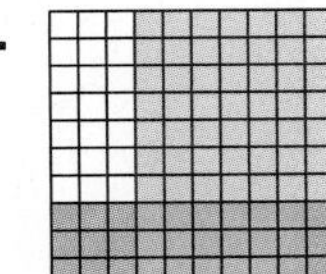

20. 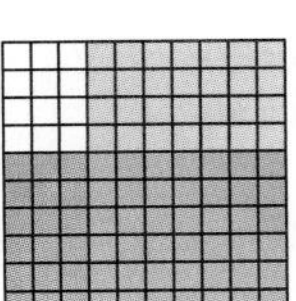

21. 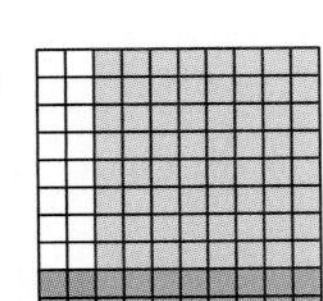

22. 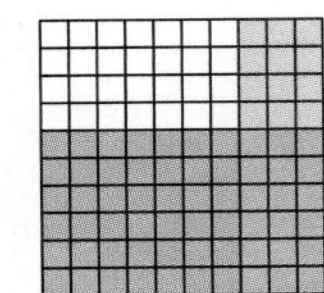

LESSON 1

Multiply Fractions and Whole Numbers

Quick Review

Write as a mixed number or whole number.

1. $\frac{8}{4}$
2. $\frac{12}{5}$
3. $\frac{10}{3}$
4. $\frac{7}{1}$
5. $\frac{9}{7}$

Learn

BAKER'S DOZEN A chef is making 3 batches of doughnuts. The recipe calls for $\frac{3}{4}$ cup of sugar for each batch. How many cups of sugar does the chef need?

Find $3 \times \frac{3}{4}$, or 3 groups of $\frac{3}{4}$.

One Way You can draw a picture.

STEP 1

Use circles to show 3 groups of $\frac{3}{4}$.

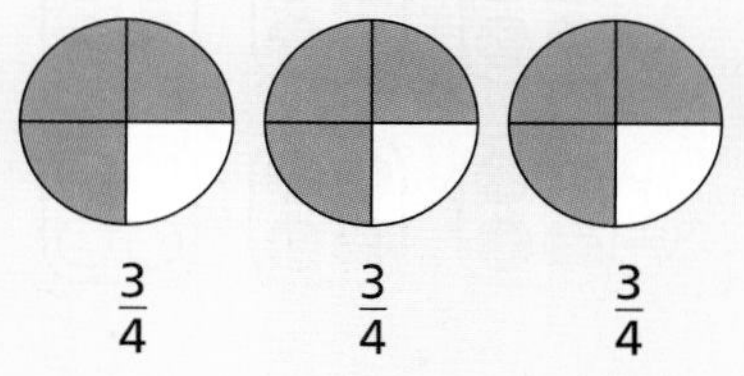

$\frac{3}{4}$ $\frac{3}{4}$ $\frac{3}{4}$

STEP 2

Count the shaded fourths.

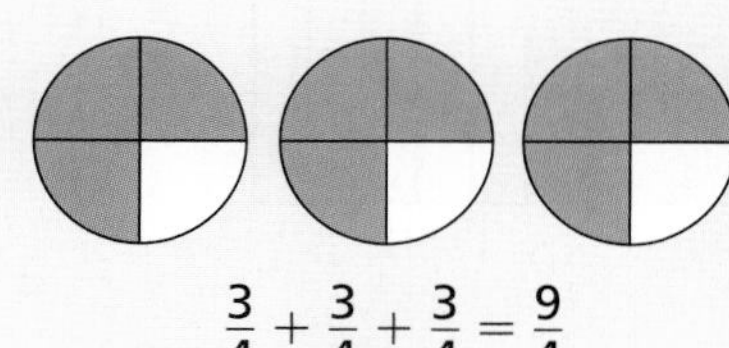

$\frac{3}{4} + \frac{3}{4} + \frac{3}{4} = \frac{9}{4}$

STEP 3

Write the answer as a mixed number.

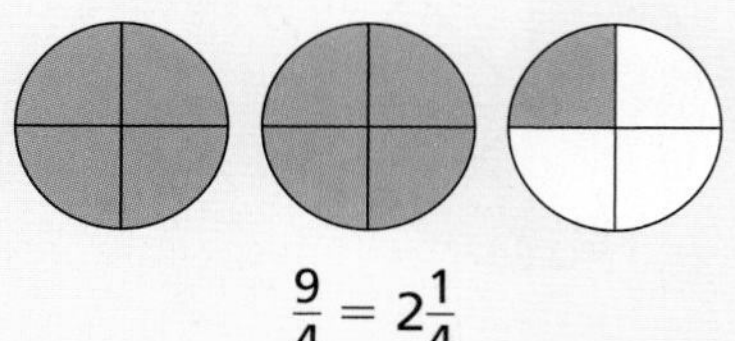

$\frac{9}{4} = 2\frac{1}{4}$

Another Way You can multiply.

STEP 1

Write the whole number as a fraction.

$\frac{3}{1} \times \frac{3}{4}$ Think: $3 = \frac{3}{1}$

STEP 2

Multiply the numerators. Then multiply the denominators.

$\frac{3 \times 3}{1 \times 4} = \frac{9}{4}$

STEP 3

Write the answer as a mixed number.

$\frac{9}{4} = 2\frac{1}{4}$

So, the chef needs $2\frac{1}{4}$ cups of sugar.

Examples

A $12 \times \frac{2}{3} = \frac{12}{1} \times \frac{2}{3}$

$= \frac{12 \times 2}{1 \times 3}$

$= \frac{24 \div 3}{3 \div 3} = \frac{8}{1} = 8$

B $\frac{2}{5} \times 3 = \frac{2}{5} \times \frac{3}{1}$

$= \frac{2 \times 3}{5 \times 1}$

$= \frac{6}{5}$, or $1\frac{1}{5}$

TECHNOLOGY LINK

More Practice: Use Mighty Math Number Heroes, *Fraction Fireworks*, Level V.

CALIFORNIA STANDARDS **NS 2.4** Understand the concept of multiplication and division of fractions. **NS 2.5** Compute and perform simple multiplication and division of fractions and apply these procedures to solving problems. **MR 2.3** Use a variety of methods, such as words, numbers, symbols, charts, graphs, tables, diagrams, and models, to explain mathematical reasoning. *also* **NS 2.0, MR 2.2, MR 3.2, MR 3.3**

Check

1. **Describe** a model you could draw to show the product $4 \times \frac{2}{3}$.

Write the number sentence each model represents.

2. 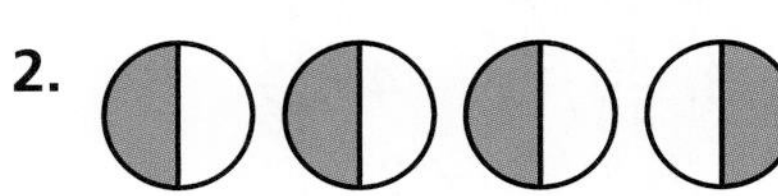

3. 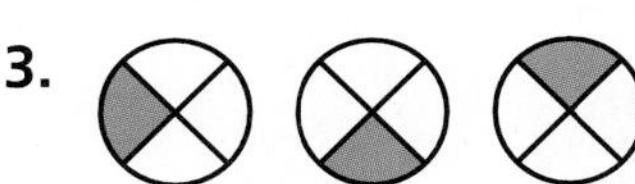

4. 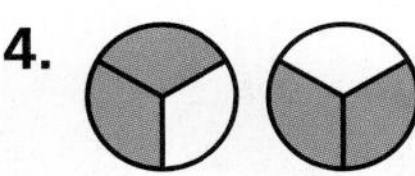

Find the product.

5. $\frac{1}{4} \times 8$
6. $6 \times \frac{2}{3}$
7. $\frac{3}{5} \times 5$
8. $\frac{1}{4} \times 15$
9. $15 \times \frac{2}{5}$

Practice and Problem Solving

Write the number sentence each model represents.

10. 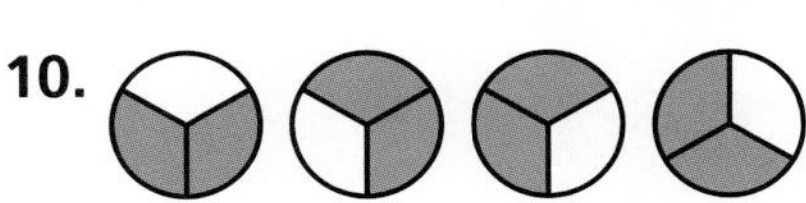

11. 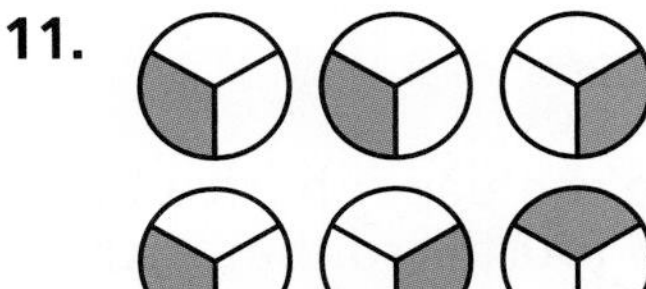

12. 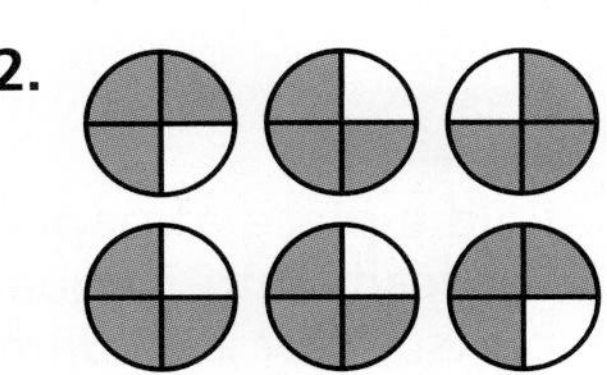

Find the product.

13. $\frac{3}{5} \times 25$
14. $\frac{2}{3} \times 15$
15. $\frac{2}{9} \times 8$
16. $16 \times \frac{7}{8}$
17. $20 \times \frac{2}{5}$
18. $30 \times \frac{5}{6}$
19. $\frac{3}{7} \times 10$
20. $14 \times \frac{5}{7}$
21. $9 \times \frac{3}{8}$
22. $\frac{2}{7} \times 35$

Algebra **Find the missing digit.**

23. $\frac{■}{2} \times 8 = 4$
24. $\frac{1}{2} \times ■ = 30$
25. $\frac{1}{■} \times 18 = 3$
26. $\frac{2}{9} \times 27 = ■$

Evaluate each expression. Then write $<$, $>$, or $=$ for each ●.

27. $\frac{1}{2} \times 14$ ● $\frac{1}{3}$ of 27
28. $\frac{2}{3} \times 21$ ● $\frac{1}{4} \times 24$
29. $\frac{3}{4} \times 16$ ● $\frac{1}{5}$ of 60

30. **What's the Question?** Cyd saved \$20 in quarters. The answer is 80 quarters.

31. **NUMBER SENSE** Without multiplying, how do you know that $4 \times \frac{1}{3} = \frac{1}{3} \times 4$?

32. **REASONING** Is the product of a fraction less than one and a whole number less than or greater than the whole number? Explain.

Mixed Review and Test Prep

33. $25.06 - 0.43$ (p. 50)

34. $93.056 - 1.8$ (p. 50)

35. Sarah has $2\frac{2}{3}$ cups of flour. She needs $2\frac{3}{4}$ cups. Does she have enough flour? Explain. (p. 286)

36. Write $\frac{4}{12}$ in simplest form. (p. 290)

37. **TEST PREP** Which expression models the height of a plant that grew $\frac{2}{5}$ inch? (p. 68)

A $\frac{2}{5} + n$ C $n \times \frac{2}{5}$
B $n \div \frac{2}{5}$ D $\frac{2}{5} \div n$

Extra Practice page H52, Set A

LESSON 2

Multiply a Fraction by a Fraction

Quick Review

Write in simplest form.

1. $\frac{12}{21}$
2. $\frac{9}{24}$
3. $\frac{5}{15}$
4. $\frac{12}{16}$
5. $\frac{18}{30}$

Learn

HIGH-STRUNG In an orchestra, $\frac{2}{3}$ of the musicians play string instruments. In the string section, $\frac{1}{4}$ of the musicians play the viola. What part of the orchestra plays the viola?

Multiply. $\frac{1}{4}$ of $\frac{2}{3}$ or $\frac{1}{4} \times \frac{2}{3}$

One Way You can make a model to find the product.

STEP 1

Fold a piece of paper vertically into 3 equal parts. Color 2 columns to represent $\frac{2}{3}$ of the whole.

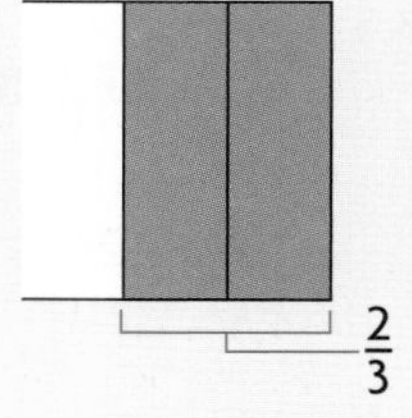

STEP 2

Fold the paper horizontally into fourths so that each of the thirds is divided into 4 equal parts.

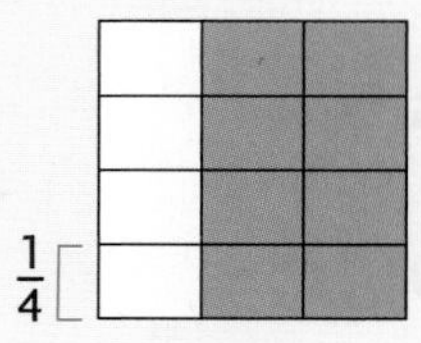

STEP 3

Use the other color to shade 1 of the fourths rows to represent $\frac{1}{4}$. The overlapped shading shows the product.

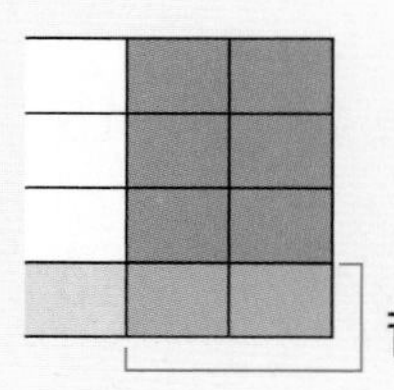

Since 2 of the 12 parts are shaded both colors, $\frac{1}{4}$ of $\frac{2}{3} = \frac{2}{12}$, or $\frac{1}{6}$.

Another Way You can compute the product $\frac{1}{4} \times \frac{2}{3}$.

STEP 1

Multiply the numerators. Then multiply the denominators.

$$\frac{1}{4} \times \frac{2}{3} = \frac{1 \times 2}{4 \times 3} = \frac{2}{12}$$

STEP 2

Write the answer in simplest form.

$$\frac{2 \div 2}{12 \div 2} = \frac{1}{6}$$

So, $\frac{1}{6}$ of the musicians in the orchestra play viola.

MATH IDEA To multiply a fraction by a fraction, multiply the numerators, and then multiply the denominators. Write the product in simplest form.

CALIFORNIA STANDARDS NS 2.4 Understand the concept of multiplication and division of fractions. **NS 2.5** Compute and perform simple multiplication and division of fractions and apply these procedures to solving problems. *also* **NS 2.0, MR 2.3, MR 2.4, MR 3.2, MR 3.3**

Check

1. **Explain** why the product of two fractions less than one is less than each of its factors.

Find the product. Write it in simplest form.

2. $\frac{1}{4} \times \frac{1}{3}$
3. $\frac{2}{3} \times \frac{1}{3}$
4. $\frac{1}{3} \times \frac{3}{4}$
5. $\frac{1}{5} \times \frac{2}{5}$
6. $\frac{1}{4} \times \frac{2}{5}$

Practice and Problem Solving

Find the product. Write it in simplest form.

7. $\frac{2}{3} \times \frac{1}{5}$
8. $\frac{1}{5} \times \frac{1}{4}$
9. $\frac{1}{3} \times \frac{3}{5}$
10. $\frac{5}{6} \times \frac{3}{4}$
11. $\frac{1}{4} \times \frac{3}{4}$
12. $\frac{3}{7} \times \frac{2}{3}$
13. $\frac{8}{9} \times \frac{5}{6}$
14. $\frac{4}{5} \times \frac{8}{11}$
15. $\frac{7}{10} \times \frac{3}{8}$
16. $\frac{6}{7} \times \frac{11}{18}$
17. $\frac{2}{3} \times \frac{3}{4}$
18. $\frac{1}{5} \times \frac{3}{4}$
19. $\frac{2}{7} \times \frac{1}{2}$
20. $\frac{7}{8} \times \frac{4}{5}$
21. $\frac{5}{6} \times \frac{1}{3}$

Write the number sentence each model represents.

22. 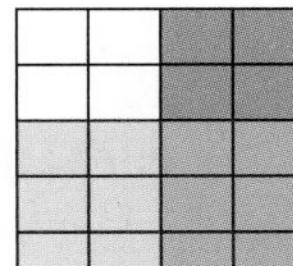

23. 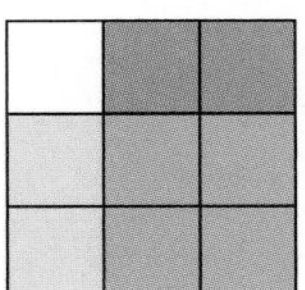

24. 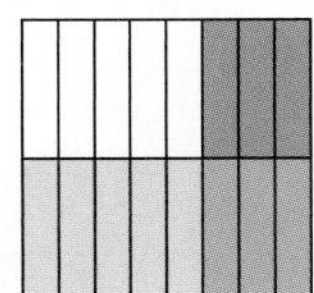

25. 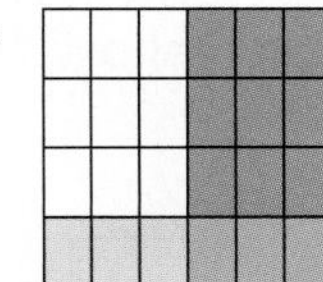

26. **What's the Error?** Mel says $\frac{2}{3} \times \frac{1}{4}$ is $\frac{3}{12}$. Describe and correct his error.

27. In a school band, $\frac{2}{5}$ of the students play wind instruments. In the woodwind section, $\frac{1}{2}$ of the students play the clarinet. What part of the band plays the clarinet?

28. **NUMBER SENSE** Without multiplying, explain how you know that $\frac{7}{5} \times \frac{1}{4}$ is greater than $\frac{1}{4}$.

29. **Write About It** Explain how to multiply $\frac{3}{8} \times \frac{1}{3} \times \frac{2}{3}$. Then find the product.

Mixed Review and Test Prep

Find the least common multiple. (p. 256)

30. 4 and 30
31. 15 and 40
32. Round 5.0699 to the nearest tenth. (p. 46)
33. Evaluate $n - 26$ if n is 50. (p. 68)
34. **TEST PREP** Which illustrates the Distributive Property? (p. 92)
   - **A** $5 \times 16 = 16 \times 5$
   - **B** $5 \times (10 + 6) = (5 \times 10) + (5 \times 6)$
   - **C** $5 \times (10 \times 6) = (5 \times 10) \times 6$
   - **D** $5 \times 16 = (1 + 1 + 1 + 1 + 1) \times 16$

Extra Practice page H52, Set B

LESSON 3

Multiply Fractions and Mixed Numbers

Quick Review

1. $3\frac{3}{5} = \frac{■}{5}$
2. $2\frac{4}{7} = \frac{■}{7}$
3. $1\frac{2}{3} = \frac{■}{3}$
4. $5\frac{1}{6} = \frac{■}{6}$
5. $6\frac{3}{4} = \frac{■}{4}$

Learn

CHEF'S GARDEN Maria's parents run a bed-and-breakfast inn. They grow fresh produce in their garden, on $1\frac{1}{3}$ acres. They planted $\frac{1}{4}$ of it with herbs. How many acres of herbs did they plant?

Multiply. $\frac{1}{4} \times 1\frac{1}{3}$

Remember
You can write a mixed number as a fraction.
$2\frac{1}{3} = 2 + \frac{1}{3} = \frac{6}{3} + \frac{1}{3} = \frac{7}{3}$

One Way Use models to multiply fractions and mixed numbers.

STEP 1

Show 2 whole squares. Divide each square into thirds. Shade 1 whole square and $\frac{1}{3}$ of the other square yellow to show $1\frac{1}{3}$, or $\frac{4}{3}$.

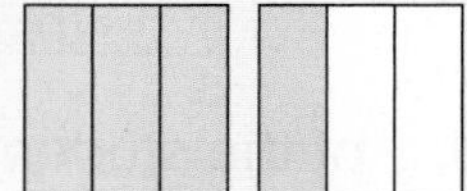

STEP 2

Divide the squares into fourths. Each of the squares is now divided into twelfths.

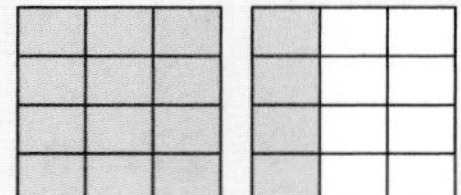

STEP 3

Shade $\frac{1}{4}$ of each whole square blue. The overlapped shading shows the product.

$\frac{3}{12} + \frac{1}{12} = \frac{4}{12}$, or $\frac{1}{3}$

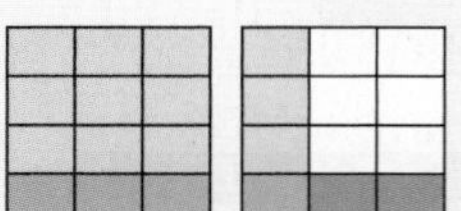

Another Way Compute the product $\frac{1}{4} \times 1\frac{1}{3}$.

STEP 1

Rename the mixed number as a fraction.

Think: $1\frac{1}{3} = \frac{4}{3}$

$\frac{1}{4} \times 1\frac{1}{3} = \frac{1}{4} \times \frac{4}{3}$

STEP 2

Multiply.

$\frac{1}{4} \times \frac{4}{3} = \frac{1 \times 4}{4 \times 3} = \frac{4}{12}$

STEP 3

Write the product in simplest form.

$\frac{4 \div 4}{12 \div 4} = \frac{1}{3}$

So, they planted $\frac{1}{3}$ acre of herbs.

Examples

A $1\frac{1}{2} \times \frac{3}{4} = \frac{3}{2} \times \frac{3}{4} = \frac{9}{8}$, or $1\frac{1}{8}$

B $\frac{2}{3} \times 3\frac{2}{5} = \frac{2}{3} \times \frac{17}{5} = \frac{34}{15}$, or $2\frac{4}{15}$

CALIFORNIA STANDARDS **NS 2.4** Understand the concept of multiplication and division of fractions. **NS 2.5** Compute and perform simple multiplication and division of fractions and apply these procedures to solving problems. *also* **NS 2.0, MR 2.3, MR 2.4, MR 3.2**

Check

1. **Explain** how multiplying a fraction and a mixed number is different from multiplying a fraction by a fraction.

TECHNOLOGY LINK
To learn more about multiplying fractions, watch the Harcourt Math Newsroom Video *Ant Wrangler.*

Find the product. Draw fraction squares as needed.

2. $\frac{1}{3} \times 1\frac{1}{5}$
3. $\frac{1}{2} \times 1\frac{2}{3}$
4. $\frac{1}{2} \times 2\frac{2}{5}$
5. $\frac{3}{5} \times 1\frac{1}{2}$

Practice and Problem Solving

Find the product. Draw fraction squares as needed.

6. $\frac{3}{4} \times 1\frac{1}{6}$
7. $\frac{2}{5} \times 3\frac{2}{3}$
8. $3\frac{1}{5} \times \frac{1}{4}$
9. $4\frac{1}{2} \times \frac{1}{9}$
10. $\frac{1}{4} \times 2\frac{1}{5}$
11. $\frac{1}{3} \times 3\frac{5}{8}$
12. $\frac{5}{12} \times 2\frac{2}{5}$
13. $6\frac{1}{4} \times \frac{3}{8}$
14. $5\frac{6}{7} \times \frac{1}{10}$
15. $\frac{1}{9} \times 3\frac{1}{2}$
16. $\frac{1}{2} \times 5\frac{3}{5}$
17. $1\frac{1}{8} \times \frac{2}{3}$
18. $\frac{1}{4} \times 6\frac{2}{7}$
19. $2\frac{5}{6} \times \frac{1}{9}$
20. $\frac{2}{5} \times 5\frac{1}{3}$
21. $3\frac{3}{4} \times \frac{1}{5}$
22. $4\frac{1}{6} \times \frac{1}{12}$
23. $\frac{3}{8} \times 2\frac{5}{6}$
24. $\frac{1}{3} \times 6\frac{3}{5}$
25. $4\frac{1}{4} \times \frac{2}{3}$
26. $\frac{2}{5} \times 4\frac{3}{4}$
27. $\frac{3}{7} \times 2\frac{1}{3}$
28. $1\frac{3}{4} \times \frac{2}{9}$
29. $3\frac{1}{5} \times \frac{1}{7}$
30. $\frac{5}{6} \times 2\frac{1}{3}$

31. Strawberries cost \$3.00 a pound. Ken bought $1\frac{1}{4}$ pounds. How much did he spend?

32. **What's the Error?** Steve says $\frac{1}{4} \times 2\frac{1}{2} = 2\frac{1}{8}$. Describe and correct his error.

33. **REASONING** How can you use decimals to find $2\frac{1}{2} \times \frac{1}{2}$?

34. **Write a problem** about finding a fractional part of $3\frac{1}{2}$ dozen apples.

Mixed Review and Test Prep

For 35–36, use the data set 95, 87, 76, 95, and 82.

35. What is the mean? (p. 102)

36. What is the median? (p. 104)

37. $908{,}056 - 32{,}569$ (p. 38)

38. $6{,}361{,}204 - 1{,}265{,}501$ (p. 38)

39. **TEST PREP** In which place would you write the first digit of the quotient for $1.2635 \div 4$? (p. 222)

A ones
B tenths
C hundredths
D thousandths

LESSON 4

Multiply with Mixed Numbers

Quick Review

1. $\frac{6 \div ■}{8 \div ■} = \frac{3}{4}$
2. $\frac{9 \div ■}{12 \div ■} = \frac{3}{4}$
3. $\frac{6 \div 6}{12 \div 6} = \frac{■}{■}$
4. $\frac{21 \div 7}{7 \div 7} = \frac{■}{■}$
5. $\frac{18 \div 6}{6 \div 6} = \frac{■}{■}$

Learn

PLAN AHEAD To multiply two mixed numbers, follow the same steps you use to multiply a fraction and a mixed number.

Multiply. $2\frac{1}{4} \times 1\frac{2}{3}$

STEP 1

Rename both mixed numbers as fractions greater than 1.

$2\frac{1}{4} \times 1\frac{2}{3} = \frac{9}{4} \times \frac{5}{3}$

STEP 2

Multiply.

$\frac{9}{4} \times \frac{5}{3} = \frac{9 \times 5}{4 \times 3} = \frac{45}{12}$

STEP 3

Write the product as a mixed number in simplest form.

$\frac{45}{12} = 3\frac{9}{12} = 3\frac{3}{4}$

Sometimes you can simplify the factors in Step 2 by dividing by the GCF of a numerator and a denominator.

Remember

The greatest common factor, or GCF, is the greatest number that is a factor of two or more numbers.

18: 1, 2, 3, 6, 9, 18
12: 1, 2, 3, 4, 6, 12

GCF = 6

Example

$2\frac{1}{7} \times 1\frac{1}{6} = \frac{\overset{5}{\cancel{15}} \times \overset{1}{\cancel{7}}}{\underset{1}{\cancel{7}} \times \underset{2}{\cancel{6}}}$ Divide 15 and 6 by their GCF, 3. Divide 7 and 7 by their GCF, 7.

$= \frac{5}{2}$, or $2\frac{1}{2}$

MATH IDEA You can divide by a GCF to simplify factors before you multiply.

Check

1. **Explain** how you can use division with a GCF before you compute $4 \times 5\frac{1}{4}$.

Copy and complete each problem. Show how to simplify before you multiply.

2. $4\frac{2}{3} \times 1\frac{3}{7} = \frac{14}{3} \times \frac{10}{7} = \frac{14 \times 10}{3 \times 7} = ■$

3. $2\frac{1}{2} \times 5\frac{1}{3} = \frac{5}{2} \times \frac{16}{3} = \frac{5 \times 16}{2 \times 3} = ■$

CALIFORNIA STANDARDS NS 2.4 Understand the concept of multiplication and division of fractions. **NS 2.5** Compute and perform simple multiplication and division of fractions and apply these procedures to solving problems. *also* **NS 2.0, MR 2.2, MR 2.3, MR 3.2, MR 3.3**

Multiply. Write the answer in simplest form.

4. $4\frac{1}{3} \times 2\frac{1}{2}$ 5. $3\frac{3}{4} \times 1\frac{1}{10}$ 6. $4\frac{1}{2} \times 1\frac{2}{3}$ 7. $6 \times 3\frac{2}{3}$ 8. $5\frac{1}{3} \times 3\frac{3}{4}$

Practice and Problem Solving

Copy and complete each problem. Show how to simplify before you multiply.

9. $2\frac{1}{4} \times 3\frac{2}{3} = \frac{9}{4} \times \frac{11}{3} = \frac{9 \times 11}{4 \times 3} = ■$

10. $3\frac{1}{3} \times 2\frac{2}{5} = \frac{10}{3} \times \frac{12}{5} = \frac{10 \times 12}{3 \times 5} = ■$

Multiply. Write the answer in simplest form.

11. $4\frac{2}{3} \times 1\frac{1}{2}$ 12. $3\frac{1}{3} \times 2\frac{2}{5}$ 13. $2\frac{3}{8} \times 1\frac{2}{5}$ 14. $\frac{1}{2} \times 2\frac{2}{3}$ 15. $\frac{4}{5} \times \frac{5}{6}$

16. $\frac{3}{8} \times 24$ 17. $1\frac{2}{3} \times \frac{1}{4}$ 18. $1\frac{1}{4} \times \frac{5}{6}$ 19. $1\frac{1}{6} \times 4\frac{2}{3}$ 20. $14 \times \frac{3}{7}$

21. $4 \times 2\frac{1}{3} \times \frac{1}{2}$ 22. $3\frac{1}{5} \times \frac{2}{3} \times \frac{1}{4}$ 23. $\frac{1}{6} \times 2\frac{2}{3} \times 1\frac{1}{3}$ 24. $7\frac{1}{4} \times \frac{1}{2} \times 2\frac{1}{2}$

Algebra Find the missing digit.

25. $2\frac{n}{2} \times \frac{1}{4} = \frac{5}{8}$ 26. $2\frac{1}{3} \times 2\frac{1}{4} = 5\frac{n}{4}$ 27. $\frac{1}{2} \times \frac{n}{4} = \frac{1}{8}$ 28. $2 \times \frac{7}{8} = 1\frac{n}{4}$

USE DATA For 29–31, use the circle graph.

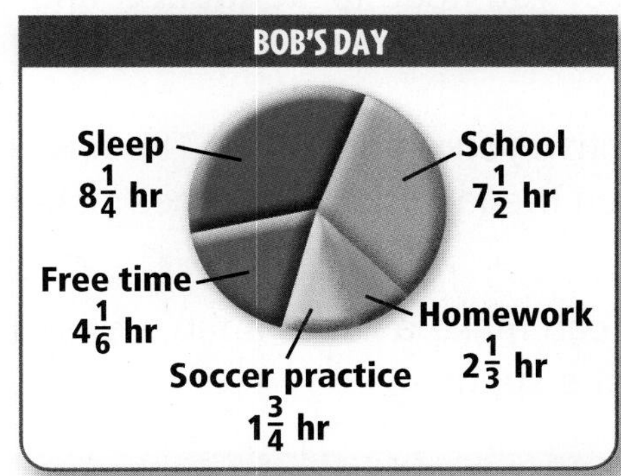

29. If Bob spent $\frac{1}{5}$ of his time at school in art class, how long was his art class?

30. Bob spent a quarter of his free time surfing the Internet. How long did he spend on the Internet?

31. How much more sleep time did he have than free time?

32. **Algebra** Find n if ♣ = 2, ♦ = 2 × ♣, ❤ = ♦ × 2, and $\frac{♣}{♦} = \frac{n}{❤}$.

33. **Write About It** How is multiplying fractions different from adding fractions?

Mixed Review and Test Prep

34. What is the least common denominator of $\frac{1}{12}$ and $\frac{7}{15}$? (p. 352)

35. Cards are sold in packs of 35 and envelopes in packs of 40. What is the least number of each pack you need to have an equal number of cards and envelopes? (p. 256)

36. What is $\frac{3}{8}$ of $\frac{4}{5}$? (p. 382)

37. What is $\frac{5}{7}$ of 6? (p. 380)

38. **TEST PREP** Which graph or plot would be best to display the number of students completing their projects on Monday, Tuesday, and Wednesday? (p. 130)

A circle graph
B bar graph
C stem-and-leaf plot
D double-bar graph

Extra Practice page H52, Set D

LESSON 5

Problem Solving Skill

Sequence and Prioritize Information

Quick Review

1. $5 ÷ 2
2. $10 ÷ 10
3. $4.20 ÷ 4
4. $3.75 ÷ 3
5. $20 \times \frac{3}{4}$

VOCABULARY

sequence prioritize

BUDGET YOUR BUCKS! Ms. James saved $\frac{1}{4}$ of her paycheck in the bank. Then she paid bills with half of what was left. She used a third of her remaining cash for tickets to a concert. After she went grocery shopping, she had $30 left. Her check was $300. How much did she spend on groceries?

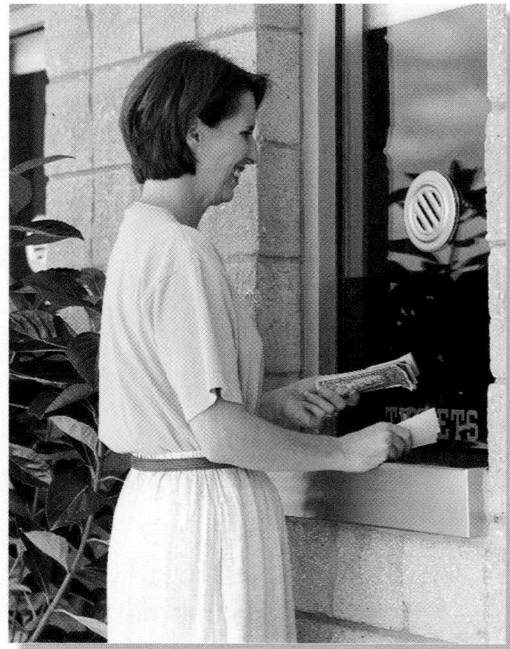

MATH IDEA Sometimes to solve a problem, you need to put events in order of time and importance—you need to **sequence** and **prioritize** information.

First, prioritize your information: The answer depends on the fact that Ms. James's check was for $300.

Then you can make a list or a table to sequence the amounts she spent.

EXPENSE TABLE		
Event	Amount Used	Balance
1. She started with $300.	0	$300
2. She saved $\frac{1}{4}$ in bank.	$300 x $\frac{1}{4}$ = $75	$300 - $75 = $225
3. She paid bills with $\frac{1}{2}$ of what was left.	$225 x $\frac{1}{2}$ = $112.50	$225 - $112.50 = $112.50
4. She spent $\frac{1}{3}$ on concert tickets.	$112.50 x $\frac{1}{3}$ = $37.50	$112.50 - $37.50 = $75
5. She bought groceries.	■	$75 - ■ = $30
6. She had $30 left.	0	$30

Look at Event 5. $75 − $45 = $30

So, Ms. James spent $45 on groceries.

- **REASONING** Why is the check amount the most important fact in the problem? Why can't you solve the problem without sequencing the events?

CALIFORNIA STANDARDS **MR 1.1** Analyze problems by identifying relationships, distinguishing relevant from irrelevant information, sequencing and prioritizing information, and observing patterns. **MR 2.0** Students use strategies, skills, and concepts in finding solutions. **MR 2.3** Use a variety of methods, such as words, numbers, symbols, charts, graphs, tables, diagrams, and models, to explain mathematical reasoning. *also* **NS 2.0, NS 2.4, NS 2.5, MR 3.2, MR 3.3**

Problem Solving Practice

Sequence and prioritize information to solve.

1. **What if** Ms. James's paycheck had been for $350? How much would she have spent on groceries?
2. The United States produced 569 movies in one year. In the same year England made $\frac{1}{2}$ the number of movies that France made and 499 fewer than the United States made. How many movies did France make that year?

The compact disc (CD) was invented 273 years after the piano. The tape recorder was invented in 1898. Thomas Edison invented the phonograph 21 years before the tape recorder and 95 years before the compact disc.

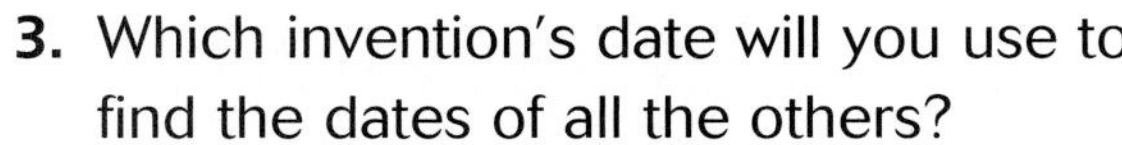

3. Which invention's date will you use to find the dates of all the others?

   A tape recorder
   B compact disc
   C piano
   D phonograph

4. Which list is in sequence from earliest invention to latest invention?

   F recorder, piano, phonograph, CD
   G piano, phonograph, CD, recorder
   H phonograph, piano, recorder, CD
   J piano, phonograph, recorder, CD

Mixed Applications

For 5–7, use the map.

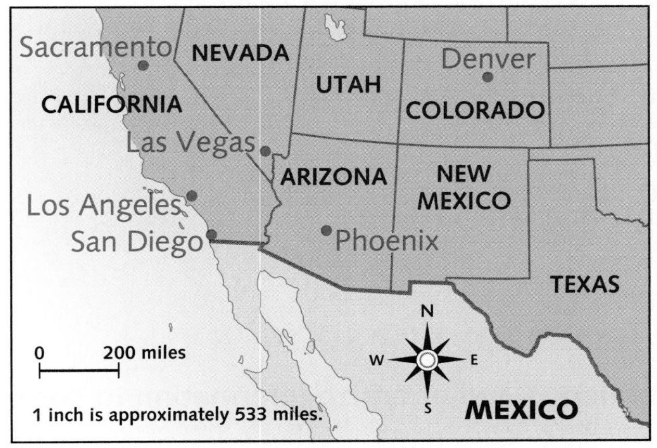

5. Alice left home and drove 124 miles to San Diego. Then she continued 358 miles to Phoenix. In what city on the map does Alice live?
6. Charlie drove from Sacramento to Denver to attend college. About how many miles did he drive?
7. Lucy visits her grandparents in Denver twice a year. If she lives in Phoenix, about how many miles does she travel altogether each year for these visits?
8. The Cineplex has 16 theaters. Each theater shows a different movie. Each movie is played 7 times a day on Saturdays and Sundays. How many shows are there in one weekend?
9. **NUMBER SENSE** Jim chose a number. Then he multiplied it by 3, added 5, and divided the sum by 2. After he subtracted 4 from the quotient, he got 6. What number did Jim choose?

CHAPTER 21

Review/Test

CHECK CONCEPTS

Write the number sentence each model represents.
(pp. 380–383)

1. 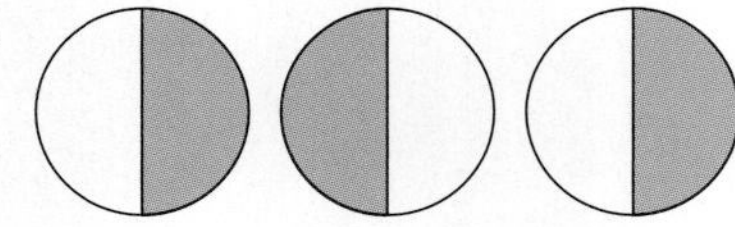

2. 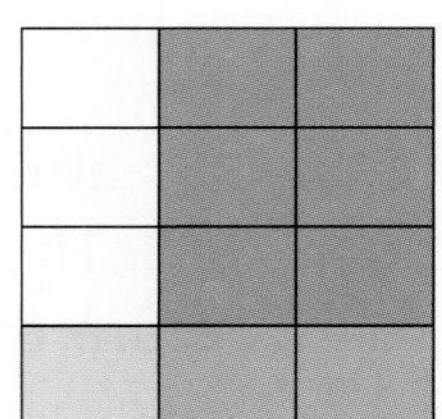

3. 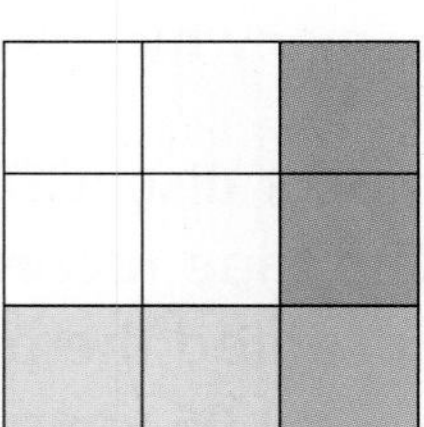

CHECK SKILLS

Find the product. Write it in simplest form. (pp. 380–387)

4. $\frac{2}{5} \times 50$
5. $6 \times \frac{4}{9}$
6. $\frac{3}{4} \times 16$
7. $2 \times \frac{3}{10}$
8. $\frac{1}{3} \times \frac{1}{4}$
9. $\frac{3}{4} \times \frac{1}{2}$
10. $\frac{1}{6} \times \frac{3}{5}$
11. $\frac{1}{7} \times \frac{2}{3}$
12. $\frac{1}{6} \times \frac{2}{7}$
13. $\frac{1}{2} \times \frac{3}{10}$
14. $\frac{3}{8} \times \frac{1}{4}$
15. $\frac{2}{9} \times \frac{1}{4}$
16. $\frac{1}{2} \times 2\frac{1}{5}$
17. $3\frac{3}{4} \times \frac{1}{3}$
18. $\frac{5}{6} \times 5\frac{1}{2}$
19. $4\frac{1}{2} \times \frac{3}{7}$
20. $1\frac{2}{3} \times 3\frac{1}{5}$
21. $4 \times 2\frac{5}{8}$
22. $3\frac{1}{4} \times 1\frac{3}{5}$
23. $2\frac{2}{5} \times 3\frac{3}{4}$

CHECK PROBLEM SOLVING

Sequence and prioritize information to solve. (pp. 388–389)

24. Colin has 2 hours before dinner is ready. He does homework for $\frac{3}{4}$ hour, plays basketball for $\frac{3}{4}$ hour, and then plays a computer game until dinner. How long does he play the computer game?

25. Karen had 100 beads. She used $\frac{3}{5}$ of the beads to make necklaces. Then she used $\frac{1}{2}$ of the leftover beads to make earrings. She used the rest to make a bracelet. How many beads are on the bracelet?

Cumulative Review

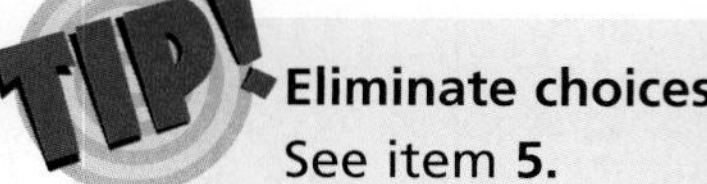

Eliminate choices.
See item **5.**

Think about a fractional part of a set. The amount will be less than the number in the whole set. Eliminate choices that are not less than the whole set.

Also see problem **5,** p. H64.

For 1–11, choose the best answer.

1. Which symbol makes this a true number sentence?

$\frac{1}{3} \times 21$ ● $\frac{1}{4} \times 28$

A $>$ **C** $<$
B $=$ **D** $+$

---

2. Kylie had 16 yards of fabric. She used $\frac{3}{4}$ of the fabric to make a dress. How many yards did she use?

F 4 **H** 12
G 8 **J** Not here

---

3. $\frac{1}{8} \times \frac{3}{4}$

A $3\frac{4}{12}$ **C** $\frac{4}{32}$
B $3\frac{3}{12}$ **D** $\frac{3}{32}$

---

4. 832,641 + 892,034

F 1,724,675 **H** 1,624,675
G 1,724,075 **J** Not here

---

5. Marsha found that $\frac{1}{4}$ of the 12 children who live on her street are too young to go to school. How many children are too young to go to school?

A 3 **C** 12
B 4 **D** 48

6. Which of the following illustrates the Distributive Property?

F $(3 \times n) \times 2 = 3 \times (n \times 2)$
G $(11 + n) \times 6 = (11 \times 6) + (n \times 6)$
H $26 \times n = n \times 26$
J $(19 + 5) + n = 19 + (5 + n)$

---

7. What is the next term in this sequence?
$\frac{1}{5}, \frac{3}{5}, 1, 1\frac{2}{5}, \underline{?}$

A $1\frac{3}{5}$ **C** 2
B $1\frac{4}{5}$ **D** $2\frac{1}{5}$

---

8. $8\frac{1}{2} \times 11$

F $19\frac{1}{2}$ **H** 88
G 44 **J** $93\frac{1}{2}$

---

9. $2\frac{5}{8} + 6\frac{1}{8}$

A $8\frac{6}{16}$ **C** $8\frac{3}{4}$
B $8\frac{1}{2}$ **D** $8\frac{7}{8}$

---

10. $1\frac{1}{4} \times \frac{3}{5}$

F $1\frac{3}{20}$ **H** $\frac{3}{5}$
G $\frac{3}{4}$ **J** $\frac{3}{10}$

---

11. LaDonna and Ted measured the length of a wall at $37\frac{1}{2}$ feet. Each will paint $\frac{1}{2}$ of the wall. How many feet will Ted paint?

A $18\frac{3}{4}$ ft **C** $17\frac{3}{4}$ ft
B $18\frac{1}{4}$ ft **D** $17\frac{3}{8}$ ft

Divide Fractions

More than two dozen types of North American birds sometimes nest in birdhouses. To attract a certain type of bird to your backyard, you must build a house to meet its needs. The table shows dimensions of houses for different birds. To attract purple martins, you need a house with many rooms. To make one room, you need 6 pieces of wood, each $\frac{1}{2}$ foot long. Suppose you have 3 feet of cedar. Would you have enough to build one room? Explain.

BIRDHOUSE DIMENSIONS			
Type of Bird	Floor of House (feet)	Height of House (feet)	Entrance Hole Diameter (inches)
Eastern bluebird	$\frac{5}{12}$ by $\frac{5}{12}$	$\frac{2}{3}$ to 1	$1\frac{1}{2}$
Chickadee	$\frac{1}{3}$ by $\frac{1}{3}$	$\frac{2}{3}$ to $\frac{5}{6}$	$1\frac{1}{8}$
Purple martin	$\frac{1}{2}$ by $\frac{1}{2}$	$\frac{1}{2}$	$2\frac{1}{4}$
Northern flicker	$\frac{7}{12}$ by $\frac{7}{12}$	$1\frac{1}{6}$ to $1\frac{1}{3}$	$2\frac{1}{2}$

Use this page to help you review and remember important skills needed for Chapter 22.

VOCABULARY

Choose the best term from the box.

1. A number that has a whole-number part and a fraction part is a __?__.
2. The greatest number that is a factor of two or more numbers is the __?__ of those numbers.
3. In the problem $21 \div 3 = 7$, 21 is the __?__, 3 is the __?__, and 7 is the __?__.

dividend
divisor
greatest common factor (GCF)
mixed number
product
quotient

FIND THE GCF (See p. H24.)

List the factors for each number.

4. 29 **5.** 24 **6.** 16 **7.** 36 **8.** 56

Write the common factors for each pair of numbers.

9. 6, 42 **10.** 15, 50 **11.** 2, 14 **12.** 24, 28 **13.** 14, 30

Write the greatest common factor for each pair of numbers.

14. 6, 15 **15.** 25, 50 **16.** 18, 45 **17.** 6, 12 **18.** 12, 28

19. 3, 6 **20.** 14, 28 **21.** 6, 21 **22.** 12, 18 **23.** 18, 27

WRITE FRACTIONS AS MIXED NUMBERS AND MIXED NUMBERS AS FRACTIONS (See p. H22.)

Write each fraction as a mixed number. Write each mixed number or whole number as a fraction.

24. $\frac{13}{3}$ **25.** $\frac{7}{4}$ **26.** $\frac{46}{5}$ **27.** $\frac{28}{8}$ **28.** $\frac{17}{6}$

29. 3 **30.** $1\frac{1}{6}$ **31.** 4 **32.** $7\frac{1}{2}$ **33.** 7

LESSON 1

HANDS ON
Divide Fractions

Quick Review

1. $\frac{3}{9} + \frac{4}{9}$
2. $\frac{5}{6} + \frac{1}{6}$
3. $\frac{1}{5} + \frac{2}{5}$
4. $\frac{1}{7} + \frac{3}{7}$
5. $\frac{1}{4} + \frac{1}{4}$

MATERIALS
fraction bars

TECHNOLOGY LINK

More Practice: Use E-Lab, *Exploring Division of Fractions.*
www.harcourtschool.com/elab2002

Explore

Kari has 2 yards of ribbon to use for a craft project. She needs to cut the ribbon into pieces that are $\frac{1}{3}$ yard long. How many pieces of ribbon will Kari have?

Divide 2 by $\frac{1}{3}$. $2 \div \frac{1}{3}$

Use fraction bars to show how many thirds are in 2.

STEP 1

Model 2 and $\frac{1}{3}$ with fraction bars.

1 | 1

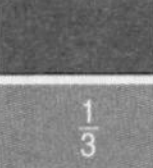

STEP 2

See how many $\frac{1}{3}$ bars are equal to 2 one-whole bars. So, six $\frac{1}{3}$ bars are equal to 2 one-whole bars.

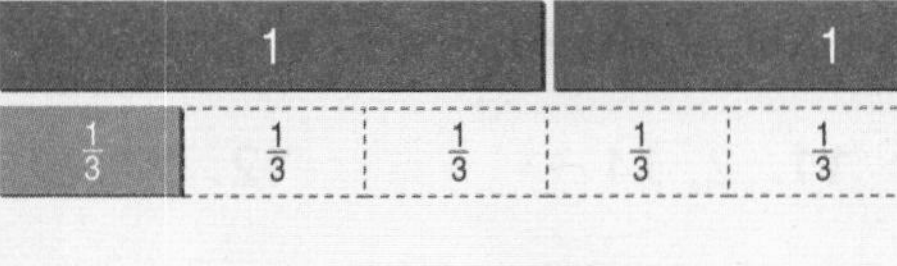

STEP 3

Record a number sentence for the model.

$2 \div \frac{1}{3} = 6$

So, Kari will have 6 pieces of ribbon.

How many $\frac{1}{4}$ bars are equal to 1 one-whole bar?

Try It

Write a number sentence for each model.

a. 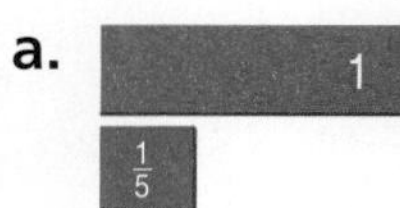

b. 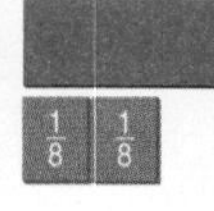

Use fraction bars to find the quotient.

c. $1 \div \frac{1}{4}$ **d.** $3 \div \frac{1}{6}$ **e.** $5 \div \frac{1}{2}$

CALIFORNIA STANDARDS NS 2.0 Students perform calculations and solve problems involving addition, subtraction, and simple multiplication and division of fractions and decimals. **NS 2.4** Understand the concept of multiplication and division of fractions. *also* **NS 1.0, NS 2.5, MR 2.0, MR 2.3, MR 3.0, MR 3.2, MR 3.3**

Connect

You can also use fraction bars to help you divide a fraction by a fraction.

Divide $\frac{1}{2}$ by $\frac{1}{8}$. $\frac{1}{2} \div \frac{1}{8}$

STEP 1

Model $\frac{1}{2}$ and $\frac{1}{8}$ with fraction bars.

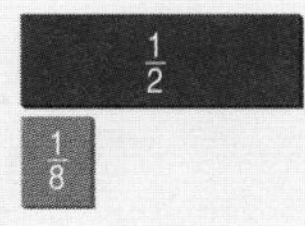

STEP 2

See how many $\frac{1}{8}$ bars are equal to a $\frac{1}{2}$ bar.

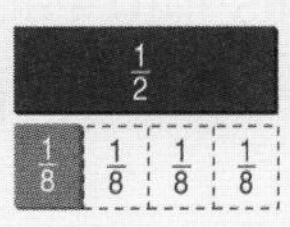

So, four $\frac{1}{8}$ bars are equal to a $\frac{1}{2}$ bar.

STEP 3

Record a number sentence for the model.

$\frac{1}{2} \div \frac{1}{8} = 4$

Practice

Write a number sentence for each model.

1.

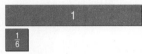

2. 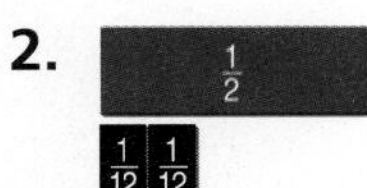

3. 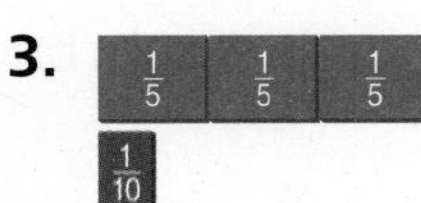

Use fraction bars to find the quotient.

4. $\frac{9}{10} \div \frac{1}{10}$
5. $1 \div \frac{1}{3}$
6. $\frac{3}{4} \div \frac{3}{12}$
7. $5 \div \frac{2}{4}$
8. $\frac{3}{4} \div \frac{1}{4}$
9. $3 \div \frac{2}{8}$
10. $\frac{1}{3} \div \frac{1}{6}$
11. $\frac{10}{2} \div \frac{1}{2}$

12. Cindy has $\frac{3}{4}$ yard of wire to use to make earrings. If each pair of earrings takes $\frac{1}{12}$ yard of wire, how many pairs of earrings can she make?

13. **REASONING** If you divide $\frac{1}{10}$ of a dollar into parts that are each $\frac{1}{100}$ of a dollar, how many parts are there? What are they called?

14. Compare the answers for **a** and **b.** What did you find?

a. $12 \times \frac{4}{3}$ **b.** $12 \div \frac{3}{4}$

Mixed Review and Test Prep

15. $\frac{7}{8} - \frac{3}{4}$ (p. 354)

16. $\frac{6}{7} - \frac{5}{14}$ (p. 354)

17. 47.04×6.1 (p. 164)

18. 82.03×2.8 (p. 164)

19. **TEST PREP** A bakery ships 24,480 loaves of bread to 60 different stores. If each store receives the same number of loaves, how many loaves does each store receive? (p. 204)

A 410 **B** 409 **C** 408 **D** 308

LESSON 2

Reciprocals

Quick Review

Write each whole number or mixed number as a fraction.

1. 6 **2.** $2\frac{2}{3}$

3. $1\frac{7}{10}$ **4.** $4\frac{3}{8}$ **5.** 2

VOCABULARY

reciprocal

Learn

WORKING FOR PEANUTS! Hanh and three friends visited the elephant exhibit at the zoo. Hanh divided a 4-pound bag of peanuts into four smaller bags. He put $\frac{1}{4}$ of the peanuts into each bag. How many pounds of peanuts did Hanh put into each bag?

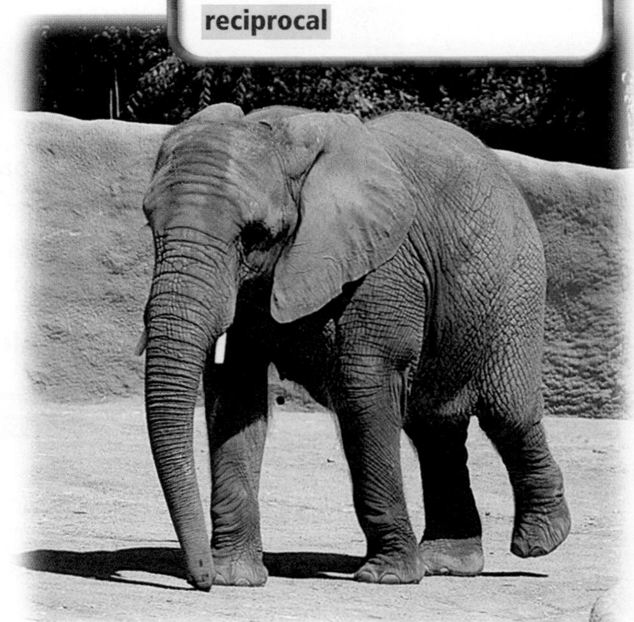

Think: $\frac{1}{4}$ of $4 \rightarrow \frac{1}{4} \times 4 = \frac{1}{4} \times \frac{4}{1}$

$= \frac{4}{4} = 1$

So, Hanh put 1 pound of peanuts into each bag.

The product of a number and its **reciprocal** is 1. $\frac{1}{4}$ and 4 are reciprocals because $\frac{1}{4} \times 4 = 1$.

MATH IDEA To write the reciprocal of a fraction, write a new fraction by exchanging the numerator and denominator.

Examples

A The reciprocal of $\frac{14}{7}$ is $\frac{7}{14}$.

$\frac{14}{7} \times \frac{7}{14} = \frac{98}{98} = 1$

B The reciprocal of 6, or $\frac{6}{1}$, is $\frac{1}{6}$.

$\frac{6}{1} \times \frac{1}{6} = \frac{6}{6} = 1$

C The reciprocal of $2\frac{2}{3}$, or $\frac{8}{3}$, is $\frac{3}{8}$.

$\frac{8}{3} \times \frac{3}{8} = \frac{24}{24} = 1$

Check

1. **Explain** how to find the reciprocal of 3 and the reciprocal of $3\frac{5}{8}$.

Are the two numbers reciprocals? Write *yes* or *no*.

2. $\frac{5}{4}$ and $\frac{4}{5}$ **3.** $\frac{1}{7}$ and $\frac{7}{3}$ **4.** $\frac{1}{6}$ and 6 **5.** $\frac{5}{7}$ and $1\frac{2}{5}$ **6.** $2\frac{3}{4}$ and $12\frac{4}{3}$

CALIFORNIA STANDARDS **NS 2.4** Understand the concept of multiplication and division of fractions. *also* **NS 1.0, NS 2.0, NS 2.5, AF 1.2, MR 2.0, MR 2.3, MR 3.0, MR 3.2, MR 3.3**

Practice

Are the two numbers reciprocals? Write *yes* or *no*.

7. $\frac{5}{9}$ and $\frac{9}{5}$ **8.** $\frac{3}{10}$ and $\frac{10}{3}$ **9.** 4 and $\frac{1}{4}$ **10.** $\frac{3}{7}$ and $1\frac{1}{3}$ **11.** $2\frac{1}{4}$ and $\frac{4}{9}$

Write the reciprocal of each number.

12. $\frac{3}{5}$ **13.** $1\frac{3}{4}$ **14.** $\frac{1}{9}$ **15.** $\frac{12}{7}$ **16.** 7

17. 12 **18.** $3\frac{1}{5}$ **19.** $\frac{23}{6}$ **20.** $\frac{1}{13}$ **21.** $\frac{2}{9}$

Algebra Find the value of *n*.

22. $\frac{1}{8} \times \frac{n}{1} = 1$ **23.** $\frac{1}{n} \times \frac{6}{1} = 1$ **24.** $\frac{8}{9} \times \frac{n}{8} = 1$ **25.** $\frac{7}{n} \times \frac{4}{7} = 1$ **26.** $8 \times \frac{1}{n} = 1$

27. $5 \times \frac{1}{n} = 1$ **28.** $\frac{9}{28} \times 3\frac{1}{n} = 1$ **29.** $\frac{1}{n} \times \frac{n}{1} = 1$ **30.** $5\frac{2}{3} \times \frac{n}{17} = 1$ **31.** $\frac{2}{n} \times 3\frac{1}{2} = 1$

Multiply. Use the Associative and Commutative Properties of Multiplication to help you.

32. $\frac{1}{5} \times \frac{5}{1} \times \frac{2}{7}$ **33.** $\frac{7}{8} \times \frac{6}{1} \times \frac{8}{7}$ **34.** $\frac{2}{5} \times \frac{5}{6} \times \frac{5}{2}$

35. $\frac{1}{5} \times \frac{11}{4} \times \frac{4}{11} \times 5$ **36.** $\frac{3}{4} \times \frac{10}{3} \times \frac{3}{10} \times 8$ **37.** $\frac{3}{2} \times \frac{2}{3} \times \frac{2}{3} \times 9$

Which is greater, the number or its reciprocal?

38. $\frac{1}{5}$ **39.** 8 **40.** $\frac{4}{7}$ **41.** $1\frac{1}{2}$ **42.** $2\frac{1}{10}$

43. **What's the Error?** Blair said that the reciprocal of $3\frac{1}{3}$ is $\frac{3}{7}$. Describe and correct her error.

44. **What's the Question?** Christy has $\frac{3}{4}$ pound of peanuts. Alex has $\frac{4}{3}$ pounds of peanuts. The answer is $\frac{7}{12}$. What is the question?

45. **REASONING** If a number is greater than 1, what can you say about its reciprocal?

46. **NUMBER SENSE** Explain how you can find the reciprocal of 0.7.

Mixed Review and Test Prep

Round to the nearest million.

47. 52,095,500 (p. 32) **48.** 496,750,000 (p. 32)

49. Brian's batting average went up by .025 points this season. He began with an average of .228. What is his average now? (p. 50)

50. Divide 49.084 by 0.7. (p. 238)

51. **TEST PREP** Which number is less than the product of 28 and 716? (p. 148)

A 24,004 **C** 20,408
B 20,804 **D** 20,008

Extra Practice page H53, Set A

LESSON 3

Divide Whole Numbers by Fractions

Quick Review

Write each fraction as a whole number or a mixed number.

1. $\frac{6}{1}$
2. $\frac{14}{3}$
3. $\frac{8}{2}$
4. $\frac{9}{4}$
5. $\frac{13}{10}$

Learn

SLOW MOTION If a giant tortoise walks $\frac{1}{5}$ mile per hour, how many hours will it take the tortoise to walk 2 miles?

Find $2 \div \frac{1}{5}$. ←**Think:** How many $\frac{1}{5}$ miles are in 2 miles?

One Way

Use a model.

1	1
$\frac{1}{5}$ $\frac{1}{5}$ $\frac{1}{5}$ $\frac{1}{5}$ $\frac{1}{5}$	$\frac{1}{5}$ $\frac{1}{5}$ $\frac{1}{5}$ $\frac{1}{5}$ $\frac{1}{5}$

Another Way

Use patterns.

$2 \times \frac{1}{1} = 2$	$2 \div 1 = 2$
$2 \times \frac{2}{1} = 4$	$2 \div \frac{1}{2} = 4$
$2 \times \frac{3}{1} = 6$	$2 \div \frac{1}{3} = 6$
$2 \times \frac{4}{1} = 8$	$2 \div \frac{1}{4} = 8$
$2 \times \frac{5}{1} = 10$	$2 \div \frac{1}{5} = 10$

So, it will take a giant tortoise 10 hours.

You can also use a reciprocal to divide by a fraction. Look at the patterns at the right. The patterns show that dividing by a number is the same as multiplying by its reciprocal.

$6 \div 2 = 3$	$6 \times \frac{1}{2} = 3$
$6 \div 1 = 6$	$6 \times \frac{1}{1} = 6$
$6 \div \frac{1}{2} = 12$	$6 \times 2 = 12$

Find $5 \div \frac{2}{3}$.

STEP 1

Write the whole number as a fraction.

$$5 \div \frac{2}{3} = \frac{5}{1} \div \frac{2}{3}$$

STEP 2

Use the reciprocal of the divisor to write a multiplication problem.

$$\frac{5}{1} \div \frac{2}{3} = \frac{5}{1} \times \frac{3}{2}$$

↑ ↑ reciprocals

STEP 3

Multiply.

$$\frac{5}{1} \times \frac{3}{2} = \frac{15}{2}, \text{ or } 7\frac{1}{2}$$

So, $5 \div \frac{2}{3} = \frac{15}{2}$, or $7\frac{1}{2}$.

MATH IDEA To divide a whole number by a fraction, write the whole number as a fraction and then multiply it by the reciprocal of the divisor.

CALIFORNIA STANDARDS NS 2.0 Students perform calculations and solve problems involving addition, subtraction, and simple multiplication and division of fractions and decimals. NS 2.4 Understand the concept of multiplication and division of fractions. NS 2.5 Compute and perform simple multiplication and division of fractions and apply these procedures to solving problems. *also* NS 1.0, 🔑 NS 2.2, MR 1.1, MR 2.3, MR 3.0, MR 3.2, MR 3.3

Check

1. **Describe** two ways you can find $3 \div \frac{1}{6}$.

Use fraction bars, patterns, or reciprocals to divide.

2. $2 \div \frac{1}{4}$
3. $3 \div \frac{1}{6}$
4. $5 \div \frac{1}{3}$
5. $1 \div \frac{1}{6}$
6. $4 \div \frac{1}{3}$

Divide.

7. $8 \div \frac{1}{2}$
8. $9 \div \frac{2}{5}$
9. $7 \div \frac{3}{4}$
10. $3 \div \frac{1}{5}$
11. $5 \div \frac{3}{10}$

Practice and Problem Solving

Use fraction bars, patterns, or reciprocals to divide.

12. $5 \div \frac{1}{2}$
13. $3 \div \frac{1}{3}$
14. $8 \div \frac{2}{3}$
15. $1 \div \frac{1}{8}$
16. $7 \div \frac{1}{4}$

Divide.

17. $9 \div \frac{3}{8}$
18. $8 \div \frac{6}{7}$
19. $4 \div \frac{4}{5}$
20. $6 \div \frac{3}{4}$
21. $1 \div \frac{2}{5}$

Find the missing number.

22. $4 \div \frac{2}{9} = \blacksquare$
23. $\blacksquare \div \frac{1}{2} = 16$
24. $5 \div \frac{1}{\blacksquare} = 30$
25. $3 \div \frac{\blacksquare}{4} = 12$
26. How many halves are in 4?
27. How many thirds are in 5?
28. How many three-fourths are in 6?
29. How many twos are in 9?
30. **What if** a tortoise travels $\frac{2}{12}$ miles per hour? How many hours will it take a tortoise to walk 3 miles?
31. **Write a problem** that has a whole number divided by a fraction, with a quotient of 4.
32. **Write About It** Show how you can use multiplication to solve $4 \div \frac{1}{2}$.
33. **What's the Error?** Describe Tony's error. Write the correct answer.

$$5 \div \frac{1}{6} = \frac{1}{5} \times \frac{1}{6} = \frac{1}{30}$$

▲ Giant tortoises have one of the longest life spans of any animal, sometimes more than 200 years.

Mixed Review and Test Prep

34. $\frac{1}{4} + \frac{1}{3}$ (p. 354)
35. $\frac{3}{5} + \frac{7}{8}$ (p. 354)
36. $\frac{3}{4} - \frac{1}{2}$ (p. 354)
37. $\frac{9}{10} - \frac{2}{5}$ (p. 354)
38. **TEST PREP** Which is 0.45 written as a fraction? (p. 280)

A $\frac{5}{20}$ **B** $\frac{9}{20}$ **C** $\frac{11}{20}$ **D** $\frac{15}{20}$

LESSON 4

Divide Fractions

Quick Review

1. $63 \div 7$
2. $64 \div 8$
3. $60 \div 6$
4. $66 \div 3$
5. $65 \div 5$

Learn

BUZZZY WORK! If one honeybee makes $\frac{1}{12}$ teaspoon of honey during its lifetime, how many honeybees are needed to make $\frac{1}{2}$ teaspoon of honey?

Find $\frac{1}{2} \div \frac{1}{12}$. ←**Think:** How many $\frac{1}{12}$ teaspoons are in $\frac{1}{2}$ teaspoon?

One Way

You can use fraction bars to help you divide fractions by fractions.

See how many $\frac{1}{12}$ bars are equal to a $\frac{1}{2}$ bar.

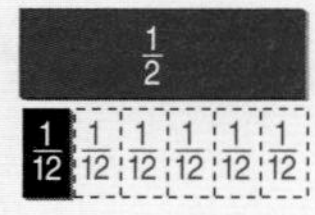

Another Way

You can find the quotient of two fractions by using what you know about reciprocals.

STEP 1

Use the reciprocal of the divisor to write a multiplication problem.

$$\frac{1}{2} \div \frac{1}{12} = \frac{1}{2} \times \frac{12}{1}$$

↑ ↑ reciprocals

STEP 2

Multiply. Write the answer in simplest form.

$$\frac{1}{2} \times \frac{12}{1} = \frac{12}{2}\text{, or } 6$$

So, $\frac{1}{2} \div \frac{1}{12} = 6$.

▲ A honeybee has to visit over 2,000,000 flowers in order to collect enough nectar to produce 1 pound of honey.

So, 6 honeybees are needed to make $\frac{1}{2}$ teaspoon of honey.

- What multiplication problem could you write for $\frac{3}{5} \div \frac{1}{5}$? What is the quotient?

MATH IDEA To find the quotient of two fractions, multiply the dividend by the reciprocal of the divisor.

CALIFORNIA STANDARDS **NS 2.0** Students perform calculations and solve problems involving addition, subtraction, and simple multiplication and division of fractions and decimals. **NS 2.4** Understand the concept of multiplication and division of fractions. **NS 2.5** Compute and perform simple multiplication and division of fractions, and apply these procedures to solving problems. *also* **NS 1.0, O━n NS 2.2, O━n AF 1.2, MR 1.0, MR 2.0, MR 2.3, MR 3.2**

Use Reciprocals to Divide

You can use reciprocals to divide mixed numbers by fractions and fractions by whole numbers.

Find $2\frac{1}{4} \div \frac{1}{3}$.

STEP 1

Write the mixed number as a fraction.

$$2\frac{1}{4} \div \frac{1}{3} = \frac{9}{4} \div \frac{1}{3}$$

STEP 2

Use the reciprocal of the divisor to write a multiplication problem.

$$\frac{9}{4} \div \frac{1}{3} = \frac{9}{4} \times \frac{3}{1}$$

↑ ↑ reciprocals

STEP 3

Multiply. Write the answer in simplest form if needed.

$$\frac{9}{4} \times \frac{3}{1} = \frac{27}{4}, \text{ or } 6\frac{3}{4}$$

So, $2\frac{1}{4} \div \frac{1}{3} = \frac{27}{4}$, or $6\frac{3}{4}$.

Examples

A Find $1\frac{2}{3} \div \frac{2}{5}$.

$1\frac{2}{3} \div \frac{2}{5} = \frac{5}{3} \div \frac{2}{5} = \frac{5}{3} \times \frac{5}{2}$

$\frac{5}{3} \times \frac{5}{2} = \frac{25}{6}$, or $4\frac{1}{6}$

B Find $1\frac{1}{4} \div 2\frac{1}{3}$.

$1\frac{1}{4} \div 2\frac{1}{3} = \frac{5}{4} \div \frac{7}{3} = \frac{5}{4} \times \frac{3}{7}$

$\frac{5}{4} \times \frac{3}{7} = \frac{15}{28}$

C Find $\frac{5}{6} \div 3$.

$\frac{5}{6} \div 3 = \frac{5}{6} \times \frac{1}{3}$

$\frac{5}{6} \times \frac{1}{3} = \frac{5}{18}$

Check

1. **REASONING** **Tell** whether the dividend is greater than or less than the divisor in Examples A, B, and C. When is the quotient greater than 1?

Write a division sentence for each model.

2. 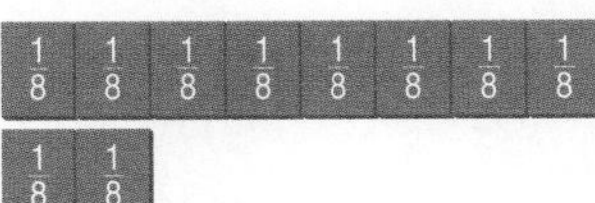

3. 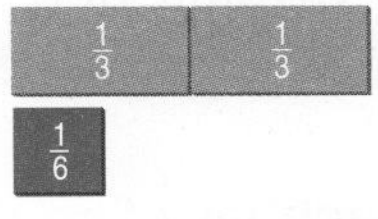

4. 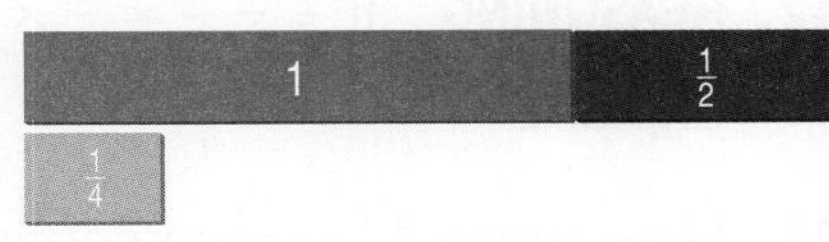

Use a reciprocal to write a multiplication problem for the division problem.

5. $\frac{3}{4} \div \frac{1}{3}$
6. $\frac{1}{3} \div \frac{1}{2}$
7. $2 \div \frac{2}{5}$
8. $3\frac{2}{3} \div \frac{3}{8}$
9. $2\frac{1}{5} \div 1\frac{1}{4}$

Divide. Write the answer in simplest form.

10. $\frac{1}{4} \div \frac{1}{2}$
11. $\frac{2}{5} \div \frac{1}{10}$
12. $\frac{1}{2} \div 2$
13. $1\frac{3}{4} \div \frac{1}{2}$
14. $1\frac{1}{3} \div 1\frac{1}{2}$

LESSON CONTINUES

Practice and Problem Solving

Write a division sentence for each model.

15. 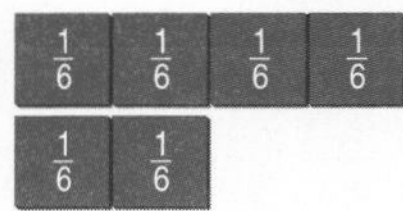

16. 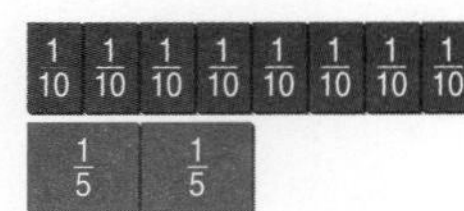

17. 

Use a reciprocal to write a multiplication problem for the division problem.

18. $\frac{3}{8} \div \frac{1}{2}$

19. $\frac{7}{8} \div \frac{1}{3}$

20. $4 \div \frac{3}{8}$

21. $1\frac{2}{5} \div \frac{1}{4}$

22. $2\frac{2}{7} \div 1\frac{1}{5}$

Divide. Write the answer in simplest form.

23. $6 \div \frac{3}{8}$

24. $7 \div \frac{6}{7}$

25. $\frac{3}{5} \div \frac{1}{5}$

26. $\frac{4}{7} \div \frac{2}{3}$

27. $1\frac{1}{4} \div \frac{3}{4}$

28. $8 \div \frac{4}{5}$

29. $\frac{5}{6} \div \frac{1}{3}$

30. $\frac{7}{12} \div \frac{2}{3}$

31. $\frac{3}{4} \div \frac{1}{8}$

32. $\frac{3}{4} \div 2\frac{1}{4}$

33. $3\frac{7}{10} \div 2\frac{4}{5}$

34. $2\frac{3}{4} \div \frac{1}{8}$

35. $\frac{4}{5} \div 8$

36. $\frac{1}{3} \div 12$

37. $2\frac{4}{5} \div 1\frac{2}{3}$

Algebra For 38–40, copy and complete each rule table.

38.

n	$n \div \frac{1}{3}$
$\frac{5}{6}$	■
$\frac{1}{2}$	■
$\frac{1}{9}$	■

39.

y	$y \div \frac{1}{4}$
$\frac{7}{8}$	■
■	2
■	$\frac{16}{5}$, or $3\frac{1}{5}$

40.

n	$n \div \frac{1}{2}$
■	$\frac{6}{8}$, or $\frac{3}{4}$
3	■
■	5

41. If a bee travels $\frac{1}{8}$ mile on each trip it makes to gather nectar, how many trips will the bee make to travel $\frac{1}{2}$ mile?

42. **REASONING** If $\frac{3}{4} \div \frac{n}{8} = \frac{3}{4} \times \frac{n}{8}$, what is the value of n? Explain.

43. How many $1\frac{1}{2}$-ounce servings are in each jar of honey on the right?

44. 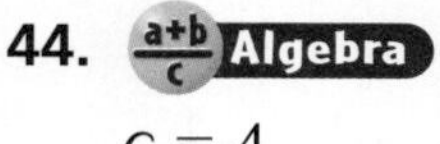 Find n if $a = 4$, $b = 6$, and $c = 4$.

$$\frac{a}{b} \div \frac{c}{n} = \frac{1}{2}$$

45. Mrs. Banak bought a $12\frac{1}{2}$-ounce box of honeycomb cereal for $3. How much does the cereal cost per ounce?

46. On average, each person in the United States eats $1\frac{1}{10}$ pounds of honey every year. At this rate, how much honey do 5 people eat in a year?

47. Scott uses $2\frac{1}{4}$ ounces of honey to make a loaf of bread. How many loaves can he make from a 12-ounce jar of honey? How much honey will be left?

Mixed Review and Test Prep

USE DATA For 48–50, use the circle graph.

48. Todd has 24 videos. How many comedy videos does he own? (p. 108)

49. What fraction of Todd's collection is either action or drama videos? (p. 108)

50. If $\frac{1}{4}$ of Todd's "Other" videos are live concerts, what fraction of his collection is live concert videos? (p. 108)

51. Write 6.9054 in word form. (p. 20)

52. $\frac{3}{4} + \frac{1}{2} + \frac{1}{4}$ (p. 354)

53. $^{+}6 + ^{-}2$ (p. 422)

54. **TEST PREP** What is the least common multiple of 16 and 12? (p. 256)

A 4 **B** 16 **C** 48 **D** 192

55. **TEST PREP** Tomas has 6 pounds of ground beef. How many quarter-pound hamburgers can he make? (p. 398)

F $1\frac{1}{2}$ **G** 12 **H** 24 **J** 96

LINKUP to Reading

STRATEGY • IDENTIFY INFORMATION A careful reader must identify the details needed to solve a word problem. Read the following problem.

The average bee collects nectar for honey within 1 mile of its hive, makes up to 25 trips each day, and can carry a load of about 0.002 ounce—about half its weight. After the first half of their lives, bees eat only honey. In fact, for every pound of honey sold in stores, 8 pounds are used by the hive. About how much does a honeybee weigh?

1. What detail do you need to solve this problem?
2. Solve the problem.

For 3–4, use the paragraph above. Write the detail you need to solve each problem. Then solve.

3. If a bee visits 75 flowers during each nectar-collecting trip, about how many flowers does it visit each day?
4. For every $\frac{1}{4}$ pound of honey sold in stores, how many pounds are used by the hive?

Extra Practice page H53, Set C

LESSON **5**

Problem Solving Strategy

Solve a Simpler Problem

Quick Review

1. 120 ÷ 3

2. 140 ÷ 2 **3.** 250 ÷ 5

4. 720 ÷ 8 **5.** $9 \div \frac{2}{3}$

PROBLEM The greatest weight ever recorded for a great white shark was about 7,000 pounds. This is $\frac{4}{25}$ the weight of the world's largest shark, the whale shark. How much does a whale shark weigh?

▲ **Great whites are the largest meat-eating sharks. They have 3,000 teeth that are triangular, razor-sharp, and up to 3 inches long.**

- What are you asked to find?
- What information will you use?

- What strategy can you use to solve the problem?

You can *solve a simpler problem* by using simpler numbers.

- How can you use the strategy to solve the problem?

Think: ■ $\times \frac{4}{25} = 7{,}000$, so $7{,}000 \div \frac{4}{25} =$ ■.

Use a simpler number instead of 7,000.
Since $7 \times 1{,}000 = 7{,}000$, use the simpler number 7 to represent the weight of the great white shark.

$$7 \div \frac{4}{25} = \frac{7}{1} \times \frac{25}{4} = \frac{175}{4}$$

To find the actual weight of a whale shark, multiply by 1,000.

$$\frac{175}{4} \times 1{,}000 = \frac{175{,}000}{4} = 43{,}750$$

So, a whale shark weighs about 43,750 pounds.

- Look back at the problem. How can you decide if your answer is correct?

CALIFORNIA STANDARDS **MR 2.2** Apply strategies and results from simpler problems to more complex problems. **MR 1.0** Students make decisions about how to approach problems. **NS 2.0** Students perform calculations and solve problems involving addition, subtraction, and simple multiplication and division of fractions and decimals. **NS 2.5** Compute and perform simple multiplication and division of fractions and apply these procedures to solving problems. *also* **NS 1.0, NS 2.4, O━┓AF 1.2, MR 2.0, MR 2.6, MR 3.0, MR 3.2**

Problem Solving Practice

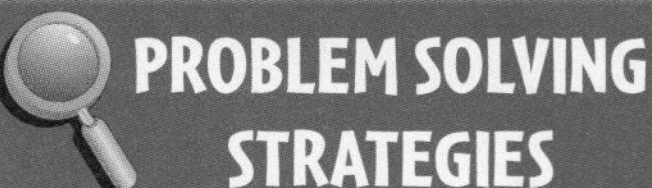

PROBLEM SOLVING STRATEGIES

- Draw a Diagram or Picture
- Make a Model or Act It Out
- Make an Organized List
- Find a Pattern
- Make a Table or Graph
- Predict and Test
- Work Backward
- Solve a Simpler Problem
- Write an Equation
- Use Logical Reasoning

Solve a simpler problem to solve.

1. **What if** a whale shark is 1,500 centimeters long and a great white shark is $\frac{2}{5}$ as long? How long is the great white shark?
2. A Pacific gray whale can weigh 70,000 pounds. This is about $\frac{7}{30}$ the weight of a blue whale. How much does a blue whale weigh?

A whale shark is 600 inches long. The world's smallest shark, the smalleye pygmy shark, is $\frac{1}{75}$ the length of a whale shark.

3. Which equation can be used to find the length of a pygmy shark?

   A $600 \times \frac{1}{75} = n$
   B $600 \times n = \frac{1}{75}$
   C $600 \div \frac{1}{75} = n$
   D $600 - \frac{1}{75} = n$

4. How long is a smalleye pygmy shark in feet?

   F 80 feet
   G 8 feet
   H $1\frac{1}{2}$ feet
   J $\frac{2}{3}$ foot

▲ Whale shark

Mixed Strategy Practice

5. Cole's dresser is $5\frac{1}{4}$ feet wide. His bookcase is $\frac{2}{3}$ as wide as his dresser. How wide is his bookcase?
6. Rachel has $4.75 in coins. She has 6 dimes and twice as many quarters. The rest are nickels. How many of each coin does she have?
7. The city bought 48,000 tulip bulbs. Workers planted $\frac{1}{8}$ in parks and $\frac{2}{3}$ along roads. How many bulbs do they have left?
8. Chloe made a square design. She colored $\frac{1}{4}$ of the square blue, $\frac{1}{8}$ yellow, and $\frac{1}{8}$ green. She left $\frac{1}{2}$ of the square white. What might her design look like?
9. **NUMBER SENSE** Oren chose a number and multiplied it by 3. Then he added 6, divided by 9, and subtracted 5. The answer was 4. What number did Oren choose?
10. **What's the Error?** Of the 398 species of sharks in the world, 199 are less than 39 inches long. Joe says the other $\frac{1}{3}$ are 39 inches or longer. Describe and correct his error.

Review/Test

CHECK VOCABULARY AND CONCEPTS

Choose the best term from the box.

denominator
divisor
fraction
numerator
reciprocal

1. To find the quotient of two fractions, multiply the dividend by the reciprocal of the __?__. (p. 400)
2. To find the reciprocal of a mixed number, first rename the mixed number as a __?__. (p. 396)
3. To divide a whole number by a fraction, first write the whole number as a fraction with a __?__ of 1. (p. 398)
4. The product of a number and its __?__ is 1. (p. 396)

CHECK SKILLS

Write the reciprocal of each number. (pp. 396–397)

5. $\frac{3}{10}$
6. $8\frac{2}{3}$
7. 14
8. $4\frac{2}{11}$
9. $\frac{2}{5}$

Divide. Write the answer in simplest form. (pp. 398–403)

10. $12 \div \frac{2}{3}$
11. $9 \div \frac{4}{5}$
12. $\frac{4}{5} \div 4$
13. $7 \div \frac{5}{9}$
14. $8 \div \frac{3}{7}$
15. $\frac{8}{9} \div 11$
16. $\frac{2}{7} \div \frac{2}{3}$
17. $\frac{5}{6} \div 1\frac{1}{9}$
18. $\frac{3}{5} \div \frac{3}{10}$
19. $1\frac{1}{2} \div \frac{1}{4}$
20. $2\frac{5}{8} \div \frac{1}{4}$
21. $2\frac{3}{4} \div 2\frac{1}{6}$

CHECK PROBLEM SOLVING

Solve. (pp. 404–405)

22. The mustard factory has 20,000 ounces of mustard to put into small packets. If each packet holds $\frac{1}{4}$ ounce, how many packets can be filled?
23. Lydia needs $\frac{3}{4}$ cup of flour to make bread. She has only a $\frac{1}{8}$-cup measuring cup. How many times will she fill the $\frac{1}{8}$ cup to get the $\frac{3}{4}$ cup she needs?
24. The Sports Card Store has 200,000 cards. If $\frac{1}{10}$ of them are baseball cards, how many cards are *not* baseball cards?
25. Gretchen has 27 feet of ribbon. She needs to cut the ribbon into $\frac{3}{4}$-foot pieces. How many pieces of ribbon can she cut?

Cumulative Review

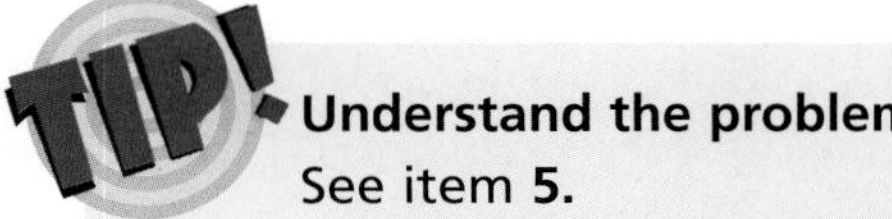

Understand the problem.
See item **5.**

The label, mph, describes speed and means the number of miles divided by the number of hours. Solve the problem by writing a number sentence with the fractions and finding the answer.

Also see problem **1,** p. H62.

For 1–11, choose the best answer.

1. $8 \div \frac{4}{5}$

A 11 **C** $\frac{32}{5}$
B 10 **D** $\frac{2}{5}$

2. Kenya has 24 math problems to do for homework. She finished $\frac{2}{3}$ of them before dinner. How many problems does she have left to do?

F 6 **H** 16
G 8 **J** 18

3. Which symbol makes this number sentence true?

$18 \div \frac{2}{3}$ ● $21 \div \frac{3}{4}$

A $>$ **C** $<$
B $=$ **D** $+$

4. $364.01 - 27.682$

F 336.328 **H** 87.19
G 243.672 **J** Not here

5. If you travel $\frac{1}{2}$ mile in $\frac{1}{10}$ of an hour, at what speed are you traveling?

A $\frac{1}{20}$ mph **C** 5 mph
B $\frac{1}{5}$ mph **D** 20 mph

6. How many $\frac{3}{4}$-inch-long pieces can you cut from a $2\frac{1}{4}$-inch-long piece of string?

F 1 **H** 3
G 2 **J** 4

7. What is the reciprocal of $2\frac{3}{5}$?

A $2\frac{5}{3}$ **C** $\frac{13}{5}$
B $2\frac{5}{13}$ **D** $\frac{5}{13}$

For 8–9, use the table.

Tunnel and Location	Length (miles)
St. Gotthard, Switzerland	$10\frac{1}{10}$
Brooklyn-Battery, New York	$1\frac{7}{10}$
Delaware Aqueduct, New York	85

8. How much longer than the Brooklyn-Battery Tunnel is the St. Gotthard Tunnel?

F $8\frac{2}{5}$ mi **H** $9\frac{3}{5}$ mi
G $8\frac{3}{5}$ mi **J** $11\frac{4}{5}$ mi

9. If there are lights every $\frac{1}{5}$ mile in the Delaware Aqueduct, how many lights are there?

A 17 **C** 405
B 80 **D** 425

10. What is 5.8651 rounded to the nearest hundredth?

F 5.86 **H** 5.87
G 5.863 **J** 5.9

11. What is the reciprocal of $2\frac{1}{2}$?

A $4\frac{1}{2}$ **C** $\frac{7}{9}$
B $\frac{2}{2}$ **D** Not here

UNIT 6

MATH DETECTIVE

You Be the Judge

For each case determine whether the solution is correct or not correct. Try not to do the actual computation, but rather base your answer on what you know about fractions and mixed numbers.

Case 1

Richard changed the fractions $\frac{3}{4}$ and $\frac{2}{3}$ to $\frac{6}{12}$ and $\frac{4}{12}$ before adding them and found a sum of $\frac{10}{12}$.

Case 2

Lynda said that to subtract the mixed number $1\frac{2}{3}$ from $2\frac{1}{3}$, you should first change them to fractions:

$2\frac{1}{3} - 1\frac{2}{3} = \frac{21}{3} - \frac{12}{3} = \frac{9}{3} = 3$

Case 3

Michael said the solution of $\frac{5}{7} - \frac{2}{3}$ is $\frac{3}{4}$.

Case 4

Susan said that to add $\frac{1}{4}$, $\frac{7}{12}$, and $\frac{2}{3}$, you could use a common denominator of 12.

Case 5

Christopher solved the problem $\frac{5}{12} + \frac{7}{12}$ and gave the answer as 1.

Case 6

David said that $4\frac{2}{3}$ can be renamed as $\frac{12}{3}$.

Think It Over!

- **Write About It** Explain the error in each case whose solution was not correct, and give the correct solution.
- **Write a problem** of your own, along with a solution, and have a classmate see if it is solved correctly.

UNIT 6

Challenge

Add Three Addends

Stephen is designing a tree house. According to his plans, he will need $4\frac{3}{4}$ yards of lumber to support the floor of the tree house, $6\frac{2}{3}$ yards for the roof, and $9\frac{1}{2}$ yards for the walls. What is the total amount of lumber Stephen will need for the tree house?

You can add three mixed numbers by first finding the least common denominator for the three fractions.

STEP 1

Find the least common multiple of the denominators.

4: 4, 8, 12, 16

3: 3, 6, 9, 12, 15

2: 2, 4, 6, 8, 10, 12, 14

The LCM is 12.

STEP 2

Write equivalent fractions.

$4\frac{3}{4} = 4\frac{9}{12}$

$6\frac{2}{3} = 6\frac{8}{12}$

$9\frac{1}{2} = 9\frac{6}{12}$

STEP 3

Add the fractions. Add the whole numbers. Write the sum in simplest form.

$$\begin{array}{r} 4\frac{9}{12} \\ 6\frac{8}{12} \\ +\ 9\frac{6}{12} \\ \hline 19\frac{23}{12} = 20\frac{11}{12} \end{array}$$

So, Stephen needs a total of $20\frac{11}{12}$ yards of lumber.

Examples

A $\frac{2}{3} + 4\frac{8}{9} + 2\frac{2}{9}$

$$\begin{array}{r} \frac{2}{3} = \frac{6}{9} \\ 4\frac{8}{9} = 4\frac{8}{9} \\ +\ 2\frac{2}{9} = 2\frac{2}{9} \\ \hline 6\frac{16}{9} = 7\frac{7}{9} \end{array}$$

B $1\frac{1}{8} + \frac{3}{4} + 1\frac{1}{2}$

$$\begin{array}{r} 1\frac{1}{8} = 1\frac{1}{8} \\ \frac{3}{4} = \frac{6}{8} \\ +\ 1\frac{1}{2} = 1\frac{4}{8} \\ \hline 2\frac{11}{8} = 3\frac{3}{8} \end{array}$$

C $2\frac{1}{3} + \frac{5}{6} + \frac{3}{4}$

$$\begin{array}{r} 2\frac{1}{3} = 2\frac{4}{12} \\ \frac{5}{6} = \frac{10}{12} \\ +\ \frac{3}{4} = \frac{9}{12} \\ \hline 2\frac{23}{12} = 3\frac{11}{12} \end{array}$$

Try It

1. $\frac{1}{12} + \frac{5}{6} + \frac{1}{3}$

2. $\frac{2}{5} + 6\frac{1}{5} + \frac{1}{2}$

3. $2\frac{1}{7} + \frac{1}{2} + 4\frac{3}{7}$

4. REASONING If I am added together three times, the total is $14\frac{1}{4}$. What number am I?

Study Guide and Review

VOCABULARY

Choose the best term from the box.

1. The least common multiple of the denominators is the __?__ for the fractions. (p. 352)
2. Two numbers with a product of 1 are called __?__. (p. 396)
3. To simplify factors before you multiply them, you can divide by the __?__ of a numerator and a denominator. (p. 386)

equivalent fractions
greatest common factor
reciprocals
rename
least common denominator

STUDY AND SOLVE

Chapter 19

Add and subtract fractions.

Find $\frac{2}{3} + \frac{1}{4}$.

Use the LCD to rename the fractions.

$$\begin{array}{rcccl} \frac{2}{3} = & \frac{2 \times 4}{3 \times 4} & = & \frac{8}{12} \\ +\frac{1}{4} = + & \frac{1 \times 3}{4 \times 3} & = + & \frac{3}{12} \\ \hline & & & \frac{11}{12} \end{array}$$

Add the numerators.

Write the sum over the denominator.

Find the sum or difference. Write the answer in simplest form. (pp. 344–357)

4. $\frac{3}{5} + \frac{2}{5}$
5. $\frac{5}{7} - \frac{3}{14}$
6. $\frac{7}{16} + \frac{3}{8}$
7. $\frac{2}{3} - \frac{5}{9}$
8. $\frac{1}{4} - \frac{1}{12}$
9. $\frac{1}{9} + \frac{5}{6}$

Chapter 20

Add and subtract mixed numbers.

$$\begin{array}{rcl} 2\frac{3}{8} = & 2\frac{3}{8} \\ -\ 1\frac{1}{4} = - & 1\frac{2}{8} \\ \hline & 1\frac{1}{8} \end{array}$$

Use the LCD to rename.

Subtract the fractions.

Then subtract the whole numbers. Simplify if possible.

Find the sum or difference. Write the answer in simplest form. (pp. 364–373)

10. $4\frac{7}{12} + 3\frac{1}{8}$
11. $3\frac{4}{6} - 2\frac{1}{6}$
12. $6\frac{2}{5} - 5\frac{3}{5}$
13. $6\frac{5}{8} + 7\frac{7}{8}$

Chapter 21

Multiply a fraction by a fraction.

Multiply. $\frac{3}{4} \times \frac{2}{5}$

$\frac{3}{4} \times \frac{2}{5} = \frac{3 \times 2}{4 \times 5}$

$= \frac{6}{20} = \frac{3}{10}$

Multiply the numerators. Multiply the denominators.

Find the product. Write the answer in simplest form. (pp. 382–383)

14. $\frac{1}{2} \times \frac{2}{7}$
15. $\frac{1}{3} \times \frac{5}{9}$
16. $\frac{4}{9} \times \frac{6}{11}$
17. $\frac{3}{8} \times \frac{1}{6}$

Multiply mixed numbers.

Multiply. $2\frac{2}{3} \times \frac{2}{3}$

$\frac{8}{3} \times \frac{2}{3} = \frac{16}{9}$, or $1\frac{7}{9}$ Rename the mixed number as a fraction. Multiply the fractions.

Find the product. Write the answer in simplest form. (pp. 384–387)

18. $\frac{5}{6} \times 3\frac{1}{2}$
19. $2\frac{1}{4} \times \frac{3}{8}$
20. $4\frac{3}{4} \times 1\frac{1}{2}$
21. $2\frac{2}{3} \times 3\frac{1}{4}$
22. $5\frac{1}{2} \times \frac{1}{4}$
23. $1\frac{5}{8} \times \frac{2}{3}$

Chapter 22

Write the reciprocal of a number.

Write the reciprocal of $\frac{2}{3}$ and $2\frac{3}{4}$.

Exchange the numerator and denominator.

$\frac{2}{3} \rightarrow \frac{3}{2}$, or $1\frac{1}{2}$ ← reciprocal

$2\frac{3}{4} = \frac{11}{4} \rightarrow \frac{4}{11}$ ← reciprocal

Write the reciprocal of each number. (pp. 396–397)

24. 19
25. $2\frac{3}{7}$
26. $\frac{7}{12}$
27. $3\frac{2}{9}$
28. $\frac{14}{31}$
29. $4\frac{1}{3}$

Divide with fractions.

Divide. $4 \div \frac{2}{3}$

$\frac{4}{1} \div \frac{2}{3} = \frac{4}{1} \times \frac{3}{2}$ Write the whole number as a fraction. Multiply using the reciprocal of the divisor.

$= \frac{12}{2} = 6$ Write the answer in simplest form.

Find the quotient. Write the answer in simplest form. (pp. 398–403)

30. $5 \div \frac{2}{3}$
31. $\frac{9}{10} \div \frac{2}{5}$
32. $12 \div \frac{5}{6}$
33. $1\frac{3}{4} \div \frac{5}{8}$

PROBLEM SOLVING PRACTICE

Solve. (pp. 374–375, 404–405)

34. Mrs. Catalanello made 3 loaves of bread. The first loaf was $10\frac{7}{8}$ inches long, the second loaf was $3\frac{1}{2}$ inches shorter than the first, and the third was $1\frac{3}{4}$ inches longer than the second. How long was the third loaf?

35. In 1998, the populations of Los Angeles, San Francisco, and San Diego totalled about 5,500,000. The population of Los Angeles was $\frac{2}{3}$ of this total. What was the population of Los Angeles in 1998?

California Connections

NATIONAL PARKS

California has eight national parks. These parks help to preserve the state's natural beauty and allow people access to wilderness areas.

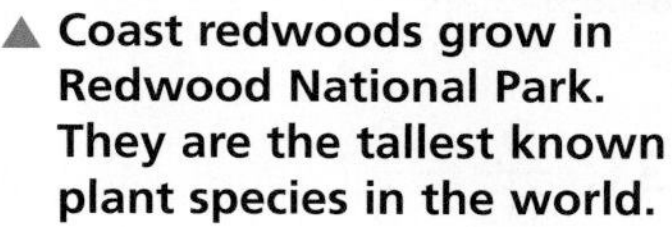

▲ Coast redwoods grow in Redwood National Park. They are the tallest known plant species in the world.

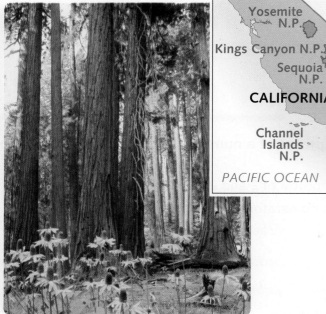

▲ Giant sequoias grow in Sequoia National Park in the Sierra Nevada at elevations from 5,000 feet to 8,000 feet.

USE DATA **For 1–5, use the table.**

1. How many more square miles are in the largest California national park than in the smallest national park?
2. Is the total area of the three smallest parks greater than or less than the area of Kings Canyon? Explain.
3. In 1994 Death Valley was upgraded from a national monument to a national park. At that time $2{,}031\frac{1}{4}$ square miles were added to its area. What was the area of Death Valley when it was a national monument?
4. Which park has about one third the area of Yosemite?
5. One half of Channel Islands National Park is under the ocean. How many square miles are under the ocean?

CALIFORNIA NATIONAL PARKS

Park	Area (in sq mi)
Channel Islands	$389\frac{5}{8}$
Death Valley	$5{,}156\frac{1}{4}$
Joshua Tree	$871\frac{7}{8}$
Kings Canyon	$721\frac{3}{4}$
Lassen Volcanic	$165\frac{5}{8}$
Redwood	$171\frac{7}{8}$
Sequoia	$628\frac{7}{8}$
Yosemite	$1{,}171\frac{7}{8}$

CHANNEL ISLANDS NATIONAL PARK

The Channel Islands are an eight-island chain lying off the coast of southern California. Five islands—San Miguel, Santa Rosa, Santa Cruz, Anacapa, and Santa Barbara—and the surrounding ocean make up the Channel Islands National Park.

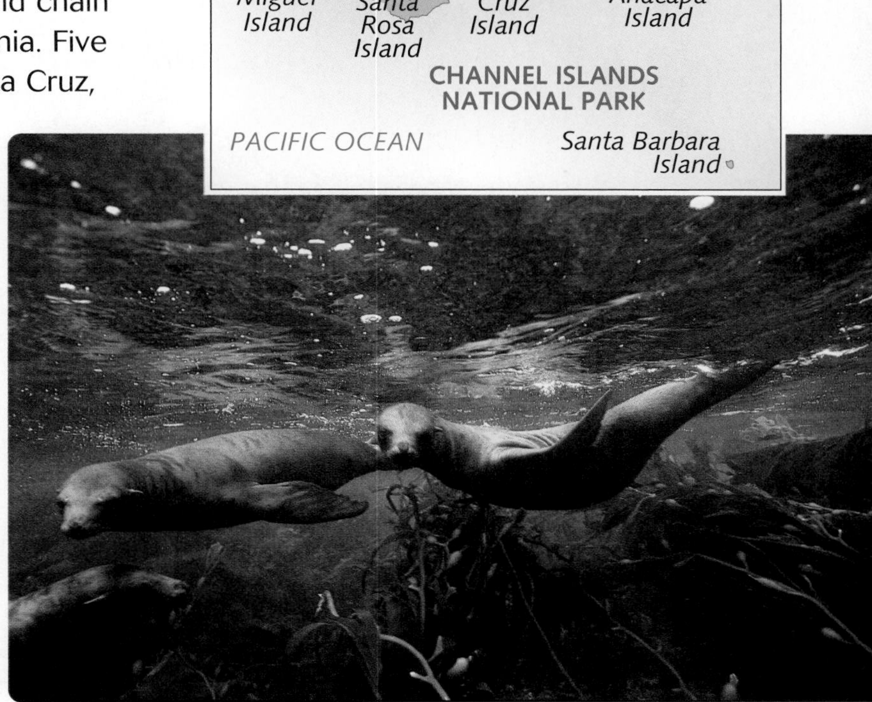

▲ **The Channel Islands National Park is home to over 2,000 types of animals and land plants, many of which are found nowhere else in the world.**

1. A boat ride from the mainland to Anacapa, the closest island to the mainland, takes $1\frac{1}{2}$ hours. How long is a round-trip boat ride between the mainland and Anacapa?

2. The Nature Conservancy owns $\frac{9}{10}$ of Santa Cruz Island. The National Park Service owns the rest of the island. What part of the island does the National Park Service own?

3. On San Miguel Island you can hike from the boat landing $2\frac{1}{4}$ miles to San Miguel Hill and continue for $5\frac{1}{4}$ miles to Point Bennett to view sea lions and seals. If you return on the same trail, how far will you have walked in all?

USE DATA The table shows the lengths of hiking trails on Santa Barbara Island. For 4–5, use the table.

4. How much longer is the Signal Peak Loop trail than the Arch Point Loop trail?

5. What if you want to hike a total of less than 3 miles? Could you walk to Elephant Seal Cove and back to the Visitor's Center? Explain.

SANTA BARBARA ISLAND TRAILS FROM THE VISITOR'S CENTER

Trail	Length (in miles)
Arch Point Loop	$1\frac{9}{10}$
Canyon View Nature	$\frac{3}{10}$
Elephant Seal Cove	$1\frac{3}{5}$
Signal Peak Loop	3

Algebra: Integers

In golf, a game that began in Scotland in the fifteenth century, the object is to get the ball into the hole in as few strokes (hits) as possible. The standard number of strokes to get the ball into the hole is called par. A golfer's score is usually shown as the number of strokes above (+) or below (–) par. The fewer strokes you take, the better your score. Use the tournament leaderboard to find whose score was below par.

2000 PEBBLE BEACH NATIONAL PRO-AM TOURNAMENT LEADERBOARD

Player	Rounds				Total Score	Par
	1	2	3	4		
Sergio Garcia	71	75	70	70	286	$^-2$
Brandt Jobe	73	73	68	76	290	$^+2$
Justin Leonard	73	70	68	70	281	$^-7$
Brian Watts	73	75	68	79	295	$^+7$
Tiger Woods	68	73	68	64	273	$^-15$

Tiger Woods

Use this page to help you review and remember important skills needed for Chapter 23.

VOCABULARY

Choose the best term from the box.

Celsius
degrees
Fahrenheit
mixed number
whole numbers

1. A number made up of a whole number and a fraction is a __?__.
2. The numbers 0, 1, 2, 3, . . . are called __?__.
3. A thermometer's scale is divided into __?__.
4. The United States Customary System uses the __?__ scale to measure temperature.

COMPARE WHOLE NUMBERS (See p. H25.)

Compare. Use <, >, or = for each ●.

5. 112 ● 111 **6.** 98 ● 99 **7.** 82 ● 87 **8.** 905 ● 950 **9.** 997 ● 999

10. 12 ● 10 **11.** 56 ● 65 **12.** 102 ● 120 **13.** 752 ● 725 **14.** 425 ● 424

15. 81 ● 88 **16.** 32 ● 320 **17.** 111 ● 101 **18.** 323 ● 423 **19.** 121 ● 120

READ A THERMOMETER (See p. H25.)

Write the Fahrenheit temperature reading for each.

20.
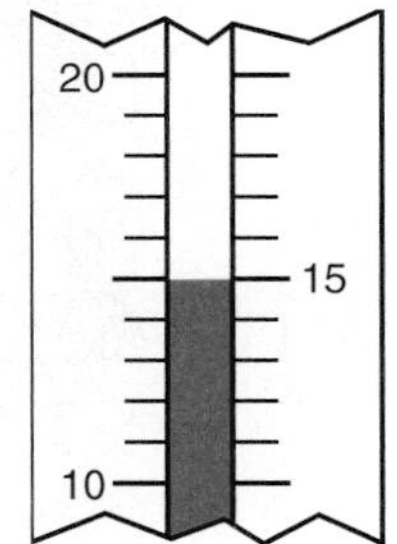

21.
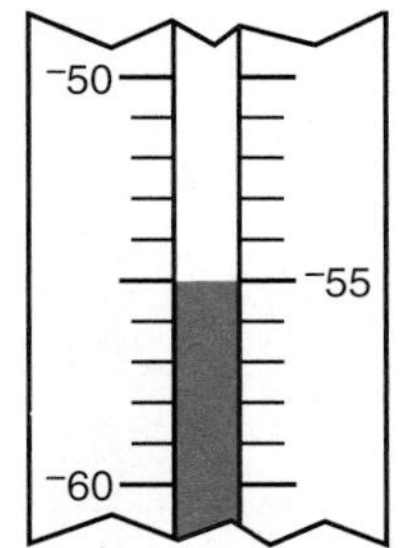

22.
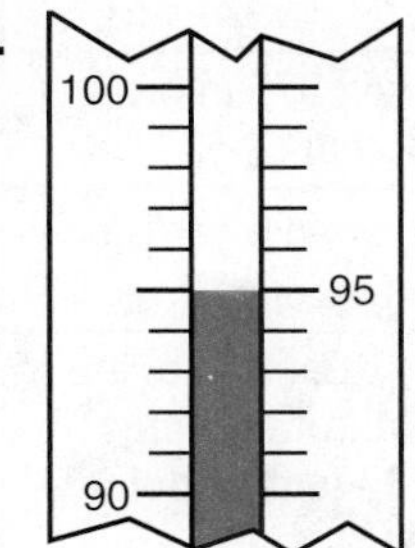

23.
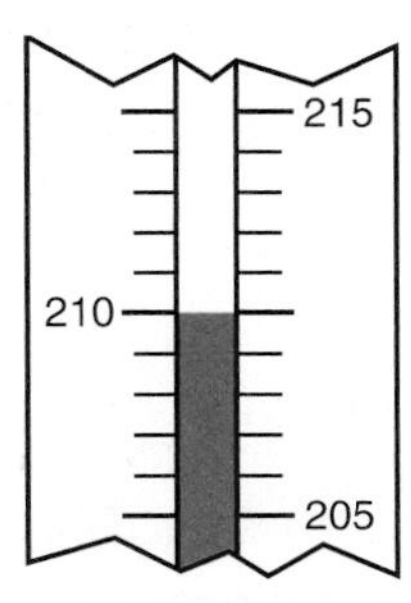

24.
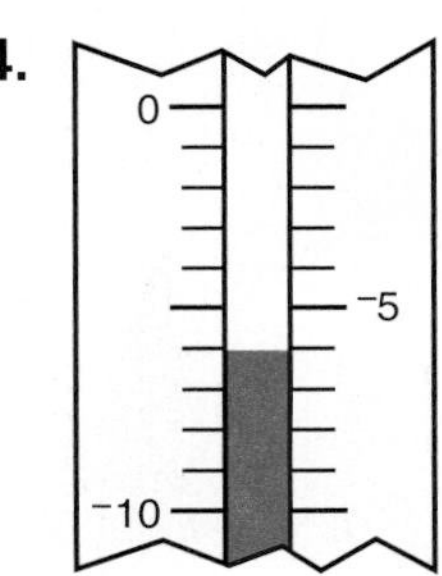

25.
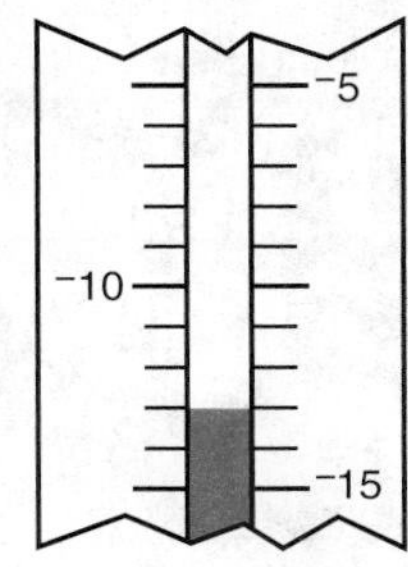

LESSON 1

Integers

Quick Review

Name the opposite.

1. up **2.** left
3. open **4.** hot
5. full

VOCABULARY

integers
negative integers
positive integers
opposites
absolute value

Learn

HIGHS AND LOWS The highest temperature in the United States, recorded in California, was $^{+}134$°F. The lowest temperature, recorded in Alaska, was $^{-}80$°F.

The numbers $^{+}134$ and $^{-}80$ are **integers**. You read $^{+}134$ as "positive one hundred thirty-four" and $^{-}80$ as "negative eighty."

Just as a thermometer shows temperatures above and below 0°, a number line shows numbers to the right and to the left of 0. Integers can be shown on a number line.

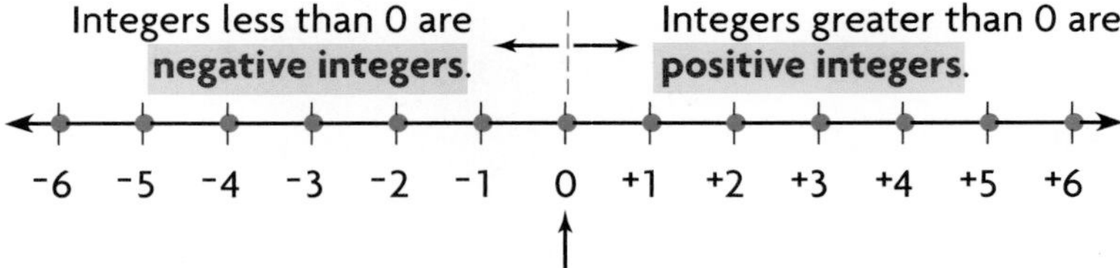
Integers less than 0 are **negative integers**.
Integers greater than 0 are **positive integers**.
⁻6 ⁻5 ⁻4 ⁻3 ⁻2 ⁻1 0 ⁺1 ⁺2 ⁺3 ⁺4 ⁺5 ⁺6

The integer 0 is neither positive nor negative.

Positive and negative integers can be used to represent word situations.

Situation	Integer
Mt. McKinley, Alaska, the highest point in the United States, is 20,320 feet above sea level.	$^{+}20{,}320$
Ken deposited $75 in his checking account.	$^{+}75$
The Lions gained 6 yards on a first down.	$^{+}6$
Death Valley, California, the lowest point in the United States, is 282 feet below sea level.	$^{-}282$
Ken withdrew $25 from his checking account.	$^{-}25$
The Bears lost 5 yards on the last play.	$^{-}5$

The polar bear's body is designed to stay warm when the air is as cold as $^{-}49$°F. Under its fur, the bear has a layer of blubber 4 inches thick to help keep it warm. ▼

CALIFORNIA STANDARDS **NS 1.5** Identify and represent on a number line, decimals, fractions, mixed numbers, and positive and negative integers. *also* **MR 1.1, MR 2.0, MR 2.3, MR 2.4, MR 3.2**

Opposites and Absolute Values

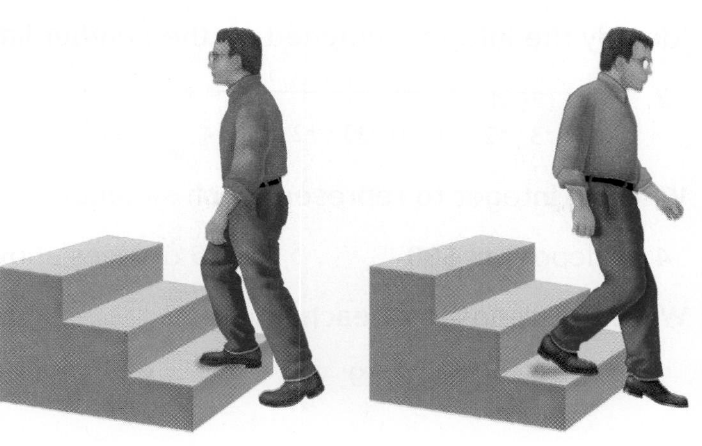

The opposite of going up 3 steps is going down 3 steps. In both cases the same distance is traveled, 3 steps.

Integers include all whole numbers and their opposites. For every positive integer, there is an opposite, negative integer. Integers that are **opposites** are the same distance from 0 on a number line, but in opposite directions.

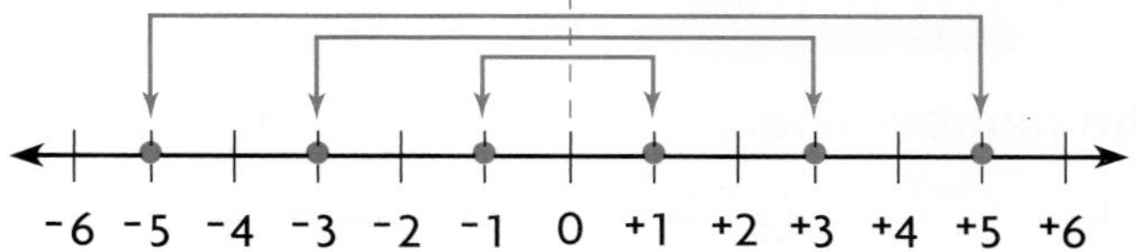

The number line above has these opposites graphed:

$^{-}1$ and $^{+}1$ $^{-}3$ and $^{+}3$ $^{-}5$ and $^{+}5$

The **absolute value** of an integer is its distance from 0.

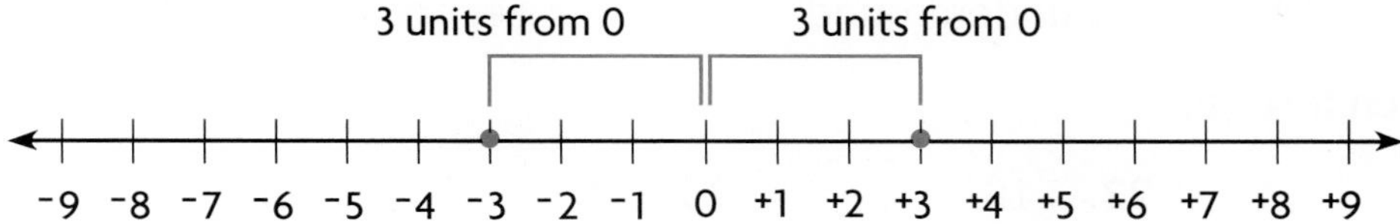

Write: $|{}^{-}3| = 3$ **Read:** The absolute value of negative three is three.

Write: $|{}^{+}3| = 3$ **Read:** The absolute value of positive three is three.

Examples

A Name the opposite of $^{-}11$.

$^{-}11 \rightarrow {}^{+}11$

B Use the number line to find $|{}^{-}4|$.

$|{}^{-}4| = 4$

C Name the opposite of $^{+}14$.

$^{+}14 \rightarrow {}^{-}14$

D Use the number line to find $|{}^{+}9|$.

$|{}^{+}9| = 9$

Check

1. **Tell** how many total points you have if you win 10 points on your first turn and then lose 10 points on your second turn. Write the integers that represent what happens on each turn. Explain how these integers are related.

LESSON CONTINUES

Identify the integers graphed on the number line.

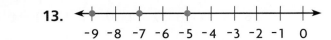

2. -4 -3 -2 -1 0 +1 +2 +3 +4

3. -6 -5 -4 -3 -2 -1 0 +1 +2

Write an integer to represent each situation.

4. a deposit of \$30
5. 85 degrees above 0
6. a loss of 15 yards

Write the opposite of each integer.

7. $^{+}7$
8. $^{-}9$
9. $^{-}2$
10. $^{+}8$
11. 0

Practice and Problem Solving

Identify the integers graphed on the number line.

12. -1 0 +1 +2 +3 +4 +5 +6 +7 +8

13. -9 -8 -7 -6 -5 -4 -3 -2 -1 0

Write an integer to represent each situation.

14. three points ahead
15. a growth of 8 inches
16. 8 feet below sea level
17. a gain of 5 pounds
18. 15 feet underground
19. 4 seconds before liftoff

Write the opposite of each integer.

20. $^{-}3$
21. $^{+}4$
22. $^{+}16$
23. $^{-}22$
24. $^{+}41$
25. $^{+}54$
26. $^{-}29$
27. $^{-}73$
28. $^{+}102$
29. $^{+}2{,}314$

Write each integer's absolute value.

30. $|^{+}25|$
31. $|^{-}10|$
32. $|^{+}1{,}000|$
33. $|^{-}1|$
34. $|0|$

35. **Algebra** What values can n have if $|n| = 6$?

USE DATA For 36–37, use the number line to locate each integer.

A B C D E F G H I J K

-5 0 +5

36. Write the letter for each integer.

a. $^{-}3$ b. $^{+}4$ c. 0 d. $^{-}1$

37. Write the letter for the opposite of each integer.

a. $^{-}2$ b. $^{+}3$ c. 0 d. $^{-}3$

38. Which of the integers $^{-}7$, $^{+}4$, 0, $^{-}4$, $^{+}2$, $^{-}2$, $^{+}7$ are positive? negative? Name the opposite pairs.

39. **REASONING** If 0 stands for today, which integers stand for yesterday and tomorrow?

40. **Write About It** Describe a situation that uses positive and negative integers. Then write each integer and explain what each integer means.

41. Israel, the hottest place in Asia, recorded a temperature of almost 129°F. That is the opposite of the coldest recorded temperature in Antarctica. How cold has it gotten in Antarctica?

42. The coldest possible temperature is called *absolute zero*. It is about four hundred fifty-nine degrees below zero Fahrenheit. Write this temperature in standard form.

Mixed Review and Test Prep

USE DATA For 43–44, use the graph.

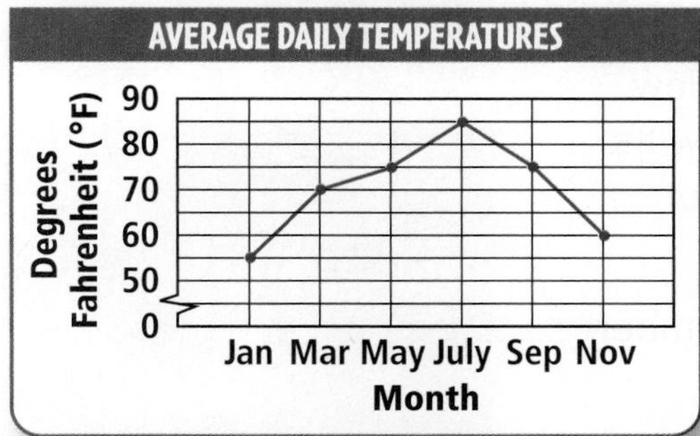

43. In which month was the temperature the highest? the lowest? (p. 108)

44. What is the mean temperature? (p. 102)

45. Identify the addition property shown. $5 + 0 = 5$ (p. 78)

A Zero C Associative
B One D Commutative

46. $1.05 + 0.509$ (p. 50)

47. $84 \times 10{,}504$ (p. 150)

48. 5.6×1.1 (p. 164)

49. 1.88×0.34 (p. 164)

50. **TEST PREP** Which is 0.13 written as a fraction? (p. 280)

F $1\frac{3}{10}$ G $\frac{13}{10}$ H $\frac{13}{100}$ J $\frac{13}{1,000}$

LINKUP to Reading

STRATEGY • USE CONTEXT To *use context,* look for words and phrases that help you understand the meanings of words, sentences, paragraphs, and situations. The words below can help you decide if an integer is negative, zero, or positive.

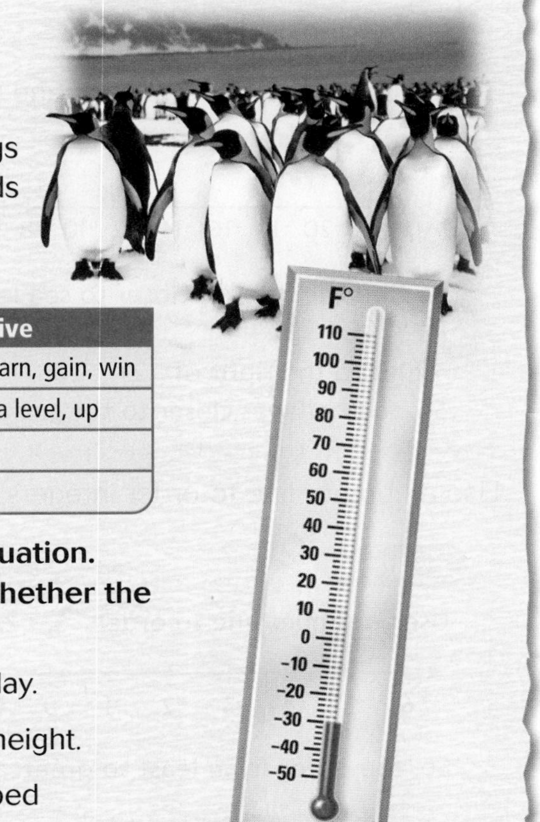

Negative	Zero	Positive
withdraw, loss, spend, lose, lost	no change, break even	deposit, profit, earn, gain, win
below, below sea level, down	sea level	above, above sea level, up
drop, behind	even	rise, ahead
before	now	after

Write an integer to represent what happens in each situation. Then write the word or phrase that helps you decide whether the integer is positive, negative, or zero.

1. Karen deposited \$20 in her savings account on Monday.
2. Between ages 16 and 17, Antoine did not change in height.
3. When the cold front moved in, the temperature dropped five degrees.

LESSON 2

Compare and Order Integers

Quick Review

Order from least to greatest.

1. 12, 0, 2
2. 9, 5, 10
3. 13, 33, 35
4. 41, 0, 14
5. 11, 20, 13

Learn

IN THE DEEP In 1718, the pirate Blackbeard's flagship, *Queen Anne's Revenge (QAR),* sank at Beaufort Inlet, North Carolina. In 1996, divers discovered a shipwreck believed to be the *QAR*. The ship's cannons were found 21 feet below the water's surface, and its bell was found 20 feet below the surface. Which was closer to the surface, the cannons or the bell?

Use a number line to compare. On a number line, a number to the right is always greater than a number to the left.

STEP 1

Name the integers that represent each situation.

cannons 21 feet below the surface: $^{-}21$

bell 20 feet below the surface: $^{-}20$

water's surface: 0

STEP 2

Graph the integers $^{-}21$, $^{-}20$, and 0 on a number line.

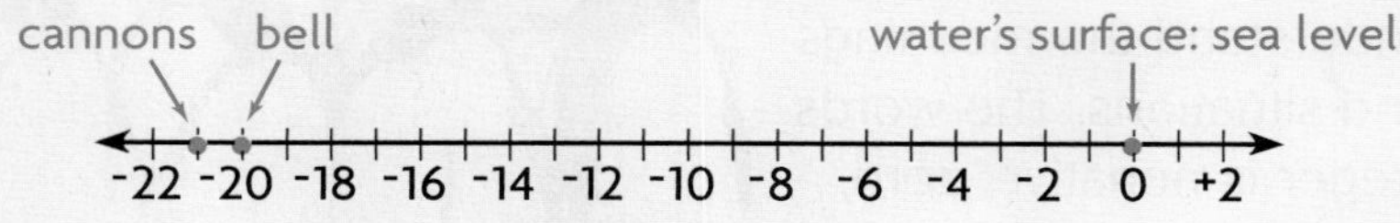

To find the integer closer to sea level, or 0, find the greater integer.

$^{-}20$ is to the right of $^{-}21$. $^{-}20 > {}^{-}21$

So, the bell was closer to the surface than the cannons.

▲ **A diver examines an anchor fluke at the QAR site. Excavation is done in 10-foot square units.**

Use a number line to order integers.

Example

Use a number line to order $^{+}5$, $^{-}2$, $^{-}4$, $^{+}1$, and $^{-}5$ from least to greatest.

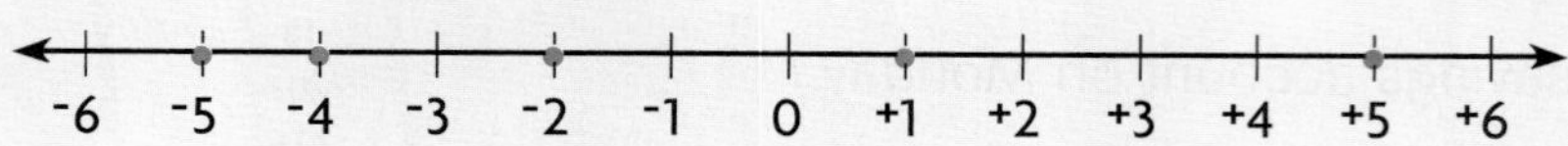

So, the order from least to greatest is $^{-}5$, $^{-}4$, $^{-}2$, $^{+}1$, and $^{+}5$.

CALIFORNIA STANDARDS **NS 1.5** Identify and represent on a number line, decimals, fractions, mixed numbers, and positive and negative integers. *also* **NS 1.0, MR 2.0, MR 2.3, MR 3.0, MR 3.2, MR 3.3**

Check

1. **Explain** why integers are easier to compare when they are graphed on a number line.

Compare. Write $<$, $>$, or $=$ for each ●.

2. $^{+}5$ ● $^{+}4$ **3.** $^{-}6$ ● $^{-}3$ **4.** $^{+}5$ ● $^{-}2$ **5.** $^{-}1$ ● 0 **6.** $^{-}2$ ● $^{-}2$

Practice and Problem Solving

Compare. Write $<$, $>$, or $=$ for each ●.

7. 0 ● $^{+}1$ **8.** $^{-}8$ ● $^{-}9$ **9.** $^{-}2$ ● $^{+}4$ **10.** $^{-}3$ ● $^{-}6$ **11.** $^{-}4$ ● $^{+}3$

Draw a number line to order each set of integers from greatest to least.

12. $^{+}1, {}^{-}2, 0, {}^{-}3$ **13.** $^{-}4, {}^{+}3, {}^{-}2, {}^{+}1$ **14.** $^{+}5, {}^{-}6, {}^{+}7, {}^{-}8$ **15.** $^{-}10, {}^{+}10, {}^{+}8, {}^{-}7$

$\frac{a+b}{c}$ **Algebra** **Name the integer that is 1 less.**

16. $^{+}8$ **17.** $^{-}4$ **18.** $^{-}1$ **19.** 0 **20.** $^{+}3$

$\frac{a+b}{c}$ **Algebra** **Name the integer that is 1 more.**

21. $^{-}4$ **22.** $^{-}1$ **23.** $^{+}6$ **24.** $^{-}8$ **25.** $^{-}10$

USE DATA **For 26–27 and 30, use the table.**

AVERAGE DEPTHS OF OCEANS	
Ocean	**Depth in feet**
Atlantic	12,880
Indian	13,002
Arctic	3,953
Pacific	13,215

26. One mile is 5,280 feet. Which ocean is less than 1 mile deep?

27. Mt. Fuji in Japan is 12,388 feet tall. Which ocean's depth is closest to being the opposite of Mt. Fuji's height?

28. **REASONING** Oarfish live at $^{-}3,000$ feet. Below $^{-}2,300$ feet, light cannot reach through water. Do oarfish live in the dark? Explain.

29. The most valuable shipwreck was found off Key West, Florida. The ship carried 36 tons of gold and silver. How many pounds is that?

30. **Write a problem** that uses the table and requires comparing or ordering integers.

Mixed Review and Test Prep

31. $5\overline{)2,455}$ (p. 188) **32.** $3\overline{)15.36}$ (p. 222)

33. Is 458 divisible by 2, 3, 4, 5, or 9? (p. 254)

34. Write $\frac{4}{5}$ as a decimal. (p. 280)

35. **TEST PREP** Which integer represents a loss of six pounds? (p. 416)

A $^{-}6$ **C** $^{+}0.6$
B $^{-}0.6$ **D** $^{+}6$

Extra Practice page H54, Set B

LESSON 3

Add Integers

Quick Review

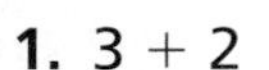

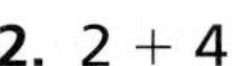

1. 3 + 2
2. 2 + 4
3. 3 + 7
4. 5 + 5
5. 2 + 9

Learn

UNDER PAR In golf, the number of strokes a golfer takes to hit a ball into a hole is compared to *par,* the standard number of strokes needed to get the ball into that hole. A score of $^{-}1$ means that the player hit the ball 1 stroke less than par, or 1 stroke under par.

For the 1997 Masters Tournament, Tiger Woods was $^{+}1$ on the first hole and $^{-}3$ on the second hole. On the first two holes, Tiger was $^{+}1 + ^{-}3$.

You can use counters to add integers. Find $^{+}1 + ^{-}3$.

On April 13, 1997, Tiger Woods became the youngest player to win the Masters Tournament. His final score was 270, which was $^{-}18$, or 18 strokes under par.

Activity

MATERIALS: two-color counters

STEP 1

Use yellow counters to represent positive integers and red counters to represent negative integers.

$= ^{+}1$

$= ^{-}3$

STEP 2

Make as many opposite pairs of one yellow and one red counter as possible. The sum of each opposite pair is 0, so take away each opposite pair.

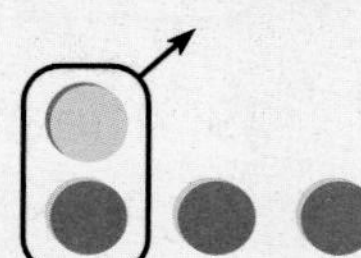

STEP 3

The number and color of unpaired counters represents the sum.

$= ^{-}2$

$^{+}1 + ^{-}3 = ^{-}2.$

Examples

A Find $^{-}3 + ^{-}2$.

$= ^{-}5$

So, $^{-}3 + ^{-}2 = ^{-}5$.

B Find $^{+}4 + ^{-}3$.

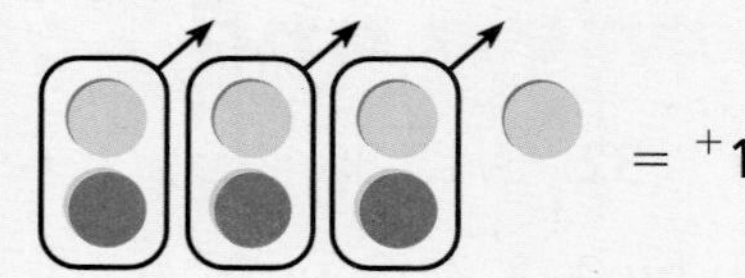

$= ^{+}1$

So, $^{+}4 + ^{-}3 = ^{+}1$.

CALIFORNIA STANDARDS **NS 2.1** Add, subtract, multiply, and divide with decimals; add with negative numbers; subtract positive integers from negative integers; and verify the reasonableness of the results. **NS 1.5** Identify and represent on a number line, decimals, fractions, mixed numbers, and positive and negative integers. *also* **NS 1.0, MR 1.0, MR 2.0, MR 2.3, MR 3.2, MR 3.3**

Use a Number Line to Add Integers

You can also use a number line to add integers. On a number line, move right for a positive integer and left for a negative integer.

Find $^{+}1 + {}^{-}3$.

STEP 1

Draw a number line. Start at 0.
Move 1 space to the right to show $^{+}1$.

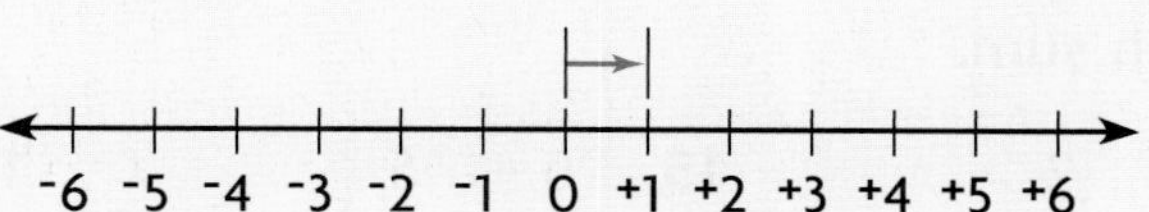

STEP 2

From $^{+}1$, move 3 spaces to the left to show $^{-}3$. This takes you to $^{-}2$.

So, $^{+}1 + {}^{-}3 = {}^{-}2$.

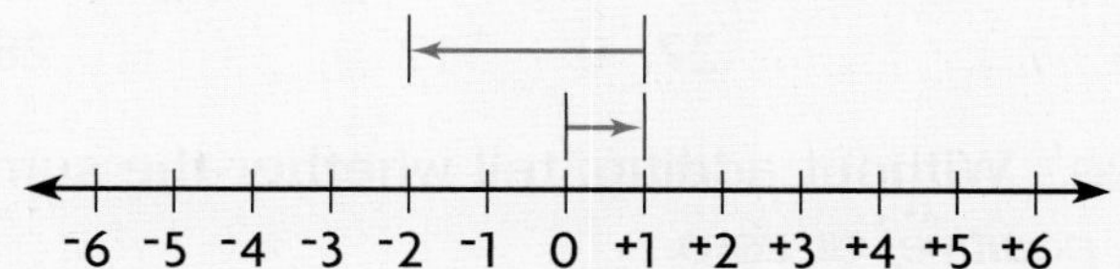

Examples

A Find $^{-}3 + {}^{-}2$.

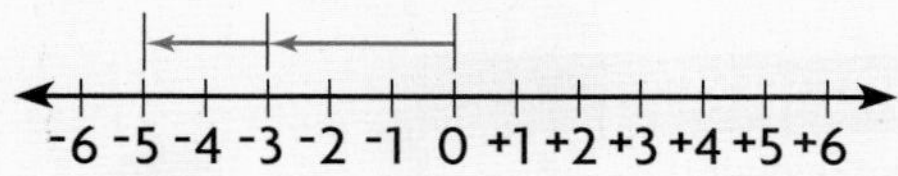

So, $^{-}3 + {}^{-}2 = {}^{-}5$.

B Find $^{+}4 + {}^{-}3$.

So, $^{+}4 + {}^{-}3 = {}^{+}1$.

- **REASONING** Will the sum of two negative integers be positive or negative? Explain.

TECHNOLOGY LINK

More Practice: Use Mighty Math Calculating Crew, ***Nautical Number Line***, Level Y.

Check

1. **Tell** the direction you would move from 0 on a number line to show $^{-}7 + {}^{+}3$.

Write the addition number sentence modeled.

2. 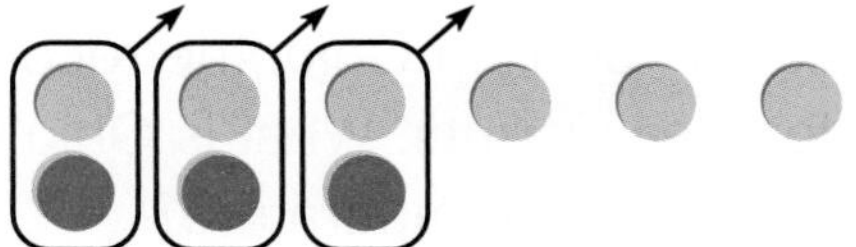

3. 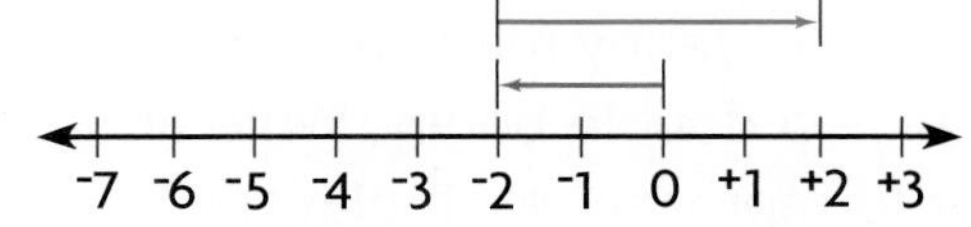

Find each sum.

4. $^{+}2 + {}^{-}3$
5. $^{-}4 + {}^{+}4$
6. $^{-}2 + {}^{-}3$
7. $^{+}7 + {}^{-}3$
8. $^{-}8 + {}^{+}9$
9. $^{-}9 + {}^{+}2$
10. $^{+}3 + {}^{+}4$
11. $^{+}2 + {}^{-}8$

LESSON CONTINUES

Practice and Problem Solving

Write the addition number sentence modeled.

12.

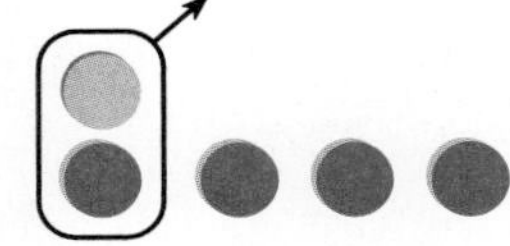

13.

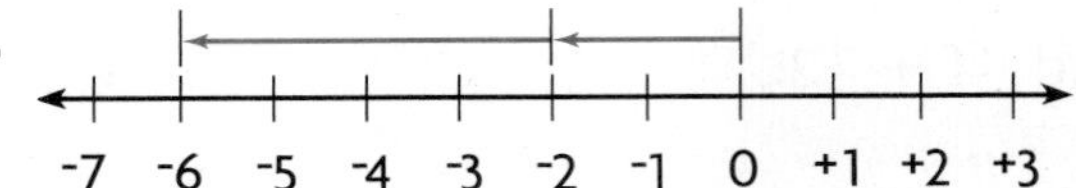

Find each sum.

14. $^{+}5 + ^{-}2$ **15.** $^{-}6 + ^{+}5$ **16.** $^{-}4 + ^{+}9$ **17.** $^{+}1 + ^{-}8$

18. $^{-}7 + ^{-}2$ **19.** $^{+}7 + ^{-}7$ **20.** $^{-}5 + ^{+}1$ **21.** $^{+}10 + ^{-}5$

22. $^{+}3 + ^{-}3$ **23.** $^{-}9 + 0$ **24.** $^{-}4 + ^{+}6 + ^{-}2$ **25.** $^{-}5 + ^{-}3 + ^{-}2$

26. $^{+}6 + ^{-}7$ **27.** $0 + ^{+}12$ **28.** $^{-}5 + ^{-}3 + 0$ **29.** $^{+}4 + ^{-}4 + ^{+}4$

REASONING **Without adding, tell whether the sum will be *negative, positive,* or *zero*.**

30. $^{-}4 + ^{-}5$ **31.** $^{+}6 + ^{-}30$ **32.** $^{-}1 + ^{+}23$ **33.** $^{+}9 + ^{+}8$

34. $^{-}17 + ^{-}3$ **35.** $^{+}10 + ^{-}15$ **36.** $^{-}200 + ^{+}200$ **37.** $^{+}350 + ^{-}60$

38. $^{+}30 + ^{-}30$ **39.** $^{-}10 + ^{+}15$ **40.** $^{+}452 + ^{-}389$ **41.** $^{-}523 + ^{-}32$

USE DATA **For 42–44, use the chart.**

GOLF TERMS
Bogey = 1 stroke over par
Double Bogey = 2 strokes over par
Birdie = 1 stroke under par
Eagle = 2 strokes under par

42. Suppose 0 represents even par for a hole. Write an integer for each of the scores listed, and order them from least to greatest.

43. Par for each round of the Masters Tournament is 72 strokes. There are 4 rounds in the Masters.

a. What is par for the tournament?

b. Tiger Woods was $^{-}18$ for the tournament. How many strokes did he hit for the 4 rounds?

44. In golf, the lowest score is the best. On the fifth hole, Jenny made an eagle and Paul made a double bogey. Write an integer to represent each player's score. Who won the hole?

45. The record for the coldest temperature in Alaska is $^{-}80$°F. In Ohio, the record is 41 degrees warmer. What is Ohio's coldest recorded temperature?

46. **Write About It** Kofi says that the sum of a positive integer and a negative integer will always be negative. Do you agree or disagree? Show an example.

47. **What's the Question?** There are 6 red counters and 4 yellow counters. The answer is $^{-}2$.

48. The deepest point in the deepest ocean is the Mariana Trench in the Pacific. It is 36,198 feet below sea level, or close to 29 times as deep as the Empire State Building is tall. About how tall is the Empire State Building?

49. To open a student savings account, Van deposited \$25 in May. In June, he withdrew \$9. In July, he deposited \$15. Write and solve an addition sentence that represents the money in his account after his deposit in July.

Mixed Review and Test Prep

50. The punch has 2 cups of orange juice for every 3 cups of cranberry juice. Write the ratio, and then write *part to whole, whole to part,* or *part to part* to describe the ratio. (p. 304)

51. **TEST PREP** Which percent is equivalent to $\frac{1}{8}$? (p. 320)

   A 0.125% **C** 12.5%
   B 1.25% **D** 125%

Write an equivalent fraction.

52. 0.4 (p. 280)

53. $\frac{3}{2}$ (p. 284)

54. Order the integers from least to greatest: $^{-}4$, $^{+}5$, $^{-}8$, $^{-}3$, $^{-}5$. (p. 420)

55. **TEST PREP** Which ratio is NOT equivalent to $\frac{2}{4}$? (p. 306)

   F 2 to 4 **H** $\frac{8}{20}$
   G 2:4 **J** 40:80

Thinker's Corner

Algebra To find $^{-}5{,}624 + {}^{-}2{,}538$ using counters it would take a long time. And you would need an enormous number line to find $^{-}50{,}329 + {}^{+}2{,}034$. Another way to add integers is to use their absolute values to find the sum.

To Add Integers with the Same Sign:

Find $^{-}948 + {}^{-}24$.

$|{}^{-}948| + |{}^{-}24| = 948 + 24 = 972$ ← Add the absolute values of the integers.

$^{-}948 + {}^{-}24 = {}^{-}972$ ← Then use the same sign as the integers.

So, $^{-}948 + {}^{-}24 = {}^{-}972$.

To Add Integers with Different Signs:

Find $^{+}157 + {}^{-}832$.

Think: $|{}^{+}157| < |{}^{-}832|$ ← Find the difference of their absolute values. Subtract the lesser absolute value from the greater absolute value.

$|{}^{-}832| - |{}^{+}157| = 832 - 157 = 675$

$^{+}157 + {}^{-}832 = {}^{-}675$ ← Then use the sign of the integer with the greater absolute value.

So, $^{+}157 + {}^{-}832 = {}^{-}675$.

Find the sum by using absolute values.

1. $^{-}42 + {}^{-}53$
2. $^{-}50{,}329 + {}^{+}2{,}304$
3. $^{-}5{,}624 + {}^{-}2{,}538$
4. $^{+}385 + {}^{-}41$

Extra Practice page H54, Set C

LESSON 4

HANDS ON
Subtract Integers

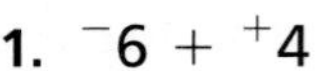

Quick Review

1. $^{-}6 + ^{+}4$
2. $^{-}8 + ^{-}2$
3. $^{+}10 + ^{-}8$
4. $^{+}1 + ^{-}3$
5. $^{-}5 + ^{-}4$

MATERIALS
two-color counters

Explore

You can use two-color counters to subtract integers. Use red counters to represent negative integers and yellow counters to represent positive integers.

Find $^{+}4 - ^{+}1$.

STEP 1

First, make a row of 4 yellow counters.

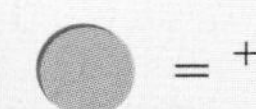

$= ^{+}4$

STEP 2

To subtract $^{+}1$, take away 1 yellow counter.

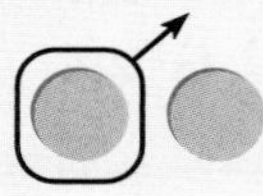
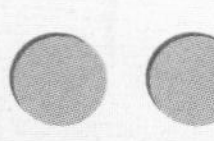

STEP 3

The number and color of counters left represents the difference.

$= ^{+}3$

So, $^{+}4 - ^{+}1 = ^{+}3$.

Find $^{-}4 - ^{+}1$.

STEP 1

First, make a row of 4 red counters.

$= ^{-}4$

STEP 2

Think: The sum of 1 red counter and 1 yellow counter is 0. In order to have 1 yellow counter to take away, add 1 yellow and 1 red counter. The model still shows $^{-}4$. Then, take away 1 yellow counter.

STEP 3

The number and color of counters left represents the difference.

$= ^{-}5$

So, $^{-}4 - ^{+}1 = ^{-}5$.

What do you do to subtract 3 yellow counters?

- What is another way to represent $^{-}4$ with red and yellow counters?

Try It

Use counters to find the difference.

a. $^{-}5 - ^{+}3$ **b.** $^{+}7 - ^{+}4$ **c.** $^{-}6 - ^{+}6$

CALIFORNIA STANDARDS **NS 2.1** Add, subtract, multiply, and divide with decimals; add with negative numbers; subtract positive integers from negative integers; and verify the reasonableness of the results. *also* **NS 1.0, MR 1.1, MR 2.3, MR 2.4, MR 3.0**

Connect

Addition and subtraction of integers are related.

Examples

A Find $^{-}6 - {}^{+}2$.

Show another way to model $^{-}6$ that includes 2 yellow counters. Then, take away 2 yellow counters.

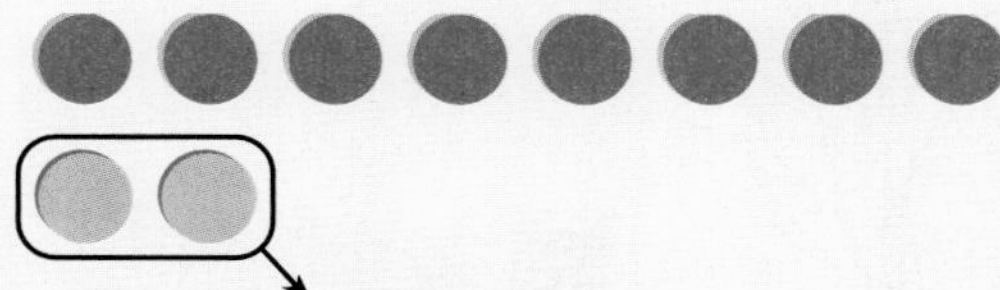

So, $^{-}6 - {}^{+}2 = {}^{-}8$.

B Find $^{-}6 + {}^{-}2$.

So, $^{-}6 + {}^{-}2 = {}^{-}8$.

Talk About It

- How does the answer to $^{-}6 - {}^{+}2$ relate to the answer to $^{-}6 + {}^{-}2$? How are $^{+}2$ and $^{-}2$ related? How are subtraction and addition related?
- Can you say $^{-}6 - {}^{+}2 = {}^{-}6 + {}^{-}2$? Explain.
- Write $^{-}7 - {}^{+}4$ as an addition problem. Then solve.

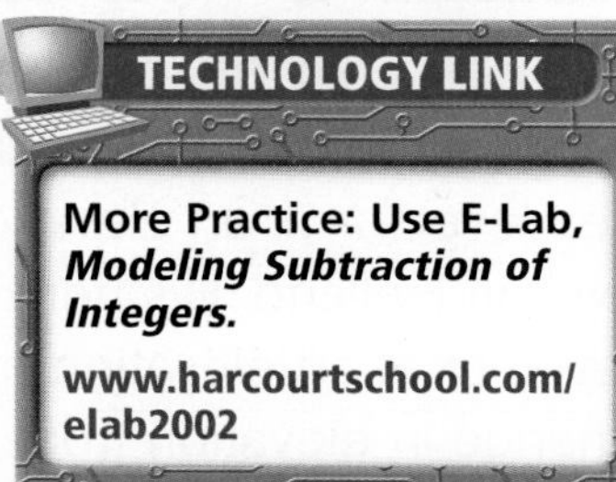

TECHNOLOGY LINK
More Practice: Use E-Lab, *Modeling Subtraction of Integers.*
www.harcourtschool.com/elab2002

Practice

Use counters to find each difference.

1. $^{+}6 - {}^{+}3$
2. $^{-}7 - {}^{+}5$
3. $^{-}8 - {}^{+}3$
4. $^{+}9 - {}^{+}4$

Algebra **Complete.**

5. $^{-}4 - {}^{+}5 = {}^{-}4 + \blacksquare$
6. $^{-}9 - {}^{+}1 = {}^{-}9 + \blacksquare$
7. $^{-}10 - {}^{+}2 = {}^{-}10 + \blacksquare$
8. One morning when Arnie left for school, the temperature outside was 4°C. By noon, the temperature had dropped to $^{-}2$°C. Find the change in temperature.
9. Write the subtraction number sentence modeled.

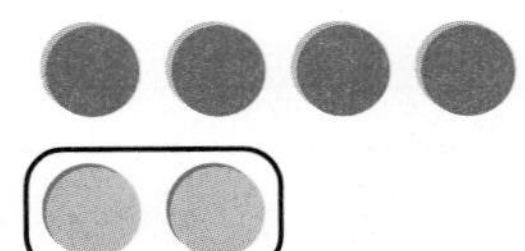

10. **What's the Error?** Correct the error in this model of $^{-}3 - {}^{+}3$.

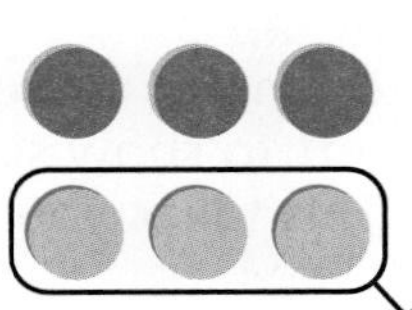

Mixed Review and Test Prep

11. $^{+}8 + {}^{-}3$ (p. 422)
12. $^{-}6 + {}^{+}4$ (p. 422)
13. Order the integers from least to greatest. $^{+}5, {}^{-}4, {}^{+}1, {}^{-}6$ (p. 420)
14. $1,569,258 + 304,852$ (p. 38)
15. **TEST PREP** Find $3\frac{4}{5} + 2\frac{7}{10}$. (p. 364)

 A $1\frac{1}{10}$ **B** $6\frac{1}{10}$ **C** $6\frac{1}{2}$ **D** $7\frac{1}{5}$

LESSON

5 Subtract Integers

Quick Review

Name the integer that is 1 less.

1. $^-6$ **2.** $^-8$ **3.** $^+10$
4. $^+1$ **5.** 0

Learn

You know that $^+6 - {}^+2 = {}^+4$. This subtraction is shown on the number line below. Notice that you move to the left to subtract a positive integer. You can use the same method to subtract a positive integer from a negative integer.

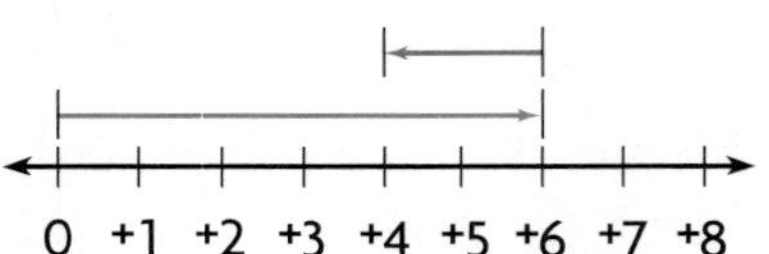

ABOVE AND BELOW When Koyi lived in Washington, D.C., his house was 1 foot above sea level, or at an elevation of $^+1$ foot. Now he lives in Holland. His house is 8 feet below sea level, or at an elevation of $^-8$ feet. Find the change in elevation from the first house to the second.

▲ **Today 2,400 kilometers of dikes protect the low, flat land of Holland—almost half of which lies below sea level—from being flooded by the North Sea.**

To find the change in elevation, subtract the starting elevation from the ending elevation.

Find $^-8 - {}^+1$.

STEP 1

Draw and label a number line. Start at 0 and move 8 spaces to the left to show $^-8$.

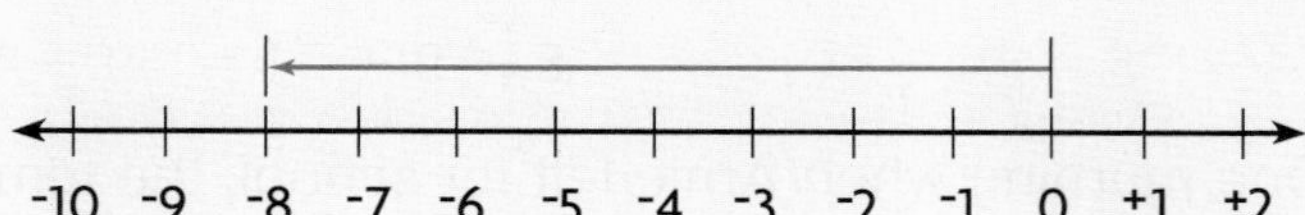

STEP 2

From $^-8$, move 1 space to the left to show subtracting $^+1$. This takes you to $^-9$.

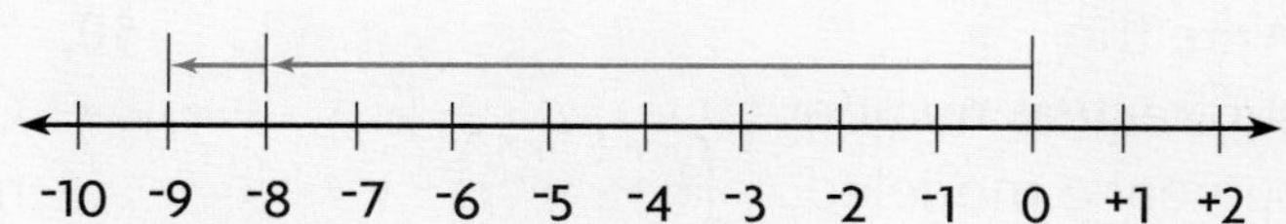

So, the change in elevation is $^-9$ feet.

- In which direction did you move from 0 to show $^-8$? In which direction did you then move to subtract $^+1$?

CALIFORNIA STANDARDS **O━NS 2.1** Add, subtract, multiply, and divide with decimals; add with negative numbers; subtract positive integers from negative integers; and verify the reasonableness of the results. **O━NS 1.5** Identify and represent on a number line, decimals, fractions, mixed numbers, and positive and negative integers. *also* **NS 1.0, MR 1.1, MR 2.3, MR 3.0, MR 3.2, MR 3.3**

Relating Addition and Subtraction

You can use a number line to show how the addition and subtraction of integers are related.

Compare the subtraction problem $^{-}2 - {}^{+}5$ to the addition problem $^{-}2 + {}^{-}5$.

Examples

A Find $^{-}2 - {}^{+}5$.

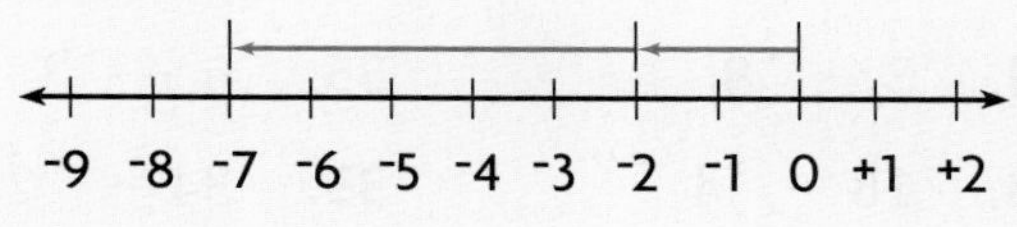

So, $^{-}2 - {}^{+}5 = {}^{-}7$.

B Find $^{-}2 + {}^{-}5$.

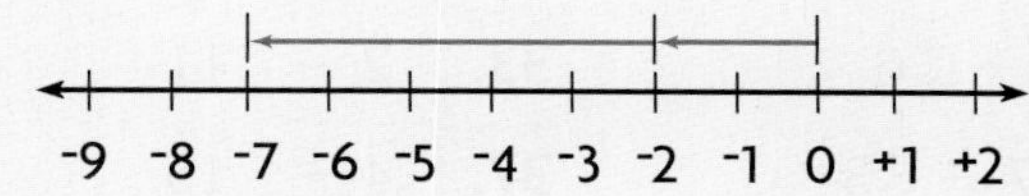

So, $^{-}2 + {}^{-}5 = {}^{-}7$.

In both examples you started at the same number on the number line, moved in the same direction the same distance, and ended up at the same number. So, $^{-}2 - {}^{+}5 = {}^{-}2 + {}^{-}5$.

Check

1. **Show** how to find $^{-}10 - {}^{+}9$ on a number line.

Draw a number line to find the difference.

2. $^{-}7 - {}^{+}5$
3. $^{+}4 - {}^{+}1$
4. $^{-}5 - {}^{+}2$
5. $^{+}8 - {}^{+}3$

Write the subtraction number sentence modeled.

6.

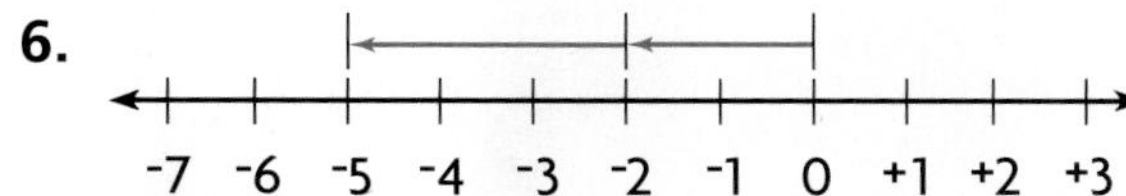

7.

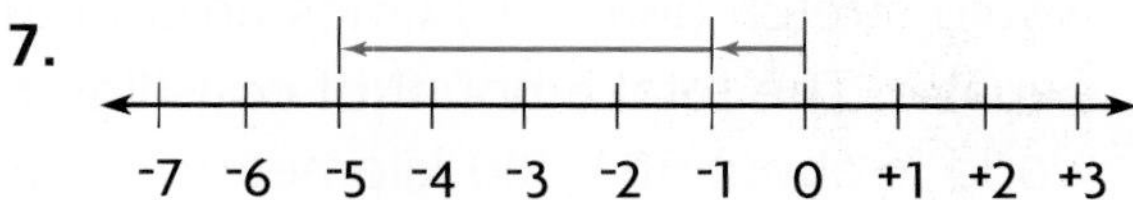

8.

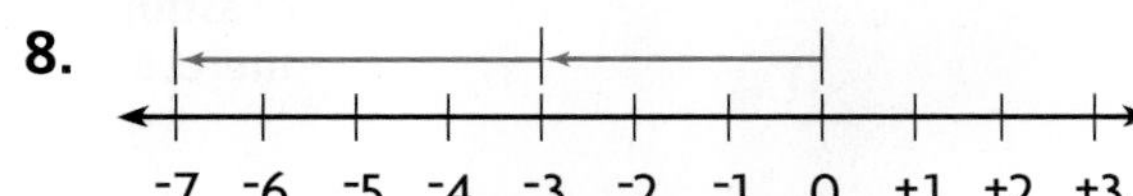

9.

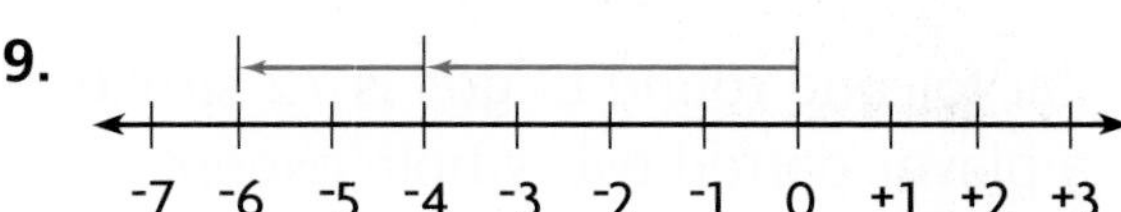

Find each difference.

10. $^{-}1 - {}^{+}3$
11. $^{-}4 - {}^{+}7$
12. $^{-}8 - {}^{+}9$
13. $^{+}7 - {}^{+}7$

Practice and Problem Solving

Draw a number line to find the difference.

14. $^{-}6 - {}^{+}5$
15. $^{-}4 - {}^{+}1$
16. $^{-}1 - {}^{+}2$
17. $^{+}5 - {}^{+}1$

LESSON CONTINUES

Write the subtraction number sentence modeled.

18.

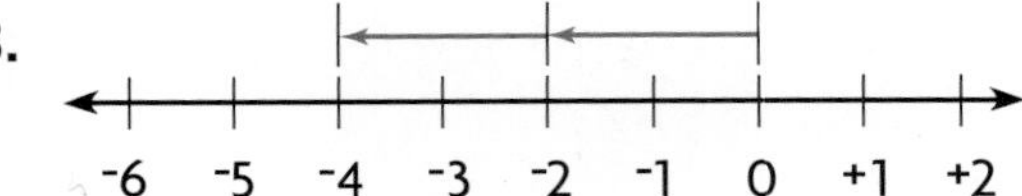

19.

-6 -5 -4 -3 -2 -1 0 +1 +2

Find each difference.

20. $^{+}5 - {}^{+}2$ **21.** $^{-}6 - {}^{+}5$ **22.** $^{-}3 - {}^{+}3$ **23.** $^{-}1 - {}^{+}8$

24. $^{-}9 - 0$ **25.** $^{-}4 - {}^{+}9$ **26.** $^{-}7 - {}^{+}2$ **27.** $0 - {}^{+}5$

28. $^{+}18 - {}^{+}7$ **29.** $^{-}4 - {}^{+}6$ **30.** $^{-}5 - {}^{+}2$ **31.** $^{-}10 - {}^{+}2$

32. $^{-}6 - {}^{+}6$ **33.** $^{-}2 - {}^{+}11$ **34.** $^{-}3 - {}^{+}8$ **35.** $^{-}6 - {}^{+}3$

36. $^{-}27 - {}^{+}3$ **37.** $^{-}4 - {}^{+}15$ **38.** $^{-}18 - {}^{+}4$ **39.** $^{-}34 - {}^{+}23$

Algebra **Complete the addition sentence.**

40. $^{-}8 - {}^{+}9 = {}^{-}8 + ■$ **41.** $^{-}2 - {}^{+}8 = {}^{-}2 + ■$

42. $^{-}1 - {}^{+}18 = {}^{-}1 + ■$ **43.** $^{-}30 - {}^{+}40 = {}^{-}30 + ■$

44. $^{-}6 - {}^{+}17 = {}^{-}6 + ■$ **45.** $^{-}25 - {}^{+}25 = {}^{-}25 + ■$

Compare. Write $<$, $>$, or $=$ for each ●.

46. $^{-}4 - {}^{+}5$ ● $^{-}4 + {}^{+}5$ **47.** $^{-}6 + {}^{+}4$ ● $^{-}4 - {}^{+}6$

48. $^{-}5 + {}^{-}3$ ● $^{-}5 - {}^{+}3$ **49.** $^{-}3 - {}^{+}9$ ● $^{-}9 + {}^{-}3$

50. $^{-}7 + {}^{+}8$ ● $^{-}15 - 0$ **51.** $^{-}24 - {}^{+}6$ ● $^{-}25 + {}^{-}5$

52. If all the coastlines of the world, excluding small bays and inlets, were straightened out, they would stretch nearly 13 times around the equator. The total amount of coastline in the world is about 504,000 kilometers. What is the distance around the equator?

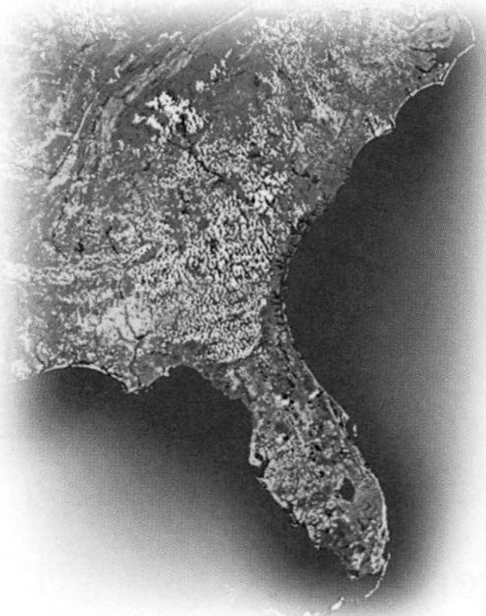

◄ Southeastern coastline of the U.S.

53. Par for one round of golf is 72 strokes. Suppose a player parred every hole except one. On the last hole, she made a birdie, or 1 stroke under par. What was her total score for the round?

54. The hottest recorded temperature in Georgia was 112°F, and the coldest recorded temperature was $^{-}17$°F. Find the change in temperature.

55. **REASONING** The sum of two integers is $^{-}4$. The difference of the two integers is $^{-}12$. What are the integers? (**HINT:** Start at $^{-}8$ on the number line.)

56. **NUMBER SENSE** Devin says the difference between two opposite integers is 0. Do you agree or disagree? Show an example.

57. **Write About It** Explain how to use a number line to model $^{-}6 - {}^{+}4$.

Mixed Review and Test Prep

For 58–60, use the data set 88, 64, 72, 98, 88, 58.

58. What are the median and mode? (p. 104)
59. What is the range? (p. 98)
60. What is the mean? (p. 102)
61. The Emperor penguin can dive 1,000 feet underwater. Name the integer that describes this situation. (p. 416)
62. TEST PREP In which place would you write the first digit of the quotient $5.426 \div 6$? (p. 222)
    - **A** ones
    - **B** tenths
    - **C** hundredths
    - **D** thousandths
63. Find the greatest common factor of 12 and 50. (p. 258)
64. What is the least common denominator of $\frac{3}{4}$ and $\frac{5}{10}$? (p. 352)
65. The submarine rose 100 feet from its position 570 feet below sea level. What is its new position? (p. 422)
66. What is $\frac{3}{4}$ and $\frac{1}{2}$? (p. 354)
67. TEST PREP The temperature was $^{-}20$°F. Then it rose by 10°F. Which number sentence models this situation? (p. 422)
    - **F** $^{+}20 + {}^{+}10 = {}^{+}30$
    - **G** $^{-}20 + {}^{+}10 = {}^{-}10$
    - **H** $^{+}20 - {}^{+}10 = {}^{+}10$
    - **J** $^{-}20 - {}^{+}10 = {}^{-}30$

LINKUP to Science

The inner planets are Mercury, Venus, Earth, and Mars. The outer planets are Jupiter, Saturn, Uranus, Neptune, and Pluto. Scientists measure the surface temperatures of Mercury, Venus, Earth, Mars, and Pluto. Since Jupiter, Saturn, Uranus, and Neptune are gas and have no surface, scientists measure the temperature of the highest layer of clouds.

AVERAGE SURFACE TEMPERATURE

Planet	Temperature
Mercury	333°F
Venus	865°F
Earth	45°F
Mars	$^{-}80$°F
Pluto	$^{-}382$°F

TEMPERATURE AT CLOUD TOPS

Planet	Temperature
Jupiter	$^{-}160$°F
Saturn	$^{-}220$°F
Uranus	$^{-}323$°F
Neptune	$^{-}330$°F

USE DATA **Use the tables.**

1. Which are the coldest and hottest inner planets?
2. Which are the coldest and hottest outer planets?
3. Order the planets from coldest to hottest.
4. Mercury is one of the hottest and coldest planets. On its light side, the surface reaches 800°F. On its dark side, temperatures can drop to $^{-}279$°F. Find the change in temperatures.

Extra Practice page H54, Set D

LESSON 6

Problem Solving Strategy
Draw a Diagram

Quick Review

1. $^{-}2 + {}^{+}5$
2. $^{-}8 - {}^{+}3$
3. $^{+}10 + {}^{-}2$
4. $^{-}4 - {}^{+}5$
5. $^{-}6 + {}^{+}1$

PROBLEM Carson parked in the deepest underground parking garage in London, the Aldersgate Car Park. It has 14 floors for 670 cars. After a long day of sightseeing, he returned to the parking garage to find his car. He forgot where he had parked. So, he entered the elevator on the ground floor and took the elevator down 8 floors and looked. Then he went up 2 floors, down 5 floors, and up another 4 floors before he found his car. His car was parked on which floor?

- What are you asked to find?
- What information will you use?
- Is there any information you will not use?

- What strategy can you use to solve the problem?

You can *draw a diagram* of the underground floors.

- How can you use this strategy to solve the problem?

Draw a vertical number line to represent the parking garage's levels. Start at the ground floor and model Carson's movements.

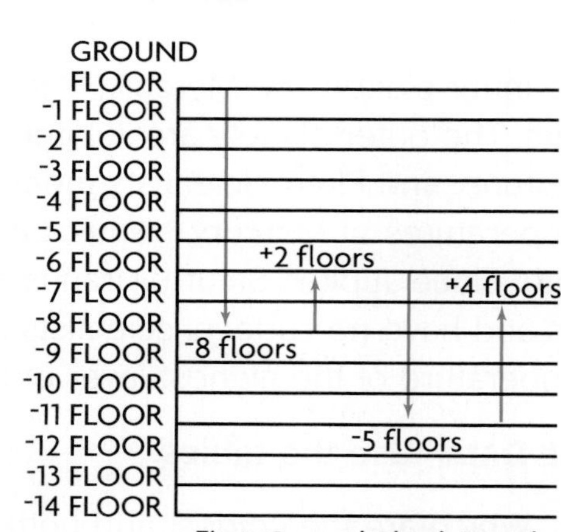

So, Carson found his car parked 7 floors underground, or on the $^{-}7$ floor.

- How can you decide if your answer is correct?

CALIFORNIA STANDARDS **MR 1.0** Students make decisions about how to approach problems. **MR 2.0** Students use strategies, skills, and concepts in finding solutions. **MR 2.3** Use a variety of methods, such as words, numbers, symbols, charts, graphs, tables, diagrams, and models, to explain mathematical reasoning. *also* **NS 1.0, O-n NS 1.5, O-n NS 2.1, MR 3.0, MR 3.2, MR 3.3**

▶ Problem Solving Practice

Draw a diagram to solve.

1. **What if** Carson entered the elevator on the ground floor and took the elevator down 9 floors and looked? Then he went up 3 floors, down 7 floors, and up another 4 floors before he found his car. His car was parked on which floor?

PROBLEM SOLVING STRATEGIES

- ▶ Draw a Diagram or Picture
- Make a Model or Act It Out
- Make an Organized List
- Find a Pattern
- Make a Table or Graph
- Predict and Test
- Work Backward
- Solve a Simpler Problem
- Write an Equation
- Use Logical Reasoning

2. At 7:00 A.M., the temperature was $^{-}3$°F. At 1:00 P.M., the temperature had risen by 11 degrees. By 11:00 P.M., it had dropped 15 degrees. What was the temperature at 11:00 P.M.?

3. A scuba diver descended to 14 feet below sea level. She then ascended 4 feet and swam 10 feet ahead. She then descended 6 feet. At what depth is she now?

4. Starting on their own 40-yard line, the Tigers gained 5 yards on their first play. Then they lost 15 yards in a penalty. What yard line are they on now?

Mixed Strategy Practice

USE DATA For 5–8, use the table. The table shows the same time in each time zone from the West coast to the East coast.

TIME ZONES IN THE UNITED STATES					
Hawaii-Aleutian	**Alaska**	**Pacific**	**Mountain**	**Central**	**Eastern**
2 P.M.	3 P.M.	4 P.M.	5 P.M.	6 P.M.	7 P.M.

5. What integer represents the change from Eastern time to Pacific time?

6. Describe the time pattern as you travel west across the time zones.

7. Suppose a pilot left the Mountain time zone at 6 P.M., flew west for 2 hours, and landed in the Pacific time zone. What time did he land?

   A 6 P.M. **C** 8 P.M.
   B 7 P.M. **D** 9 P.M.

8. Suppose a pilot left the Pacific time zone at 3 P.M., flew east for 3 hours, and landed in the Central time zone. What time did she land?

   F 6 P.M. **H** 8 P.M.
   G 7 P.M. **J** 9 P.M.

9. On May 6, 1992, a U.S. diver went underwater in a module and stayed there for a record 69 days. On what date did he come up?

10. **Write a problem** using integers that you can solve by drawing a diagram. Draw your diagram.

CHAPTER 23

Review/Test

CHECK VOCABULARY AND CONCEPTS

Choose the best term from the box.

absolute value
integers
negative integer
opposites
positive integer

1. The distance that an integer is from 0 is the __?__ of the integer. (p. 417)
2. An integer less than 0 is a __?__. (p. 416)
3. Integers that are the same distance from 0 on a number line, but in opposite directions, are called __?__. (p. 417)
4. Whole numbers and their opposites are called __?__. (p. 417)

CHECK SKILLS

Name an integer to represent each situation. (pp. 416–419)

5. a profit of $50
6. 20 degrees below 0
7. a loss of 10 yards

Write the opposite of each integer. (pp. 416–419)

8. $^{+}14$
9. $^{-}6$
10. $^{-}5$
11. $^{+}22$
12. $^{-}7$

Order each set of integers from least to greatest. (pp. 420–421)

13. $^{+}2$, $^{-}3$, $^{+}8$, $^{-}8$
14. $^{+}1$, $^{-}4$, 0, $^{-}5$
15. $^{-}7$, $^{+}3$, $^{-}6$, $^{+}2$
16. $^{+}5$, $^{-}5$, $^{+}3$, $^{+}2$

Find each sum or difference. (pp. 422–431)

17. $^{+}3 + {}^{-}2$
18. $^{-}11 - {}^{+}1$
19. $^{-}8 + {}^{-}3$
20. $^{-}4 - {}^{+}11$
21. $^{-}7 + {}^{+}2$
22. $^{+}8 + {}^{-}4$
23. $^{-}9 - {}^{+}4$
24. $^{-}7 + {}^{+}7$
25. $^{-}10 - {}^{+}1$
26. $^{+}9 + {}^{-}10$
27. $^{+}6 - {}^{+}3$
28. $^{-}7 + {}^{+}3$
29. $^{-}6 - {}^{+}2$
30. $^{+}5 + {}^{-}4$
31. $^{-}1 - {}^{+}3$

CHECK PROBLEM SOLVING

Draw a diagram to solve. (pp. 432–433)

32. Tara works in the mail room on the second floor. For her first two deliveries she went up 4 floors and then down 3 floors. Then she went down 4 floors for her last delivery. On what floor was her last delivery?

33. The temperature at 6 A.M. was $^{-}2$°F. By 10 A.M., it had risen 5°. By 10 P.M., the temperature had dropped 3°. What was the temperature at 10 P.M.?

Cumulative Review

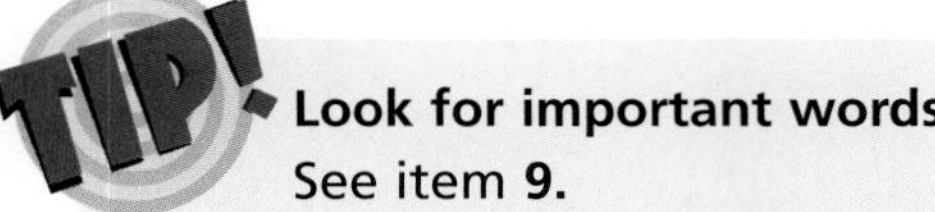

Look for important words.
See item **9**.

Important words are *lost yards* and *gained yards.* Write a number sentence to show where the ball is after two plays.

Also see problem **2**, p. H62.

For 1–10, choose the best answer.

1. $^{-}16 + {}^{+}14$

A $^{+}30$ **C** $^{-}2$
B $^{+}2$ **D** $^{-}30$

For 2–4, use the table.

HOTTEST AND COLDEST DAYS IN THE U.S.	
Place/Date	**Temperature °F**
California/July 10, 1913	$^{+}134°$
Arizona/June 29, 1994	$^{+}128°$
Nevada/June 29, 1994	$^{+}125°$
Alaska/Jan. 23, 1971	$^{-}80°$
Montana/Jan. 20, 1934	$^{-}70°$
Utah/Feb. 1, 1985	$^{-}69°$

2. In which location was the temperature 198° less than the record high temperature in Arizona?

F Alaska **H** Montana
G Utah **J** Nevada

3. How many degrees are between the hottest and coldest temperatures in the chart?

A 54° **C** 203°
B 65° **D** 214°

4. The record low temperature for which state is closest to 0°F?

F Nevada **H** Montana
G Alaska **J** Utah

5. Which is the best estimate for this difference?

$$\begin{array}{r} 8{,}652{,}108 \\ -\ 3{,}651{,}900 \\ \hline \end{array}$$

A 4,000,000 **C** 6,000,000
B 5,000,000 **D** 7,000,000

6. $^{-}25 - {}^{+}7$

F $^{+}32$ **H** $^{-}18$
G $^{+}18$ **J** $^{-}32$

7. Which multiplication property does the equation illustrate?

$(3 \times 4) \times 6 = 3 \times (4 \times 6)$

A Commutative Property
B Associative Property
C Property of One
D Distributive Property

8. What is 3.2089 rounded to the nearest thousandth?

F 3.209 **H** 3.2
G 3.21 **J** Not here

9. Dr. Fox parked his car on the 5th floor of the garage. Then he took the elevator down 5 floors to the basement, went through the walkway to the hospital, and took the elevator up 3 floors to his office. On what floor is his office?

A 5th floor **C** 3rd floor
B 4th floor **D** 2nd floor

10. Find $|{}^{-}14|$.

F 14 **H** $^{-}14$
G 0 **J** Not here

Geometry and the Coordinate Plane

Many cities are planned before they are built. When this occurs, a grid is sometimes used to lay out the streets. The first city to be laid out in a grid was central Philadelphia, by William Penn in 1682. Every street was laid out at right angles with another except one, Dock Street. Use the map grid to find the landmark, Independence Hall. Where is it located?

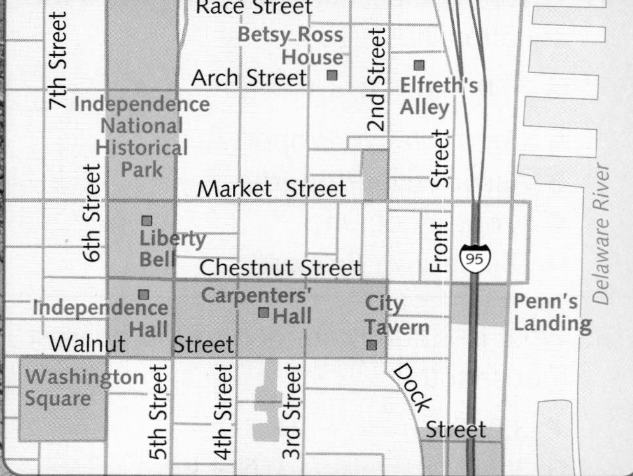

Independence Hall, known as the birthplace of the United States, is where the Declaration of Independence and the Constitution of the United States were adopted.

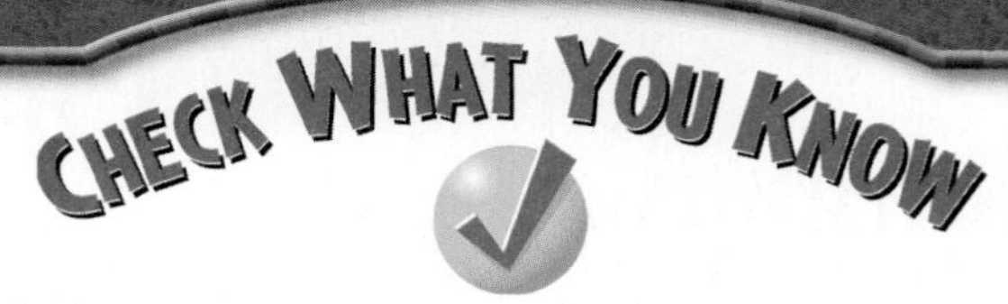

Use this page to help you review and remember important skills needed for Chapter 24.

VOCABULARY

Choose the best term from the box.

coordinate grid
ordered pair
horizontal
vertical
rule

1. A line that goes straight across is a __?__ line.
2. "Multiply x by 3" is an example of a __?__ for a function table.
3. A pair of numbers used to locate a point on a grid is called an __?__.
4. A line that goes straight up and down is a __?__ line.

IDENTIFY POINTS ON A GRID (See p. H10.)

Use the ordered pairs to name each point on the grid.

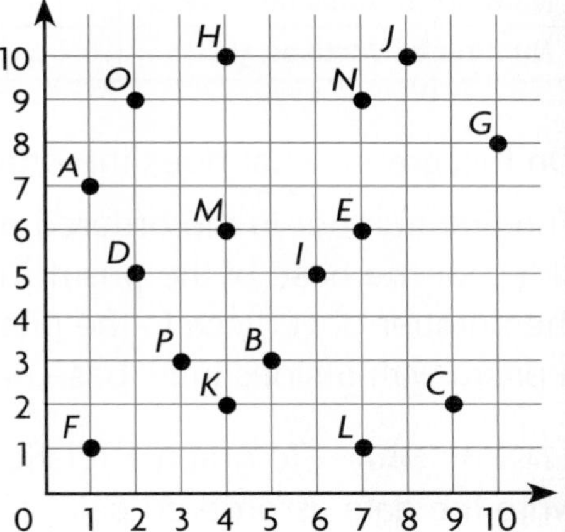

5. (5,3)	**6.** (1,1)	**7.** (7,6)
8. (1,7)	**9.** (2,9)	**10.** (7,1)
11. (6,5)	**12.** (7,9)	**13.** (4,2)
14. (4,10)	**15.** (4,6)	**16.** (3,3)
17. (10,8)	**18.** (8,10)	**19.** (9,2)

FUNCTION TABLES (See p. H26.)

Write a rule for each function table.

20.

x	y
1	4
2	5
3	6
4	7

21.

x	y
20	15
19	14
18	13
17	12

22.

x	y
4	8
5	10
6	12
7	14

23.

x	y
1	9
2	8
3	7
4	6

24.

x	y
5	9
8	12
4	8
3	7

25.

x	y
12	4
9	3
6	2
3	1

LESSON 1

Graph Relationships

Quick Review

Write each ordered pair so that the second number is 2 more than the first number.

1. (2,▒) **2.** (3,▒)
3. (4,▒) **4.** (5,▒)
5. (6,▒)

VOCABULARY

x-axis *y*-axis

Learn

PRISM PATTERNS Karen is making models of prisms. She uses gumdrops for the vertices and straws for the edges. She made a table and a graph to show the relationship between the number of sides on the prism's base and the number of vertices on the prism. On the graph, the horizontal line is the **x-axis** and the vertical line is the **y-axis**. Each point on the graph can be represented by an ordered pair, (x,y).

Number of Sides on Base, *x*	3	4	5	6
Number of Vertices, *y*	6	8	10	12

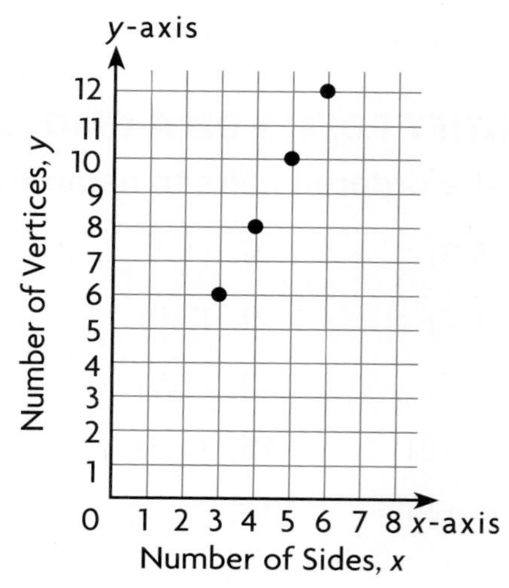

On this graph, what does the ordered pair (5,10) mean?

The first number in the ordered pair tells the number of sides for the base of the prism. The second number tells the number of vertices in the prism. So, (5,10) means that a prism with 5 sides in its base has 10 vertices.

MATH IDEA To graph a relationship shown in a table, write the data as ordered pairs and then graph the ordered pairs.

Example

Graph the relationship shown in the table.

Number of Sides on Base, *x*	3	4	5	6
Number of Faces, *y*	5	6	7	8

- Write ordered pairs for the data.
  (3,5), (4,6), (5,7), (6,8)
- Graph the ordered pairs.

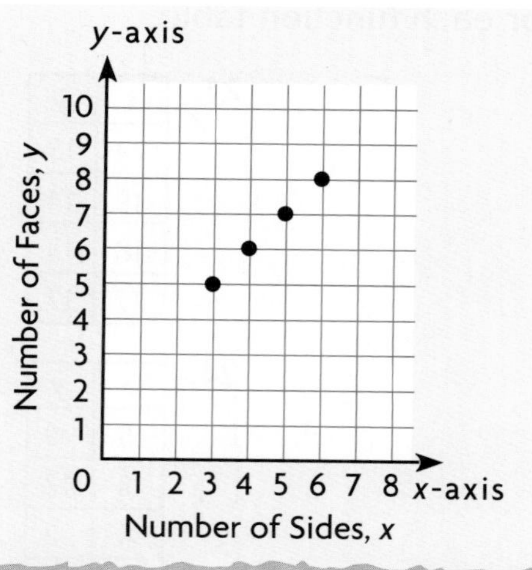

CALIFORNIA STANDARDS **SDAP 1.4** Identify ordered pairs of data from a graph and interpret the meaning of the data in terms of the situation depicted by the graph. **SDAP 1.5** Know how to write ordered pairs correctly; for example, (*x*,*y*). *also* **AF 1.1, AF 1.5, SDAP 1.0, MR 1.1, MR 2.0, MR 2.3, MR 2.4, MR 3.3**

Check

1. **Tell** what the point (5,7) means in the second graph on page 438.

Write the ordered pairs. Then graph them.

2.

Input, x	1	2	3	4
Output, y	3	6	9	12

3.

Input, x	4	5	6	7
Output, y	2	3	4	5

Practice and Problem Solving

Write the ordered pairs. Then graph them.

4.

Input, x	0	1	2	3
Output, y	3	4	5	6

5.

Input, x	7	8	9	10
Output, y	3	4	5	6

6.

Input, x	1	2	3	4
Output, y	4	8	12	16

7.

Input, x	1	2	3	4
Output, y	2	4	6	8

USE DATA **For 8–11, use the table.**

Number of Sides on Base, x	3	4	5	6
Number of Vertices on Pyramid, y	4	5	6	7

8. Write the ordered pairs. Then graph them.

9. **REASONING** What does (4,5) mean in the graph for the table?

10. **REASONING** How can you use your graph to find the number of vertices in an octagonal pyramid?

11. **Write About It** What kind of pyramid is represented by the ordered pair (5,6)? Explain.

Mixed Review and Test Prep

Find each sum. (p. 423)

12. $^{+}8 + ^{-}5$

13. $^{-}7 + ^{-}21$

14. Find the mean of the data set 28, 35, 67, 29, 31. (p. 102)

15. Write 3,869,472 in word form. (p. 4)

16. **TEST PREP** Dante won 80% of his 25 games. Kevin won $\frac{3}{4}$ of his 20 games. How many more games did Dante win than Kevin? (p. 324)

A 2
B 5
C 8
D 10

LESSON 2

Graph Integers on the Coordinate Plane

Quick Review

Write the integer that is 1 less than the given integer.

1. $^{-}5$ **2.** $^{-}1$

3. $^{+}3$ **4.** 0

5. $^{-}6$

VOCABULARY

coordinate plane **origin**

coordinates

Learn

MAPPING HISTORY Archaeologists record the locations of artifacts and features that they find in a dig site by graphing them on a coordinate plane.

A **coordinate plane** is formed by two intersecting and perpendicular number lines. The point where the two lines intersect is called the **origin**, or (0,0).

The numbers to the left of the origin on the *x*-axis and below the origin on the *y*-axis are negative.

Start at the origin. Move 4 units to the left on the *x*-axis and 2 units up on the *y*-axis. The **coordinates**, or numbers in the ordered pair, are $(^{-}4, ^{+}2)$.

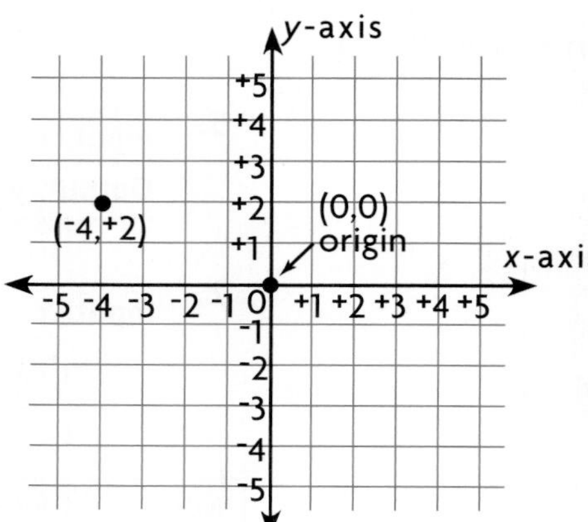

▲ Christopher Wolfe, at the age of 7, co-discovered the oldest horned ceratopsian dinosaur.

Activity

MATERIALS: coordinate plane

- Graph the ordered pair $(^{+}4, ^{-}5)$. Start at the origin. Move **right** 4 units and then **down** 5 units. Plot and label the point, *A*.
- Graph the ordered pair $(^{-}3, ^{-}2)$. Start at the origin. Move **left** 3 units and then **down** 2 units. Plot and label the point, *B*.
- In which direction and how far would you move to graph $(^{-}4, ^{+}5)$?

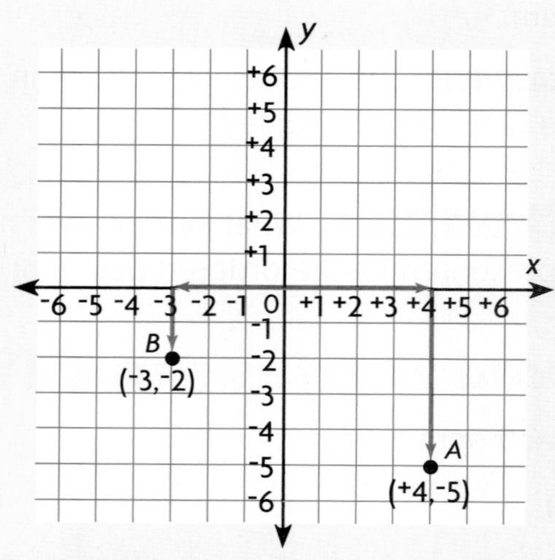

MATH IDEA The coordinates of a point tell you how far and in which direction to move first horizontally and then vertically on the coordinate plane.

CALIFORNIA STANDARDS **O━SDAP 1.4** Identify ordered pairs of data from a graph and interpret the meaning of the data in terms of the situation depicted by the graph. **O━SDAP 1.5** Know how to write ordered pairs correctly; for example, (*x*,*y*). **O━AF 1.4** Identify and graph ordered pairs in the four quadrants of the coordinate plane. *also* **O━AF 1.5, MR 1.0, MR 1.1, MR 2.0, MR 2.3, MR 3.2**

Graphing Ordered Pairs

Examples

A To graph ($^-2$,$^+3$), start at the origin. Move 2 units to the ***left*** and 3 units ***up.***

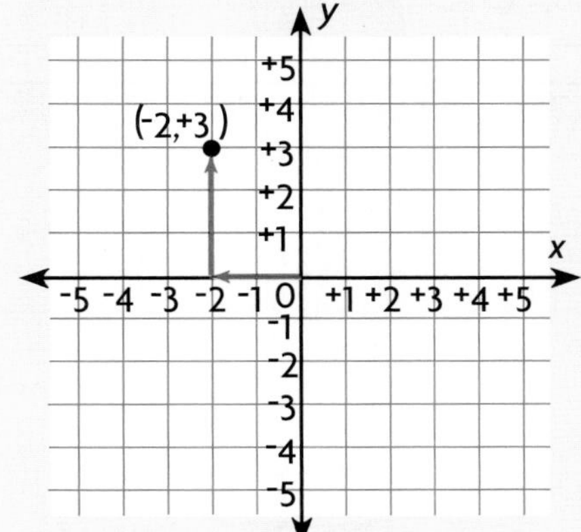

B To graph ($^+5$,$^-4$), start at the origin. Move 5 units to the ***right*** and 4 units ***down.***

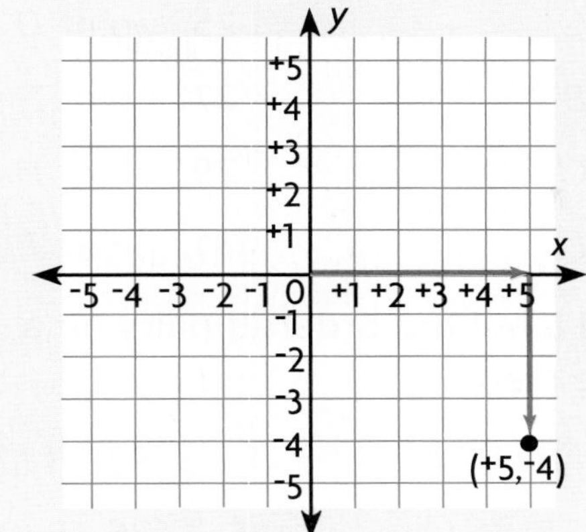

Check

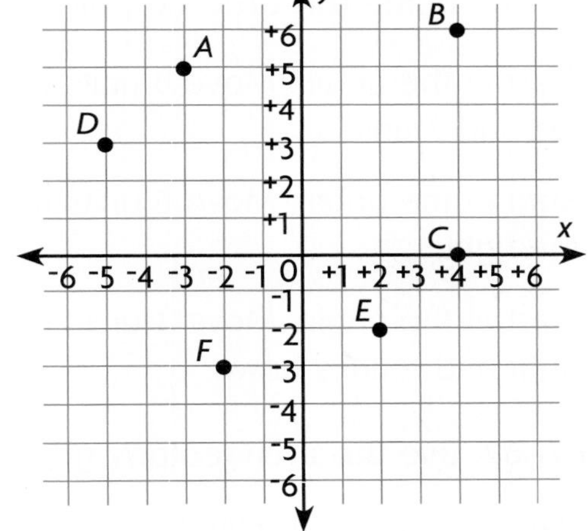

1. **Describe** how to graph ($^+4$,$^+2$) on a coordinate plane.

For 2–7, identify the ordered pair for each point.

2. point A
3. point B
4. point C
5. point D
6. point E
7. point F

Graph and label the ordered pairs on a coordinate plane.

8. A ($^-5$,$^+8$)
9. B ($^-7$,$^-4$)
10. C ($^+4$,$^+4$)
11. D ($^+1$,$^-2$)
12. E ($^-6$,$^+5$)
13. F ($^-2$,$^+8$)
14. G ($^+2$,$^+6$)
15. H ($^+3$,$^-3$)

Name the ordered pair that is described.

16. Start at the origin. Move 2 units to the right and 3 units down.
17. Start at the origin. Move 3 units to the left and 6 units up.
18. Start at the origin. Move 5 units to the left and 5 units down.
19. Start at the origin. Move 1 unit to the right and 3 units up.
20. Start at the origin. Move 3 units up.
21. Start at the origin. Move 4 units to the left.

LESSON CONTINUES 

Practice and Problem Solving

For 22–31, identify the ordered pair for each point.

22. point A
23. point B
24. point C
25. point D
26. point E
27. point F
28. point G
29. point H
30. point I
31. point J

Graph and label the ordered pairs on a coordinate plane.

32. $A\ (^-1, ^+3)$
33. $B\ (^-3, ^-7)$
34. $C\ (^+6, ^+7)$
35. $D\ (^+1, ^-5)$
36. $E\ (0, ^+8)$
37. $F\ (^+2, ^+3)$
38. $G\ (^+9, ^-2)$
39. $H\ (^-3, ^+3)$
40. $J\ (^-8, ^+1)$

For 41–46, name the ordered pair that is described.

41. Start at the origin. Move 8 units to the left and 7 units up.

42. Start at the origin. Move 3 units to the right and 9 units up.

43. Start at the origin. Move 6 units to the right.

44. Start at the origin. Move 5 units to the left and 6 units down.

45. Start at the origin. Move 4 units to the right and 6 units down.

46. Start at the origin. Move 5 units down.

For 47–52, use the archaeology grid.

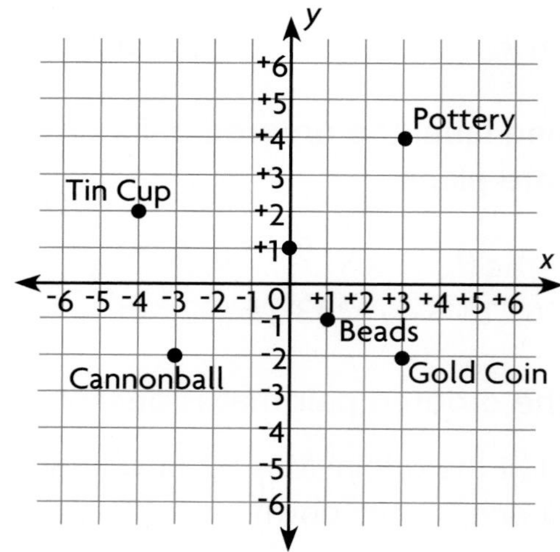

47. Which artifact is 3 units to the right of the origin and 4 units up from the origin?

48. Which two artifacts have negative x-coordinates?

49. Which two artifacts have opposite x-coordinates and the same y-coordinates?

50. What are the coordinates for the gold coin?

51. **What's the Error?** Tamara says the location of the beads is $(^-1, ^+1)$. Describe and correct her error.

52. **Write a problem** that uses the archaeology grid to solve.

53. **What's the Question?** Mark placed a coordinate grid over a map of his neighborhood. He labeled his house point *H*. The answer is $(^+4,^-3)$.

54. The temperature at 6 P.M. was 12°F. It fell 17° during the night. What was the low temperature?

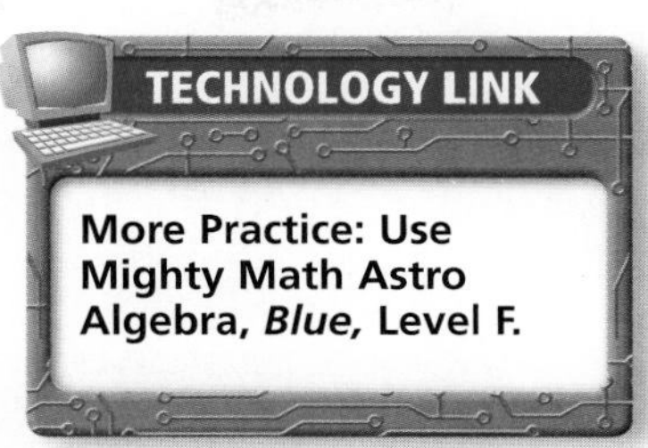

Mixed Review and Test Prep

For 55–57, write the integer that is 1 more than the given integer. (p. 416)

55. $^+5$ **56.** $^-3$ **57.** $^-1$

58. $\begin{array}{r} 210{,}875 \\ -\ \ 89{,}286 \\ \hline \end{array}$ (p. 38) **59.** $\begin{array}{r} 800{,}301 \\ -654{,}692 \\ \hline \end{array}$ (p. 38)

60. **TEST PREP** Which is the sum $^-7 + {}^+2$? (p. 422)

A $^-9$ **B** $^-5$ **C** $^+5$ **D** $^+9$

For 61–63, write the opposite of each integer. (p. 416)

61. $^-32$ **62.** $^-83$ **63.** $^+101$

64. **TEST PREP** Which is the greatest common factor of 16 and 24? (p. 258)

F 2 **H** 8
G 4 **J** 16

LINKUP to Geometry

Miguel is rearranging the furniture in his room. He drew a diagram of the furniture on a grid.

Use a coordinate grid to draw Miguel's furniture in the new locations given below. Tell how each piece was moved on the grid: by *rotating* (turning); by *reflecting* (flipping); or by *translating* (sliding).

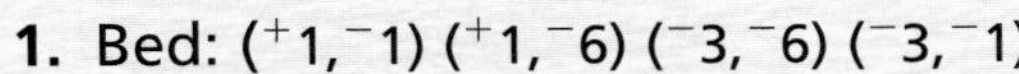

1. Bed: $(^+1,^-1)$ $(^+1,^-6)$ $(^-3,^-6)$ $(^-3,^-1)$
2. Dresser: $(^+2,^+4)$ $(^+2,^+6)$ $(^-3,^+6)$ $(^-3,^+4)$
3. Chair: $(^+4,^-2)$ $(^+6,^-4)$ $(^+4,^-6)$ $(^+2,^-4)$
4. Desk: $(^-5,^+2)$ $(^-5,^+5)$ $(^-6,^+5)$ $(^-6,^+2)$

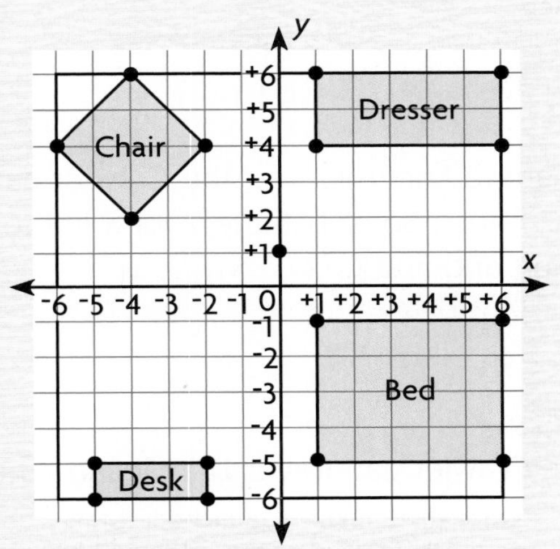

LESSON

3 Use an Equation to Graph

Quick Review

Graph each point on a coordinate grid.

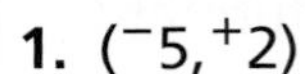

1. $(^{-}5, ^{+}2)$ **2.** $(^{+}1, ^{+}3)$

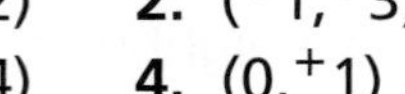

3. $(^{-}4, ^{-}4)$ **4.** $(0, ^{+}1)$

5. $(^{-}3, ^{-}1)$

Learn

SLOW START The transporter that takes the space shuttle to its launch pad is called the Crawler. It got its name from its slow speed of 2 miles per hour. How many miles can the Crawler go in 7 hours?

Speed is a relationship between time and the distance traveled in that amount of time. You can show this relationship with x and y values in a function table.

Time in hours, x	1	2	3	4	5
Distance in miles, y	2	4	6		

Remember
A positive number can be written without a positive sign.

$^{+}3 = 3$

The table shows that the miles the Crawler travels is twice the hours it takes.

Rule: Multiply x by 2.

You can use the rule to find each value for y in the table, to write the ordered pairs, and make a graph.

You can also use the rule to write an equation for this relationship.

miles traveled	equals	speed	times	hours traveled
↓	↓	↓	↓	↓
y	$=$	2	$\times$	x

You want to know how far the Crawler can go in 7 hours. So, substitute 7 for x in the equation and find y.

$y = 2x$
$y = 2 \times 7$
$y = 14$

So, the Crawler can travel 14 miles in 7 hours.

Once the Shuttle is launched, it takes about 8 minutes to reach a speed of more than 17,000 mph. ▶

CALIFORNIA STANDARDS **AF 1.5** Solve problems involving linear functions with integer values; write the equation; and graph the resulting ordered pairs of integers on a grid. **SDAP 1.5** Know how to write ordered pairs correctly; for example, (x,y). **AF 1.4** Identify and graph ordered pairs in the four quadrants of the coordinate plane. *also* **AF 1.0, AF 1.1, SDAP 1.0, SDAP 1.4, MR 1.0, MR 1.1, MR 2.0, MR 2.3, MR 2.4, MR 3.3**

More Graphing

Graph the relationship $y = 6x$.

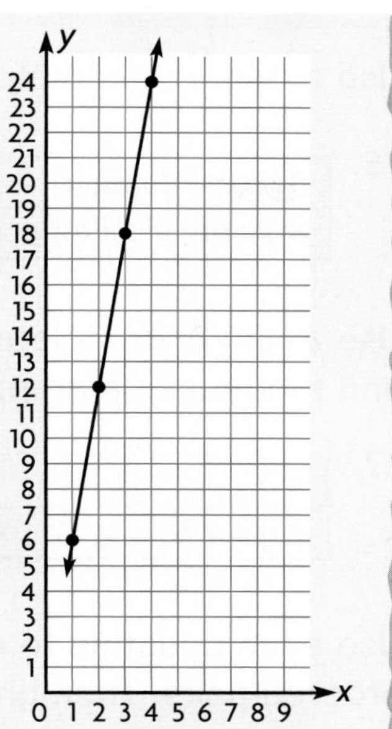

STEP 1

Make a table for the relationship. List at least four values for x. For each value of x, use the equation to find y.

Time in hours, x	1	2	3	4
Distance in miles, y	6	12	18	24

STEP 2

Write the data in the table as ordered pairs. (1,6), (2,12), (3,18), (4,24)

STEP 3

Graph the ordered pairs and draw a line connecting the points.

MATH IDEA You can use an equation to find ordered pairs and use the ordered pairs to graph the relationship.

Examples

A Graph the relationship $y = 2x + {}^{-}3$.

x	0	1	2	3
y	$^{-}3$	$^{-}1$	1	3

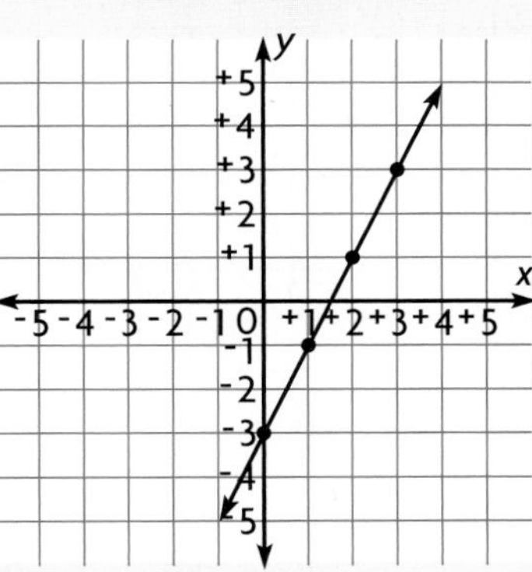

B Graph the relationship $y = x - 4$.

x	$^{-}2$	$^{-}1$	1	2
y	$^{-}6$	$^{-}5$	$^{-}3$	$^{-}2$

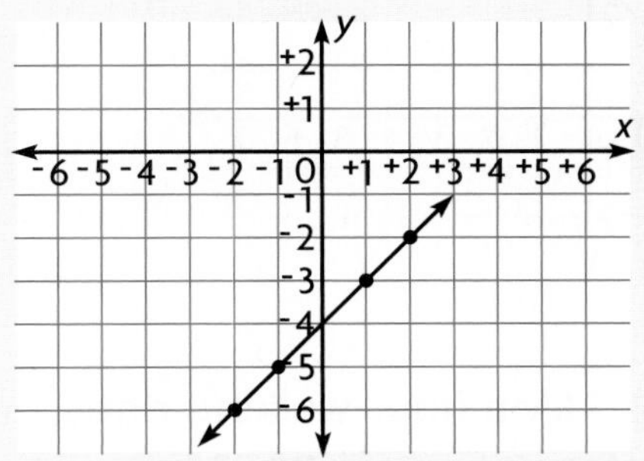

Check

Use a rule to complete the table. Then complete the equation.

1.

Hours, x	1	2	3	4
Miles, y	50	100	150	■

$y = ■ \times x$

2.

Feet, x	15	12	9	6
Yards, y	5	4	3	■

$y = x \div ■$

3.

Sales, x	\$10	\$9	\$8	\$7
Profits, y	\$4	\$3	\$2	■

$y = x - ■$

4.

Inches, x	24	36	48	60
Feet, y	2	3	4	■

$y = x \div ■$

LESSON CONTINUES

▶ Practice and Problem Solving

Use a rule to complete the table. Then write the equation.

5.

Gallons of gas, x	1	2	3	4	5
Distance in miles, y	25	50	75	100	■

6.

Hours, x	1	2	3	4
Earnings, y	\$5	\$10	\$15	■

Use a rule to complete the table. Write the ordered pairs and then make a graph.

7.

x	1	2	3	4	5
y	5	6	7	■	■

8.

x	$^-2$	$^-3$	$^-4$	$^-5$	$^-6$
y	$^-4$	$^-5$	$^-6$	■	■

9.

x	2	4	6	8	10
y	1	2	3	■	■

Use each equation to make a table with at least 4 ordered pairs. Then graph.

10. $y = x + 3$ **11.** $y = 2x + {}^-1$ **12.** $y = 4x$ **13.** $y = x \div 3$

14. $y = x$ **15.** $y = x + 5$ **16.** $y = {}^-2 + x$ **17.** $y = x + 0$

Algebra Write an equation for the relationship shown in each graph. HINT: First make a table to show the ordered pairs.

18.

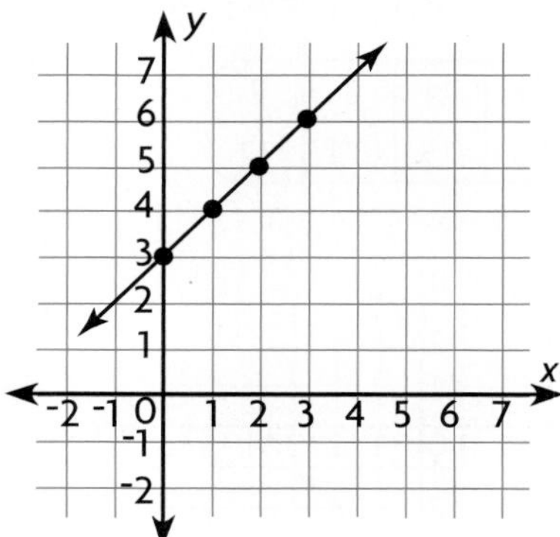

19.

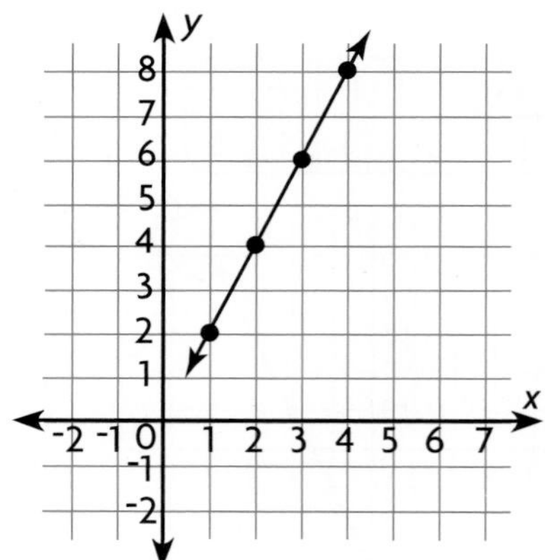

Sam earns \$6 an hour walking dogs. Use this information for 20–25.

20. Let x = number of hours worked and y = money earned. Write an equation for Sam's earnings.

21. Use the equation to make a table of ordered pairs, and graph the relationship.

22. Does the graph show an increase or decrease in the amount earned? Explain.

23. **REASONING** Explain why it does not make sense to use negative integers for x to find Sam's earnings.

24. How much money will Sam earn in 12 hours?

25. Explain what the ordered pair (3,18) means in the graph.

26. **REASONING** How can you use the equation $y = 3x$ to find the value of x when $y = 24$?

27. **Write About It** Is the ordered pair (1,2) on the graph of $y = x + 2$? Explain.

Mixed Review and Test Prep

For 28–31, write the integer that is one less than the given integer. (p. 416)

28. 5 **29.** $^{-}3$ **30.** $^{-}8$ **31.** 0

32. **TEST PREP** Find the product 0.32×8.61. (p. 162)

A 2.7522 **C** 27.52
B 2.7552 **D** 27.552

For 33–35, graph and label the ordered pairs on a coordinate plane. (p. 441)

33. $(^{+}3,^{-}6)$ **34.** $(^{-}5,^{+}1)$ **35.** $(^{-}2,^{-}4)$

36. **TEST PREP** Which is the least common multiple of 3 and 15? (p. 256)

F 3 **H** 15
G 5 **J** 30

LINKUP to Algebra

What two numbers have a sum of 7 and a difference of 1?

- Write the two equations. Let x = one number and y = the other number. $x + y = 7$, and $x - y = 1$
- Make a table for each equation's ordered pairs.

$x + y = 7$

x	1	2	3	4	5
y	6	5	4	3	2

$x - y = 1$

x	1	2	3	4	5
y	0	1	2	3	4

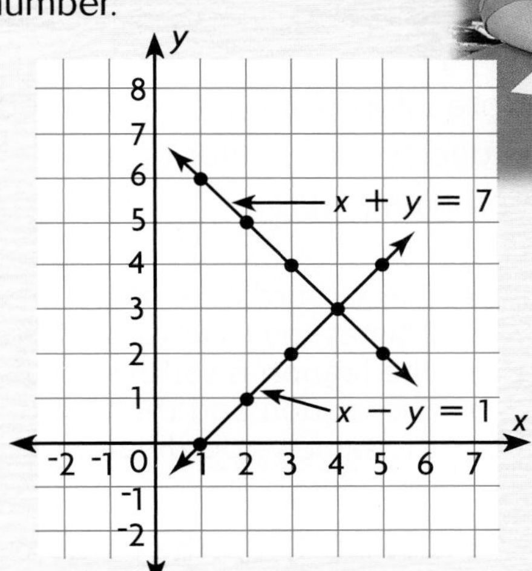

- Graph each set of ordered pairs on the same coordinate plane. Connect the points of each equation with a line.
- The point where the two lines intersect (4,3) is the solution to the problem. So, the two numbers are 4 and 3.

Write and graph two equations to solve each problem.

1. What two numbers have a sum of 10 and a difference of 4?
2. What two numbers have a sum of 4 and a difference of 2?

LESSON 4

Problem Solving Skill
Relevant or Irrelevant Information

> **Quick Review**
>
> **Name the *x*- or *y*-coordinate that is given by each direction from the origin.**
>
> **1.** 2 units up
> **2.** 1 unit left
> **3.** 4 units down
> **4.** 3 units right
> **5.** 2 units left

TREASURE HUNT Jeff found this old map with instructions for finding a buried treasure. Where is the treasure?

The x-coordinate of the treasure is 2 times the cave's x-coordinate. The lagoon is southwest of the cave. The lagoon and the treasure have the same y-coordinate.

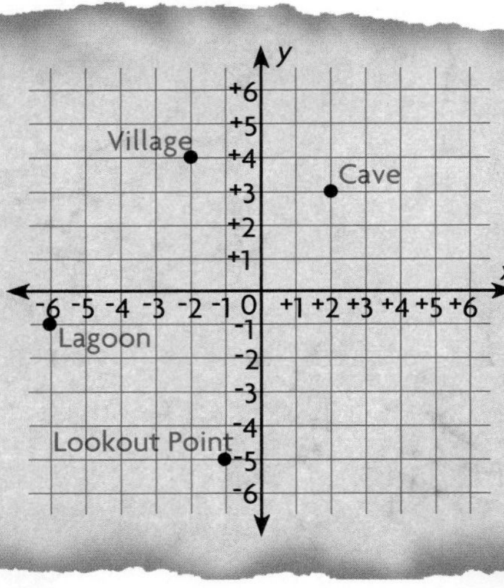

Sometimes a problem contains more information than you need to answer the question. You must decide which information is relevant, or needed, to solve the problem.

Read each fact and decide whether it is relevant or irrelevant to solving the problem.	• The *x*-coordinate of the treasure is 2 times the cave's *x*-coordinate.	relevant
	• The lagoon is southwest of the cave.	irrelevant
	• The lagoon and the treasure have the same *y*-coordinate.	relevant
Use the relevant information to solve the problem.	• Cave: $(^+2, ^+3)$. The treasure's *x*-coordinate is $2 \times 2 = 4$.	
	• Lagoon: $(^-6, ^-1)$. The treasure's *y*-coordinate is $^-1$.	

So, the coordinates of the treasure are $(^+4, ^-1)$.

Talk About It

- Which of the locations on the map are irrelevant to finding the treasure?

CALIFORNIA STANDARDS **SDAP 1.4** Identify ordered pairs of data from a graph and interpret the meaning of the data in terms of the situation depicted by the graph. **MR 1.1** Analyze problems by identifying relationships, distinguishing relevant from irrelevant information, sequencing and prioritizing information, and observing patterns. *also* **AF 1.1, AF 1.4, AF 1.5, SDAP 1.5, MR 1.0, MR 2.0, MR 2.1, MR 3.3**

Problem Solving Practice

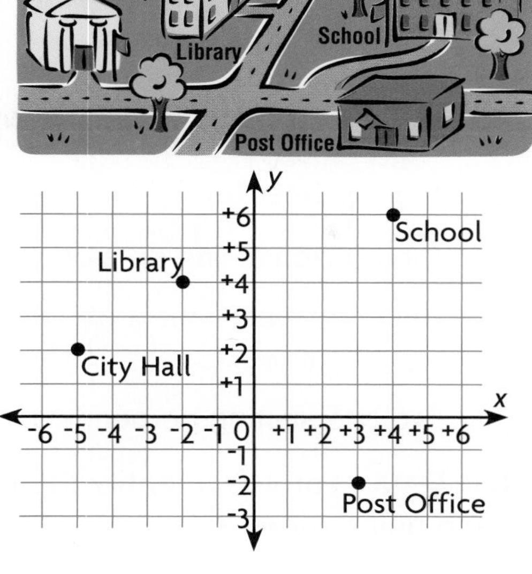

For 1–2, use the map. Tell the relevant information and solve.

1. The park and the library have the same *x*-coordinate. The library is 5 units left and 6 units up from the post office. The *y*-coordinate of the school is 3 more than the park's *y*-coordinate. Where is the park?

2. The Mall is south of the school and east of City Hall. If you go down 3 units and right 4 units from the library, you will find the Mall. Where is the Mall?

At 1 P.M., Amy left her house and rode her bike north 3 miles to Jeb's house. Next she rode east 2 miles and then turned south and rode 3 miles to Ann's house, where they had lunch. From there, she rode west back to her house. What shape did her trip make?

3. Which information is irrelevant to solving the problem?

   A Amy rode north to Jeb's house.
   B Amy and Ann had lunch at Ann's house.
   C Amy rode east and then south from Jeb's to Ann's house.
   D Amy rode west back home.

4. Which question cannot be answered with the given information?

   F In what direction did Amy ride to Jeb's house?
   G What shape did her trip make?
   H How long did it take Amy to get to Ann's house?
   J How far is it from Amy's house to Ann's house?

Mixed Applications

5. An original copy of Chaucer's *Canterbury Tales* sold in 1998 for $7,394,400. This was 9 times as much as the expected price. What was the expected price?

6. The most valuable violin in the world is the Kreutzer, created in Italy in 1727. It was sold at auction for $1,516,000 in England in 1998. How old was the violin when it was sold?

7. Kenya has two dogs. The sum of their ages is 7 years, and the difference in their ages is 3 years. How old are her two dogs?

8. **Patterns** Draw the next figure in this sequence.

9. **What's the Question?** Sue's triangle has vertices (1,2) and (1,5). The answer is ($^-2$,$^-2$).

10. **Write a problem** that includes relevant and irrelevant information. Exchange problems with a partner. Identify the relevant and irrelevant information and solve.

Review/Test

CHECK VOCABULARY AND CONCEPTS

Choose the best term from the box.

coordinate plane
x-axis
y-axis
origin

1. The horizontal line on a coordinate plane is the __?__. (p. 438)
2. The point where the *x*-axis and the *y*-axis intersect is the __?__. (p. 440)
3. Two intersecting perpendicular number lines form a __?__. (p. 440)

USE DATA For 4–9, identify the ordered pair for each point. (pp. 440–443)

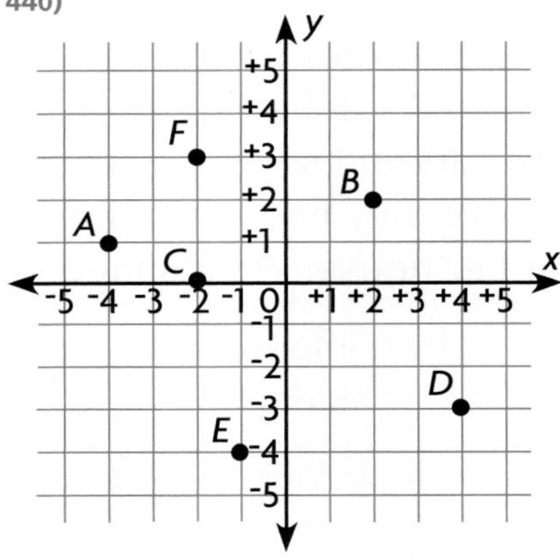

4. point A
5. point B
6. point E
7. point D
8. point C
9. point F

CHECK SKILLS

Graph and label the ordered pairs on a coordinate plane. (pp. 440–443)

10. $A\ (^-5, ^+8)$
11. $B\ (^-7, ^-4)$
12. $C\ (^+4, ^+4)$

Write the ordered pairs. Then graph them. (pp. 438–439)

13.

Input, *x*	2	4	6	8
Output, *y*	5	7	9	11

14.

Cost of Items, *x*	1	2	3	4
Amount, *y*	$6	$12	$18	$24

Use each equation to make a table. Then graph. (pp. 444–447)

15. $y = x + 1$
16. $y = 2x + {}^-3$
17. $y = 3x$
18. $y = x \div 2$

CHECK PROBLEM SOLVING

For 19–20, use the grid. Tell the relevant information and solve. (pp. 448–449)

19. Mei and Bobby are on the same *y*-coordinate. Bobby is south of Terrell. Karen's *x*-coordinate is 1 more than Bobby's. Where is Bobby?
20. Sam is between Mei and Terrell. If you go up 6 units and right 3 units from Anita, you will find Sam. Where is Sam?

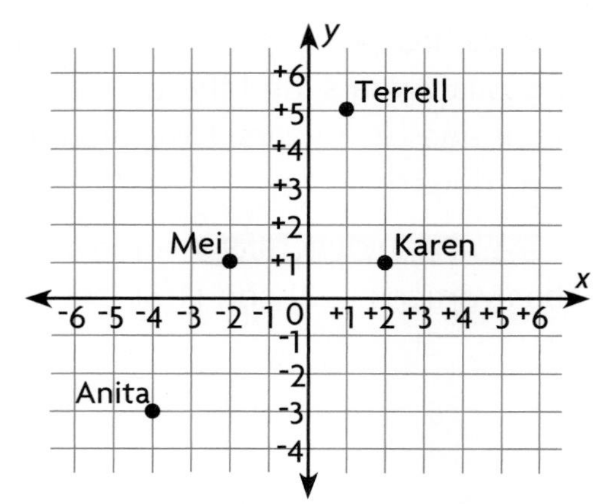

Cumulative Review

Understand the problem.
See item **3.**

Two points that are *opposite* are the same distance from the origin and reflected over an axis. The *y*-coordinate is the second number in an ordered pair.

Also see problem **1**, p. H62.

For 1–9, choose the best answer.

1. $^{-}38 + ^{-}23$

A 61 **C** $^{-}15$
B 15 **D** $^{-}61$

For 2–4, use the coordinate plane.

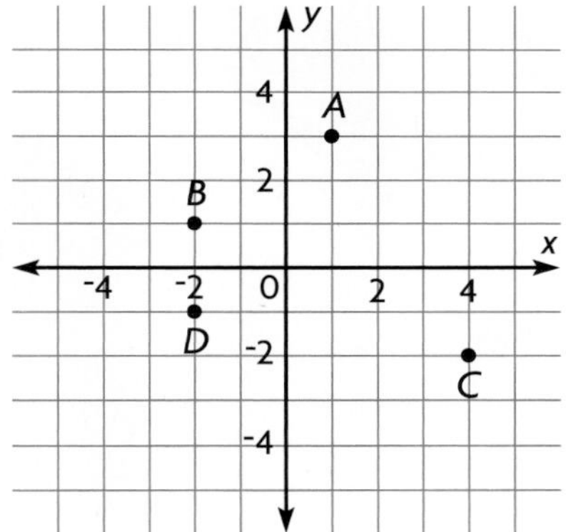

2. What are the coordinates of Point *C*?

F $(4,^{-}2)$ **H** $(^{-}4,2)$
G $(^{-}2,4)$ **J** $(1,3)$

3. Which two points have opposite *y*-coordinates?

A *A* and *B* **C** *B* and *D*
B *A* and *C* **D** *C* and *D*

4. Which point is located 2 units to the left of the origin and 1 unit up?

F Point *A* **H** Point *C*
G Point *B* **J** Point *D*

5. Michael used 2 rolls of film to take pictures on his vacation. Each roll made 24 pictures. He used the flash on 9 of the pictures. It cost \$0.12 to develop each picture. What information is not needed to find the total cost to develop Michael's pictures?

A He used the flash on 9 pictures.
B He used 2 rolls of film.
C Each roll made 24 pictures.
D It cost \$0.12 to develop each picture.

---

6. $45{,}600 \div 48$

F 95 **H** 9,500
G 950 **J** Not here

---

7. What equation matches this graph?

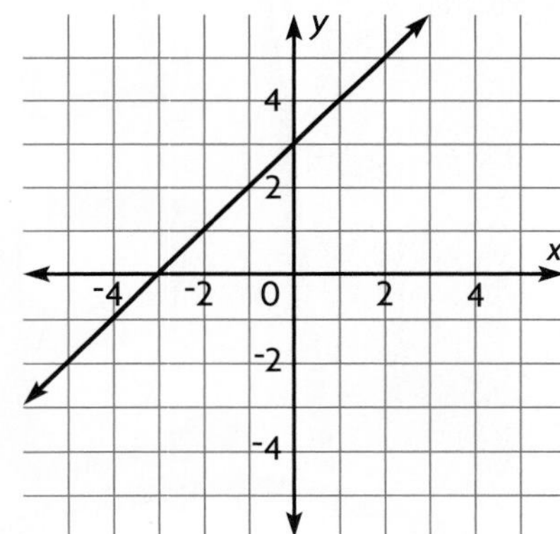

A $x + 3 = y$ **C** $x + 2 = y$
B $x + 1 = y$ **D** $x - 1 = y$

---

8. Which ratio is equivalent to $\frac{3}{8}$?

F 1:8 **H** 4:9
G 6:24 **J** 9:24

---

9. What is 5% of 600?

A 3,000 **C** 30
B 300 **D** Not here

Plane Figures

Traffic signs are placed along or above a walkway, roadway, or highway. They regulate the flow of traffic for motor vehicles, bicycles, and pedestrians. The signs give information through their shape, color, messages, and placement. Name the shape for each type of sign.

STANDARD SIGN SHAPES

Shape	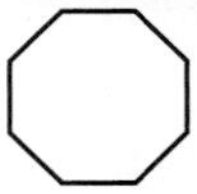	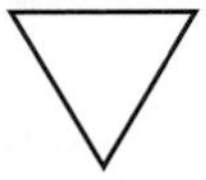	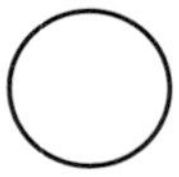	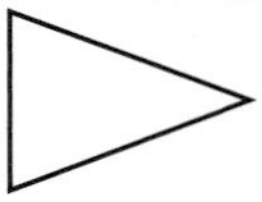		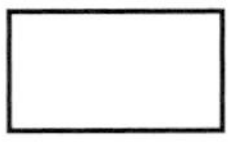	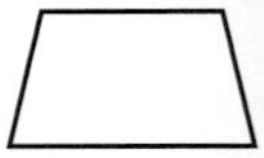
Message	Stop	Yield	Railroad Warning	No Passing Zone	Warning Signs	Guide Signs	Recreational Area Signs

CHECK WHAT YOU KNOW

Use this page to help you review and remember important skills needed for Chapter 25.

VOCABULARY

Choose the best term from the box.

quadrilateral
acute angle
right angle
plane figure
polygon

1. A figure that lies in one plane is a __?__.
2. A closed plane figure with straight sides that is named by the number of its sides and angles is a __?__.
3. A polygon with four angles and four sides is a __?__.
4. An angle that forms a square corner is a __?__.

CLASSIFY PLANE FIGURES (See p. H26.)

Is the figure a *closed* figure?

5. 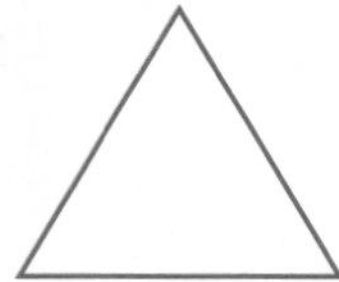
6. 
7. 
8. 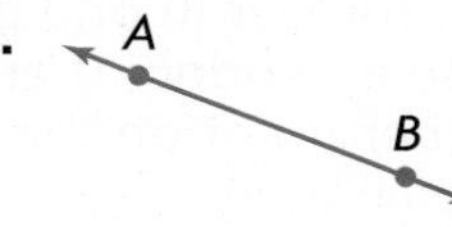

9. 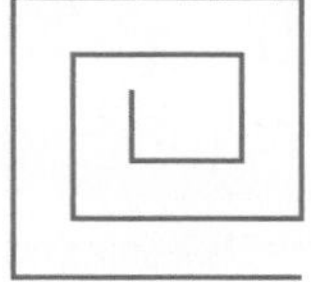
10. 
11. 
12. 

NAME POLYGONS (See p. H27.)

Name each polygon. Tell the number of sides and angles.

13. 
14. 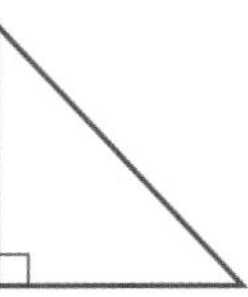
15. 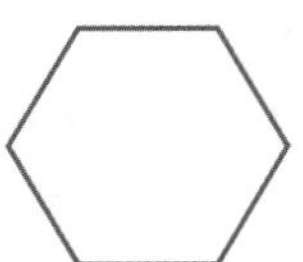
16. 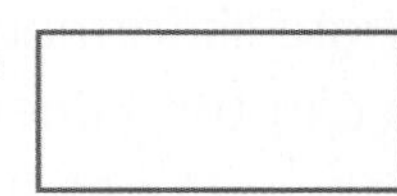
17. 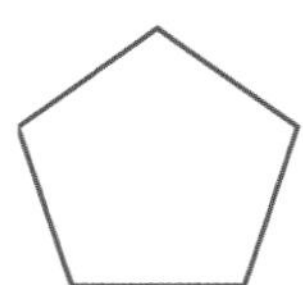
18. 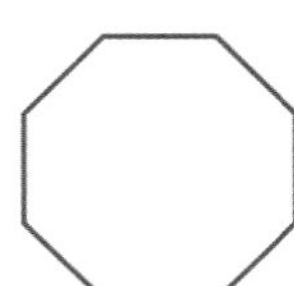
19. 
20. 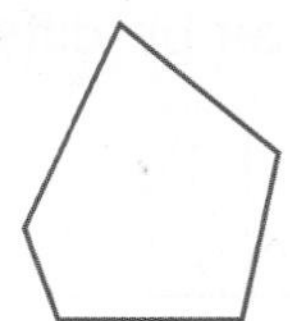

LESSON 1

Lines and Angles

Quick Review

1. 7 × 8 **2.** 72 ÷ 9
3. 10 × 90 **4.** 240 ÷ 80
5. 130 × n = 2,600

VOCABULARY

point, line, ray, plane, angle, line segment, parallel lines, intersecting lines, perpendicular lines

Learn

WHAT'S THE ANGLE? You can see basic geometric figures all around you. Some geometric figures have special names and symbols.

A **point** marks an exact location in space. Use a letter to name a point.	• A point A
A **line** is an endless straight path. It has no endpoints. Use two points on the line to name the line.	line AB or $\overleftrightarrow{AB}$ or line BA or $\overleftrightarrow{BA}$
A **ray** is part of a line that has one endpoint and goes on forever in one direction. Use an endpoint and one other point on the ray to name a ray.	ray KL or $\overrightarrow{KL}$
A **line segment** is part of a line between two endpoints. Use the two endpoints to name the line segment.	line segment CD or $\overline{CD}$ or line segment DC or $\overline{DC}$
A **plane** is an endless flat surface. Use three points that are not on a line to name the plane.	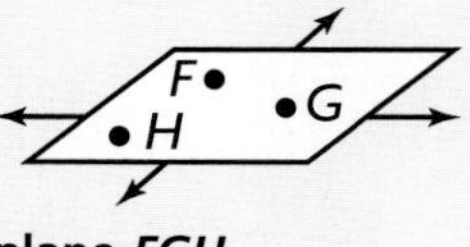 plane FGH

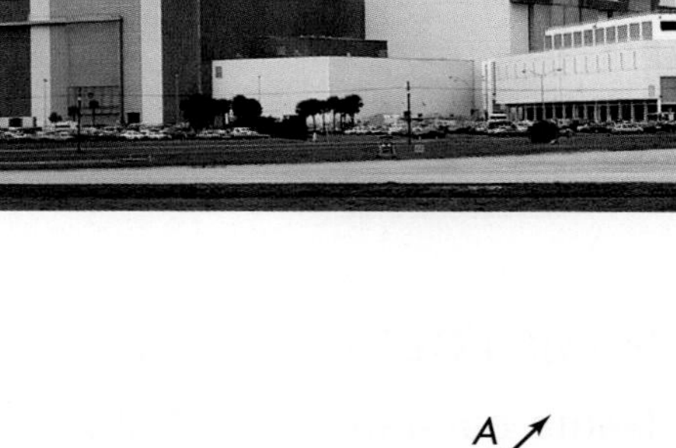

An **angle** is formed by two rays with the same endpoint. An angle can be named by three letters—a point from each side and the vertex as the middle letter. It can also be named by a single letter, its vertex.

vertex

$\angle ABC$, $\angle CBA$, or $\angle B$

Angles can be different sizes.

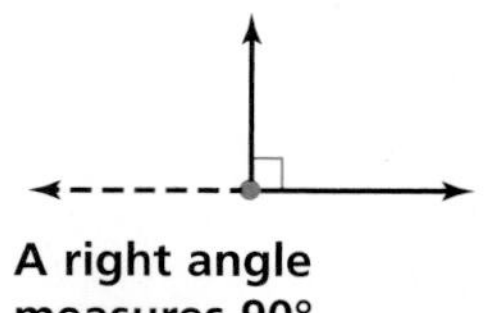

A right angle measures 90°.

An acute angle is greater than 0° and less than 90°.

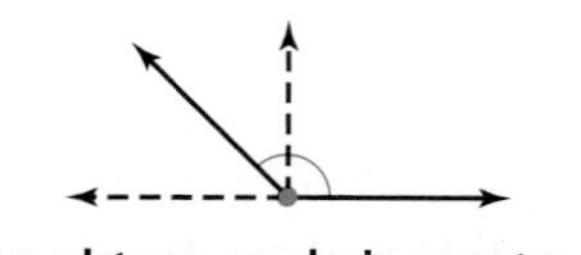

An obtuse angle is greater than 90° and less than 180°.

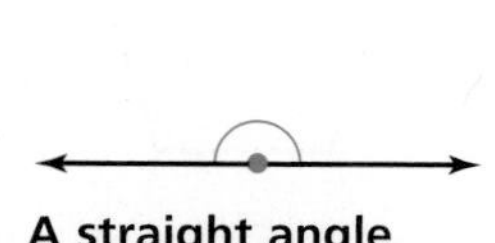

A straight angle measures 180°.

CALIFORNIA STANDARDS **MG 2.1** Measure, identify, and draw angles, perpendicular and parallel lines, rectangles and triangles by using appropriate tools. **MR 2.3** Use a variety of methods, such as words, numbers, symbols, charts, graphs, tables, diagrams, and models, to explain mathematical reasoning. *also* **MG 2.0, MR 2.3, MR 3.0, MR 3.2**

Line Relationships

Within a plane, lines can have different relationships.

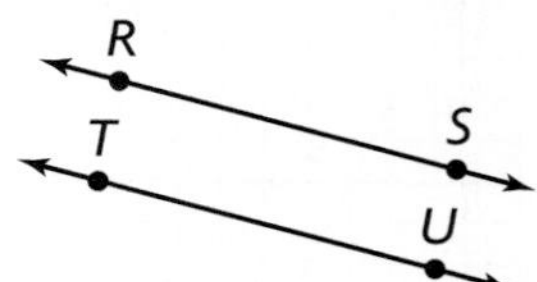

Lines in a plane that never intersect and are the same distance apart at every point are **parallel lines**. $\overleftrightarrow{RS} \parallel \overleftrightarrow{TU}$

HINT: Remember that $\parallel$ means "is parallel to" and $\perp$ means "is perpendicular to."

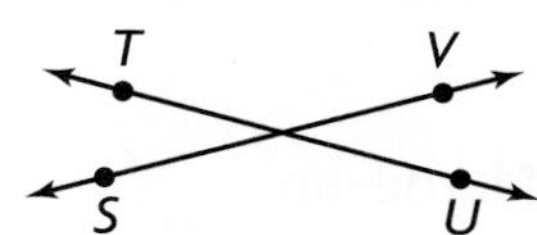

Lines that cross at one point are **intersecting lines**.

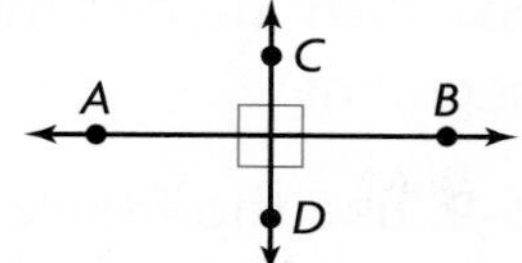

Lines that intersect to form four right angles are **perpendicular lines**. $\overleftrightarrow{AB} \perp \overleftrightarrow{CD}$

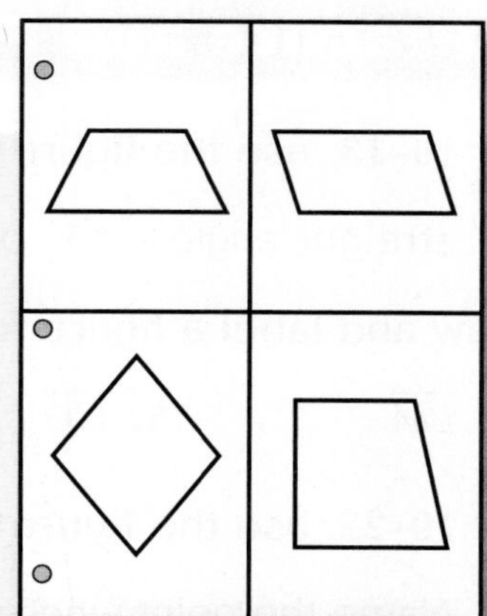

Activity

On the worksheet, draw and label parallel, perpendicular, and intersecting lines on each figure.

MATERIALS: ruler, polygons worksheet

STEP 1

Label the vertices of your figure *A, B, C,* and *D.* Draw a line through points *A* and *D.* Draw arrowheads to show that the line is endless. Repeat through points *B* and *C* and points *D* and *C.* You have drawn $\overleftrightarrow{AD}$, $\overleftrightarrow{DC}$, and $\overleftrightarrow{BC}$. Now, draw a line from point *A* to point *B.*

STEP 2

Classify the lines. Examples:

Parallel: $\overleftrightarrow{AB}$ is parallel to $\overleftrightarrow{DC}$, or $\overleftrightarrow{AB} \parallel \overleftrightarrow{DC}$

Perpendicular: $\overleftrightarrow{AD}$ is perpendicular to $\overleftrightarrow{DC}$ or $\overleftrightarrow{AD} \perp \overleftrightarrow{DC}$.

Intersecting: $\overleftrightarrow{DC}$ intersects $\overleftrightarrow{BC}$.

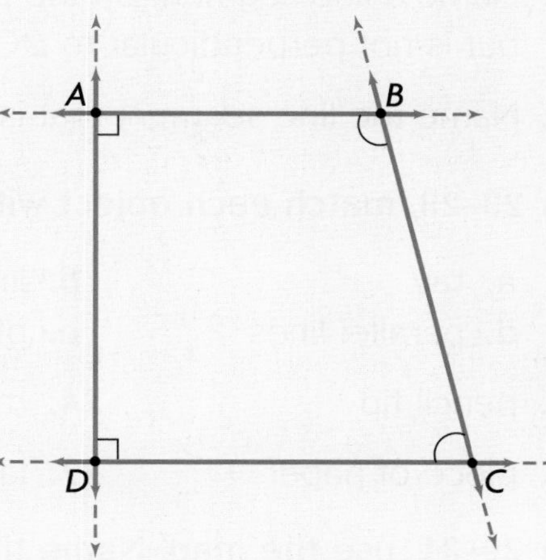

STEP 3

Classify the angles. Examples:

Right: $\angle A$ and $\angle D$ **Obtuse:** $\angle B$ **Acute:** $\angle C$

- If $\overleftrightarrow{AD}$ and $\overleftrightarrow{BC}$ were extended, would they be parallel, perpendicular, or intersecting? Explain.
- Repeat Steps 1–3 for each figure on the worksheet.

LESSON CONTINUES

Check

1. **Explain** the difference between a line and a line segment.

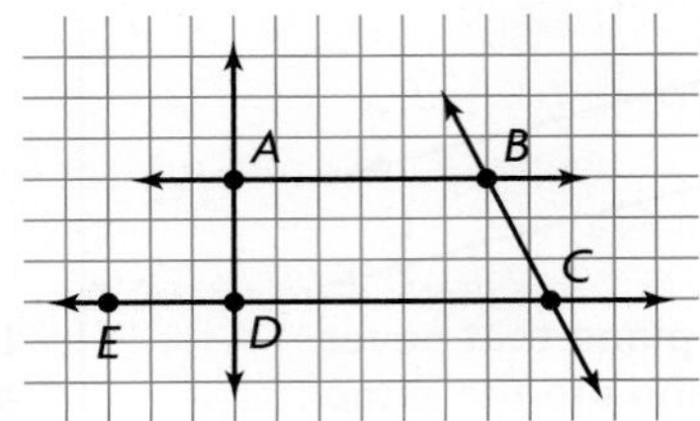

For 2–9, use the figure at the right. Name an example of each term.

2. acute angle
3. parallel lines
4. line segment
5. plane
6. ray
7. line
8. right angle
9. perpendicular lines

Practice and Problem Solving

For 10–13, use the figure above. Name an example of each.

10. straight angle
11. obtuse angle
12. point
13. intersecting lines

Draw and label a figure for each.

14. $\overline{LM}$
15. $\overrightarrow{CD}$
16. $\overleftrightarrow{AB}$
17. $\angle BDE$
18. point P

For 19–22, use the figure at the right.

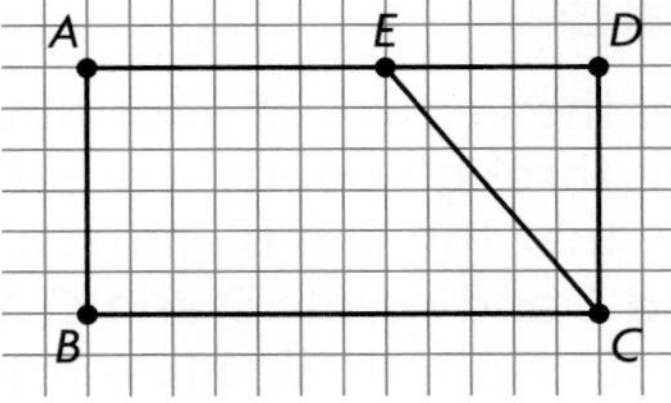

19. Name the point where $\overline{AB}$ and $\overline{BC}$ intersect.
20. Name a line segment on plane ABC that is perpendicular to $\overline{AD}$
21. Name a line segment on the plane that intersects but is not perpendicular to $\overline{DC}$.
22. Name the line segment that is parallel to $\overline{BC}$.

For 23–28, match each object with a term.

a. ray
b. line segment
c. intersecting lines
d. parallel lines
e. plane
f. point

23. pencil tip
24. railroad tracks
25. the letter X
26. piece of paper
27. laser beam
28. flagpole

For 29–31, use the map. Name the geometric figure suggested by

29. each of the three towns.
30. Interstate 18 and Interstate 287.
31. Maple Avenue and Interstate 287.
32. **Write About It** Explain why parallel lines never meet.

33. Mark two points on your paper. Label them *C* and *D*. How many different lines can you draw that go through both points *C* and *D*?

34. Mark one point on your paper. Label it *P*. How many lines can you draw through point *P*?

35. Copy these four points on your paper. What is the greatest number of line segments you can draw joining two of these points?

36. Copy the points on your paper. Connect the points to make a closed figure. Find the perimeter and area of the figure.

Mixed Review and Test Prep

37. Name the addition property shown. $(5 + 2) + 8 = 5 + (2 + 8)$ (p. 78)

38. Write the number that is 0.05 greater than 0.125 (p. 50)

39. $193{,}023 \times 24$ (p. 150)

40. $218{,}384 \times 13$ (p. 150)

41. **TEST PREP** Which number is 0.25 times 900? (p. 164)

A 150 **C** 225
B 215 **D** 250

USE DATA For 42–44, use the table.

WEEK'S SALES						
M	T	W	Th	F	Sa	Su
$150	$100	$160	$200	$250	$300	$275

42. What is the range of data? (p. 99)

43. Between which pairs of days did sales decrease? (p. 12)

44. **TEST PREP** Choose the kind of graph that would best display these data. (p. 130)

F line graph **H** circle graph
G bar graph **J** pictograph

LINKUP to Science

Astronomers use star charts such as this one of the northern constellations to plot locations of stars.

1. What geometric figure represents each star?
2. What geometric figure represents the distance between two stars in a constellation?
3. Name an acute angle and an obtuse angle in Cepheus.
4. In the diagram of Cepheus, is $\overline{CD}$ parallel to $\overline{ED}$ or does it intersect $\overline{ED}$ at point *D*?
5. **Write a problem** using the star chart.

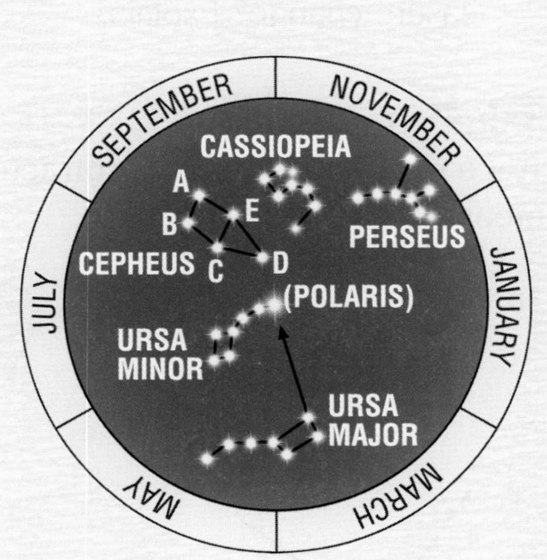

▲ To use the chart, face north soon after sunset and hold the chart in front of you with the current month at the top.

LESSON 2

HANDS ON

Measure and Draw Angles

Quick Review

1. (110 + 70) − 90
2. 120 + 240
3. 270 − 180
4. 4 × 90
5. 360 − 90

VOCABULARY

protractor degree

MATERIALS

protractor

Explore

A **protractor** is a tool used to measure angles. The unit to measure an angle is called a **degree** (°). You can use a protractor to measure ∠*ABC*.

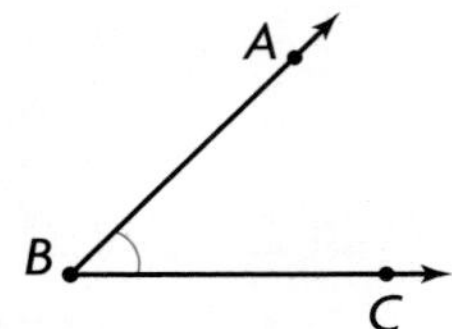

STEP 1

Place the center of the protractor on the vertex of the angle.

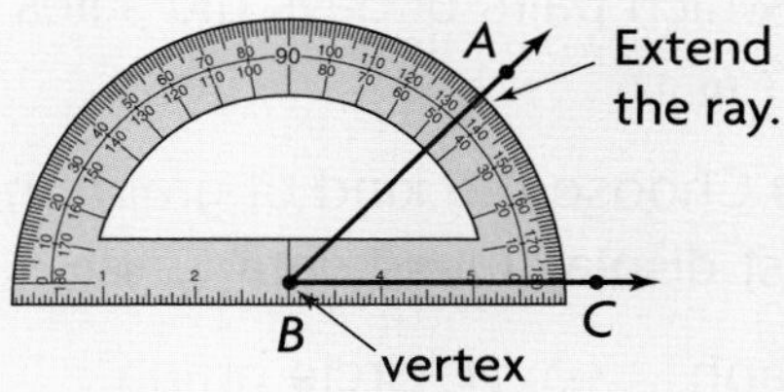

STEP 2

Line up the center point and the 0° mark on the protractor with one ray of the angle.

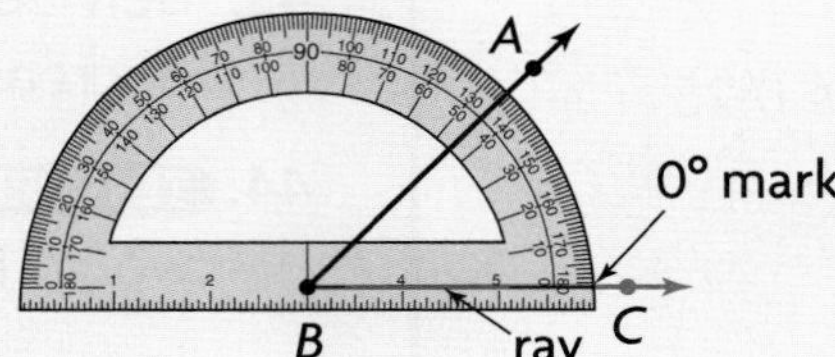

STEP 3

Read the measure of the angle where the other ray passes through the scale.

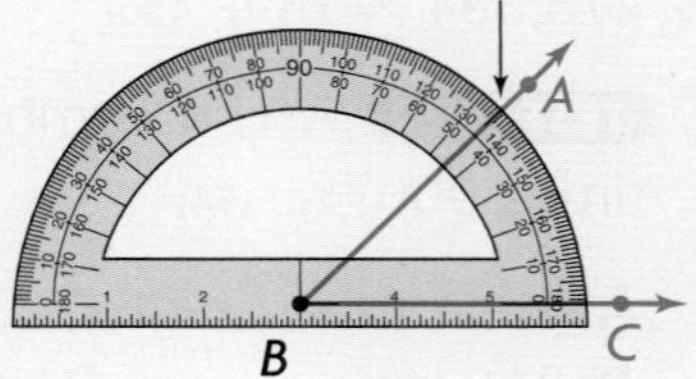

The measure of ∠*ABC* is 45°.

- If the rays are extended, will the angle measure change? Explain.

Which scale on the protractor do you read to find the measure of an angle?

Try It

Trace each angle. Use a protractor to measure each angle.

a. 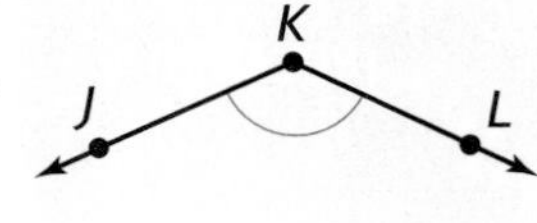

b. 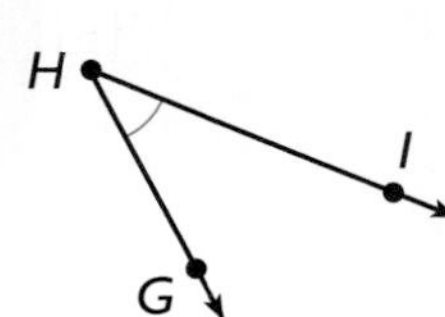

c. 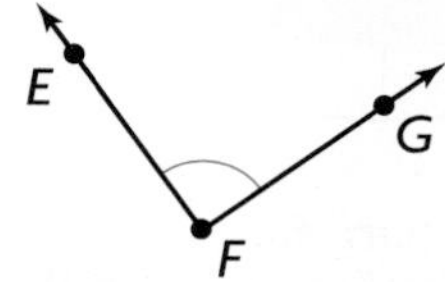

CALIFORNIA STANDARDS MG 2.0 Students identify, describe, and classify the properties of, and relationships between, plane and solid figures. 🔑MG 2.1 Measure, identify, and draw angles, perpendicular and parallel lines, rectangles and triangles by using appropriate tools (e.g. protractor). *also* **MR 1.1, MR 2.3, MR 2.4**

Connect

You can use a protractor to draw a right angle. A right angle is 90°.

STEP 1

Draw and label a ray. Place the center point of the protractor on the endpoint of the ray. Then line up the ray with the 0° mark on the protractor.

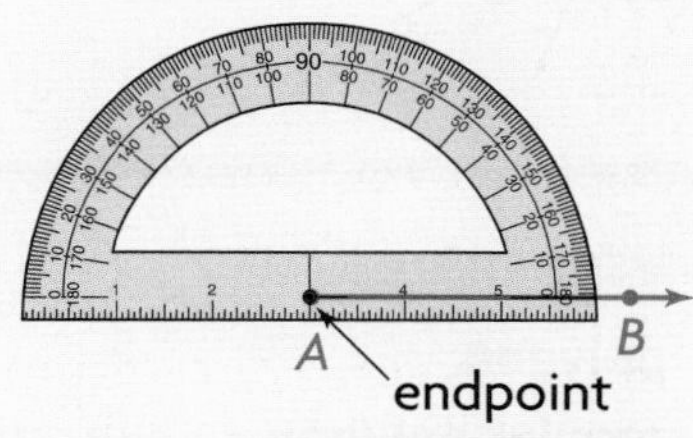

STEP 2

Find 90° on the protractor scale. Mark a point on your paper at 90°.

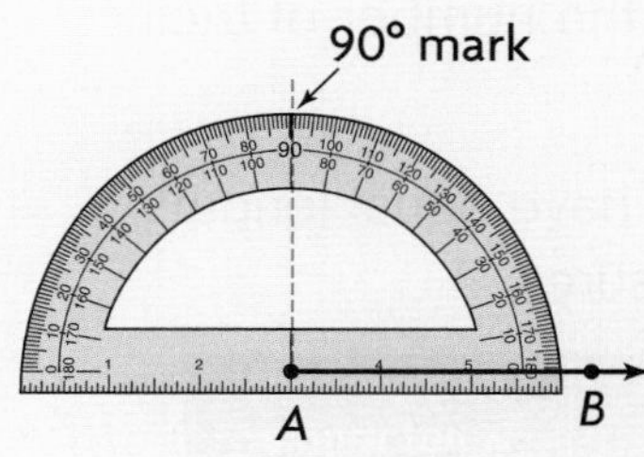

STEP 3

Draw the ray connecting the endpoint and your 90° mark. Label one point on this ray.

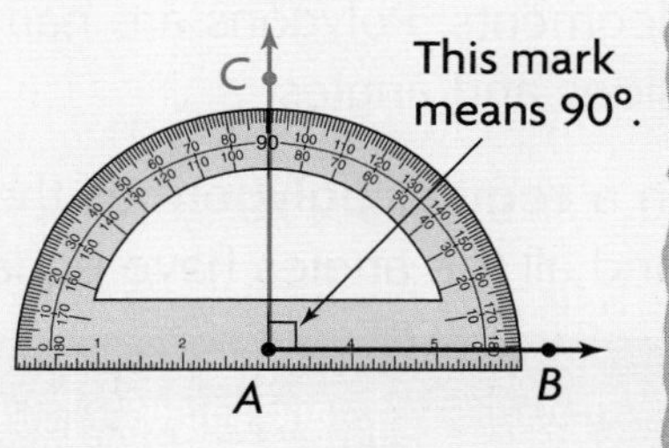

- A right angle measures 90°. What can you conclude about the measures of acute and obtuse angles?

Practice

For 1–8, use the figure at the right. Copy each angle. Then use a protractor to measure and classify each angle.

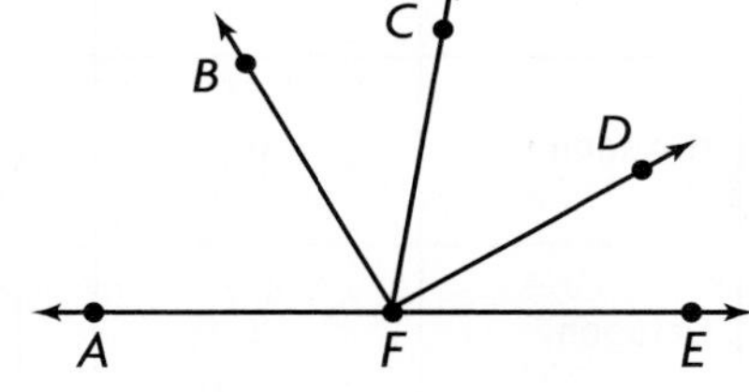

1. ∠*AFB*
2. ∠*BFE*
3. ∠*AFE*
4. ∠*AFD*
5. ∠*AFC*
6. ∠*CFD*
7. ∠*DFE*
8. ∠*BFD*

Use a protractor to draw each angle. Then write *acute, right,* or *obtuse* for each angle.

9. 40°
10. 90°
11. 120°
12. 15°
13. 95°

14. **What's the Error?** Ben measured a 50° angle as 130°. Explain how Ben made this error.

15. The fastest sailing vessel in the world can sail at 46 knots. One knot equals 1.15 mph. How fast, in miles per hour, is the fastest sailing vessel?

16. **REASONING** What fraction of a circle is 1°? 90°? 180°? HINT: There are 360° in a circle.

Mixed Review and Test Prep

17. $^{+}6 - {}^{+}3$ (p. 428)
18. $^{-}9 - {}^{+}1$ (p. 428)
19. $\frac{3}{7} + \frac{1}{7}$ (p. 344)
20. $\frac{5}{8} - \frac{3}{8}$ (p. 344)
21. **TEST PREP** Which number is 20% of 1,200? (p. 324)
    A 220 **B** 240 **C** 250 **D** 350

LESSON 3

Angles and Polygons

Learn

SIGNS OF THE TIMES When you recognize a stop sign by its shape, you are classifying a polygon. A **polygon** is a closed plane figure formed by three or more line segments. Polygons are named by the number of their sides and angles.

In a **regular polygon**, all the sides have equal lengths and all the angles have equal measures.

	Sides and angles	Regular polygon	Polygon that is not regular
Triangle	3		
Quadrilateral	4		
Pentagon	5		
Hexagon	6		
Octagon	8		

Quick Review

Write *acute, obtuse,* or *right* for each angle.

1. 
2. 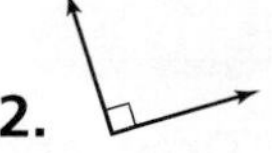
3. 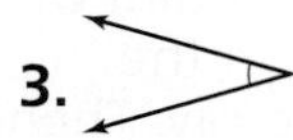
4. 
5. 

VOCABULARY

polygon
regular polygon

Activity 1

MATERIALS: square and triangle dot paper, ruler

STEP 1

Draw a regular polygon.

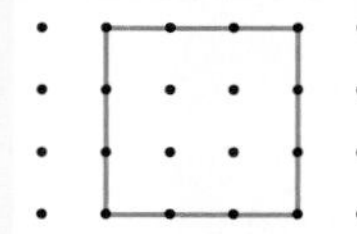

STEP 2

Draw a polygon that is not regular.

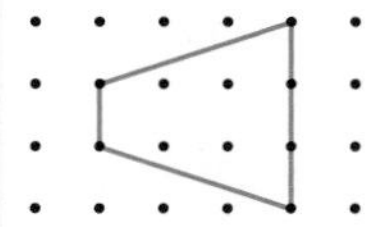

- Now, use dot paper to draw two 6-sided polygons, one regular and one that is not regular. Name the figures and explain how they are alike and how they are different.
- What kind of polygon is a rectangle? Is it regular? Explain.

CALIFORNIA STANDARDS **MG 2.1** Measure, identify, and draw angles, perpendicular and parallel lines, rectangles and triangles by using appropriate tools. **MG 2.2** Know that the sum of the angles of any triangle is 180° and the sum of the angles of any quadrilateral is 360° and use this information to solve problems. *also* **MR 1.1, MR 2.2, MR 2.4, MR 3.2, MR 3.3**

Angles of Polygons

A straight angle measures 180°. You can use this fact to find the sums of the angles of polygons.

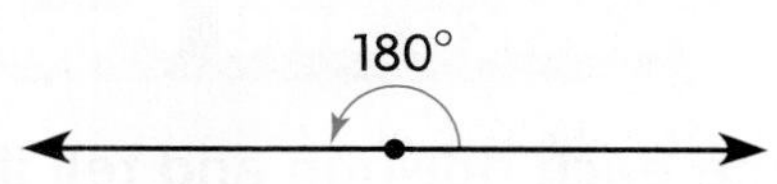

Activity 2

Find the sum of the angles in a triangle.

STEP 1

Draw a right triangle. Label each angle. Cut out the triangle.

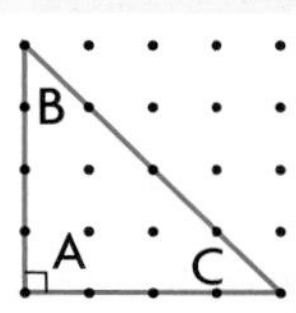

STEP 2

Cut off the angles as shown.

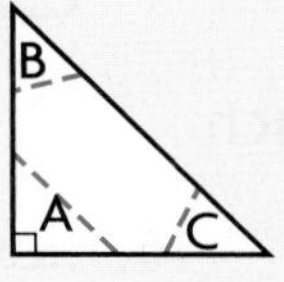

STEP 3

The angles placed together at a point form a straight angle as shown.

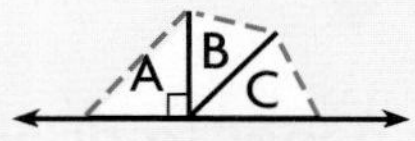

The sum of the angles is 180°.

- Repeat the steps with two different-shaped triangles. Write a rule about the sum of the angles in a triangle.

Find the sum of the angles in a quadrilateral.

STEP 1

Draw a quadrilateral on dot paper. Label each angle. Cut out the quadrilateral.

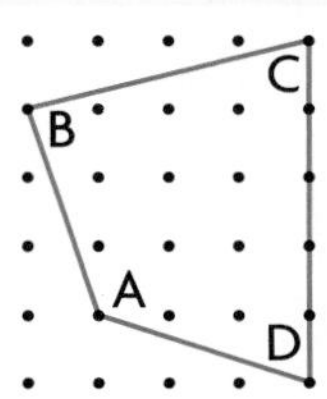

STEP 2

Cut off the angles as shown.

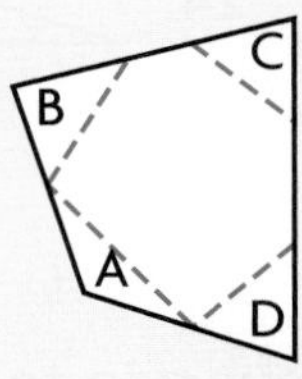

STEP 3

Place the labeled angles together at a point as shown.

The angles completely surround a point. So, the sum of the angles is 360°.

- Repeat the steps with two different-shaped quadrilaterals. Write a rule about the sum of the angles in a quadrilateral.

MATH IDEA The sum of the angles in a triangle is 180°, and the sum of the angles in a quadrilateral is 360°.

Check

1. **Explain** another way to check that the sum of the angles in a quadrilateral is 360°.

Name each polygon and tell if it is *regular* or *not regular*.

2. 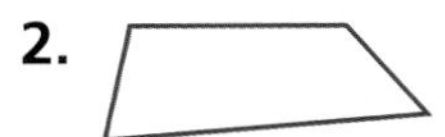

3. 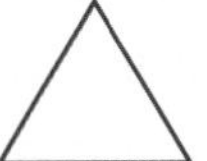

4. 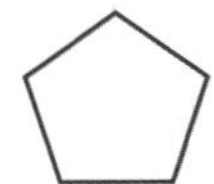

5. 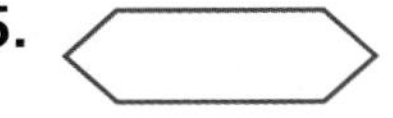

LESSON CONTINUES

Practice and Problem Solving

Name each polygon and tell if it is *regular* or *not regular*.

6. 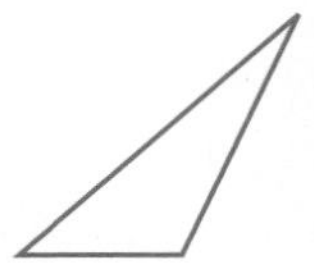

7. 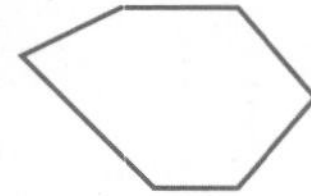

8. 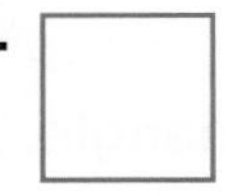

9. 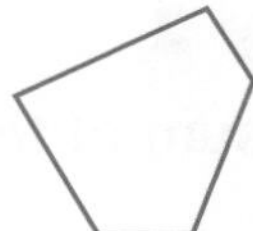

10. 

11. 

12. 

13. 

Use dot paper to draw an example of each.

14. regular triangle

15. regular octagon

16. hexagon that is not regular

17. quadrilateral that is not regular

Find the unknown angle measure.

18. 

19. 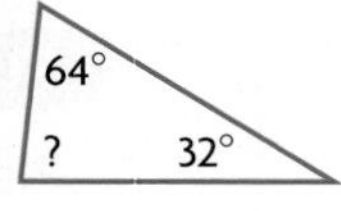

20. 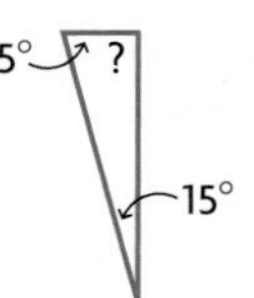

21. 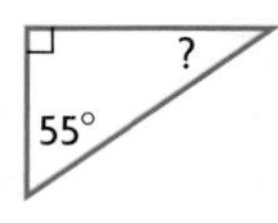

22. 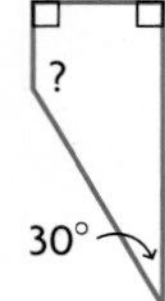

23. 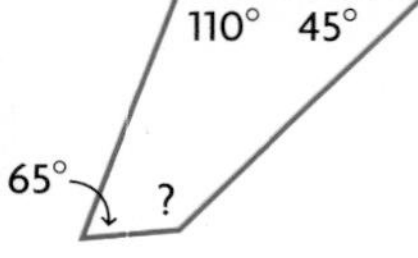

24. 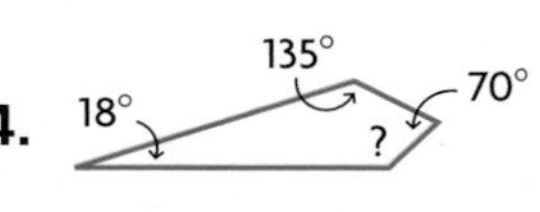

25. 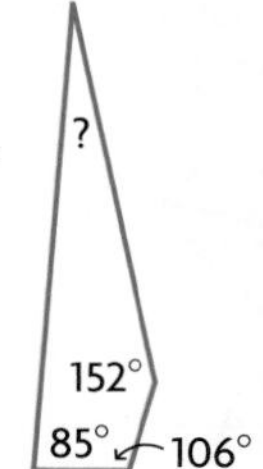

Find the pattern. Then write a rule. Use your rule to draw the next figure in the pattern.

26. 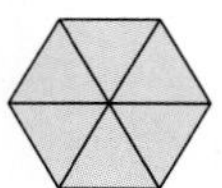 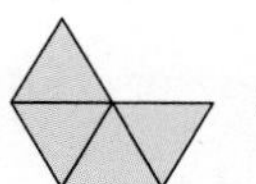

27. 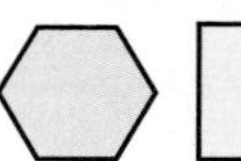  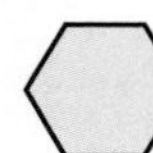 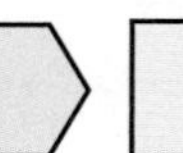 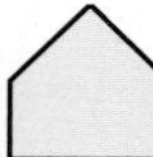

28. **Reasoning** Copy the figure at the right and cut out $\angle A$ and $\angle B$. How does the sum of the measures of $\angle A$ and $\angle B$ compare to the measure of $\angle BCD$?

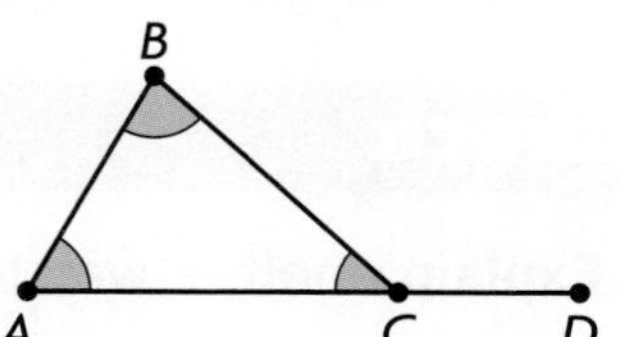

29. How many triangles are in this figure?

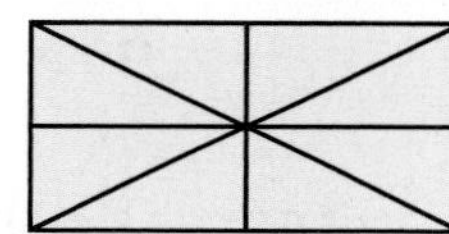

30. **What's the Error?** Rhonda says all the angles in a regular hexagon are greater than 90°, so they are acute. Describe and correct her error.

31. PATTERNS A nonagon has 9 sides. Follow the pattern in the table of polygons on page 460 to find how many angles a nonagon has.

32. REASONING Bob and Rosa are walking down parallel streets. If they are 100 feet apart when they start, how far will they have to walk until their paths cross?

Mixed Review and Test Prep

For 33–34, tell whether the angle is *acute, obtuse,* or *right.* (p. 454)

33. 95°

34. 53°

35. 1.825×3.06 (p. 164)

36. 20.66×2.7 (p. 164)

37. TEST PREP Which is a possible sum of a right angle and an acute angle? (p. 454)

A 45°
B 60°
C 90°
D 95°

USE DATA For 38–39, use the graph. (p. 108)

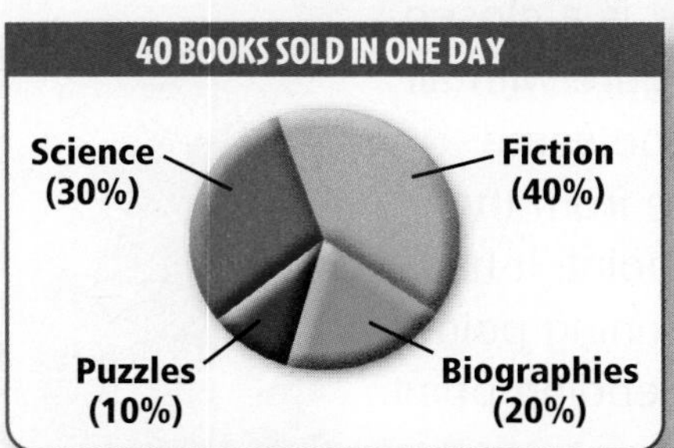

38. How many of each type of book were sold?

39. TEST PREP What fraction of the graph is fiction books?

F $\frac{1}{9}$ **G** $\frac{1}{3}$ **H** $\frac{2}{5}$ **J** $\frac{3}{10}$

Thinker's Corner

SUM IT UP! You can divide any polygon into triangles to find the sum of that polygon's angles.

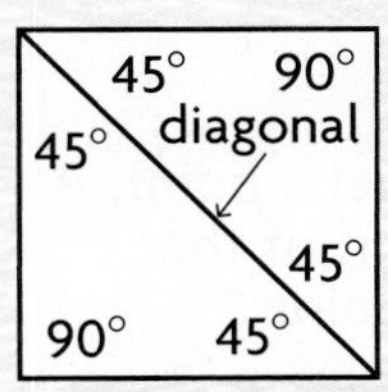

◀ **The sum of the angles in a triangle is 180°.**

Find the sum of the angles in a pentagon.

STEP 1 Draw a pentagon.

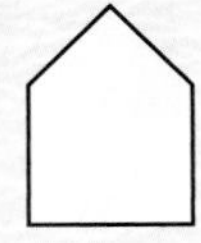

STEP 2 Draw diagonals.

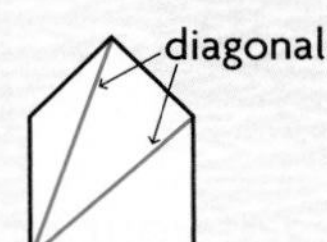

STEP 3 Count the triangles.

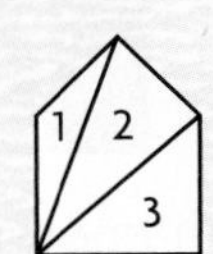

STEP 4 Multiply:
3 × 180° = 540°
So, the sum of the angles in any pentagon is 540°.

Find the sum of the angles in each figure.

1. quadrilateral

2. hexagon

3. octagon

LESSON 4

Circles

Quick Review

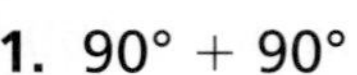

1. 90° + 90°
2. 180° + 180°
3. 120° + 120° + 120°
4. 60° + 240° + 60°
5. 90° + 90° + 90° + 90°

VOCABULARY

circle, chord, radius, compass, diameter, central angle

Learn

ROUND AND ROUND The largest Ferris wheel in the world, the London Eye, is shaped like a giant circle.

A **circle** is a closed plane figure with all points the same distance from the center point. It has no beginning point and no ending point.

A line segment that connects the center with a point on the circle is called a radius (plural: *radii*). $\overline{BA}$ is a radius.

A chord that passes through the center of the circle is called a diameter. $\overline{CD}$ is a diameter.

A line segment that connects any two points on the circle is called a chord. $\overline{EF}$ is a chord.

Activity 1

MATERIALS: compass, centimeter ruler

A **compass** is a tool for constructing circles. You can use a compass to construct a circle with a radius of 7 centimeters to model the London Eye Ferris Wheel.

STEP 1

Draw and label a point, *P*. Place the point of the compass on point *P*.

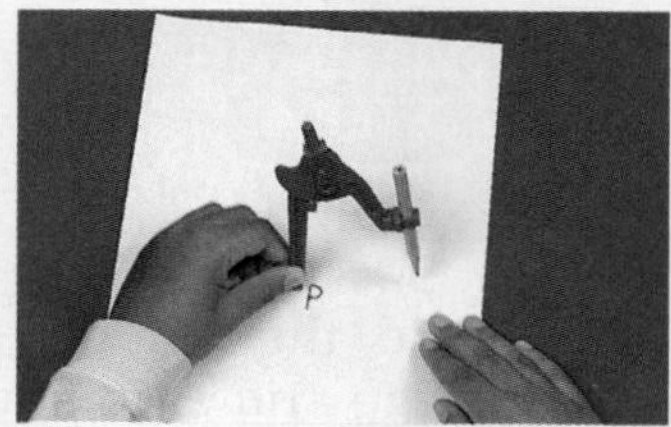

STEP 2

Open the compass 7 cm, which is the length of the radius.

STEP 3

Use the compass to make the circle.

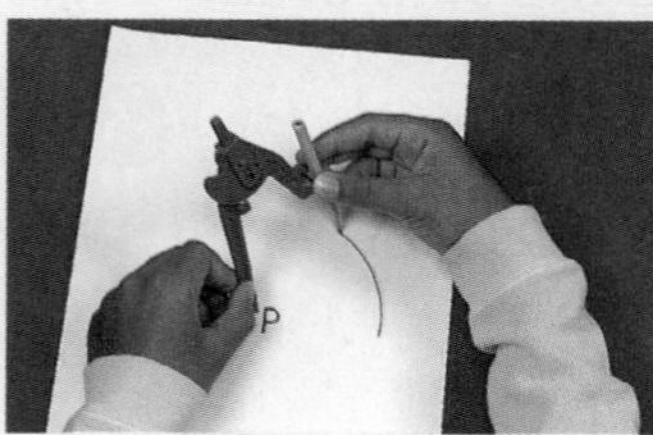

- Draw and label a chord, a diameter, and a radius of your circle *P*. Write the length of each.
- Draw three more circles of different sizes. Measure the lengths of each circle's radius and diameter. Record the measurements in a table.
- What relationship do you notice between the radius and diameter of a circle?

CALIFORNIA STANDARDS MG 2.0 Students identify, describe, and classify the properties of, and relationships between, plane and solid figures. **MG 2.1** Measure, identify, and draw angles, perpendicular and parallel lines, rectangles and triangles by using appropriate tools. *also* **MR 1.1, MR 2.0, MR 2.3, MR 2.4, MR 3.0**

Central Angles

Activity 2

When two radii of a circle meet at its center, they form a **central angle**. The measure of a central angle is less than or equal to 180°.

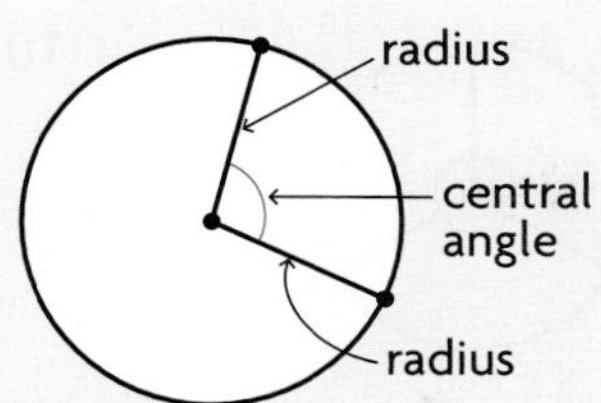

What is the sum of all the central angles in a circle?

STEP 1

Use a compass to draw a circle. Draw 4 radii of the circle to show 4 central angles.

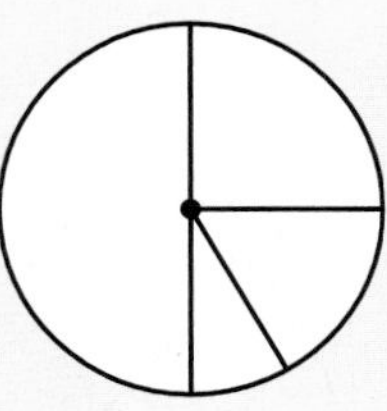

STEP 2

Use the protractor to measure each central angle.

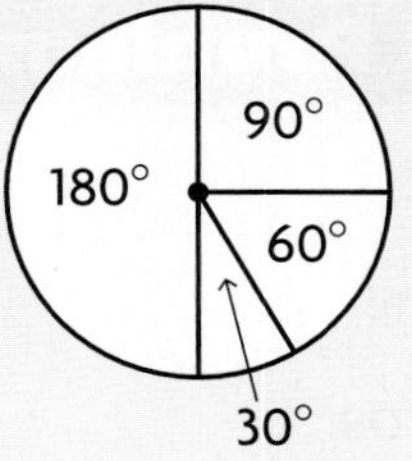

STEP 3

Find the sum of the angles in the circle.

180° + 90° + 60° + 30° = 360°

- Trace Figures A and B. Use a protractor to measure each central angle. Then find the sum of the angles in each circle.

A

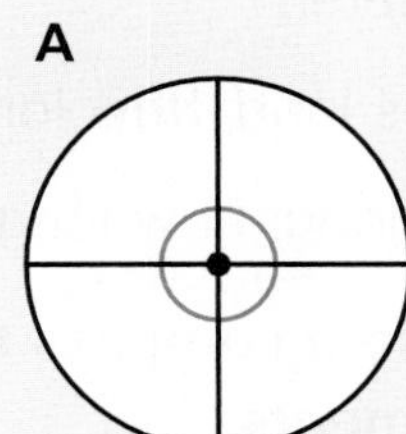

B

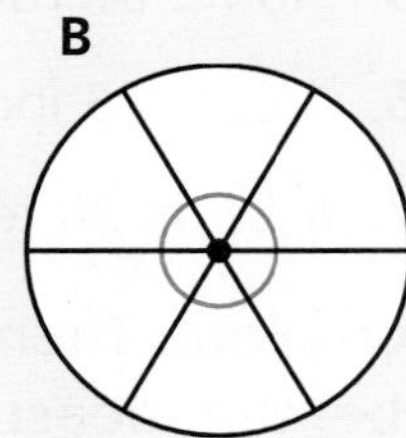

So, the sum of the angles in any circle is 360°.

- If a circle is divided into 4 sections and you know the measures of 3 of the angles, how can you find the measure of the fourth angle?

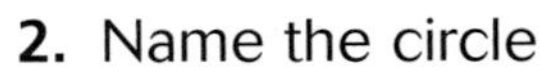

Check

1. **Explain** how you can find the length of the diameter of the London Eye Ferris Wheel. HINT: The London Eye Ferris Wheel is about 1,000 times as large as your model.

Use the circle for 2–5.

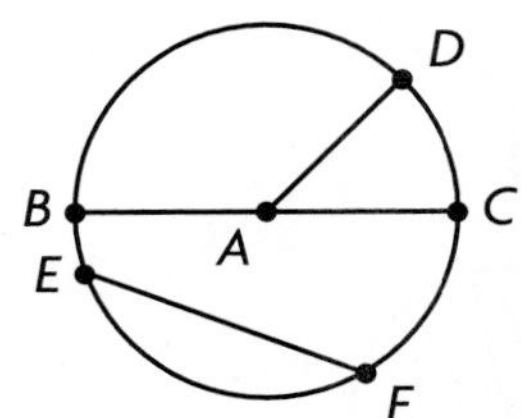

2. Name the circle.
3. Name a radius of the circle.
4. Name a diameter of the circle.
5. Name a chord of the circle.

Complete. Then use a compass to draw each circle. Draw and label the measurements.

6. radius = 4 cm diameter = ■
7. radius = ■ diameter = 6 cm
8. radius = 5 cm diameter = ■

LESSON CONTINUES

Use a protractor to measure each central angle.

9. 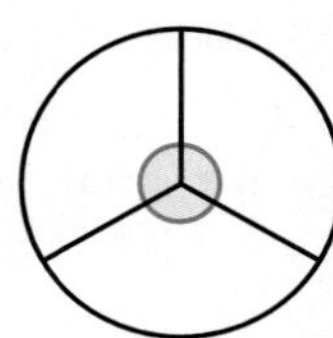

10. 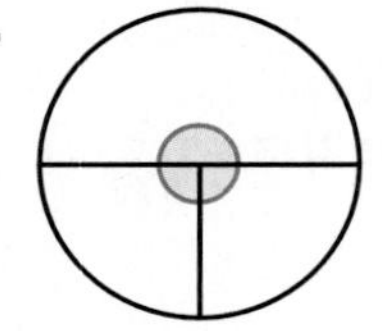

Find the unknown angle measure.

11. 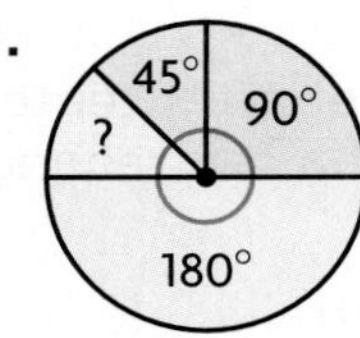

12. 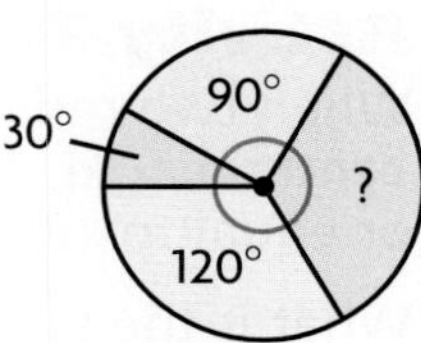

▶ Practice and Problem Solving

For 13–17, use circle *C*.

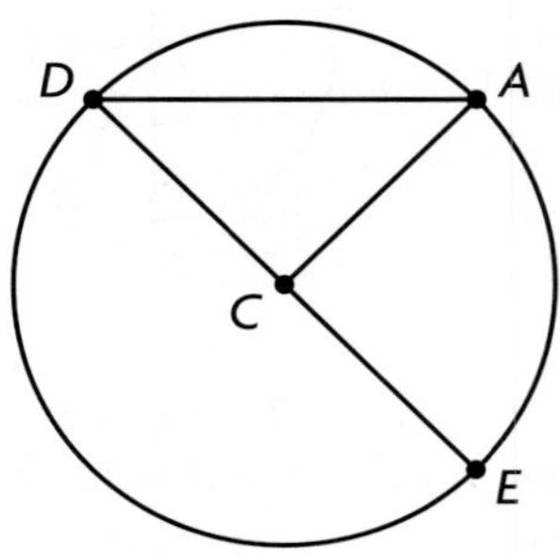

13. Name the three radii.
14. Name the two chords.
15. Name the diameter.
16. If $\overline{CD}$ is 5 inches long, how long is $\overline{DE}$?
17. If $\overline{DE}$ is 20 feet long, how long is $\overline{AC}$?

Complete. Then use a compass to draw each circle. Draw and label the measurements.

18. radius = 2.5 cm
diameter = ■

19. radius = ■
diameter = 9 cm

20. radius = 1.5 cm
diameter = ■

21. radius = ■
diameter = 7 cm

22. radius = 3 cm
diameter = ■

23. radius = ■
diameter = 11 cm

Use a protractor to measure each central angle.

24. 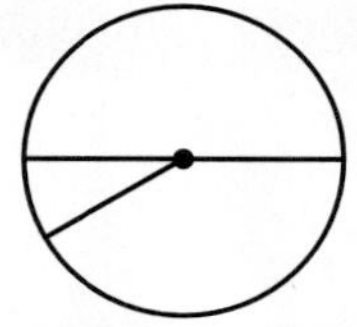

25. 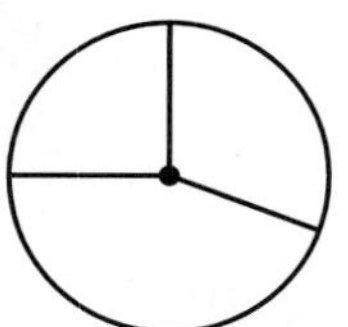

26. 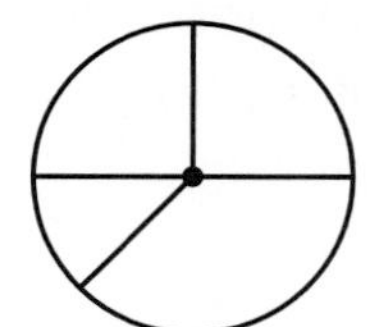

27. 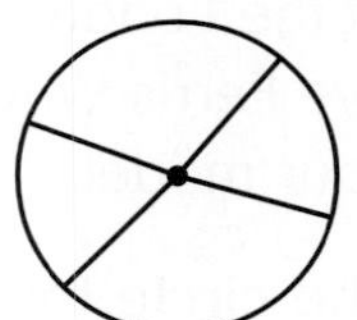

Find the unknown angle measure.

28. 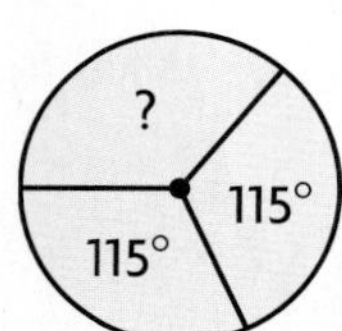

29. 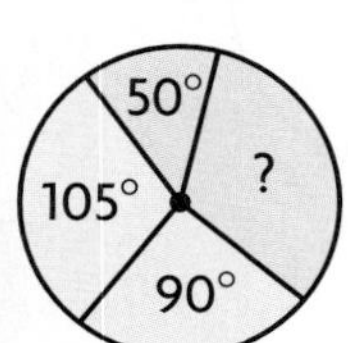

30. 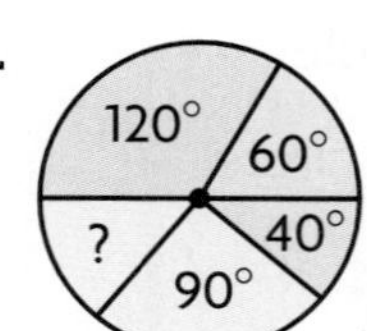

31. 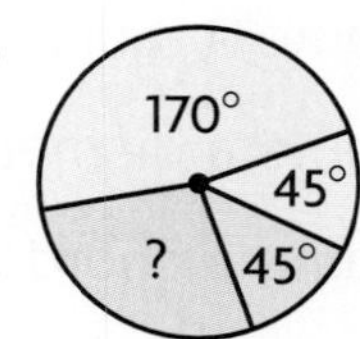

32. **What's the Error?** The radius of Ken's circle is 12 cm. The diameter of Tom's circle is 24 cm. Tom says his circle is bigger. Describe his error.

33. What line segments in a circle determine the size of the circle?

34. How many degrees are there in $\frac{1}{2}$ of a circle? $\frac{1}{4}$ of a circle? $\frac{1}{8}$ of a circle?

35. **REASONING** How many degrees are between the 2 and the 3 on a clock?

36. The second largest Ferris wheel in the world is in Japan. Its radius is about 42 meters. What is its diameter?

37. The largest Ferris wheel in the United States is the Texas Star in Dallas. Its diameter is about 212 feet. What is its radius?

Mixed Review and Test Prep

38. 6×12 (p. 146)

39. 9×12 (p. 146)

40. Two lines intersect to form right angles. What kind of lines are they? (p. 454)

41. The temperature at 5 A.M. was $^{-}7$°F. By 2 P.M. it had risen 18°F. What was the temperature at 2 P.M.? (p. 422)

Write the fraction in simplest form. (p. 290)

42. $\frac{8}{10}$

43. $\frac{24}{30}$

44. **TEST PREP** Which of these angles is acute? (p. 454)

A 180°   C 36°
B 90°   D 120°

For 45–47, use the figures. (p. 460)

a.

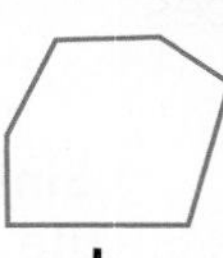
b.

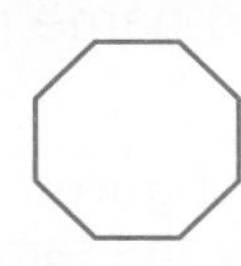
c.

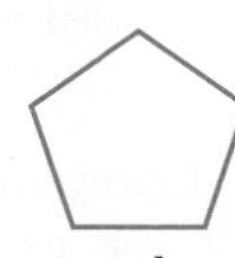
d.

45. Which of the polygons is not regular? What kind of polygon is it?

46. The sum of the angles of which figure is 360°? What kind of polygon is it?

47. **TEST PREP** Which is the sum of the angles in figure C? (p. 467)

F 540°   H 900°
G 720°   J 1,080°

Thinker's Corner

YOU BE THE JUDGE!

1. Which is longer, $\overline{AB}$ or $\overline{BC}$?

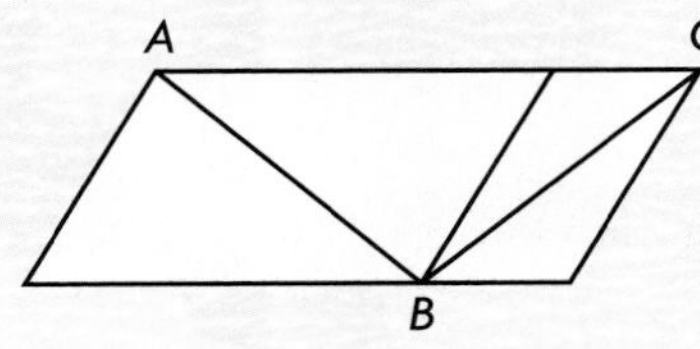

2. Which of the lines is the most curved?

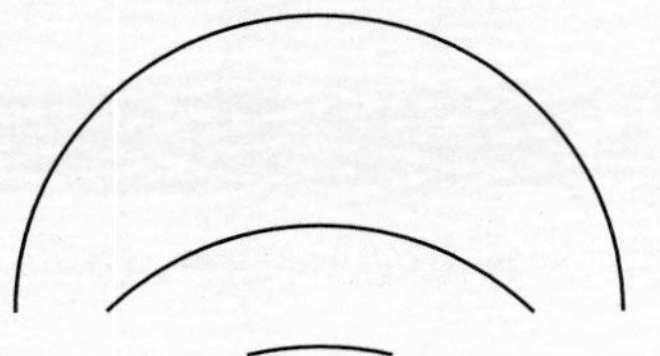

3. What happened to the ruler? Are the sides parallel?

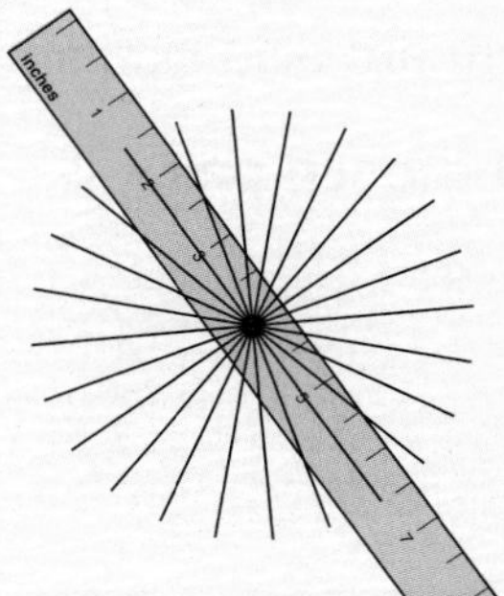

4. Look at the heavy lines in the figures. Which is longer?

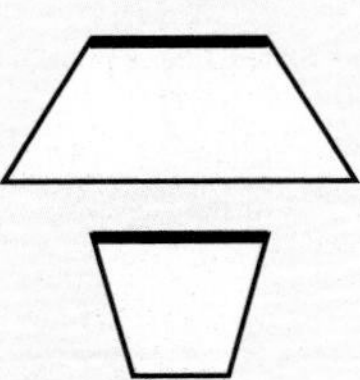

LESSON

5 Congruent and Similar Figures

Learn

DOUBLE TAKE! On the plaza of Philadelphia's City Hall, there are several giant sculptures of game pieces. All of these sculptures are similar to actual-sized game pieces.

Congruent figures have exactly the same size and shape.

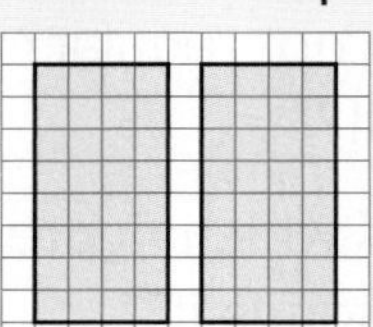

Similar figures have exactly the same shape but do not have to be the same size.

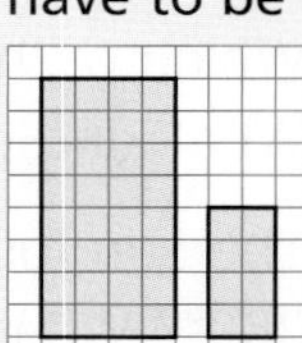

- Are a real domino and the statue of a domino in Philadelphia *congruent, similar,* or *neither*? Explain.

Examples **Write *similar*, *congruent*, or *neither*.**

A

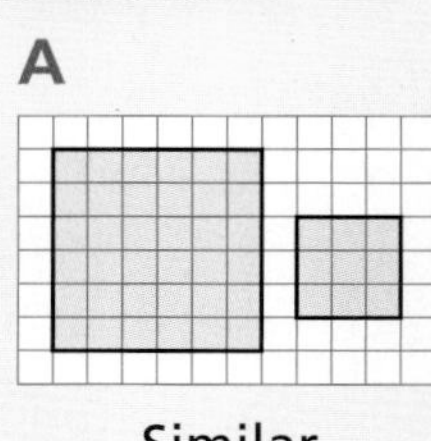

Similar

B

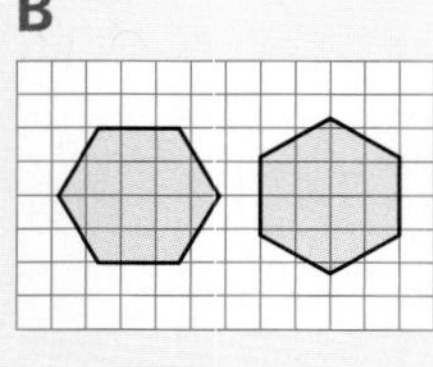

Congruent

C

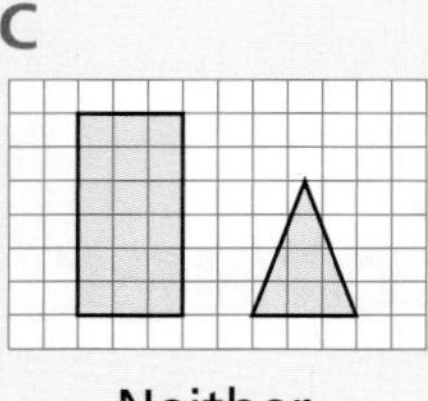

Neither

Quick Review

Name the polygon.

1. 
2. 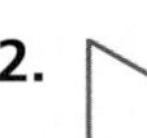
3. 
4. 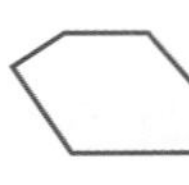
5. 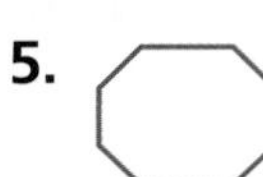

VOCABULARY

congruent similar

Check

1. **Explain** why two figures that are congruent are also similar.

Write *similar*, *congruent*, or *neither* to describe each pair.

2. 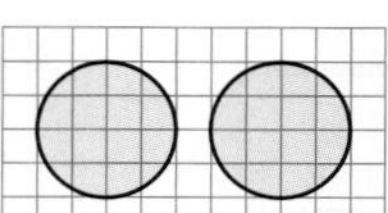
3. 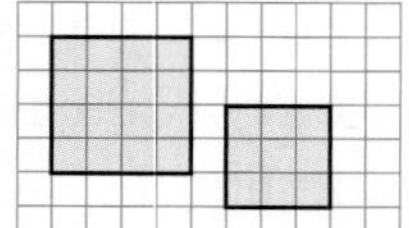
4. 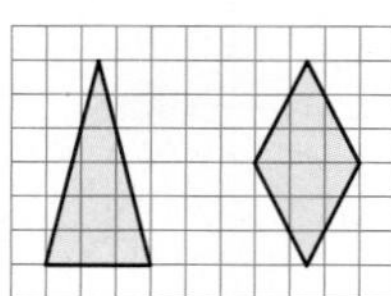

CALIFORNIA STANDARDS MG 2.0 Students identify, describe, and classify the properties of, and relationships between, plane and solid figures. **MR 2.3** Use a variety of methods, such as words, numbers, symbols, charts, graphs, tables, diagrams, and models, to explain mathematical reasoning. *also* **MR 1.1, MR 2.0, MR 3.2, MR 3.3**

Practice and Problem Solving

Write *similar, congruent,* or *neither* to describe each pair.

5. 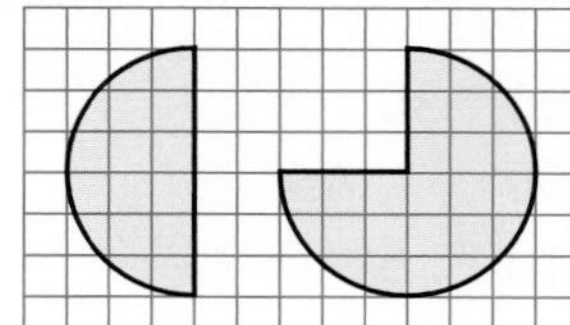

6. 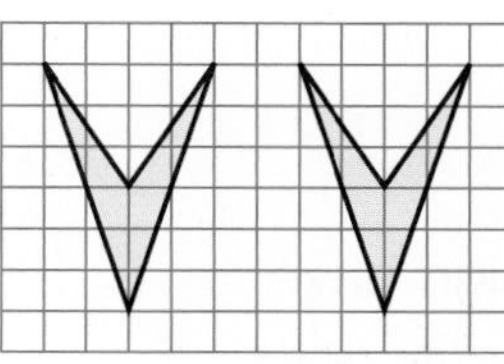

7. 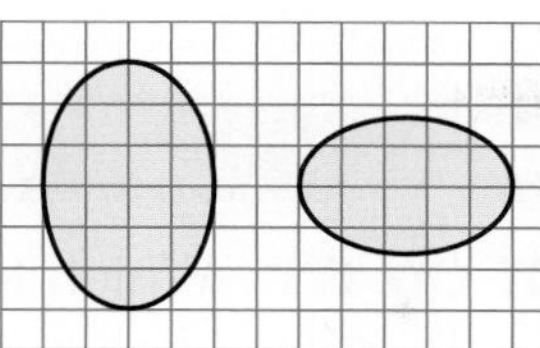

8. 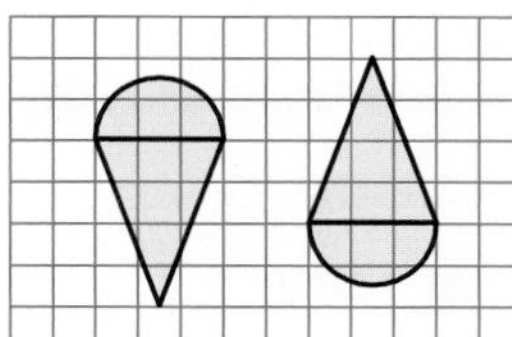

9. 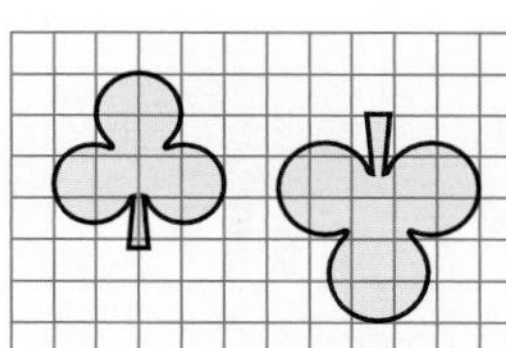

10. 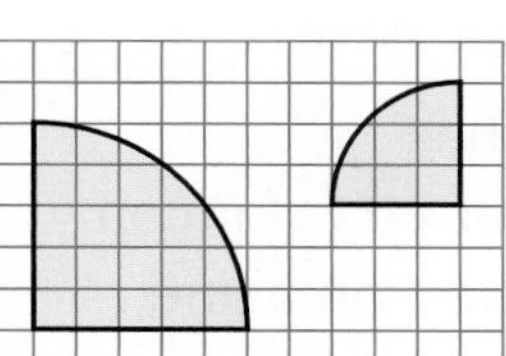

For 11–14, use the figures below.

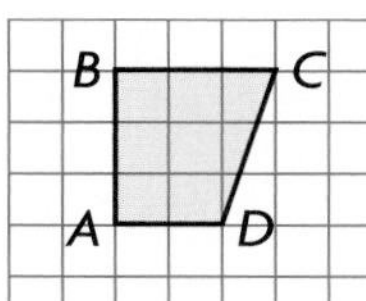

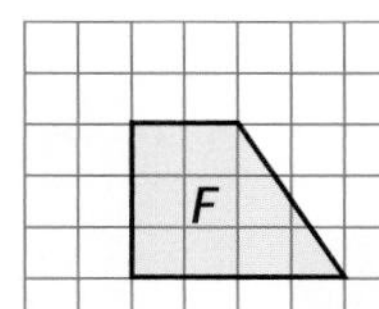

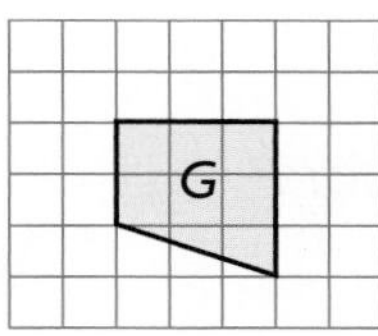

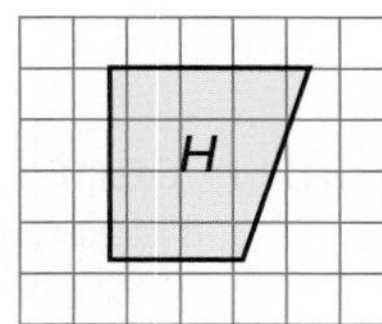

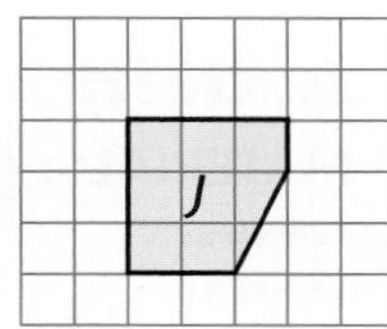

11. Write the letter of the one figure that is congruent to *ABCD*.

12. Write the letter of the one figure that is similar but not congruent to *ABCD*.

13. Write the letter of the two figures that are neither congruent nor similar to *ABCD*.

14. What is the same and what is different about figures *G* and *H*?

15. Mary says that all rectangles are similar. Is this statement correct? Explain.

16. Two sculptures are similar. The height of one sculpture is four times as great as that of the other sculpture. The smaller sculpture is 2.5 feet tall. How tall is the larger sculpture?

17. **REASONING** Square *ABCD* is congruent to square *EFGH*. The perimeter of square *ABCD* is 14 inches. What is the length of $\overline{EF}$?

Mixed Review and Test Prep

Find the radius of each circle with the given diameter. (p. 464)

18. 12 in.

19. 16.5 cm

20. Write 10,000,000 + 5,000,000 + 20,000 + 700 + 2 in standard form. (p. 4)

21. Solve for n. $5 + n = 94$ (p. 74)

22. **TEST PREP** Which of the following is NOT part of a line? (p. 454)

A point
B angle
C line segment
D ray

Extra Practice page H56, Set D

LESSON

Symmetric Figures

Quick Review

Find the missing angle measure for each triangle.

1. 90°, 30°, ■
2. 90°, ■, 45°
3. 100°, 5°, ■
4. ■, 75°, 75°
5. 50°, 120°, ■

VOCABULARY

line symmetry
rotational symmetry

Learn

REFLECT ON IT! Which letters in this sign have line symmetry? How many lines of symmetry does each letter have?

A figure has **line symmetry** if it can be reflected on a line so that the two parts are congruent.

Activity

MATERIALS: grid paper, pencil, scissors

STEP 1

Copy the H on grid paper by shading the squares as shown. Cut out the H.

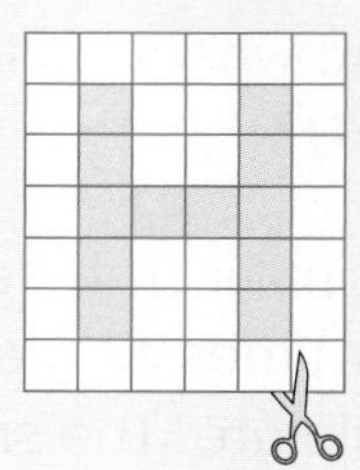

STEP 2

Fold the H in half so the two halves match exactly.

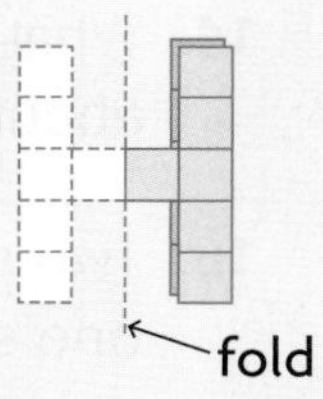

STEP 3

Fold the H in half in different ways. If the halves match, then the fold is another line of symmetry.

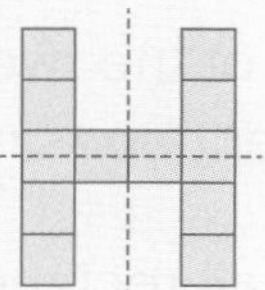

So, the H has 2 lines of symmetry. Repeat with the other letters.

The H and 0 each have 2 lines of symmetry, and the Y, W, and D each have 1 line of symmetry. The L does not have line symmetry.

Another type of symmetry involves turning the figure instead of folding it. A figure has **rotational symmetry** if it can be rotated less than 360° around a central point and still match the original figure.

Examples

A $\frac{1}{4}$ turn, or 90°

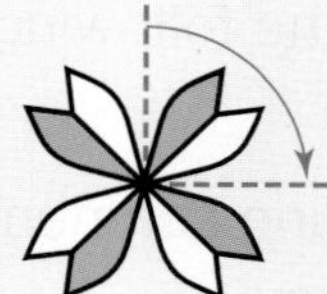

B $\frac{1}{3}$ turn, or 120°

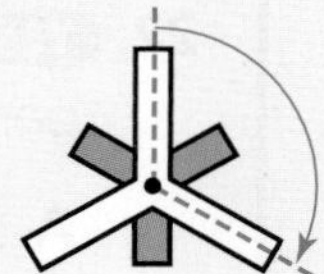

C $\frac{1}{2}$ turn, or 180°

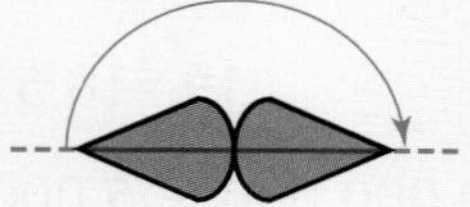

CALIFORNIA STANDARDS MG 2.0 Students identify, describe, and classify the properties of, and relationships between, plane and solid figures. **MR 2.3** Use a variety of methods, such as words, numbers, symbols, charts, graphs, tables, diagrams, and models, to explain mathematical reasoning. *also* **MR 2.0, MR 3.0, MR 3.2, MR 3.3**

Check

1. **Explain** how to find out how many lines of symmetry a regular pentagon has.

Trace each figure. Draw the lines of symmetry. Tell whether each figure has rotational symmetry. Write *yes* or *no*.

2. 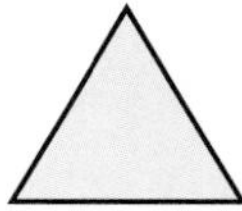

3. 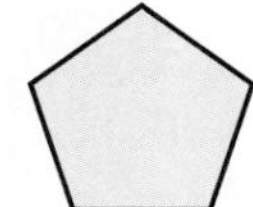

4. 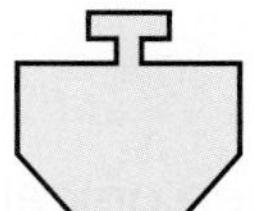

5. 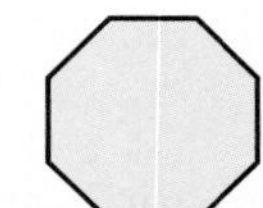

6. 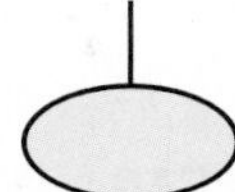

Practice and Problem Solving

Trace each figure. Draw the lines of symmetry. Tell whether each figure has rotational symmetry. Write *yes* or *no*.

7. 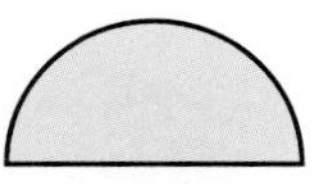

8. 

9. 

TECHNOLOGY LINK

More Practice: Use **Mighty Math Number Heroes,** ***Geoboard,*** Level L.

10. 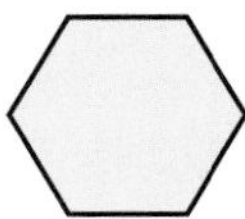

11. 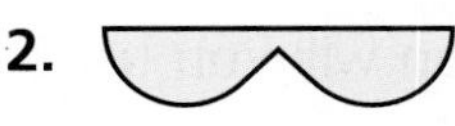

12.

Each figure has rotational symmetry. Tell the fraction and angle measure of each turn.

13. 

14. 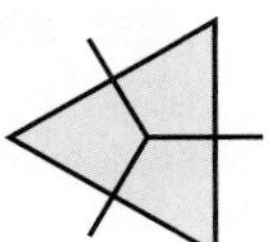

15. 

Trace each figure and line of symmetry on grid paper. Then draw the other half of the figure.

16. 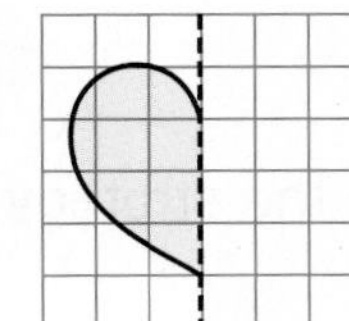

17. 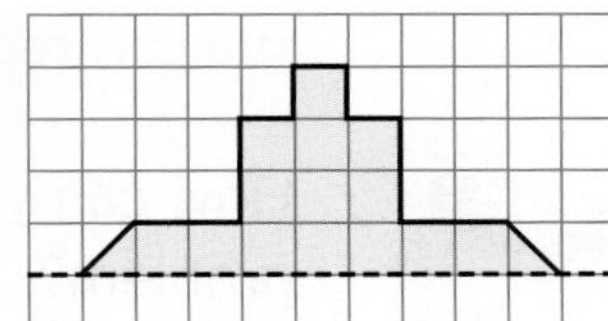

18. **Write About It** Describe how you could use a mirror to test a figure's line of symmetry.

19. Look again at the Hollywood sign. What fraction of the letters are consonants?

Mixed Review and Test Prep

20. Is 71 prime or composite? (p. 264)

21. 12×135 (p. 148)

22. $9{,}005 - 3{,}538$ (p. 36)

23. $13{,}272 \div 24$ (p. 188)

24. **TEST PREP** Which of the following is divisible by 4? (p. 254)

A 13
B 56
C 105
D 182

Extra Practice page H56, Set E

LESSON 7

Problem Solving Strategy

Find a Pattern

Quick Review

Write the rule. Then find the missing number.

1. 4, 8, 12, ■

2. 2, 4, 12, 48, ■, 1,440

3. 30, 24, ■, 12

4. $\frac{1}{2}, \frac{3}{2}, \frac{5}{2}$, ■

5. 0.5, 1.6, 2.7, 3.8, ■

PROBLEM While volunteering at an archaeological dig, Kelly found this piece of painted pottery. She wants to copy the design onto a pot she is making. What should the rest of the design look like?

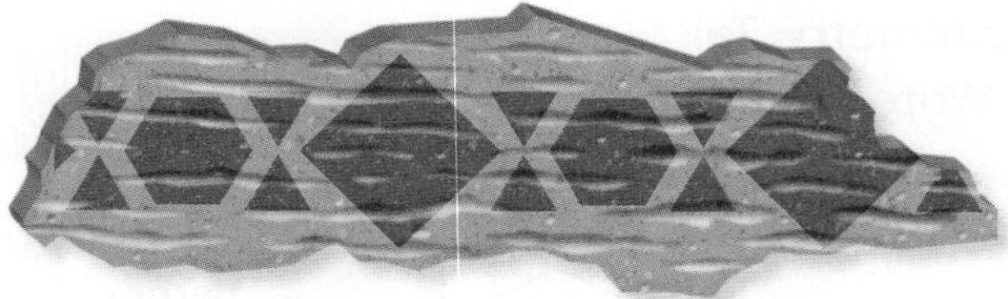

- What are you asked to find?
- What information will you use?
- Is there any information you will not use?

- What strategy can you use to solve the problem?
  You can *find a pattern* in the design. Patterns are details that repeat in the same order over and over again. If Kelly can find the pattern, she can repeat it.

- How can you use the strategy to solve the problem?

  Find the part of the design that repeats.

So, the repeated pattern of the design is 1 hexagon, 2 triangles, 1 square, 2 triangles. The next hexagon is the beginning of the repeating pattern.

- What shape should Kelly paint next to continue from the broken edge of the pottery's design?

CALIFORNIA STANDARDS MR 1.1 Analyze problems by identifying relationships, distinguishing relevant from irrelevant information, sequencing and prioritizing information, and observing patterns. *also* **MG 2.0, MR 1.0, MR 1.1 MR 2.0, MR 2.2, MR 3.3**

Problem Solving Practice

PROBLEM SOLVING STRATEGIES

- Draw a Diagram or Picture
- Make a Model or Act It Out
- Make an Organized List
- Find a Pattern
- Make a Table or Graph
- Predict and Test
- Work Backward
- Solve a Simpler Problem
- Write an Equation
- Use Logical Reasoning

Find a pattern to solve. Describe the pattern.

1. **What if** there were two hexagons next to each other instead of one hexagon on the pottery's design? What pattern should Kelly repeat on her pot?

2. Pablo painted half of the border below on the kitchen walls. Tony wants to continue the border. Draw the shapes and colors Sam should paint next.

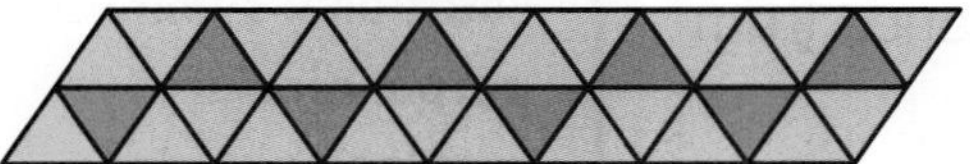

Ronnie painted the bricks along the driveway to his house.

3. What color should the next two bricks be?

   A red, green
   B red, gray
   C red, blue
   D red, red

4. What color will the fiftieth brick in the border be?

   F red
   G green
   H gray
   J blue

Mixed Strategy Practice

For 5–6, use the figures at the right.

5. Draw the next figure in the pattern.

6. Write the number that goes with each figure in the pattern. What is the eighth number in the sequence?

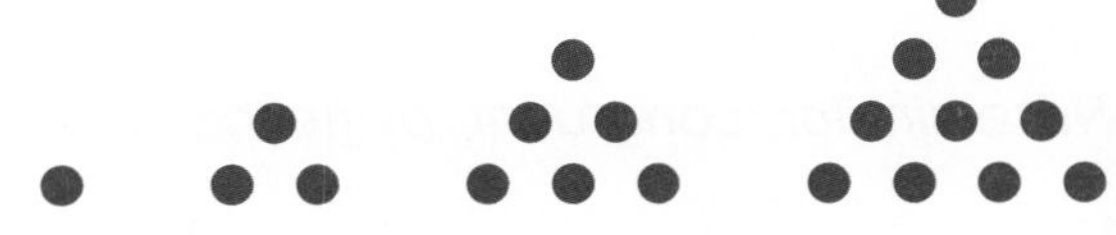

7. Ted's quilt is 5 ft $2\frac{1}{2}$ in. wide and 10 ft 2 in. long. He wants to put ribbon around the edge. How much ribbon does he need?

8. Anne spent half of her money on CDs. After she spent half of what was left on stickers, she had $9.50 left. How much did she have to start?

9. The daily temperatures for one week in Atlanta were 78°, 80°, 78°, 81°, 80°, 82°, and 82°. Make a graph to display this data. Explain your choice of graph.

10. Danny, Brenda, Ari, and Laura won the first four prizes in a contest. Danny won second prize. Laura did not win third prize. Ari won fourth prize. What prize did Brenda win?

Review/Test

CHECK VOCABULARY AND CONCEPTS

Choose the best term from the box.

circle
congruent
line symmetry
rotational symmetry
perpendicular
regular polygon
similar

1. If a figure can be rotated less than 360° around a central point and still match the original figure, it has __?__. (p. 470)
2. A closed plane figure with all points the same distance from the center point is a __?__. (p. 464)
3. Figures that have exactly the same size and shape are __?__. (p. 468)

CHECK SKILLS

Draw and label a figure for each. (pp. 454–457)

4. $\angle CLS$
5. $\overline{SO}$
6. $\overrightarrow{KA}$
7. $\overleftrightarrow{AC}$

Use a protractor to draw each angle. Then write *acute, right,* or *obtuse* for each angle. (pp. 458–459)

8. 210°
9. 30°
10. 145°
11. 95°

Name each polygon and tell if it is *regular* or *not regular*. (pp. 460–463)

12. 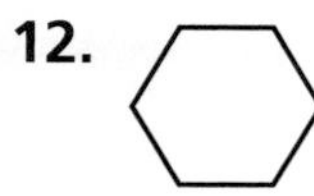

13. 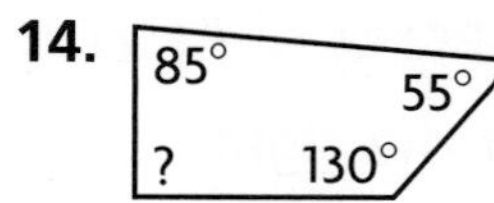

Find the unknown angle measure. (pp. 460–463)

14. 85°, 55°, ?, 130°

15. 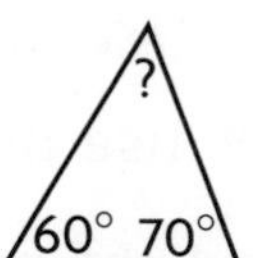

Write *similar, congruent,* or *neither* to describe each pair. (pp. 468–469)

16. 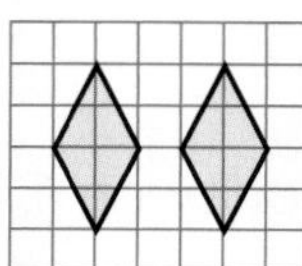

17. 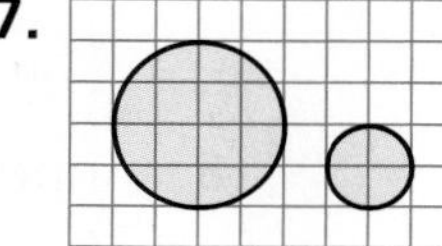

18. 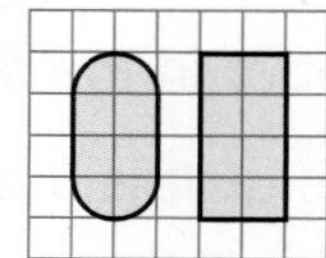

CHECK PROBLEM SOLVING

Find a pattern to solve. (pp. 472–473)

19. What shape and color should you draw next to continue the design?

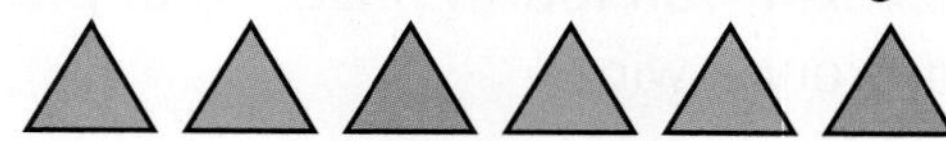

20. The school bell rang at 8:30 A.M., 9:15 A.M., 10:00 A.M., and 10:45 A.M. If the pattern continues, what is the next time the school bell will ring?

Cumulative Review

Check your work.
See item **5.**
Be sure you added the two given angle measures correctly. To find the measure of the third angle, you need to subtract the sum from the total degree measure of a triangle, which is 180.

Also see problem **7,** p. H65.

For 1–10, choose the best answer.

1. Which could be the measure of an acute angle?

   A 85° **C** 95°
   B 90° **D** 180°

2. What is true about these figures?

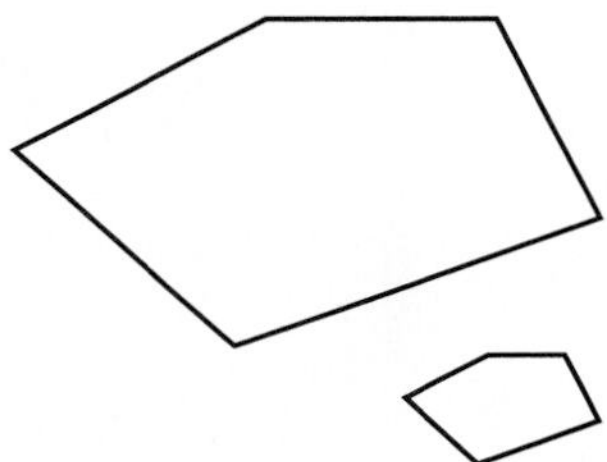

   F They are congruent.
   G They are similar.
   H They are hexagons.
   J They are regular pentagons.

3. Which of the following does *not* name a line?

   A line *AB* **C** line *BA*
   B $\overleftrightarrow{AB}$ **D** *BA*

4. Which is the best estimate for this difference?

$$\begin{array}{r} 8{,}365{,}251 \\ -2{,}185{,}204 \\ \hline \end{array}$$

   F 11,000,000 **H** 6,000,000
   G 7,000,000 **J** 5,000,000

5. If two angles of a triangle measure 45° and 60°, what is the measure of the third angle?

   A 75° **C** 105°
   B 90° **D** Not here

6. Which kind of lines intersect to form right angles?

   F parallel **H** horizontal
   G perpendicular **J** vertical

For 7–9, use the bar graph.

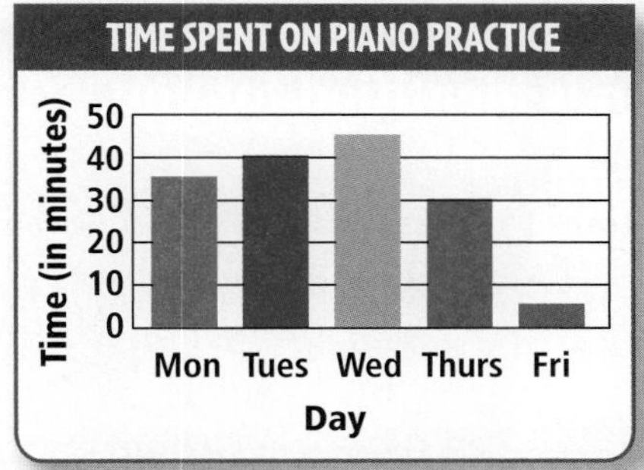

7. On which day did Miguel spend the most time on practice?

   A Monday **C** Wednesday
   B Tuesday **D** Thursday

8. On which day did Miguel spend 35 minutes on practice?

   F Monday **H** Wednesday
   G Tuesday **J** Thursday

9. For the week, what is the mean amount of time Miguel practiced each day?

   A 40 minutes **C** 35 minutes
   B 37.5 minutes **D** 31 minutes

10. 3.16×8.2

   F 24.812 **H** 25.912
   G 25.812 **J** Not here

CHAPTER 26

Classify Plane and Solid Figures

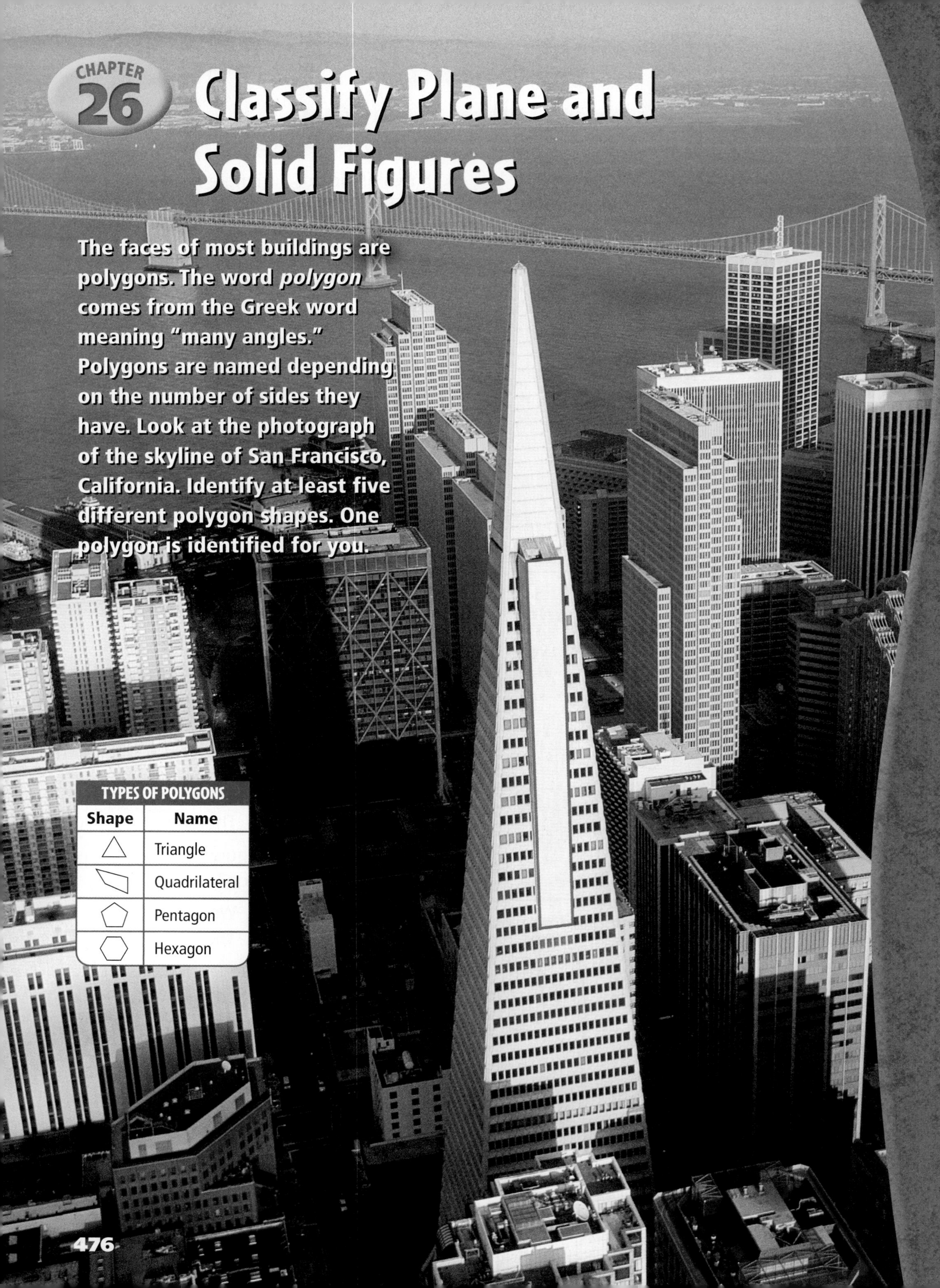

The faces of most buildings are polygons. The word *polygon* comes from the Greek word meaning "many angles." Polygons are named depending on the number of sides they have. Look at the photograph of the skyline of San Francisco, California. Identify at least five different polygon shapes. One polygon is identified for you.

TYPES OF POLYGONS

Shape	Name
△	Triangle
▱	Quadrilateral
⬠	Pentagon
⬡	Hexagon

Use this page to help you review and remember important skills needed for Chapter 26.

VOCABULARY

Choose the best term from the box.

solid figure
triangle
congruent
perpendicular
parallel

1. Lines in a plane that are the same distance apart are called __?__ lines.
2. A figure with three dimensions is called a __?__.
3. Figures that have the same shape and the same size are __?__.
4. The sum of the measures of the angles of a __?__ is 180°.

CLASSIFY ANGLES (See p. H27.)

Classify the angle. Write *acute, right,* or *obtuse*.

5. 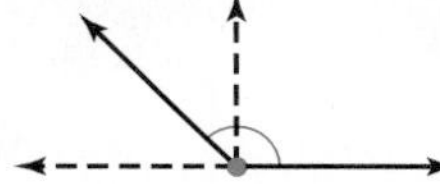

6. 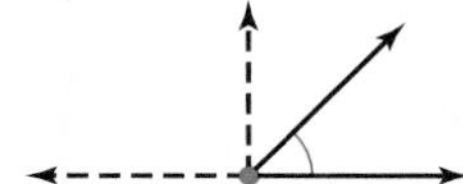

7. 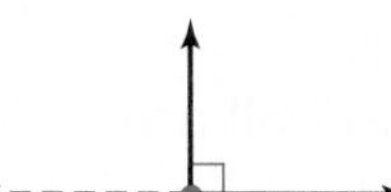

8. 

FACES OF SOLID FIGURES (See p. H28.)

Name the plane figure that is the shaded face of the solid figure.

9. 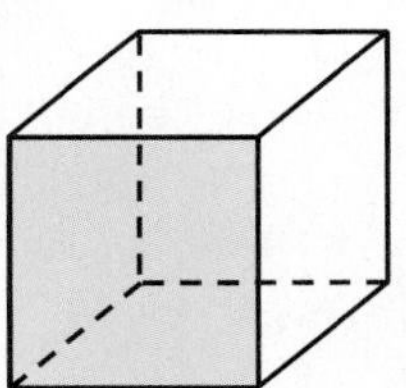

10. 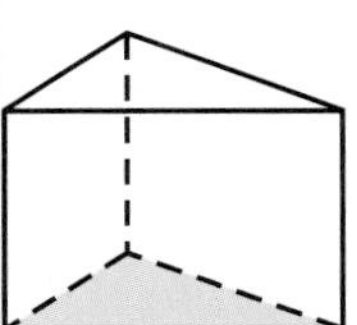

11. 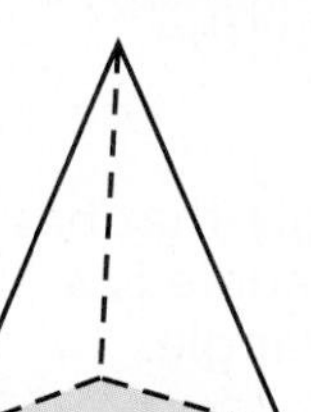

12. 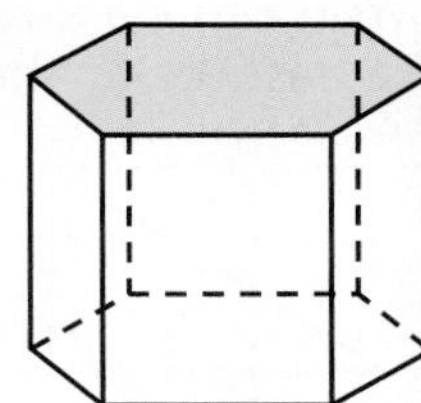

13. 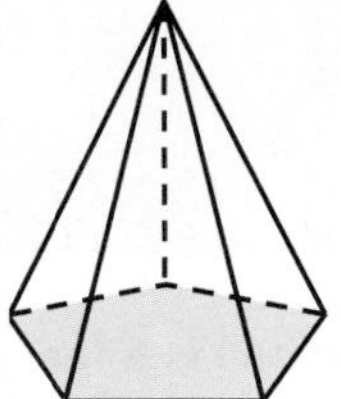

14. 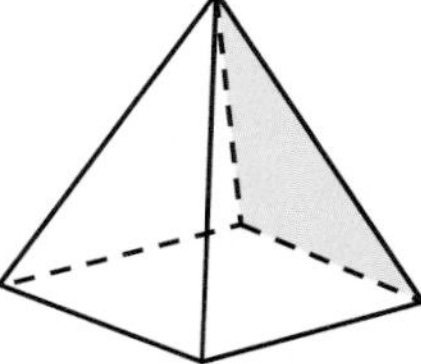

15. 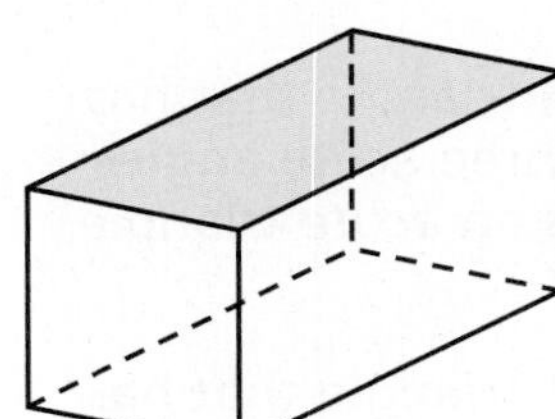

16. 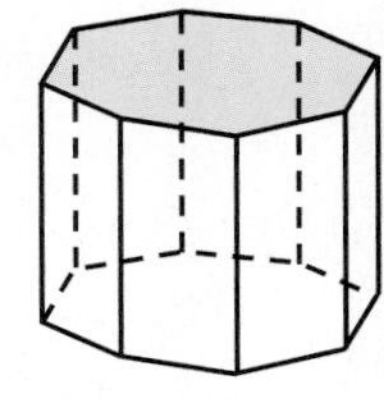

LESSON 

Triangles

Learn

SIDE BY SIDE The crystal ball that was dropped in Times Square for New Year's Eve 2000 contains 504 equilateral triangular pieces. Why are the triangles called equilateral?

You can classify triangles by the lengths of their sides.

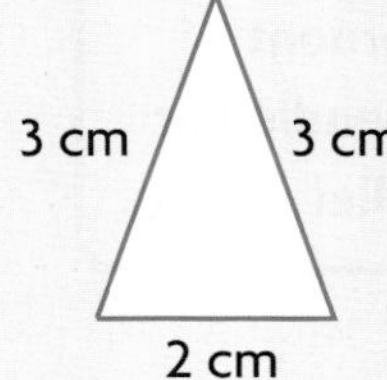

A triangle with exactly two congruent sides is an **isosceles triangle**.

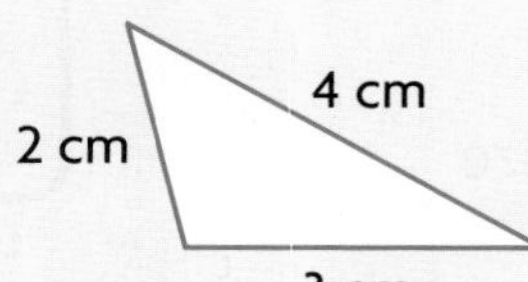

A triangle with no congruent sides is a **scalene triangle**.

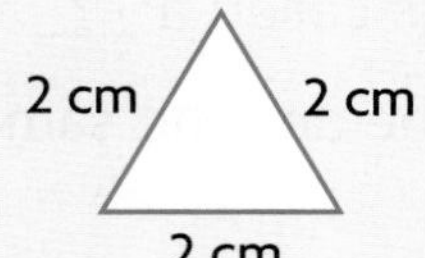

A triangle with all sides congruent is an **equilateral triangle**.

So, the triangles of the crystal ball are called equilateral triangles because they each have three congruent sides.

- A regular polygon has congruent sides and angles. Which triangle is a regular polygon?

You can classify triangles by the measures of their angles.

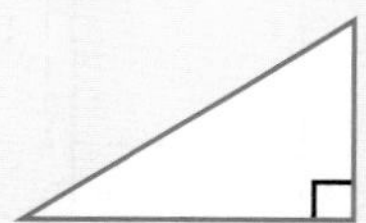

A triangle that has a right angle is a *right* triangle.

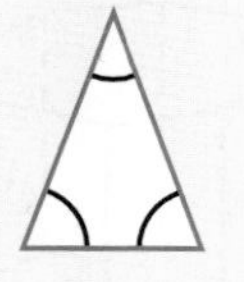

A triangle that has three acute angles is an *acute* triangle.

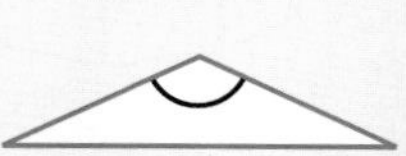

A triangle that has one obtuse angle is an *obtuse* triangle.

- Without measuring the angles, how can you tell that the triangles on the crystal ball are acute?

Quick Review

Classify each angle as right, acute, or obtuse.

1.  **2.** 

3. **4.**

5.

VOCABULARY

isosceles triangle
scalene triangle
equilateral triangle
figurate numbers
triangular numbers

▲ **The crystal ball is 6 feet in diameter and contains 600 light bulbs, 96 strobe lights, and 92 pyramid mirrors.**

CALIFORNIA STANDARDS **O‑MG 2.1** Measure, identify, and draw angles, perpendicular and parallel lines, rectangles, and triangles by using appropriate tools. **O‑MG 2.2** Know that the sum of the angles of any triangle is 180° and the sum of the angles of any quadrilateral is 360°, and use this information to solve problems. *also* **MG 2.0, MR 1.0, MR 1.1, MR 2.0, MR 2.3, MR 3.3**

Draw Triangles

Activity

MATERIALS: ruler, protractor

Draw a triangle with angles measuring 30°, 60°, and 90°. Make the side between the 30° angle and the 60° angle 2 inches long.

> **Remember**
> The sum of the measures of the angles in a triangle is always 180°.

STEP 1
Use the protractor to draw a 30° angle. Label the vertex *A*.

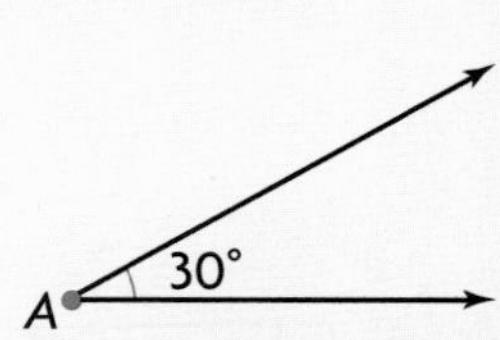

STEP 2
Make a dot 2 inches from *A* along one of the rays. Label the point *B*.

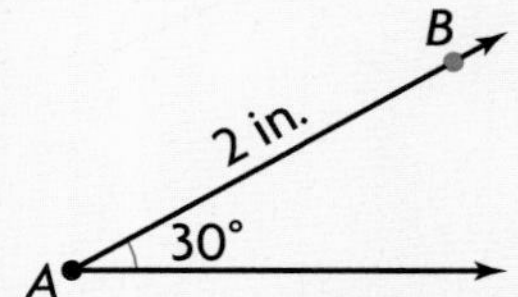

STEP 3
Draw a 60° angle with vertex at *B*. Use $\overrightarrow{BA}$ as one of the rays.

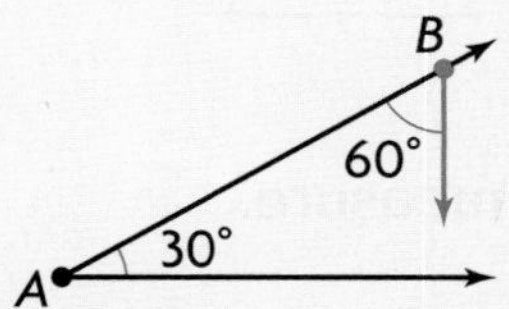

STEP 4
Extend the rays of the 60° angle and the 30° angle until they intersect. Label the point of intersection *C*.

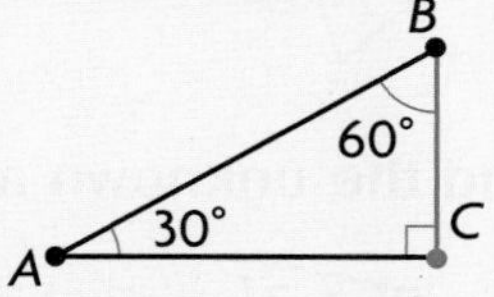

- What is the measure of ∠*C*? What kind of triangle is triangle *ABC*? Explain.

> **TECHNOLOGY LINK**
> To learn more about triangles, watch Harcourt Math Newsroom Video *Georgia Dome Architecture.*

Check

1. **Explain** why the other two angles of a triangle must be acute if one angle is a right angle.

Classify each triangle. Write *isosceles, scalene,* or *equilateral*.

2. 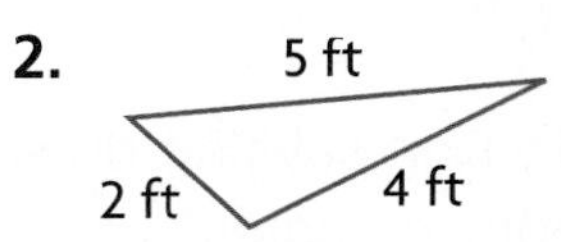

3. 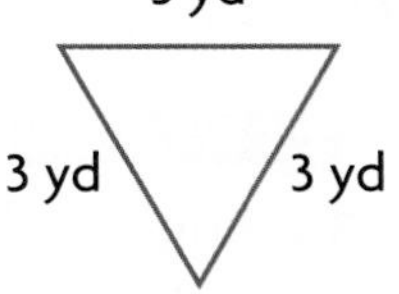

4. 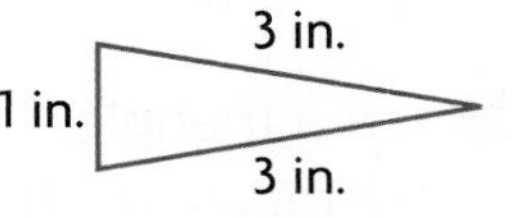
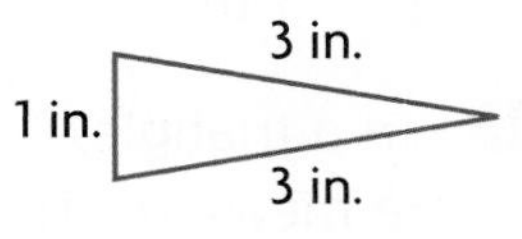

5. 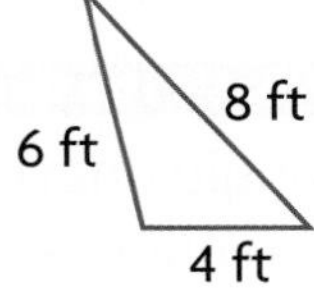

Classify each triangle. Write *right, acute,* or *obtuse*.

6. 

7. 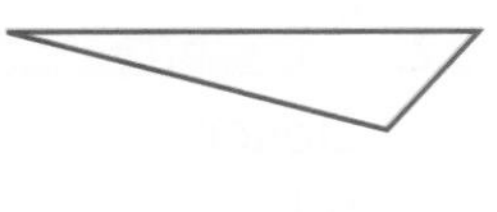

8. 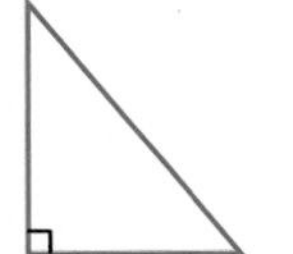

9. 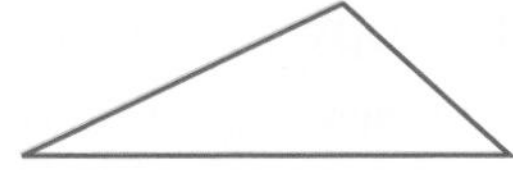

10. 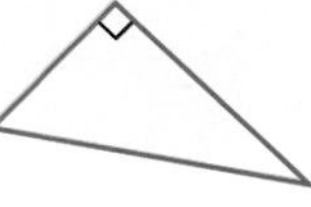

11. 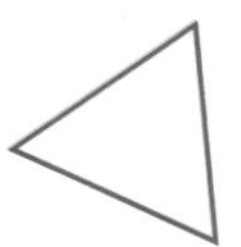

12. 

LESSON CONTINUES 

Practice and Problem Solving

Classify each triangle. Write *isosceles, scalene,* or *equilateral*.

13. 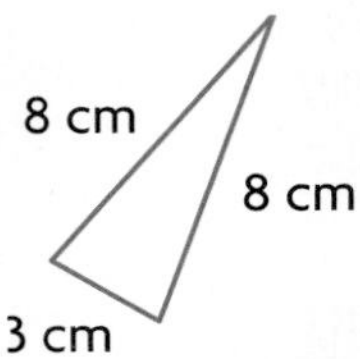

14. 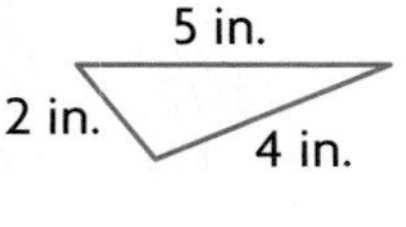

15. 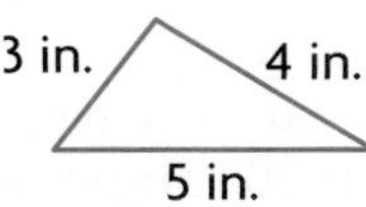

16. 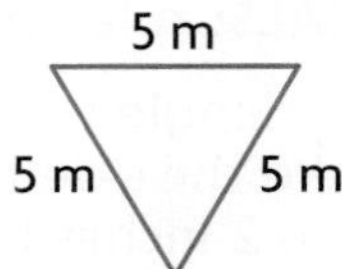

Classify each triangle. Write *right, acute,* or *obtuse*.

17. 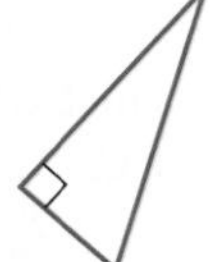

18. 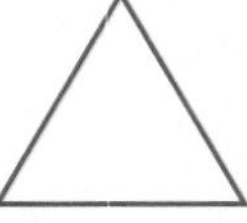

19. 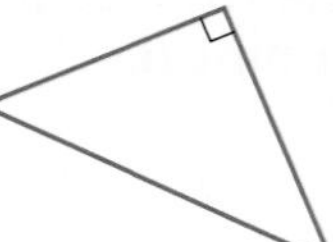

20. 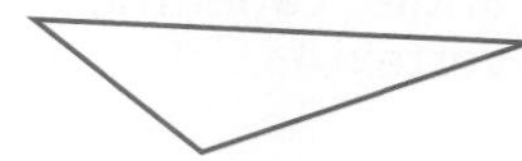

Find the unknown angle measure.

21. 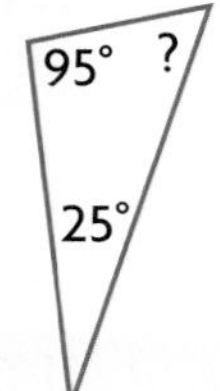

22. 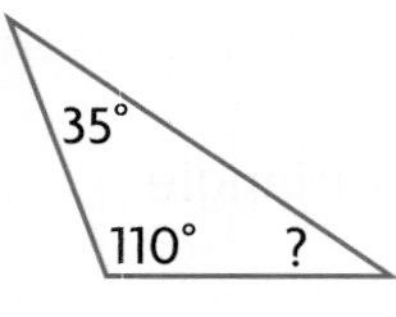

23. 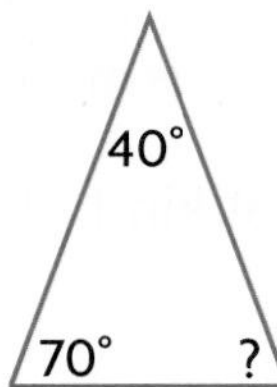

24. 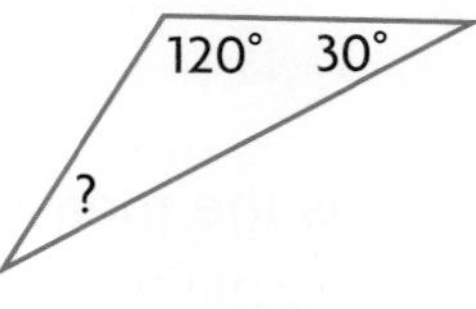

Use a protractor and ruler to draw triangle *ABC* according to the given measurements. Classify the triangle by its sides and by its angles. Then find the measure of the third angle.

25. $\angle A = 60°$
$\angle B = 60°$
$\overline{AB} = 2$ in.

26. $\angle C = 90°$
$\angle B = 55°$
$\overline{CB} = 3$ in.

27. $\angle A = 65°$
$\angle C = 65°$
$\overline{AC} = 4$ in.

28. $\angle C = 50°$
$\angle B = 20°$
$\overline{CB} = 2\frac{1}{2}$ in.

29. **What's the Question?** In a triangle, two angles have the same measure. The third angle measure is equal to the sum of the other two angle measures. The answer is right triangle.

30. **What's the Error?** Dan says he drew a triangle with two obtuse angles. Describe his error.

31. Copy this design onto your paper. See how many congruent triangles you can draw within the square array, using only three dots. Each vertex must touch a dot. One triangle is drawn for you.

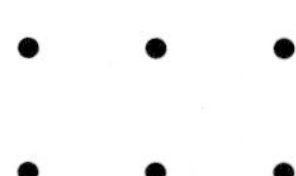

32. **Write About It** Describe a relationship shown in this Venn diagram.

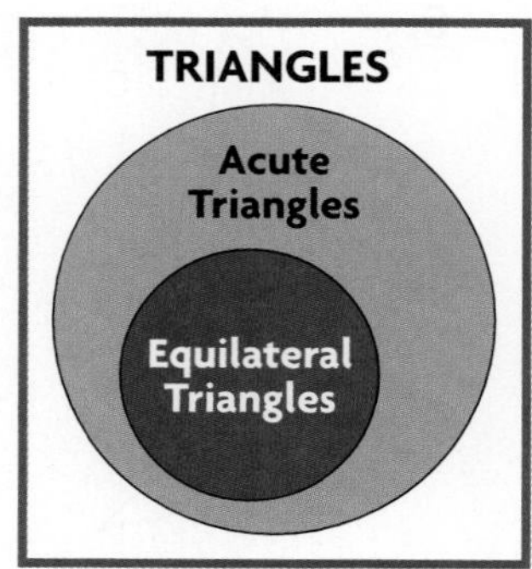

33. Keith drew an equilateral triangle. By drawing one line, how can he divide the triangle into two right triangles?

34. Draw a square and one of its diagonals. Classify the triangles by their angles and by their sides.

Mixed Review and Test Prep

USE DATA **For 35–37, use the graph.**

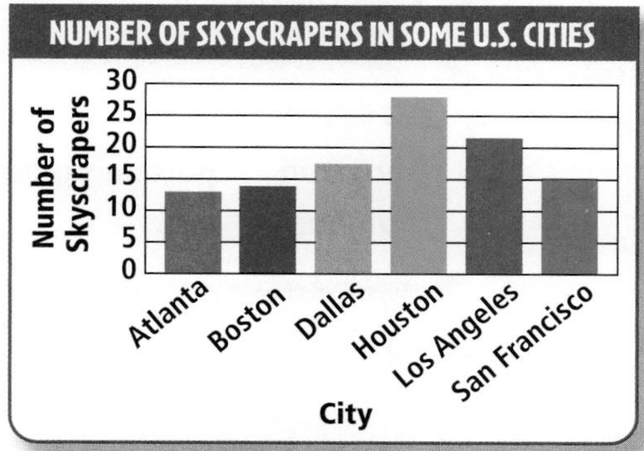

35. Which cities have more than 20 skyscrapers? (p. 108)

36. Which cities have fewer than 15 skyscrapers? (p. 108)

37. What other kind of graph would be appropriate to display this data? (p. 130)

38. **TEST PREP** Which multiplication property is illustrated in $5 \times 1 = 5$? (p. 90)
   - **A** Property of One
   - **B** Associative Property
   - **C** Commutative Property
   - **D** Distributive Property

39. **TEST PREP** Which shows the numbers ordered from greatest to least? (p. 24)
   - **F** 0.25, 1.2, 1.5, 1.05, 2.5
   - **G** 1.05, 2.5, 0.25, 1.5, 1.2
   - **H** 2.5, 0.25, 1.5, 1.2, 1.05
   - **J** 2.5, 1.5, 1.2, 1.05, 0.25

Thinker's Corner

ALGEBRA **Figurate numbers** are numbers that can be represented by geometric patterns. The number of dots used to make each of the triangular arrays below show the first four **triangular numbers**.

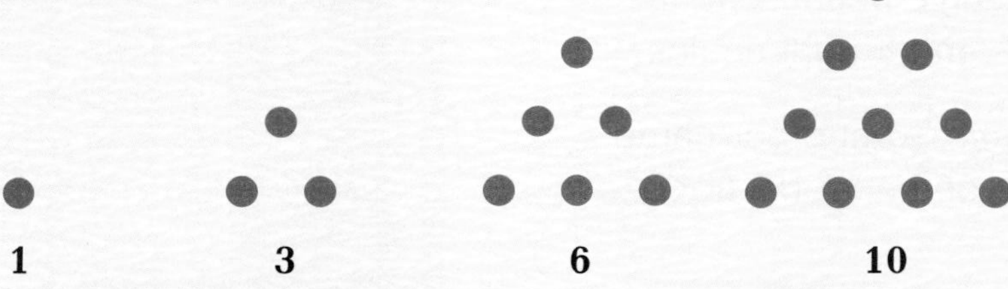

▲ **More than 2,500 years ago, the Greek mathematician Pythagoras first studied figurate numbers.**

1. What kind of triangle is used to show these triangular numbers? Explain.

2. Draw the pattern of dots for the next three triangular numbers. How many dots did you draw for each triangular number?

3. **REASONING** Explain how the number of dots added for each new triangular number is related to the number of dots added to the triangular number just before it.

LESSON 2

Quadrilaterals

Learn

FOUR BY FOUR The faces of the World Trade Center's two towers in New York City are quadrilaterals because they have four sides and four angles. Why can the faces also be called rectangles?

Just as there are many types of triangles, there are many types of quadrilaterals.

general quadrilateral	**trapezoid**	**parallelogram**
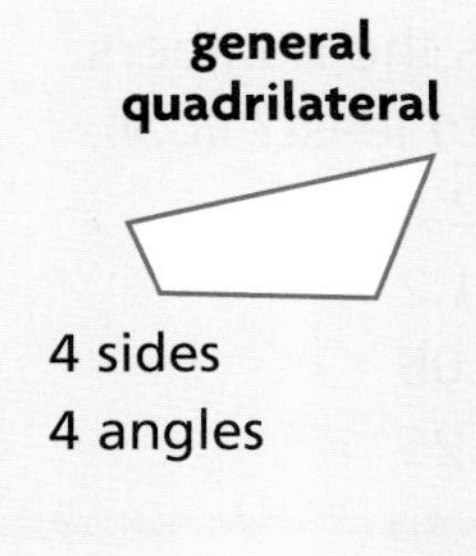	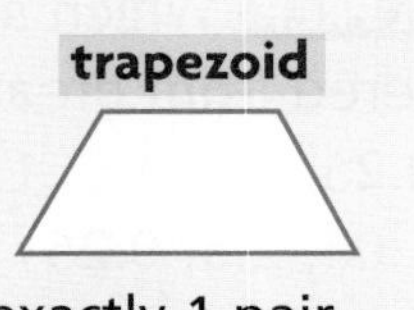	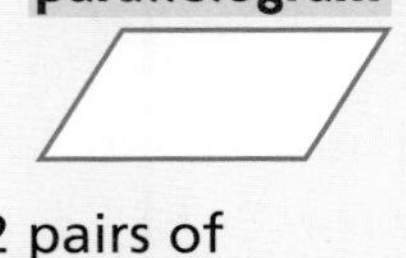
4 sides 4 angles	exactly 1 pair of parallel sides	2 pairs of congruent sides 2 pairs of parallel sides
rectangle	**rhombus**	**square**
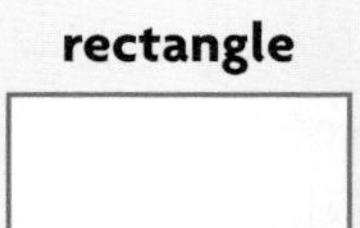 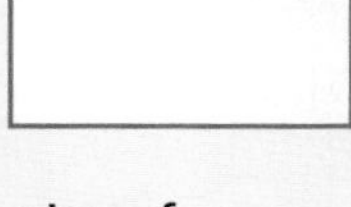		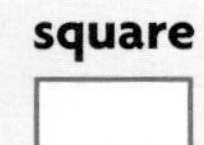
2 pairs of congruent sides 4 right angles	4 congruent sides 2 pairs of congruent angles	4 congruent sides 4 right angles

So, each face can be called a rectangle because it has 2 different pairs of congruent sides and 4 right angles.

- Besides *quadrilateral* and *rectangle,* what other name can you use to classify the faces of the World Trade Center's two towers? Explain.

MATH IDEA You can classify quadrilaterals by the characteristics of their sides and angles.

Check

1. **Explain** how a square and a rectangle are alike and how they are different.

Quick Review

Write the number of sides for each polygon.

1. square
2. triangle
3. rectangle
4. pentagon
5. octagon

VOCABULARY

trapezoid **parallelogram** **rhombus**

Remember

Two figures are *congruent* if they have the same size and shape.

Two lines are *parallel* if they never intersect and are the same distance apart at every point.

▲ **Although the two towers of the World Trade Center look congruent, the North Tower is 4 feet taller than the South Tower.**

CALIFORNIA STANDARDS MG 2.0 Students identify, describe, and classify the properties of, and the relationships between, plane and solid geometric figures. **MG 2.1** Measure, identify, and draw angles, perpendicular and parallel lines, rectangles, and triangles by using appropriate tools. *also* **MR 1.1, MR 2.0, MR 2.3, MR 3.3**

Classify each figure in as many ways as possible. Write *quadrilateral, parallelogram, square, rectangle, rhombus,* or *trapezoid.*

2. 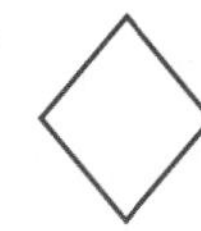

3. 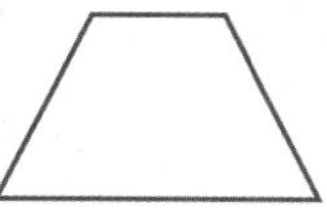

4. 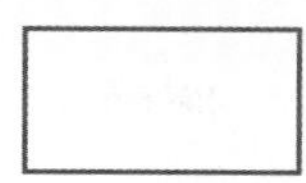

5. 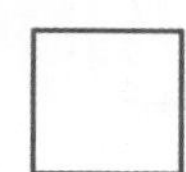

Practice and Problem Solving

Classify each figure in as many ways as possible. Write *quadrilateral, parallelogram, square, rectangle, rhombus,* or *trapezoid.*

6. 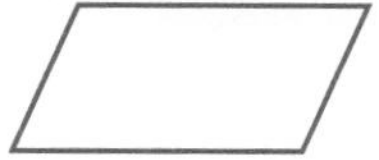

7. 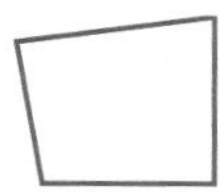

8. 

9. 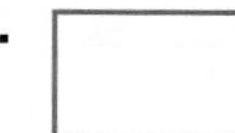

Write all of the names for each figure. Then write the names that do *not* name the figure. Choose from *trapezoid, parallelogram, rhombus, rectangle,* and *square.*

10. 

11. 

12. 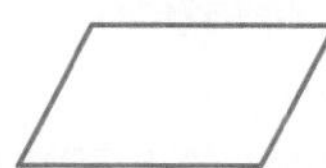

13. 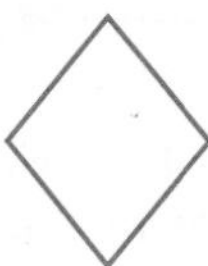

For 14–17, draw and classify each quadrilateral described.

14. a parallelogram that has all sides congruent

15. a parallelogram with 4 right angles

16. a figure that is both a rhombus and a rectangle

17. a figure that has only 1 pair of parallel lines

18. The North Tower of the World Trade Center has 110 stories and is 1,378 feet tall. About how tall is each story?

19. **REASONING** Troy says that all parallelograms are rectangles. Do you agree? Explain.

20. I have 540 rectangles and trapezoids. I have 120 more trapezoids than rectangles. How many of each do I have?

21. **REASONING** Are all squares also rectangles? Explain.

Mixed Review and Test Prep

22. $654{,}821 \times 22$ (p. 148)

23. $78{,}975 \times 43$ (p. 148)

24. $\frac{2}{3} + \frac{7}{9}$ (p. 354)

25. $\frac{1}{5} + \frac{3}{8}$ (p. 354)

26. **TEST PREP** Two angles in a triangle measure 48° and 67°. What is the measure of the third angle? (p. 460)

A 48° **B** 65° **C** 67° **D** 180°

Extra Practice page H57, Set B

LESSON 3

HANDS ON
Algebra: Transformations

Quick Review

How many congruent sides must each quadrilateral have?

1. square **2.** rectangle
3. rhombus **4.** trapezoid
5. parallelogram

VOCABULARY

transformation **translation**
reflection **rotation**

MATERIALS

coordinate planes, straightedge, scissors, colored pencils

Explore

A rigid **transformation**, or movement, of a figure does not change its size or shape. Here are examples of three types of transformations.

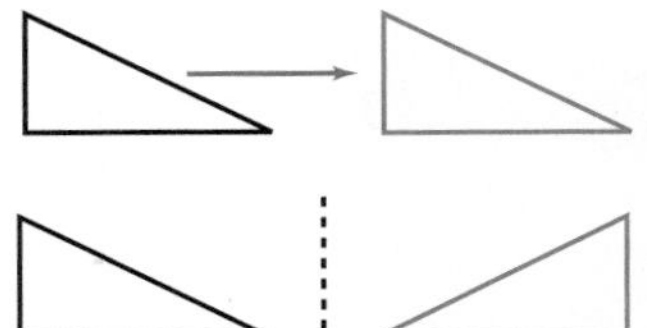

Sliding a figure in a straight line is called a **translation**.

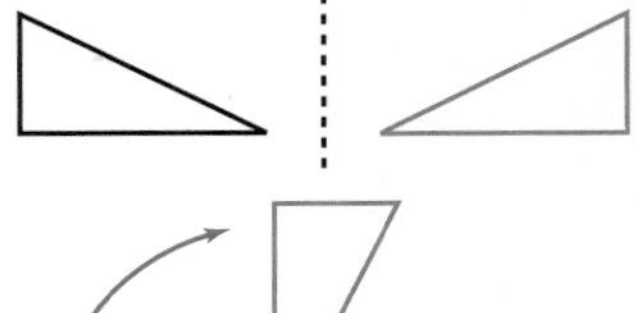

Flipping a figure over a line is called a **reflection**.

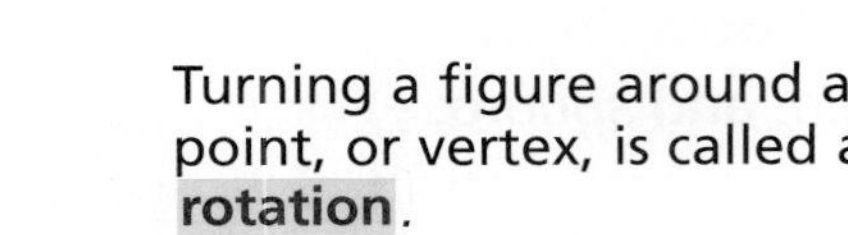

Turning a figure around a point, or vertex, is called a **rotation**.

Graph a triangle with vertices ($^+1$,$^+2$), ($^+4$,$^+2$), and ($^+4$,$^+6$). Trace it on a sheet of paper. Color and cut it out. Place it on the triangle on the coordinate plane. Transform the triangle according to the directions. Name the ordered pairs for the new vertices.

Reflection

Reflect the triangle across the *y*-axis.

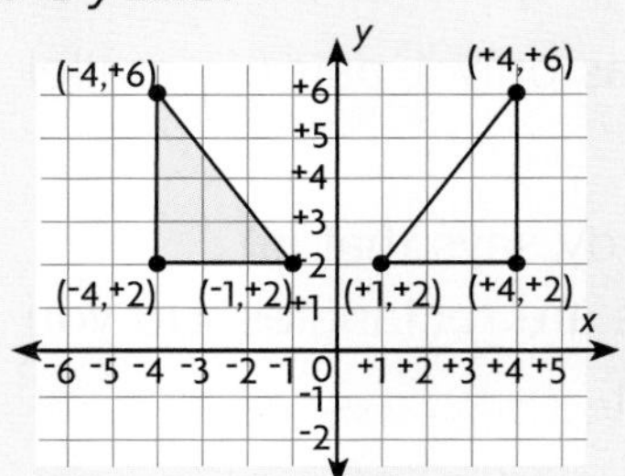

New ordered pairs:
($^-1$,$^+2$), ($^-4$,$^+2$), ($^-4$,$^+6$)

Rotation

Rotate the triangle 90° around the origin.

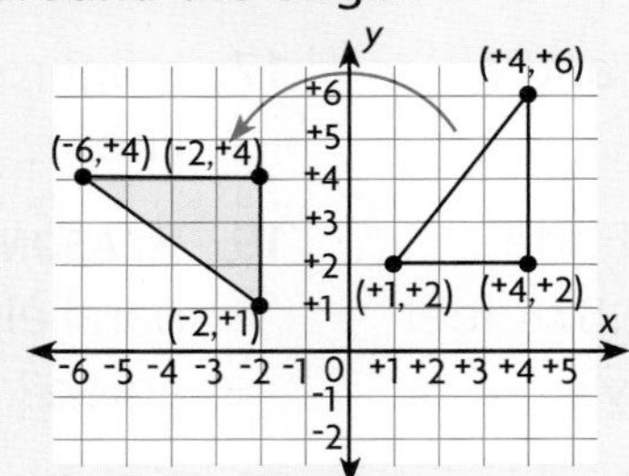

New ordered pairs:
($^-2$,$^+1$), ($^-2$,$^+4$), ($^-6$,$^+4$)

Translation

Translate each vertex 2 units left and 1 unit down.

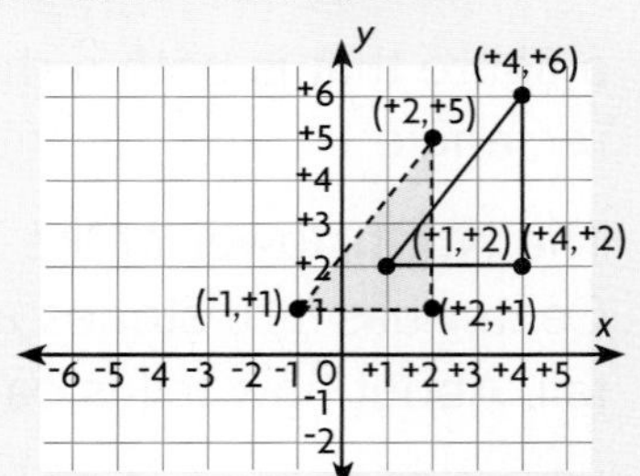

New ordered pairs:
($^-1$,$^+1$), ($^+2$,$^+1$), ($^+2$,$^+5$)

Try It

Graph the triangle with vertices (1,1), (3,1), and (1,4). Then transform the triangle to the new given vertices. Write *translation, reflection,* or *rotation* to describe the move.

a. (3,3), (3,1), and (6,3) **b.** (4,$^-1$), (6,$^-1$), and (4,2)

TECHNOLOGY LINK

More Practice: Use E-Lab, *Transformations on the Coordinate Grid.*

www.harcourtschool.com/elab2002

CALIFORNIA STANDARDS **AF 1.4** Identify and graph ordered pairs in the four quadrants of the coordinate plane. **MG 2.1** Measure, identify, and draw angles, perpendicular and parallel lines, rectangles, and triangles by using appropriate tools. **SDAP 1.5** Know how to write ordered pairs correctly. *also* **MG 2.0, MR 1.0, MR 1.1, MR 2.0, MR 3.0, MR 3.3**

Connect

When you translate a figure, you move all of its vertices the same number of spaces in the same directions. So, all of its ordered pairs change in the same way. The table shows how an ordered pair changes when you translate that point.

Translation of (2,4)	Change to Ordered Pair	New Ordered Pair
Move 3 spaces to the *right.*	*Add* 3 to the first number in the pair.	(2+3,4) → (5,4)
Move 2 spaces to the *left.*	*Subtract* 2 from the first number in the pair.	(2−2,4) → (0,4)
Move 4 spaces *up.*	*Add* 4 to the second number in the pair.	(2,4+4) → (2,8)
Move 3 spaces *down.*	*Subtract* 3 from the second number in the pair.	(2,4−3) → (2,1)

- **REASONING** If a point at (4,7) is moved 4 spaces to the right and 9 spaces down, what will its new ordered pair be?

Practice

Graph the triangle with vertices (2,2), (2,4), and (6,2). Then transform the triangle to the new given vertices. Write *translation, reflection,* or *rotation* to describe the move.

1. (4,0), (4,2), and (8,0)
2. (6,6), (8,6), and (6,2)
3. (2,2), (2,4), and ($^-$2,2)

For 4–5, graph the triangle with vertices (1,2), (5,2), and (1,4).

4. Translate the triangle 4 spaces to the right and 1 space up. What are the ordered pairs for the vertices of the triangle's new location?
5. Reflect the triangle across the horizontal line through the point (1,4). What are the ordered pairs for the vertices of the triangle's new location?

Mixed Review and Test Prep

6. $\frac{3}{5} - \frac{1}{8}$ (p. 354)
7. $\frac{4}{5} - \frac{1}{4}$ (p. 354)
8. What quadrilateral has 4 congruent sides and 2 pairs of congruent angles? (p. 482)
9. Write 3.695 in word form. (p. 20)
10. **TEST PREP** In what kind of triangle does each side always have a different length? (p. 478)
    - **A** scalene
    - **B** isosceles
    - **C** equilateral
    - **D** right

LESSON 4

Solid Figures

Quick Review

Draw an example of each polygon.

1. rhombus
2. isosceles triangle
3. parallelogram
4. right triangle
5. trapezoid

VOCABULARY

polyhedron **prism**
pyramid **base**

Learn

BOX, BALL, OR CAN? The design of the Rock and Roll Hall of Fame and Museum in Cleveland, Ohio, suggests several solid figures.

A solid figure with faces that are polygons is called a **polyhedron**.

All of the flat surfaces are called *faces*.

The line where two faces come together is an *edge*.

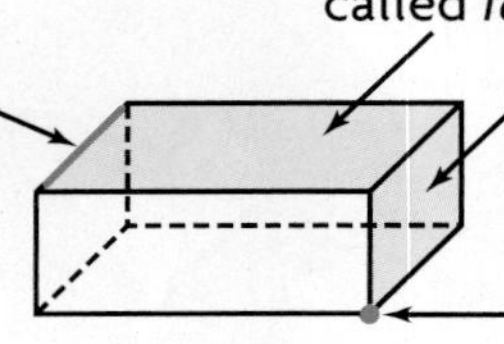

The point where several edges come together is a *vertex*.

The main tower suggests a rectangular prism. A **prism** is a polyhedron that has two congruent **bases**. All other faces of a prism are rectangles. Polyhedrons are named by the polygons that form their bases.

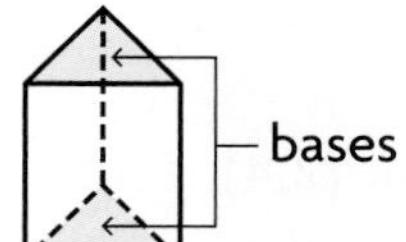

triangular prism

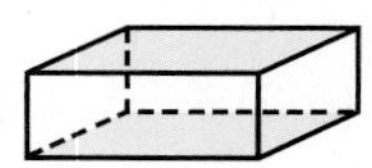

rectangular prism

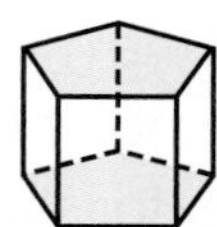

pentagonal prism

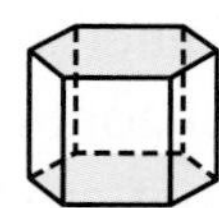

hexagonal prism

▲ **I. M. Pei, the architect of the Rock and Roll Hall of Fame, is known for combining pyramids and other solids in his designs.**

The glass structure next to the museum's tower suggests a triangular pyramid. A **pyramid** is a polyhedron with only one base. All other faces are triangles that meet at the same vertex.

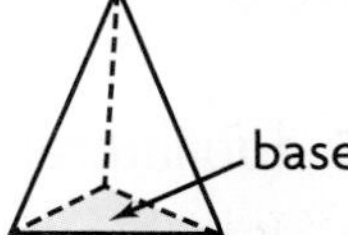

triangular pyramid

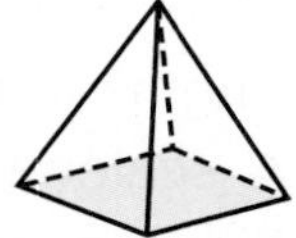

square pyramid

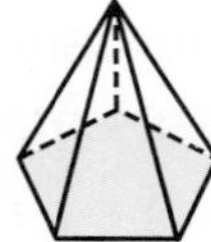

pentagonal pyramid

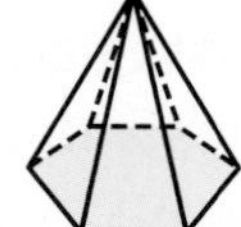

hexagonal pyramid

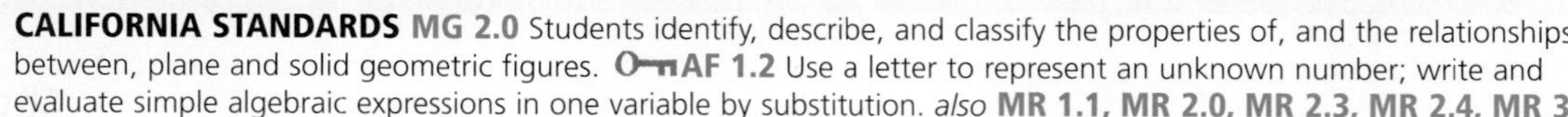

CALIFORNIA STANDARDS MG 2.0 Students identify, describe, and classify the properties of, and the relationships between, plane and solid geometric figures. **AF 1.2** Use a letter to represent an unknown number; write and evaluate simple algebraic expressions in one variable by substitution. *also* **MR 1.1, MR 2.0, MR 2.3, MR 2.4, MR 3.3**

Classify Solid Figures

Solids with curved surfaces are *not* polyhedrons.
The column next to the museum's tower suggests a cylinder.

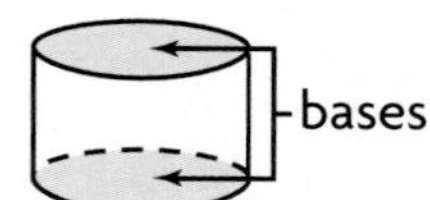

cylinder
A cylinder has 2 flat circular bases and 1 curved surface.

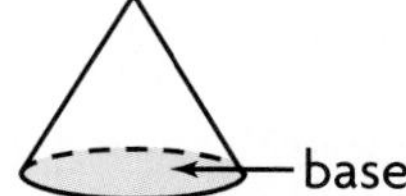

cone
A cone has 1 flat circular base and 1 curved surface.

sphere
A sphere has no bases and 1 curved surface.

MATH IDEA Solid figures can be classified by the shape and number of their bases, faces, vertices, and edges.

Examples

Classify the solid figure suggested by each building. If the figure is a polyhedron, tell the number of faces, vertices, and edges it has.

A Flatiron Building New York City, NY

It has 2 triangular bases.
Faces: **5**
Vertices: **6**
Edges: **9**

Classify: triangular prism

B Grande Louvre Paris, France

It has 1 square base.
Faces: **5**
Vertices: **5** Edges: **8**

Classify: square pyramid

C Renaissance Center Detroit, MI

It has 2 circular bases.
A cylinder is not a polyhedron.

Classify: cylinder

Check

1. **Explain** why a cylinder is *not* a polyhedron.

Classify each solid figure. Write *prism, pyramid, cone, cylinder,* or *sphere*.

2. 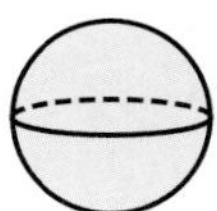

3. 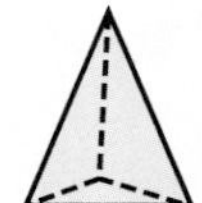

4. 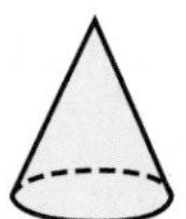

5. 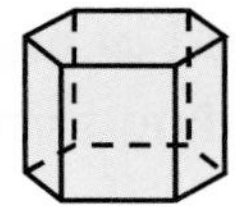

Classify the solid figure. Then, write the number of faces, vertices, and edges.

6. 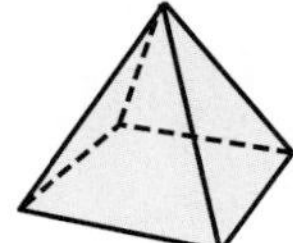

7. 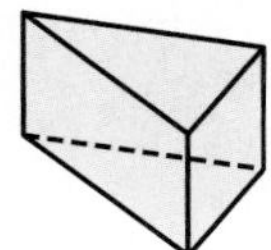

8. 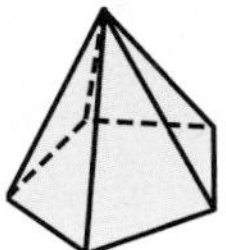

9. 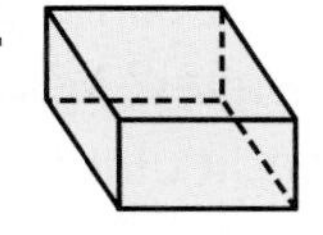

LESSON CONTINUES 

▶ Practice and Problem Solving

Classify each solid figure. Write *prism, pyramid, cone, cylinder,* or *sphere*.

10. 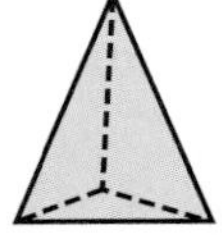

11. 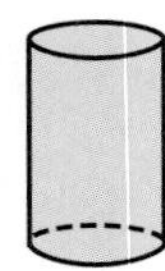

12. 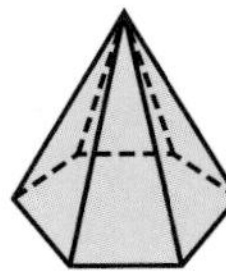

13. 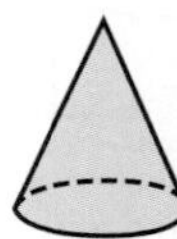

Classify the solid figure. Then, write the number of faces, vertices, and edges.

14. 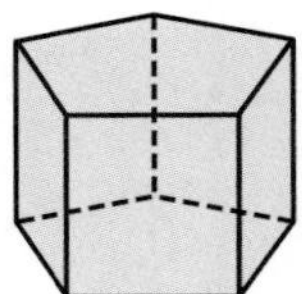

15. 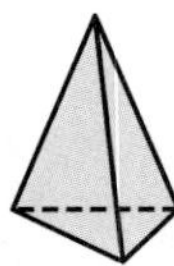

16. 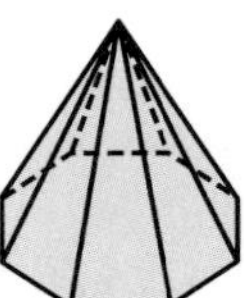

17. 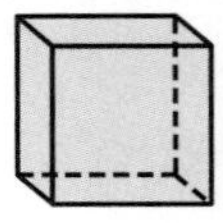

Draw and classify each solid figure described.

18. I have a base with 4 congruent sides. My other faces are 4 triangles.

19. I have 2 bases that are congruent circles.

20. I have 2 congruent triangles for bases and 3 rectangular faces.

21. All of my 4 faces are congruent triangles.

Copy and complete the table of solid figures.

	Name of Solid Figure	Number and Name of Bases	Number of Faces	Number of Vertices	Number of Edges
22.	rectangular prism	2 rectangles	6	■	■
23.	triangular pyramid	1 triangle	■	■	6
24.	pentagonal prism	■	7	10	■
25.	pentagonal pyramid	■	■	■	10

For 26–28, use the picture at the right of the Octagon House in Camillus, NY.

▼ **Octagon House, Camillus, NY**

26. What shape is the floor of the building?

27. What kind of angle does each corner of the building form?

28. When the Octagon House was built, each of its sides was 17 feet long. What was the perimeter of the house?

29. Which faces on a rectangular prism are parallel? Which faces are perpendicular?

30. Yoko walked around the perimeter of a rectangular building. The length of the building was 110 ft and the width was 82 ft. How many yards did Yoko walk?

31. The Miller family spent 4 days on a camping trip. They spent $\frac{1}{6}$ of the time hiking. How many hours did they spend hiking?

Mixed Review and Test Prep

32. 6.25×3.1 (p. 164)

33. 106.54×5.8 (p. 164)

For 34–37, name the polygon with the given number of sides. (p. 460)

34. 5 35. 3 36. 8 37. 6

Evaluate. (p. 268)

38. 3^3 39. 5^2 40. 2^4 41. 10^3

42. **TEST PREP** What is the least common multiple of 15 and 20? (p. 256)

A 5 C 20
B 15 D 60

For 43–45, use the bar graph. (p. 108)

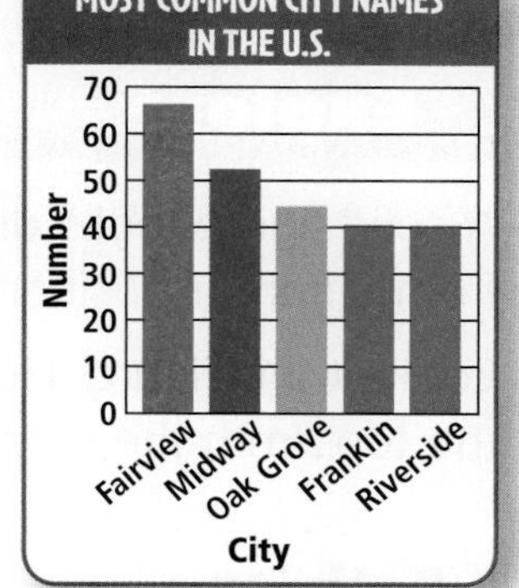

43. Which two names appear the same number of times?

44. What is the most common U.S. city name?

45. **TEST PREP** Which is the third most common U.S. city name?

F Fairview H Oak Grove
G Midway J Franklin

Thinker's Corner

ALGEBRA EXPRESS! You can write and evaluate expressions to find the number of faces, vertices, and edges for prisms and pyramids.

Prisms	Pyramids
Let n = the number of sides on the base	Let n = the number of sides on the base
$n + 2$ = number of faces	$n + 1$ = number of faces
$n \times 2$ = number of vertices	$n + 1$ = number of vertices
$n \times 3$ = number of edges	$n \times 2$ = number of edges

▲ A 16th century Swiss mathematician named Leonhard Euler discovered a relationship among the number of vertices, edges, and faces of certain solid figures such as prisms and pyramids.

hexagonal prism

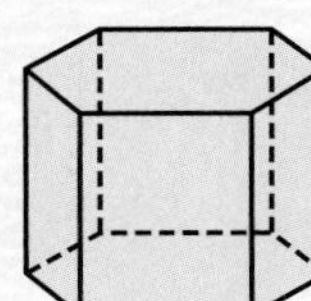

The number of sides on the base is 6.
So, $n = 6$.
number of faces: $n + 2 = 6 + 2 = 8$
number of vertices: $n \times 2 = 6 \times 2 = 12$
number of edges: $n \times 3 = 6 \times 3 = 18$

For 1–3, tell how many faces, vertices, and edges each figure has.

1. triangular prism
2. hexagonal pyramid
3. octagonal prism

LESSON 5

HANDS ON

Draw Solid Figures from Different Views

Quick Review

Identify the solid that has the base or bases described.

1. 2 squares **2.** 1 circle
3. 1 rectangle **4.** 2 circles
5. 2 hexagons

MATERIALS
connecting cubes, grid paper

Explore

A solid figure looks different when it is viewed from different positions.

Look at this figure. How would it look from the top, from the side, and from the front?

Activity

Use connecting cubes to build the solid above. Draw three pictures on grid paper to show how the figure looks from the top, from the side, and from the front.

Top View

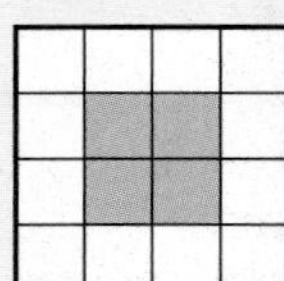

Side View

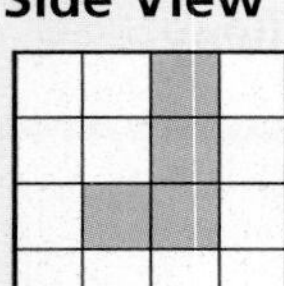

Front View

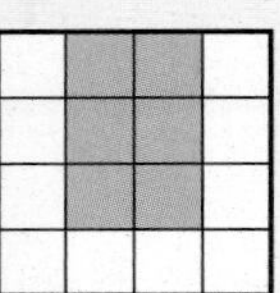

Which squares should you shade to show the side view?

Try It

Use cubes to build each figure. On grid paper, draw each figure from the top, from the side, and from the front.

a.

b.

c.

CALIFORNIA STANDARDS MG 2.3 Visualize and draw two-dimensional views of three-dimensional objects made from rectangular solids. MG 2.0 Students identify, describe, and classify the properties of, and the relationships between, plane and solid geometric figures. *also* **MG 2.1, MR 2.0, MR 3.0, MR 3.2, MR 3.3**

Connect

You can identify a solid by the way it looks from different views.

Which solid figure is shown by these three views?

Top View
The top view shows that the base is a circle and that the top comes to a point.

Front View
From the front the figure looks like a triangle.

Side View
From the side the figure looks like a triangle.

So, a solid figure with these views is a cone.

- What solid figure has squares in all of its views?

TECHNOLOGY LINK

More Practice: Use E-Lab, *Solid Figures from Different Views.*
www.harcourtschool.com/elab2002

Practice

Use centimeter cubes to build each figure. On grid paper, draw each figure from the top, from the side, and from the front.

1. 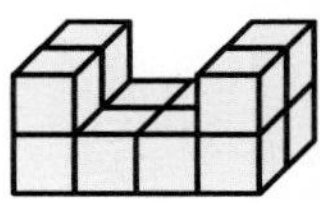
2. 
3. 
4. 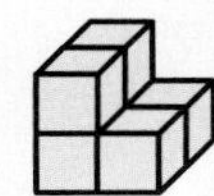

Identify the solid figure that has the given views.

5. 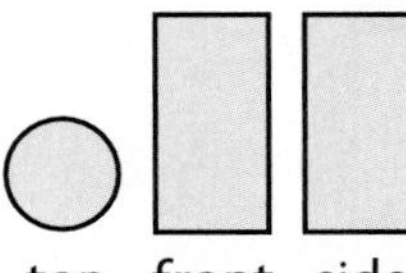

6. 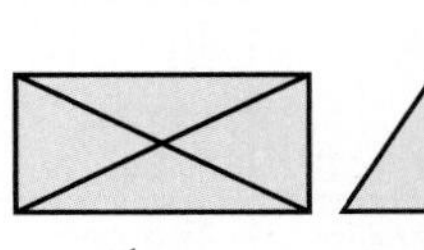

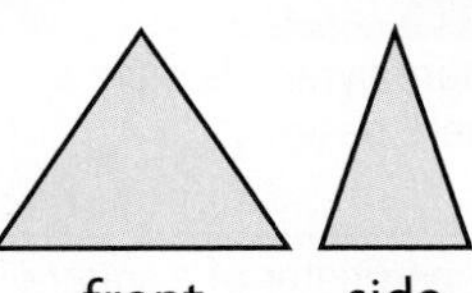

7. 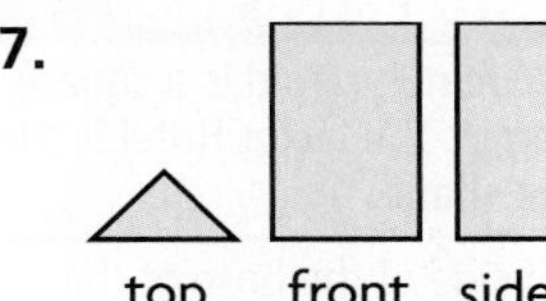

8. Which solids have circles in some of their views?

9. **Write About It** Name a solid figure and describe it from two views.

Mixed Review and Test Prep

10. $\frac{3}{5} + \frac{1}{2}$ (p. 354)

11. $\frac{1}{3} + \frac{1}{7}$ (p. 354)

12. $3\overline{)6,852}$ (p. 188)

13. $6\overline{)22,692}$ (p. 188)

14. **TEST PREP** Which figure has 4 faces, 4 vertices, and 6 edges? (p. 486)
   A cube
   B sphere
   C cone
   D triangular pyramid

LESSON 6

Problem Solving Skill

Make Generalizations

Quick Review

Find the perimeter of each regular polygon.

1. 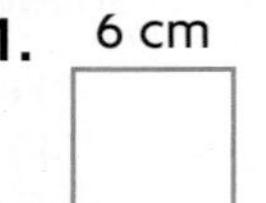

2. 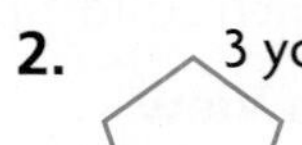

3. 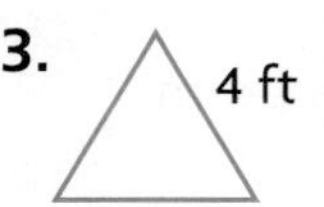

4. 

5. 

GENERALLY SPEAKING . . . The Luxor Hotel in Las Vegas, Nevada, and the Great Pyramid in Egypt are the same shape. The Great Pyramid is a square pyramid. Each side of its base is 756 feet long. The perimeter of the base of the hotel is 600 feet less than the perimeter of the base of the Great Pyramid. What is the perimeter of the base of the hotel?

Sometimes you need to *make generalizations* to solve a problem. When you generalize, you make a statement that is true about a whole group of similar situations or objects.

▲ The Great Pyramid in Giza, Egypt

▲ The Luxor Hotel in Las Vegas, Nevada

What You Know	Generalization	Conclusion
The Great Pyramid is a square pyramid. The Luxor Hotel is the same shape.	Square pyramids have a square base.	The hotel has a square base.
Each side of the base of the Great Pyramid is 756 feet long.	The perimeter of a square is 4 × length of one side.	The perimeter of the base of the Great Pyramid is 4 × 756 ft, or 3,024 ft.
The perimeter of the base of the hotel is 600 feet less than the perimeter of the base of the Great Pyramid.	To find an amount less than a given amount, you subtract.	The perimeter of the base of the hotel is 3,024 ft − 600 ft, or 2,424 ft.

So, the perimeter of the base of the hotel is 2,424 feet.

Talk About It

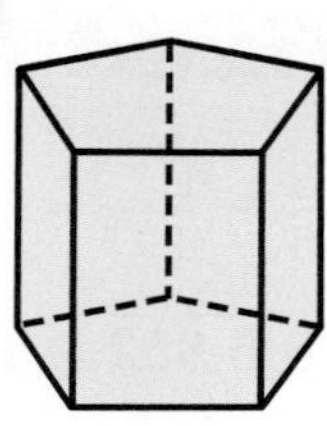

- What are some generalizations you can make about a building that is the same shape as the model at the right?

CALIFORNIA STANDARDS **MR 3.3** Develop generalizations of the results obtained and apply them in other circumstances. **MR 3.0** Students move beyond a particular problem by generalizing to other situations. **MG 2.0** Students identify, describe, and classify the properties of, and the relationships between, plane and solid geometric figures. *also* **MR 1.0, MR 2.0, MR 3.2**

Problem Solving Practice

Make generalizations to solve.

1. Two brands of soda are packaged in congruent cylinders. The first brand contains 12 ounces of soda in each can. How many ounces are in a 6-pack of the second brand?

2. The Pyramid of Khafre is the second largest pyramid in Giza. It is the same shape as the Great Pyramid. The perimeter of its base is 2,816 feet. How long is each side of its base?

A plane figure has 4 congruent sides.

3. The perimeter of the figure is 6 centimeters. What is the length of each side?

   A 24 cm **C** 3 cm
   B 12 cm **D** 1.5 cm

4. One of the angles of the figure has a measure of 110°. What is the shape of the figure?

   F rectangle **H** circle
   G square **J** rhombus

Mixed Applications

USE DATA For 5–7, use the bar graph.

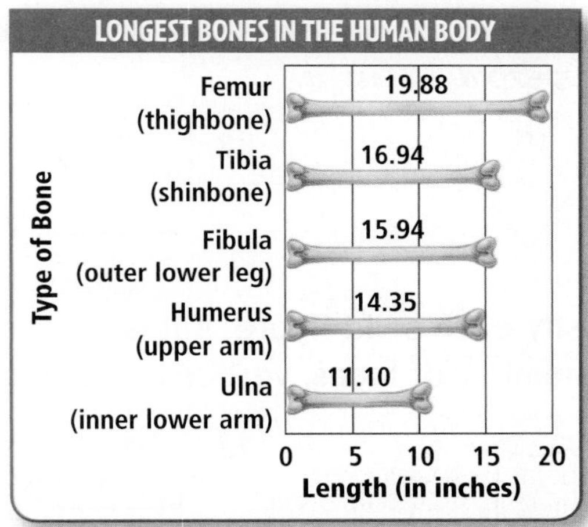

5. What is the difference in length between the longest leg bone and longest arm bone?

6. **REASONING** This bone's length has a 1 in the tens place and a 9 in the tenths place. It is not the second longest bone. Which bone is it?

7. What is the range of the data displayed on the graph?

8. About 2,300,000 stone blocks were used to build the Great Pyramid. Each stone weighs 2.5 tons. About how much do they weigh altogether?

9. **REASONING** Keyshawn's tree house is 5 feet by 7 feet. His circular table has a radius of 3 feet. Will the table fit in his tree house? Explain.

10. I have a circular base. From the side I look like a square. What figure am I?

11. My two bases are congruent. I have 3 other faces. What figure am I?

12. The printer estimated it would cost $0.14 per brochure to print 1,000 one-color brochures. What is the total cost to print 1,000 brochures?

13. **Write a problem** using the data in the table above.

Review/Test

CHECK VOCABULARY AND CONCEPTS

Choose the best term from the box.

rhombus
bases
prism
trapezoid

1. Prisms and pyramids are named by the polygons that form their __?__. (p. 486)
2. A quadrilateral with 4 congruent sides and 2 pairs of congruent angles is a __?__. (p. 482)

CHECK SKILLS

Classify each triangle by its sides and angles. (pp. 478–481)

3. 3 ft, 3 ft, 2 ft
4. 4 in., 3 in., 5 in.
5. 2 cm, 2 cm, 2 cm
6. 6 yd, 10 yd, 8 yd

Classify each figure in as many ways as possible. Write *quadrilateral, parallelogram, square, rectangle, rhombus,* or *trapezoid.* (pp. 482–483)

7. 
8. 
9.
10.

Classify each solid figure. If it is a prism or pyramid, write the number of faces, vertices, and edges. (pp. 486–489)

11. 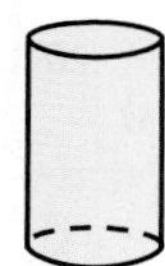
12. 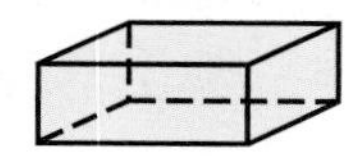
13. 
14. 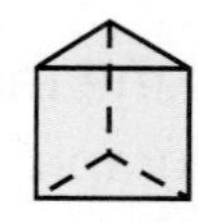

Identify the solid figure that has the given views. (pp. 490–491)

15. 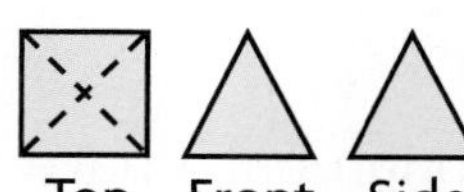

16. 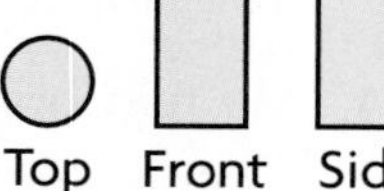

17. 

18. 

CHECK PROBLEM SOLVING

For 19–20, the figure has one right angle and a total of three angles. (pp. 492–493)

19. One angle measures 45°. What is the third angle measure?
20. What is the shape of the figure? Can all 3 sides on the figure be the same length?

Cumulative Review

Decide on a plan.
See item **2**.
Look at the relationships of faces, edges, and vertices in the drawing. Picture the top view of each solid in your mind to find the same relationships.

Also see problem **4**, p. H63.

For 1–10, choose the best answer.

1. What is the value of the 6 in 56,894.103?

   A 6,000,000   C 60,000
   B 600,000   D 6,000

2. This drawing could be a top view of which solid?

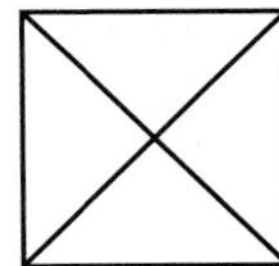

   F hexagonal prism
   G triangular prism
   H square pyramid
   J triangular pyramid

3. Which term describes a triangle that has angles measuring 85°, 30°, and 65°?

   A right   C obtuse
   B acute   D isosceles

4. What is the value of $16 \times n$ for $n = 40$?

   F 240   H 440
   G 280   J Not here

5. $$\begin{array}{r} 6{,}672{,}054 \\ +4{,}821{,}368 \\ \hline \end{array}$$

   A 11,493,422   C 10,493,322
   B 10,493,412   D Not here

6. Which of the following is **not** a solid figure?

   F cone   H circle
   G prism   J pyramid

For 7–9, use the figures.

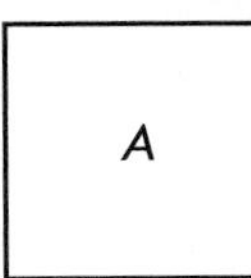

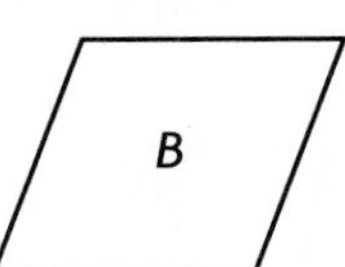

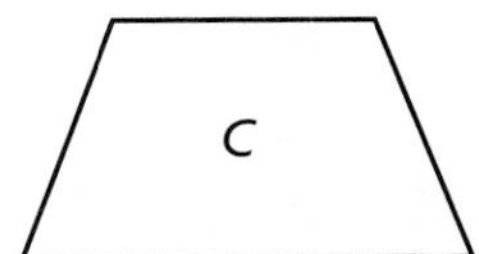

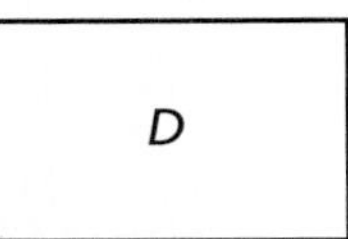

7. In which figure are all the angles and all the sides congruent?

   A Figure A   C Figure C
   B Figure B   D Figure D

8. Which figure is **not** a parallelogram?

   F Figure A   H Figure C
   G Figure B   J Figure D

9. Which figure is the shape of some of the faces in all rectangular prisms?

   A Figure A   C Figure C
   B Figure B   D Figure D

10. Which is a name for the figure below?

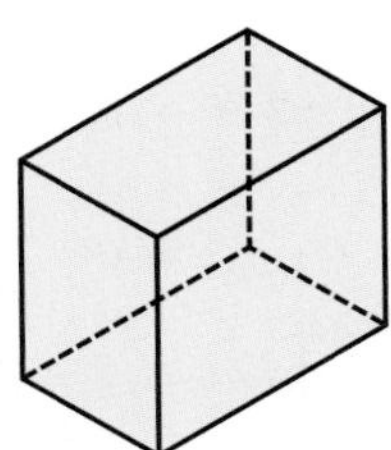

   F rectangular prism
   G triangular prism
   H rectangular pyramid
   J triangular pyramid

UNIT 7

MATH DETECTIVE

Get the Point!

Use the algebraic clues given to find the coordinates of the missing points.

Find the values of x and y in the ordered pair (x,y) that satisfy both clues at the same time. For each case, there is only one value for x and one value for y.

Case 1

$x + y = {}^{+}5$

$x - y = {}^{+}1$

Case 2

$x + y = {}^{+}17$

$x - y = {}^{+}3$

Case 3

$x + y = 0$

$x + {}^{+}1 = {}^{+}5$

Case 4

$x + y = {}^{-}5$

$y + {}^{+}4 = {}^{-}4$

Case 5

$x - y = {}^{+}1$

$x - {}^{+}1 = {}^{+}2$

Case 6

$y - x = {}^{-}9$

$y + x = {}^{+}5$

Case 7

$x + y = {}^{+}5$

$x - {}^{+}2 = {}^{+}2$

Case 8

$y + x = {}^{+}1$

$y + {}^{+}2 = {}^{+}2$

Think It Over!

- What figure is formed by plotting on a coordinate plane and connecting the points found in Cases 1–3? in Cases 5–7?
- What common point do these two figures share?

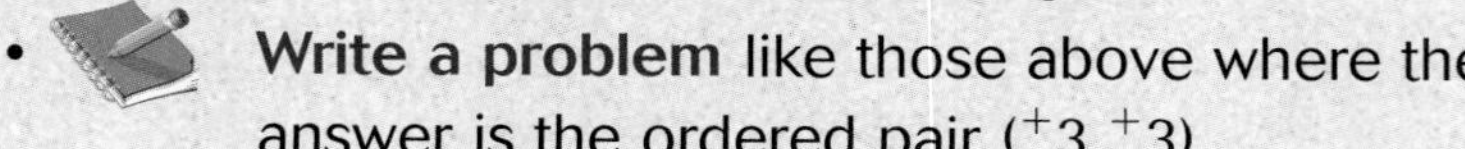

- **Write a problem** like those above where the answer is the ordered pair $({}^{+}3, {}^{+}3)$.

UNIT 7

Challenge

Tessellations

When closed figures are arranged in a repeating pattern that covers a surface so that there are no gaps or overlaps, the pattern is called a **tessellation**.

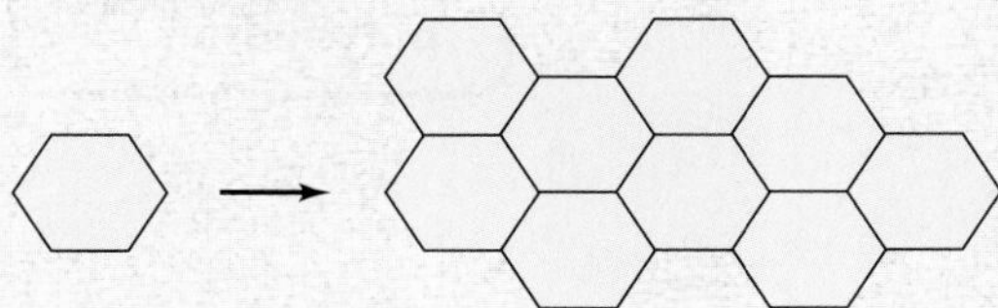

- How do you know these are tessellations?

Some tessellations use two or more different figures.

Example

Using the figures at the right, make a tessellation.

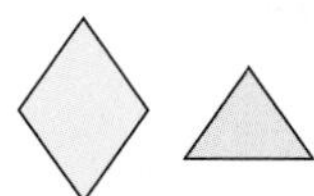

STEP 1

Trace the two figures and cut out several of each.

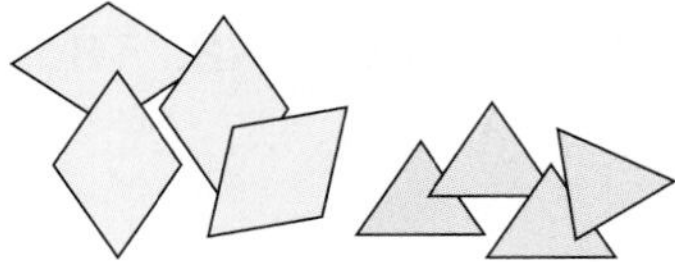

STEP 2

Find a way in which the two figures can fit together with no gaps or overlaps to form a tessellation.

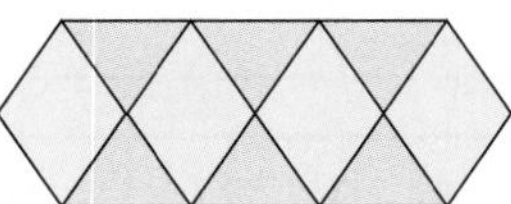

Try It

Trace each figure. Write *yes* or *no* to tell whether each figure can be used to make a tessellation.

1. 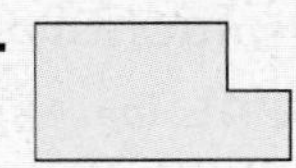

2. 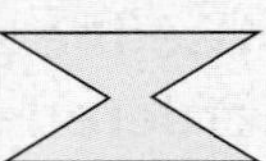

3. 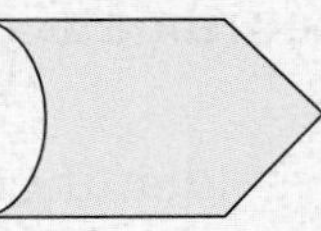

4. 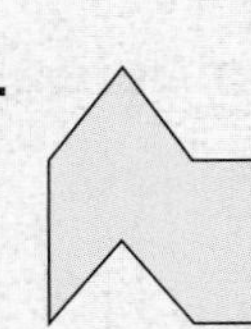

5. 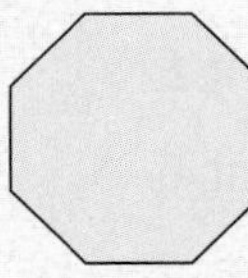

6. 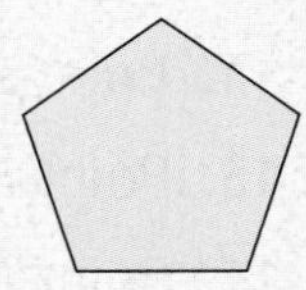

7. 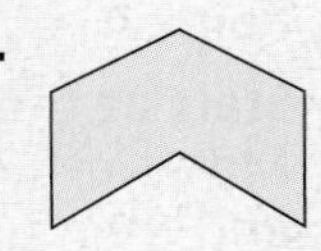

8. 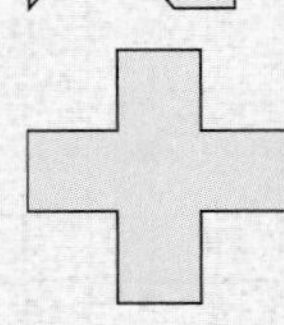

9. Shelley wants to cover her kitchen floor with square and octagonal tiles. Draw what the design might look like.

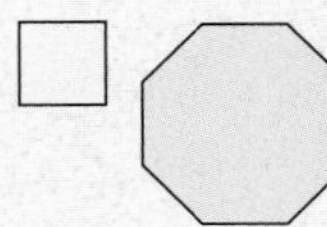

10. Leticia has pattern blocks that are equilateral triangles, squares, and hexagons. All blocks have sides that are 1 inch long. Which two of these blocks can be used together for a tessellation? Draw the design.

Study Guide and Review

VOCABULARY

Choose the best term from the box.

1. The distance of an integer from 0 is the __?__ of the integer. (p. 417)
2. A closed plane figure formed by three or more line segments is a(n) __?__. (p. 460)
3. A solid figure with faces that are polygons is called a(n) __?__. (p. 486)

absolute value
opposite
polygon
sphere
polyhedron

STUDY AND SOLVE

Chapter 23

Add and subtract integers.

Subtract. $^{-}3 - {}^{+}2$

On a number line, start at 0 and move 3 spaces to the left to show $^{-}3$.

From $^{-}3$, move 2 spaces to the left to show subtracting 2. This takes you to $^{-}5$.

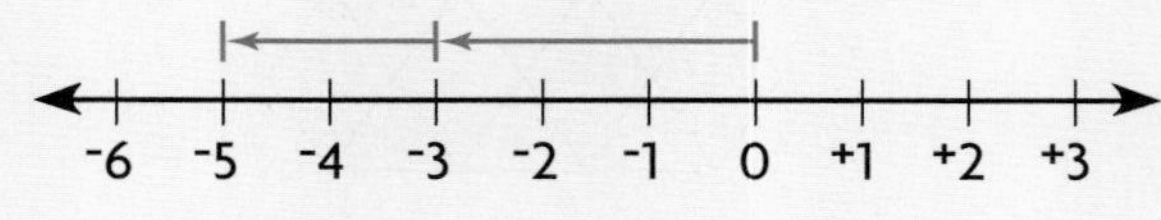

So, $^{-}3 - {}^{+}2 = {}^{-}5$.

Find each sum or difference. (pp. 422–431)

1. $^{+}3 + {}^{-}2$
2. $^{-}4 - {}^{+}3$
3. $^{-}2 - {}^{+}5$
4. $^{-}4 + {}^{-}4$
5. $^{-}2 - {}^{+}7$
6. $^{+}3 - {}^{+}6$
7. $^{+}1 + {}^{-}6$
8. $^{-}8 - {}^{+}7$

Chapter 24

Identify ordered pairs on a coordinate plane.

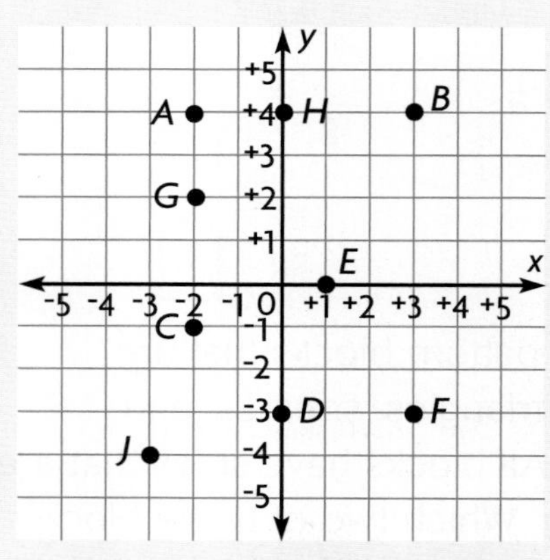

Identify the ordered pair for Point *A*.

Start at the origin. Move left 2 units and then up 4 units. Plot and label the point.

So, Point *A* is at $(^{-}2, {}^{+}4)$.

Use the graph at the left. Identify the ordered pair for each point. (pp. 440–443)

9. Point *B*
10. Point *C*
11. Point *D*
12. Point *E*
13. Point *F*
14. Point *G*
15. Point *H*
16. Point *J*

Chapter 25

Identify congruent and similar figures.

Tell whether the figures are similar, congruent, or neither.

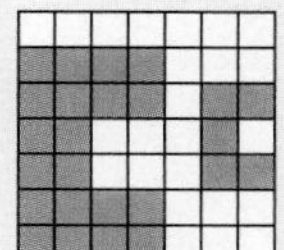

similar

Write *similar, congruent,* or *neither* to describe each pair. (pp. 468–469)

17. 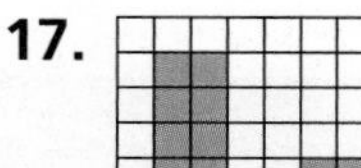

18. 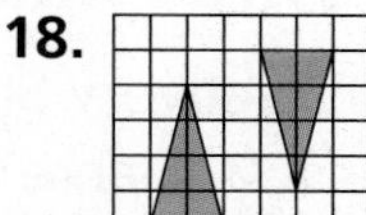

Chapter 26

Identify and classify triangles and quadrilaterals.

Classify the figure in as many ways as possible.

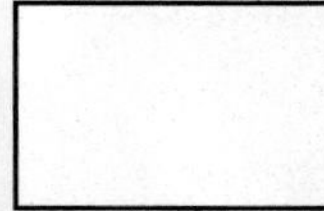

This is a quadrilateral, a parallelogram, and a rectangle.

Classify each figure in as many ways as possible. For 19, write *isosceles, scalene,* or *equilateral* and *right, acute,* or *obtuse*. For 20, write *quadrilateral, parallelogram, square, rectangle, rhombus,* or *trapezoid*. (pp. 478–483)

19. 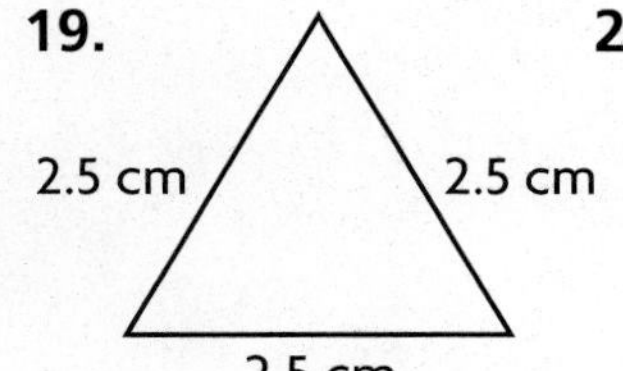

20. 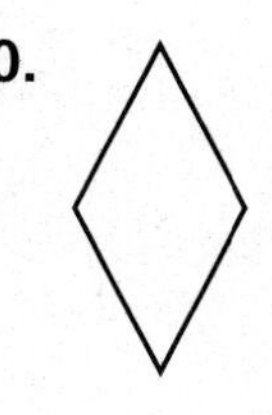

Identify solid figures.

Identify the solid figure.

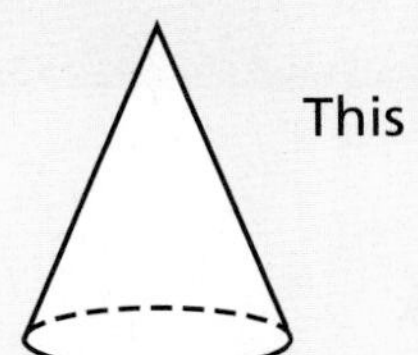

This is a cone.

Classify each solid figure. Write *prism, pyramid, cone, cylinder,* or *sphere*. (pp. 486–489)

21. 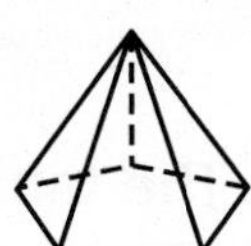

22. 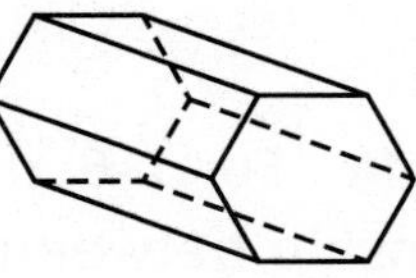

PROBLEM SOLVING PRACTICE

Solve. (pp. 432–433, 472–473)

23. A scuba diver was 22 ft below the surface of the water when he saw 3 starfish and 4 sea urchins. He then swam up 10 ft and down 7 ft. How far below the surface did he finish?

24.      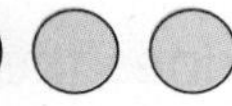

What color are the next three circles in the pattern?

California Connections

MARSHALL GOLD DISCOVERY STATE HISTORIC PARK

Marshall Gold Discovery State Historic Park is a 274-acre park in Coloma, California. At this location, James W. Marshall found gold flecks at a sawmill he was building for himself and John Sutter. This discovery changed the course of California history. The park is dedicated to protecting and preserving the place where the California Gold Rush began in 1848.

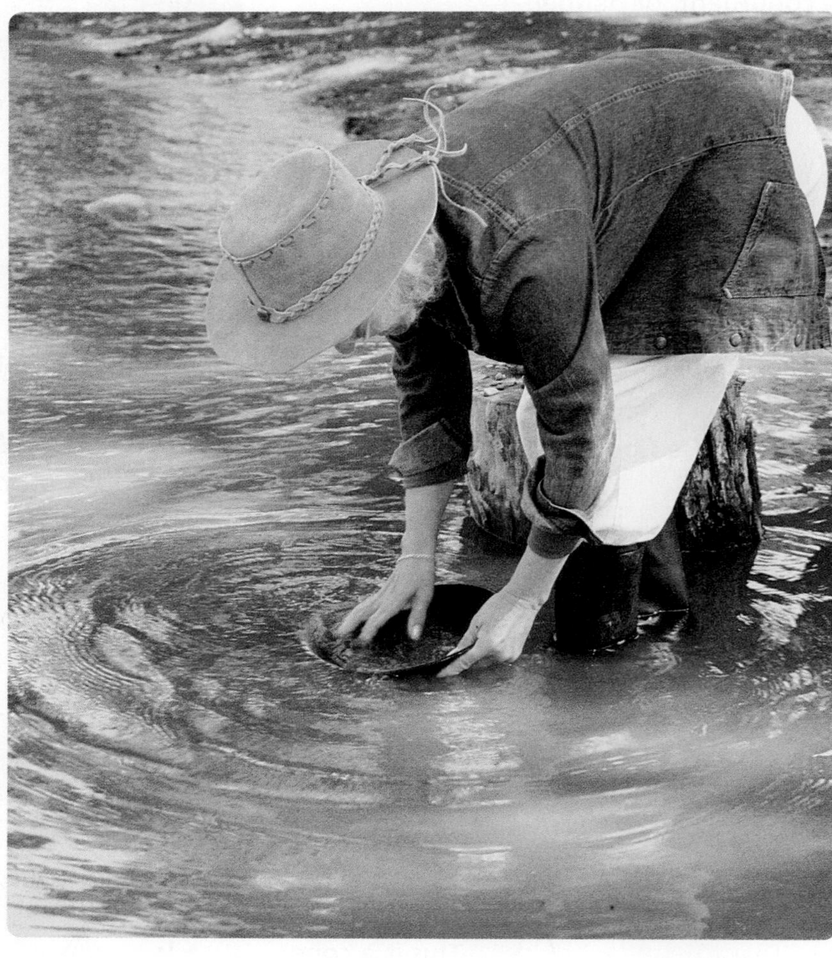

Panning for gold is a popular recreational pastime in "Gold Country." ▶

USE DATA **For 1–6, use the table. The table shows the relationship between the number of cars that enter the park and the number of dollars collected for admittance.**

Number of cars, x	1	2	3	4	5	6
Number of dollars, y	5	10	15	20	■	■

1. What does the ordered pair (1,5) mean?
2. Copy and complete the table.
3. Write the ordered pairs.
4. Graph the relationship shown in the table.
5. Write a rule for the relationship.
6. Write an equation for the relationship.

SUTTER'S MILL

An outstanding attraction at the park is the full-sized replica of Sutter's Mill. The original mill was abandoned and disappeared in the floods of the 1850's. The replica was completed in 1968.

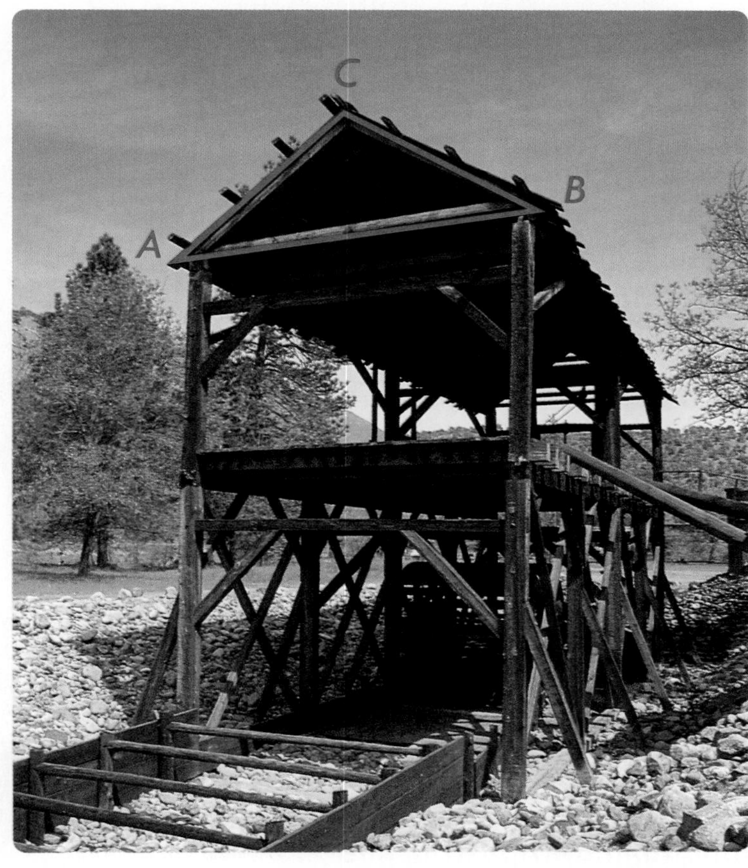

This replica of the original sawmill was built with hand tools and is operated regularly for park visitors. ▶

For 1–4, use the photo of Sutter's Mill.

1. Write three sentences to describe lines in the photo. Write one sentence for parallel lines, one for perpendicular lines, and one for lines that are intersecting but not perpendicular.

2. Use a protractor and a ruler to measure the angles and the lengths of the sides of triangle *ABC*. Classify the triangle in two ways. Write *isosceles, scalene,* or *equilateral.* Then write *right, acute,* or *obtuse.*

3. Describe two shapes in the photo that appear to be congruent.

4. Describe a part of the mill that appears to be symmetrical.

5. The diagram shows the base of the mill.

   a. Write the ordered pairs for the vertices of the base.

   b. What if you translate each vertex 2 units left and 3 units down? Name the ordered pairs for the new location of the base.

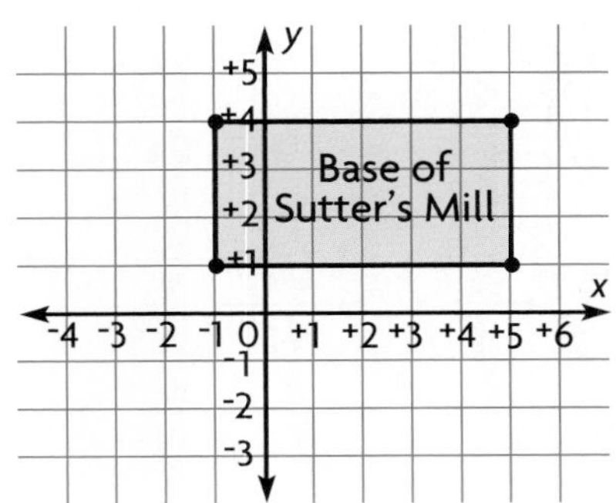

CHAPTER 27 Customary and Metric Systems

Climate is an average of weather conditions over a long time period. The major factors that determine a location's climate are temperature and precipitation. Most of the United States has a temperate climate—one that has cool winters and warm summers. The graph below shows the average July temperatures for five U.S. cities. Which city's July temperature is higher than 76°F and lower than 85°F?

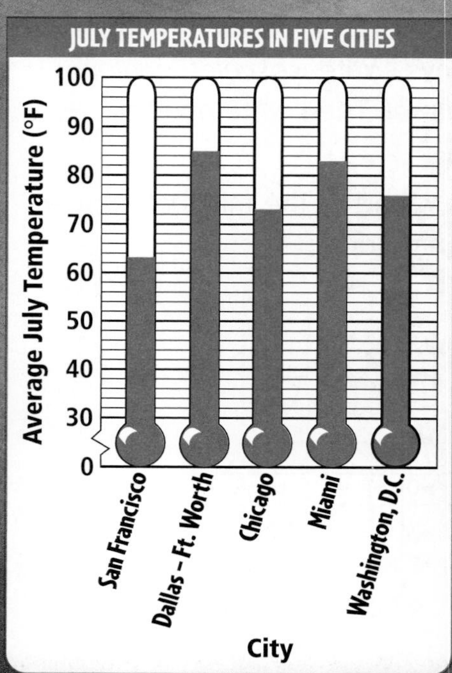

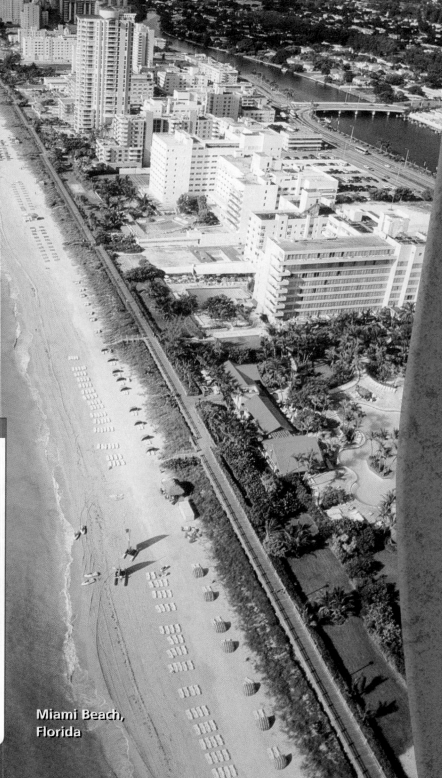

Miami Beach, Florida

CHECK WHAT YOU KNOW

Use this page to help you review and remember important skills needed for Chapter 27.

VOCABULARY

Choose the best term from the box.

centimeters
inches
feet
yards

1. There are 12 __?__ in 1 foot.
2. There are 3 __?__ in 1 yard.
3. There are 100 __?__ in 1 meter.

CUSTOMARY UNITS AND TOOLS (See p. H28.)

Choose the tool you would use to measure each.

4. the length of the chalkboard
5. the amount of flour needed to make a cake
6. the weight of a stapler

a. 

b. 

c.

Choose the unit you would use to measure each.

7. the length of your foot
   a. inches
   b. yards
   c. miles

8. the weight of a compact car
   a. ounces
   b. pounds
   c. tons

9. the amount of water in a pool
   a. cups
   b. quarts
   c. gallons

METRIC UNITS AND TOOLS (See p. H29.)

Choose the tool you would use to measure each.

10. the temperature outside
11. the mass of a bracelet
12. the distance across the classroom

a. 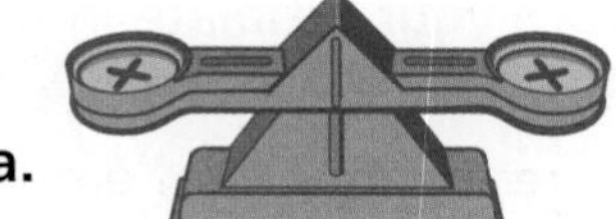
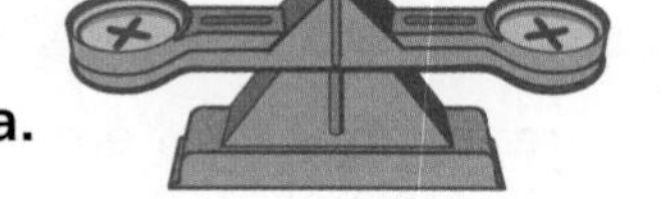

b. 

c.

Choose the unit you would use to measure each.

13. the distance run in a track meet
    a. centimeters
    b. kilometers

14. the mass of a big dog
    a. grams
    b. kilograms

15. a large bottle of soda
    a. liters
    b. milliliters

LESSON 1

HANDS ON
Customary Length

Quick Review

Compare. Write <, >, or = for each ●.

1. $\frac{2}{16}$ ● $\frac{3}{16}$
2. $\frac{1}{4}$ ● $\frac{4}{16}$
3. $\frac{5}{16}$ ● $\frac{1}{2}$
4. $\frac{8}{16}$ ● $\frac{1}{2}$
5. $\frac{3}{4}$ ● $\frac{15}{16}$

VOCABULARY

precision

MATERIALS

ruler with markings to $\frac{1}{16}$ inch; objects such as marker, paper clip, pencil

Explore

The **precision** of a measurement is related to the unit of measure you choose. The smaller the unit, the more precise the measurement will be.

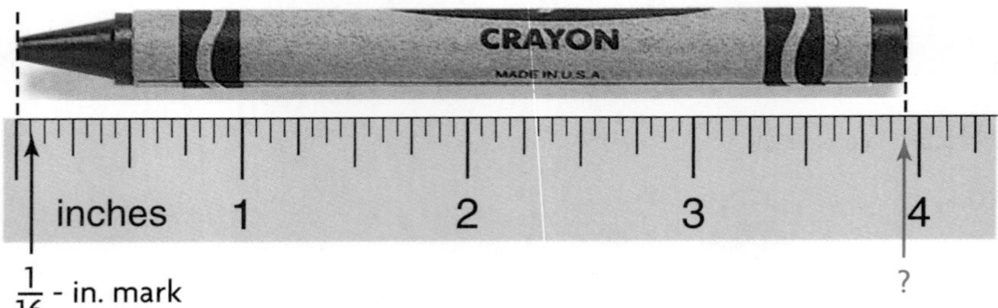

To the nearest inch, the crayon's length is 4 inches. A more precise measurement is $3\frac{15}{16}$ inches.

Estimate and measure the length of each object.

STEP 1 Copy the table.

Object	Estimate (nearest inch)	Measurement (nearest $\frac{1}{8}$ inch)	Measurement (nearest $\frac{1}{16}$ inch)
marker			
paper clip			
pencil			

STEP 2 Estimate the length of each object to the nearest inch. Record your estimate in the table.

STEP 3 Use a ruler. Measure the length of each object to the nearest $\frac{1}{8}$ inch. Measure the length of each object to the nearest $\frac{1}{16}$ inch. Record your measurements in the table.

How can you tell the measurement to the nearest $\frac{1}{16}$ inch?

Talk About It

- Why is measuring to the nearest $\frac{1}{16}$ inch more precise than measuring to the nearest $\frac{1}{8}$ inch?

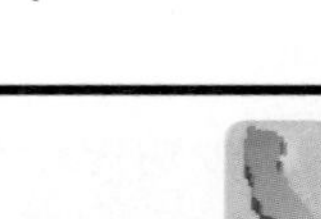

CALIFORNIA STANDARDS MR 3.2 Note the method of deriving the solution, and demonstrate a conceptual understanding of the derivation by solving similar problems. **MR 2.5** Indicate the relative advantages of exact and approximate solutions to problems, and give answers to a specified degree of accuracy. *also* **MR 2.0, MR 2.3**

Connect

You can draw a line segment to the nearest $\frac{1}{16}$ inch. Follow these steps to draw a line segment that is $3\frac{5}{16}$ inches:

TECHNOLOGY LINK
More Practice: Use E-Lab, *Precise Measurements.*
www.harcourtschool.com/elab 2002

- Line up your pencil with the end of the ruler. Count 3 inches on the ruler. Count five $\frac{1}{16}$ marks to the right of the 3-inch mark.
- Draw a line segment along the ruler until you reach $3\frac{5}{16}$ inches.

Practice

Estimate the length in inches. Then measure to the nearest $\frac{1}{16}$ inch.

1. 

2. 

Estimate the length in inches. Then measure to the nearest $\frac{1}{8}$ inch.

3. 

4. 

Draw a line segment to the given length.

5. $4\frac{1}{16}$ inches

6. $2\frac{1}{4}$ inches

7. $3\frac{7}{16}$ inches

8. $6\frac{3}{4}$ inches

9. **Write About It** Explain how you would measure your pencil to the nearest $\frac{1}{8}$ in.

10. Trenton's ruler has only $\frac{1}{4}$-inch marks. How can he measure $3\frac{12}{16}$ inches of string?

Mixed Review and Test Prep

Write each fraction as a decimal. (p. 280)

11. $\frac{7}{8}$

12. $\frac{3}{40}$

13. $\frac{11}{25}$

14. List the prime numbers between 5 and 30. (p. 264)

15. **TEST PREP** Which number is less than the product of 35 and 2.58? (p. 164)

A 90.03 **C** 90.33
B 90.3 **D** 93

LESSON 2

HANDS ON
Metric Length

Quick Review

Write each missing number in the number line.

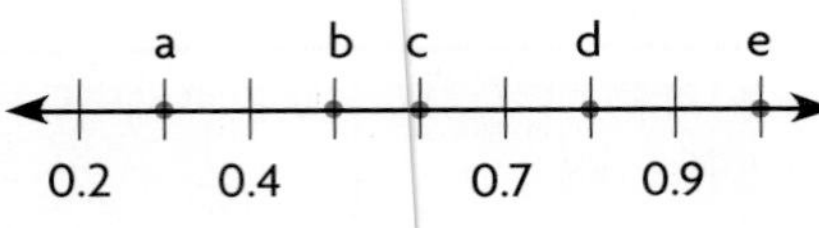

1. a **2.** b **3.** c **4.** d **5.** e

VOCABULARY

centimeter (cm)
millimeter (mm)

MATERIALS

centimeter ruler, paper clip, other objects

Explore

You can use metric units to measure length or height.

A **centimeter (cm)** is about the width of your index finger.

A **millimeter (mm)** is about the thickness of a dime.

The height of the plant to the nearest centimeter is 3 cm and to the nearest millimeter is 3 cm 4 mm, or 34 mm.

Estimate and measure the length of 3 objects in your classroom to the nearest centimeter and millimeter.

STEP 1

Make a table to show the objects, estimated lengths, measurement to the nearest centimeter, and measurement to the nearest millimeter.

STEP 2

Estimate the length of each object to the nearest centimeter. Record your estimate in the table.

STEP 3

Use a centimeter ruler. Measure the length of each object to the nearest centimeter. Measure the length of each object to the nearest millimeter. Record your measurements.

The length of your little finger is nearest to what centimeter mark?

Talk About It

- Joe's eraser is 4 cm 6 mm. What is the measurement to the nearest centimeter?

CALIFORNIA STANDARDS MR 3.2 Note the method of deriving the solution and demonstrate a conceptual understanding of the derivation by solving similar problems. **MR 2.5** Indicate the relative advantages of exact and approximate solutions to problems and give answers to a specified degree of accuracy. *also* **MR 2.0, MR 2.3**

Connect

TECHNOLOGY LINK
To learn more about measurement, watch the Harcourt Math Newsroom Video *River Forecasters*.

You can draw a line segment to the nearest millimeter, or 0.1 cm.

Follow these steps to draw a line segment that is 3.4 cm long. 3.4 cm is the same as 3 cm 4 mm.

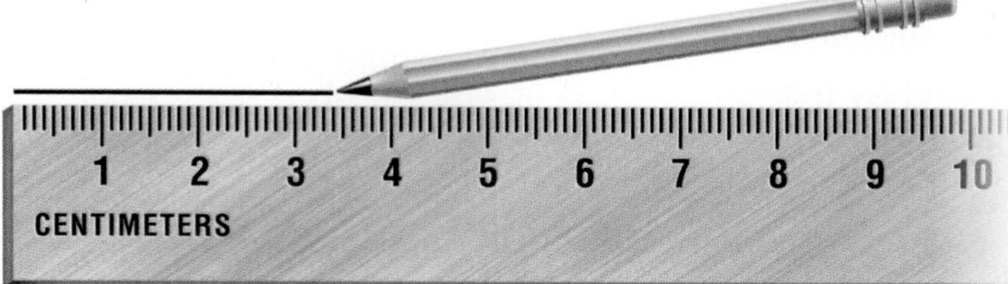

- Line up your pencil with the end of the ruler. Find 3 cm on the ruler. Count 4 millimeter marks to the right of 3 cm.
- Draw a line segment along the ruler until you reach 3.4 cm.

Practice

Estimate the length in centimeters. Measure to the nearest centimeter and then to the nearest millimeter.

1.

2. 

3. 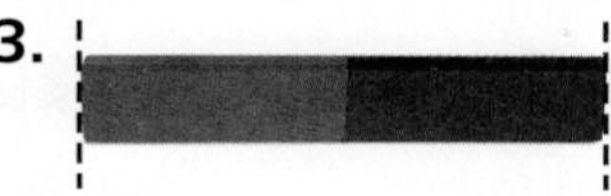

4. 

5. 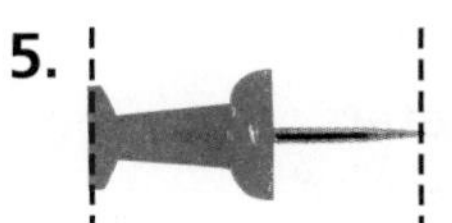

6. 

Draw a line segment to the given length.

7. 5 cm 2 mm

8. 2.6 cm

9. 12 mm

10. 8.8 cm

11. **GEOMETRY** A rectangular frame has a perimeter of 50.6 cm. If the length is 8.3 cm, what is the width?

12. **NUMBER SENSE** Anne says 0.2 cm is the same as 2 mm. Do you agree? Explain.

Mixed Review and Test Prep

Write each in simplest form. (p. 290)

13. $\frac{3}{12}$

14. $\frac{16}{24}$

15. $\frac{15}{35}$

16. $\frac{6}{18}$

17. **TEST PREP** Find $\frac{3}{8}$ of 16. (p. 380)

A $\frac{3}{16}$

B $\frac{48}{128}$

C 6

D 16

LESSON 3

Change Linear Units

Quick Review

1. 2×12
2. $60 \div 5$
3. $84 \div 3$
4. $1,760 \times 2$
5. 25×3

Learn

CUSTOM TOWELS Jenna is making personalized beach towels for each member of her family. She needs 30 feet of fabric. Since fabric is sold by the yard, how many yards of fabric does Jenna need?

Think: 30 feet = ■ yards

To change smaller units to larger units, divide.

number of feet	÷	number of feet in 1 yard	=	total yards
↓		↓		↓
30	÷	3	=	10

So, Jenna needs 10 yards of fabric.

CUSTOMARY UNITS OF LENGTH

12 inches (in.) = 1 foot (ft)
3 feet = 1 yard (yd)
5,280 feet = 1 mile (mi)
1,760 yards = 1 mile

METRIC UNITS OF LENGTH

10 millimeters (mm) = 1 centimeter (cm)
100 centimeters = 1 meter (m)
1,000 meters = 1 kilometer (km)

Examples

A How many millimeters are in 230 centimeters?

Think: 230 cm = ■ mm

To change larger units to smaller units, multiply.

number of cm	×	number of mm in 1 cm	=	number of mm
↓		↓		↓
230	×	10	=	2,300 mm

There are 2,300 millimeters in 230 centimeters.

B How many feet are in 168 inches?

Think: 168 inches = ■ feet

To change smaller units to larger units, divide.

number of inches	÷	number of inches in 1 foot	=	number of feet
↓		↓		↓
168	÷	12	=	14

There are 14 feet in 168 inches.

- Would you multiply or divide to find how many kilometers are in 3,400 meters? Explain.

CALIFORNIA STANDARDS MR 1.0 Students make decisions about how to approach problems. **MR 1.1** Analyze problems by identifying relationships, distinguishing relevant from irrelevant information, sequencing and prioritizing information, and observing patterns. **MR 2.0** Students use strategies, skills, and concepts in finding solutions. *also* **MR 2.3, MR 3.0, MR 3.2**

Add and Subtract Measurements

You may need to change units to add and subtract measurements.

Jenna needs 5 feet 7 inches of fabric for her mother's towel and 4 feet 6 inches for her little brother's towel. How much fabric does she need for both towels?

Find 5 ft 7 in. + 4 ft 6 in.

STEP 1

Add each kind of unit.

$$\begin{array}{rr} 5 \text{ ft} & 7 \text{ in.} \\ +4 \text{ ft} & 6 \text{ in.} \\ \hline 9 \text{ ft} & 13 \text{ in.} \end{array}$$

STEP 2

Think: 12 in. = 1 ft
Since 13 in. is more than 1 ft, rename 13 in. as 1 ft + 1 in.

9 ft 13 in. = 9 ft + (1 ft + 1 in.)

STEP 3

Combine like units.

9 ft + (1 ft + 1 in.) =
(9 ft + 1 ft) + 1 in. = 10 ft 1 in.

So, she needs 10 feet 1 inch of fabric.

Garth made a towel bar with a piece of wooden dowel that was 60.5 cm long. If he used 48.9 cm of the dowel, how much of it remained?

Find 60.5 cm − 48.9 cm.

STEP 1

Line up the decimal points. Subtract the tenths. Regroup if needed.

$$\begin{array}{r} \scriptstyle 9 \\ \scriptstyle 5\ 10\ 15 \\ \not{6}\not{0}.\not{5} \\ -48.9 \\ \hline .6 \end{array}$$

STEP 2

Subtract the ones.

$$\begin{array}{r} \scriptstyle 9 \\ \scriptstyle 5\ 10\ 15 \\ \not{6}\not{0}.\not{5} \\ -48.9 \\ \hline 1.6 \end{array}$$

STEP 3

Subtract the tens.

$$\begin{array}{r} \scriptstyle 9 \\ \scriptstyle 5\ 10\ 15 \\ \not{6}\not{0}.\not{5} \\ -48.9 \\ \hline 11.6 \end{array}$$

So, 11.6 cm of wooden dowel was left.

Check

1. **Tell** how you would rename units to subtract 4 feet 6 inches from 7 feet 2 inches.

Change the unit.

2. 3 ft = ■ in.
3. 5 yd = ■ ft
4. 60 in. = ■ ft
5. 24 ft = ■ yd
6. 5 m = ■ cm
7. 20 mm = ■ cm
8. 2 km = ■ m
9. 4,000 m = ■ km

Complete.

10. 4 ft 3 in. = 3 ft ■ in.
11. 5,360 ft = 1 mi ■ ft
12. 10 yd 1 ft = 9 yd ■ ft

LESSON CONTINUES

▶ Practice and Problem Solving

Change the unit.

13. 24 in. = ■ ft **14.** 30 ft = ■ yd **15.** 13 yd = ■ ft **16.** 8 ft = ■ in.

17. 4 mi = ■ yd **18.** 48 in. = ■ ft **19.** 6 m = ■ cm **20.** 50 mm = ■ cm

21. 7 km = ■ m **22.** 65 cm = ■ mm **23.** 8 m = ■ cm **24.** 700 cm = ■ m

Complete.

25. 2 ft = 1 ft ■ in. **26.** 12 yd 2 ft = 11 yd ■ ft **27.** 3 mi 27 ft = 2 mi ■ ft

Find the sum or difference.

28. 2 ft 2 in. + 3 ft 7 in.

29. 3 yd 2 ft + 8 yd 1 ft

30. 7 ft 4 in. − 5 ft 1 in.

31. 14 yd − 5 yd 2 ft

32. 15 ft 4 in. + 12 ft 6 in.

33. 3 mi 345 ft + 1 mi 39 ft

34. 8 ft 6 in. − 3 ft 10 in.

35. 12 yd 2 ft 3 in. − 3 yd 1 ft 11 in.

36. 5 yd 2 ft + 2 yd 2 ft

37. 3 mi 820 ft + 7 mi 4,460 ft

38. 8 ft − 4 ft 7 in.

39. 6 yd 1 ft − 2 yd 2 ft

40. 4.5 cm + 3.2 cm **41.** 12.5 km + 2.8 km **42.** 6.7 cm − 3.9 cm

43. 88 cm − 69 cm **44.** 50 km + 6.8 km **45.** 8.3 cm − 4.6 cm

46. 13.4 km − 2.8 km **47.** 10.3 cm + 7.8 cm **48.** 15.6 cm − 6.4 cm

49. **What's the Error?** Jason wanted to know how many feet are in 27 yards. He wrote 27 yards ÷ 3 = 9 feet. Describe his error. Give the correct number of feet.

50. **ESTIMATION** Estimate how many yards long your classroom is. Then use a yardstick to check. Record your estimate and the actual measurement.

51. **NUMBER SENSE** Describe how you could find how many centimeters are in a kilometer.

52. **NUMBER SENSE** Joel says that 40 cm is the same as 0.4 m. Do you agree? Explain.

53. Mr. Gomez had 15 ft of cable. He used 12 ft 6 in. to hook up his television. How long was the remaining piece?

54. Grace has 13 feet of fabric. Her pattern calls for $4\frac{1}{4}$ yards of fabric. Does she have enough? Explain.

55. Use the diagram on the right. By car, how many kilometers is it from Bill's house to the library following the path shown?

Mixed Review and Test Prep

56. Write the reciprocal. $\frac{7}{12}$ (p. 396)

For 57–58, round to the hundredths place. (p. 46)

57. 55.073

58. 172.6053

59. TEST PREP Which number is greater than the product of 28 and 163? (p. 148)

A 4,744
B 4,453
C 3,352
D 3,176

60. $6\overline{)3.6}$ (p. 222)

61. $5\overline{)4.0}$ (p. 222)

62. Solve for *n*. $n \times 15 = 105$ (p. 88)

63. An angle that is less than 90° is called an __?__ angle. (p. 454)

64. TEST PREP Find $\frac{4}{21} \times \frac{7}{9}$. (p. 382)

F $\frac{1}{8}$
G $\frac{4}{27}$
H $\frac{28}{32}$
J $\frac{1}{3}$

LINKUP to Reading

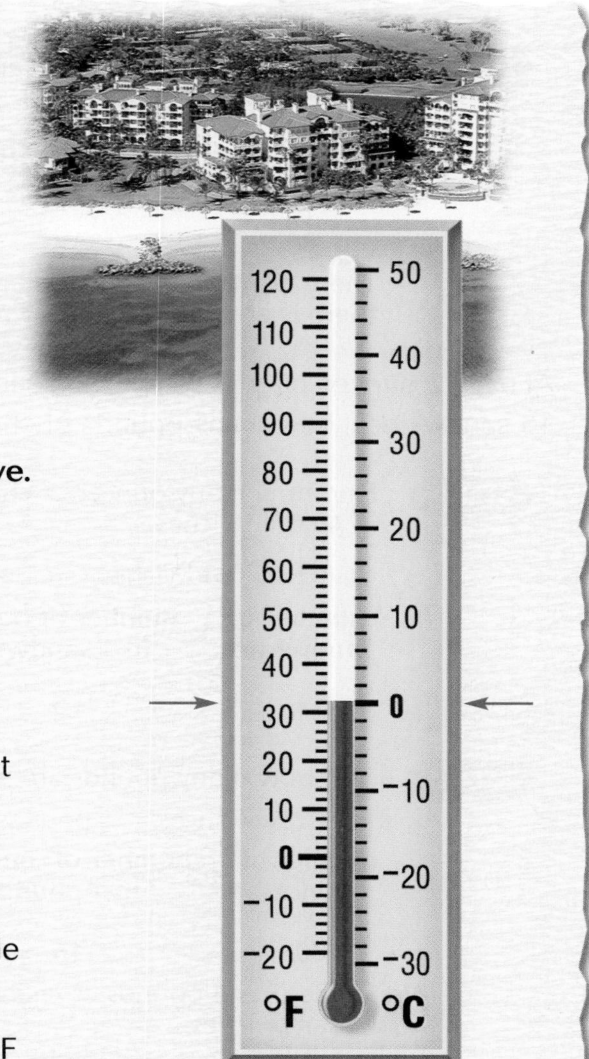

STRATEGY • USE GRAPHIC AIDS Sometimes using pictures can help you find the answer to a problem more easily. When it is 0°C, what is the temperature in degrees Fahrenheit? Read the thermometer. Look on the right-hand side to find 0°C. Then look on the left to find degrees Fahrenheit. So, 0°C is the same as 32°F.

For 1–3, use the picture of the thermometer to solve.

1. On a warm day the temperature might be 86°F. What is that in degrees Celsius?
2. On a cold day the temperature might be 1°C. What is that in degrees Fahrenheit?
3. Evan's family keeps the temperature of their house at 68°F. Lucy's family keeps their house at 25°C. Which family's house has the higher temperature?
4. Algebra To estimate degrees Celsius, use this rule: °C is about $0.5 \times (°F - 30)$. Use the rule to estimate the temperature in degrees Celsius.

   a. 90°F **b.** 50°F **c.** 35°F

Extra Practice page H58, Set A

LESSON 4

Customary Capacity and Weight

Quick Review

1. 8×20
2. 16×6
3. 4×32
4. $2{,}000 \times 3$
5. 8×16

Learn

ALL SHOOK UP! Al's Deli uses 1 pint of milk in each milkshake. There are 15 milkshakes ordered. If Al has 2 gallons of milk, does he have enough for all the milkshakes?

CUSTOMARY UNITS OF CAPACITY

8 fluid ounces (fl oz) = 1 cup (c)
2 cups = 1 pint (pt)
2 pints = 1 quart (qt)
4 cups = 1 quart
4 quarts = 1 gallon (gal)

CUSTOMARY UNITS OF WEIGHT

16 ounces (oz) = 1 pound (lb)
2,000 pounds = 1 ton (T)

STEP 1 First, find how many quarts are in 2 gallons.

Think: 2 gallons = ■ quarts

number of gallons	×	number of quarts in 1 gallon	=	total quarts
↓		↓		↓
2	×	4	=	8

STEP 2 Now, find how many pints are in 8 quarts.

Think: 8 quarts = ■ pints

number of quarts	×	number of pints in 1 quart	=	total pints
↓		↓		↓
8	×	2	=	16

There are 16 pints in 2 gallons. Since $15 < 16$, there is enough milk for 15 milkshakes.

Al uses 2 ounces of cheese for each sandwich. If he makes 40 sandwiches, how many pounds of cheese does he use?

STEP 1 Find how many ounces of cheese are used in 40 sandwiches.

Think: 40 sandwiches = ■ ounces of cheese

number of sandwiches	×	number of ounces in 1 sandwich	=	total ounces
↓		↓		↓
40	×	2	=	80

STEP 2 Find how many pounds are in 80 ounces.

Think: 80 ounces = ■ pounds

number of ounces	÷	number of ounces in 1 pound	=	total pounds
↓		↓		↓
80	÷	16	=	5

So, he uses 5 pounds of cheese to make 40 sandwiches.

CALIFORNIA STANDARDS MR 1.0 Students make decisions about how to approach problems. **MR 1.1** Analyze problems by identifying relationships, distinguishing relevant from irrelevant information, sequencing and prioritizing information, and observing patterns. **MR 1.2** Determine when and how to break a problem into simpler parts. *also* **MR 2.0, MR 3.0, MR 3.2**

Check

1. **Describe** how you could find the number of pounds in 160 ounces.

Change the unit.

2. 3 gal = ■ qt
3. 32 c = ■ qt
4. 4 gal = ■ c
5. 2 pt = ■ c
6. 48 oz = ■ lb
7. 2 lb = ■ oz
8. 1 T = ■ oz
9. 8,000 lb = ■ T

Practice and Problem Solving

Change the unit.

10. 10 pt = ■ qt
11. 5 c = ■ fl oz
12. 3 gal = ■ pt
13. 16 fl oz = ■ pt
14. 2 T = ■ lb
15. 6 lb = ■ oz
16. 96 oz = ■ lb
17. 1.5 T = ■ lb

Choose the best estimate.

18. A dog weighs
    a. 40 oz.
    b. 40 lb.
    c. 40 T.
19. A bucket holds
    a. 2 cups.
    b. 2 pints.
    c. 2 gallons.
20. A car weighs
    a. 1 oz.
    b. 1 lb.
    c. 1 T.
21. A mug holds
    a. 1 cup.
    b. 1 quart.
    c. 1 gallon.

22. **What if** Al had 23 milkshake orders and 3 gallons of milk? Would he have enough milk for 1 pint in each milkshake? Explain.

23. Use the Frosty Freeze ad. Toby needs 1 gallon of ice cream for a party. What is the best buy? Explain.

24. Rochelle bought 8 pounds of apples for pies. If each apple weighs 4 ounces, how many apples did she buy?

FROSTY FREEZE
ICE CREAM SALE!
Two 1-quart containers for $1.00!
1 gallon for $3.79!

25. **Write About It** Describe three items that are measured by capacity. Describe three items that are measured by weight.

26. Antonia's room is 10 feet 4 inches long. Her brother's room is 9 feet 8 inches long. How much longer is Antonia's room?

Mixed Review and Test Prep

27. A 90° angle is called a __?__ angle. (p. 454)
28. A 120° angle is called an __?__ angle. (p. 454)
29. 56.3×2.1 (p. 164)
30. 18.1×0.5 (p. 164)
31. **TEST PREP** In which set are all numbers factors of 180? (p. 264)
    A 2, 5, 7
    B 2, 3, 5, 13
    C 2, 3, 5
    D 2, 3, 5, 7

Extra Practice page H58, Set B

LESSON

5 Metric Capacity and Mass

Quick Review

1. 1,000 ÷ 100
2. 1,000 ÷ 10
3. 1,000 × 5
4. 500 ÷ 2
5. 250 × 4

Learn

PUNCH LINE Ms. Allison needs 4 liters of juice to make fruit punch. The cafeteria sends her 12 metric cups of juice. Is this enough juice to make the punch?

Think: 12 metric cups = ■ liters

To change smaller units to larger units, divide.

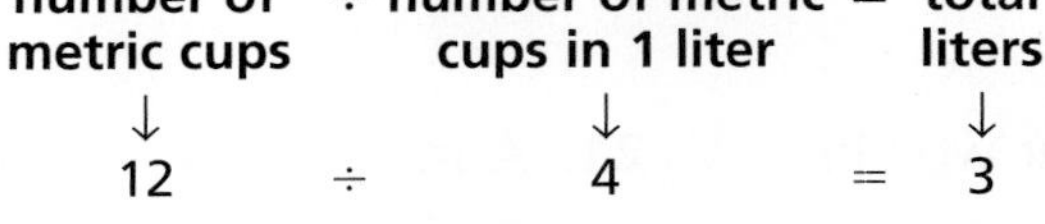

number of metric cups	÷	**number of metric cups in 1 liter**	=	**total liters**
↓		↓		↓
12	÷	4	=	3

There are 3 liters in 12 metric cups. Since $3 < 4$, Ms. Allison does not have enough juice.

250 milliliters (mL), or 1 metric cup

1 liter

Ms. Allison needs to make 50 sandwiches for the fifth grade. If she uses two 1-kilogram jars of peanut butter, how many grams of peanut butter will she use for each sandwich?

STEP 1 Find how many grams are in 2 kilograms.

Think: 2 kilograms = ■ grams

To change larger units to smaller units, multiply.

number of kilograms	×	**number of grams in 1 kilogram**	=	**total grams**
↓		↓		↓
2	×	1,000	=	2,000

So, there are 2,000 grams in 2 kilograms.

STEP 2 To find how many grams will be used for each sandwich, divide 2,000 grams into 50 equal groups.

number of grams	÷	**number of sandwiches**	=	**grams per sandwiches**
↓		↓		↓
2,000	÷	50	=	40

So, Ms. Allison will use 50 grams of peanut butter for each sandwich.

METRIC UNITS OF CAPACITY

1,000 milliliters (mL) = 1 liter (L)
250 milliliters = 1 metric cup
4 metric cups = 1 liter
1,000 liters = 1 kiloliter (kL)

A large paper clip is about 1 gram.

A large container of peanut butter is about 1 kilogram.

METRIC UNITS OF MASS

1,000 milligrams (mg) = 1 gram (g)
1,000 grams = 1 kilogram (kg)

CALIFORNIA STANDARDS **MR 1.0** Students make decisions about how to approach problems. **MR 1.1** Analyze problems by identifying relationships, distinguishing relevant from irrelevant information, sequencing and prioritizing information, and observing patterns. *also* **MR 2.0, MR 3.0, MR 3.2**

Check

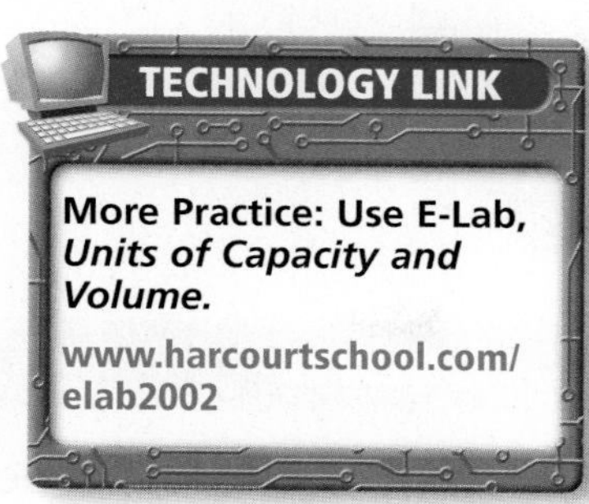

TECHNOLOGY LINK
More Practice: Use E-Lab, *Units of Capacity and Volume.*
www.harcourtschool.com/elab2002

1. **Explain** whether you multiply or divide to find how many milliliters are in 5 liters.

Change the unit.

2. 4,000 L = ■ kL
3. 8 metric cups = ■ L
4. 6 kg = ■ g
5. 2,000 mg = ■ g

Practice and Problem Solving

Change the unit.

6. 16 metric cups = ■ L
7. 8 kL = ■ L
8. 5 L = ■ mL
9. 1.5 kg = ■ g
10. 1,000 g = ■ kg
11. 3 g = ■ mg

Choose the best estimate.

12. The mass of a strawberry is
    a. 6 mg.
    b. 6 g.
    c. 6 kg.

13. The dropper holds
    a. 3 mL.
    b. 3 L.
    c. 3 kL.

14. The mass of a cat is
    a. 6 mg.
    b. 6 g.
    c. 6 kg.

15. The container holds
    a. 1 mL.
    b. 1 L.
    c. 1 kL.

16. The 4 members of We-Care Lawn Mowers drank a total of 2 L of water. If they all drank an equal amount of water, how many mL did each member drink?

17. A can of soda holds 355 mL. Which holds more, a 2-L bottle of soda or a 6-pack of soda cans? Explain.

18. **What's the Question?** Monique had a 1-kg jar of peanut butter for sandwiches. She used 750 g of peanut butter in one week. The answer is 250 g.

19. **ESTIMATION** 150 pennies have a mass of about 0.5 kg. Ramon has $3 in pennies. About how many grams is this?

Mixed Review and Test Prep

20. Divide 345 by 20. (p. 204)
21. Find 4.30 − 1.32. (p. 50)
22. 37.5 ÷ 1.5 (p. 238)
23. 3.6 ÷ 0.3 (p. 238)
24. **TEST PREP** Which shows the simplest form of $\frac{12}{32}$? (p. 290)
    A $\frac{3}{8}$ **B** $\frac{8}{32}$ **C** $\frac{24}{64}$ **D** $\frac{2}{3}$

Extra Practice page H58, Set C

LESSON 6

Time

Quick Review

1. 2 hr = ■ min
2. 3 wk = ■ days
3. 30 mo = ■ yr
4. 1 wk = ■ hr
5. ■ hr = 3,600 sec

Learn

BEAT THE CLOCK Noemi wants to run 5 miles in less than $1\frac{3}{4}$ hours. If she starts her run at 10:47 A.M. and runs 5 miles by 12:26 P.M., has she reached her goal?

One Way **Use a clock.**

Count forward on the clock from the starting time to the ending time.

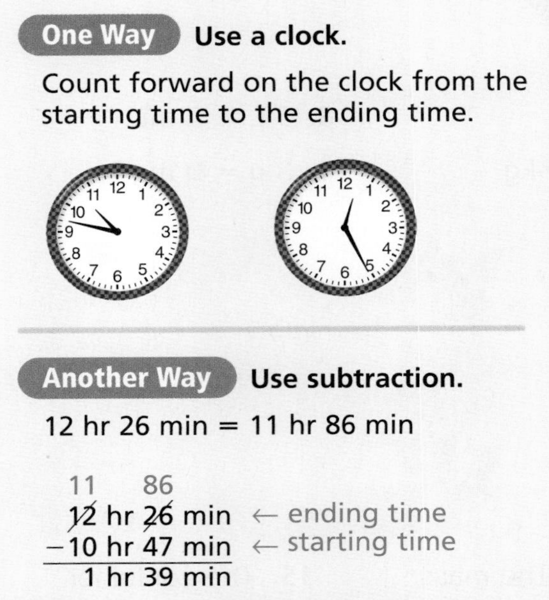

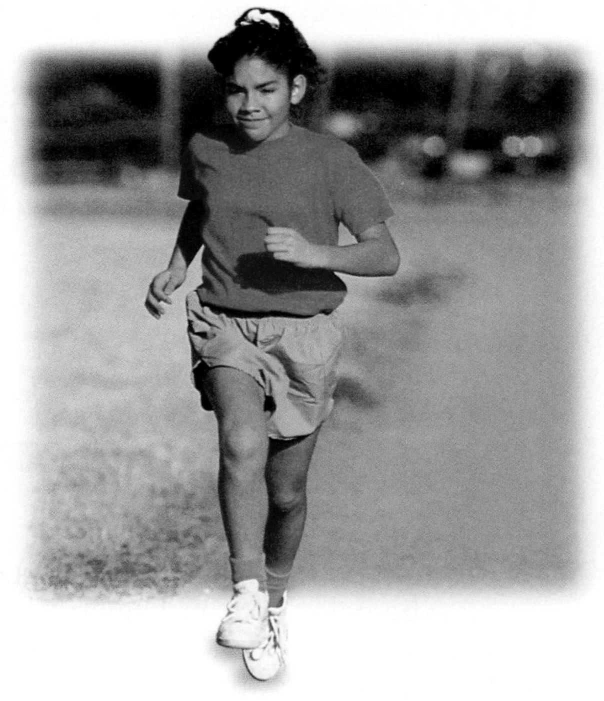

Another Way **Use subtraction.**

12 hr 26 min = 11 hr 86 min

$$\begin{array}{r} \overset{11}{\cancel{12}} \text{ hr } \overset{86}{\cancel{26}} \text{ min} \\ -10 \text{ hr } 47 \text{ min} \\ \hline 1 \text{ hr } 39 \text{ min} \end{array}$$

← ending time
← starting time

So, Noemi ran for 1 hr 39 min. Since 1 hr 39 min is less than $1\frac{3}{4}$ hr, or 1 hr 45 min, she has reached her goal.

You can use a calendar to find elapsed time.

Examples

A Count the whole days from 10 P.M. September 16 to 10 P.M. September 18.

September 2002
1 2 3 4 5 6 7
8 9 10 11 12 13 14
15 16 17 18 19 20 21
22 23 24 25 26 27 28
29 30

2 days

B Count the weeks from 2:00 A.M. April 3 to 2:00 A.M. April 24.

April 2002
1 2 3 4 5 6
7 8 9 10 11 12 13
14 15 16 17 18 19 20
21 22 23 24 25 26 27
28 29 30

3 weeks

UNITS OF TIME

60 seconds (sec) = 1 minute (min)
60 minutes = 1 hour (hr)
24 hours = 1 day
7 days = 1 week (wk)
about 52 weeks = 1 year (yr)
365 days = 1 year
366 days = 1 leap year
10 years = 1 decade
100 years = 1 century
1,000 years = 1 millennium

CALIFORNIA STANDARDS MR 2.0 Students use strategies, skills, and concepts in finding solutions. *also* MR 3.0, MR 3.2, MR 3.3

Check

1. **Describe** how you use a clock to find the elapsed time between Wednesday, 3:00 P.M. and Thursday, 1:00 A.M.

Write the time for each.

2. Start: 4:30 P.M.
   45 min elapsed time
   End: ■

3. Start: ■
   25 min elapsed time
   End: 8:00 P.M.

4. Start: June 11, 9:00 P.M.
   ■ time has elapsed.
   End: June 16, 5:30 A.M.

Add or subtract.

5. 5 hr 10 min + 2 hr 45 min

6. 8 hr 11 min − 3 hr 24 min

7. 5 hr 15 min − 3 hr 35 min

8. 1 hr 50 min + 3 hr 22 min

Practice and Problem Solving

Write the time for each.

9. Start: ■
   25 min elapsed time
   End: 4:15 P.M.

10. Start: 1:45 A.M.
    3 hr 15 min elapsed time
    End: ■

11. Start: May 1, 2:00 P.M.
    3 days, 6 hr, 10 min elapsed time
    End: ■

Add or subtract.

12. 8 hr 36 min + 3 hr 55 min

13. 16 hr 9 min − 10 hr 15 min

14. 6 hr 23 min − 2 hr 47 min

15. 24 hr 49 min + 12 hr 33 min

16. Hiromi began practicing his tuba at 11:20 A.M. If he needs to practice for 45 minutes, at what time should he stop?

17. Leeza began doing her chores at 3:30 P.M. She finished at 4:45 P.M. How many minutes did it take her to finish her chores?

18. Pablo read 90 pages of a book in 3 hours. On average, how long did it take him to read each page?

19. The game ended at 1:45 P.M. It began $2\frac{1}{2}$ hours earlier. At what time did it begin?

20. **Write a problem** about how much time it might take to do your homework or bake cookies. Include beginning and ending times.

21. How many centuries are in a millennium?

Mixed Review and Test Prep

22. Write 10,000 by using an exponent. (p. 266)

23. 493 × 64 (p. 148)

24. 704 × 88 (p. 148)

25. 108 × 22 (p. 148)

26. TEST PREP What is the least common denominator for $\frac{6}{7}$ and $\frac{3}{4}$? (p. 352)

    A 24 **B** 28 **C** 30 **D** 36

Extra Practice page H58, Set D

LESSON 7

Problem Solving Strategy

Make a Table

Quick Review

Begin at 3:20 P.M. Find the elapsed time for each time given.

1. 4:25 P.M. **2.** 5:00 P.M.
3. 3:55 P.M. **4.** 6:05 P.M.
5. 5:15 P.M.

PROBLEM You and your aunt are planning to take a train to visit City Museum. Your aunt will pick you up at 10:30 A.M. It takes 20 minutes to drive to the Smithfield train station. You must be home by 5:00 P.M. How long can you stay at the museum?

Departure	Arrival
Smithfield	*City Museum*
9:50 A.M.	10:20 A.M.
11:10 A.M.	11:40 A.M.
12:30 P.M.	1:00 P.M.
City Museum	*Smithfield*
2:00 P.M.	2:30 P.M.
3:30 P.M.	4:00 P.M.
4:45 P.M.	5:15 P.M.

- What are you asked to find?
- What information will you use?

- What strategy can you use to solve the problem?

You can use the train schedule and the information in the problem to *make a table* to show the elapsed time.

- How can you use the table to solve the problem?

You can use the table to plan your schedule.

You'll get to the train station at 10:50 A.M.

The next train is at 11:10 A.M.

If you take the 4:45 P.M. train home, you won't be home by 5:00 P.M. So, take the 3:30 P.M. train.

You can stay at the museum from 11:40 A.M. to 3:30 P.M. So, you can spend 3 hr 50 min at the museum.

Part of Trip	Start	End
Drive to train station	10:30 A.M.	10:50 A.M.
Train to Museum	11:10 A.M.	11:40 A.M.
Visit Museum	11:40 A.M.	?
Train from Museum	3:30 P.M.	4:00 P.M.
Drive home from train station	4:00 P.M.	4:20 P.M.

Check

- How can you decide if your answer is reasonable?

CALLIFORNIA STANDARDS **MR 1.0** Students make decisions about how to approach problems. **MR 1.2** Determine when and how to break a problem into simpler parts. **MR 2.0** Students use strategies, skills, and concepts in finding solutions. *also* **MR 2.3, MR 2.4, MR 2.6, MR 3.2, MR 3.3**

Problem Solving Practice

- Draw a Diagram or Picture
- Make a Model or Act It Out
- Make an Organized List
- Find a Pattern
- ▶ Make a Table or Graph
- Predict and Test
- Work Backward
- Solve a Simpler Problem
- Write an Equation
- Use Logical Reasoning

For 1–2, use the train schedule on page 518.

1. **What if** your aunt arrives to pick you up at 11:00 A.M., instead of 10:30? How long can you stay at the museum?
2. Conner wants to take the train to the museum. He can walk to the train station in 10 minutes. He must be home by 2:45 P.M. If he wants to spend at least 3 hours at the museum, by what time should he leave for the train station?
3. It takes 25 minutes for each batch of biscuits to bake. It is 2:00 P.M. How many batches of biscuits can Greg bake before he must leave for soccer practice at 3:45 P.M.?

On Wednesday, the third graders will use the library for half an hour, beginning at 10:45 A.M. Then the fifth graders will use the library for 50 minutes. At 1:45 P.M., the fourth graders will come to the library to work on their research papers.

4. How long is the library free after the fifth grade leaves?

   A 1 hr **C** 1 hr 40 min
   B 1 hr 30 min **D** 2 hr

5. When might the sixth graders use the library to watch a 90-minute video on Wednesday?

   F 8:30 A.M. **H** 10:30 A.M.
   G 9:30 A.M. **J** 11:30 A.M.

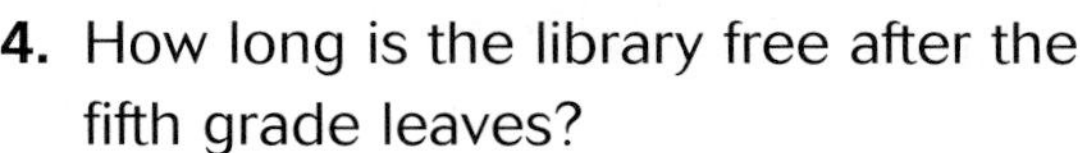

Mixed Strategy Practice

6. At Awards Night, there were 225 students and 328 parents. The chairs were set up in 20 rows of 25 chairs each. Did everyone get a seat? Explain.
7. Jorge had 30 meters of rope. He cut 14 centimeters off both ends of the rope. How many centimeters long is the rope now?
8. Cari is 3 times as old as Jen. Together, their ages add to 24. Use the diagram to find Jen's age.

   Cari ☐☐☐
   Jen ☐

9. The school band sold 120 boxes of cookies. The trombone section sold $\frac{1}{4}$ of the total number of boxes. How many boxes did the trombone section sell?

CHAPTER 27

Review/Test

CHECK VOCABULARY AND CONCEPTS

Choose the correct term from the box.

centimeter
liters
millimeters

1. There are 10 __?__ in 1 centimeter. (p. 506)
2. A __?__ is about the width of your index finger. (p. 506)

CHECK SKILLS

Estimate the length in inches. Then measure to the nearest $\frac{1}{16}$ inch. (pp. 504–505)

3.

4. Presenting
6 CHILD PRICE $5.50
*AUSTRALIAN ADVENTURE
Auditorium 6
7.30 pm Friday
West Maple Theater

Draw a line segment to the given length. (pp. 506–507)

5. 3 cm 6 mm
6. 10.2 cm
7. 13 mm
8. 6 cm
9. 3.8 cm

Change the unit. (pp. 508–515)

10. 2 mi = ■ yd
11. 900 yd = ■ ft
12. 200 cm = ■ m
13. 20 mm = ■ cm
14. 5 T = ■ lb
15. 8 pt = ■ gal
16. 15 g = ■ mg
17. 20 metric cups = ■ L
18. 2.5 kg = ■ g

Find the sum or difference. (pp. 508–511)

19. 8 yd 2 ft − 5 yd 1 ft
20. 6.7 cm + 2.9 cm
21. 4 ft 6 in. − 2 ft 7 in.
22. 3.6 cm − 1.9 cm

Write the time for each. (pp. 516–517)

23. Start: March 18, 10:05 A.M.
5 days, 5 hr, 40 min elapsed time
End: ■

24. Start: 5:10 A.M.
■ time has elapsed.
End: 12:49 P.M.

CHECK PROBLEM SOLVING

Use the schedule to solve. (pp. 518–519)

Departure	Arrival
Green Street	Shoppers' Plaza
10:30 A.M.	11:20 A.M.
12:45 P.M.	1:35 P.M.
Shoppers' Plaza	Green Street
1:35 P.M.	2:25 P.M.
4:00 P.M.	4:50 P.M.

25. John wants to go to the arcade at Shoppers' Plaza. It takes him 20 minutes to walk to the Green Street train station. He must be home by 3:00 P.M. If he leaves for the train station at 10:00 A.M., how much time can he spend at the arcade?

Cumulative Review

Look for important words.
See item **9.**

Important words are *good estimate* and *Celsius.* Notice in the table that the water temperature of the habitat is given in degrees Fahrenheit. You need to find a reasonable Celsius approximation for the Fahrenheit temperature given in the table.

Also see problem **2,** p. H62.

For 1–12, choose the best answer.

1. How many grams are in 6.5 kilograms?

A 65,000 **C** 650
B 6,500 **D** 65

2. The average length of a giraffe's tongue is 53.34 centimeters. How many millimeters is that?

F 53,340 **H** 533.4
G 5,334 **J** Not here

3. Which term describes a triangle that has sides measuring 5 cm, 7 cm, and 8 cm?

A isosceles **C** scalene
B right **D** equilateral

4. $2.7 \div 0.03$

F 90 **H** 0.9
G 9 **J** Not here

5. $3.654 \div 9$

A 0.406 **C** 4.006
B 0.46 **D** 4.06

6. What is 65,934 rounded to the nearest hundred?

F 65,000 **H** 65,980
G 65,900 **J** 66,000

7. 6.9 cm + 4.4 cm

A 10 cm **C** 11 cm
B 10.3 cm **D** 11.3 cm

8. Which is a name for this figure?

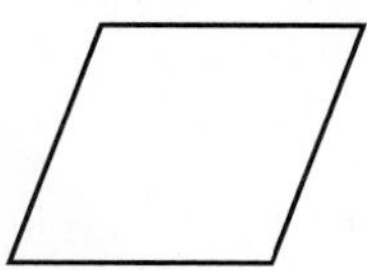

F square **H** prism
G rectangle **J** rhombus

For 9–11, use the table.

BLUE WHALE FACTS	
Average length	110 feet
Average weight	209 tons
Average water temperature of habitat	55°F

9. What is a good estimate of the average temperature in degrees Celsius of the whale habitat?

A 0°C **C** 32°C
B 12°C **D** 142°C

10. About how many yards long is the average blue whale?

F about 1,320 **H** about 37
G about 330 **J** about 9

11. How many pounds does an average blue whale weigh?

A 418,000 **C** 41,800
B 209,000 **D** 2,090

12. How long is a class that starts at 11:15 A.M. and ends at 1:55 P.M.?

F 1 hr 30 min **H** 2 hr 30 min
G 1 hr 40 min **J** 2 hr 40 min

CHAPTER 28

Perimeter and Area

Central Park, located in the center of Manhattan Island, New York, was one of the first public parks laid out by landscape architects. It includes lakes, open fields, a zoo, an art museum, a skating rink, and an open-air theater. The park is shaped like a rectangle that is 2.5 miles by 0.5 mile. What are the perimeter and area of Central Park? How does the area of Central Park compare to the areas of the parks in the graph?

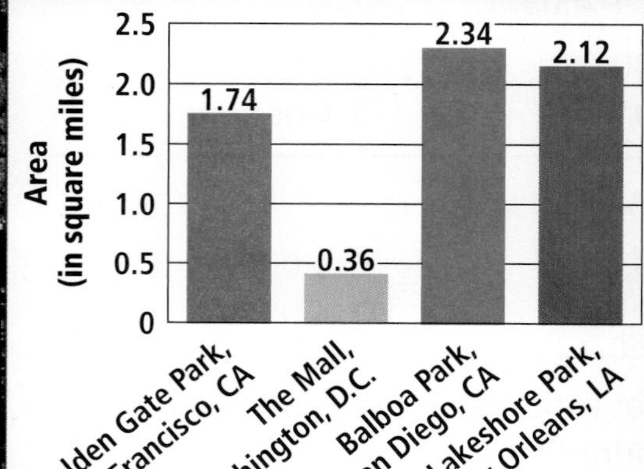

CHECK WHAT YOU KNOW

Use this page to help you review and remember important skills needed for Chapter 28.

VOCABULARY

Choose the best term from the box.

parallelogram right triangle perimeter

1. A quadrilateral with two pairs of congruent and parallel sides is a __?__.
2. A triangle that has a right angle is a __?__.

PERIMETER AND AREA (See p. H29.)

Find the area and perimeter of each figure.

3. 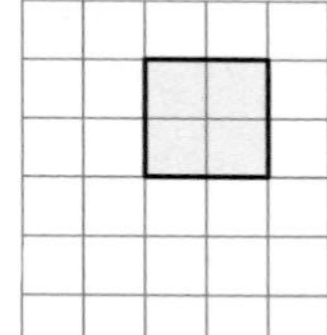
4. 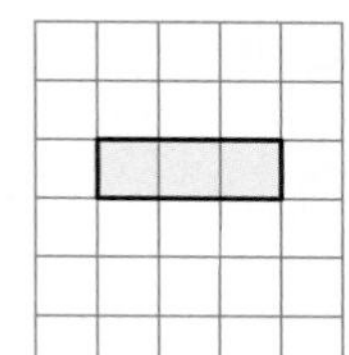
5. 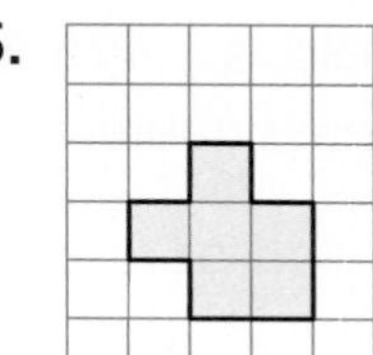
6. 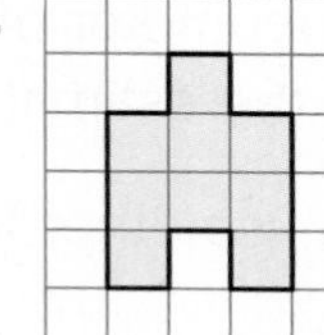
7. 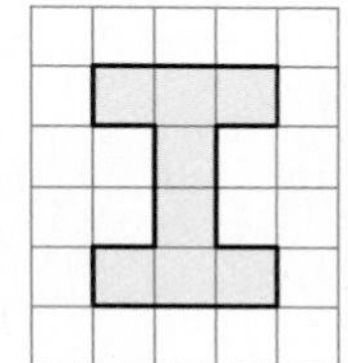
8. 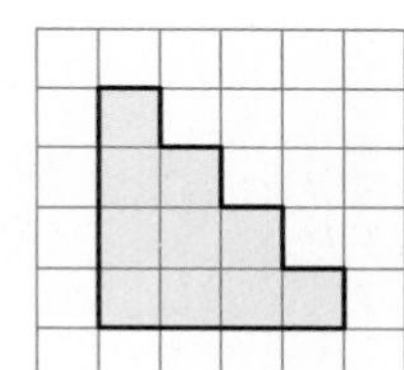
9. 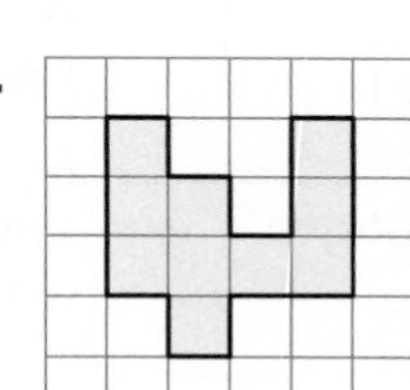
10. 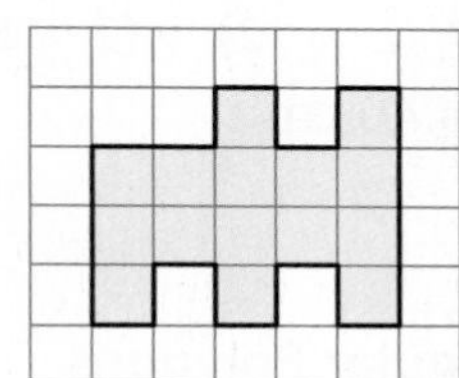

MULTIPLICATION (See p. H30.)

Find the product.

11. 9×6
12. 8×7
13. 12×5
14. 19×6
15. 42×4
16. 29×12
17. 35×21
18. 81×42
19. 1.29×6
20. 2.35×8
21. 4.6×1.5
22. 2.8×2.8
23. 3.24×12
24. 4.12×1.4
25. 8.5×8.5

LESSON 

Perimeter

Quick Review

Name the number of sides for each figure.

1. triangle
2. rectangle
3. square
4. pentagon
5. octagon

VOCABULARY

perimeter

Learn

IT ALL ADDS UP! Dan drew a diagram of the playing field at Wrigley Field in Chicago. What is the perimeter of the playing field?

The **perimeter** is the distance around a figure.

MATH IDEA You can find the perimeter of any polygon by adding the lengths of its sides.

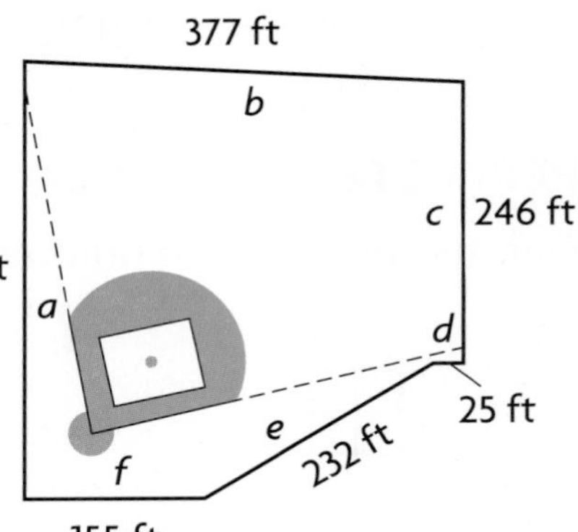

Estimate. $400 + 400 + 200 + 0 + 200 + 200 = 1{,}400$

$a + b + c + d + e + f = P$

$\downarrow \quad \downarrow \quad \downarrow \quad \downarrow \quad \downarrow \quad \downarrow \quad \downarrow$

Find the sum. $390 + 377 + 246 + 25 + 232 + 155 = 1{,}425$

So, the perimeter of Wrigley Field is 1,425 ft. The answer is close to the estimate, so it is reasonable.

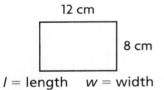

You can use other formulas to find perimeter.

Rectangles

$P = (2 \times l) + (2 \times w)$

$P = (2 \times 12) + (2 \times 8)$

$P = 40$ cm

12 cm

8 cm

l = length w = width

Regular Polygons

$P =$ (number of sides) $\times s$

$P = 6 \times s$

$P = 6 \times 2$

$P = 12$ ft

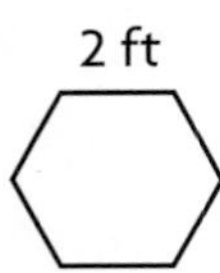

s = side

Check

1. **Write** three formulas you could use to find the perimeter of a park measuring 150 ft by 150 ft.

Find the perimeter of each polygon.

2. 

3. 4 cm, 5 cm, 3 cm

4. 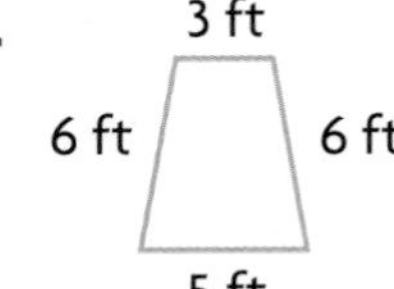

5. 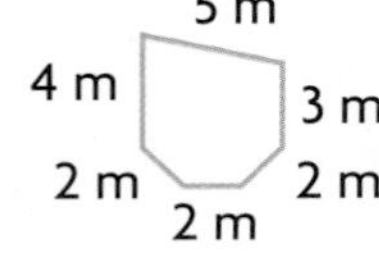

CALIFORNIA STANDARDS MG 1.4 Differentiate between, and use appropriate units of measures for, two- and three-dimensional objects (e.g., find the perimeter, area, volume). **AF 1.2** Use a letter to represent an unknown number: write and evaluate simple algebraic expressions in one variable by substitution. *also* **AF 1.0, MR 2.0, MR 2.1, MR 2.4, MR 3.2, MR 3.3**

Practice and Problem Solving

Find the perimeter of each polygon.

6. 10 cm; 3.5 cm

7. 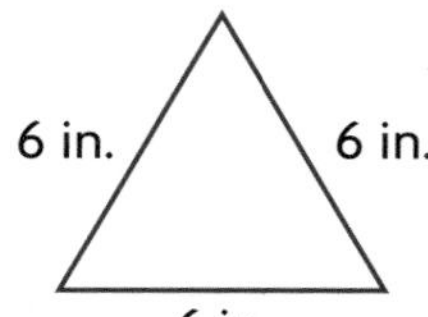

8. 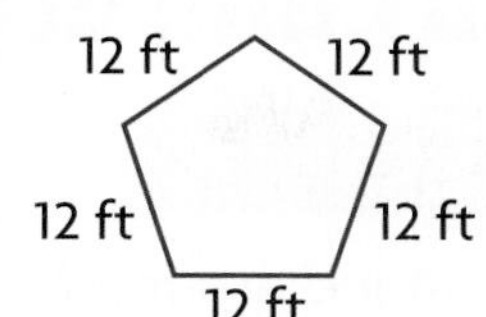

9. 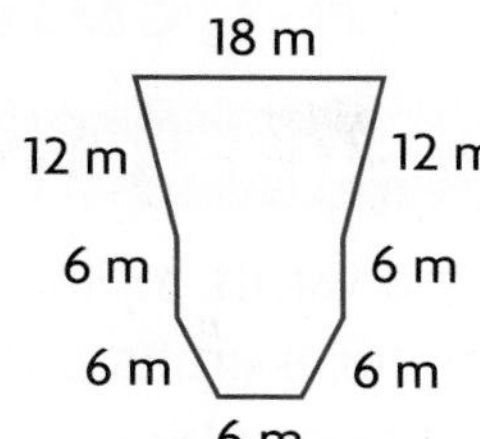

10. 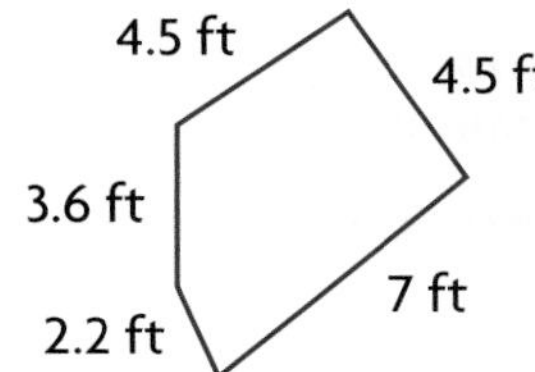

11. 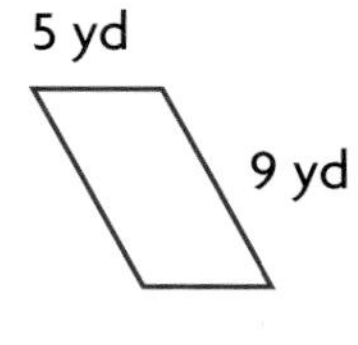

12. 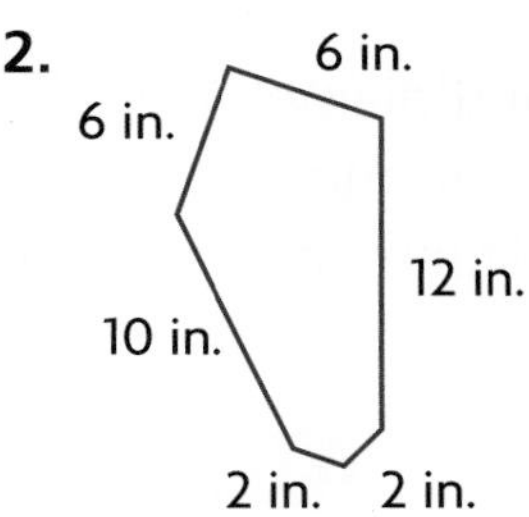

13. 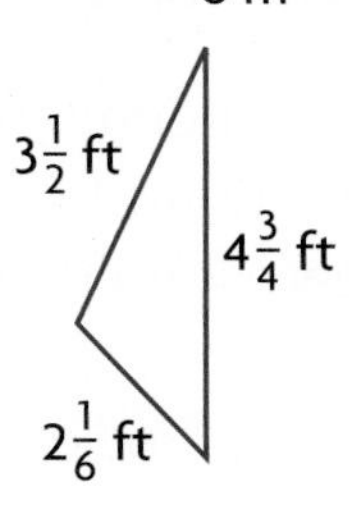

USE DATA For 14–15, use the table.

14. What is the perimeter of each regulation field?

15. Which has a greater perimeter, a sand volleyball court or a basketball court?

SIZE OF REGULATION FIELDS

Field/Court	Dimensions
Basketball court	94 ft × 50 ft
Baseball diamond	90 ft × 90 ft
Soccer field	110 m × 75 m
Sand volleyball court	29 ft 6 in. × 59 ft
Ice hockey rink	58 m × 26 m

Algebra For 16–20, complete the table.

	Regular Polygon	Number of Equal Sides	Formula for Perimeter
16.	triangle	■	$P = 3 \times s$
17.	■	4	$P = 4 \times s$
18.	pentagon	5	■
19.	hexagon	■	$P = 6 \times s$
20.	■	■	$P = 8 \times s$

21. **Write About It** The perimeter of a triangle is 30 cm, and two of its sides are 10 cm and 7 cm long. Explain how to find the length of its third side.

22. **REASONING** A square and a regular pentagon have the same perimeter. Which polygon has longer sides? Explain.

Mixed Review and Test Prep

Find the sum or difference. (p. 38)

23. $\begin{array}{r} 5{,}605{,}361 \\ +1{,}222{,}604 \\ \hline \end{array}$

24. $\begin{array}{r} 8{,}125{,}605 \\ -1{,}854{,}627 \\ \hline \end{array}$

25. Sam drove 225 miles at an average speed of 54 miles per hour. Did he make the trip in less than 4 hours? (p. 204)

26. Express 0.3 + 0.5 as a fraction in simplest form. (p. 280)

27. **TEST PREP** Joan has half a pizza. If she gives $\frac{1}{5}$ of it to Bob, what fraction of a pizza will she have left? (p. 354)

A $\frac{9}{10}$ **B** $\frac{4}{5}$ **C** $\frac{2}{5}$ **D** $\frac{1}{10}$

LESSON 2

HANDS ON

Algebra: Circumference

Quick Review

1. 1.2 ÷ 3
2. 2.4 ÷ 8
3. 3.5 ÷ 7
4. 0.81 ÷ 9
5. 0.56 ÷ 8

VOCABULARY

circumference

MATERIALS

centimeter ruler, string, cans

Explore

Carla wants to find the distance around a can for a crafts project. The distance around a circular object is its **circumference**.

Find the circumference and diameter of the can. Round your measurements to the nearest tenth of a centimeter.

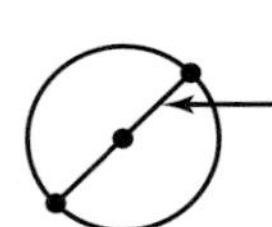

STEP 1

Wrap the string around a can.

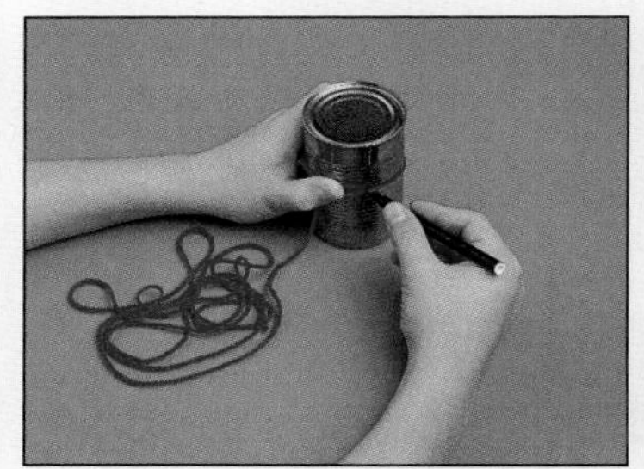

STEP 2

Use the ruler to measure the length of the string. This is the circumference (*C*) of the can.

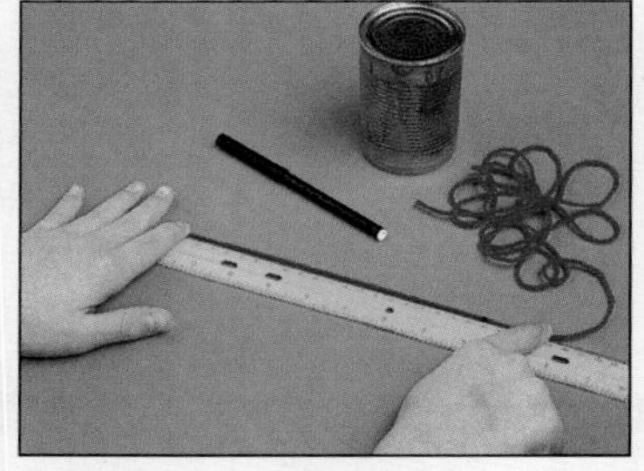

STEP 3

Trace the base of the can, and measure the diameter (*d*) of the circle.

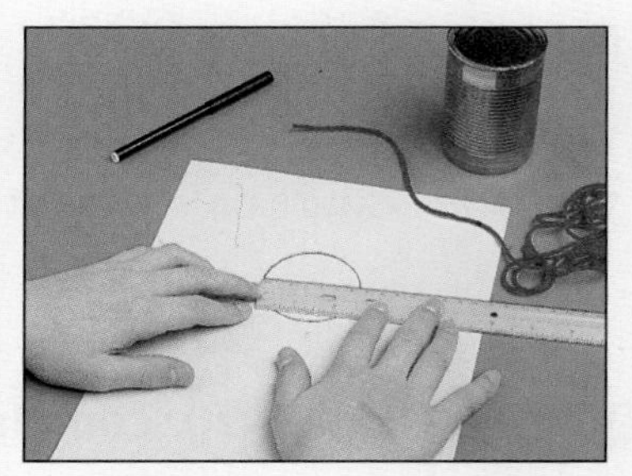

- Divide the circumference by the diameter. How many times as long as the diameter is the circumference?

Try It

Use three different size cans. Follow the steps above. Copy and complete the table. Round each quotient to the nearest hundredth of a centimeter.

Circle	Circumference (C)	Diameter (d)	C÷d
Example	15.7 cm	5 cm	
A			
B			

- Use your results to describe the ratio between the circumference and the diameter of a circle.

What do you do next to find the relationship between the circumference and the diameter?
a
b
c

CALIFORNIA STANDARDS MG 1.4 Differentiate between, and use appropriate units of measures for, two- and three-dimensional objects. **AF 1.2** Use a letter to represent an unknown number: write and evaluate simple algebraic expressions in one variable by substitution. *also* **AF 1.0, MR 1.1, MR 2.0, MR 2.3, MR 2.5, MR 3.3**

Connect

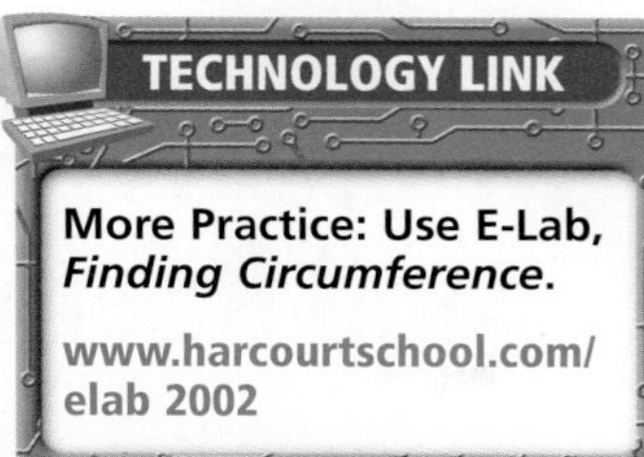

TECHNOLOGY LINK
More Practice: Use E-Lab, *Finding Circumference*.
www.harcourtschool.com/elab 2002

The ratio of the circumference to the diameter of a circle, $C : d$, is called pi (π). The approximate decimal value of π is 3.14.

So, when you know the diameter, use the formula at the right to find the circumference.

$$C = \pi \times d$$

circumference ≈ 3.14 × diameter

≈ means "is approximately equal to"

Examples Find the circumference.

A

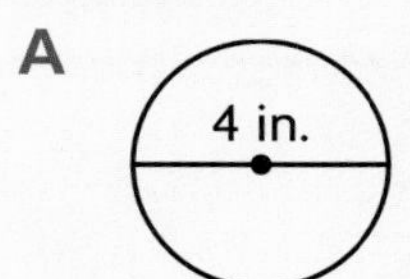

$C = \pi \times d$
$C \approx 3.14 \times 4$
$C \approx 12.56$
The circumference is 12.56 in.

B

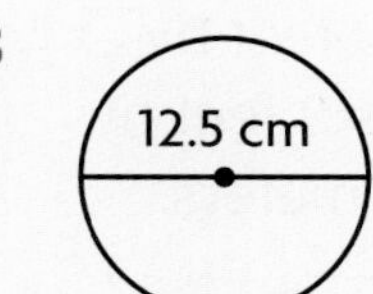

$C = \pi \times d$
$C \approx 3.14 \times 12.5$
$C \approx 39.25$
The circumference is 39.25 cm.

Practice

For 1–3, complete the table.

	Object	C	d	$C \div d$
1.	spool	9.4 cm	3 cm	■
2.	mug	28.3 cm	■	3.14
3.	CD	■	12 cm	3.14

4. Marta is putting lace trim around a can for her project. The diameter of the can is 7.5 cm. How many centimeters of lace trim does she need?

Find the circumference of a circle that has

5. a diameter of 10 cm.

6. a diameter of 17.8 ft.

7. a radius of 6 in.

8. The diameter of Bobby's can is 8 cm. He wants to put two rows of red tape around the can. How many centimeters of tape does he need?

9. The diameter of the dinner plate is $10\frac{7}{8}$ in. The salad plate's diameter is $8\frac{1}{4}$ in. How much greater is the dinner plate's diameter?

10. **Write About It** Explain to a fourth grader how to find the circumference of a circular object.

11. **REASONING** The radius of the Earth is about 3,960 miles. About how long is the equator?

Mixed Review and Test Prep

12. 2,368,561 + 3,826,692 (p. 38)

13. 10,008 − 8,976 (p. 38)

14. Write $3\frac{6}{4}$ in simplest form. (p. 290)

15. Aaron has seventeen quarters. Express this in dollars and cents. (p. 158)

16. **TEST PREP** Which expression models this statement? Mrs. Ray took 2 vitamins 3 times a day for a week. (p. 86)

A $2 + 3$ **C** 2×3
B $2 \times 3 \times 7$ **D** $2 + 3 + 7$

LESSON 3

Algebra: Area of Squares and Rectangles

Quick Review

Find the value of *n*.

1. $2^2 = n$
2. $3^2 = n$
3. $4^2 = n$
4. $5^2 = n$
5. $6^2 = n$

VOCABULARY

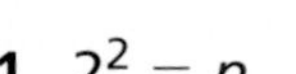

area

Learn

MAKING PLANS! In art class, Holly and Jake are designing floor plans for a house. In her floor plan, Holly designs her room to be 10 feet by 8 feet. In Jake's floor plan, his room is 9 feet by 9 feet. The rooms have the same perimeters. Are the areas the same?

Find the  **area** of each room, or the number of square units needed to cover each room's floor space.

Activity

MATERIALS: grid paper

You can use grid paper to find the area of squares and rectangles.

STEP 1

Let each square on the grid represent 1 square foot. Draw a rectangle 8 squares by 10 squares and shade it. Draw a square 9 squares by 9 squares and shade it.

STEP 2

Count the number of shaded grid squares in your rectangle and square. Compare the areas.

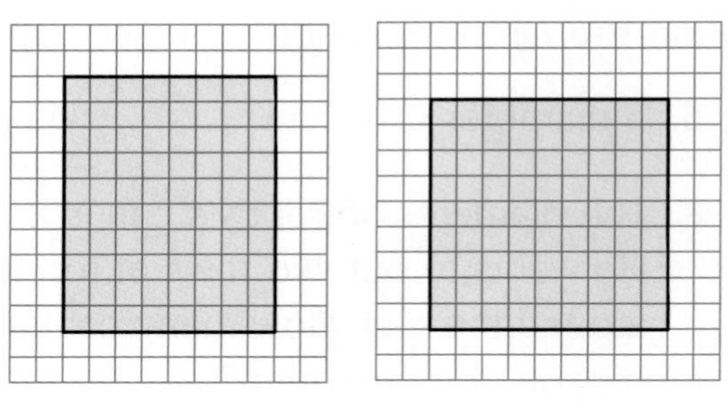

80 sq ft < 81 sq ft

No, the areas are not the same. Jake's room has the greater area.

You can also use formulas to find the area of rectangles and squares.

Area of a rectangle = length × width

$A = l \times w$

$A = 7 \times 3.5$

$A = 24.5$

The area is 24.5 sq m, or 24.5 m^2.

Area of a square = side × side

$A = s \times s$

$A = 8.2 \times 8.2$

$A = 67.24$

The area is 67.24 sq cm, or 67.24 cm^2.

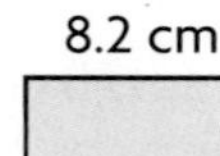

CALIFORNIA STANDARDS MG 1.4 Differentiate between, and use appropriate units of measures for, two- and three-dimensional objects (e.g., find the perimeter, area, volume). **AF 1.2** Use a letter to represent an unknown number: write and evaluate simple algebraic expressions in one variable by substitution. *also* **AF 1.0, MG 1.0, MR 1.0, MR 2.0, MR 2.3, MR 3.2, MR 3.3**

Check

1. **Explain** how the formula for the area of a rectangle can be used to find the area of a square.

Find the area of each figure.

2. 6 in. / 6 in.

3. 13 m / 8.5 m

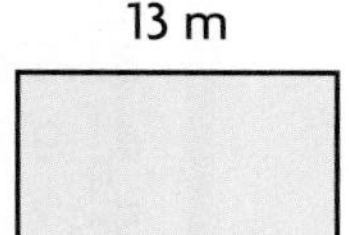

4. 22.3 cm / 10 cm

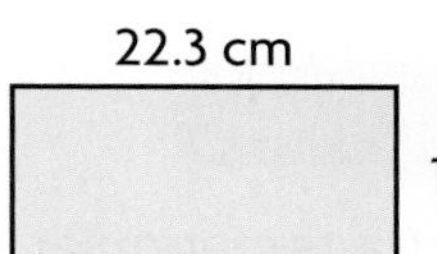

5. 54 mi / 54 mi

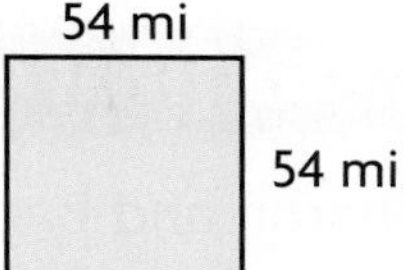

Practice and Problem Solving

Find the area of each figure.

6. 12 in. / 12 in.

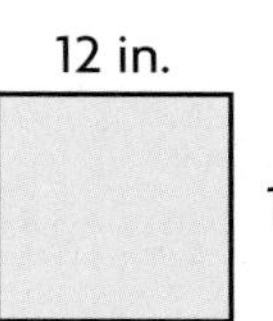

7. $3\frac{1}{4}$ ft / $2\frac{1}{2}$ ft

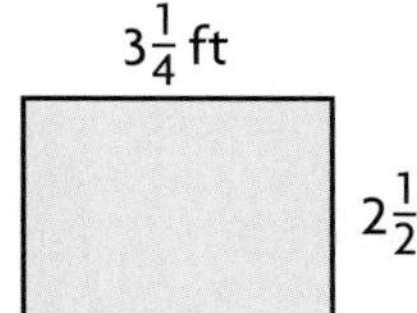

8. 15.5 km / 12 km

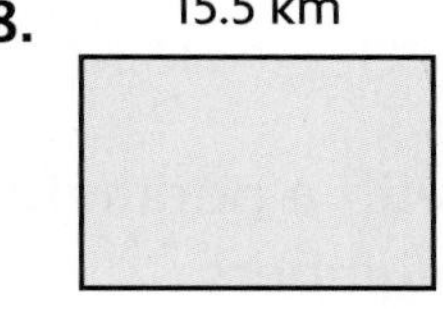

9. 32 mi / 32 mi

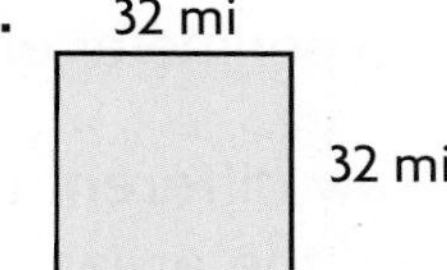

Find each missing measurement.

10. $s = 4\frac{1}{2}$ yd
 $A = ■$

11. $s = 3.5$ km
 $A = ■$

12. $s = 7$ ft
 $A = ■$

13. $s = 1.25$ m
 $A = ■$

14. $l = 3.2$ m
 $w = 4$ m
 $A = ■$

15. $l = ■$
 $w = 7.2$ cm
 $A = 28.8 \text{ cm}^2$

16. $l = 2\frac{2}{3}$ ft
 $w = 4\frac{1}{4}$ ft
 $A = ■$

17. $P = 24$ yd
 $l = 8$ yd
 $w = ■$
 $A = ■$

18. **Write About It** How can the formula for the area of a square be used to find the area of a rectangle that is 12 ft by 24 ft? Explain.

19. **REASONING** If you double the length of a rectangle, how does that affect its area? **What if** you double the width and the length?

20. How many 1-ft^2 tiles are needed to cover a 15 ft × 25 ft patio?

21. **What's the Question?** A garden is 6 yards by 8 yards. The answer is 28 yards.

Mixed Review and Test Prep

22. $\begin{array}{r} 7.3 \\ \times 4.8 \\ \hline \end{array}$ (p. 170)

23. $\begin{array}{r} 4.8 \\ \times 6.3 \\ \hline \end{array}$ (p. 170)

24. What are the factors of 56? (p. 258)

25. Order $^{+}3$, $^{-}4$, $^{+}7$, $^{-}7$ from least to greatest. (p. 420)

26. **TEST PREP** Which are the least and the greatest fractions in this list? (p. 286)

$\frac{7}{4}, \frac{8}{7}, \frac{6}{3}, \frac{11}{9}, \frac{5}{5}, \frac{7}{2}, \frac{2}{8}, \frac{9}{6}$

A $\frac{2}{8}, \frac{7}{2}$

B $\frac{6}{3}, \frac{11}{9}$

C $\frac{2}{8}, \frac{8}{7}$

D $\frac{2}{8}, \frac{9}{6}$

LESSON 

HANDS ON
Relate Perimeter and Area

Quick Review

Find the area of each rectangle.

1. 3 in. × 4 in.
2. 2 ft × 8 ft
3. 6 cm × 6 cm
4. 9 in. × 5 in.
5. 7 m × 11 m

MATERIALS
grid paper

Explore

Sharon and Brad are planning a rectangular sandbox for the preschool class. They have 20 feet of wood to make its sides. If they want their sandbox to have the greatest possible area, what dimensions should they use?

You can use grid paper to model the relationship between perimeter and area. Find the greatest area of a rectangle with a perimeter of 20 feet.

Draw different rectangles with a perimeter of 20 feet. Find the areas. Use whole numbers.

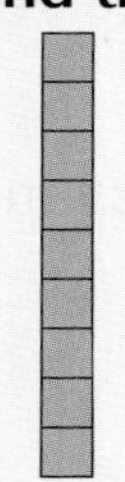

$P = 20$ ft
$A = l \times w$
$A = 1 \times 9$
$A = 9$ ft^2

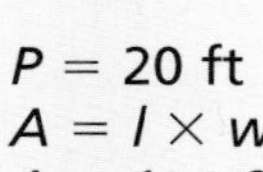

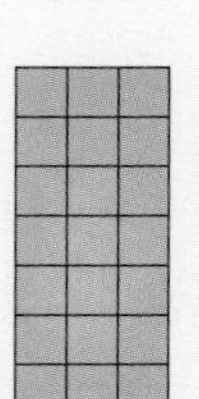

$P = 20$ ft
$A = l \times w$
$A = 2 \times 8$
$A = 16$ ft^2

$P = 20$ ft
$A = l \times w$
$A = 3 \times 7$
$A = 21$ ft^2

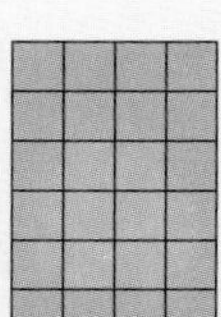

$P = 20$ ft
$A = l \times w$
$A = 4 \times 6$
$A = 24$ ft^2

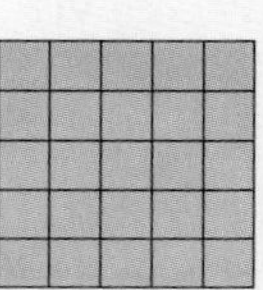

$P = 20$ ft
$A = l \times w$
$A = 5 \times 5$
$A = 25$ ft^2

So, Sharon and Brad should build a 5-ft × 5-ft sandbox.

- What dimensions gave the least area? What kind of rectangle gave the greatest area?

What other rectangles can you draw with a perimeter of 30 cm? Which one has the greatest area?

Try It

Use grid paper to draw rectangles for the given perimeter. Name the length and width of the rectangle with the greatest area. (Use whole numbers.)

a. $P = 30$ cm **b.** $P = 24$ cm

c. $P = 10$ cm **d.** $P = 36$ cm

CALIFORNIA STANDARDS MG 1.4 Differentiate between, and use appropriate units of measures for, two- and three-dimensional objects. **MR 1.1** Analyze problems by identifying relationships, distinguishing relevant from irrelevant information, sequencing and prioritizing information, and observing patterns. *also* **AF 1.0, AF 1.2, MG 1.0, MR 1.0, MR 2.0, MR 2.6, MR 3.0, MR 3.1**

Connect

Now, find the rectangle with the least perimeter that has an area of 24 ft^2. Use whole numbers only. Record your data in a table. (HINT: To find all the possible lengths and widths, find all the factors of 24.)

Length	Width	Perimeter	Area
1	24	50 ft	24 ft^2
2	12	28 ft	24 ft^2
3	8	22 ft	24 ft^2
4	6	20 ft	24 ft^2

So, the rectangle with the least perimeter that has an area of 24 ft^2 is 4 ft $\times$ 6 ft, or 6 ft $\times$ 4 ft.

REASONING As a rectangle gets closer to being a square, what happens to its perimeter?

MATH IDEA You can change the area of a figure without changing its perimeter, and you can change the perimeter of a figure without changing its area.

Practice

Use grid paper to draw rectangles for the given perimeter. Name the length and width of the rectangle with the greatest area. (Use whole numbers.)

1. $P = 16$ cm **2.** $P = 14$ cm **3.** $P = 32$ cm **4.** $P = 42$ cm **5.** $P = 60$ cm

Find the dimensions of the rectangle with the least perimeter for the given area. (Use whole numbers.)

6. $A = 8$ cm^2 **7.** $A = 25$ cm^2 **8.** $A = 32$ cm^2 **9.** $A = 64$ cm^2 **10.** $A = 100$ cm^2

11. $A = 15$ cm^2 **12.** $A = 36$ cm^2 **13.** $A = 49$ cm^2 **14.** $A = 56$ cm^2 **15.** $A = 81$ cm^2

16. **REASONING** If you increase the length of a rectangle with a perimeter of 20 ft, what has to happen to its width to keep the perimeter the same?

17. **What's the Error?** Bud says that with a given perimeter, the rectangle with the greatest length has the greatest area. Describe his error.

18. Using 48 feet of fencing, what is the greatest area that can be fenced? the least area? (Use whole numbers.)

19. **Write a problem** about a soccer field with dimensions of 110 m by 75 m.

Mixed Review and Test Prep

20. $4\overline{)3.68}$ (p. 222) **21.** $8\overline{)8.64}$ (p. 222)

22. What kind of angle measures greater than 90 degrees? (p. 454)

23. $\frac{1}{3} + \frac{3}{8}$ (p. 354)

24. **TEST PREP** Which of the following is NOT a measurement of length? (pp. 504, 506)

A meter **C** foot
B liter **D** kilometer

LESSON 5

Algebra: Area of Triangles

Quick Review

1. $\frac{1}{2} \times 10$ 2. $\frac{1}{2} \times 16$

3. $\frac{1}{2} \times 24$ 4. $\frac{1}{2} \times 2.4$

5. $\frac{1}{2} \times (3 \times 4)$

VOCABULARY

base **height**

Learn

SURF THE WIND

Dion is a windsurfer. He is making a model of his sailboard. How much fabric does he need for the sail?

Find the *area* of the triangular sail.

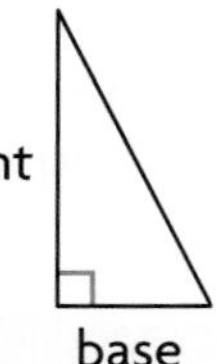

The **height** is the length of a line segment perpendicular to the **base** of the triangle.

Activity 1

MATERIALS: grid paper

Use grid paper and what you know about the area of a rectangle to find the area of a triangle.

STEP 1

Draw and shade a model of the triangular sail on grid paper.

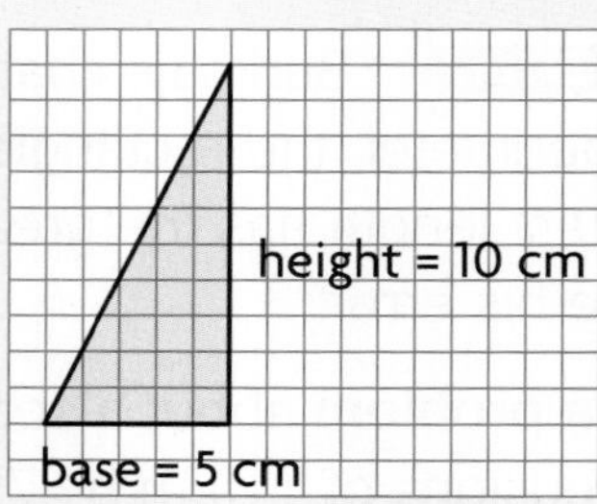

STEP 2

Draw a rectangle around the triangle as shown. Find the area of the rectangle.

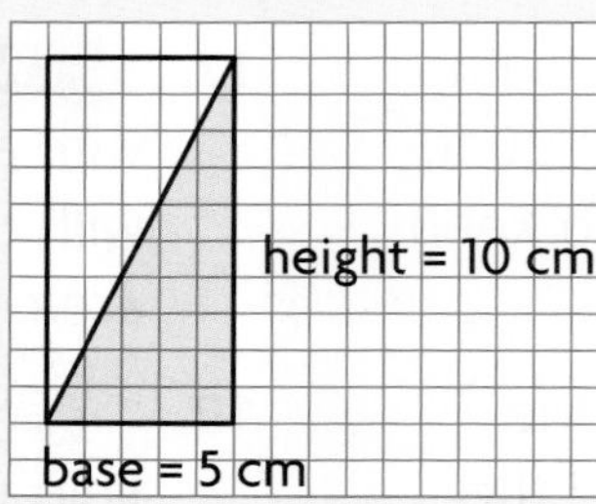

Rectangle:

$A = b$ (base) $\times\ h$ (height)

$A = 10 \times 5$

$A = 50$

STEP 3

Cut out the rectangle. Cut it in half to make two triangles. Notice that they are congruent.

So, the area of the triangle is half the area of the rectangle.

Triangle:

$A = \frac{1}{2} \times (b \times h)$

$A = \frac{1}{2} \times 50$

$A = 25$

The area of the triangle is 25 cm^2.

So, Dion needs 25 cm^2 of fabric to make the sail.

- How do the base and height of the sail relate to the length and width of the rectangle in Step 2?

CALIFORNIA STANDARDS **MG 1.1** Derive and use the formula for the area of a triangle and of a parallelogram by comparing it with the formula for the area of a rectangle. **MG 1.4** Differentiate between, and use appropriate units of measures for, two- and three-dimensional objects. *also* **NS 2.5, MG 1.0, MG 2.0, AF 1.0, AF 1.2, MR 1.0, MR 1.1, MR 3.0, MR 3.2, MR 3.3**

Other Triangles

Activity 2

MATERIALS: grid paper

Some triangles are not right triangles. Find the area of the triangle below.

STEP 1

Draw and shade a model of the triangle inside a rectangle.

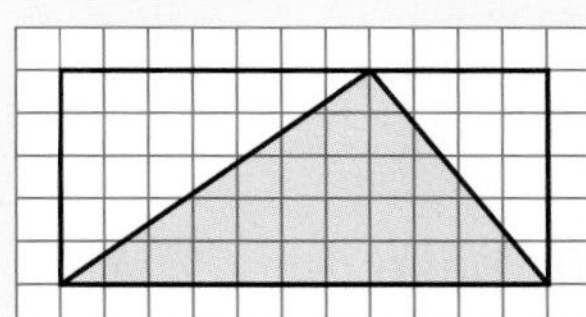

STEP 2

Cut out the rectangle and then the shaded triangle.

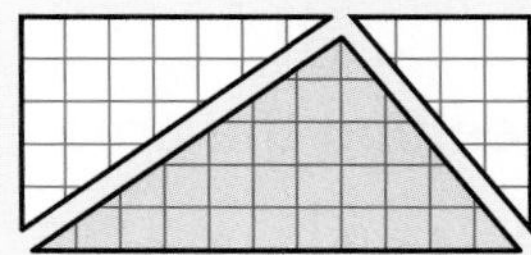

STEP 3

The unshaded parts of the rectangle fit exactly over the shaded triangle.

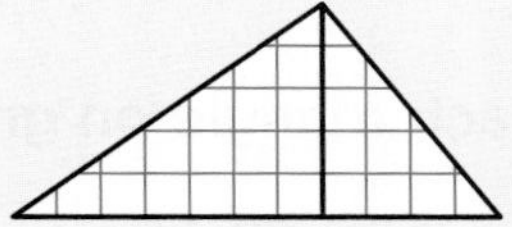

So, the area of the triangle is half the area of the rectangle.

MATH IDEA You can use the formula $A = \frac{1}{2} \times (b \times h)$ to find the area of any triangle.

Examples

A Find the area.

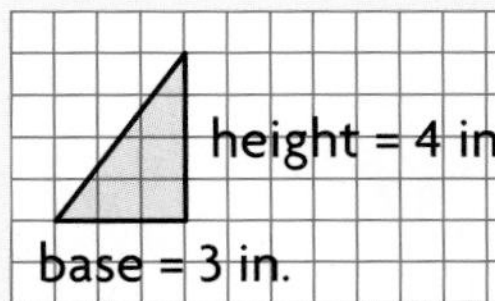

$A = \frac{1}{2} \times b \times h$

$A = \frac{1}{2} \times 3 \times 4$

$A = 6$

The area is 6 in.^{2}

B Find the area.

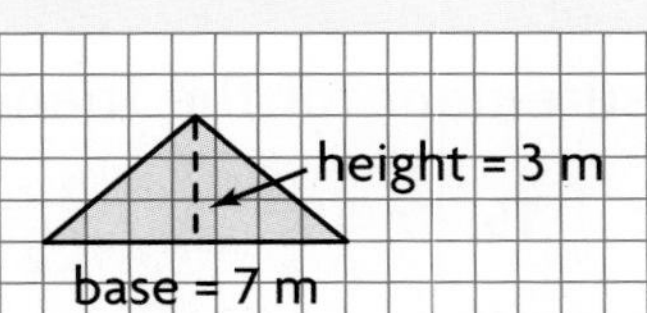

$A = \frac{1}{2} \times b \times h$

$A = \frac{1}{2} \times 7 \times 3$

$A = 10.5$

The area is 10.5 m^2.

Check

1. **Explain** the relationship in Activity 1 between the area of the rectangle and the area of the triangle.

Find the area of each triangle.

2. 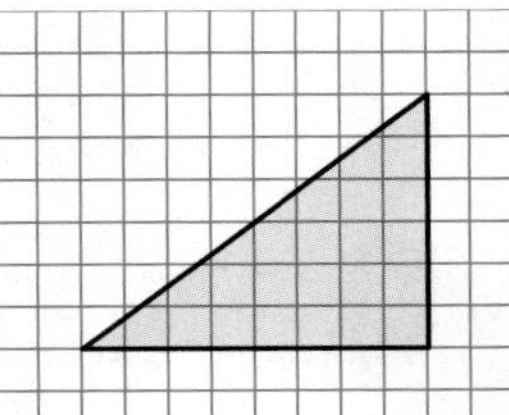

3. 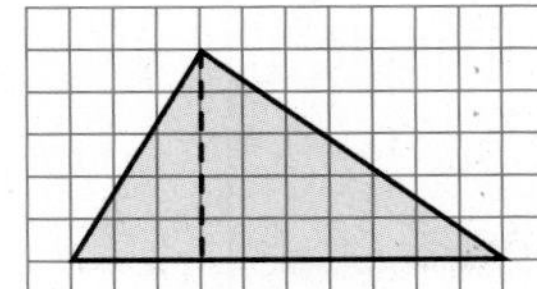

4. 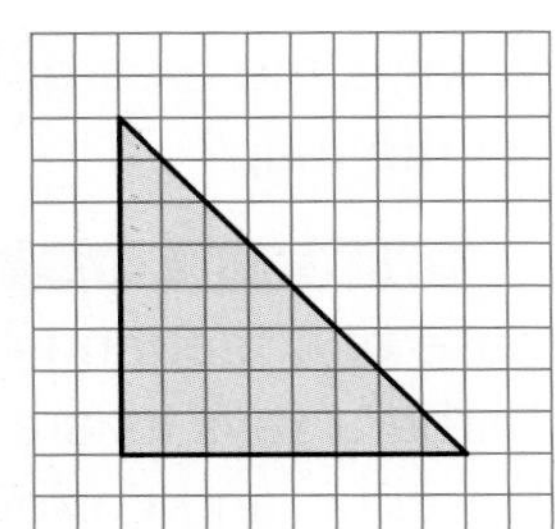

LESSON CONTINUES

Practice and Problem Solving

Find the area of each triangle.

5. 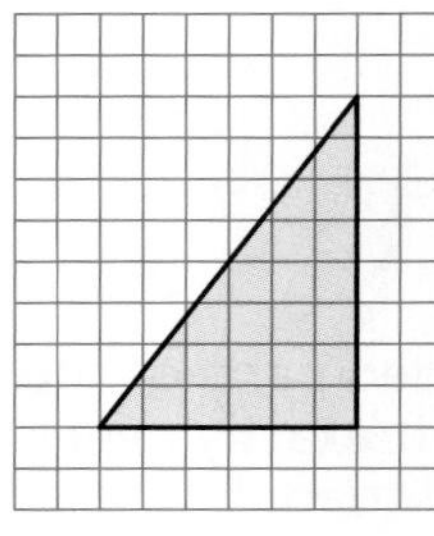

6. 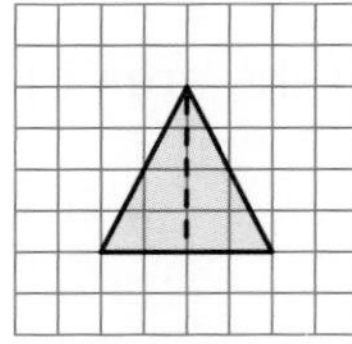

7. 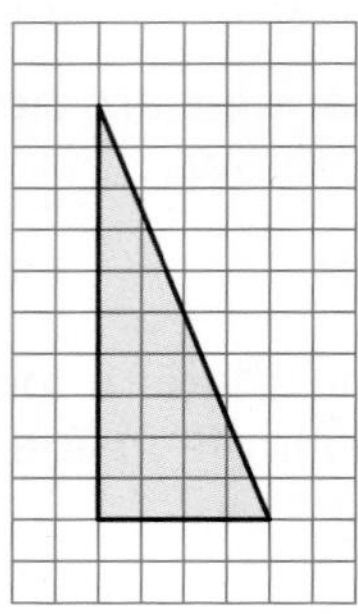

Draw each triangle on grid paper. Then find the area.

8. base $(b) = 10$ cm
height $(h) = 4$ cm
Area $(A) = ■$

9. base $(b) = 6$ m
height $(h) = 6$ m
Area $(A) = ■$

10. base $(b) = 7$ yd
height $(h) = 11$ yd
Area $(A) = ■$

Find the area of each triangle.

11. 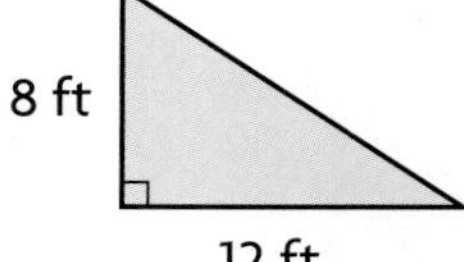

12. 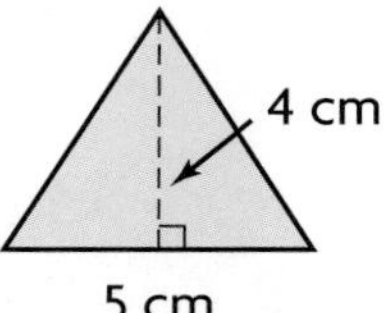

13. 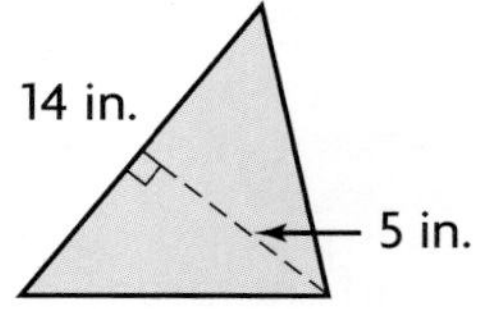

Find the missing measurement for the triangle.

14. $b = 2.6$ cm
$h = 4.7$ cm
$A = ■$

15. $b = ■$
$h = 10$ in.
$A = 40$ in.^{2}

16. $b = 5$ ft
$h = ■$
$A = 15$ ft^2

17. $b = 3\frac{1}{2}$ yd
$h = 2$ yd
$A = ■$

For 18–21, use the diagram of the Bermuda Triangle.

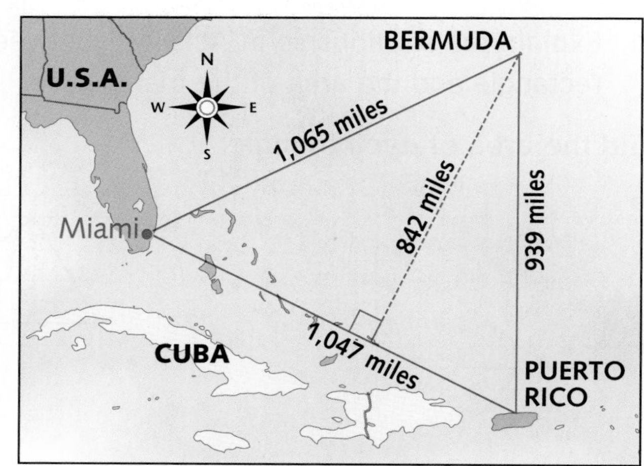

18. What is the perimeter of the Bermuda Triangle?

19. What are the base and height of the Bermuda Triangle?

20. What is the area of the Bermuda Triangle on the map?

21. **What's the Error?** A triangle has a base of 4 feet and a height of 7 feet. Barbara says its area is 28 ft^2. Describe and correct her error.

22. **Write About It** Explain the difference between an inch and a square inch. Draw an example of both.

23. **REASONING** Why do you need to multiply the area of a rectangle by $\frac{1}{2}$ when finding the area of a triangle?

Mixed Review and Test Prep

For 24–26, use the bar graph. (p. 108)

24. How many people chose vanilla?

25. How many more people chose chocolate than strawberry?

26. **TEST PREP** How many people voted in all?

A 28 people
B 29 people
C 30 people
D 31 people

Find the perimeter and area of each rectangle. (p. 528)

27. $s = 2.5$ ft

28. $l = \frac{1}{2}$ in.; $w = 3\frac{1}{2}$ in.

29. What is $\frac{2}{3}$ of 4? (p. 380)

30. **TEST PREP** The distance around a circle is called the __?__. (p. 526)

F area
G diameter
H circumference
J radius

Thinker's Corner

VISUAL THINKING To solve the problems, use what you know about finding the area of geometric figures.

1. How many triangles can you find in the puzzle?
2. Find the area of square *ABCD*.
3. Find the area of triangle *ABC*.
4. The small square that is inside square *ABCD* is $\frac{1}{4}$ the area of the large square. What is the area of each blue triangle?

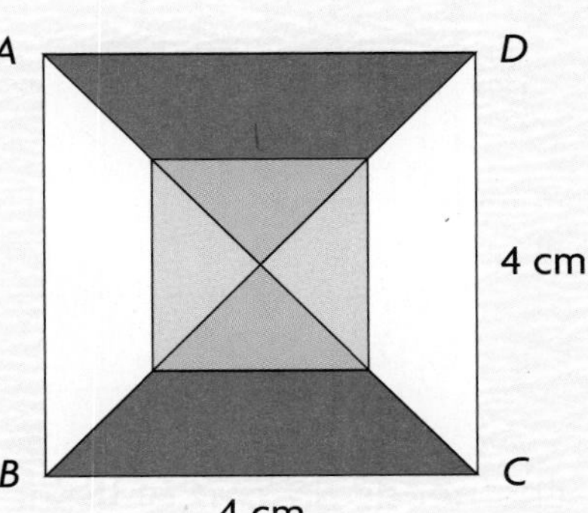

LESSON 6

Algebra: Area of Parallelograms

Learn

CATCH THE SUN Mecha is making a model of a stained-glass window out of colored tissue paper. How much paper does she need for each small parallelogram?

Find the area of a parallelogram.

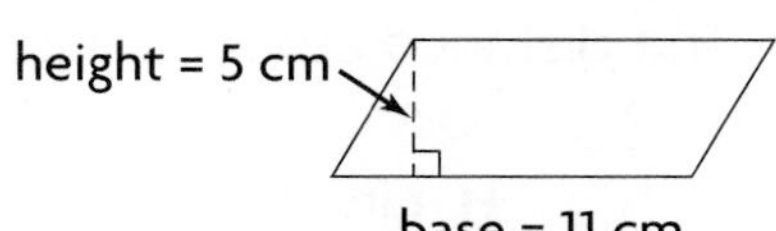

Quick Review

Find the area of each figure.

1. 7 ft by 7 ft square
2. 10 in. by 12 in. rectangle
3. 12 ft by 12 ft square
4. circle, 10 ft diameter
5. right triangle, 4 in., 3 in.

Remember
A *parallelogram* is a quadrilateral with parallel and congruent opposite sides.

Activity

MATERIALS: grid paper, scissors

Use grid paper and what you know about the area of rectangles to find the area of parallelograms.

STEP 1

Copy the parallelogram and cut it out. Draw a line segment to form a right triangle as shown.

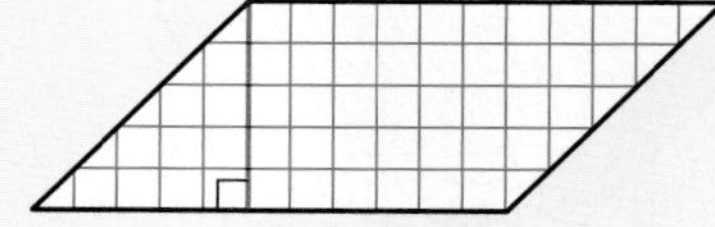

STEP 2

Cut out the right triangle on the left and move it to the right of the parallelogram to form a rectangle.

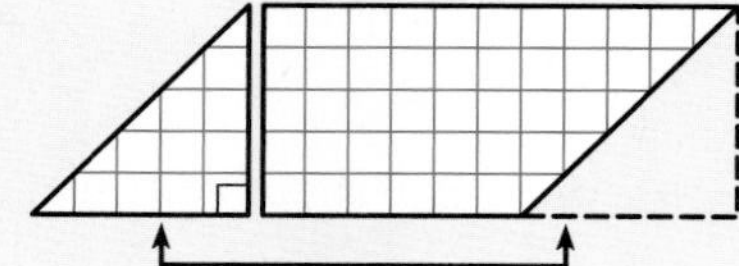

STEP 3

The rectangle has the same area as the parallelogram. Count the grid squares to find its area.

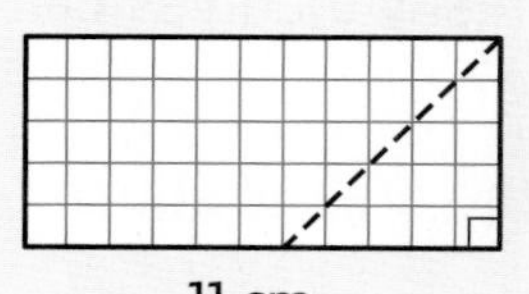

$A = 11 \times 5$, or 55

The area is 55 cm^2.

So, Mecha needs 55 cm^2 of tissue paper to make each small parallelogram.

- How do the base and height of the parallelogram in **Step 1** relate to the length and width of the rectangle in **Step 3**?

CALIFORNIA STANDARDS **MG 1.1** Derive and use the formula for the area of a triangle and of a parallelogram by comparing it with the formula for the area of a rectangle. **MG 1.4** Differentiate between, and use appropriate units of measures for, two- and three-dimensional objects. *also* **MG 1.0, MG 2.0, AF 1.0, AF 1.2, MR 1.1, MR 2.3, MR 2.4, MR 3.0, MR 3.3**

Use a Formula

MATH IDEA The area of a parallelogram is equal to the area of a rectangle with the same base (length) and height (width).

Area of a rectangle	=	**length**	×	**width**	$A = l \times w$
Area of a parallelogram	=	**base**	×	**height**	$A = b \times h$

Examples

A Find the area.

6 cm

9 cm

$A = b \times h$

$A = 9 \times 6$

$A = 54$

The area is 54 cm^2.

B Find the area.

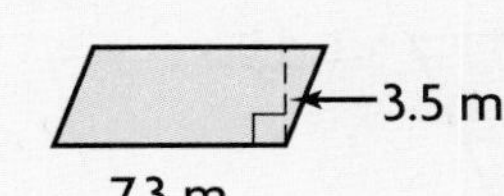

$A = b \times h$

$A = 3.5 \times 7.3$

$A = 25.55$

The area is 25.55 m^2.

Check

1. **Compare** the areas of a rectangle with a length of 6 cm and a width of 4 cm and a parallelogram with a base of 6 cm and a height of 4 cm.

Write the base and the height of each figure.

2. 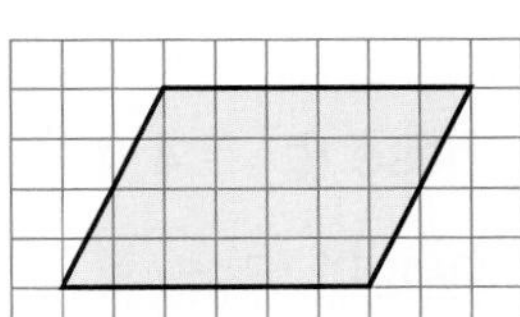

3. 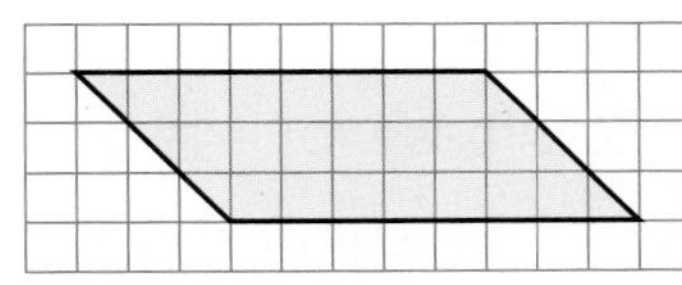

4. 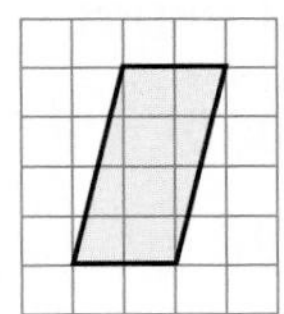

Algebra **Find the missing dimension for each parallelogram. Use grid paper if necessary.**

5. base (b) = ■ cm

   height (h) = 5 cm

   Area (A) = 20 cm^2

6. base (b) = 11 ft

   height (h) = ■ ft

   Area (A) = 88 ft^2

7. base (b) = ■ in.

   height (h) = 12 in.

   Area (A) = 36 $in.^2$

Find the area of each parallelogram.

8. 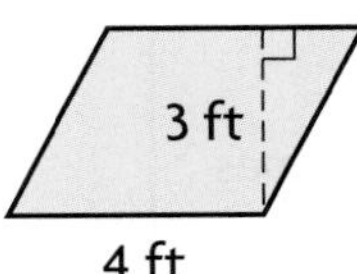

9. 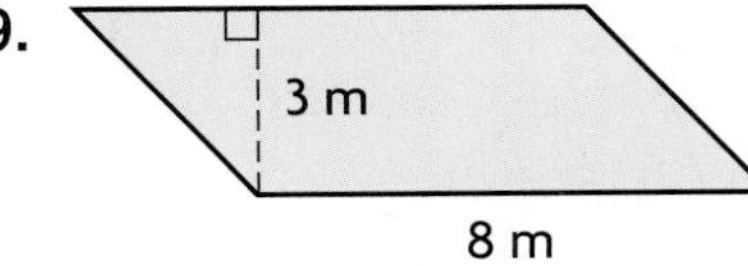

10. 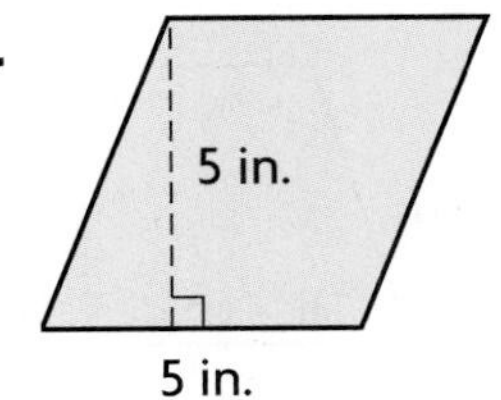

LESSON CONTINUES 

Practice and Problem Solving

Write the base and the height of each figure.

11. 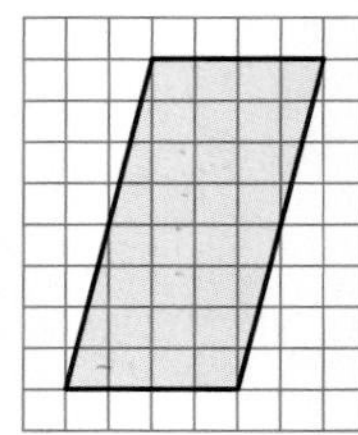

12. 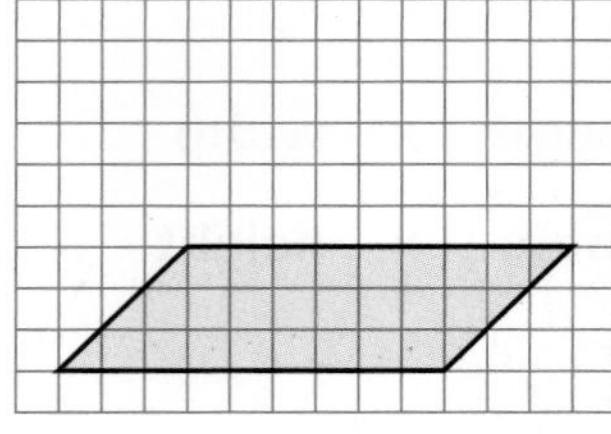

13. 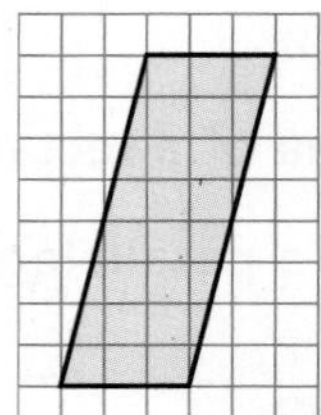

Find the area of each parallelogram.

14. 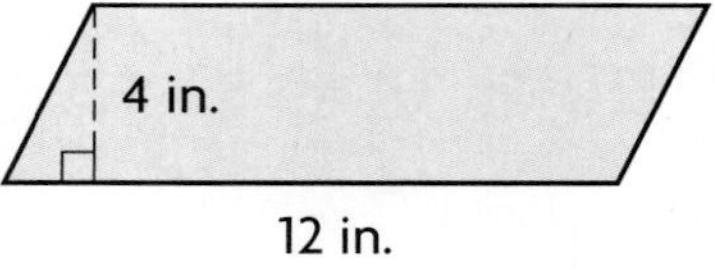

15. 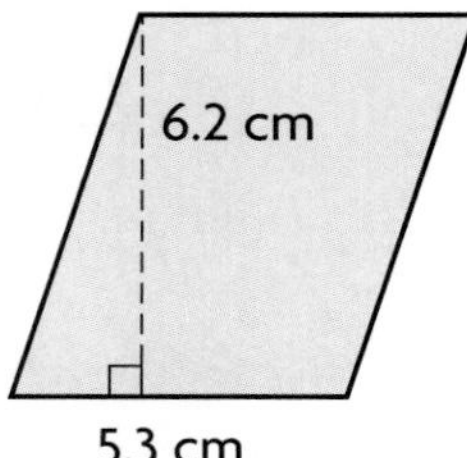

16. 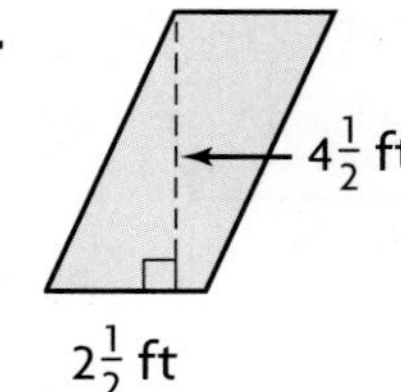

17. 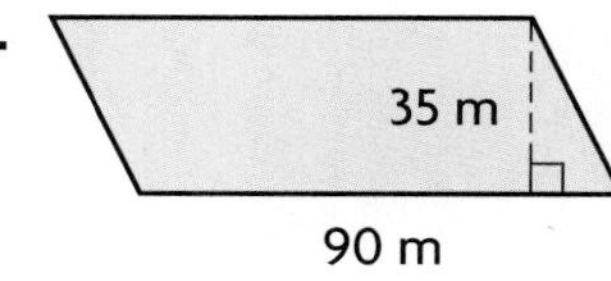

18. 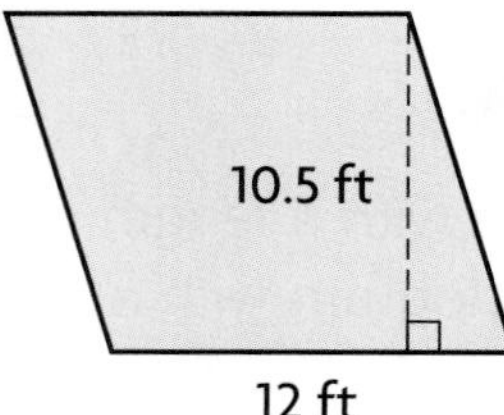

19. 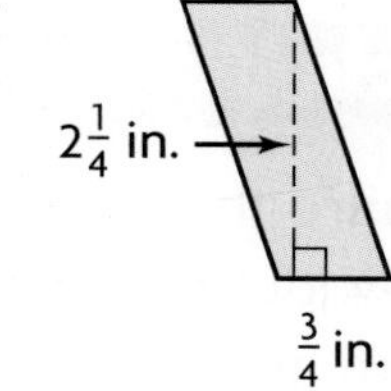

Find the missing dimension for each parallelogram. Use grid paper if necessary.

20. base $(b) = 3.5$ cm
height $(h) = 8.2$ cm
Area $(A) = ■$

21. base $(b) = ■$
height $(h) = 12$ in.
Area $(A) = 24$ in.^{2}

22. base $(b) = 4$ ft
height $(h) = ■$
Area $(A) = 28$ ft^2

23. **What's the Error?** The base of a parallelogram is 7 cm and its height is 5 cm. Ken says its area is 35 cm. Describe and correct his error.

24. **What's the Question?** The base of a parallelogram is 8 ft. The area is 32 ft^2. The answer is 4 ft.

25. **REASONING** Can you find the area of this parallelogram with the given measurements? Explain.

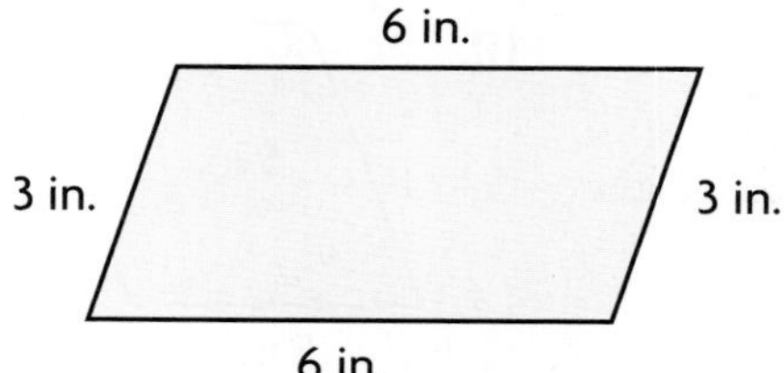

26. **Write About It** Describe the relationships shown in this Venn diagram.

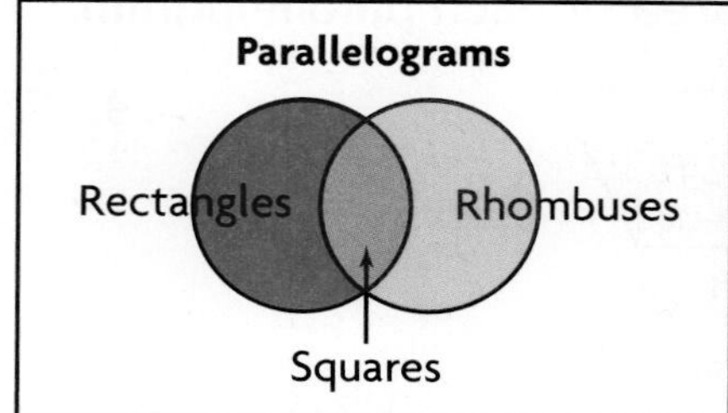

27. Candy's stained-glass window model is 18 inches by 12 inches. How many square inches of tissue paper does she need?

28. **REASONING** The base of a parallelogram is two times its height. If the base is 10 in., what is the area?

Mixed Review and Test Prep

For 29–31, use the pictograph. (p. 108)

HOURS WATCHING TV EACH DAY

Japan	
U.S.	
U.K.	
France	

Each [TV] = 2 hours.

29. How many hours does the average American watch TV each day?

30. How many more hours a day does a person in Japan watch TV than a person in France?

31. **TEST PREP** Which is the average number of hours spent watching TV over 2 weeks by a person in France?

A 7 hr **B** 14 hr **C** 28 hr **D** 42 hr

Find the area of each triangle. (p. 532)

32. base = 4 cm; height = 3 cm

33. base = $\frac{1}{2}$ in.; height = $\frac{1}{2}$ in.

34. Round 1.999 to the nearest tenth. (p. 46)

35. $0.5\overline{)0.875}$ (p. 238)

36. **TEST PREP** The distance around a square is called the __?__. (p. 524)

F area **G** perimeter **H** diameter **J** side

Thinker's Corner

DOUBLE IT You can use the area of a triangle to find the area of a parallelogram.

- Draw two of these triangles on grid paper. Cut them out.
- Put them together to make a parallelogram.

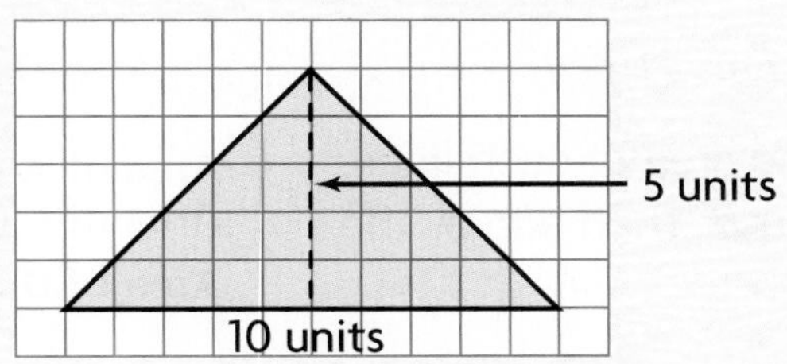

1. Find the area of one triangle.

2. How is the area of the triangle related to the area of the parallelogram?

3. Write a formula for the area of a parallelogram.

LESSON 7

Problem Solving Strategy

Solve a Simpler Problem

Quick Review

Name the figure for each formula.

1. $P = 4 \times s$ **2.** $A = l \times w$

3. $A = b \times h$ **4.** $A = s \times s$

5. $A = \frac{1}{2} \times b \times h$

PROBLEM Mike and Stephanie are designing a flat backdrop for their school play. How many square feet of backdrop will they use to make the house shown at the right?

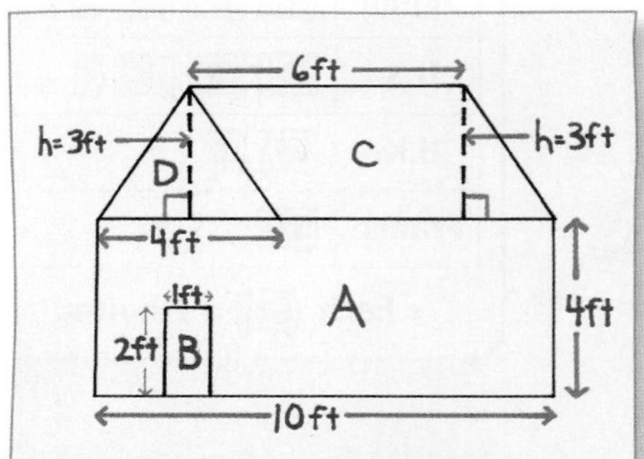

- What are you asked to find?
- What information will you use?
- Is there any information you will not use? Explain.

- What strategy can you use to solve the problem?

You can *solve a simpler problem* by breaking the diagram of the house into simpler figures.

- How can you use the strategy to solve the problem?

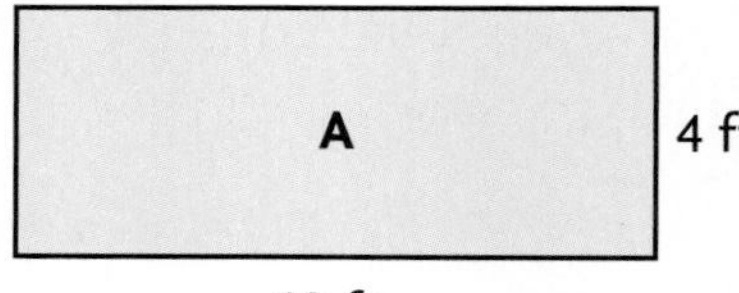

Area = $l \times w$

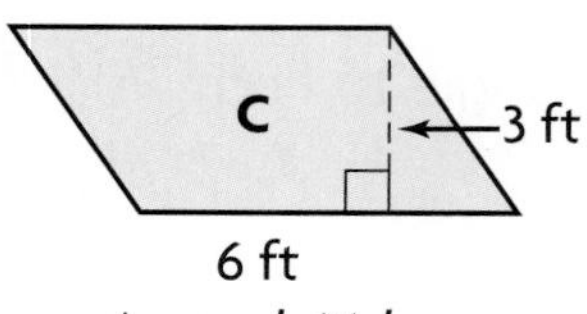

Area = $b \times h$

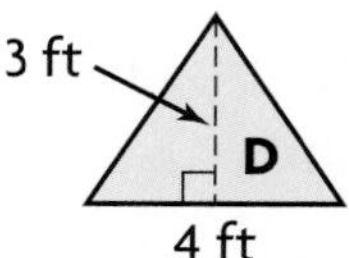

Area = $\frac{1}{2} \times b \times h$

Find the area of each figure in the house. Then add all the areas.

Area of A (rectangle)	$10 \times 4 =$	40
Area of C (parallelogram)	$6 \times 3 =$	18
Area of D (triangle)	$\frac{1}{2} \times (4 \times 3) =$	+ 6
		64

So, they need 64 ft^2 of backdrop to make the house.

- How can you estimate to check your answer?

CALIFORNIA STANDARDS MR 3.2 Note the method of deriving the solution and demonstrate a conceptual understanding of the derivation by solving similar problems. **O-n AF 1.2** Use a letter to represent an unknown number: write and evaluate simple algebraic expressions in one variable by substitution. *also* **NS 2.5, AF 1.0, MG 1.0, O-n MG 1.1, MG 1.4, MG 2.0, MR 1.0, MR 1.1, MR 2.0, MR 3.0, MR 3.3**

Problem Solving Practice

Solve a simpler problem to solve.

1. **What if** Vince cuts a 1 ft by 3 ft window from the house? What will the area be?
2. Antoine made this design by wrapping string around nails on a piece of wood. How much string did he use?

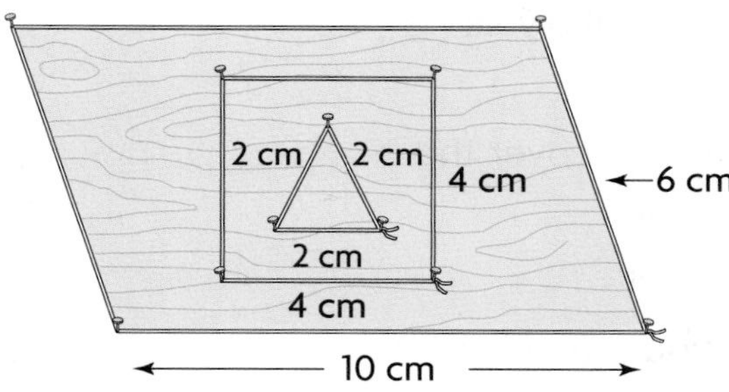

- Draw a Diagram or Picture
- Make a Model or Act It Out
- Make an Organized List
- Find a Pattern
- Make a Table or Graph
- Predict and Test
- Work Backward
- Solve a Simpler Problem
- Write an Equation
- Use Logical Reasoning

3. The cartons of computer parts are labeled on the top and the front of each carton. How many labels are showing if there are 8 rows of 3 boxes each?

   A 24 **C** 32
   B 27 **D** 36

4. What is the area of the figure?

   F 28 sq units
   G 30 sq units
   H 32 sq units
   J 36 sq units

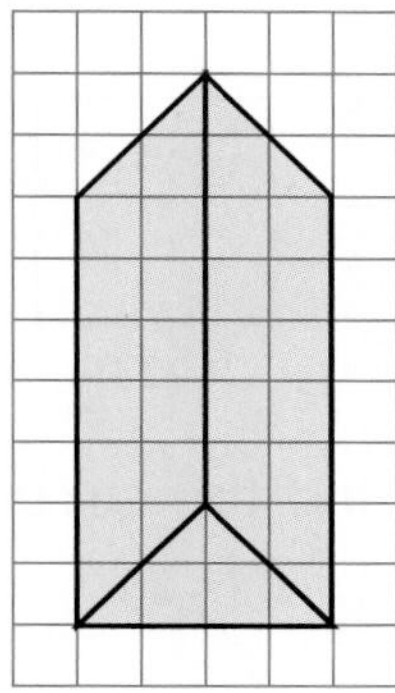

Mixed Strategy Practice

5. Mr. Bell's rectangular garden is 15 ft by 10 ft. Fencing costs $2.50 a foot. How much will he spend to fence in his entire garden?
6. Each can of paint covers 50 ft^2. How many cans does Vince need to paint a 12 ft by 16 ft backdrop for the play?
7. The diameter of the moon is about 3,500 kilometers. The diameter of the sun is 400 times as great as the diameter of the moon. What is the circumference of the sun? the moon? (Use $\pi = 3.14$.)
8. Mr. Kelly shopped for art supplies. He bought 3 canvases for $10.98 each, brushes for $6.50, and 8 bottles of paint for $5.69 each. He paid with five $20 bills. How much change did he receive?
9. Two equilateral triangles are put together to form a quadrilateral. Draw a diagram to find what kind of quadrilateral is formed.
10. Describe one way you could break apart a pentagon to find its area.

Review/Test

CHECK VOCABULARY AND CONCEPTS

Choose the best term from the box.

area
base
circumference
height
perimeter

1. The distance around a figure is the __?__. (p. 524)
2. The distance around a circle is its __?__. (p. 526)
3. To find the area of a parallelogram, multiply its __?__ by its __?__. (p. 536)
4. The number of square units needed to cover the surface of an object is the __?__. (p. 528)

CHECK SKILLS

Find the perimeter of each polygon. (pp. 524–525)

5. 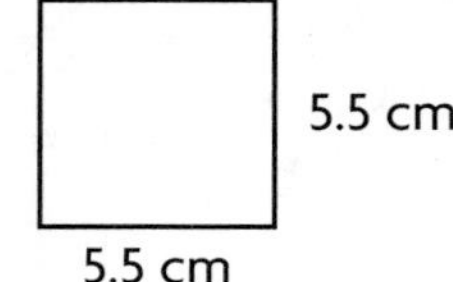

6. 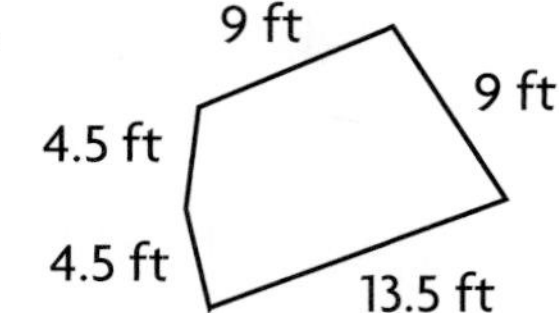

7. 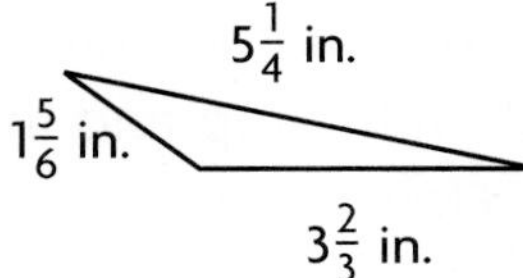

Find the circumference of each circle. Use $\pi = 3.14$. (pp. 526–527)

8. 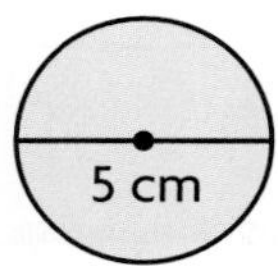

9. 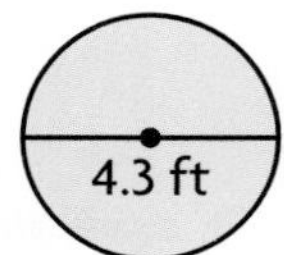

10. 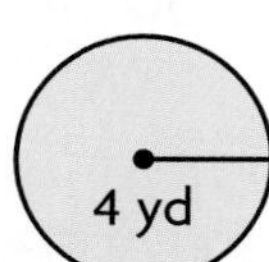

Find the area of each figure. (pp. 528–537)

11. 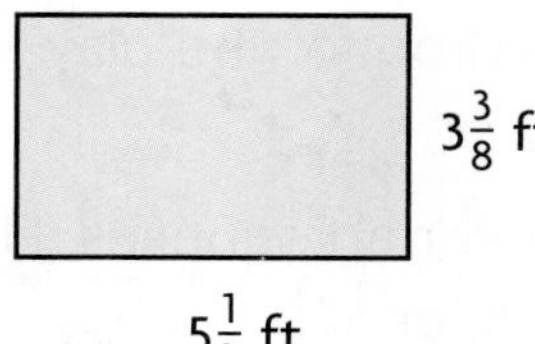

12. 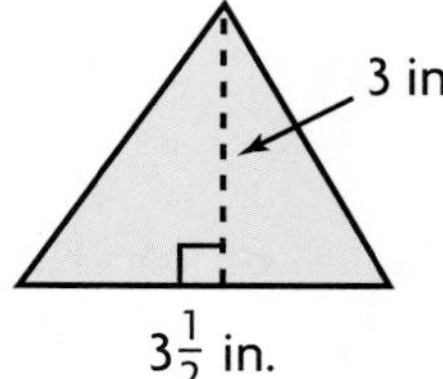

13. 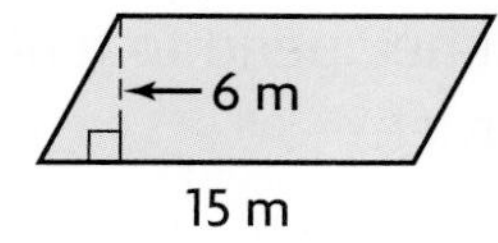

CHECK PROBLEM SOLVING

Solve a simpler problem. (pp. 540–541)

14. Ben planted a garden with this shape. What is the perimeter of the garden to the nearest foot?

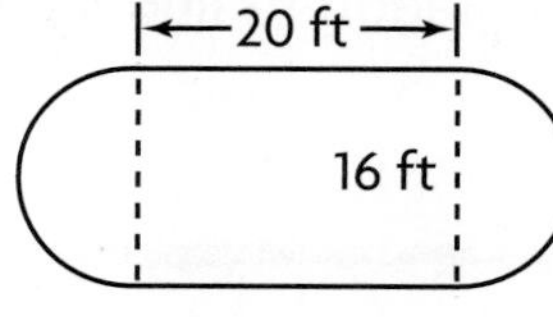

15. Bill's Sign Shop makes road signs in this shape. What is the area of each sign?

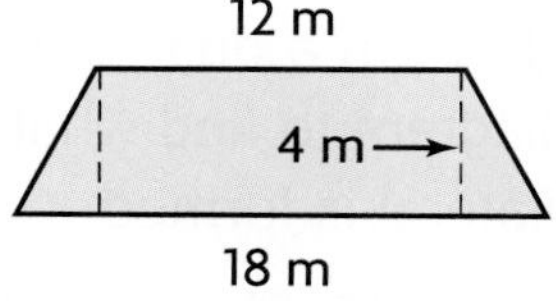

Cumulative Review

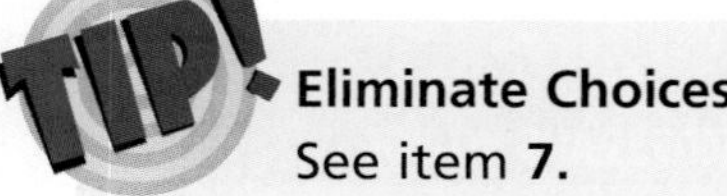

Eliminate Choices.
See item **7**.

Round the decimal to the nearest whole number. Multiplying with the rounded number will give you an estimated area. Eliminate any choices that are not reasonable based on your estimate.

Also see problem **5**, p. H64.

For 1–10, choose the best answer.

1. How many quarts are in 2 gallons?

A $\frac{1}{4}$ quarts **C** 8 quarts
B 4 quarts **D** 16 quarts

2. What is the circumference of a circle with a radius of 3 centimeters? (Use 3.14 for the value of π.)

F 18.84 cm **H** 28.26 cm
G 18.84 cm^2 **J** 28.26 cm^2

3. What is the perimeter of a square with sides that measure $8\frac{1}{5}$ inches?

A $32\frac{4}{5}$ in. **C** 32 in.
B $32\frac{4}{5}$ in.^{2} **D** 32 in.^{2}

4. $12.8 \div 8$

F 0.16 **H** 16
G 1.6 **J** Not here

5. What is the area of this triangle?

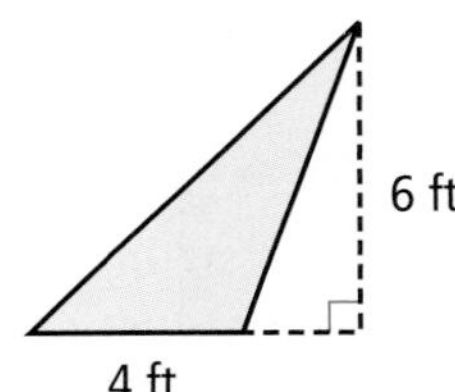

A 10 ft^2 **C** 24 ft^2
B 12 ft^2 **D** Not here

6. Mulan swam 4,000 meters at practice. How many kilometers did she swim?

F 400 km **H** 4 km
G 40 km **J** 0.4 km

7. What is the area of a rhombus with a base of 4 cm and a height of 2.1 cm?

A 6.1 cm^2 **C** 8.4 cm^2
B 8 cm^2 **D** 14.2 cm^2

For 8–10, use the table.

SIZES OF FAMOUS PAINTINGS

Title	Artist	Dimensions of Painting (in inches)
Mona Lisa	da Vinci	30 × 21
Water Lily Pond	Monet	35 × 36
Apples and Oranges	Cézanne	29 × 37
Dancers in Blue	Degas	34 × 30

8. What is the perimeter of the *Mona Lisa*?

F 51 inches **H** 102 inches
G 81 inches **J** 630 inches

9. Which artist's painting has the greatest area?

A da Vinci **C** Cézanne
B Monet **D** Degas

10. Which painting measures $2\frac{5}{6}$ feet by $2\frac{1}{2}$ feet?

F *Mona Lisa*
G *Water Lily Pond*
H *Apples and Oranges*
J *Dancers in Blue*

CHAPTER 29

Surface Area and Volume

PACKAGING OF GYROSCOPES AND YO-YOS		
Item	**Size of Item**	**Size of Carton**
Gyroscope	length: $2\frac{3}{4}$ inches	length: 11 inches
	width: $2\frac{3}{4}$ inches	width: $8\frac{1}{4}$ inches
	height: $2\frac{3}{4}$ inches	height: $5\frac{1}{2}$ inches
Yo-Yo	length: $2\frac{1}{4}$ inches	length: 9 inches
	width: $2\frac{1}{4}$ inches	width: 9 inches
	height: $1\frac{1}{2}$ inches	height: $4\frac{1}{2}$ inches

Many products, like the toys in this photo, are packaged in cardboard cartons and shipped to stores worldwide. A store needs to know how much space the various-size cartons occupy and how many items are in each carton, so it can calculate how much of a product to order. Use the data in the table to find how many gyroscopes can fit in the first carton and how many yo-yos can fit in the second carton. Then find how many cubic feet of space 10 cartons of yo-yos would occupy.

The yo-yo is over 3,000 years old and is the second-oldest known toy in the world. Since 1930, over half a billion yo-yos have been sold in the United States.

Use this page to help you review and remember important skills needed for Chapter 29.

VOCABULARY AND CONCEPTS

Choose the best term from the box.

area bases faces polygon vertices

1. Prisms and pyramids are named by the polygons that form their _?_.
2. A closed plane figure with straight sides that is named by the number of its sides and angles is a _?_.
3. The number of square units needed to cover a surface is its _?_.
4. The flat surfaces of a prism or pyramid are its _?_.

FACES, EDGES, AND VERTICES (See p. H30.)

Copy and complete the following table.

		Name of Figure	Number of Faces	Number of Edges	Number of Vertices
5.		Cube	6	12	8
6.		?	■	■	■
7.		?	■	■	■
8.		?	■	■	■
9.		?	■	■	■

EXPRESSIONS WITH EXPONENTS (See p. H31.)

Write each expression by using an exponent, and find its value.

10. 4×4 **11.** $9 \times 9 \times 9$ **12.** $(2 \times 2) + (6 \times 6)$ **13.** $2 \times 2 \times 2 \times 3 \times 3$

14. $2 \times 2 \times 5 \times 5 \times 5$ **15.** $(3 \times 3 \times 3) + (7 \times 7)$ **16.** $8 \times 8 \times 8 \times 8 \times 8 \times 8$

Find the value of each expression.

17. 8^2 **18.** 3^3 **19.** $5^2 \times 2^3$ **20.** $5^2 \times 3^2$

LESSON 1

Nets for Solid Figures

Quick Review

Name each polygon.

1. 

2. 

3.

4. 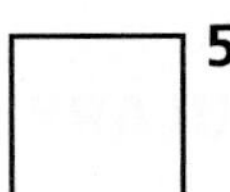

5.

VOCABULARY

net

Learn

BOXED IN Boxes that contain products you buy in stores are made from patterns, or nets. A **net** is a two-dimensional pattern that can be folded into a three-dimensional prism or pyramid.

Here is one net for a cube-shaped box. A cube is formed by 6 congruent square faces: 1 for the top, 1 for the bottom, and 4 for the sides.

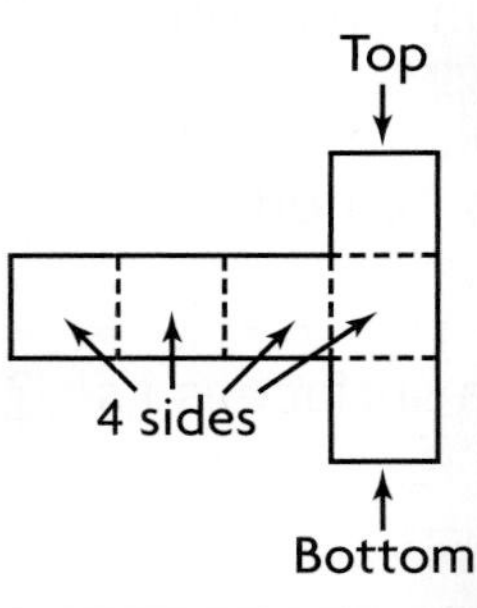

Activity

MATERIALS: 1-inch grid paper, scissors, tape

Use the net to make a cube-shaped box.

STEP 1

Copy the net onto grid paper and cut it out.

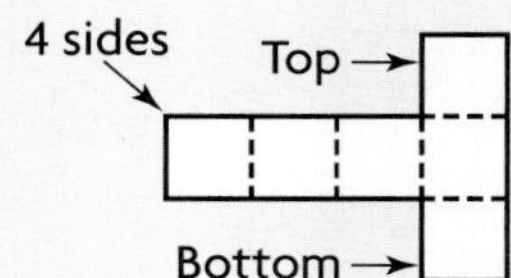

STEP 2

Fold on the dashed lines.

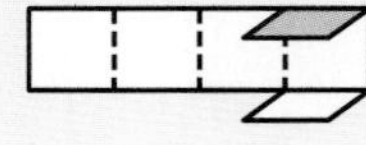

STEP 3

Tape the edges together. Be sure there are no gaps and that none of the sides overlap.

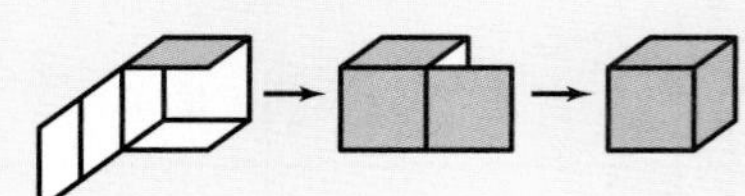

Save this cube to use in Lesson 2.

Examples

A

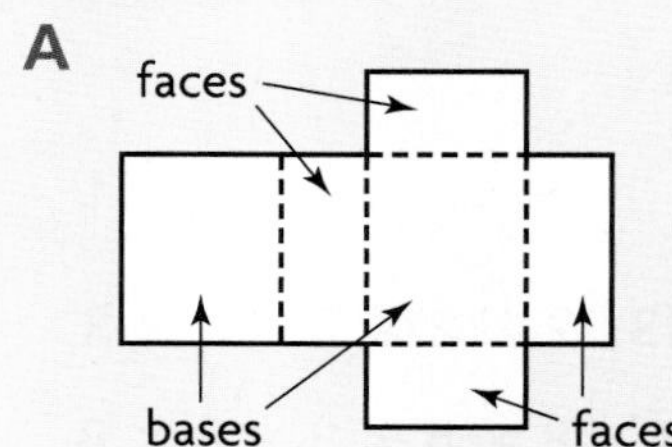

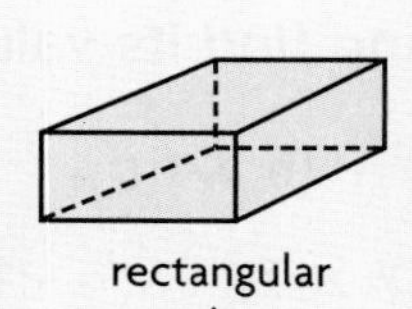

rectangular prism

B

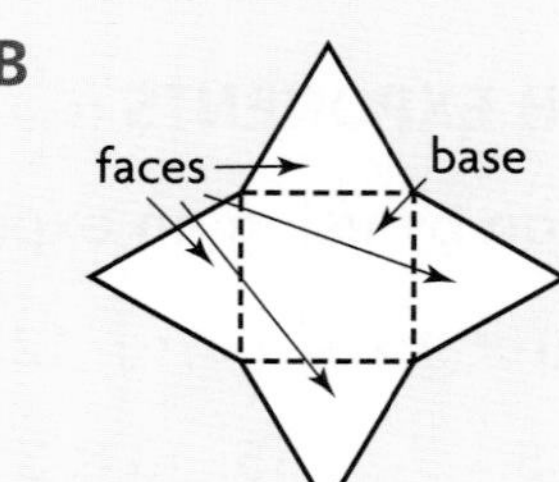

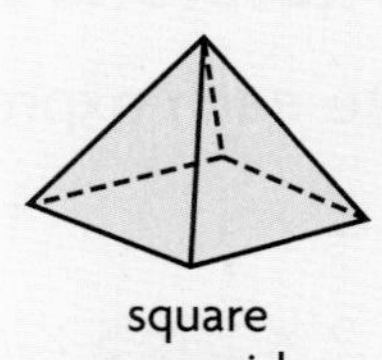

square pyramid

CALIFORNIA STANDARDS O—n MG 1.2 Construct a cube and rectangular box from two-dimensional patterns and use these patterns to compute the surface area for these objects. **MG 2.3** Visualize and draw two-dimensional views of three-dimensional objects. *also* **MG 1.0, MG 2.0, MR 2.0, MR 2.3, MR 2.4, MR 3.2, MR 3.3**

Check

1. **Explain** why a cube is a prism.

Match each solid figure with its net. Write *a, b, c,* or *d*.

2. 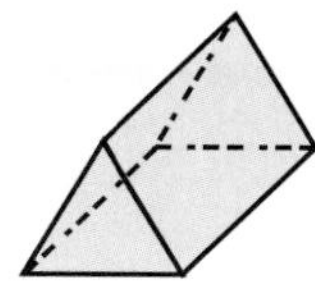

3. 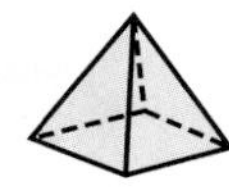

4. 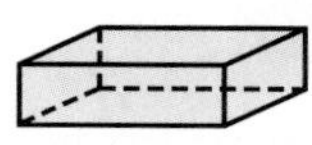

5. 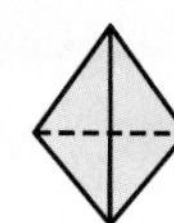

a. 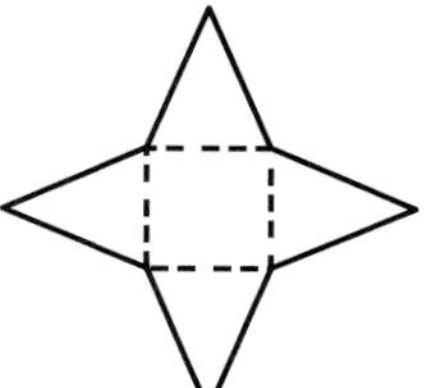

b. 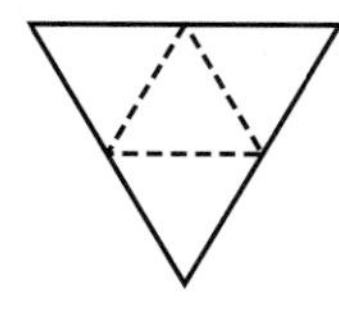

c. 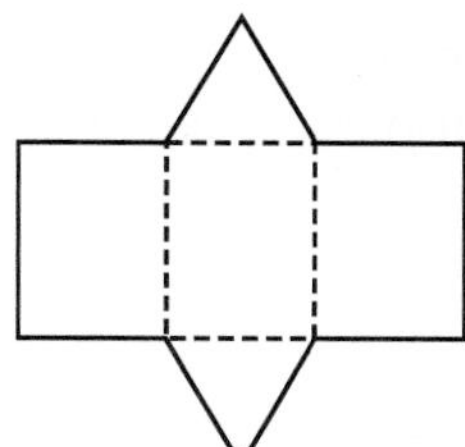

d. 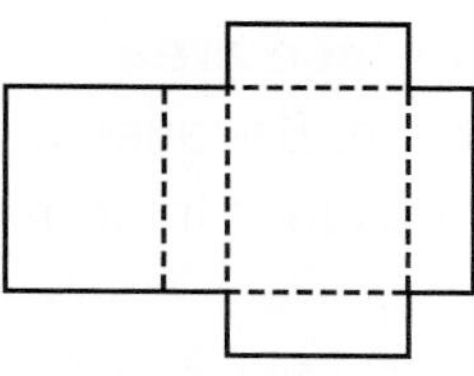

Practice and Problem Solving

Match each solid figure with its net. Write *a, b, c,* or *d*.

6. 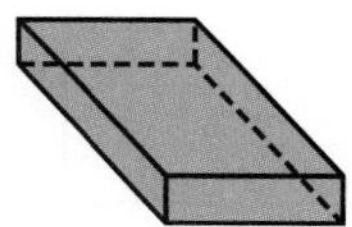

7. 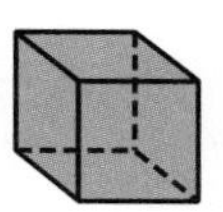

8. 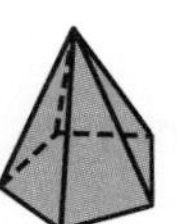

9. 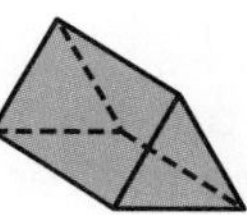

a. 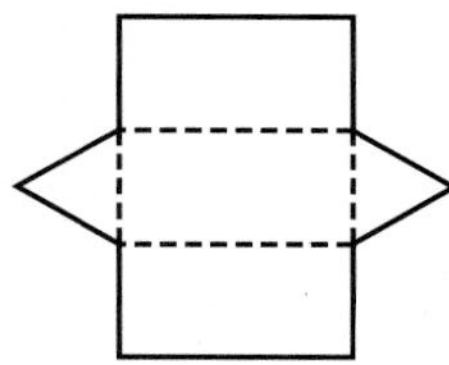

b. 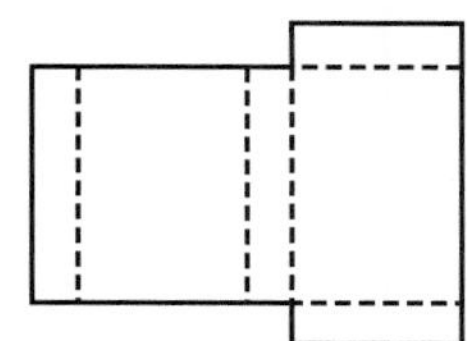

c. 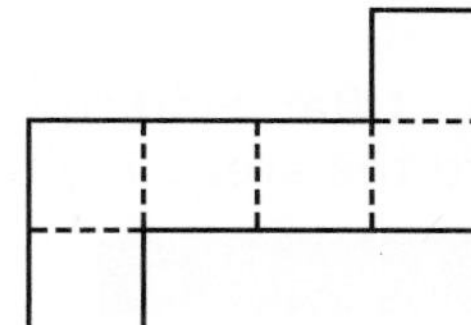

d. 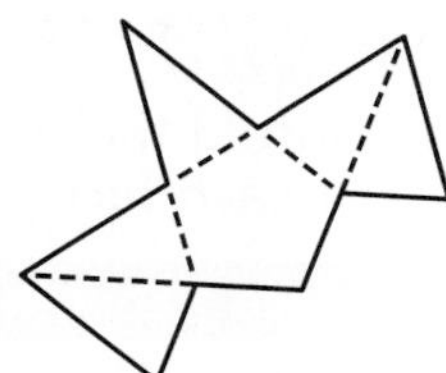

10. Find the area of one of the 5-in. × 7-in. bases of a rectangular prism.

11. **REASONING** Which pyramid or prism requires the least number of faces to make its net?

12. **Write About It** Explain how you would make a net for a square pyramid.

Mixed Review and Test Prep

13. $3 \times \frac{1}{3}$ (p. 380)

14. $2\frac{2}{3} \times \frac{1}{2}$ (p. 384)

15. Evaluate 3^4. (p. 268)

16. What is the least common denominator of $\frac{1}{3}$ and $\frac{3}{4}$? (p. 352)

17. **TEST PREP** Which is the value for n if $3\frac{1}{2} + n = 6$? (p. 364)

A $n = 3\frac{1}{2}$
B $n = 2\frac{4}{5}$
C $n = 2\frac{1}{2}$
D $n = 1\frac{1}{2}$

Extra Practice page H60, Set A

LESSON 2

Surface Area

Quick Review

1. 9×7
2. $2(4 \times 5)$
3. $2(3 \times 1.5)$
4. $2(1\frac{1}{2} \times \frac{3}{4})$
5. $2(0.75 \times 4)$

VOCABULARY

surface area

Learn

WRAPPED UP! You can use what you know about finding the area of a rectangle to find the surface area of a rectangular prism.

Surface area is the sum of the areas of the faces of a solid figure. Because surface area measures two dimensions, it is measured in square units.

Activity 1

MATERIALS: centimeter grid paper, scissors, tape

To find the surface area of a rectangular prism, use its net.

STEP 1

Copy the net onto centimeter grid paper. Label the faces *A–F* as shown.

Cut out the net and fold on the lines to make the rectangular prism.

STEP 3

Unfold the prism and lay it flat. Find the area of each face, *A–F*. Record the areas in a table.

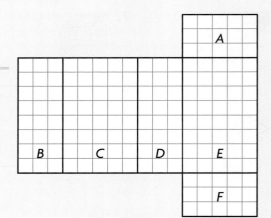

Face	Length	Width	Area = $l \times w$
A	5 cm	3 cm	15 cm^2
B	■	■	■
C	■	■	■
D	■	■	■
E	■	■	■
F	■	■	■

STEP 4

Add the areas of the faces to find the surface area.

15 + 24 + 40 + 24 + 40 + 15 = 158

So, the surface area is 158 cm^2.

- Which faces have the same area? Why?

CALIFORNIA STANDARDS O━━ **MG 1.2** Construct a cube and rectangular box from two-dimensional patterns and use these patterns to compute the surface area for these objects. **MG 1.0** Students understand and compute the volumes and areas of simple objects. *also* **MG 2.0, MR 1.2, MR 2.3, MR 3.3**

Compute Surface Area

You can also compute the areas of opposite faces to find the surface area of a rectangular prism.

Find the surface area of the figure.

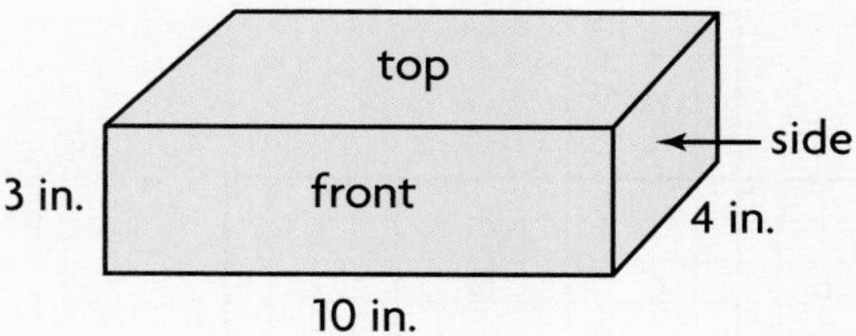

- The top and bottom have the same area. So, their combined area is: $2 \times (10 \times 4) = 80$
- The front and back have the same area. So, their combined area is: $2 \times (10 \times 3) = 60$
- The left and right sides have the same area. So, their combined area is: $2 \times (3 \times 4) = 24$
- The areas for opposite faces of the rectangular prism are $80 + 60 + 24 = 164$.

So, the surface area is 164 in.^{2}

Activity 2

- Use your cube net from Lesson 1. Find its surface area if the length of each side is 1 unit.

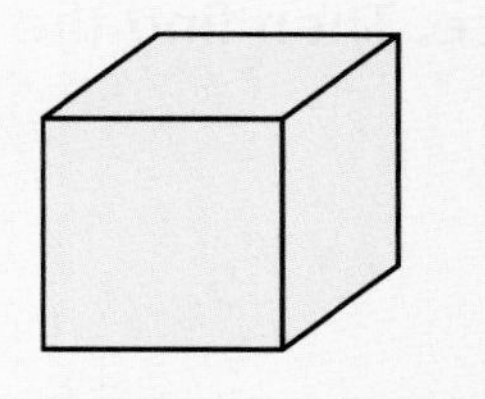

Remember

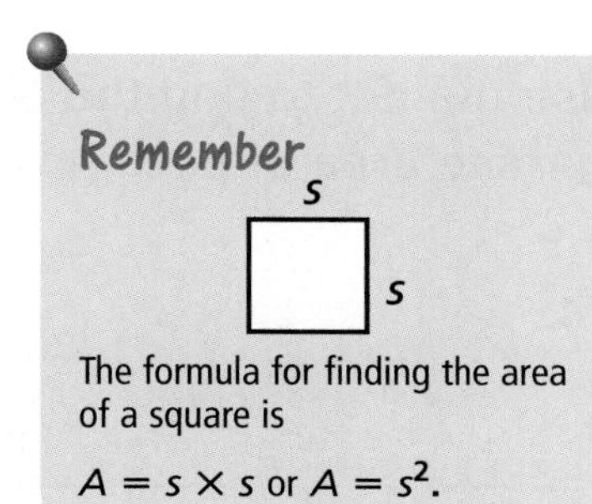

The formula for finding the area of a square is

$A = s \times s$ or $A = s^2$.

REASONING A formula for the surface area of a cube using s, the side measurement, is $6s^2$. Explain why this formula works for a cube.

Examples

A Find the surface area of this figure.

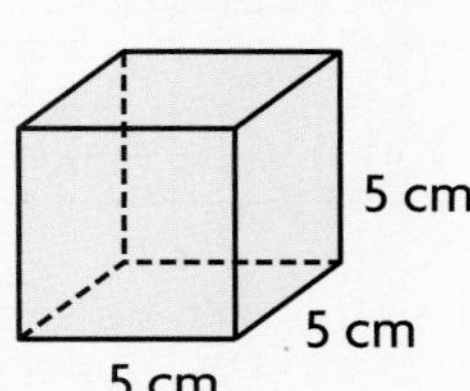

$6 \times (5 \times 5) = 150 \text{ cm}^2$

B Find the surface area of this figure.

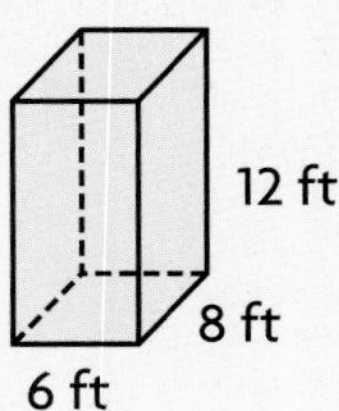

$2 \times (6 \times 12) + 2 \times (8 \times 12) + 2 \times (6 \times 8) = 432 \text{ ft}^2$

MATH IDEA To find the surface area of a three-dimensional figure, find the area of each face. Then add these to find the total area.

Check

1. **Explain** why fewer steps are needed to find the surface area of a cube than of a rectangular prism.

Use the net to find the area of each face. Then find the surface area of each prism.

2. 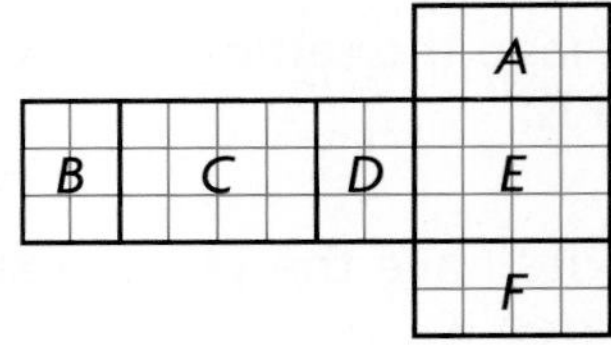

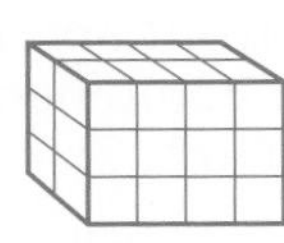

3. 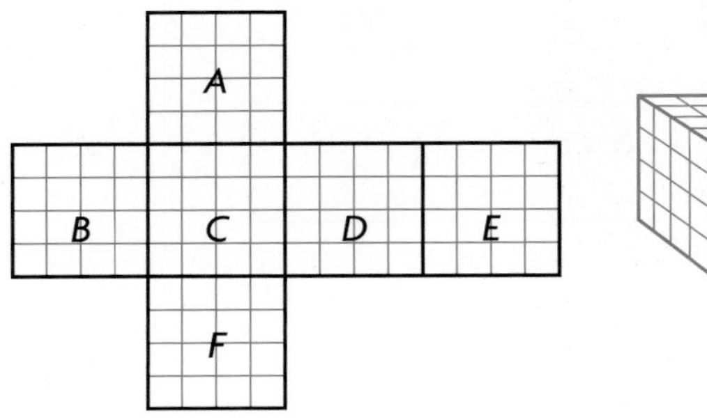

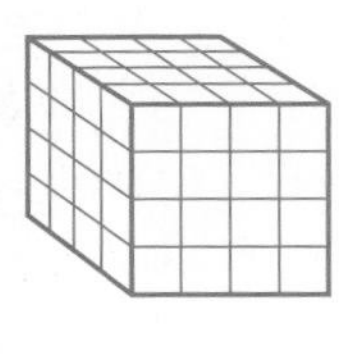

Practice and Problem Solving

Use the net to find the area of each face. Then find the surface area of each prism.

4. 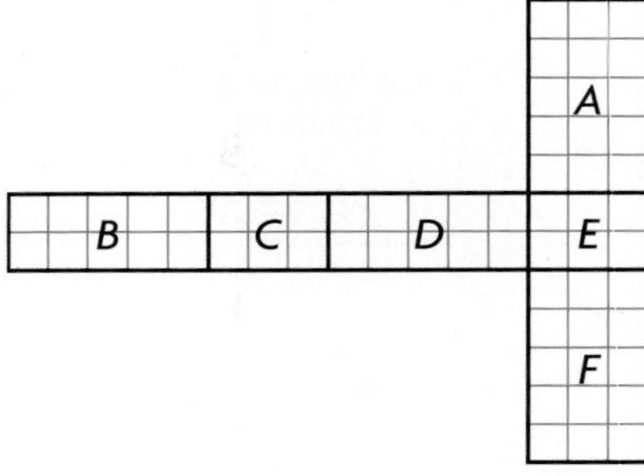

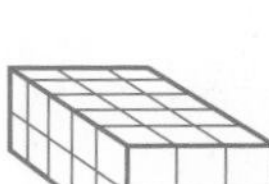

5. 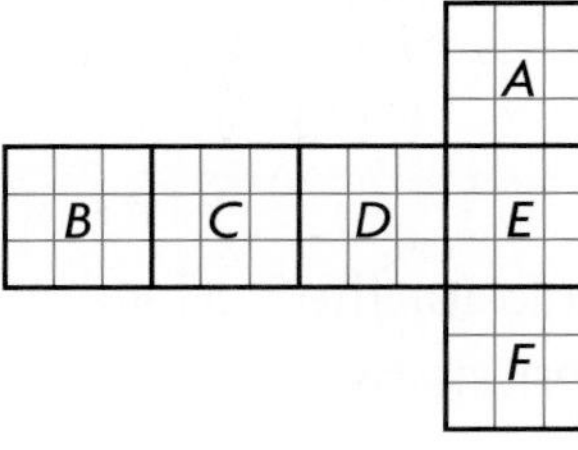

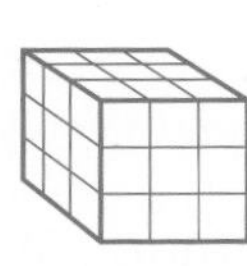

Find the surface area in cm^2. You may want to make the net.

6. 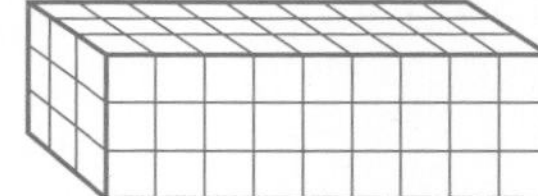

7. 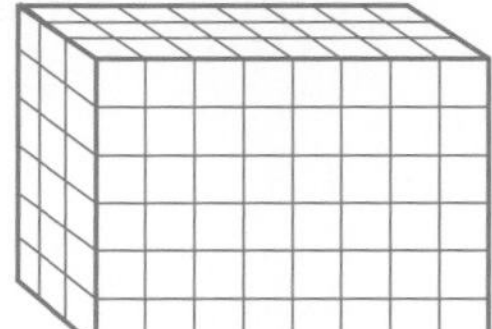

8. 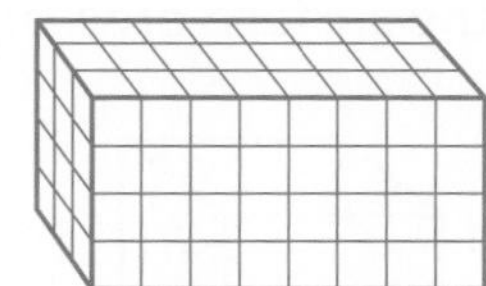

9. What is the surface area of a box 5 feet long, 3 feet wide and 7 feet high?

10. What is the surface area of a box 11 cm long, 10 cm wide, and 12 cm high?

11. **REASONING** Ellie wants to build a doghouse in the shape of a cube. Each face will have an area of 9 ft^2. What will be the surface area of the doghouse?

12. Edward is painting his room, which is 12 ft long, 10 ft wide, and 10 ft high. He will not paint the ceiling or the floor. How much surface area is he painting?

Nicole wants to decorate a square prism to use as a pencil holder. She wants to cover the entire surface except the top.

13. What shapes will she need to cut out to do this?

14. What dimensions will she need to know to find the surface area?

Mixed Review and Test Prep

15. 6×0.5 (p. 158)

16. $\frac{4}{5} \div \frac{2}{5}$ (p. 400)

17. Write a decimal equivalent to 0.40. (p. 22)

18. Evaluate 2^3 and 3^2. (p. 266)

19. $5\frac{3}{4} - 2\frac{2}{3}$ (p. 368)

20. **TEST PREP** Which fraction is equivalent to $\frac{3}{4}$? (p. 284)

A $\frac{6}{9}$ **B** $\frac{6}{8}$ **C** $\frac{9}{10}$ **D** $\frac{10}{12}$

21. Find the least common multiple (LCM) of 9 and 15. (p. 256)

22. Evaluate $n \div 12$ if $n = 60$. (p. 86)

23. Change $\frac{1}{4}$ to a decimal. (p. 280)

24. **TEST PREP** Which of the following is *not* a quadrilateral? (p. 460)

F square **G** hexagon **H** trapezoid **J** rectangle

LINKUP to Reading

STRATEGY • FOLLOW INSTRUCTIONS
Read the instructions below carefully to make an origami whirlybird. Look for the underlined and **boldfaced** words to help you.

First, draw an 11 in. by $4\frac{1}{2}$ in. rectangle and cut it out. **Next,** cut a slit down the middle and another slit in each side as shown in Figure A. **Then,** fold the two lower flap-sides in toward the center as in Figure B and tape them together. **Now,** fold up 1 in. of the bottom and secure it with two paper clips. **Finally,** fold one copter blade down toward you and one down away from you as in Figure C. Hold the whirlybird out, let it drop, and watch it spin.

Figure A

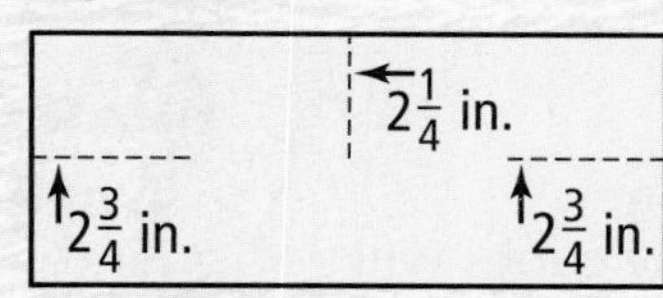

Figure B

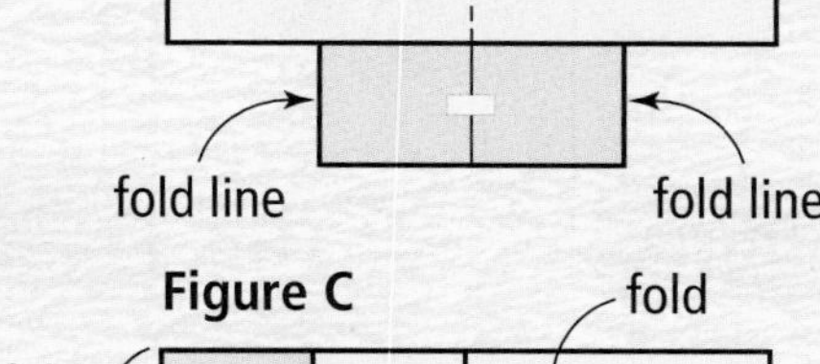

Figure C

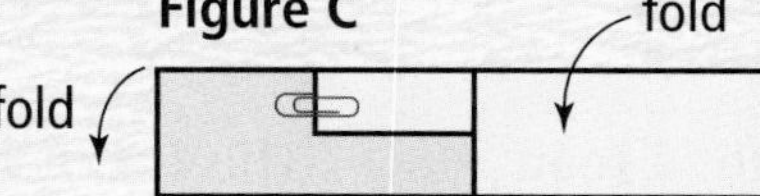

1. One of the steps could not be done correctly without looking at one of the figures. Which step was it? Explain.

2. How did the underlined words and boldfaced words help you follow the instructions correctly?

LESSON 3

Algebra: Volume

Quick Review

Find the area of each rectangle in square inches.

1. $l = 3$, $w = 4$
2. $l = 4$, $w = 5$
3. $l = 5$, $w = 5$
4. $l = 3$, $w = 6$
5. $l = 10$, $w = 9$

VOCABULARY

volume

Learn

GOAL KEEPER Antonio's soccer team stores their team uniforms and equipment in their locker room in a cabinet that is 4 ft × 2 ft × 3 ft. How many 1-cubic-foot boxes can be stored in the cabinet?

Antonio needs to find the volume of the cabinet. **Volume** is the amount of space a solid figure occupies. Volume is measured in *cubic units.*

Activity

MATERIALS: connecting cubes

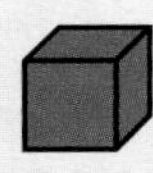

Make a rectangular prism in the same shape as Antonio's cabinet to find its volume. Let 1 connecting cube represent 1 cubic foot (ft^3).

STEP 1

One Dimension
Find how many cubes will make 1 row.

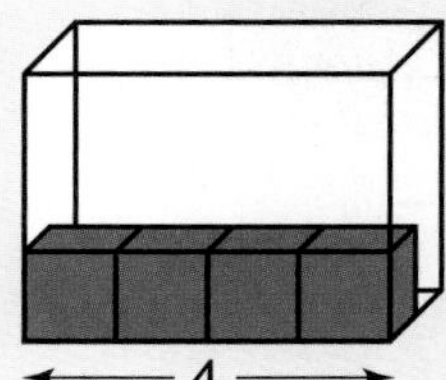

This is the length: 4 feet.

STEP 2

Two Dimensions
Find how many rows of 4 cubes will make 1 layer.

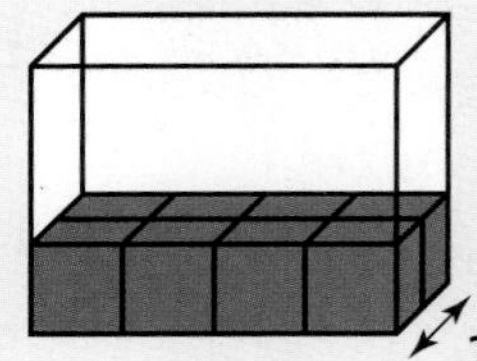

This is the width: 2 feet.
Count the number of cubes in each layer.
8, or 2 × 4

STEP 3

Three Dimensions
Find how many layers of 8 cubes will complete the prism.

This is the height: 3 feet.
Count the total number of cubes that make the prism.
24, or 3 × 2 × 4

So, 24 boxes can be stored in the cabinet.

- What are the dimensions that are multiplied to find the volume of the rectangular prism?

CALIFORNIA STANDARDS **MG 1.3** Understand the concept of volume and use appropriate units in common measuring systems to compute the volume of rectangular solids. **MG 1.0** Students understand and compute the volumes and areas of simple objects. *also* **MR 2.0, MR 2.2, MR 2.3, MR 3.3**

Use a Formula

Instead of counting cubes, you can use a formula to find the volume of a rectangular prism:

Volume = length × width × height, or $V = l \times w \times h$

Examples

A

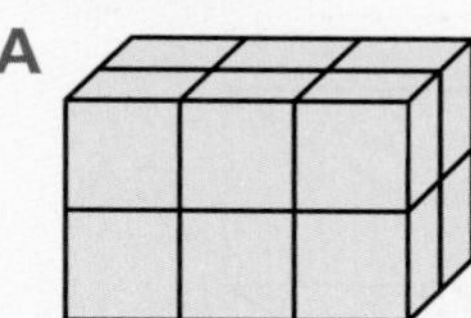

$V = l \times w \times h$
$V = 3 \times 2 \times 2$
$V = 12$
Volume is 12 units^3.

B

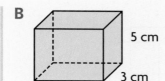

5 cm
3 cm
6 cm

$V = l \times w \times h$
$V = 6 \times 3 \times 5$
$V = 90$
Volume is 90 cm^3.

C

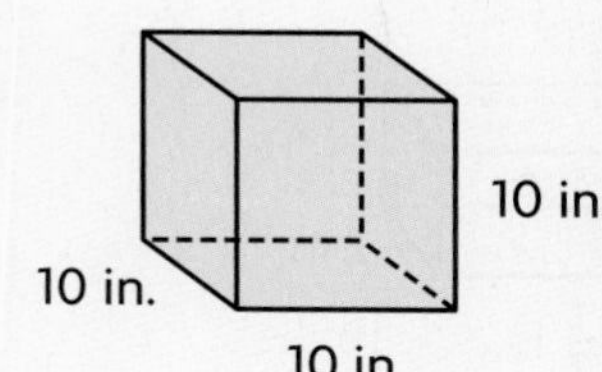

$V = l \times w \times h$
$V = 10 \times 10 \times 10$
$V = 1{,}000$
Volume is 1,000 in.^{3}

If you know the volume of a rectangular prism and two of its dimensions, you can use the formula to find the unknown dimension.

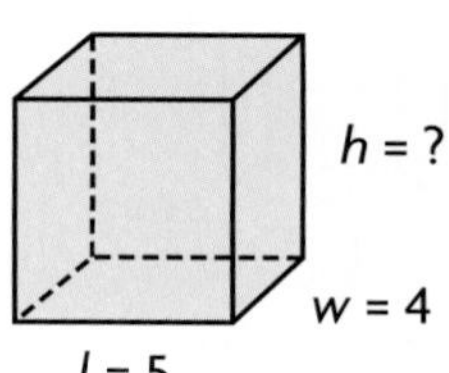

Volume: 120 in.^{2}

$$V = l \times w \times h$$
$$120 = 5 \times 4 \times h$$
$$120 = 20 \times h$$
$$\frac{120}{20} = h$$ The inverse of multiplication is division, so divide.
$$6 = h$$ The height is 6 in.

Check

1. **Explain** what is different about the rectangular prism in Example C compared to the prisms in Examples A and B.

Find the volume of each rectangular prism.

2. 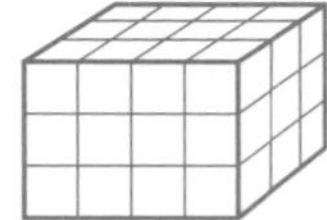

3. 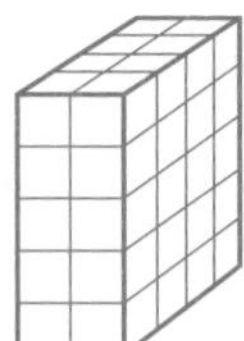

4. 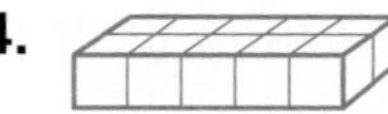

5. 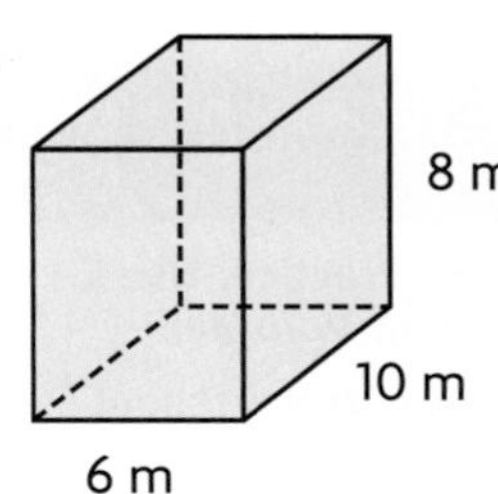

6. 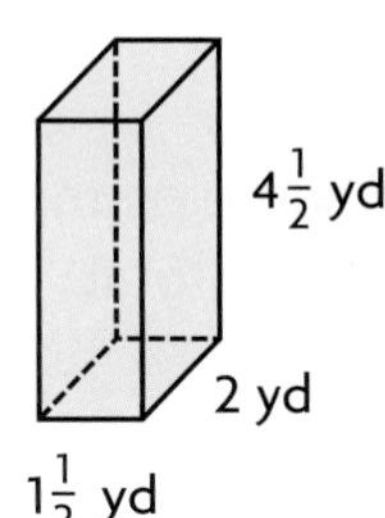

7. 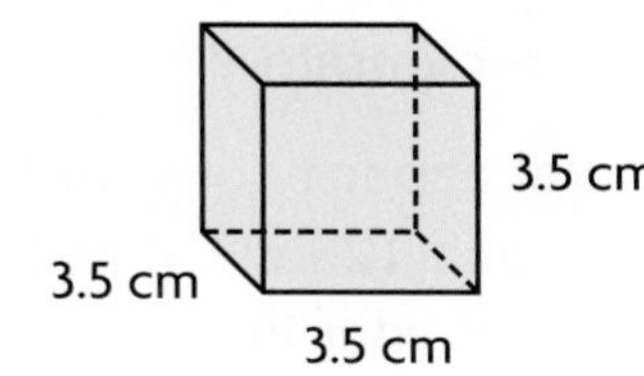

LESSON CONTINUES 

Practice and Problem Solving

Find the volume of each rectangular prism.

8.

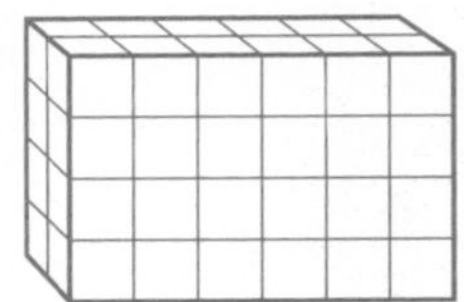

9.

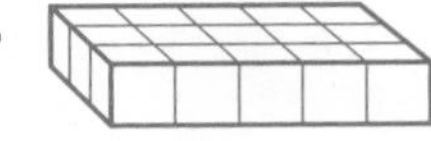

10.

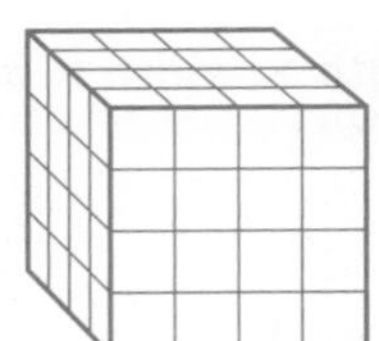

11.

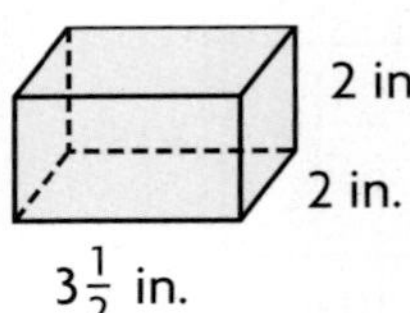

12.

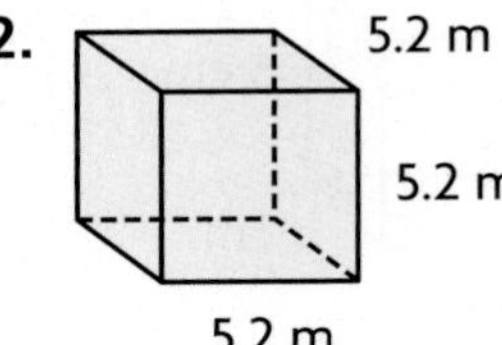

13.

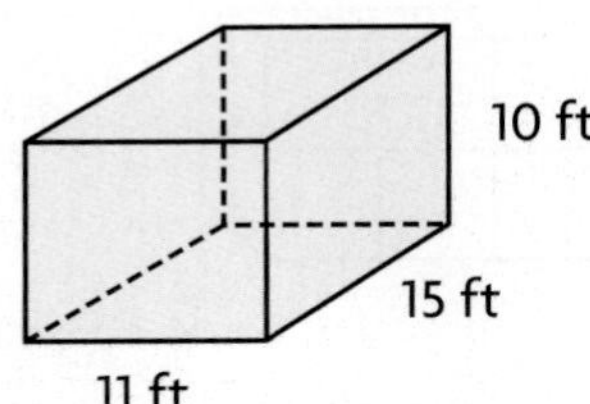

Algebra **Find the unknown dimension.**

14. $V = 60 \text{ ft}^3$

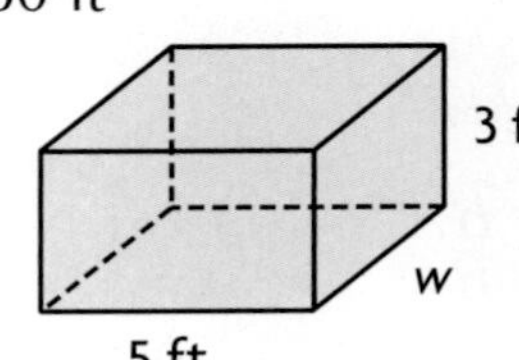

15. $V = 64 \text{ m}^3$

8 m

2 m

l

16. $V = 144 \text{ in.}^3$

h

6 in.

12 in.

17. $l = 3$ in.
$w = 2$ in.
$h = 10$ in.
$V = ■$

18. $l = 6$ yd
$w = 4$ yd
$h = ■$
$V = 48 \text{ yd}^3$

19. $l = 12$ ft
$w = ■$
$h = 2$ ft
$V = 192 \text{ ft}^3$

20. $l = ■$
$w = 7$ m
$h = 5$ m
$V = 350 \text{ m}^3$

For 21–23, use the drawings.

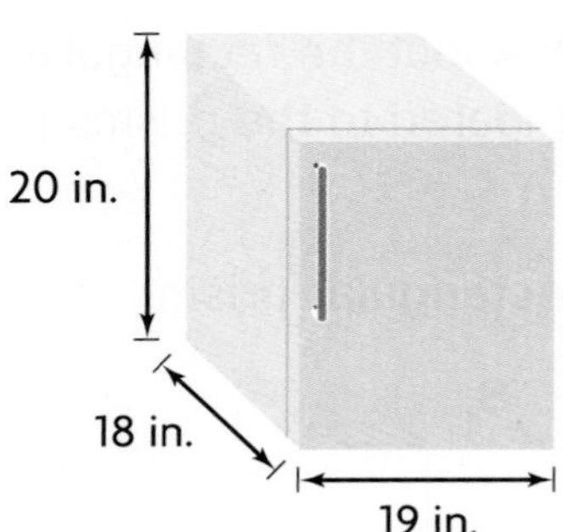

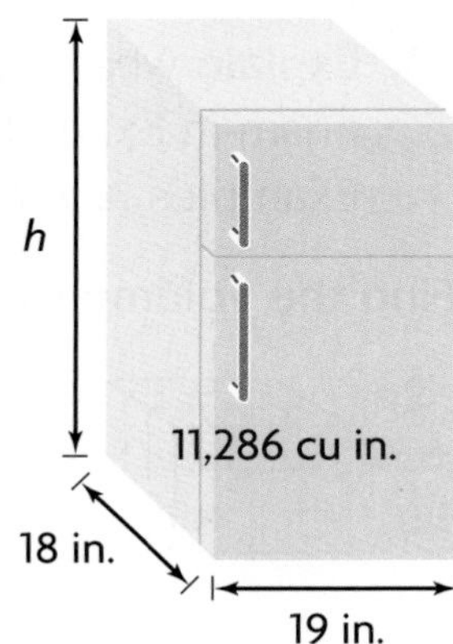

21. What is the volume of the small refrigerator? the height of the larger one?

22. How much area of the floor does each refrigerator cover?

23. Alan bought 2 small refrigerators. What is the difference between their combined volume and the volume of 1 large refrigerator?

24. The total capacity of your lungs is about the same as the volume of a 12 in. × 8 in. × 5 in. box. What is the total capacity of your lungs?

25.  **Write a problem** that requires finding volume to solve.

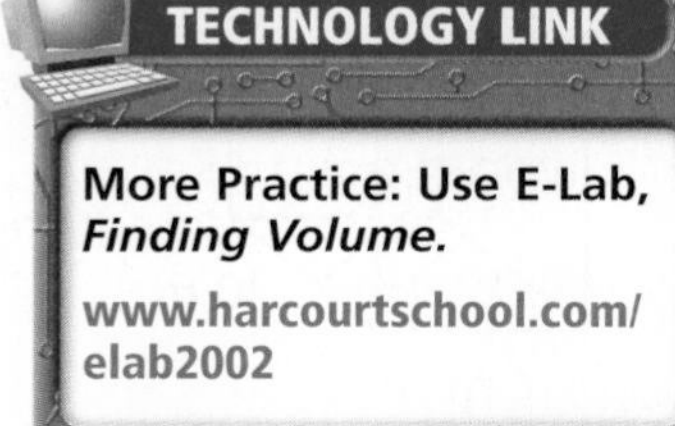

26. REASONING Explain the difference between an inch, a square inch, and a cubic inch. Make a model of each.

27. Juan says the formula for the volume of a cube, where s is the length of one side, is $V = s^3$. Is he correct? Explain.

Mixed Review and Test Prep

USE DATA **For 28–30, use the graph.** (p. 124)

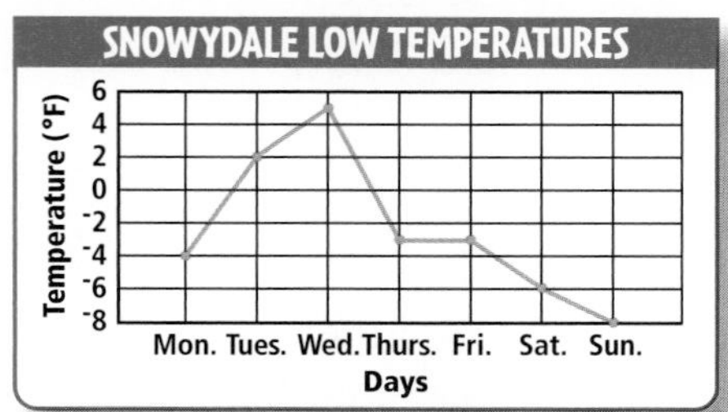

28. On which day was the lowest temperature recorded?

29. **TEST PREP** What was the average low temperature on Wednesday?

A 6°F
B 5°F
C 0°F
D ⁻5°F

30. What was the difference between the low temperature on Monday and on Tuesday?

31. $\frac{1}{4} + \frac{5}{6} + \frac{3}{8}$ (p. 346)

32. Write and evaluate an expression for this situation: Joe had 20 pens. He gave 2 students 3 pens each. How many pens are left? (p. 86)

33. **TEST PREP** A triangle with only two congruent sides is __?__. (p. 478)

F isosceles
G scalene
H parallel
J equilateral

LINKUP to Science

More than 10 million households in the United States have pet fish. For the best environment, a general rule to follow is 1 gallon of water per inch of fish. So, if your fish is 4 in. long, you need 4 gal of water in its tank. The volume of 1 gal of water is about 231 in.^{3} So, if the volume of a fish tank is 1,000 in.^{3}, the tank will hold 1,000 in.^{3} $\div$ 231 in.^{3} $\approx$ 4 gal of water.

1. What is the volume of this fish tank?
2. How many gallons of water will it hold? (Round your answer to the nearest whole gallon.)
3. How many 1-inch-long fish can you keep in the tank?
4. How many 3-inch-long goldfish can you keep in this tank?
5. What is the volume of a 50-gallon tank? a 100-gallon tank?

LESSON 4

Measure Perimeter, Area, and Volume

Quick Review

1. 24 in. = ■ ft
2. 27 ft^2 = ■ yd^2
3. 2 yd = ■ in.
4. 3 m = ■ cm
5. 35 cm = ■ m

Learn

COVER UP Keisha's grandmother is making a table by covering the outside of a wooden cube with fabric that matches her bedspread. The cube is 18 inches on each edge. What units will she use to express the amount of fabric needed? How much fabric will she need?

Geometric figures can be measured in one, two, or three dimensions. The unit you choose depends upon the number of dimensions being used.

Perimeter is the distance around a figure.

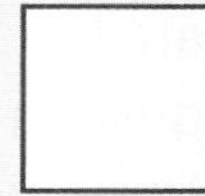

Its measure is in one dimension, *length.* Use linear units such as in., ft, yd, mi, cm, m, or km.

Area is the measure of the flat surface of a figure.

Its measure is in two dimensions, *length* and *width.* Use square units such as $in.^2$, ft^2, yd^2, mi^2, cm^2, m^2, or km^2.

Volume is the measure of the space a figure occupies.

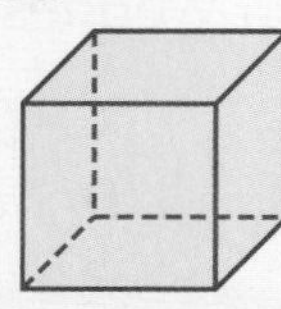

Its measure is in three dimensions, *length, width,* and *height.* Use cubic units such as $in.^3$, ft^3, yd^3, mi^3, cm^3, m^3, or km^3.

Since she is finding area, she should use square inches.

1 face: 18 in. × 18 in. = 324 $in.^2$
6 faces: 6 × 324 $in.^2$ = 1,944 $in.^2$

So, Keisha's grandmother needs 1,944 $in.^2$ of fabric.

Check

1. **Explain** how to find the amount of wood trim needed to go around a window that measures 48 in. by 30 in.

Tell the appropriate units for measuring each. Write *units, square units,* or *cubic units.*

2. surface area
3. volume
4. perimeter
5. area

CALIFORNIA STANDARDS MG 1.4 Differentiate between, and use appropriate units of measure for, two- and three-dimensional objects (e.g., find the perimeter, area, volume). **MG 1.0** Students understand and compute the volumes and areas of simple objects. *also* **MG 1.3, MG 2.0, MR 1.0, MR 1.1, MR 2.0, MR 2.2, MR 2.3, MR 3.3**

Practice and Problem Solving

Tell the appropriate units for measuring each. Write *units, square units,* or *cubic units*.

6. fence for a garden
7. tile for a floor
8. paper to cover a box
9. space in a refrigerator

Write the units you would use to measure each.

10. area of this rectangle

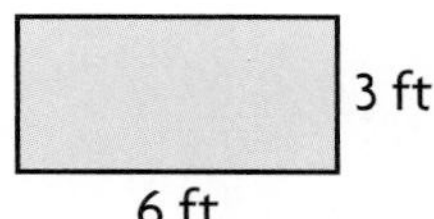

11. volume of this prism

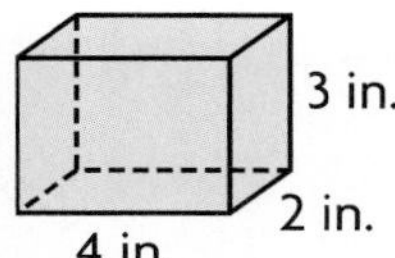

12. perimeter of this figure

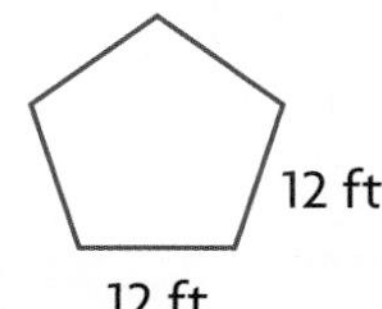

13. surface area of this figure

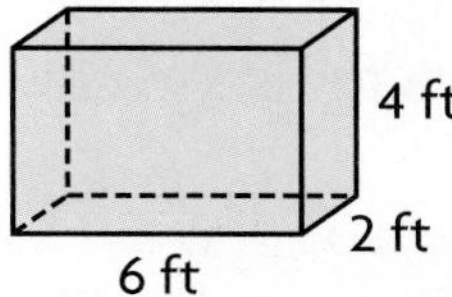

14. 8 in. by 10 in. picture frame
15. fabric covered box, 5 cm × 3 cm × 2 cm
16. a 9 ft by 10 ft rug
17. space inside a box, 2 ft × 8 ft × 4 ft

Tom and his dad are making a toy box for Tom's little sister. Use the picture for 18–20.

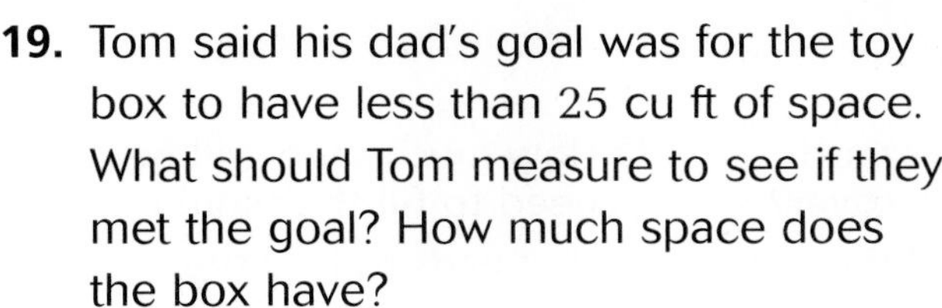

18. Tom used tape to put two stripes around the toy box. What did he measure to find out how much tape he needed? How much tape did Tom use?
19. Tom said his dad's goal was for the toy box to have less than 25 cu ft of space. What should Tom measure to see if they met the goal? How much space does the box have?

20. A can of paint for the toy box covers 20 sq ft. What did Tom measure to find out how much paint he needed? What units did he use for this measurement?
21. **What's the Error?** Joe's lunchbox is 10 in. long, 8 in. wide, and 3 in. high. He estimates the volume to be 240 in.^{2} Explain his error.

Mixed Review and Test Prep

22. Find the sum. $3.01 + 4.15$ (p. 50)
23. 84×37 (p. 148)
24. 209×43 (p. 148)
25. $7\frac{5}{6} - 3\frac{2}{3}$ (p. 368)
26. **TEST PREP** Evaluate $n \div 25$ if $n = 750$. (p. 190)

A 20 **B** 25 **C** 30 **D** 35

Extra Practice page H60, Set D

LESSON 5

Problem Solving Skill
Use a Formula

Quick Review

1. $\frac{3}{8} \times 8$
2. $8 \times 6 \times 4$
3. $36 + 40 + 36$
4. 6.3×4.5
5. Find the value of n.
$15 + (2 \times 5) - n = 5$

FILL IT UP! Mr. Lee plans to build in his backyard a sandbox that is 5 feet long, 4 feet wide, and $\frac{1}{2}$ foot high.

Use this table to help you decide which formula to use to solve each problem below.

RECTANGLE FORMULAS

Want To Know	Find	Formula
Distance around an object	Perimeter	$P = (2 \times l) + (2 \times w)$
Number of square units needed to cover a flat surface	Area	$A = l \times w$
Space occupied by a solid figure	Volume	$V = l \times w \times h$

MATH IDEA To use a formula to solve a problem, first decide what you want to know. Then put your information in the formula and do the calculations.

A What length of 6-in. boards will Mr. Lee need to **surround** the sandbox?

He needs to find the *perimeter* of the sandbox.

Use a formula:

$P = (2 \times l) + (2 \times w)$

$P = (2 \times 5) + (2 \times 4)$

$P = 10 + 8 = 18$

He needs 18 feet of 6-in. boards to make the sandbox.

B How much of his yard will the sandbox **cover**?

He needs to find the *area* the sandbox will cover.

Use a formula:

$A = l \times w$

$A = 5 \times 4$

$A = 20$

The sandbox will cover 20 square feet of his yard.

C How much sand will he need to **fill** the sandbox?

He needs to find the *volume* of the sandbox.

Use a formula:

$V = l \times w \times h$

$V = 5 \times 4 \times \frac{1}{2}$

$V = 10$

He needs 10 cubic feet of sand to fill the sandbox.

- If you did not use formulas, how could you answer A–C?

CALIFORNIA STANDARDS MR 2.3 Use a variety of methods, such as words, numbers, symbols, charts, graphs, tables, diagrams, and models, to explain mathematical reasoning. **MG 1.4** Differentiate between, and use appropriate units of measure for, two- and three-dimensional objects. **MG 1.0** Students understand and compute the volumes and areas of simple objects. *also* **MG 1.2, O-n MG 1.3, O-n MG 2.0, MR 1.0, MR 1.1, MR 2.0, MR 2.2, MR 3.3**

Problem Solving Practice

Use a formula to solve.

1. **What if** Mr. Lee decides to make the sandbox $\frac{3}{4}$ foot deep? How much sand will he need to fill the sandbox?

2. Rachel's bedroom is 15 ft long and 12 ft wide. She is having wall-to-wall carpet installed. How many square yards of carpet does she need? (HINT: Change feet to yards.)

3. A playground area that is 66 ft long and 42 ft wide needs to be fenced. How much fencing will be needed?

4. Mrs. Lopez wants to store some boxes of winter clothes in a cabinet that measures 5 ft by 2 ft by 6 ft. Each box is 1 ft $\times$ 2 ft $\times$ 2 ft. How many boxes can she fit into the cabinet?

A regulation American football field is 120 yards long, including the end zones, and $53\frac{1}{3}$ yards wide. Each end zone is 10 yards deep and $53\frac{1}{3}$ yards wide.

5. What is the perimeter of the field?

   A $5{,}033\frac{1}{3}$ yd^2 **C** $346\frac{2}{3}$ yd^2

   B $5{,}033\frac{1}{3}$ yd **D** $346\frac{2}{3}$ yd

6. What is the area of each end zone?

   F $306\frac{2}{3}$ yd^2 **H** $5{,}033\frac{1}{3}$ yd

   G $533\frac{1}{3}$ yd^2 **J** $53{,}033\frac{2}{3}$ yd^2

Mixed Applications

7. The veterinarian puts bird seed in a bin 15 in. $\times$ 5 in. $\times$ 6 in. How much bird seed can the bin hold?

8. A bag of fertilizer will cover an area of 100 ft^2. Will 4 bags be enough to cover a garden 25 ft $\times$ 18 ft?

9. Janine's book shelf is $4\frac{2}{3}$ ft tall. Her desk is $\frac{3}{4}$ as high as the book shelf. How high is Janine's desk?

10. **Write About It** Explain how you know whether the perimeter or the area formula should be used in a problem.

11. **What's the Question?** Mr. Lyle needs 66 square yards of ceiling tile for his basement. Each box of tiles contains 30 square yards of tile. The answer is 3 boxes.

12. Mr. Chu is unpacking some books from a box. There are 4 layers of books. Each layer has 3 rows with 3 books in each row. How many books are in the box?

CHAPTER 29

Review/Test

CHECK VOCABULARY AND CONCEPTS

Choose the best term from the box.

net
base
surface area
volume

1. The measure of the space a solid figure occupies is _?_. (p. 552)
2. A two-dimensional pattern for a three-dimensional prism or pyramid is a _?_. (p. 546)
3. The sum of the areas of the faces of a solid figure is the measure of _?_. (p. 548)

Match each solid figure with its net. Write *a, b,* or *c*. (pp. 546–547)

4. 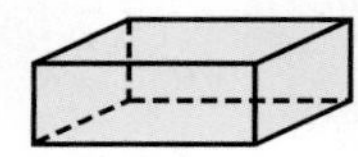

5. 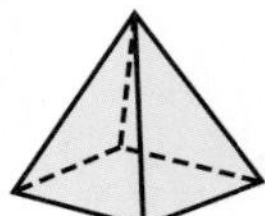

6. 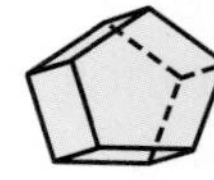

a. 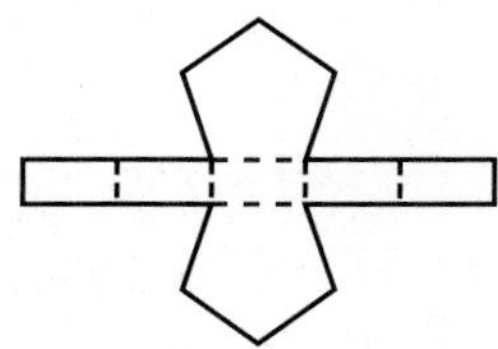

b. 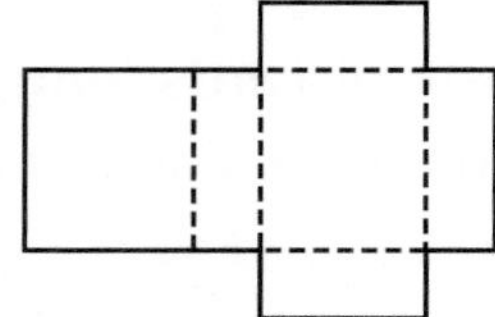

c. 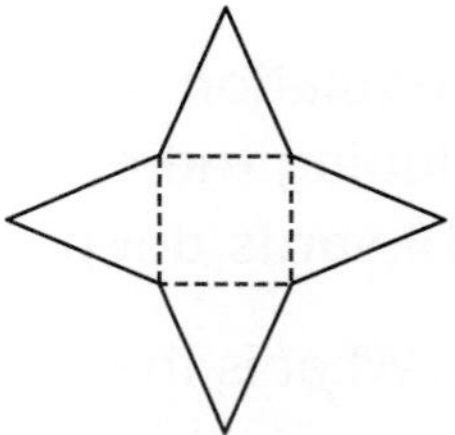

CHECK SKILLS

Find the area of each face in cm^2. Then find the surface area. (pp. 548–551)

7. 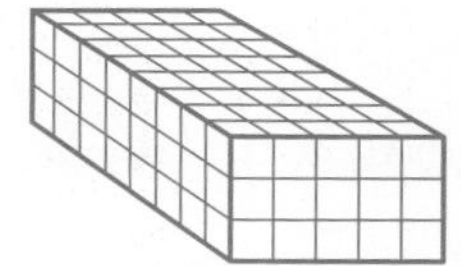

8. 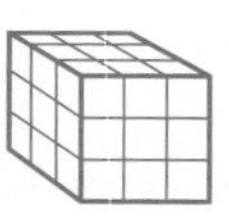

Find the unknown dimension. (pp. 522–525)

9. length = 3 yd
   width = 10 yd
   height = 17 yd
   volume = ■

10. length = 9 ft
    width = ■
    height = 8 ft
    volume = 504 cu ft

Tell the appropriate units for measuring each. Write *units, square units,* or *cubic units*. (pp. 556–557)

11. wood for a picture frame
12. fabric to cover a triangular prism
13. space in a closet

CHECK PROBLEM SOLVING

Solve. (pp. 558–559)

14. The stereo packaging box is 3 ft long and 2.5 ft wide. What is the area of the base of the box?
15. The storage locker is 5 yd long, 3.5 yd high, and 6 yd wide. How much space is in the locker?

Cumulative Review

Choose the answer.
See item **4.**

Surface area is related to area and uses the same units of measure. Think about the number of faces on a cube when finding surface area.

Also see problem **6,** p. H65.

For 1–11, choose the best answer.

1. Which shows the net for this solid?

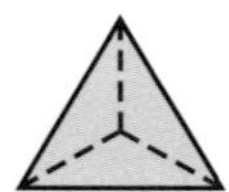

A 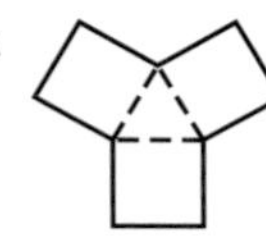

C 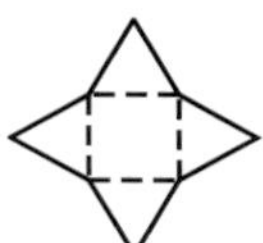

B 

D 

2. What is the volume of this toy box?

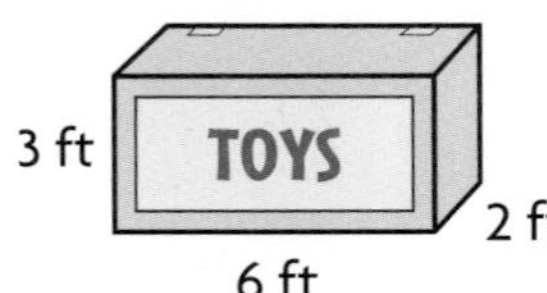

F 11 ft **H** 36 ft^2
G 36 ft^3 **J** 150 ft

3. Which decimal is equivalent to $\frac{3}{5}$?

A 6 **C** 0.3
B 0.6 **D** 0.06

4. What is the surface area of a cube that measures 5 centimeters on each edge?

F 125 cm^3 **H** 150 cm^2
G 150 cm^3 **J** 150 cm

5. Two angles of a triangle measure 78° and 33°. What is the measure of the third angle?

A 69° **C** 111°
B 71° **D** 249°

6. 3.6 ÷ 4

F 90 **H** 0.9
G 9 **J** Not here

7. What is the value of the 9 in 31.6592?

A 0.9 **C** 0.009
B 0.09 **D** 0.0009

8. The volume of a trailer is 2,640 ft^3. It measures 11 feet high and 8 feet wide. What is the length of this trailer?

F 2,651 ft **H** 330 ft
G 2,552 ft **J** 30 ft

9. The area of a rectangular rug is 84 ft^2. It measures 7 feet long. What is the width of this rug?

A 12 ft **C** 77 ft
B 28 ft **D** 91 ft

10. Which expression shows a way to find the value of pi ?

F $r \div C$ **H** $d \div C$
G $d \div r$ **J** $C \div d$

11. What is the perimeter of a rectangle that measures 6.2 centimeters by 3.4 centimeters?

A 21.08 cm^2 **C** 19.2 cm
B 21.08 cm **D** 19.2 cm^2

UNIT 8

MATH DETECTIVE

Match It

In each set, three of the members are related to each other, while the other member does not share that relationship. Use what you know about measurement to figure out which member doesn't go with the others.

1.	kilometer	yard	foot	mile
2.	32 fl oz	1 qt	16 pt	4 c
3.	2,500 mm	0.025 cm	250 cm	2.5 m
4.	fluid ounces	gallons	pounds	cups
5.	ounces	liters	tons	pounds
6.	4 hr 15 min	255 min	2 hr 135 min	275 min
7.	3 ft	36 in.	1 mi	1 yd
8.	1 day	43,200 sec	12 hr	720 min
9.	4.5 lb	72 oz	3 lb 24 oz	4 lb 5 oz
10.	millimeter	inch	meter	kilometer

Think It Over!

- **Write About It** State the relationship that is shared by the three members in each group.
- **Write a problem** like the ones above.

CASE CLOSED

UNIT 8

Challenge

Area of a Circle

To find the area of a circle, use the formula $A = \pi \times r^2$. The r in the formula represents the length of the radius of the circle. The approximate decimal value of π (pi) is 3.14.

To find the area of the circle, replace r with 5 and π with 3.14.

$A = \pi \times r^2$ $\quad r = 5$

$A \approx 3.14 \times (5)^2$

$A \approx 3.14 \times 25$

$A \approx 78.5 \text{ in.}^2$

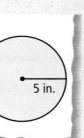

5 in.

The units in the answer are squared because area is measured in square units.

Examples

A $A = \pi \times r^2$ $\quad r = 7$

$A \approx 3.14 \times (7)^2$

$A \approx 3.14 \times 49$

$A \approx 153.86 \text{ cm}^2$

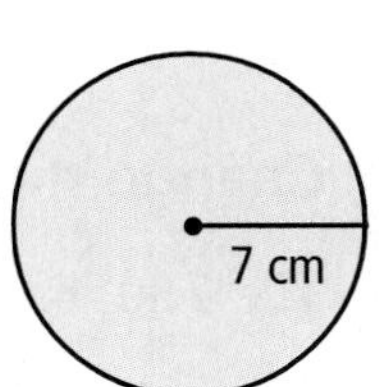

B $A = \pi \times r^2$ $\quad r = 9$

$A \approx 3.14 \times (9)^2$

$A \approx 3.14 \times 81$

$A \approx 254.34 \text{ yd}^2$

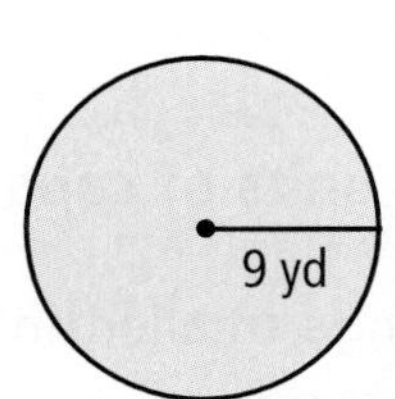

Talk About It

- How do you determine what unit to use to express the area of a circle?

Try It

Find the area of the circle with the given radius. Use $\pi = 3.14$.

1. 4 in. **2.** 6 cm **3.** 8.5 ft **4.** 10 mi

Find the radius of the circle with the given area.

5. 28.26 ft^2 **6.** 19.625 m^2 **7.** 254.34 in.^2 **8.** 379.94 cm^2

Study Guide and Review

VOCABULARY

Choose the best term from the box.

circumference
net
volume
radius
area

1. The distance around a circle is the _?_. (p. 526)
2. A two-dimensional pattern that can be folded into a three-dimensional prism or pyramid is a _?_. (p. 546)
3. The amount of space a solid figure occupies is its _?_. (p. 552)

STUDY AND SOLVE

Chapter 27

Change linear units.

To change smaller units to larger units, divide.
200 cm = ■ m
200 ÷ 100 = 2 ← 100 cm = 1 m
So, 200 cm = 2 m.

To change larger units to smaller units, multiply.
3 ft = ■ in.
3 × 12 = 36 ← 1 ft = 12 in.
So, 3 ft = 36 in.

Change the unit. (pp. 508–511)

4. 5 ft = ■ in.
5. 15 yd = ■ ft
6. 6,160 yd = ■ mi
7. 7 m = ■ cm
8. 2.5 km = ■ m
9. 45 mm = ■ cm

Change units of capacity and mass.

To change smaller units to larger units, divide.
8 qt = ■ gal
8 ÷ 4 = 2 ← 4 qt = 1 gal
So, 8 qt = 2 gal.

To change larger units to smaller units, multiply.
2 kg = ■ g
2 × 1,000 = 2,000 ← 1 kg = 1,000 g
So, 2 kg = 2,000 g.

Change the unit. (pp. 512–515)

10. 7 kL = ■ L
11. 5 qt = ■ pt
12. 64 oz = ■ lb
13. 4 kg = ■ g
14. 300 mL = ■ L
15. 6 qt = ■ gal

Chapter 28

Find the perimeter of a polygon.

Find the perimeter.
$P = (2 \times l) + (2 \times w)$
$P = (2 \times 5.6) + (2 \times 3)$
$P = 11.2 + 6$
$P = 17.2$ cm

3 cm
5.6 cm

Find the perimeter of each polygon. (pp. 524–525)

16. 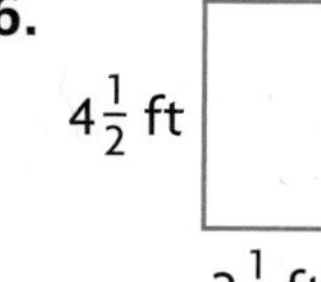

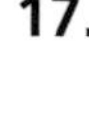
17. 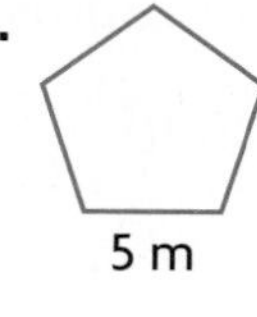

Find the area of a polygon.

Find the area.

$A = l \times w$
$A = 6.4 \times 4.5$
$A = 28.8$

4.5 cm

6.4 cm

So, the area is 28.8 cm^2.

Find the area of each figure. (pp. 528–539)

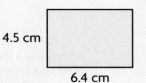

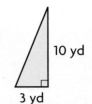

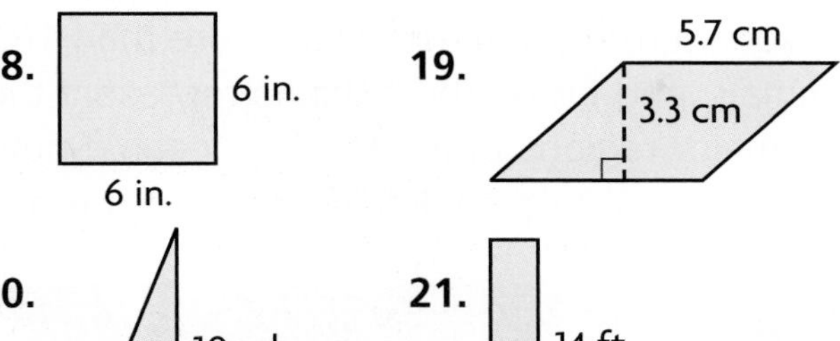

Chapter 29

Find the surface area.

The surface area of a solid figure is the sum of the areas of all the faces.

Find the surface area.

$2 \times (3 \times 4) = 24$
$2 \times (2 \times 3) = 12$
$2 \times (2 \times 4) = 16$
$24 + 12 + 16 = 52$

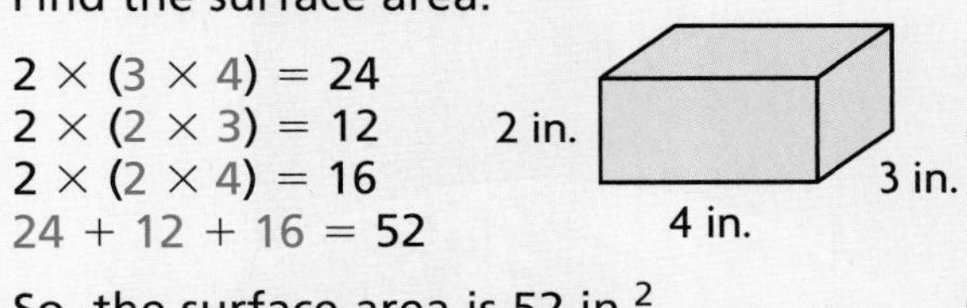

So, the surface area is 52 $in.^2$

Find the surface area. (pp. 548–551)

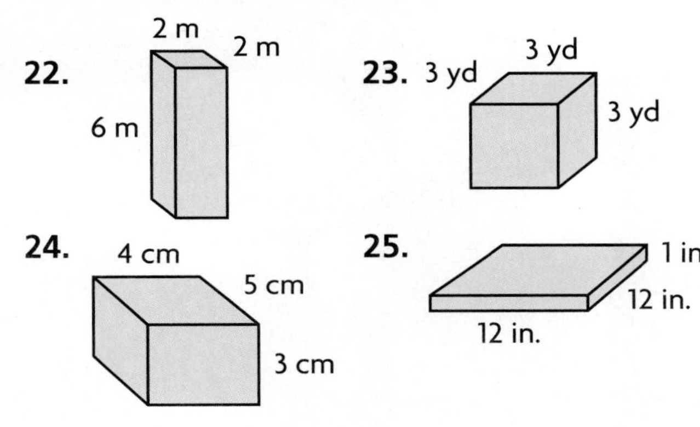

Find the volume of a rectangular prism.

Find the volume.

$V = l \times w \times h$
$V = 6 \times 2 \times 10$
$V = 120$

So, the volume is 120 $in.^3$

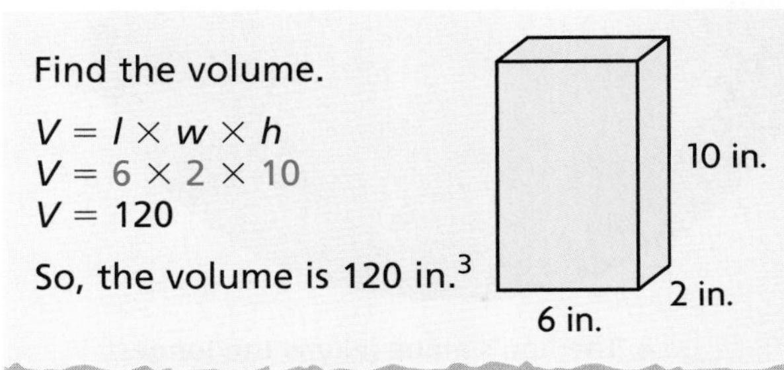

Find the volume of each rectangular prism. (pp. 552–557)

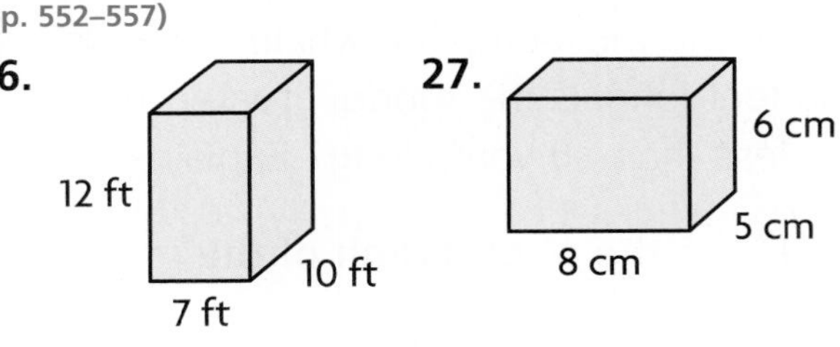

PROBLEM SOLVING PRACTICE

Solve. (pp. 518–519, 558–559)

28. The first gym class starts at 9:45 and lasts 50 minutes. The second gym class begins at 11:30. How long is the gym free between classes?

29. Marie is using construction paper to cover a box 1 ft long, 3 ft wide, and 2 ft high. How much construction paper does she need?

California Connections

LONG BEACH AQUARIUM OF THE PACIFIC

The Aquarium of the Pacific has more than 10,000 fish, mammals, and birds. The exhibits represent the three major regions of the Pacific Ocean: Southern California/Baja, Tropical Pacific, and Northern Pacific.

▼ **The blue whale is the largest animal on the planet.**

▲ **The lion's mane jelly is the longest animal in the world.**

1. A full-scale model of a blue whale floats high above the heads of visitors as they enter the museum. The whale model is 88 feet long. Is the model greater than or less than 30 yards long? Explain.

For 2–4, use the photograph of the lion's mane jelly.

2. Measure the diameter of the bell-shaped body to the nearest $\frac{1}{8}$ inch. Measure to the nearest millimeter.

3. The actual diameter of a lion's mane jelly can be as much as 70 times the diameter shown in the photo. About how many feet can the actual diameter be? Explain how you found the answer.

4. The length of the tentacles of the lion's mane jelly can be as much as 12.5 times its diameter. About how many feet long can the tentacles be? Compare this length to the length of your classroom.

THE MONTEREY BAY AQUARIUM

About 360 miles up the coast from the Aquarium of the Pacific is the Monterey Bay Aquarium. The exhibits in this aquarium are devoted to the rich marine life of Monterey Bay.

The viewing window of the Outer Bay Waters exhibit is 54 feet long and 15 feet tall.

1. What is the perimeter of the window?
2. What is the area of the window?
3. **What if** you wanted to make a window with the same perimeter but with the greatest possible area? What dimensions would you use? Explain.
4. The window is about 1.1 feet thick. What is the volume of the acrylic material in the window?

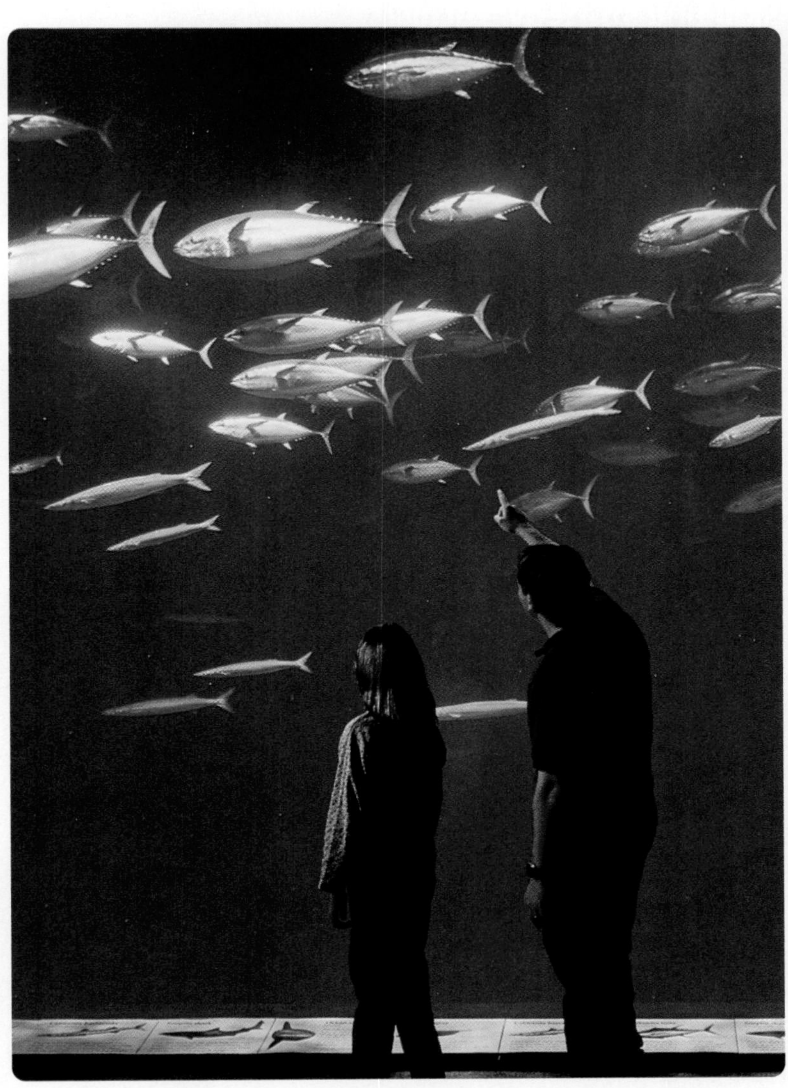

▲ The Outer Bay exhibit, which opened in 1996, has a million-gallon tank with the world's largest window.

Suppose the aquarium had a holding tank that was 10 ft long, 8 ft wide, and 6 ft deep. The tank has a base and sides, but no top.

5. What is the surface area of the tank?
6. What is the volume of the tank?
7. **What if** the length of the tank is doubled? How does the volume change?

CHAPTER 30 Probability

If you like to play games, you know that sometimes winning depends on chance, and sometimes it depends on skill and strategy. Pachisi is a game of chance that is the national game of India. The players toss cowrie shells and move pieces around a gameboard according to how the shells land. An updated version of this game exists today. Look at the games in the photos, and tell whether each is a game of chance or one of skill and strategy.

- 4000 B.C. — Babylonian board game played that was ancestor of chess and checkers
- 3000 B.C. — Backgammon played in ancient Sumeria
- 2000 B.C. — Egyptians played game like modern-day checkers
- 1000 B.C.
- B.C./A.D.
- Pachisi played in India
- Mah-Jongg first played in China
- A.D. 1000
- A modern version of Pachisi first played in 1959
- A.D. 2000 — Computer games first played in 1970's

Pachisi was first played in India 1,500 years ago. The name *Pachisi* comes from the Indian word for 25, which is the highest score a player can get with one throw of the shells.

Use this page to help you review and remember important skills needed for Chapter 30.

VOCABULARY

Choose the best term from the box.

certain
impossible
likely
unlikely

1. An event that has more chances to happen than not to happen is __?__ to happen.
2. An event is __?__ if it will always happen.
3. An event is __?__ if it will never happen.

IMPOSSIBLE, CERTAIN, LIKELY, UNLIKELY (See p. H31.)

Tell whether each event is *certain* or *impossible*.

4. going to the planet Jupiter on a family vacation
5. becoming warmer in front of a fire
6. spinning an odd number on a spinner labeled 1, 3, 5, and 7
7. pulling a red counter from a bag containing only blue counters

Tell whether each event is *likely* or *unlikely*.

8. rolling a 3 on a cube numbered 1, 1, 1, 2, 2, 3
9. pulling a yellow marble from a bag of 9 yellow and 2 green marbles
10. spinning an even number on a spinner with equally-divided sections numbered 1, 3, 5, 6, 7, and 9
11. pulling a blue or red marble from a bag of 3 blue, 5 red, and 2 green marbles

For 12–15, use the spinner. Classify each event as *impossible, certain, likely*, or *unlikely*.

12. spinning purple
13. spinning yellow
14. spinning red or yellow
15. spinning red

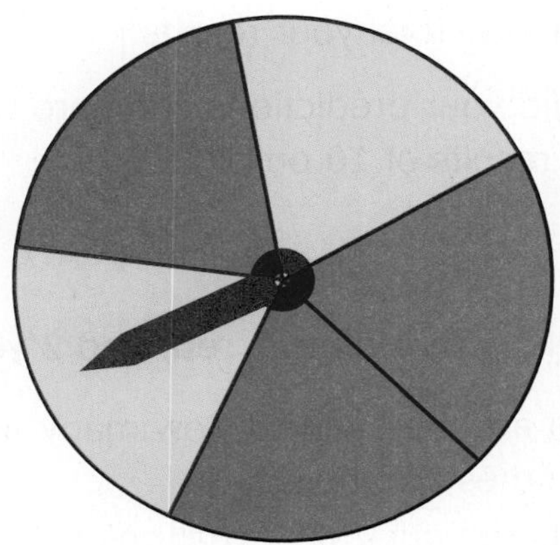

LESSON 1

HANDS ON
Probability Experiments

Quick Review

Write the fraction of each circle that is shaded.

1. 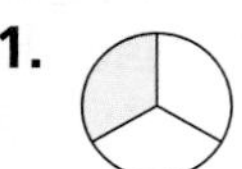 **2.** 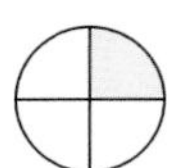

3. 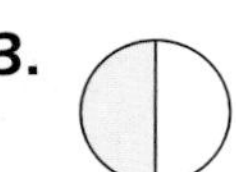 **4.** 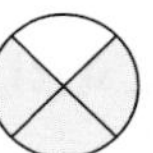

5. 

VOCABULARY

possible outcomes

MATERIALS
color tiles, paper bag

Explore

Mark conducted an experiment. He put 5 red, 3 green, and 2 yellow tiles into a bag and then pulled out one tile.

Possible outcomes have an equal chance of happening in an experiment. Each tile has an equal chance of being pulled. There are 10 possible outcomes in this experiment.

The three possible colors in this experiment are events:

- pull red
- pull green
- pull yellow

Conduct the same experiment Mark did. Place 5 red, 3 green, and 2 yellow tiles in the bag. Predict the number of times out of 10 you will pull red.

COLOR-TILE EXPERIMENT			
Events	**Yellow**	**Red**	**Green**
Predicted Frequency			
Actual Frequency			
Total			

- Make a table like the one shown. Record your predictions for all three events.
- Pull a tile 10 times without looking. Replace the tile after each pull. Record your pulls with tally marks. Total your results.
- How did your predictions compare with the actual results of 10 pulls?

If you pull out 1 tile, what color do you think it will be?

Try It

Use a bag with 6 blue, 2 red, and 2 yellow tiles.

a. What are the events? How many possible outcomes are there?

b. If you pull out 1 tile, what color do you think it will be? Explain.

CALIFORNIA STANDARDS **MR 3.3** Develop generalizations of results obtained and apply them in other circumstances. *also* **MR 2.3, MR 2.4, MR 3.0, MR 3.2**

Connect

Look back at Mark's experiment. Predict how many times out of 100 he would pull red. Since half of the tiles are red, a good prediction is 50 pulls.

Mark and nine other students recorded their results.

TECHNOLOGY LINK

More Practice: Use E-Lab, *Probability Experiments.*

www.harcourtschool.com/elab2002

COLOR-TILE EXPERIMENT (100 Pulls)											
Students	**A**	**B**	**C**	**D**	**E**	**F**	**G**	**H**	**I**	**J**	**Total**
Red Pulls	4	6	4	7	5	4	6	7	5	4	52
Yellow Pulls	2	2	3	1	2	2	1	1	3	2	19
Green Pulls	4	2	3	2	3	4	3	2	2	4	29

With nine classmates, make a table for 10 pulls, and find the totals.

- How do your group's actual results compare to the prediction for pulling red? To the results of other groups?

MATH IDEA The more times you repeat a probability experiment, the closer you will get to the expected, or predicted, result.

Practice

1. Write the possible events for pulling a marble from a bag with 3 green marbles, 2 pink marbles, and 7 blue marbles.

2. Keith has a bag of 312 marbles. He has equal numbers of 6 colors. How many of each color does he have?

3. Felicia has a bag filled with 5 red and 5 blue tiles. She pulls and replaces a tile 20 times. Predict how many times she will pull a red tile.

4. Write the possible events for rolling a number cube labeled 1 to 6.

5. **What's the Error?** June has a bag of tiles. There are 99 blue tiles and 1 red tile in the bag. June says that it is certain she will pull a blue tile from the bag. Describe her error. Write the correct answer.

Mixed Review and Test Prep

Complete each pattern. (p. 218)

6. $0.072 \div 8 = 0.009$
   $0.72 \div 8 = 0.09$
   $7.2 \div 8 = \blacksquare$

7. $30 \div 5 = 6$
   $3 \div 5 = 0.6$
   $0.3 \div 5 = \blacksquare$

8. $13\overline{)78}$ (p. 204)

9. 9×21 (p. 146)

10. **TEST PREP** Which is a prime number? (p. 264)

    A 39 **C** 53
    B 45 **D** 76

LESSON 2

Outcomes

Quick Review

Evaluate each expression.

1. $2 \times 5 \times 3$ **2.** $4 \times 2 \times 3$

3. $5 \times 4 \times 2$ **4.** $6 \times 3 \times 4$

5. $3 \times 8 \times 2$

VOCABULARY

tree diagram

Learn

LOOKING GOOD! Shawn is choosing an outfit to wear. He has tan, blue, and black pants. Shawn can choose a blue, green, white, or red shirt. How many different choices of pants and shirt outfits does Shawn have?

One Way

You can use a tree diagram to organize and show all the possible outcomes.

Pants	Shirts	Choices
tan	blue	tan pants with a blue shirt
	green	tan pants with a green shirt
	white	tan pants with a white shirt
	red	tan pants with a red shirt
blue	blue	blue pants with a blue shirt
	green	blue pants with a green shirt
	white	blue pants with a white shirt
	red	blue pants with a red shirt
black	blue	black pants with a blue shirt
	green	black pants with a green shirt
	white	black pants with a white shirt
	red	black pants with a red shirt

So, Shawn can choose from 12 different outfits.

Another Way

You can find the number of choices by multiplying.

Number of Pants		Number of Shirts		Number of Choices
↓		↓		↓
3	×	4	=	12

MATH IDEA You can use a tree diagram to organize and show a list of all possible choices. You can multiply to find the number of choices.

CALIFORNIA STANDARDS **MR 2.3** Use a variety of methods, such as words, numbers, symbols, charts, graphs, tables, diagrams, and models, to explain mathematical reasoning. *also* **MR 1.0, MR 2.4, MR 3.2, MR 3.3**

Check

1. **Explain** how it would change the outcomes if Shawn could choose from only 3 shirts.

For 2–3, make a tree diagram to show the possible choices.

2. **School Trips**
   Place: museum, park, theater
   Transportation: car, bus, subway

3. **Room Decorating Choices**
   Paint: white, blue, pink
   Borders: wallpaper, stencils
   Windows: blinds, curtains

Practice and Problem Solving

For 4–7, make a tree diagram to show the possible choices. Solve.

4. Lisa is making a sandwich. She can use either smooth or chunky peanut butter. She can choose from grape, strawberry, peach, or apricot jelly. How many choices does Lisa have? How many choices include grape jelly?

5. Tom can buy only one book. His choices are mystery, biography, or science fiction. He has the choice of hardback or paperback. How many choices does Tom have?

6. Janice is choosing a summer Enrichment Class. She can choose Foreign Language, Computers, Art, Math, or Physical Fitness in either the first or second session. How many choices does Janice have?

7. Matt is ordering dinner. He has a choice of salad or soup. Then he has a choice among spaghetti, meatloaf, or fried chicken. How many dinner choices does Matt have?

8. **Algebra** You have x number of choices for shirts and y number of choices for pants. Write an expression to show how many outfits you can make.

9. **Write About It** Explain how a tree diagram can help you organize your lunch choices with 4 different kinds of sandwiches and 4 types of fruit.

Mixed Review and Test Prep

10. $3{,}289 \times 43$ (p. 148)

11. $72{,}439 \times 82$ (p. 148)

Round to the nearest hundred thousand. (p. 32)

12. 7,809,432

13. 57,345,987

14. **TEST PREP** Which is the product of $\frac{3}{4} \times \frac{2}{3}$? (p. 382)

   A $\frac{5}{12}$
   B $\frac{1}{2}$
   C $\frac{5}{7}$
   D $\frac{6}{7}$

Extra Practice page H61, Set A

LESSON 3

Probability Expressed as a Fraction

Quick Review

Write in simplest form.

1. $\frac{10}{12}$
2. $\frac{9}{15}$
3. $\frac{8}{24}$
4. $\frac{14}{18}$
5. $\frac{12}{16}$

VOCABULARY

probability
equally likely

Learn

CLASS PET Mr. Gonzalez will choose one student to care for the class pet, a rabbit named Buffy. There are 25 students in his class. He writes each student's name on an index card, places the cards in a box, and pulls one name without looking.

Ellen wants to be Buffy's caretaker. What is the probability that Ellen's name will be pulled?

Probability is the chance that an event will happen.

Each outcome is **equally likely**, or has the same chance of happening.

$$\text{Probability of an event} = \frac{\text{number of favorable outcomes}}{\text{total number of possible outcomes}}$$

$$\text{Probability of Ellen being chosen} = \frac{\text{number of cards with Ellen's name}}{\text{total number of cards with students' names}} = \frac{1}{25}$$

So, the probability that Ellen's name will be pulled is $\frac{1}{25}$.

Ellen hopes that if she is not chosen, her friend Rosa will be. What is the probability that Ellen's or Rosa's name will be pulled?

Instead of only one name, there are two different names, Ellen or Rosa, that can be chosen.

$$\text{Probability} = \frac{\text{number of cards with Ellen's or Rosa's name}}{\text{total number of cards with students' names}} = \frac{2}{25}$$

The probability of choosing Ellen or Rosa is $\frac{2}{25}$.

MATH IDEA The probability of an event can be written as a fraction comparing the number of favorable outcomes to the total number of possible equally likely outcomes.

CALIFORNIA STANDARDS **MR 2.3** Use a variety of methods, such as words, numbers, symbols, charts, graphs, tables, diagrams, and models, to explain mathematical reasoning. **MR 2.4** Express the solution clearly and logically by using the appropriate mathematical notation and terms and clear language; support solutions with evidence in both verbal and symbolic work. *also* **MR 2.0, MR 2.2, MR 3.2**

Check

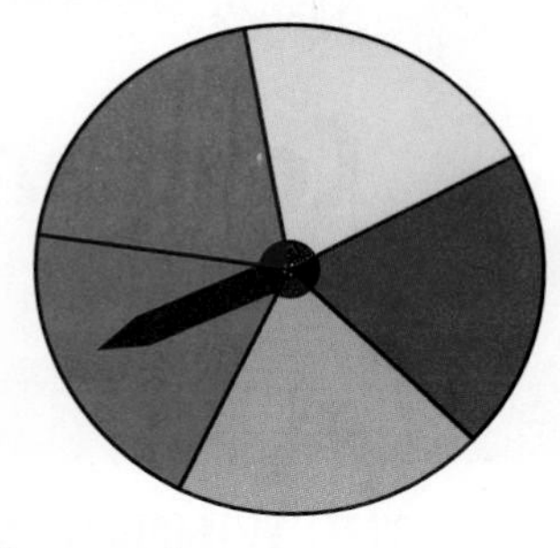

1. **Tell** the probability that Mr. Gonzalez will pull a name of one of the students in the class. Explain.

Write a fraction for the probability of the pointer of the spinner landing on each color.

2. blue or yellow
3. red
4. not green
5. purple

Practice and Problem Solving

Write the probability of the pointer landing on red.

6. 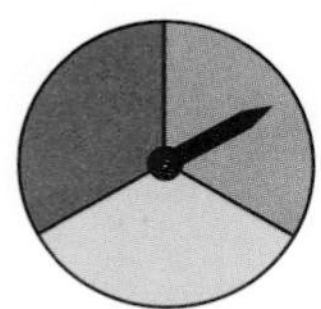
7. 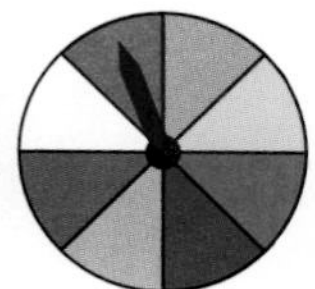
8. 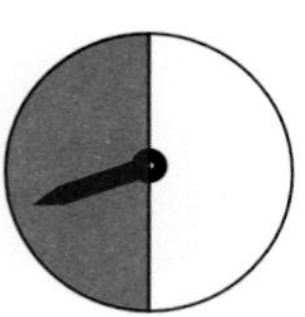
9. 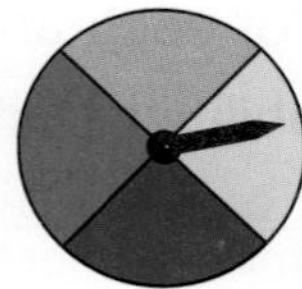

Write a fraction for the probability of rolling each number on a number cube labeled 1 to 6.

10. 4
11. 2 or 3
12. not 1
13. an even number
14. not 8

For 15–19, one of the marbles below is drawn. Write a fraction for the probability of each event.

15. blue
16. yellow or red
17. not orange
18. not blue
19. black

20. Jimmy tosses a coin. Write a fraction that expresses the probability of the coin showing heads. Express this probability as a decimal.
21. **REASONING** If you toss three coins at once, how many outcomes are possible? What are the outcomes? What is the probability of three coins showing heads?
22. **What's the Question?** Janet rolls a number cube labeled 1 to 6. The probability is $\frac{1}{2}$.

Mixed Review and Test Prep

23. Write an equivalent decimal for 0.70. (p. 22)
24. Round 0.2134 to the nearest thousandth. (p. 46)
25. $3,159 \div 81$ (p. 204)
26. Write the reciprocal of $7\frac{1}{7}$. (p. 396)
27. **TEST PREP** The sum of two numbers is 49. Their difference is less than 10. Which is a possible pair? (p. 72)

   A 30 and 19
   B 24 and 25
   C 40 and 9
   D 35 and 14

LESSON 4

Compare Probabilities

Quick Review

Compare. Write <, >, or = for each ●.

1. $\frac{3}{5}$ ● $\frac{1}{2}$ **2.** $\frac{4}{7}$ ● $\frac{3}{4}$ **3.** $\frac{2}{3}$ ● $\frac{5}{6}$ **4.** $\frac{1}{4}$ ● $\frac{1}{5}$ **5.** $\frac{7}{10}$ ● $\frac{1}{2}$

Learn

AND THE WINNER IS . . . Kelsey and Josh are playing a game. The first player to score 25 points wins. Kelsey earns a point when the pointer on the spinner lands on red. Josh earns a point when it lands on blue. Who is more likely to win the game?

To determine if one event is more likely than another, find the probability of each and compare fractions.

Probability of red

$$\frac{\text{number of red sections}}{\text{total number of sections}} = \frac{5}{9}$$

Probability of blue

$$\frac{\text{number of blue sections}}{\text{total number of sections}} = \frac{4}{9}$$

Since $\frac{5}{9} > \frac{4}{9}$, the probability of spinning red is greater than that of spinning blue, so Kelsey is more likely to win the game.

The probability of an event occurring is always expressed as 0, 1, or a fraction between 0 and 1.

MATH IDEA You can compare probabilities to decide whether one event is more likely than another.

Check

1. **Use** Kelsey and Josh's spinner. Express as a fraction the probability of the pointer landing on green. What is the probability of it landing on red or blue?

TECHNOLOGY LINK

More Practice: Use Mighty Math Number Heroes, *Probability,* Levels E and J.

Use the spinner. Write each probability as a fraction. Tell which event is more likely.

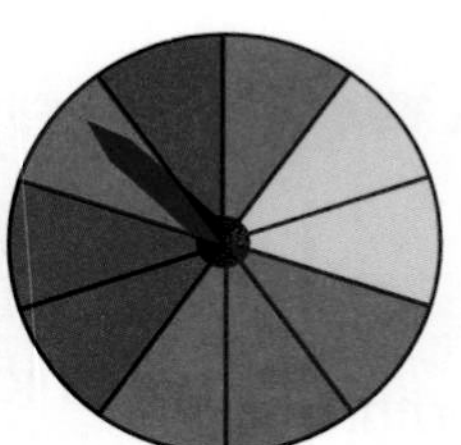

2. The pointer will land on blue; the pointer will land on red.
3. The pointer will land on either blue or green; the pointer will land on either red or yellow.

CALIFORNIA STANDARDS **MR 2.3** Use a variety of methods, such as words, numbers, symbols, charts, graphs, tables, diagrams, and models, to explain mathematical reasoning. **MR 2.4** Express the solution clearly and logically by using the appropriate mathematical notation and terms and clear language; support solutions with evidence in both verbal and symbolic work. *also* **MR 1.0, MR 1.1, MR 2.0**

Practice and Problem Solving

For 4–9, use the bag of marbles. Write each probability as a fraction. Tell which event is more likely.

4. You pull a red marble; you pull a blue marble.
5. You pull a yellow marble; you pull a green marble.
6. You pull a red or blue marble; you pull a green or yellow marble.
7. You pull a yellow marble; you pull a purple marble.
8. You pull a marble that is not red; you pull a marble that is not yellow.
9. You pull a marble that is not blue; you pull a marble that is not green.

For 10–13, use the spinner at the right. Write each probability as a fraction. Tell which event is more likely.

10. You spin an even number; you spin a prime number.
11. You spin a number greater than 3; you spin a number greater than or equal to 3.
12. You spin a prime number; you spin a prime number less than or equal to 5.
13. In a game with a number cube labeled 1 to 6, Sue gets a point for an even number and Julie gets a point for an odd number. Is the game fair? Explain.

USE DATA **For 14–15, use the bar graph.**

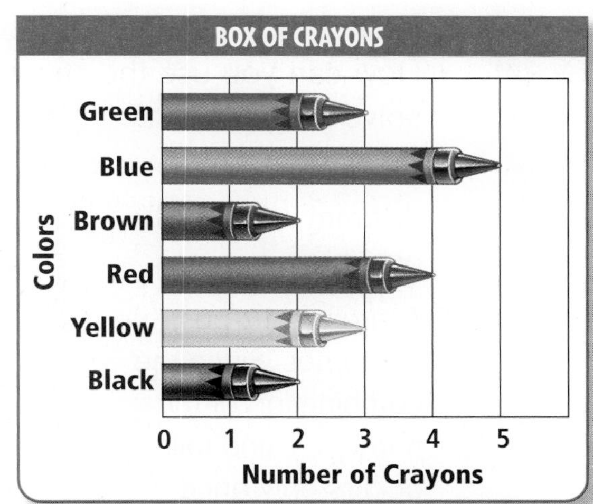

14. How many of each colored crayon are in the box?
15. **Write a problem** in which you compare two probabilities.

Mixed Review and Test Prep

16. Order 23,546; 24,546; 23,456; and 23,564 from least to greatest. (p. 10)
17. $3{,}456 \times 24$ (p. 150)
18. $72{,}891 \times 58$ (p. 150)
19. Write 0.25 as a fraction. (p. 280)
20. **TEST PREP** For lunch, Bob can have a ham or turkey sandwich and a soda, juice, or milk. How many choices does he have for lunch? (p. 572)

A 2 **C** 6
B 5 **D** 8

Extra Practice page H61, Set C

LESSON 5

Problem Solving Strategy

Make an Organized List

Quick Review

Write each probability as a fraction.

1. 2 out of 7 **2.** 4 out of 9
3. 1 out of 6 **4.** 2 out of 8
5. 3 out of 5

PROBLEM Selena is conducting a probability experiment. She will roll a number cube labeled 1 to 6 and toss a coin. What is the probability of rolling a 1 on the number cube and having the coin land tails up?

- What are you asked to find?
- What information will you use?

- What strategy can you use to solve the problem?

You can *make an organized list* of all the possible outcomes.

- How can you use the strategy to solve the problem?

You can make a tree diagram to list all the possible outcomes.

There are 12 possible outcomes. Since the combination of the number 1 and tails occurs only once, the probability is 1 out of 12, or $\frac{1}{12}$.

cube	coin	outcome
1	heads	1 and heads
	tails	1 and tails
2	heads	2 and heads
	tails	2 and tails
3	heads	3 and heads
	tails	3 and tails
4	heads	4 and heads
	tails	4 and tails
5	heads	5 and heads
	tails	5 and tails
6	heads	6 and heads
	tails	6 and tails

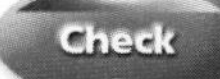

- Look back at the problem. Does the answer make sense?

CALIFORNIA STANDARDS MR 2.3 Use a variety of methods, such as words, numbers, symbols, charts, graphs, tables, diagrams, and models, to explain mathematical reasoning. **MR 1.1** Analyze problems by identifying relationships, distinguishing relevant from irrelevant information, sequencing and prioritizing information, and observing patterns. *also* **MR 1.0, MR 2.6, MR 3.2, MR 3.3**

▶ Problem Solving Practice

Make an organized list to solve.

1. **What if** Selena added a spinner that was half red and half yellow to her experiment? What probability would she have of getting 1, tails, and red?

2. Diana has a spinner, divided into 5 equal sections numbered 1 to 5, along with a coin. If she spins the pointer and tosses the coin, what is the probability that the pointer will land on 4 and the coin will be heads?

- Draw a Diagram or Picture
- Make a Model or Act It Out
- ▶ Make an Organized List
- Find a Pattern
- Make a Table or Graph
- Predict and Test
- Work Backward
- Solve a Simpler Problem
- Write an Equation
- Use Logical Reasoning

Marcus is conducting a probability experiment by tossing a coin and taking 1 marble from the bag below. Marcus will replace the marble after each turn.

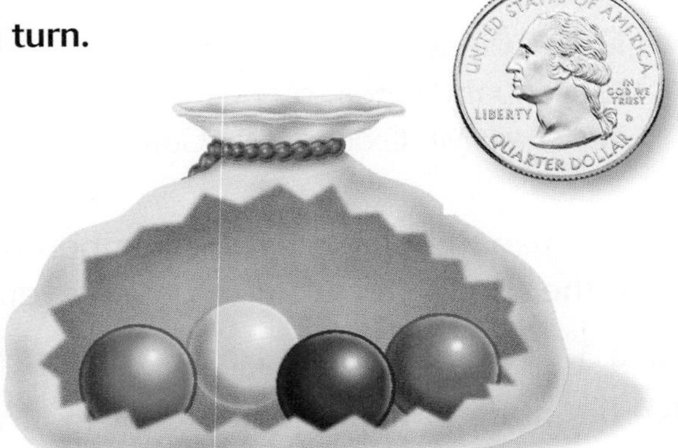

3. How many possible outcomes are there for this experiment?

   A 4 **B** 6 **C** 8 **D** 16

4. What is the probability that Marcus will choose a red marble and toss heads?

   F $\frac{1}{2}$ **G** $\frac{1}{4}$ **H** $\frac{1}{6}$ **J** $\frac{1}{8}$

Mixed Strategy Practice

5. Sam stacks 4 cubes, one on top of another, to make a plant stand. He paints the outside of the stand blue, but not the bottom. How many faces of the original cubes are painted?

6. After school, Li rode his bike to the library. The ride took 10 minutes. He stayed there for 2 hours and 10 minutes. Riding home took another 35 minutes. If Li got home at 5:15 P.M., at what time did he leave school?

7. Alexandra has \$0.30 more than Barbara. Together they have \$4.50. How much money does each girl have?

8. Marshal has 15¢. How many combinations of coins could he have? What are they?

9. Ben is making a sandbox for his brother. He has 12 ft of wood for the sides. He wants the sandbox to have the greatest area possible. What are the dimensions and area of the sandbox?

10. The drawing shows 1-square-foot tiles around a fish pond. What is the area that will be tiled?

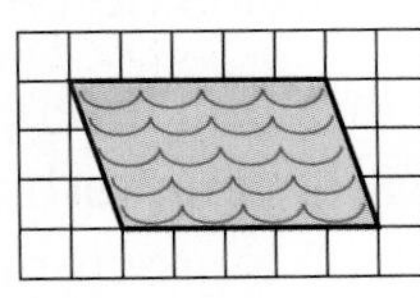

Let 1 square = 1 ft^2

CHAPTER 30

Review/Test

CHECK VOCABULARY AND CONCEPTS

Choose the best term from the box.

equally likely
possible outcomes
probability
tree diagram

1. The chance that an event will happen is _?_. (p. 574)
2. You can use a _?_ to show all the possible outcomes of an event. (p. 572)
3. All the results that have an equal chance of happening are _?_. (p. 570)

CHECK SKILLS

Find the number of choices. (pp. 572–573)

4. **Music Class Choices**
   Class: band, choir, orchestra
   Teacher: Mr. Gibson, Ms. Garcia

5. **Pizza Choices** (1 topping)
   Crust: thick, thin
   Toppings: mushrooms, pepperoni, onions, anchovies

For 6–8, use the number cube numbered 1 to 6. Write the probability of rolling each event. (pp. 574–575)

6. a 5
7. either 1 or 6
8. an even number

Use the number cube numbered 1 to 6. Write the probability of each event. Tell which event is more likely. (pp. 574–577)

9. a number less than 3; an odd number
10. a number greater than 1; a number less than 6
11. a prime number; a composite number

CHECK PROBLEM SOLVING

Solve. (pp. 578–579)

12. Ben has 25¢, but he does not have any pennies. Find the possible combinations of coins he has.
13. Adam tosses 2 coins. What are the possible outcomes? Determine the probability of tossing exactly 2 heads.
14. In a bag, there are 5 blue, 2 purple, and 3 green marbles. Are you more likely to pull a green marble or a blue one? Explain.
15. A bag contains 20 marbles. There are red marbles and blue marbles. The probability of pulling a red marble is $\frac{3}{5}$. Are you more likely to pull a red marble or a blue marble? Explain.

Cumulative Review

Get the information you need.
See item **7**.

Think about the meaning of *most likely.* Compare the numbers on the cards with each answer choice to find which choice is most likely to occur when one of these cards is chosen.

Also see problem **6**, p. H65.

For 1–9, choose the best answer.

1. What is the value of 3^4?

   A 9   C 27
   B 12   D 81

2. Which outcome is *certain* using this spinner?

   F landing on 1 or 5
   G landing on 1, 3, or 5
   H landing on an odd number
   J landing on 3 or 7

3. What is the volume of this box?

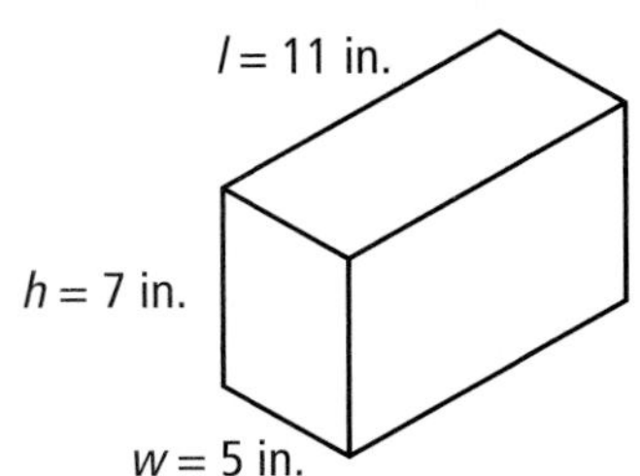

   A 23 in.^{3} C 77 in.^{3}
   B 35 in.^{3} D 385 in.^{3}

4. A jar contains 22 pink, 18 green, 14 blue, and 20 purple jelly beans. Which color of jelly bean is most likely to be pulled out of the jar at random?

   F pink   H blue
   G green   J purple

5. What is the probability of picking a red marble out of a bag that contains 6 red, 4 blue, and 10 yellow marbles?

   A $\frac{3}{10}$ C $\frac{6}{14}$
   B $\frac{3}{7}$ D $\frac{2}{10}$

6. $6.417 \div 2.3$

   F 0.279   H 2.8
   G 2.79   J Not here

7. The cards shown are placed face down.

2	7	13	20	52	37

   If you choose one of these cards, which outcome is most likely?

   A It has a number greater than 20.
   B It has a number less than 20.
   C It has a two-digit number.
   D It has a one-digit number.

8. How many different outfits of one shirt and one pair of pants can Ted make if he packs 4 shirts and 2 pairs of pants?

   F 8   H 6
   G 7   J Not here

9. What is the probability that a fair coin will land with tails up when it is tossed?

   A $\frac{1}{5}$ C $\frac{1}{2}$
   B $\frac{2}{5}$ D $\frac{7}{10}$

UNIT 9

MATH DETECTIVE

Side by Side

Use what you know about probability to determine how the event on the left compares to the event on the right. Is the probability of the event on the left greater than, less than, or equal to the probability of the event on the right? Use your detective skills and mental math skills rather than paper and pencil to determine the answers.

A number cube labeled 1 to 6.

	LEFT	RIGHT
1.	rolling an even number	rolling an odd number
2.	rolling a prime number	rolling a composite number
3.	rolling a 5 or 6	rolling a 1, 2, 3, or 4

	LEFT	RIGHT
4.	pulling a yellow marble	pulling a red or blue marble
5.	not pulling a red marble	not pulling a blue marble
6.	pulling a blue marble	pulling a green marble

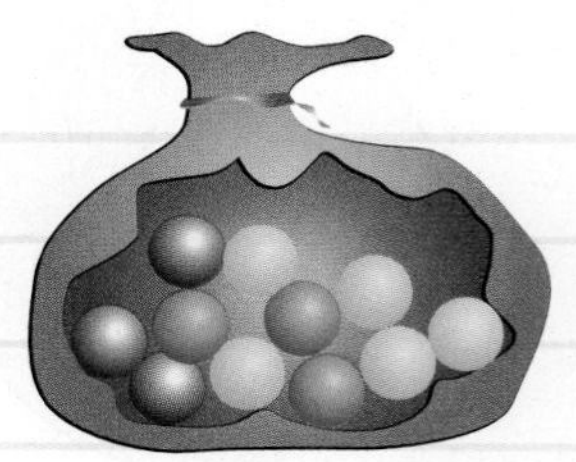

	LEFT	RIGHT
7.	landing on blue	landing on orange
8.	landing on blue or orange	landing on white or green
9.	not landing on green	not landing on white

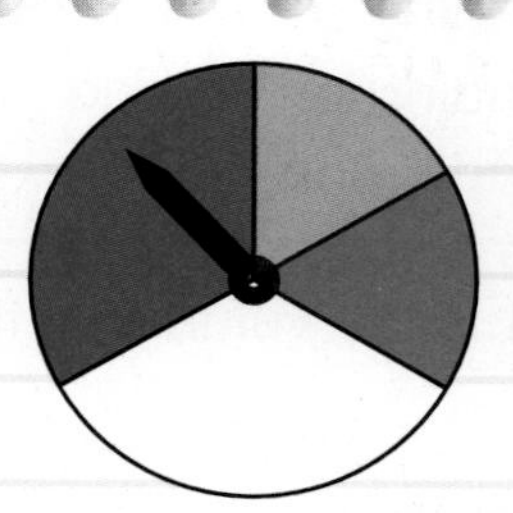

Think It Over!

- What is the probability of randomly selecting a problem from the ones above in which ">" is the answer? in which "<" is the answer? in which "=" is the answer?
- **Write a problem** like those above comparing the probability of two outcomes of a roll of the number cube.

UNIT 9

Challenge

Fairness

Number Fun is a two-player game played with a number cube labeled 1 to 6. Each player scores a point when his or her type of number is rolled.

Rules

Player 1 scores 1 point if the number is even.

Player 2 scores 1 point if the number is a multiple of 3.

Is this game fair?

You can use what you know about probability to see if this game is fair. A game is **fair** if each player has the same chance of winning.

Example

Player 1 scores if the number is even.

Even numbers on the number cube are 2, 4, and 6.

Probability of tossing an even number $= \frac{3}{6}$.

Player 2 scores if the number is a multiple of 3.

Multiples of 3 on the number cube are 3 and 6.

Probability of tossing a multiple of 3 $= \frac{2}{6}$.

Since $\frac{3}{6} > \frac{2}{6}$, the game is unfair. Player 1's chances of winning are greater than Player 2's.

Talk About It

- In order for a game to be fair, how should the probabilities of winning compare?

Try It

Use the same number cube but different rules. Tell whether the game is *fair* or *unfair*. Explain your answer.

1. Player 1 wins if the number is even.
   Player 2 wins if the number is odd.
2. Player 1 wins if the number is a multiple of 4.
   Player 2 wins if the number is prime.

Study Guide and Review

VOCABULARY

Choose the best term from the box.

1. A diagram showing all the possible outcomes is a __?__ diagram. (p. 572)
2. If each outcome has the same chance of happening, these outcomes are __?__. (p. 574)
3. The chance that an event will happen is the __?__. (p. 574)

probability
tree
certainty
unlikely
Venn
equally likely

STUDY AND SOLVE

Chapter 30

Predict outcomes of probability experiments.

What are the possible outcomes for tossing a coin?

heads and tails

Write the possible outcomes. (pp. 570–571)

4. Rolling a number cube labeled 1 to 6
5. Pulling a marble from a bag containing 10 yellow marbles and 3 red marbles
6. Randomly selecting a day of the week

Predict number of outcomes of probability experiments.

For summer vacation, Kerri's family can go to the beach, the mountains, or a theme park. They can travel by plane, train, or automobile. How many different trips can they take?

To find the number of choices, make a tree diagram or multiply the number of locations by the number of ways of travel.

locations × ways of travel = number of choices

↓ ↓ ↓

3 × 3 = 9 choices

So, they can take 9 different trips.

Find the number of choices. (pp. 572–573)

7. **Main course:** steak, chicken, fish
   Dessert: pie, cake
8. **Shirt:** blue, white, gray, beige
   Tie: striped, solid, print
9. Two theaters are each showing a western, a drama, and a comedy at 5:00, 7:00, and 9:00.

Write a probability as a fraction.

A bag has 3 green, 1 red, and 2 blue marbles. Find the probability of picking a red marble from the bag.

The probability of an event = $\frac{\text{number of favorable outcomes}}{\text{total number of possible outcomes}}$

Probability of red = $\frac{\text{1 red marble}}{\text{6 total marbles}} = \frac{1}{6}$

Compare probabilities.

Which event is more likely? You pull a yellow marble; you pull a green marble.

Probability of yellow = $\frac{2}{20} = \frac{1}{10}$

Probability of green = $\frac{5}{20} = \frac{1}{4}$

Compare. $\frac{1}{4} > \frac{1}{10}$

So, pulling a green marble is more likely.

Write a fraction for the probability of the pointer landing on blue. (pp. 574–575)

10. 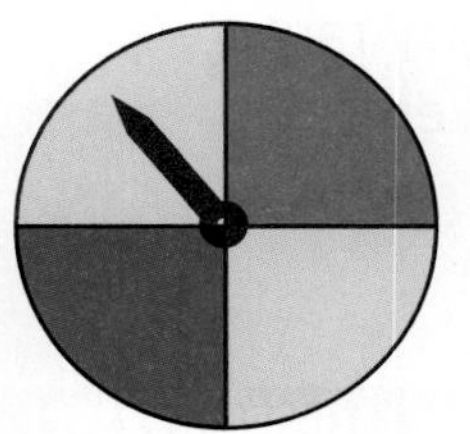

11. 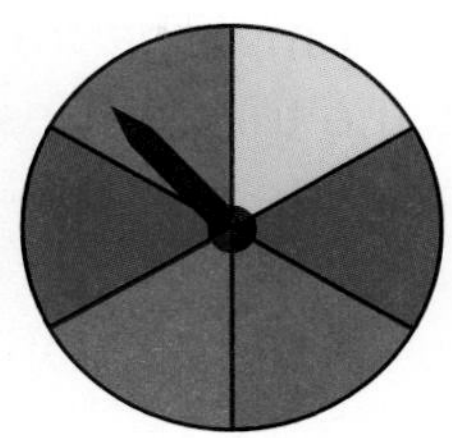

Write a fraction for the probability of rolling each number on a number cube labeled 1 to 6. (pp. 574–575)

12. an even number

13. a number less than 5

For 14–17, use the bag of marbles. Write each probability as a fraction. Tell which event is more likely. (pp. 576–577)

14. You pull a green marble; you pull a blue marble.

15. You pull a red marble; you pull an orange marble.

16. You pull a green or blue marble; you pull a marble that is not blue, yellow, or red.

17. You pull a red or an orange marble; you pull a blue marble.

PROBLEM SOLVING PRACTICE

Solve. (pp. 578–579)

18. Trey is going to spin the pointer in problem 10 and toss a coin. What is the probability that the pointer will land on blue and the coin will land heads up?

19. Trevor has 3 coins in his pocket. Each is either a nickel or a dime. What are the different amounts of money that he could have with these 3 coins?

California Connections

TRAVEL

Tourism is important to California's economy. The travel industry ranks as the third-largest employer in the state.

California is a favorite travel destination for both business travelers and vacationers.▼

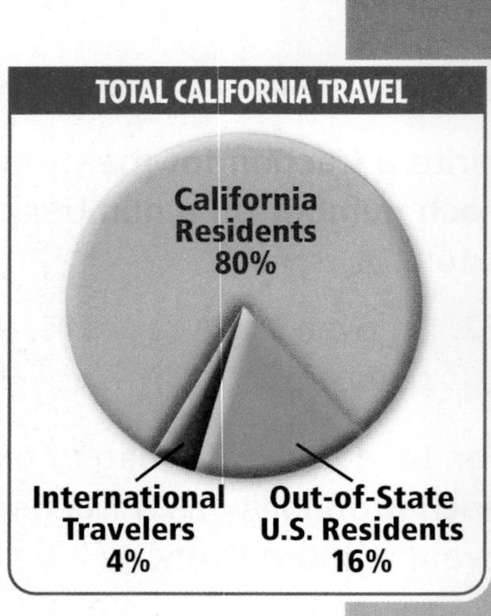

USE DATA For 1–3, use the circle graph. Suppose a person on a trip in California is chosen at random. Classify each event as *impossible, certain, likely,* or *unlikely*.

1. The person is from California.
2. The person is an international traveler.
3. The person is either a United States traveler or an international traveler.
4. **What if** a family visiting California decides to choose one favorite adult's activity and one favorite children's activity? Make a tree diagram to show the possible choices.
5. People travel to California for leisure and for business. The probability that a randomly selected trip is for business is $\frac{3}{10}$. What is the probability that the trip is for leisure?

FAVORITE ACTIVITIES OF TRAVELERS*

Adults	Children
Shopping	Theme Parks
Outdoor Activities	Beach
Museums	Water Parks

*based on California Tourism Research

VISITING SAN DIEGO AND LOS ANGELES

San Diego and Los Angeles, two of the most visited cities in California, are only 125 miles apart. It is possible to visit an attraction in San Diego and an attraction in Los Angeles in the same day!

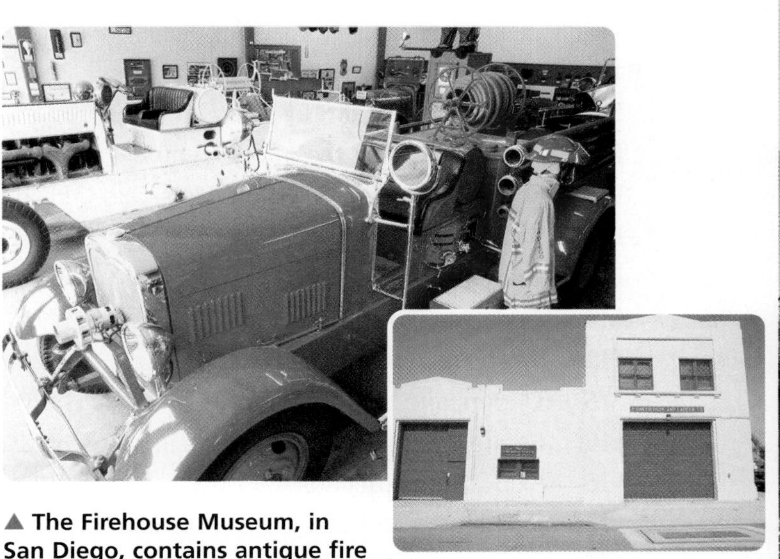

▲ The Firehouse Museum, in San Diego, contains antique fire equipment, an old steamer, helmets, and other items from around the world.

The Angels Flight Railway is an inclined cable railway that provides a good view of downtown Los Angeles. ▼

Suppose each member in a family of six writes the name of a favorite Los Angeles attraction on a slip of paper.

USE DATA For 1–2, use the slips of paper at the right. Write each probability as a fraction. Tell which event is more likely.

1. You pull a slip with Hollywood. You pull a slip with Music Center.
2. You pull a slip with California Science Center. You pull a slip that does not have California Science Center.
3. **What if** each member of a travel club writes either Zoo or Firehouse Museum as the sight he or she wants to see in San Diego? The members put the slips in a bag. The probability of pulling Zoo is $\frac{3}{4}$. Are they more likely to pull Zoo or Firehouse Museum? Explain.
4. Suppose you decide to visit one of the Los Angeles attractions and one of the San Diego attractions suggested in Exercises 1–3. How many possible choices do you have? Explain.

STUDENT HANDBOOK

PREREQUISITE SKILLS REVIEW Do you have the math skills needed to start a new chapter? Use this list of skills to review and remember your skills from last year.

TROUBLESHOOTING

PLACE VALUE

You can use a place-value chart to understand the value of each digit.

THOUSANDS			ONES		
Hundreds	Tens	Ones	Hundreds	Tens	Ones
5	2	0,	6	4	9

Examples

A What is the value of the digit 6 in the place-value chart?

The 6 is in the hundreds place.

So, its value is 6 × 100, or 600.

B What is the value of the digit 2 in the place-value chart?

The 2 is in the ten thousands place.

So, its value is 2 × 10,000, or 20,000.

▶ Practice

Write the value of the blue digit.

1. 340 **2.** 1,522 **3.** 75,811 **4.** 291,063

5. 872 **6.** 9,548 **7.** 15,433 **8.** 680,472

READ AND WRITE WHOLE NUMBERS

Numbers can be written in three forms: standard form, expanded form, and word form.

Example

Write the number in standard form, expanded form, and word form.

THOUSANDS			ONES		
Hundreds	Tens	Ones	Hundreds	Tens	Ones
4	0	5,	6	2	7

Standard form: 405,627

Expanded form: 400,000 + 5,000 + 600 + 20 + 7

Word form: four hundred five thousand, six hundred twenty-seven

▶ Practice

Write the number in standard form.

1. three hundred fifty thousand, four **2.** 30,000 + 600 + 90 + 7

Write the number in word form.

3. 211 **4.** 9,403 **5.** 105,078 **6.** 78,050

Write the number in expanded form.

7. 326,047 **8.** 93,117 **9.** 600,489 **10.** 10,320

11. 123,006 **12.** 47,021 **13.** 216,313 **14.** 523,410

READ AND WRITE DECIMALS

A decimal is a number with one or more digits to the right of the decimal point. A place-value chart can help you read and write decimals.

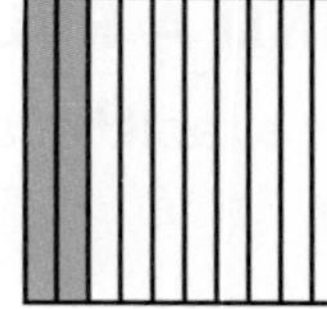

Write: 0.2
Read: two tenths

Hundreds	Tens	Ones	Tenths	Hundredths
3	1	2	5	4

The decimal point is read as "and."

Standard form: 312.54

Expanded form: 300 + 10 + 2 + 0.5 + 0.04

Word form: three hundred twelve and fifty-four hundredths

► Practice

Write as a decimal.

1. 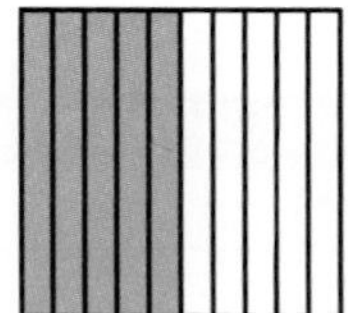

2. 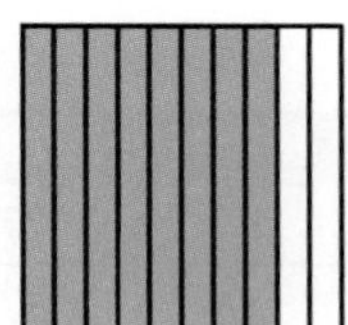

3. 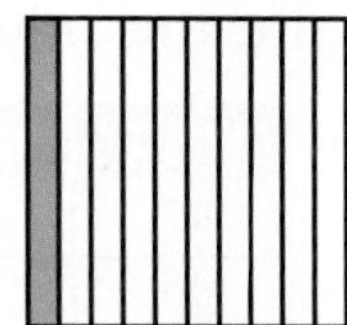

Write the number in standard form.

4. seventy-five and eight hundredths

5. 100 + 20 + 6 + 0.5 + 0.03

ROUNDING WHOLE NUMBERS

You can round numbers to the nearest 10 or 100.

Example

Round 8,356 to the nearest hundred.

You can use the rounding rules.

Think: The digit in the rounding place is 3.
The digit to its right is 5.

rounding place

8,356 Since 5 = 5 or more, the hundreds digit increases by 1.

So, 8,356 rounds to 8,400.

> ***REMEMBER***
>
> To round to a given place value, look at the digit to the right of the place.
>
> - If the digit is less than 5, the digit in the rounding place stays the same.
> - If the digit is 5 or more, the digit in the rounding place increases by 1.

► Practice

Round each number to the nearest ten.

1. 74 **2.** 318 **3.** 29 **4.** 531 **5.** 5,738

Round each number to the nearest hundred.

6. 328 **7.** 6,905 **8.** 21,280 **9.** 4,093 **10.** 15,650

TROUBLESHOOTING

MENTAL MATH: FIND A RULE

You can find a rule for a table of input and output values.

Examples

A

INPUT	OUTPUT
50	45
40	35
30	25
20	15

Think: The output value is 5 less than the input value.

So, the rule is subtract 5.

B

INPUT	OUTPUT
10	22
20	32
30	42
40	52

Think: The output value is 12 more than the input value.

So, the rule is add 12.

▶ Practice

Write the rule. Use mental math.

1.

INPUT	OUTPUT
12	8
22	18
32	28
42	38

2.

INPUT	OUTPUT
17	27
27	37
37	47
47	57

3.

INPUT	OUTPUT
45	70
50	75
55	80
60	85

SUBTRACTION ACROSS ZEROS

When you subtract, you may need to regroup more than one place value.

Subtract. 600 − 337

STEP 1

Look at the ones place. Since 7 > 0, regroup.

```
  6 0 0
− 3 3 7
```

STEP 2

Regroup 6 hundreds as 5 hundreds 9 tens 10 ones. Subtract the ones.

```
    9
  5 10 10
  6̸ 0̸ 0̸
− 3 3 7
      3
```

STEP 3

Subtract the tens and the hundreds.

```
    9
  5 10 10
  6̸ 0̸ 0̸
− 3 3 7
  2 6 3
```

▶ Practice

Subtract. Regroup when necessary.

1. 50 − 33

2. 600 − 462

3. 400 − 105

4. 1,500 − 281

5. 5,300 − 2,453

MENTAL MATH STRATEGIES: DECIMALS

You can add and subtract decimals mentally. Look for decimal sums that are whole numbers.

Examples

A Add. 5.7 + 0.3

Think: 0.7 + 0.3 = 1

So, 5.7 + 0.3 = 5 + 1, or 6.0.

B Subtract. 2 − 0.6

Think: What number added to 0.6 is 1? Since 0.6 + 0.4 = 1, 1 − 0.6 = 0.4.

So, 2 − 0.6 = 1.4.

▶ Practice

Add or subtract. Use mental math.

1. 2.5 + 0.5 **2.** 4 − 0.6 **3.** 4 + 0.4 **4.** 8 − 0.4

5. 5.6 + 0.4 **6.** 0.8 + 7.2 **7.** 6 − 0.8 **8.** 2 − 0.9

ADD AND SUBTRACT DECIMALS

You can use decimal squares divided into tenths to help you add and subtract decimals.

Examples

A Find the sum. 1.6 + 0.3

Shade 1.6 decimal squares red. Then, shade 0.3 of the second square blue. Count the shaded parts to find the sum.

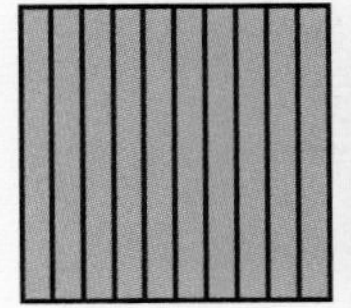

Record:

$$\begin{array}{r} 1.6 \\ +\ 0.3 \\ \hline 1.9 \end{array}$$

So, 1.6 + 0.3 = 1.9.

B Find the difference. 2.0 − 0.7

Shade 2 decimal squares red. Then, take 0.7 away. Count to find what is left.

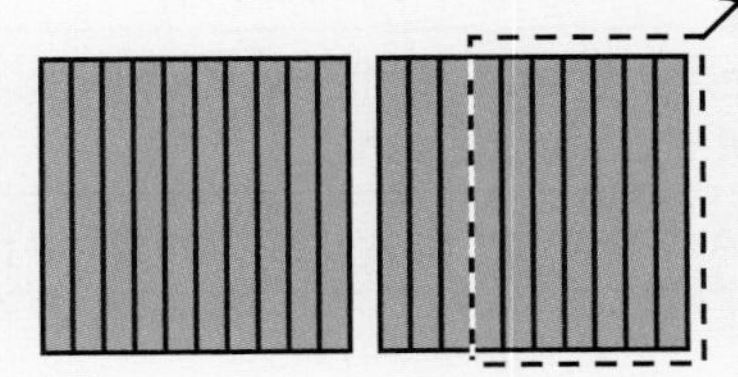

Record:

$$\begin{array}{r} 2.0 \\ -\ 0.7 \\ \hline 1.3 \end{array}$$

So, 2.0 − 0.7 = 1.3.

▶ Practice

Add or subtract. Regroup when necessary.

1. $\begin{array}{r} 3.4 \\ +\ 0.5 \\ \hline \end{array}$ **2.** $\begin{array}{r} 4.8 \\ +\ 2.5 \\ \hline \end{array}$ **3.** $\begin{array}{r} 6.0 \\ -\ 4.8 \\ \hline \end{array}$ **4.** $\begin{array}{r} 7.0 \\ -\ 2.1 \\ \hline \end{array}$ **5.** $\begin{array}{r} 2.3 \\ +\ 5.9 \\ \hline \end{array}$

6. $\begin{array}{r} 7.8 \\ -\ 0.5 \\ \hline \end{array}$ **7.** $\begin{array}{r} 20.0 \\ -\ 8.7 \\ \hline \end{array}$ **8.** $\begin{array}{r} 12.6 \\ +\ 5.7 \\ \hline \end{array}$ **9.** $\begin{array}{r} 17.7 \\ -\ 2.6 \\ \hline \end{array}$ **10.** $\begin{array}{r} 13.0 \\ -\ 8.2 \\ \hline \end{array}$

TROUBLESHOOTING

ADDITION AND SUBTRACTION

You can use n for a missing number in an equation. To solve these equations, find the value of n by finding the sum or difference.

Examples

Write the value of n.

A $15 + 6 = n$

$$\begin{array}{r} 15 \\ +\ 6 \\ \hline 21 \end{array}$$

So, $n = 21$.

B $n = 44 - 24$

$$\begin{array}{r} 44 \\ -\ 24 \\ \hline 20 \end{array}$$

So, $n = 20$.

▶ Practice

Write the value of n.

1. $15 + 3 = n$
2. $n = 9 + 12$
3. $7 + 6 + 9 = n$
4. $4 + 12 + 8 = n$
5. $32 + 18 = n$
6. $n = 51 + 17$
7. $n = 19 - 7$
8. $28 - 11 = n$
9. $45 - 13 = n$
10. $n = 70 - 15$
11. $36 - 13 = n$
12. $n = 62 - 12$

ADDITION PROPERTIES

The Addition Properties can help you find a sum mentally.

PROPERTY	EXAMPLE
A. **Order Property:** Changing the order in which you add two numbers does not change the sum.	$7 + 4 = 11$ and $4 + 7 = 11$
B. **Grouping Property:** Changing the way addends are grouped does not change the sum.	$(2 + 5) + 6 = 7 + 6 = 13$ and $2 + (5 + 6) = 2 + 11 = 13$
C. **Zero Property:** Any number plus 0 is that same number.	$8 + 0 = 8$ and $0 + 8 = 8$

▶ Practice

Write the letter of the addition property used in each equation.

1. $0 + 12 = 12$
2. $15 + 4 = 4 + 15$
3. $35 + 0 = 35$
4. $(4 + 5) + 9 = 4 + (5 + 9)$
5. $22 + 6 = 6 + 22$
6. $7 + (4 + 1) = (7 + 4) + 1$
7. $9 + 2 = 2 + 9$
8. $0 + 100 = 100$
9. $(5 + 2) + 8 = 5 + (2 + 8)$
10. $2 + (6 + 4) = (2 + 6) + 4$
11. $10 + 6 = 6 + 10$
12. $23 + 0 = 23$

MENTAL MATH: FUNCTION TABLES

To complete a function table, use the rule and the input number.

Example

Complete the function table.

Rule: Subtract 5.

Input	12	14	16	18
Output	7	9	11	■

Think: $18 - 5 = 13$

So, the output for 18 is 13.

▶ Practice

Copy and complete the table. Use mental math.

1. Rule: Add 11.

Input	2	4	6	8	10
Output	13	15	■	■	■

2. Rule: Subtract 6.

Input	13	18	23	28	33
Output	7	12	■	■	■

3. Rule: Subtract 9.

Input	50	40	30	20	10
Output	41	31	■	■	■

4. Rule: Add 5.

Input	10	15	20	25	30
Output	15	20	■	■	■

MULTIPLICATION FACTS THROUGH 12

Learning to skip-count can help you remember the multiplication facts. The number line below shows skip-counting by twelves.

Example

Find the value of n.

$5 \times 12 = n$

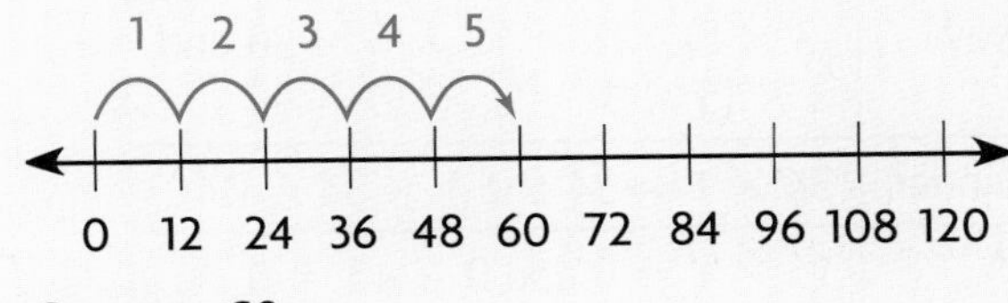

So, $n = 60$.

▶ Practice

Find the value of n.

1. $8 \times 11 = n$ **2.** $9 \times 8 = n$ **3.** $n = 4 \times 7$ **4.** $6 \times 12 = n$

5. $4 \times 9 = n$ **6.** $n = 8 \times 7$ **7.** $12 \times 8 = n$ **8.** $7 \times 7 = n$

9. $n = 10 \times 11$ **10.** $n = 9 \times 12$ **11.** $6 \times 9 = n$ **12.** $12 \times 12 = n$

TROUBLESHOOTING

USE PARENTHESES

Parentheses show you what operations to do first.

Examples

Choose the expression that has a value of 20.

A $(5 \times 12) - 8$ Multiply first.

$60 - 8$ Then subtract.

52

B $5 \times (12 - 8)$ Subtract first.

5×4 Then multiply.

20

So, the expression in Example B has a value of 20.

▶ Practice

Choose the expression that has the given value.

1. 38
- **a.** $7 + (12 \times 2)$
- **b.** $(7 + 12) \times 2$

2. 9
- **a.** $(12 - 9) \div 3$
- **b.** $12 - (9 \div 3)$

3. 13
- **a.** $25 - (2 \times 6)$
- **b.** $(25 - 2) \times 6$

4. 20
- **a.** $(15 + 25) \div 5$
- **b.** $15 + (25 \div 5)$

5. 44
- **a.** $(21 \times 2) + 2$
- **b.** $21 \times (2 + 2)$

6. 6
- **a.** $18 \div (6 - 3)$
- **b.** $(18 \div 6) - 3$

7. 160
- **a.** $45 - (5 \times 4)$
- **b.** $(45 - 5) \times 4$

8. 23
- **a.** $9 + (7 \times 2)$
- **b.** $(9 + 7) \times 2$

MULTIPLICATION PROPERTIES

The Multiplication Properties can help you find a product mentally.

PROPERTY	EXAMPLE
A. Order Property: Factors can be multiplied in any order without changing the product.	$6 \times 4 = 24$ and $4 \times 6 = 24$
B. Grouping Property: Factors can be grouped in any way without changing the product.	$(3 \times 5) \times 6 = 15 \times 6 = 90$ and $3 \times (5 \times 6) = 3 \times 30 = 90$
C. Property of One: Any number times 1 equals that number.	$34 \times 1 = 34$ and $1 \times 34 = 34$
D. Zero Property: Any number times 0 is 0.	$15 \times 0 = 0$ and $0 \times 15 = 0$

▶ Practice

Write the letter of the multiplication property used in each equation.

1. $(4 \times 5) \times 7 = 4 \times (5 \times 7)$ **2.** $78 \times 0 = 0$ **3.** $39 \times 6 = 6 \times 39$

4. $12 \times 5 = 5 \times 12$ **5.** $25 = 1 \times 25$ **6.** $2 \times (3 \times 9) = (2 \times 3) \times 9$

7. $47 \times 1 = 47$ **8.** $0 \times 32 = 0$ **9.** $14 \times 7 = 7 \times 14$

FREQUENCY TABLES

A frequency table uses numbers to show how often something happens. This table shows how often different students at Lakewood School joined teams. Each student is on only one team.

To read a frequency table, read the heading for each column. Then for each group, look at the number to find the number of times each event happened.

LAKEWOOD SCHOOL	
Team	**Frequency**
Tennis	16
Swimming	40
Football	35
Basketball	15

▶ Practice

For 1–4, use the frequency table.

1. How many students joined the football team?
2. How many students joined either the tennis or the football team?
3. What is the total number of students who joined a team?
4. Which teams had fewer than 20 students?

READ GRAPHS

Bar graphs use bars to stand for data. To read a bar graph, find the length of each bar on the scale.

Example

This bar graph tells you the types of sandwiches sold at a deli. It also tells how many of each type were sold.

▶ Practice

For 1–4, use the bar graph.

1. How many types of sandwiches did the deli sell?
2. Which sandwich did the deli sell the most of?
3. How many more salami sandwiches were sold than roast beef?
4. A ham sandwich costs $3.50. How much money did the deli earn by selling ham sandwiches?

TROUBLESHOOTING

SKIP-COUNT ON A NUMBER LINE

You can use a number line to skip-count or to complete a pattern.

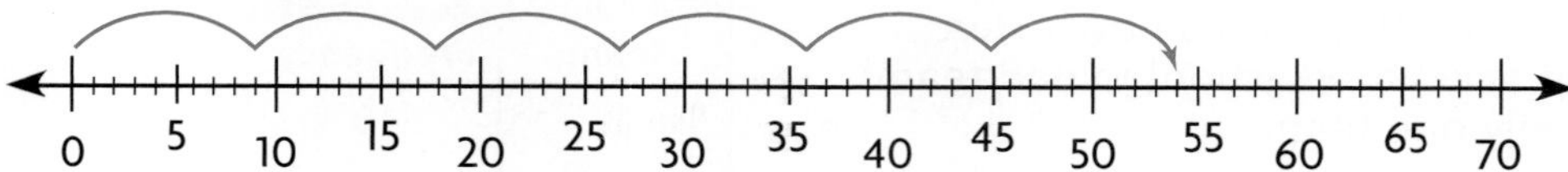

Example

Complete the pattern.

0, 9, 18, 27, ■, ■, 54

Think: Use the number line.
Skip-count by nines.

So, 36 and 45 complete the pattern.

▶ Practice

Complete the pattern.

1. 0, 5, 10, 15, 20, ■, ■, ■, 40
2. 0, 11, 22, ■, ■, 55, ■

List the numbers you would count.

3. Count by twelves to 60.
4. Count by threes to 30.

GRAPH ORDERED PAIRS

You can use an ordered pair to locate points on a grid.
The first number tells you how far to move horizontally.
The second number tells you how far to move vertically.

Example

- Point *A* names the ordered pair (1,4).
- Point *B* names the ordered pair (5,3).
- Point *C* names the ordered pair (2,0).

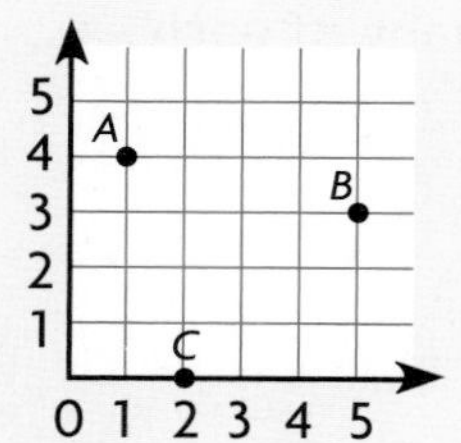

▶ Practice

Use the ordered pair to name the point on the grid.

1. (8,8)
2. (5,0)
3. (0,9)
4. (4,3)
5. (3,7)
6. (7,3)
7. (1,4)
8. (5,10)

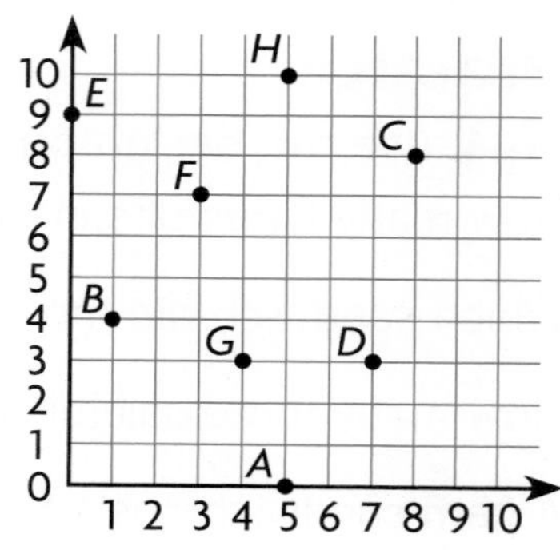

MULTIPLY BY TENS AND HUNDREDS

You can use basic facts and a pattern to multiply by tens and hundreds.

Examples

Find the product.

A $6 \times 8 = 48$
$6 \times 80 = \blacksquare$
$6 \times 800 = 4{,}800$
So, the product is 480.

B $7 \times 9 = 63$
$7 \times 90 = 630$
$7 \times 900 = \blacksquare$
So, the product is 6,300.

▶ Practice

Find the product.

1. 3×10 **2.** 6×10 **3.** 4×100 **4.** 8×100

5. 4×20 **6.** 3×30 **7.** 2×300 **8.** 5×200

9. 8×40 **10.** 7×60 **11.** 4×900 **12.** 8×800

MULTIPLY 3 DIGITS BY 1 DIGIT

When multiplying a 3-digit number by a 1-digit number, remember to multiply ones, tens, and then hundreds.

Find the product. 6×247

STEP 1

Multiply the ones.
6×7 ones $= 42$ ones
Regroup 42 ones as
4 tens 2 ones.

$$\begin{array}{r} {}^{4} \\ 247 \\ \times\ \ 6 \\ \hline 2 \end{array}$$

STEP 2

Multiply the tens.
6×4 tens $= 24$ tens
Add the regrouped tens.
Regroup 28 tens as
2 hundreds 8 tens.

$$\begin{array}{r} {}^{2\,4} \\ 247 \\ \times\ \ 6 \\ \hline 82 \end{array}$$

STEP 3

Multiply the hundreds.
6×2 hundreds $= 12$ hundreds
Add the regrouped hundreds.
Regroup 14 hundreds as
1 thousand 4 hundreds.

$$\begin{array}{r} {}^{2\,4} \\ 247 \\ \times\ \ 6 \\ \hline 1{,}482 \end{array}$$

▶ Practice

Find the product.

1. $\begin{array}{r} 128 \\ \times\ 5 \\ \hline \end{array}$ **2.** $\begin{array}{r} 263 \\ \times\ 4 \\ \hline \end{array}$ **3.** $\begin{array}{r} 526 \\ \times\ 8 \\ \hline \end{array}$ **4.** $\begin{array}{r} 814 \\ \times\ 9 \\ \hline \end{array}$ **5.** $\begin{array}{r} 681 \\ \times\ 7 \\ \hline \end{array}$

6. $\begin{array}{r} 499 \\ \times\ 6 \\ \hline \end{array}$ **7.** $\begin{array}{r} 322 \\ \times\ 7 \\ \hline \end{array}$ **8.** $\begin{array}{r} 711 \\ \times\ 6 \\ \hline \end{array}$ **9.** $\begin{array}{r} 187 \\ \times\ 4 \\ \hline \end{array}$ **10.** $\begin{array}{r} 503 \\ \times\ 2 \\ \hline \end{array}$

TROUBLESHOOTING

REPEATED ADDITION OF DECIMALS

You can use skip-counting on a number line to find the sum of repeated decimals.

Example

Add. 0.4 + 0.4 + 0.4 + 0.4

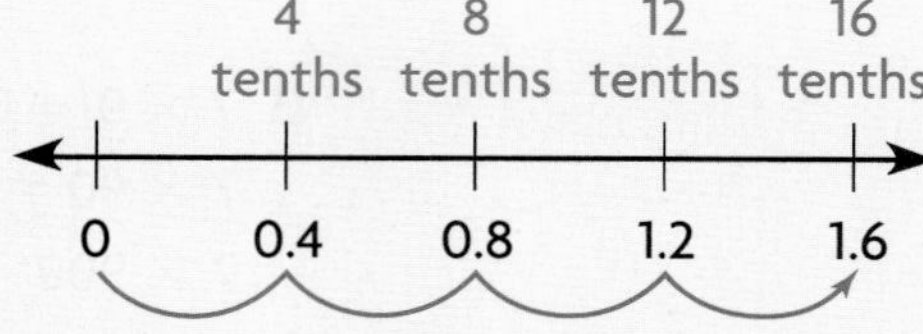

So, 4 tenths + 4 tenths + 4 tenths + 4 tenths = 16 tenths, or 1.6.

▶ Practice

Find the sum.

1. 0.3 + 0.3 **2.** 0.9 + 0.9 + 0.9 **3.** 0.5 + 0.5 + 0.5

4. 0.2 + 0.2 + 0.2 **5.** 0.8 + 0.8 **6.** 0.4 + 0.4 + 0.4

7. 1.1 + 1.1 **8.** 1.2 + 1.2 + 1.2 **9.** 0.8 + 0.8 + 0.8

10. 0.7 + 0.7 **11.** 0.6 + 0.6 + 0.6 **12.** 0.7 + 0.7 + 0.7

MULTIPLY MONEY

To multiply money, you could model by using coins.

Example

Multiply. 5 × $0.05

ONE WAY

Count 5 nickels.

$0.05 $0.10 $0.15 $0.20 $0.25

ANOTHER WAY

$$\begin{array}{r} {}^{2} \\ \$0.05 \\ \times \quad 5 \\ \hline \$0.25 \end{array}$$

▶ Practice

Find the product.

1. 4 × $0.10 **2.** 4 × $0.05 **3.** 10 × $0.25 **4.** 4 × $0.50

5. 12 × $0.01 **6.** 5 × $0.25 **7.** 7 × $0.10 **8.** 8 × $0.05

9. $\begin{array}{r} \$0.05 \\ \times \quad 12 \\ \hline \end{array}$ **10.** $\begin{array}{r} \$0.50 \\ \times \quad 7 \\ \hline \end{array}$ **11.** $\begin{array}{r} \$0.10 \\ \times \quad 5 \\ \hline \end{array}$ **12.** $\begin{array}{r} \$0.25 \\ \times \quad 12 \\ \hline \end{array}$

DIVIDE 2 DIGITS BY 1 DIGIT

You can use basic facts to divide greater numbers.

Divide. $6\overline{)79}$

STEP 1

Divide the 7 tens.

$$\begin{array}{r} 1 \\ 6\overline{)79} \\ -\ 6\downarrow \\ \hline 19 \end{array}$$

Divide. $7 \div 6$
Multiply. 6×1
Subtract. $7 - 6$
Compare. $1 < 6$
Bring down the 9 ones.

STEP 2

Divide the 19 ones.

$$\begin{array}{r} 13 \\ 6\overline{)79} \\ -\ 6 \\ \hline 19 \\ -\ 18 \\ \hline 1 \end{array}$$

Divide. $19 \div 6$
Multiply. 6×3
Subtract. $19 - 18$
Compare. $1 < 6$

STEP 3

Write the quotient and remainder.

$$\begin{array}{r} 13 \text{ r1} \\ 6\overline{)79} \\ -\ 6 \\ \hline 19 \\ -\ 18 \\ \hline 1 \end{array}$$

▶ Practice

Divide.

1. $5\overline{)29}$
2. $4\overline{)90}$
3. $9\overline{)48}$
4. $4\overline{)97}$
5. $7\overline{)84}$
6. $4\overline{)29}$
7. $4\overline{)76}$
8. $3\overline{)31}$
9. $5\overline{)54}$
10. $4\overline{)81}$

CHECK DIVISION

You can check that a quotient is correct by multiplying.

Example

Check that the quotient is correct. $\begin{array}{r} 13 \text{ r1} \\ 5\overline{)66} \end{array}$

Think: Does 66 equal 5 groups of 13 with 1 left over?

$5 \times 13 = 65$ Multiply the divisor by the quotient.

$65 + 1 = 66$ Add the remainder.

Since 66 equals the dividend, the answer checks.

▶ Practice

Check the division by using multiplication.

1. $\begin{array}{r} 19 \\ 3\overline{)57} \end{array}$
2. $\begin{array}{r} 13 \text{ r2} \\ 4\overline{)54} \end{array}$
3. $\begin{array}{r} 15 \text{ r5} \\ 6\overline{)95} \end{array}$
4. $\begin{array}{r} 6 \text{ r3} \\ 5\overline{)33} \end{array}$
5. $\begin{array}{r} 11 \text{ r4} \\ 8\overline{)92} \end{array}$
6. $\begin{array}{r} 7 \text{ r5} \\ 9\overline{)68} \end{array}$
7. $\begin{array}{r} 22 \text{ r3} \\ 4\overline{)91} \end{array}$
8. $\begin{array}{r} 8 \text{ r4} \\ 7\overline{)60} \end{array}$
9. $\begin{array}{r} 7 \text{ r6} \\ 8\overline{)62} \end{array}$
10. $\begin{array}{r} 11 \text{ r4} \\ 6\overline{)70} \end{array}$

TROUBLESHOOTING

MULTIPLY BY 2-DIGIT NUMBERS

An estimate can help you check that a product is reasonable.

Find the product. 23 × 58

Estimate. 20 × 60 = 1,200

STEP 1

Multiply by the ones.

$$\begin{array}{r} {}^{2} \\ 58 \\ \times\ 23 \\ \hline 174 \end{array} \leftarrow 3 \times 58$$

STEP 2

Multiply by the tens.

$$\begin{array}{r} \not{2}^{1} \\ 58 \\ \times\ 23 \\ \hline 174 \\ 1{,}160 \end{array} \leftarrow 20 \times 58$$

STEP 3

Add the products.

$$\begin{array}{r} \not{2}^{1} \\ 58 \\ \times\ 23 \\ \hline 174 \\ +\ 1{,}160 \\ \hline 1{,}334 \end{array}$$

Since 1,334 is close to 1,200, the product is reasonable.

▶ Practice

Find the product. Estimate to check.

1. $\begin{array}{r} 18 \\ \times\ 92 \\ \hline \end{array}$ **2.** $\begin{array}{r} 189 \\ \times\ 51 \\ \hline \end{array}$ **3.** $\begin{array}{r} 447 \\ \times\ 26 \\ \hline \end{array}$ **4.** $\begin{array}{r} 64 \\ \times\ 75 \\ \hline \end{array}$

5. $\begin{array}{r} 1{,}250 \\ \times\ 23 \\ \hline \end{array}$ **6.** $\begin{array}{r} 949 \\ \times\ 54 \\ \hline \end{array}$ **7.** $\begin{array}{r} 3{,}726 \\ \times\ 62 \\ \hline \end{array}$ **8.** $\begin{array}{r} 8{,}409 \\ \times\ 31 \\ \hline \end{array}$

DIVIDE BY 10

To divide any number by 10, look for a pattern.

Examples

Find the quotient.

A 50 ÷ 10 = 5
500 ÷ 10 = 50
5,000 ÷ 10 = ■
So, the quotient is 500.

B 180 ÷ 10 = 18
1,800 ÷ 10 = ■
18,000 ÷ 10 = 1,800
So, the quotient is 180.

▶ Practice

Divide.

1. 50 ÷ 10 **2.** 90 ÷ 10 **3.** 240 ÷ 10 **4.** 190 ÷ 10 **5.** 6,000 ÷ 10

6. 980 ÷ 10 **7.** 3,600 ÷ 10 **8.** 6,420 ÷ 10 **9.** 3,370 ÷ 10 **10.** 4,020 ÷ 10

DIVIDE BY 1-DIGIT NUMBERS

In division problems, use place value to place the first digit in the quotient.

Divide. $108 \div 6$

STEP 1

Decide where to place the first digit in the quotient.

$6\overline{)108}$ $1 < 6$, so look at the tens.

$6\overline{)108}$ $10 > 6$, so place the first digit in the tens place.

STEP 2

Divide the tens.

$$\begin{array}{r} 1 \\ 6\overline{)108} \\ -\ 6 \\ \hline 4 \end{array}$$

Divide. $6\overline{)10}$
Multiply. 6×1
Subtract. $10 - 6$
Compare. $4 < 6$

STEP 3

Bring down the 8 ones.

$$\begin{array}{r} 18 \\ 6\overline{)108} \\ -\ 6\downarrow \\ \hline 48 \\ -\ 48 \\ \hline 0 \end{array}$$

Divide. $6\overline{)48}$
Multiply. 6×8
Subtract. $48 - 48$
Compare. $0 < 6$

▶ Practice

Divide.

1. $5\overline{)25}$
2. $4\overline{)144}$
3. $6\overline{)192}$
4. $3\overline{)75}$
5. $8\overline{)176}$
6. $126 \div 6$
7. $186 \div 3$
8. $322 \div 7$
9. $95 \div 5$
10. $212 \div 4$
11. $732 \div 3$
12. $530 \div 5$
13. $968 \div 8$
14. $810 \div 9$
15. $496 \div 8$

DIVISION PATTERNS

You can use basic facts and patterns to divide by multiples of 10.

Example

$48{,}000 \div 40 = n$

$$\begin{aligned} 48 \div 4 &= 12 \\ 480 \div 40 &= 12 \\ 4{,}800 \div 40 &= 120 \\ 48{,}000 \div 40 &= 1{,}200 \end{aligned}$$

Think: Use the basic fact. As the number of zeros in the dividend increases, the number of zeros in the quotient increases.

So, $n = 1{,}200$.

▶ Practice

Complete the pattern.

1. $160 \div 20 = 8$
   $1{,}600 \div 20 = 80$
   $16{,}000 \div 20 = n$
2. $720 \div 60 = n$
   $7{,}200 \div 60 = 120$
   $72{,}000 \div 60 = 1{,}200$
3. $450 \div 50 = 9$
   $4{,}500 \div 50 = n$
   $45{,}000 \div 50 = 900$
4. $560 \div 80 = 7$
   $5{,}600 \div 80 = 70$
   $56{,}000 \div 80 = n$
5. $810 \div 90 = n$
   $8{,}100 \div 90 = 90$
   $81{,}000 \div 90 = 900$
6. $210 \div 70 = 3$
   $2{,}100 \div 70 = n$
   $21{,}000 \div 70 = 300$

TROUBLESHOOTING

EQUIVALENT DECIMALS

Equivalent decimals are decimals that name the same number or amount. You can write zeros to the right of the last digit after the decimal point. This does not change the value.

Example

0.3

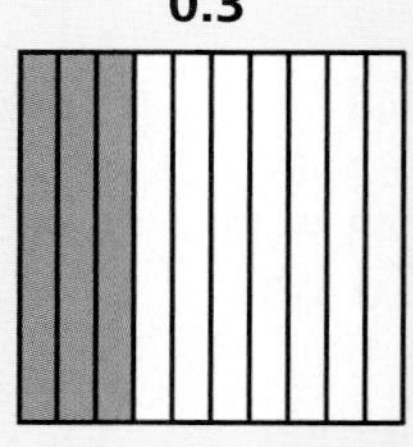

three tenths

3 out of 10

0.30

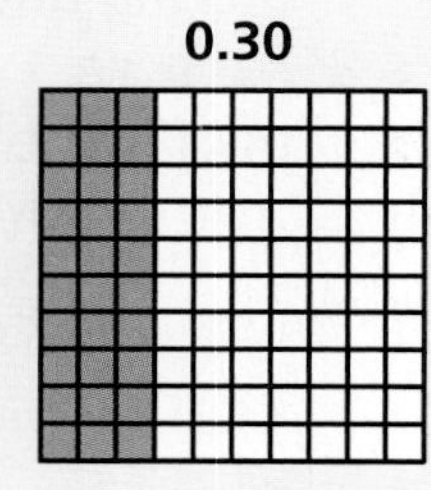

thirty hundredths

30 out of 100

▶ Practice

Write two equivalent decimals for each.

1. 0.4 **2.** 7.8 **3.** 2.50 **4.** 0.95 **5.** 4.08

6. 0.75 **7.** 0.3 **8.** 1.7 **9.** 12.4 **10.** 9.06

RELATED FACTS

The fact families for multiplication and division can have either two or four related number sentences.

A multiplication family with two different factors has four related facts. This fact family is for 7, 4, and 28.

$7 \times 4 = 28$

$4 \times 7 = 28$

$28 \div 4 = 7$

$28 \div 7 = 4$

A multiplication family with two factors that are the same has only two related facts. This fact family is for 12, 12, and 144.

$12 \times 12 = 144$

$144 \div 12 = 12$

▶ Practice

For each multiplication fact, write a related division fact.

1. $6 \times 8 = 48$ **2.** $5 \times 4 = 20$ **3.** $6 \times 6 = 36$ **4.** $8 \times 7 = 56$ **5.** $11 \times 7 = 77$

6. $9 \times 9 = 81$ **7.** $6 \times 12 = 72$ **8.** $9 \times 11 = 99$ **9.** $4 \times 6 = 24$ **10.** $4 \times 4 = 16$

MULTIPLY BY 10 AND BY 100

To multiply a decimal by 10 or 100, think about patterns when you multiply whole numbers by 10 or 100.

Look for patterns in the placement of the decimal point.

Examples Find the product.

A $5.2 \times 1 = 5.2$
$5.2 \times 10 = 52$
$5.2 \times 100 = 520$

B $0.12 \times 1 = 0.12$
$0.12 \times 10 = 1.2$
$0.12 \times 100 = 12$

▶ Practice

Find the product.

1. 4×10
4×100

2. 2.7×10
2.7×100

3. 9.22×10
9.22×100

4. 53.5×10
53.5×100

5. 70×10
70×100

6. 0.96×10
0.96×100

7. 0.1×10
0.1×100

8. 0.04×10
0.04×100

DIVIDE DECIMALS BY WHOLE NUMBERS

You can use an estimate to help you place the decimal point in a quotient.

Divide. $56.4 \div 6$

STEP 1

Estimate the quotient.
Use compatible numbers.

Think:
$56.4 \div 6$ → $54 \div 6 = 9$
$56.4 \div 6$ → $60 \div 6 = 10$

So, $56.4 \div 9$ is between 9 and 10.

STEP 2

Divide as with whole numbers.

```
    94
 6)56.4
 - 54
    2 4
  - 2 4
      0
```

STEP 3

Use the estimate to place the decimal point.

```
   9.4
 6)56.4
 - 54
    2 4
  - 2 4
      0
```

▶ Practice

Find the quotient.

1. $8\overline{)27.2}$

2. $7\overline{)3.15}$

3. $4\overline{)14.08}$

4. $3\overline{)19.23}$

5. $37.0 \div 5$

6. $5.04 \div 9$

7. $6.18 \div 6$

8. $419.2 \div 8$

9. $14.48 \div 4$

10. $4.30 \div 5$

11. $43.4 \div 7$

12. $120.6 \div 6$

TROUBLESHOOTING

MULTIPLICATION AND DIVISION FACTS

You can use arrays to help you remember multiplication and division facts.

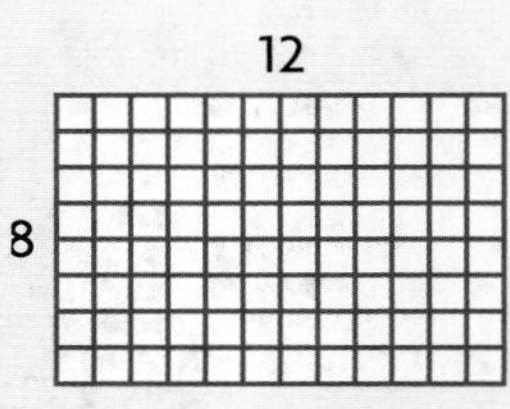

$8 \times 12 = 96$ **Think:** 8 rows of 12

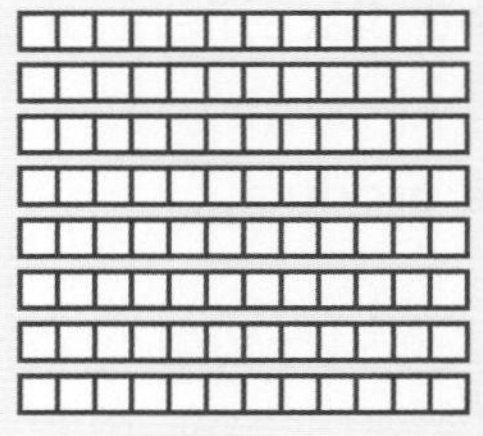

$96 \div 12 = 8$ **Think:** Divide 96 into 8 rows of 12 in each row.

▶ Practice

Find the product.

1. 11×6 **2.** 8×9 **3.** 12×7 **4.** 12×5 **5.** 8×8

6. 8×7 **7.** 9×6 **8.** 9×10 **9.** 9×7 **10.** 12×9

Find the quotient.

11. $60 \div 12$ **12.** $70 \div 10$ **13.** $56 \div 8$ **14.** $54 \div 9$ **15.** $72 \div 8$

16. $110 \div 11$ **17.** $99 \div 11$ **18.** $45 \div 9$ **19.** $48 \div 8$ **20.** $120 \div 10$

FACTORS

A **factor** is a number multiplied by another number to find a product. To find all the factors of a number, divide it by each of the numbers 1, 2, 3, ... Stop when you find repeated factors.

Example

Find all the factors of 40.

$40 = 1 \times 40$
$\quad 2 \times 20$
$\quad$ 3 is not a factor.
$\quad 4 \times 10$
$\quad 5 \times 8$
$\quad$ 6 is not a factor.
$\quad$ 7 is not a factor.
$\quad 8 \times 5$ repeated factors

So, all the factors of 40 are: 1, 2, 4, 5, 8, 10, 20, and 40.

▶ Practice

Write all the factors for each number.

1. 6 **2.** 16 **3.** 20 **4.** 33 **5.** 49

6. 15 **7.** 28 **8.** 24 **9.** 42 **10.** 30

11. 12 **12.** 9 **13.** 36 **14.** 25 **15.** 45

UNDERSTAND FRACTIONS

In the model, 3 out of 5 equal parts are shaded. The fraction $\frac{3}{5}$ represents the part of the figure that is shaded.

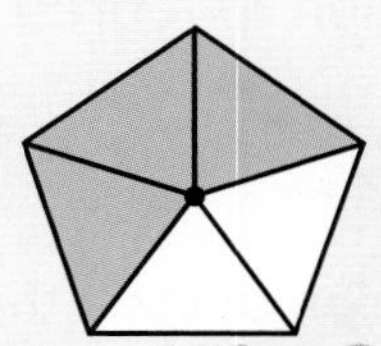

Write: $\frac{3}{5}$

Read: three fifths, or three out of five

▶ Practice

Write the fraction for the shaded part of each model.

1. 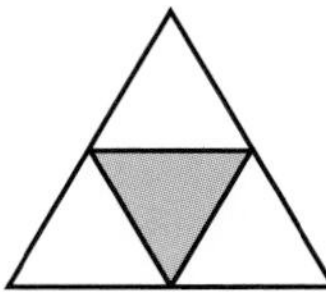

2. 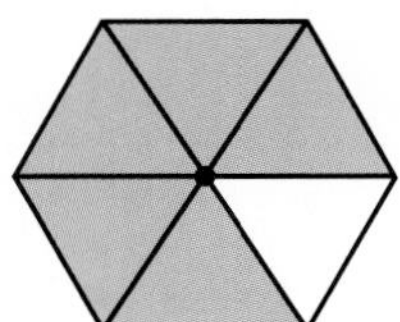

3. 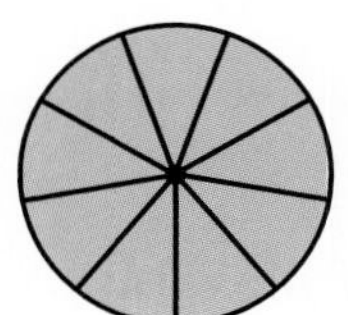

4. 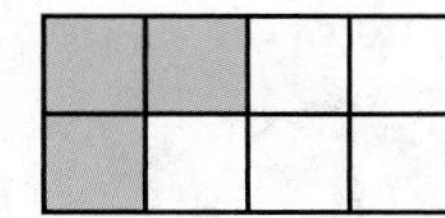

Write in words.

5. $\frac{1}{5}$
6. $\frac{3}{4}$
7. $\frac{3}{6}$
8. $\frac{7}{8}$
9. $\frac{4}{9}$

COMPARE FRACTIONS

You can use fractions bars to compare fractions.

Compare $\frac{1}{2}$ and $\frac{1}{3}$.

STEP 1

Start with the bar for 1. Line up the fraction bar for $\frac{1}{2}$.

STEP 2

Line up the fraction bar for $\frac{1}{3}$. Compare the fraction bars for $\frac{1}{2}$ and $\frac{1}{3}$. The longer bar represents the greater fraction.

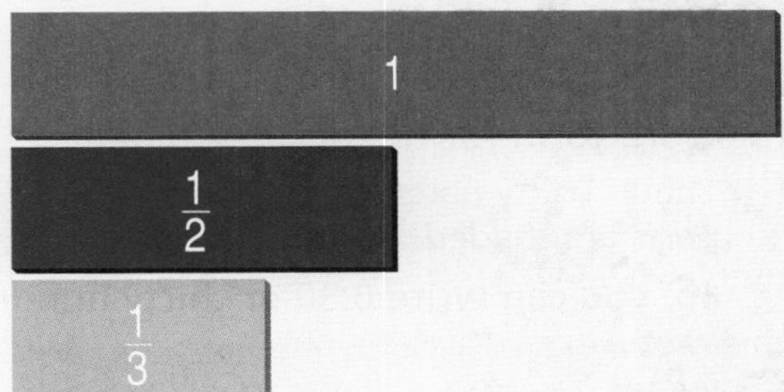

The fraction bar for $\frac{1}{2}$ is longer than the fraction bar for $\frac{1}{3}$. So, $\frac{1}{2} > \frac{1}{3}$.

▶ Practice

Compare the fractions. Write <, >, or = for each ●.

1. $\frac{4}{6}$ ● $\frac{2}{3}$
2. $\frac{1}{6}$ ● $\frac{1}{8}$
3. $\frac{3}{4}$ ● $\frac{4}{6}$
4. $\frac{1}{3}$ ● $\frac{7}{10}$
5. $\frac{6}{12}$ ● $\frac{1}{2}$
6. $\frac{3}{8}$ ● $\frac{1}{3}$
7. $\frac{7}{8}$ ● $\frac{8}{10}$
8. $\frac{1}{4}$ ● $\frac{1}{3}$

TROUBLESHOOTING

WRITE FRACTIONS

You can write a fraction to represent a part of a group.

Example

At the right, the fraction of stars that are red is $\frac{12}{18}$.

$\frac{12}{18} \leftarrow \frac{\text{number of red stars}}{\text{total number of stars}}$

You can write this fraction in simplest form. Divide the numerator and denominator by 6, the greatest common factor of 12 and 18.

$\frac{12}{18} = \frac{12 \div 6}{18 \div 6} = \frac{2}{3}$

▶ Practice

Write a fraction for the following.

1. 

What fraction of the squares are *not* green?

2. What fraction of the circles are red?

3. What fraction of the triangles are either red or yellow?

Write the fraction in simplest form.

4. $\frac{2}{16}$ **5.** $\frac{15}{30}$ **6.** $\frac{8}{48}$ **7.** $\frac{25}{55}$ **8.** $\frac{9}{9}$ **9.** $\frac{18}{27}$

UNDERSTAND HUNDREDTHS

You can use a decimal model divided into 100 equal parts to understand hundredths.

Examples

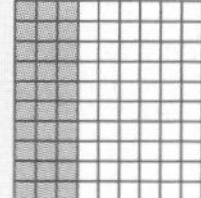

A Write the standard form and the word form for the decimal model.

Think: Thirty parts out of 100 equal parts are shaded.

So, you can write 0.30 or thirty hundredths.

B Write a fraction for the decimal model.

Think: The decimal 0.30 is 30 parts out of 100 parts.

So, $0.30 = \frac{30}{100}$.

▶ Practice

Write the standard form and the word form for the decimal model.

1. 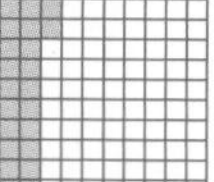

2. 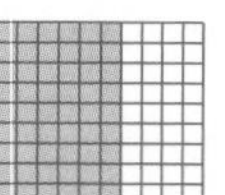

3. 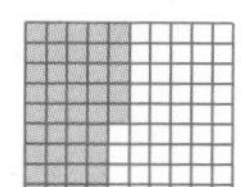

4. 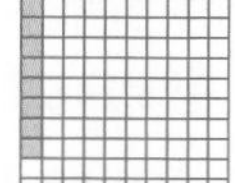

Write as a decimal and as a fraction.

5. seventeen hundredths

6. sixty-two hundredths

RELATE FRACTIONS AND DECIMALS

Fractions and decimals tell what part of a whole is being used.

Example

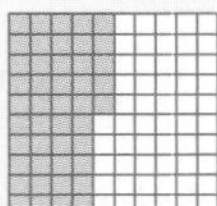

Write a decimal and a fraction for the model.

Think: Forty-five parts out of 100 equal parts are shaded.

This is written as 0.45 or $\frac{45}{100}$.

▶ Practice

Write a decimal and a fraction or a mixed number for each model.

1. 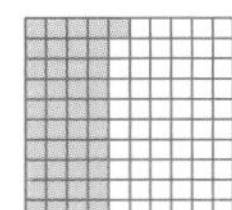

2. 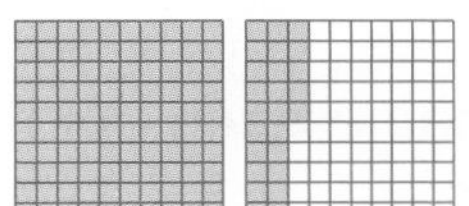

3. 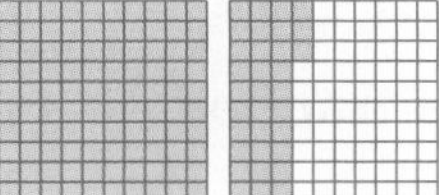

SKIP-COUNT FRACTIONS

You can use a number line to help you skip-count fractions.

Complete the pattern and write a rule for $2\frac{1}{6}$, ■, ■, $4\frac{1}{6}$, $4\frac{5}{6}$, ■.

STEP 1

Draw a number line from 2 to 6 with tick marks every $\frac{1}{6}$. Graph a point for $2\frac{1}{6}$, $4\frac{1}{6}$, and $4\frac{5}{6}$.

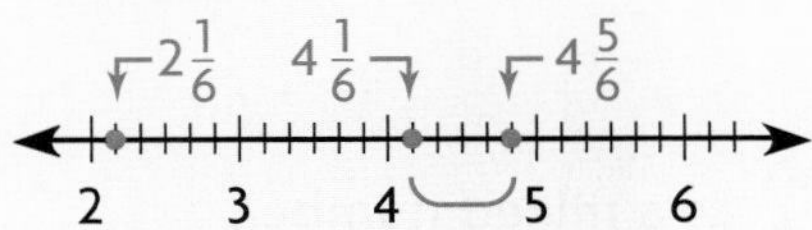

Then, look for a pattern.

Show a skip from $4\frac{1}{6}$ to $4\frac{5}{6}$.

Think: To skip from $4\frac{1}{6}$ to $4\frac{5}{6}$, you add $\frac{4}{6}$.

STEP 2

Starting with $2\frac{1}{6}$, show a skip by adding $\frac{4}{6}$.

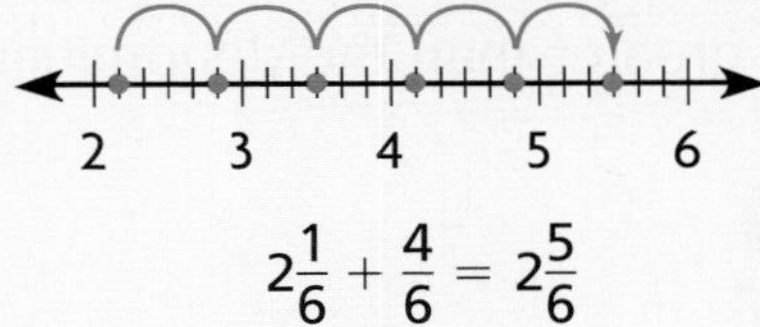

$$2\frac{1}{6} + \frac{4}{6} = 2\frac{5}{6}$$

Continue until you complete the pattern.

$2\frac{5}{6} + \frac{4}{6} = 3\frac{3}{6}$ $\quad$ $4\frac{5}{6} + \frac{4}{6} = 5\frac{3}{6}$

So, the missing numbers are $2\frac{5}{6}$, $3\frac{3}{6}$, and $5\frac{3}{6}$.

The rule is add $\frac{4}{6}$.

▶ Practice

Complete the pattern and write a rule.

1. $\frac{4}{5}$, 1, ■, ■, ■, $1\frac{4}{5}$

2. $3\frac{3}{8}$, $3\frac{4}{8}$, ■, ■, $3\frac{7}{8}$, ■

3. $1\frac{3}{10}$, ■, ■, ■, $1\frac{7}{10}$, $1\frac{8}{10}$

4. $2\frac{2}{9}$, ■, ■, ■, $2\frac{6}{9}$, $2\frac{7}{9}$

TROUBLESHOOTING

FRACTIONS ON A RULER

To name a fraction on a ruler, first decide whether the tick marks represent halves, fourths, or eighths.

Example

Write a fraction or a mixed number to name point *P*.

Point *P* is at the seventh tick mark after 1. So, the mixed number is $1\frac{7}{8}$.

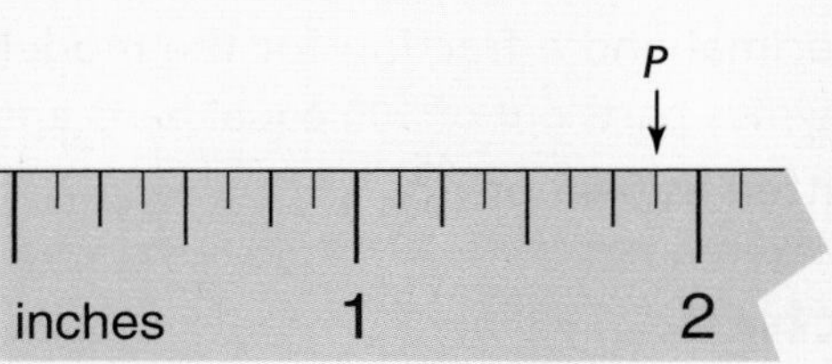

▶ Practice

Write a fraction or a mixed number to name each point.

1.

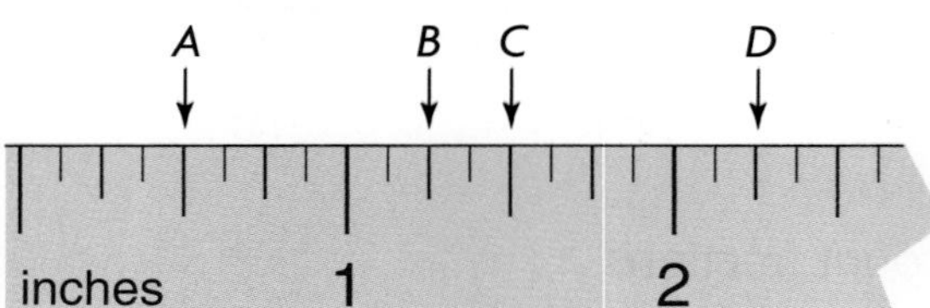

2.

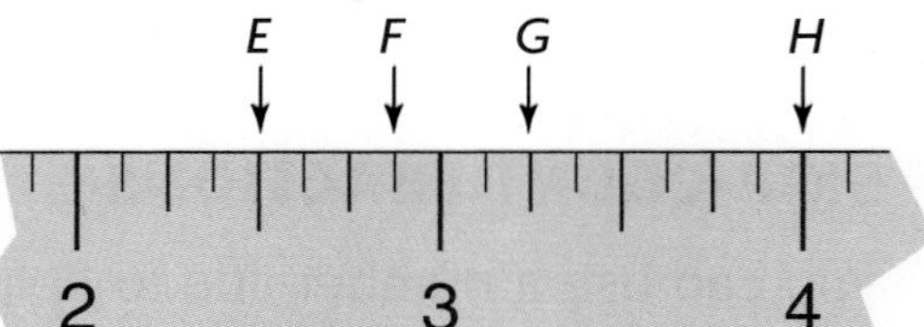

UNDERSTAND MIXED NUMBERS

You can write a mixed number as a fraction. You can also write a fraction as a mixed number when the numerator is greater than the denominator.

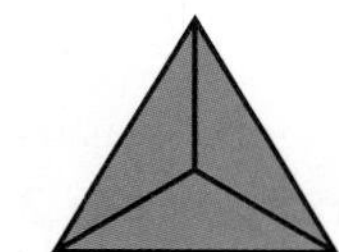
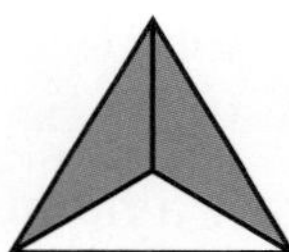

Read: one and two thirds **Write:** $1\frac{2}{3}$

Examples

A Rename $1\frac{1}{5}$ as a fraction.

Think: A whole divided into fifths has 5 equal parts.

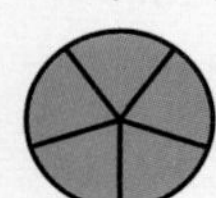
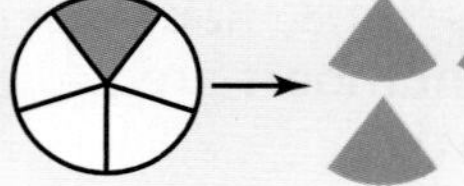
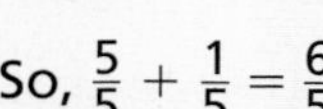

So, $\frac{5}{5} + \frac{1}{5} = \frac{6}{5}$.

B Rename $\frac{11}{6}$ as a mixed number.

Think: There are $\frac{6}{6}$ in one whole.

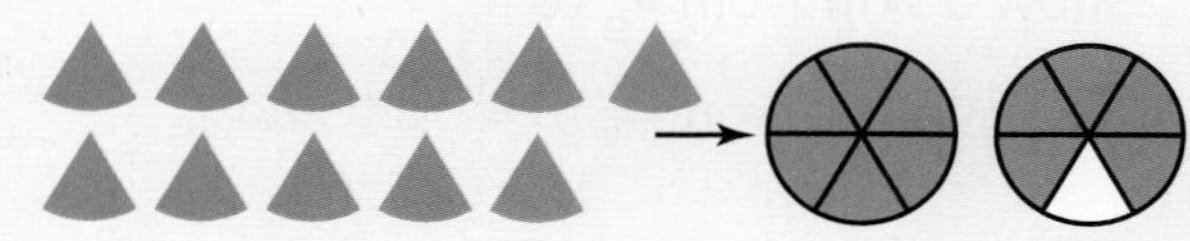

So, $\frac{6}{6} + \frac{5}{6} = 1\frac{5}{6}$.

▶ Practice

Rename each fraction as a mixed number.

1. $\frac{7}{2}$ **2.** $\frac{11}{8}$ **3.** $\frac{17}{6}$ **4.** $\frac{15}{4}$ **5.** $\frac{21}{10}$

Rename each mixed number as a fraction.

6. $2\frac{1}{3}$ **7.** $4\frac{3}{5}$ **8.** $3\frac{3}{7}$ **9.** $5\frac{5}{8}$ **10.** $1\frac{5}{12}$

ADD AND SUBTRACT FRACTIONS

To add and subtract fractions, you can use the least common denominator to change the fractions to like fractions.

Subtract. $\frac{3}{4} - \frac{1}{6}$

STEP 1

Find the least common denominator.

$$\begin{array}{r} \frac{3}{4} \\ -\frac{1}{6} \\ \hline \end{array}$$

Think: The LCD of fourths and sixths is twelfths.

STEP 2

Write equivalent fractions.

$$\begin{array}{rr} \frac{3 \times 3}{4 \times 3} = & \frac{9}{12} \\ -\frac{1 \times 2}{6 \times 2} = & -\frac{2}{12} \\ \hline \end{array}$$

STEP 3

Subtract. Simplify if necessary.

$$\begin{array}{rr} \frac{3 \times 3}{4 \times 3} = & \frac{9}{12} \\ -\frac{1 \times 2}{6 \times 2} = & -\frac{2}{12} \\ \hline & \frac{7}{12} \end{array}$$

▶ Practice

Find the sum or difference. Write the answer in simplest form.

1. $\frac{5}{8} - \frac{3}{8}$
2. $\frac{2}{3} - \frac{1}{2}$
3. $\frac{2}{5} + \frac{7}{10}$
4. $\frac{1}{4} + \frac{5}{12}$
5. $\frac{5}{12} - \frac{1}{6}$
6. $\frac{1}{2} - \frac{3}{8}$
7. $\frac{6}{7} - \frac{1}{3}$
8. $\frac{2}{3} + \frac{5}{9}$
9. $\frac{1}{3} + \frac{3}{4}$
10. $\frac{5}{6} - \frac{3}{4}$

FRACTIONS OF A WHOLE OR A GROUP

You can use fractions to describe parts of a whole or parts of a group.

Examples

A The circle is divided into eighths. Three parts are shaded.

So, the fraction for the shaded part is $\frac{3}{8}$.

B 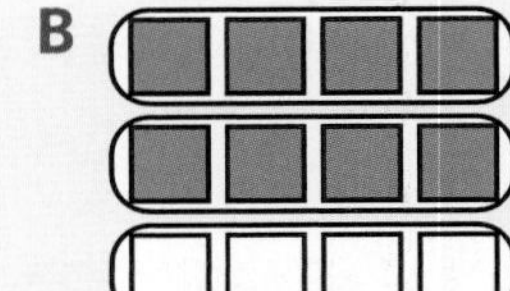

The group is divided into thirds. Two rows are shaded.

So, the fraction for the shaded part is $\frac{2}{3}$.

▶ Practice

Write a fraction to represent the shaded part.

1. 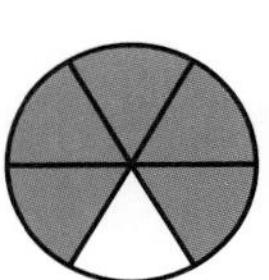
2. 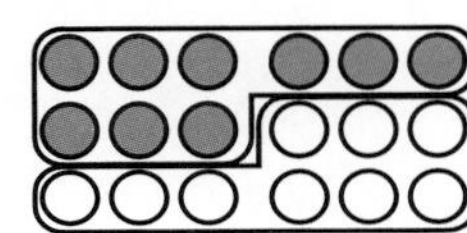
3. 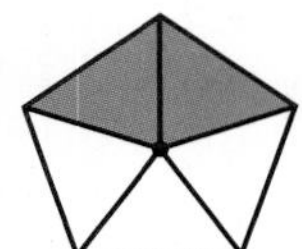
4. 
5. 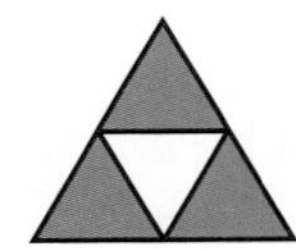
6. 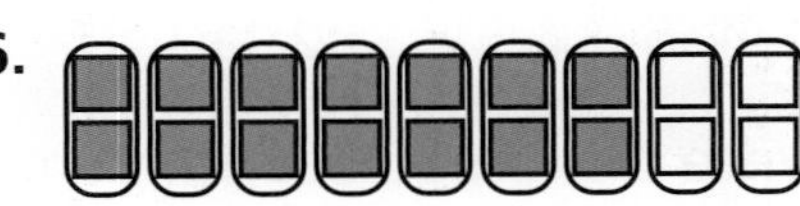

TROUBLESHOOTING

MODEL DECIMAL MULTIPLICATION

You can model decimal multiplication using decimal squares divided into hundredths.

Model the product. 0.3 × 0.2

STEP 1

Shade 3 rows yellow for 0.3.

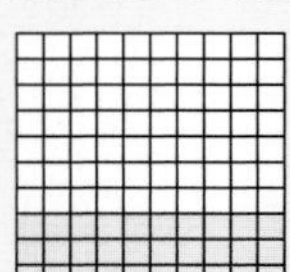

STEP 2

Shade 2 columns blue for 0.2.

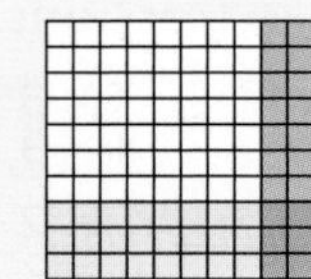

The area in which the shading overlaps shows the product, 0.06.

So, 0.3 × 0.2 = 0.06.

▶ Practice

Write the number sentence each model represents.

1. 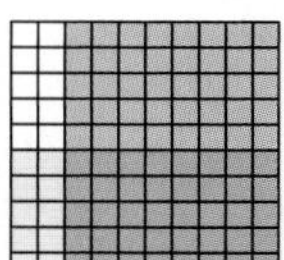

2. 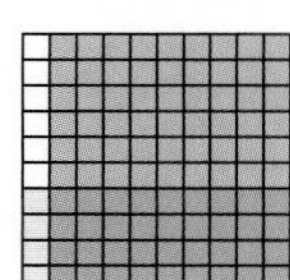

3. 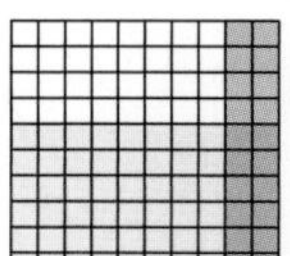

FIND THE GCF

The GCF of two numbers is their greatest common factor. This is the greatest number that is a factor of two or more numbers. Use these steps to find the GCF of two numbers.

Find the GCF of 6 and 8.

STEP 1

List the factors of each number.

Factors of 6: 1, 2, 3, 6

Factors of 8: 1, 2, 4, 8

STEP 2

Find the common factors, or the factors in both lists.

(1)	(2)	3	6
(1)	(2)	4	8

STEP 3

The greatest of the common factors is the GCF.

So, the GCF is 2.

▶ Practice

Write the common factors for each pair of numbers.

1. 4, 6 **2.** 12, 15 **3.** 12, 24 **4.** 10, 30 **5.** 15, 25

Write the greatest common factor for each pair of numbers.

6. 6, 10 **7.** 12, 16 **8.** 15, 18 **9.** 12, 30 **10.** 10, 15

COMPARE WHOLE NUMBERS

Compare whole numbers by using a number line or by comparing place values.

Examples

ONE WAY

Compare. 80 ● 88

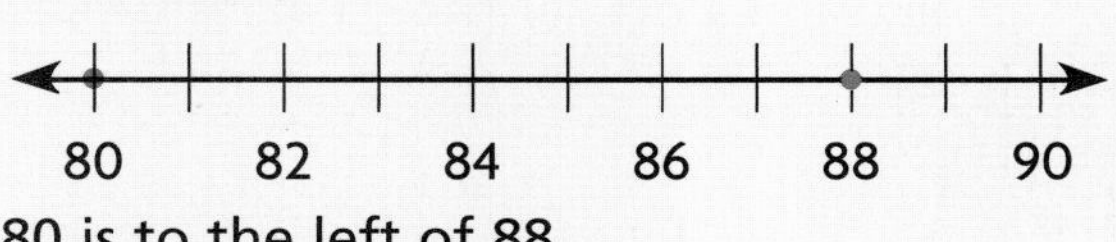

80 is to the left of 88.

So, $80 < 88$.

ANOTHER WAY

Compare. 225 ● 220 Use $<$, $>$, or $=$ for ●.

Compare the hundreds. 225 220
Hundreds are equal.

Compare the tens. 225 220
Tens are equal.

Compare the ones. 225 220
5 ones $>$ 0 ones.

So, $225 > 220$.

▶ Practice

Compare. Use $<$, $>$, or $=$ for each ●.

1. 48 ● 47 **2.** 7 ● 72 **3.** 39 ● 30 **4.** 201 ● 210 **5.** 111 ● 110

6. 67 ● 76 **7.** 24 ● 20 **8.** 92 ● 920 **9.** 425 ● 452 **10.** 162 ● 621

READ A THERMOMETER

A scale on a thermometer has both positive and negative numbers.

Example

Write the Fahrenheit temperature reading.

The red on the thermometer ends at the 15°F mark.

So, the temperature is 15°F.

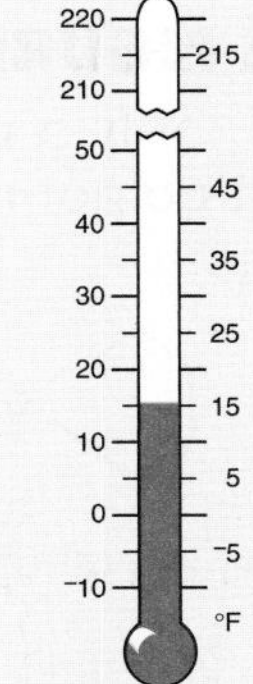

▶ Practice

Write the Fahrenheit temperature reading.

1.

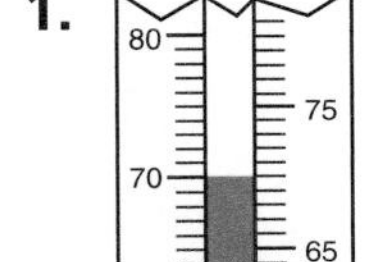

2.

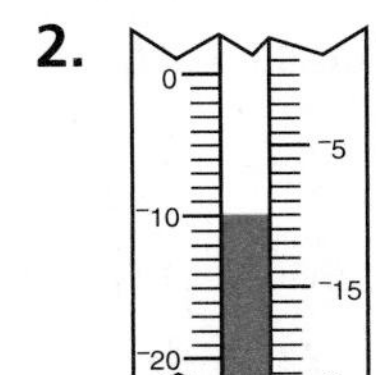

3.

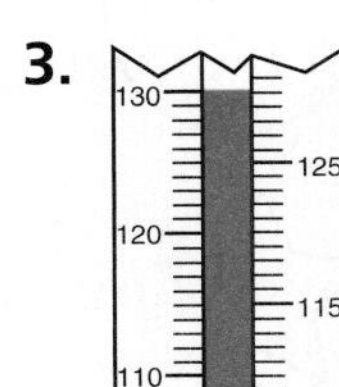

TROUBLESHOOTING

FUNCTION TABLES

To write a rule for a function table, look for the operation that is used with each input to get the output.

- When the output is greater than the input, try a rule with addition or multiplication.
- When the output is less than the input, try a rule with subtraction or division.

Example

Write a rule for the function table.

INPUT, x	OUTPUT, y
12	6
11	5
10	4
9	3

Look for a pattern.

Think: Each output is less than the input. Try a rule with subtraction or division.

Each output is the input minus 6.

So, the rule is: Subtract 6.

▶ Practice

Write a rule for each function table.

1.

x	y
5	16
6	17
7	18
8	19

2.

x	y
2	8
3	12
4	16
5	20

3.

x	y
32	4
24	3
16	2
8	1

CLASSIFY PLANE FIGURES

Figures with ends that do not meet are **open** figures.

Open figures

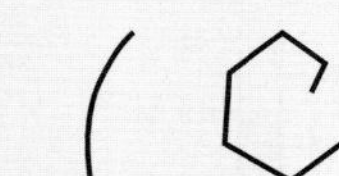

Figures that begin and end at the same point are **closed** figures.

Closed figures

▶ Practice

Is the figure a *closed* figure or an *open* figure?

1. 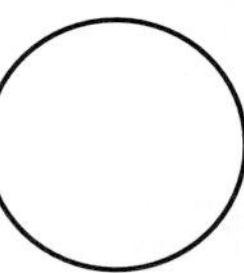

2. 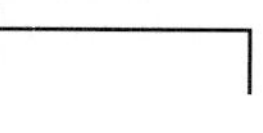

3. 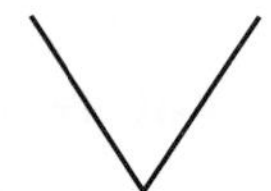

4. 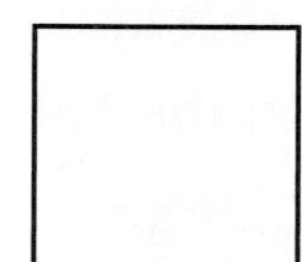

5. 

6. 

7. 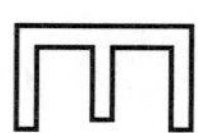

8. 

NAME POLYGONS

Polygons are closed figures with straight sides. Polygons are named by the number of their sides and angles.

Examples

TRIANGLE	QUADRILATERAL	PENTAGON	HEXAGON	OCTAGON
3 sides, 3 angles	4 sides, 4 angles	5 sides, 5 angles	6 sides, 6 angles	8 sides, 8 angles

▶ Practice

Name each polygon. Tell the number of sides and angles.

1.

2. 

3. 

4. 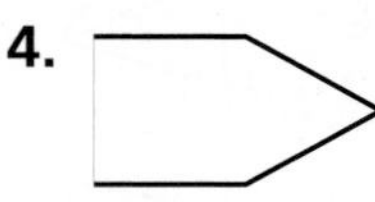

5.

6. 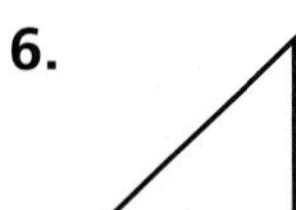

7. 

8. 

CLASSIFY ANGLES

Angles can be classified by their measures. An angle is a right angle, or it is greater than or less than a right angle.

Examples

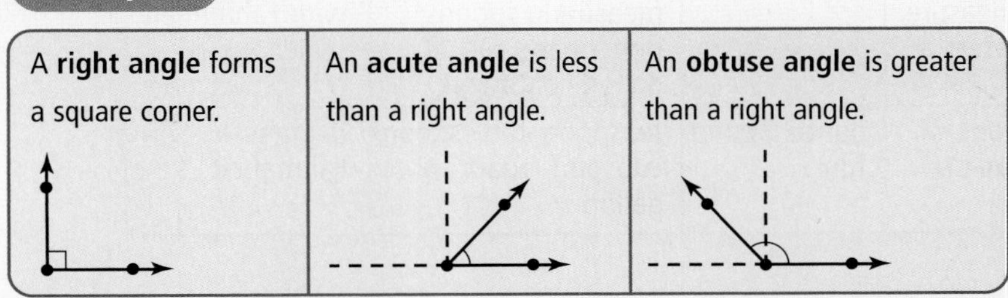

A right angle forms a square corner.
An acute angle is less than a right angle.
An obtuse angle is greater than a right angle.

▶ Practice

Classify the angle. Write *acute, right,* or *obtuse*.

1.

2. 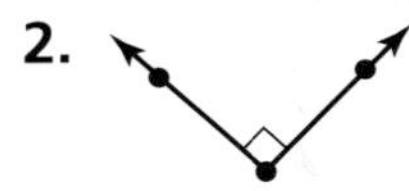

3. 

4. 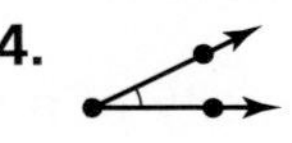

TROUBLESHOOTING

FACES OF SOLID FIGURES

A solid figure is a three-dimensional figure. The sides of a solid figure are called faces. The faces are plane figures.

Example

The solid figure at the right is a hexagonal prism.
The two shaded faces are hexagons.
The other faces are rectangles.

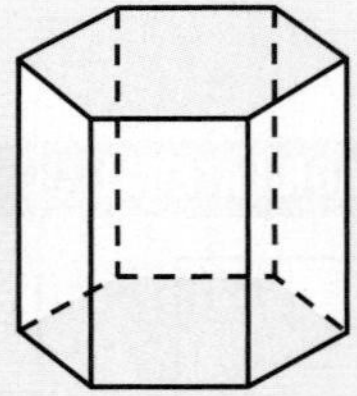

▶ Practice

Name the plane figure that is the shaded face of the solid figure.

1. 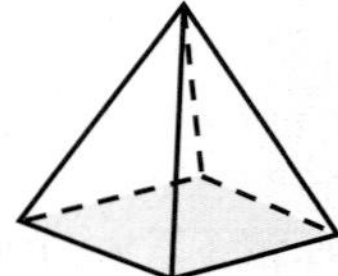

2. 

3. 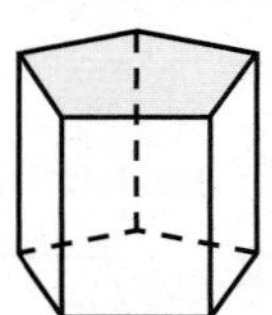

4. 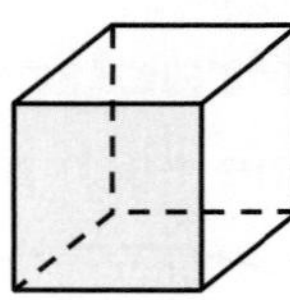

5. 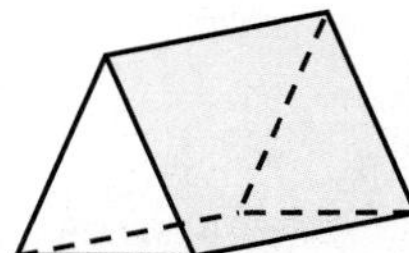

6. 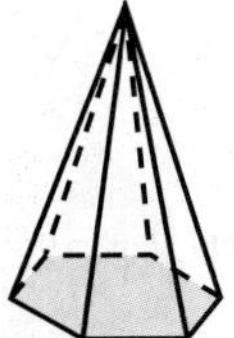

7. 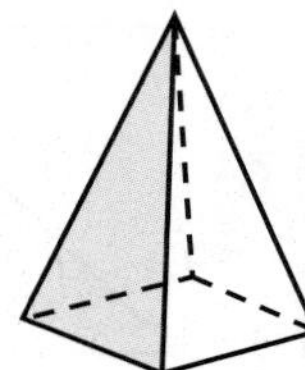

8. 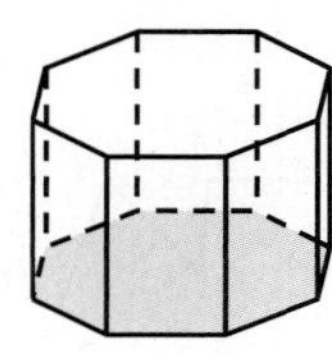

CUSTOMARY UNITS AND TOOLS

Customary units of measure are used every day in the United States. The table lists the units and the tools used to measure them.

	LENGTH	WEIGHT	CAPACITY	TEMPERATURE
Tools for measuring	ruler, yardstick, tape measure, odometer	spring scale	measuring cups; measuring spoons; pint, quart, and gallon containers	thermometer with Fahrenheit scale
Units of measure	inch, foot, yard, mile	ounce, pound, ton	teaspoon, tablespoon, cup, pint, quart, gallon	degrees Fahrenheit

▶ Practice

Choose the customary measuring tool and unit you would use to measure each.

1. width of a finger
2. weight of a pencil
3. oven temperature
4. capacity of a saucepan
5. distance between two cities
6. amount of salt to put into a recipe

METRIC UNITS AND TOOLS

The table lists metric units and tools.

	LENGTH	MASS	CAPACITY	TEMPERATURE
Tools for measuring	meterstick, odometer	balance scale	measuring cup or other milliliter or liter containers	thermometer with Celsius scale
Units of measure	millimeter, centimeter, decimeter, meter, kilometer	grams, kilograms	milliliter, liter	degrees Celsius

▶ Practice

Choose the metric measuring tool and unit you would use to measure each.

1. height of a ladder
2. mass of a nail
3. freezer temperature
4. volume of a drinking glass
5. volume of a large carton of milk
6. length of a bridge across a wide river

PERIMETER AND AREA

Perimeter is the distance around a figure. To find the perimeter of a rectangle, add the lengths of the sides. **Area** is the number of square units needed to cover a given surface. To find the area of a rectangle, count the number of square units or multiply the length and width, $A = l \times w$.

Examples

A Find the perimeter of the rectangle above.
Add the lengths of the sides.
$P = 5 + 4 + 5 + 4 = 18$
So, the perimeter is 18 units.

B Find the area of the rectangle above.
Count the number of square units, or multiply the length and width.
$A = 5 \times 4 = 20$
So, the area is 20 square units.

▶ Practice

Find the perimeter and area.

1. 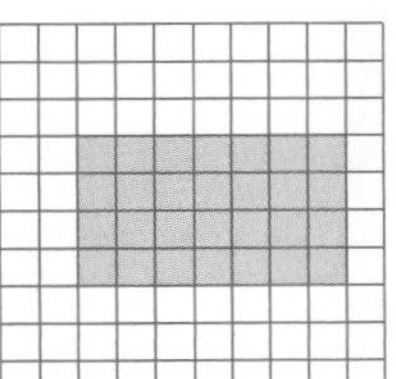
2. 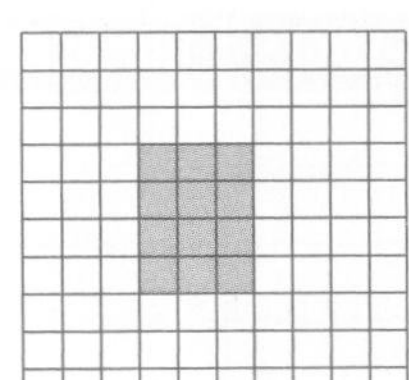
3. 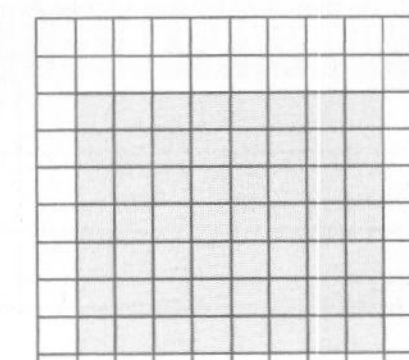
4. 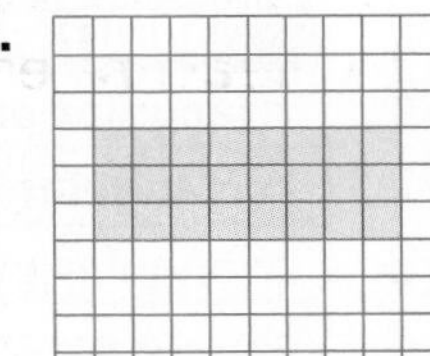

TROUBLESHOOTING

MULTIPLICATION

An estimate can help you check the reasonableness of your answer. It can also help you place the decimal point in a product.

Multiply. 1.2×2.3

STEP 1

Estimate the product. Round each factor.

$$\begin{array}{c} 1.2 \times 2.3 \\ \downarrow \quad \downarrow \\ 1 \times 2 = 2 \end{array}$$

STEP 2

Multiply as with whole numbers.

$$\begin{array}{r} 2.3 \\ \times\ 1.2 \\ \hline 46 \\ +\ 230 \\ \hline 276 \end{array}$$

STEP 3

Use the estimate to place the decimal point in the product.

$$\begin{array}{r} 2.3 \\ \times\ 1.2 \\ \hline 46 \\ +\ 230 \\ \hline 2.76 \end{array}$$

Since the estimate is 2, place the decimal point after the 2.

▶ Practice

Find the product. Estimate to check.

1. 87×5 **2.** 62×9 **3.** 38×12 **4.** 103×7 **5.** 74×15

6. 5.4×6 **7.** 7.2×2 **8.** 4.7×1.2 **9.** 4.4×7 **10.** 8.2×3.1

FACES, EDGES, AND VERTICES

Prisms and pyramids have faces, edges, and vertices. A **face** is a polygon that forms a surface of a solid figure. An **edge** is a line segment formed where two faces meet. A **vertex** is a point where three or more edges meet.

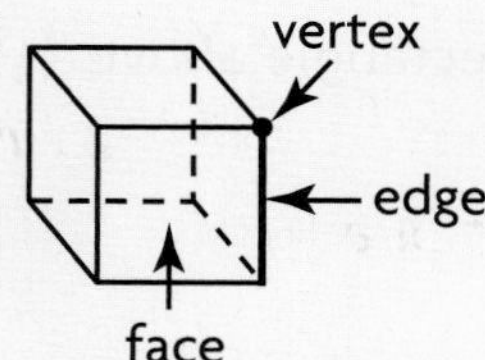

A cube has 6 faces, 12 edges, and 8 vertices.

▶ Practice

Copy and complete the table.

	SOLID FIGURE	NUMBER OF FACES	NUMBER OF EDGES	NUMBER OF VERTICES
1.	triangular prism	5	■	■
2.	rectangular prism	■	12	■
3.	triangular pyramid	■	■	4
4.	square pyramid	5	■	■

EXPRESSIONS WITH EXPONENTS

An **exponent** tells how many times a number is used as a factor. The **base** is the number being used as a factor.

$$10 \times 10 \times 10 \times 10 = 10^4$$

↓ exponent (4); ↑ base (10)

▶ Practice

Write the expression by using an exponent, and find its value.

1. 6×6 **2.** $7 \times 7 \times 7$ **3.** $10 \times 10 \times 10$ **4.** $1 \times 1 \times 1 \times 1 \times 1$

5. 5×5 **6.** $2 \times 2 \times 3 \times 3$ **7.** $(9 \times 9) + (2 \times 2)$ **8.** $(3 \times 3) + (6 \times 6)$

IMPOSSIBLE, CERTAIN, LIKELY, UNLIKELY

An event is **certain** if it always happens. An event is **impossible** if it never happens. Some events are neither certain nor impossible. These events are **likely** if they usually happen and **unlikely** if they usually do not happen.

Examples

A Tell if the event is *certain* or *impossible*. From a bag that holds only pennies, an object you will pull out will be a penny.

A penny is the only object you can pull from the bag. This event is certain.

B Tell if the event is *likely* or *unlikely*. At a shopping mall, you will see a shopper.

A mall has shoppers during most hours of the day. This event is likely.

▶ Practice

Tell whether each event is *certain* or *impossible*.

1. pulling a green marble out of a bag of all yellow marbles

2. getting heads or tails when you toss a quarter

3. pulling a red counter from a bag holding only red counters

4. finding a blue marker in a box holding only green and red markers

Tell whether each event is *likely* or *unlikely*.

5. feeling hungry after two days of not eating

6. feeling tired after sleeping soundly

7. getting heads at least once when you toss a penny 10 times

8. getting caught in a snowstorm in the desert

CHAPTER 1

EXTRA PRACTICE

Set A (pp. 4–7)

Write the value of the blue digit.

1. 14,350
2. 54,079
3. 635,017
4. 9,450,180
5. 47,329,052
6. 8,506,332,189

Write each number in standard form.

7. 3,000,000 + 40,000 + 9,000 + 200 + 7
8. 50,000,000 + 1,000,000 + 400,000 + 10,000 + 300 + 20 + 5
9. forty-three million, three hundred ten thousand, five hundred forty-seven
10. six billion, five hundred six million, thirteen thousand, four hundred twenty

Write each number in word form.

11. 17,642,007
12. 202,062,311
13. 4,501,078
14. In 1999, there were 82,079,454 people in Germany. What is this number in word form?
15. Write the greatest 9-digit whole number without repeating a digit. What is the value of the 2?

Set B (pp. 8–9)

Start at the left. Name the first place-value position where the digits differ. Name the greater number.

1. 4,778; 4,596
2. 10,015; 17,051
3. 432,759; 432,764

Compare. Write <, >, or = for each ●.

4. 18,276 ● 18,287
5. 513,640 ● 513,450
6. 845,418 ● 845,418
7. 173,090 ● 172,900
8. 2,722,250 ● 2,722,250
9. 50,967,300 ● 5,967,021
10. Write a 7-digit number that is greater than 9,895,999 and has a 4 in the thousands place.

Set C (pp. 10–11)

Order from greatest to least.

1. 12,945; 12,693; 12,990
2. 10,235,561; 1,235,561; 10,253,561
3. 280,056; 280,605; 280,560

Order from least to greatest.

4. 436,042,303; 463,054,119; 64,332,989
5. 1,525,202; 1,252,202; 1,522,202
6. 186,452; 180,642; 186,522

Set A (pp. 18–19)

Write as a decimal and a fraction or a mixed number.

1. $4 + 0.07$ **2.** $10 + 0.3 + 0.09$ **3.** $8 + 0.8 + 0.01$

4. thirteen and four hundredths **5.** fifty-seven hundredths

6. Mike rode his bike three and four hundredths of a mile. Write this number in standard form.

Set B (pp. 20–21)

Write in expanded form and in standard form.

1. two hundred sixteen ten-thousandths **2.** five and five thousandths **3.** two and seven hundred two thousandths

Write in word form.

4. 44.009 **5.** 2.0189 **6.** 1.2002 **7.** 0.505

8. Eric wrote $78\frac{106}{1,000}$ as seventy-eight and one hundred six hundredths. Did he write this correctly? Explain.

Set C (pp. 22–23)

Write *equivalent* or *not equivalent* to describe each pair of decimals.

1. 03.45 and 3.450 **2.** 0.097 and 0.970 **3.** 23.504 and 23.50

Write an equivalent decimal for each number.

4. 5.2 **5.** 9.320 **6.** 87.0800 **7.** 2.02

Set D (pp. 24–25)

Order from greatest to least.

1. 54.453, 54.59, 54.811 **2.** 7.564, 17.45, 11.94

Order from least to greatest.

3. 31.104, 31.05, 31.94 **4.** 6.309, 6.42, 6.341

5. Leslie has $4.68 in her pocket. Which of the following items can she buy?

Notebook	$2.30	Box of Pencils	$4.67
Backpack	$8.53	Markers	$4.69

CHAPTER 3

EXTRA PRACTICE

Set A (pp. 32–33)

Round each number to the place of the blue digit.

1. 2,517 2. 450,339 3. 724,950 4. 3,620,740 5. 7,852,060

Round 8,056,231 to the place named.

6. thousands 7. hundred thousands 8. millions

Name the place to which each number was rounded.

9. 3,128 to 3,100 10. 13,098 to 10,000 11. 4,509,176 to 5,000,000

Set B (pp. 34–35)

Estimate by rounding.

1. 48,502 + 37,154
2. 518,273 + 284,043
3. 482,054 + 375,407
4. 927,306 + 92,517
5. 259,345 + 109,254
6. 4,513,932 − 2,189,603
7. 3,028,458 + 687,543

Set C (pp. 36–37)

Find the sum or difference. Estimate to check.

1. 9,002 − 7,784
2. 16,456 + 9,023
3. 31,216 − 18,927
4. 48,898 + 61,124
5. 77,991 − 56,178
6. 59,233 + 71,099
7. 67,938 − 30,556
8. 99,043 + 87,914
9. There were 11,561 fans at Game 1 and 15,907 fans at Game 2. How many more fans attended Game 2?

Set D (pp. 38–39)

Find the sum or difference. Estimate to check.

1. 1,658,350 − 1,374,035
2. 6,340,583 + 2,394,051
3. 8,429,083 − 672,039
4. 2,600,898 + 452,917
5. 2,567,790 − 1,408,196
6. 7,112,095 + 8,011,403
7. 4,567,805 − 832,509
8. 11,503,288 + 1,629,917
9. 403,687 + 280,321
10. 8,311,298 − 6,901,283
11. 13,674,950 + 3,560,528

Set A (pp. 46–47)

Round each number to the place of the blue digit.

1. 2.0647 **2.** 1.3686 **3.** 34.0685 **4.** 7.3168 **5.** 79.5063

Name the place to which each number was rounded.

6. 4.638 to 4.6 **7.** 6.0571 to 6.06 **8.** 10.4937 to 10.494 **9.** 8.084 to 8.08

Set B (pp. 48–49)

Estimate the sum or difference to the nearest whole number or dollar.

1. $5.83 + 3.74

2. 6.57 − 3.45

3. 45.731 + 21.07

4. 19.76 − 11.37

5. $367.09 + 14.52

Estimate the sum or difference to the nearest tenth.

6. 0.385 + 0.284

7. 7.251 − 4.951

8. 8.402 + 6.917

9. 7.829 − 2.136

10. 11.673 − 7.492

Set C (pp. 50–53)

Find the sum or difference. Estimate to check.

1. 9.2 + 2.6

2. 17.04 + 9.5

3. 43.04 + 22.6

4. 126.73 + 91.5

5. 28.01 + 40.7

6. 6.4 − 3.9

7. 35.28 − 31.9

8. 78.2 − 71.34

9. 567.25 − 435.87

10. 410.4 − 32.56

11. Nicole cut 4.25 yards of ribbon but used only 3.7 yards of it. How much did she have left?

12. Joe measured 6.8 yards of wire and then 11.45 more yards. How much did he measure in all?

Set D (pp. 54–55)

Find the difference.

1. 7.2 − 5.62 **2.** 4 − 1.08 **3.** 12.23 − 3.254 **4.** 15.04 − 4.835

5. Gerry's new green shirt cost $17.86 and his new yellow one cost $15.25. How much more did the green one cost?

6. Amanda hiked 7.2 kilometers on Saturday. How far does she have to go to finish the 15-kilometer course?

CHAPTER 5

EXTRA PRACTICE

Set A (pp. 68–71)

Write an expression. Explain what the variable represents.

1. Eric had some baseball cards. Tracy gave him 5 more.
2. Rebecca gathered 18 pinecones. On the way home, she dropped some.
3. Josh had some oranges. He gave 3 to Nikki and 9 to James.
4. Some girls came to the first practice, and 12 more came to the second practice.

Find the value of the expression.

5. $7 + n$ if n is 9
6. $(n - 4) + 6$ if n is 12
7. $19 - (n + 3)$ if n is 6

Set B (pp. 72–73)

Write an equation. Explain what the variable represents.

1. Jesse had 5 goldfish, and his friend gave him some more. Now Jesse has 12 goldfish. How many did his friend give him?
2. A store had 83 lawn mowers for sale. The store sold some on Monday and had 60 left. How many did the store sell?

Write a problem for the equation. State what the variable *n* represents.

3. $n + 17 = 25$
4. $(n - 2) + 7 = 16$

Set C (pp. 74–77)

Solve the equation. Check your solution.

1. $9 + n = 19$
2. $52 - 14 = n$
3. $12 + 2 = n - 8$
4. $18 + 32 = n$
5. $n + 6 = 6$
6. $15 + n = 11 + 8$
7. $n - 12 = 40$
8. $n + 55 = 123$
9. $16 + n = 38$
10. $81 - n = 60$
11. $2.5 + 1.5 + n = 7$
12. $87 - 20 = n + 7$

Set D (pp. 78–79)

Find the value of *n*. Identify the addition property used.

1. $8 + 13 = 13 + n$
2. $12.4 + (9.3 + 0.7) = (12.4 + n) + 0.7$
3. $0 + n = 322$
4. $n + 16.9 = 16.9 + 14.5$
5. $9 + (5 + 6 + 3) = (9 + 5 + 6) + n$
6. $9.2 + 4.6 + 8.2 = n + 9.2 + 4.6$

Set A (pp. 86–87)

Write an expression. If you use a variable, tell what it represents.

1. Stan had 8 board games. He got 3 more for his birthday.
2. Danielle bought 4 pounds of apples. Each pound contained 3 apples.
3. George did the same number of sit-ups each day for 5 days.

Evaluate each expression.

4. $30 + n$ if $n = 18$
5. $(8 \times n) - 15$ if $n = 4$
6. $3 \times (n + 12)$ if $n = 22$
7. $13 \times n$ if $n = 17$
8. $(n \times 8) + 20$ if $n = 3$
9. $4 \times (n - 25)$ if $n = 75$

Set B (pp. 90–91)

Solve the equation. Identify the property used.

1. $y \times 35 = 0$
2. $n \times 10 = 10$
3. $4 \times 2 = z \times 4$
4. $3 \times 2 = 2 \times n$
5. $(12 \times 2) \times 10 = n \times (2 \times 10)$
6. $4 \times (3 \times n) = (4 \times 3) \times 6$
7. $1 \times 0 = n$
8. Charles says that $5 \times 16 = 16 \times 5$. Is he correct? Explain.
9. On their farm, Tyler and Antoni take care of an equal number of animals. Tyler takes care of 6 pens with 4 rabbits in each. Antoni takes care of 4 pens with an equal number of guinea pigs in each. How many guinea pigs are in each pen?

Set C (pp. 92–93)

Use the Distributive Property to restate each expression. Find the product.

1. 33×4
2. 8×17
3. 22×6
4. 5×24
5. 4×83
6. 3×94
7. 55×8
8. 63×9

Restate the expression, using the Distributive Property. Then find the value of the expression.

9. $8 \times (n + 5)$ if $n = 30$
10. $n \times (10 + 9)$ if $n = 3$
11. $n \times (20 + 7)$ if $n = 9$
12. $4 \times (60 + n)$ if $n = 7$
13. $7 \times (n + 9)$ if $n = 50$
14. $(70 + 2) \times n$ if $n = 6$
15. Warren says the value of $3 \times (n + 5)$ is 35 if $n = 10$. Ellen says the value is 45. Who is correct? Explain.

CHAPTER 7

EXTRA PRACTICE

Set A (pp. 98–101)

For 1–4, use the frequency table.

1. How many miles did Dan run on Day 1? Day 4?
2. How many total miles did Dan run by the end of Day 3?
3. What is the range of miles Dan ran each day?
4. How many total miles did Dan run by the end of Day 5?

DAN'S RUNNING LOG

Day	Frequency (Number of Miles)	Cumulative Frequency
1	6	6
2	3	9
3	4	13
4	2	15
5	5	20

Set B (pp. 102–103)

Find the mean for each set of data.

1. 8, 21, 18, 17
2. 45, 31, 20, 27, 52
3. 15, 64, 33, 18, 22, 40

For 4–5, use the line plot.

4. What is the mean of the students' ages?
5. What is the range of the students' ages?

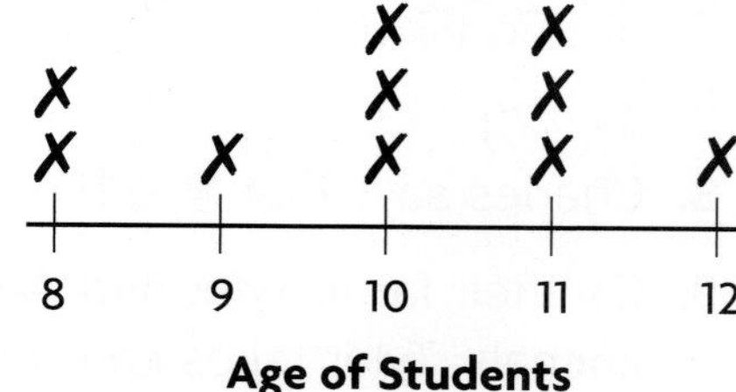

Set C (pp. 104–105)

Find the median and the mode for each set of data.

1.

SAVINGS

Weeks	1	2	3	4	5	6	7	8	9
Amount	$10	$14	$8	$7	$10	$12	$10	$12	$12

2.

POINTS SCORED

Game	1	2	3	4	5	6	7
Points	23	17	28	20	24	30	20

3. 16, 24, 11, 29, 20
4. 25, 50, 75, 25, 80
5. 26, 28, 28, 36, 40, 28

Set D (pp. 108–111)

For 1–3, use the bar graph.

1. How many students voted for cake? ice cream?
2. Which two foods got the same number of total votes?
3. How many more students voted for cake than for popcorn?

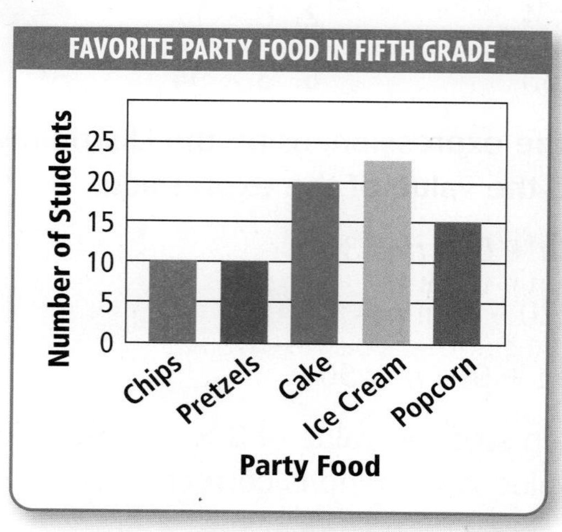

Set A (pp. 116–119)

Choose 5, 10, 15, or 20 as the most reasonable interval for the data.

1. 30, 16, 60, 45, 75
2. 3, 6, 10, 14, 22
3. 10, 30, 40, 60, 20

Choose the more reasonable scale for the data.

4.

FAVORITE ANIMAL	
Animal	**Number**
Cat	20
Dog	40
Horse	30
Bird	10
Rabbit	9

a. 50, 40, 30, 20, 10, 0

b. 25, 20, 15, 10, 5, 0

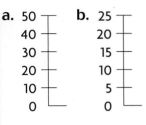

5.

FAVORITE COLOR	
Color	**Number**
Red	6
Blue	8
Green	2
Purple	5
Yellow	4

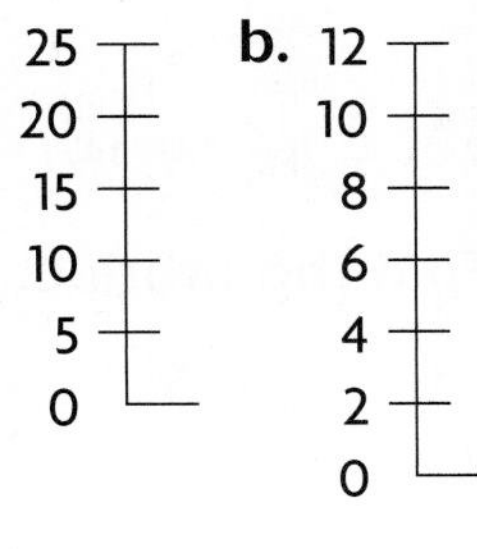

Set B (pp. 122–123)

Graph and label the following points on a coordinate grid.

1. *A* (0,6)
2. *B* (1,3)
3. *C* (2,2)
4. *D* (3,4)
5. *E* (4,1)
6. *F* (5,3)
7. *G* (6,5)
8. *H* (7,2)

Set C (pp. 124–127)

USE DATA For 1–4, use the table.

TIME MARIA SPENDS ON HOMEWORK					
Day	Mon	Tue	Wed	Thu	Fri
Time	20 min	45 min	60 min	75 min	0 min

1. Make a line graph for the data in the table.
2. Between which days does Maria increase her homework time the most?
3. On your graph, what does the ordered pair (Mon,20) tell you?
4. How many more minutes does Maria spend on homework on Thursday than on Monday?

Set D (pp. 130–133)

Choose the best type of graph or plot for the data. Explain your choice.

1. how you spend your allowance
2. results of a survey to find the most popular running shoes
3. the number and types of sports cards
4. store earnings during a month

CHAPTER 9 EXTRA PRACTICE

Set A (pp. 144–145)

Estimate the product.

1. 73×403
2. $6 \times 7{,}834$
3. $32 \times 8{,}561$
4. 93×468
5. 702×69
6. 855×28
7. 276×82
8. 807×51
9. $1{,}231 \times 28$

Set B (pp. 146–147)

Find the product. Estimate to check.

1. $2{,}335 \times 6$
2. $3{,}584 \times 4$
3. $5{,}824 \times 5$
4. $4{,}854 \times 3$
5. $530{,}944 \times 4$
6. $90{,}256 \times 3$
7. $6{,}993{,}028 \times 7$
8. $3{,}511{,}824 \times 8$
9. A theater complex has 9 theaters. Each theater seats 1,275 people. How many people can be seated at one time?

Set C (pp. 148–149)

Find the product. Estimate to check.

1. 343×29
2. 89×37
3. 612×52
4. $4{,}261 \times 83$
5. 872×21
6. $4{,}002 \times 43$
7. 860×80
8. $9{,}089 \times 67$
9. 27×63
10. 312×74
11. 92×859
12. 532×84

Set D (pp. 150–151)

Find the product. Estimate to check.

1. 592×534
2. $239{,}518 \times 68$
3. $214{,}093 \times 406$
4. $2{,}742{,}521 \times 687$
5. $3{,}853{,}632 \times 790$
6. $8{,}989{,}898 \times 898$
7. $4{,}612{,}811 \times 416$
8. $6{,}399{,}017 \times 696$
9. $8{,}300{,}605 \times 208$

Set A (pp. 160–161)

Multiply each number by 10, by 100, and by 1,000.

1. 0.7
2. 0.32
3. 0.003
4. 0.152

Algebra Find the value of n.

5. $100 \times 0.1 = n$
6. $10 \times 5.643 = n$
7. $1{,}000 \times 0.9023 = n$
8. $10 \times n = 0.06$
9. $n \times 10 = 0.27$
10. $100 \times n = 0.45$
11. In 1910, it cost about $0.80 per week to educate a public school student in the United States. By 1920, the weekly cost had doubled. How much did it cost to educate a student for 1 week in 1920?

Set B (pp. 162–163)

Make a model to find the product.

1. 0.6×0.6
2. 0.1×0.8
3. 0.5×0.3
4. 0.7×0.3

Find the product.

5. 0.6×0.8
6. 0.3×0.8
7. 0.9×0.5
8. 0.4×0.9
9. 0.2×0.9
10. 0.5×0.4
11. 0.2×0.3
12. 0.2×0.2

Set C (pp. 164–167)

Find the product. Estimate to check.

1. 0.3×14
2. $0.5 \times 1{,}206$
3. 6.8×4.5
4. 7.25×3.8
5. $\begin{array}{r} 0.67 \\ \times \quad 8 \\ \hline \end{array}$
6. $\begin{array}{r} 3.94 \\ \times 0.04 \\ \hline \end{array}$
7. $\begin{array}{r} 0.53 \\ \times \ 58 \\ \hline \end{array}$
8. $\begin{array}{r} 5.06 \\ \times \ 28 \\ \hline \end{array}$
9. Gustavo has a pumpkin with a mass of 4.8 kilograms. If 0.9 of its mass is water, how much of its mass is water? How much is not water?
10. Bernadette bought 2 loaves of bread at $0.97 each and a carton of milk for $1.18. How much did she spend?

Set D (pp. 168–169)

Find the product. Estimate to check.

1. 8×0.006
2. 0.03×2
3. 0.02×4
4. 7×0.004
5. 0.7×0.07
6. 0.025×0.6
7. 0.075×0.9
8. 0.5×0.024

CHAPTER 11

EXTRA PRACTICE

Set A (pp. 182–183)

Estimate the quotient. Tell what compatible numbers you used.

1. $8\overline{)902}$ **2.** $7\overline{)2{,}936}$ **3.** $9\overline{)1{,}989}$ **4.** $7\overline{)51{,}409}$ **5.** $7\overline{)313{,}084}$

Estimate the quotient, using two sets of compatible numbers.

6. $2{,}958 \div 3$ **7.** $46{,}133 \div 5$ **8.** $172{,}506 \div 8$ **9.** $217{,}028 \div 2$

Set B (pp. 184–185)

Name the position of the first digit of the quotient.

1. $2\overline{)417}$ **2.** $5\overline{)213}$ **3.** $3\overline{)697}$ **4.** $6\overline{)489}$ **5.** $2\overline{)526}$

Divide.

6. $4\overline{)448}$ **7.** $5\overline{)615}$ **8.** $6\overline{)212}$ **9.** $7\overline{)488}$ **10.** $8\overline{)568}$

11. Bailey and his family went on a 108-mile hiking trip in the mountains. They hiked 9 miles each day. How many days did they hike?

Set C (pp. 186–187)

Divide. Estimate to check.

1. $7\overline{)714}$ **2.** $5\overline{)54}$ **3.** $9\overline{)936}$ **4.** $9\overline{)725}$ **5.** $8\overline{)167}$

6. $3\overline{)92}$ **7.** $2\overline{)815}$ **8.** $6\overline{)628}$ **9.** $8\overline{)105}$ **10.** $7\overline{)425}$

Set D (pp. 188–189)

Divide. Estimate to check.

1. $7\overline{)4{,}102}$ **2.** $3\overline{)4{,}005}$ **3.** $8\overline{)32{,}580}$ **4.** $5\overline{)26{,}253}$

5. $7\overline{)239{,}305}$ **6.** $3\overline{)875{,}249}$ **7.** $2\overline{)3{,}269{,}325}$ **8.** $4\overline{)8{,}469{,}631}$

Set E (pp. 190–193)

Evaluate the expression $n \div 4$ for each value of n.

1. $n = 88$ **2.** $n = 456$ **3.** $n = 212$ **4.** $n = 6{,}344$

Solve the equation. Then check the solution.

5. $99 \div n = 9$ **6.** $672 \div n = 84$ **7.** $n \times 8 = 336$ **8.** $n \div 6 = 17$

9. $36 \div n = 12$ **10.** $64 \div p = 8$ **11.** $m \div 8 = 9$ **12.** $n \div 7 = 12$

Set A (pp. 200–201)

Use mental math to complete. Write the basic fact you use.

1. $200 \div 40 = 5$
   $2{,}000 \div 40 = 50$
   $20{,}000 \div 40 = \blacksquare$
2. $160 \div 40 = 4$
   $1{,}600 \div 40 = \blacksquare$
   $16{,}000 \div 40 = 400$
3. $180 \div 30 = \blacksquare$
   $1{,}800 \div 30 = 60$
   $18{,}000 \div 30 = \blacksquare$
4. A tour group traveled 6,000 miles in 30 days. How many miles a day did it average?
5. If you average 50 miles per hour for 550 miles, how many hours will you travel?

Set B (pp. 202–203)

Write two pairs of compatible numbers for each. Give two possible estimates.

1. $198 \div 28$
2. $336 \div 46$
3. $1{,}344 \div 87$

Estimate the quotient.

4. $22\overline{)215}$
5. $33\overline{)605}$
6. $42\overline{)1{,}378}$
7. $51\overline{)9{,}750}$
8. $41\overline{)8{,}261}$

Set C (pp. 204–207)

Divide.

1. $842 \div 27$
2. $2{,}946 \div 76$
3. $8{,}888 \div 62$
4. $15{,}432 \div 18$
5. $23\overline{)532}$
6. $31\overline{)8{,}461}$
7. $18\overline{)2{,}468}$
8. $27\overline{)16{,}798}$
9. $23\overline{)86{,}698}$

Set D (pp. 208–209)

Write *too high, too low,* or *just right* for each estimate.

1. $34\overline{)\overset{800}{26{,}163}}$
2. $41\overline{)\overset{6}{325}}$
3. $29\overline{)\overset{800}{23{,}007}}$
4. $69\overline{)\overset{10}{698}}$
5. $76\overline{)\overset{20}{1{,}826}}$

Choose the better estimate to use for the quotient. Write *a* or *b*.

6. $14\overline{)92}$ **a.** 6 **b.** 5
7. $18\overline{)138}$ **a.** 6 **b.** 7
8. $61\overline{)192}$ **a.** 3 **b.** 4

Set E (pp. 210–211)

Divide.

1. $12\overline{)58}$
2. $18\overline{)79}$
3. $24\overline{)9{,}168}$
4. $33\overline{)234}$
5. $84\overline{)31{,}164}$
6. $36\overline{)24{,}469}$
7. $42\overline{)523{,}965}$
8. $26\overline{)635{,}204}$

CHAPTER 13

EXTRA PRACTICE

Set A (pp. 218–219)

Copy and complete each pattern.

1. $300 \div 6 = ■$
$30 \div 6 = ■$
$3 \div 6 = ■$

2. $200 \div 5 = ■$
$20 \div 5 = ■$
$2 \div 5 = ■$

3. $600 \div 5 = ■$
$60 \div 5 = ■$
$6 \div 5 = ■$

4. $100 \div 4 = ■$
$10 \div 4 = ■$
$1 \div 4 = ■$

5. $400 \div 8 = ■$
$40 \div 8 = ■$
$4 \div 8 = ■$

6. $500 \div 2 = ■$
$50 \div 2 = ■$
$5 \div 2 = ■$

7. $900 \div 10 = ■$
$90 \div 10 = ■$
$9 \div 10 = ■$

8. $100 \div 5 = ■$
$10 \div 5 = ■$
$1 \div 5 = ■$

9. $200 \div 4 = ■$
$20 \div 4 = ■$
$2 \div 4 = ■$

Set B (pp. 222–225)

Copy the quotient and correctly place the decimal point.

1. $2\overline{)13.70}$ quotient 685
2. $3\overline{)34.29}$ quotient 1143
3. $4\overline{)2.348}$ quotient 0587
4. $5\overline{)0.075}$ quotient 0015
5. $6\overline{)57.384}$ quotient 9564
6. $7\overline{)0.8638}$ quotient 01234
7. $8\overline{)71.104}$ quotient 8888
8. $9\overline{)190.08}$ quotient 2112

Find the quotient. Check by multiplying.

9. $5\overline{)\$4.45}$
10. $4\overline{)\$5.84}$
11. $3\overline{)1.92}$
12. $7\overline{)10.5}$
13. $7.92 \div 6$
14. $3.29 \div 7$
15. $\$0.32 \div 8$
16. $75.36 \div 3$
17. $\$0.64 \div 4$
18. $10.92 \div 7$
19. $0.265 \div 5$
20. $51.56 \div 4$

Set C (pp. 228–229)

Write as a decimal.

1. $\frac{7}{8}$
2. $\frac{2}{5}$
3. $\frac{9}{16}$
4. $\frac{7}{20}$
5. $\frac{9}{10}$
6. $\frac{17}{20}$
7. $\frac{19}{20}$
8. $\frac{6}{10}$
9. $\frac{9}{12}$
10. $\frac{6}{16}$
11. $\frac{13}{20}$
12. $\frac{5}{10}$
13. $\frac{3}{10}$
14. $\frac{18}{24}$
15. $\frac{3}{8}$
16. $\frac{3}{16}$
17. $\frac{5}{8}$
18. $\frac{3}{12}$

19. Carol ate $\frac{2}{5}$ of the cake, and Evelyn ate $\frac{1}{5}$ of the cake. Write as a decimal the part of the cake they ate.

20. Brett ate $\frac{3}{4}$ of a pizza. Write as a decimal the part of the pie that is left.

Set A (pp. 234–235)

Complete each division pattern.

1. $720 \div 12 = 60$
   $72 \div 1.2 = \blacksquare$
   $7.2 \div 0.12 = \blacksquare$
2. $63 \div 7 = 9$
   $6.3 \div 0.7 = \blacksquare$
   $0.63 \div 0.07 = \blacksquare$
3. $240 \div 30 = 8$
   $24 \div 3.0 = \blacksquare$
   $2.4 \div 0.30 = \blacksquare$
4. $132 \div 11 = 12$
   $13.2 \div 1.1 = \blacksquare$
   $1.32 \div \blacksquare = 12$
5. $96 \div 24 = 4$
   $9.6 \div \blacksquare = 4$
   $0.96 \div \blacksquare = 4$
6. $150 \div 6 = 25$
   $15 \div \blacksquare = 25$
   $1.5 \div \blacksquare = 25$

Use a basic fact and patterns to solve for *n*.

7. $0.36 \div 0.04 = n$
8. $0.049 \div 0.07 = n$
9. $0.16 \div n = 8$
10. $4.5 \div 1.5 = n$
11. $6.9 \div n = 3$
12. $0.018 \div 0.06 = n$
13. Shannon earned $12.30 in dimes. How many dimes did Shannon earn?
14. Shantay won $23.85 in nickels. How many nickels did Shantay win?

Set B (pp. 238–241)

Copy the problem. Place the decimal point in the quotient. Draw arrows to help you.

1. $\begin{array}{r} 75 \\ 0.5\overline{)3.75} \end{array}$
2. $\begin{array}{r} 8 \\ 0.12\overline{)0.96} \end{array}$
3. $\begin{array}{r} 1172 \\ 0.7\overline{)8.204} \end{array}$
4. $\begin{array}{r} 064 \\ 0.09\overline{)0.0576} \end{array}$
5. $\begin{array}{r} 55 \\ 0.5\overline{)2.75} \end{array}$
6. $\begin{array}{r} 16 \\ 2.8\overline{)4.48} \end{array}$
7. $\begin{array}{r} 63 \\ 0.2\overline{)1.26} \end{array}$
8. $\begin{array}{r} 196 \\ 0.23\overline{)45.08} \end{array}$

Divide.

9. $0.4\overline{)2.48}$
10. $1.3\overline{)22.555}$
11. $\$0.07\overline{)\$6.58}$
12. $2.1\overline{)23.961}$
13. $0.05\overline{)0.515}$
14. $0.6\overline{)5.22}$
15. $\$0.14\overline{)\$0.42}$
16. $0.08\overline{)4}$
17. $26 \div 0.013$
18. $1.755 \div 2$
19. $33.12 \div 9.2$
20. $4.34 \div 7$
21. $\$5.75 \div \$0.25$
22. $\$6.96 \div \$0.08$
23. $\$6.00 \div \$0.75$
24. $\$3.15 \div \$0.05$
25. Natasha bought a bag of muffins for $4.25. Each muffin cost $0.85. How many muffins did she buy?
26. Dwight had 1.23 meters of ribbon. He cut the ribbon into 0.41-meter pieces. How many pieces does he have now?

CHAPTER 15

EXTRA PRACTICE

Set A (pp. 254–255)

Tell if each number is divisible by **2, 3, 4, 5, 6, 9,** or **10**.

1. 15 **2.** 26 **3.** 60 **4.** 132 **5.** 725

6. 1,410 **7.** 135 **8.** 109 **9.** 15,324 **10.** 378

Set B (pp. 258–261)

List the factors for each number.

1. 3 **2.** 17 **3.** 32 **4.** 75 **5.** 100

Write the common factors for each set of numbers.

6. 17, 18 **7.** 12, 32 **8.** 10, 50 **9.** 35, 75 **10.** 60, 100

Write the greatest common factor for each set of numbers.

11. 2, 12 **12.** 6, 21 **13.** 14, 28 **14.** 30, 36 **15.** 9, 27

Set C (pp. 266–267)

Write in exponent form.

1. 10,000 **2.** 1,000,000 **3.** 10,000,000 **4.** 100 **5.** 100,000

Find the value.

6. 10^6 **7.** 10^8 **8.** 10^1 **9.** 10^9 **10.** 10^{10}

Set D (pp. 268–271)

Write each expression by using an exponent.

1. $5 \times 5 \times 5 \times 5 \times 5$ **2.** $8 \times 8 \times 8 \times 8 \times 8 \times 8$ **3.** $10 \times 10 \times 10 \times 10$

Find the value.

4. 2^4 **5.** 4^5 **6.** 3^6 **7.** 7^2 **8.** 5^5

Set E (pp. 272–275)

Rewrite the prime factorization by using exponents.

1. $3 \times 3 \times 5$ **2.** $2 \times 2 \times 2 \times 2$ **3.** $2 \times 3 \times 5 \times 5$ **4.** $2 \times 7 \times 7 \times 7$

Find the prime factorization of the number. Use exponents when possible.

5. 49 **6.** 54 **7.** 23 **8.** 36

9. 84 **10.** 25 **11.** 52 **12.** 60

Set A (pp. 280–283)

Write a fraction for each decimal.

1. 0.55 2. 0.7 3. 0.26 4. 0.751 5. 0.2 6. 0.305

Write a decimal for each fraction.

7. $\frac{1}{4}$ 8. $\frac{17}{20}$ 9. $\frac{3}{8}$ 10. $\frac{6}{10}$ 11. $\frac{35}{100}$ 12. $\frac{1}{8}$

Set B (pp. 284–285)

Write an equivalent fraction.

1. $\frac{2}{4}$ 2. $\frac{3}{8}$ 3. $\frac{4}{5}$ 4. $\frac{6}{9}$ 5. $\frac{12}{15}$ 6. $\frac{22}{33}$

7. $\frac{8}{24}$ 8. $\frac{2}{8}$ 9. $\frac{1}{2}$ 10. $\frac{6}{7}$ 11. $\frac{2}{12}$ 12. $\frac{2}{3}$

Set C (pp. 286–289)

Rename the fractions, using the LCM, and compare.
Write <, >, or = for each ●.

1. $\frac{1}{2}$ ● $\frac{2}{6}$ 2. $\frac{3}{4}$ ● $\frac{1}{6}$ 3. $\frac{2}{3}$ ● $\frac{4}{6}$ 4. $\frac{5}{7}$ ● $\frac{3}{5}$ 5. $\frac{5}{6}$ ● $\frac{2}{4}$ 6. $\frac{4}{8}$ ● $\frac{2}{4}$

Write in order from least to greatest.

7. $\frac{1}{6}, \frac{2}{5}, \frac{1}{3}$ 8. $\frac{4}{8}, \frac{2}{3}, \frac{1}{6}$ 9. $\frac{4}{5}, \frac{2}{10}, \frac{1}{2}$ 10. $\frac{1}{4}, \frac{2}{12}, \frac{4}{6}$ 11. $\frac{4}{6}, \frac{9}{11}, \frac{7}{12}$ 12. $\frac{8}{9}, \frac{2}{5}, \frac{12}{18}$

Set D (pp. 290–293)

Write each fraction in simplest form.

1. $\frac{4}{16}$ 2. $\frac{6}{18}$ 3. $\frac{12}{18}$ 4. $\frac{6}{10}$ 5. $\frac{6}{36}$ 6. $\frac{15}{20}$

7. $\frac{19}{38}$ 8. $\frac{8}{24}$ 9. $\frac{18}{72}$ 10. $\frac{18}{24}$ 11. $\frac{30}{45}$ 12. $\frac{25}{125}$

Set E (pp. 294–295)

Write each fraction as a mixed number. Write each mixed number as a fraction.

1. $\frac{9}{7}$ 2. $5\frac{2}{3}$ 3. $\frac{13}{4}$ 4. $6\frac{4}{5}$ 5. $\frac{16}{6}$ 6. $6\frac{3}{4}$

7. $4\frac{1}{2}$ 8. $\frac{8}{3}$ 9. $\frac{25}{8}$ 10. $5\frac{1}{4}$ 11. $\frac{8}{5}$ 12. $4\frac{2}{7}$

CHAPTER 17 EXTRA PRACTICE

Set A (pp. 304–305)

Write *a* or *b* to show which fraction represents the ratio.

1. 5 to 3
a. $\frac{3}{5}$ **b.** $\frac{5}{3}$

2. 4:9
a. $\frac{4}{9}$ **b.** $\frac{9}{4}$

3. 5 to 5
a. $\frac{5}{5}$ **b.** $\frac{1}{5}$

4. 7:2
a. $\frac{7}{1}$ **b.** $\frac{7}{2}$

5. 1 to 6
a. $\frac{1}{6}$ **b.** $\frac{6}{1}$

For 6–8, use the graph.

6. What is the ratio of guards to forwards in the league?

7. What is the ratio of forwards to the total number of players in the league?

8. If 2 more guards join the league, what will be the ratio of guards to total players in the league?

BASKETBALL LEAGUE PLAYERS

Centers	
Forwards	
Guards	

= 2 players

Set B (pp. 306–307)

Tell whether the ratios are equivalent. Write *yes* or *no*.

1. $\frac{8}{12}$ and $\frac{12}{24}$

2. 6:6 and 8:8

3. 2 to 4 and 5 to 10

4. 7 to 6 and 6 to 5

5. 3:5 and $\frac{6}{10}$

6. 27 to 3 and $\frac{9}{1}$

Write three ratios that are equivalent to the given ratio.

7. 1 to 3

8. 4:5

9. $\frac{10}{7}$

10. 1 to 1

11. Twelve members of the drama club acted in the school play, and the remaining 19 members were stagehands. What is the ratio of actors to the total club membership?

12. The science club has 40 members, but only 30 members went on the trip to the aquarium. What is the ratio of members who went on the trip to those who did not?

Set C (pp. 308–309)

Copy and complete the map scale ratio table.

1.

Scale Length	1 in.	■ in.	5 in.	■ in.
Actual Length	5 mi	15 mi	■ mi	35 mi

2.

Scale Length	1 cm	8 cm	■ cm	12 cm
Actual Length	4.5 km	■ km	45 km	■ km

Set A (pp. 318–319)

Write the number as a decimal and a percent.

1. six tenths
2. two hundredths
3. nineteen hundredths

Write each decimal as a percent and each percent as a decimal.

4. 64%
5. 0.37
6. 1.21
7. 60%

Set B (pp. 320–323)

For 1–4, copy and complete the tables. Write each fraction in simplest form.

	FRACTION	DECIMAL	PERCENT
1.	$\frac{3}{5}$	■	60%
3.	■	■	10%

	FRACTION	DECIMAL	PERCENT
2.	■	■	40%
4.	$\frac{1}{20}$	■	■

5. Shawn scored 30% of the points. Mike scored 30 points out of 100. Jeff scored $\frac{2}{5}$ of the points. Who scored the most points?

Set C (pp. 324–327)

Find the percent of the number.

1. 10% of 10
2. 12% of 75
3. 150% of 130
4. 70% of 65
5. Mark got 70% out of 80 questions correct and Richard got 80% out of 70 questions correct. Who got more correct?
6. Julie found a quarter, 2 dimes, and 4 pennies in her coat pocket. What percent of a dollar did she find?

Set D (pp. 328–329)

Use mental math to find the percent of each number.

1. 10% of 400
2. 35% of 40
3. 85% of 160
4. 60% of 70
5. 25% of 40
6. 30% of 900
7. 40% of 90
8. 70% of 280

Set E (pp. 332–333)

For 1–2, use the table.

1. In which survey did the water park get more votes?
2. In which survey did the water park get a greater percent of the vote?

FIELD TRIP CHOICES

	Water Park	Zoo	Planetarium	Museum
Survey 1 (25 students)	40%	20%	20%	20%
Survey 2 (50 students)	30%	20%	20%	30%

CHAPTER 19 EXTRA PRACTICE

Set A (pp. 344–345)

Find the sum or difference in simplest form.

1. $\frac{3}{6} + \frac{1}{6}$
2. $\frac{4}{10} + \frac{1}{10}$
3. $\frac{1}{3} + \frac{2}{3}$
4. $\frac{3}{8} + \frac{4}{8}$
5. $\frac{1}{4} + \frac{2}{4}$
6. $\frac{4}{5} - \frac{1}{5}$
7. $\frac{5}{8} - \frac{3}{8}$
8. $\frac{5}{12} + \frac{2}{12}$

Set B (pp. 350–351)

Estimate each sum or difference.

1. $\frac{5}{6} - \frac{1}{8}$
2. $\frac{3}{7} + \frac{7}{9}$
3. $\frac{10}{11} - \frac{2}{4}$
4. $\frac{2}{3} + \frac{7}{8}$
5. $\frac{6}{10} - \frac{1}{5}$

Estimate to compare. Write < or > for each ●.

6. $\frac{2}{3} + \frac{3}{4}$ ● $\frac{4}{5} + \frac{1}{10}$
7. $\frac{5}{8} - \frac{1}{9}$ ● $\frac{4}{10} + \frac{5}{12}$
8. $\frac{5}{6} - \frac{3}{7}$ ● 0

Set C (pp. 352–353)

Name the LCD. Find the sum or difference in simplest form.

1. $\frac{1}{3} + \frac{1}{6}$
2. $\frac{1}{5} + \frac{4}{10}$
3. $\frac{7}{8} - \frac{1}{4}$
4. $\frac{2}{3} + \frac{1}{4}$
5. $\frac{1}{3} + \frac{1}{4}$
6. $\frac{5}{8} - \frac{1}{2}$
7. $\frac{8}{9} - \frac{1}{3}$
8. $\frac{4}{5} - \frac{3}{4}$

Set D (pp. 354–357)

Find the sum or difference in simplest form.

1. $\frac{2}{3} + \frac{2}{9}$
2. $\frac{1}{4} + \frac{3}{8}$
3. $\frac{1}{3} + \frac{2}{4}$
4. $\frac{4}{5} - \frac{1}{2}$
5. $\frac{8}{9} - \frac{2}{3}$
6. $\frac{1}{3} + \frac{2}{9}$
7. $\frac{7}{8} - \frac{1}{4}$
8. $1 - \frac{2}{7}$
9. $\frac{3}{10} + \frac{2}{5}$
10. $\frac{7}{12} - \frac{1}{3}$
11. $\frac{6}{7} - \frac{1}{2}$
12. $\frac{1}{4} + \frac{1}{6}$
13. $\frac{1}{3} + \frac{2}{5}$
14. $\frac{3}{4} - \frac{5}{12}$
15. $\frac{2}{3} + \frac{1}{6}$
16. Eva used $\frac{3}{4}$ yard, $\frac{7}{8}$ yard, and $\frac{1}{2}$ yard of ribbon to make a hair bow. How much ribbon did she use?
17. The sauce has $\frac{3}{4}$ teaspoon of basil, $\frac{1}{2}$ teaspoon of oregano, and $\frac{3}{4}$ teaspoon of parsley. How many teaspoons of herbs does the sauce have in all?

Set A (pp. 364–367)

Find the sum in simplest form. Estimate to check.

1. $2\frac{1}{2} + \frac{1}{6}$
2. $3\frac{2}{5} + 1\frac{1}{10}$
3. $4\frac{1}{3} + 2\frac{5}{9}$
4. $1\frac{5}{8} + 2\frac{1}{2}$
5. $6\frac{2}{3} + 4\frac{1}{6}$
6. $5\frac{1}{4} + 2\frac{1}{2}$
7. $7\frac{1}{4} + 1\frac{4}{5}$
8. $9\frac{1}{3} + 1\frac{1}{4}$
9. $10\frac{2}{5} + 6\frac{1}{2}$
10. Dana put $1\frac{1}{2}$ quarts of oil in her car on March 3rd and $1\frac{1}{4}$ quarts on March 28th. How much oil did Dana put in her car in March?
11. Vic and his teammates had $1\frac{1}{10}$ hours to practice drills and run laps. If they ran for $\frac{2}{5}$ hour, how long did they practice drills?

Set B (pp. 368–369)

Find the difference in simplest form. Estimate to check.

1. $3\frac{5}{7} - 2\frac{3}{14}$
2. $10\frac{4}{5} - 6\frac{3}{10}$
3. $8\frac{1}{3} - 3\frac{5}{6}$
4. $5\frac{1}{2} - 2\frac{7}{12}$
5. $5\frac{1}{2} - 2\frac{1}{3}$
6. $7\frac{5}{6} - 6\frac{2}{3}$
7. $3\frac{3}{4} - 1\frac{1}{2}$
8. $10\frac{7}{8} - 6\frac{1}{2}$
9. $12\frac{2}{3} - 4\frac{5}{6}$

Set C (pp. 372–373)

Add or subtract. Write the answer in simplest form. Estimate to check.

1. $4\frac{2}{3} + 5\frac{5}{6}$
2. $12\frac{1}{3} - 6\frac{5}{9}$
3. $7\frac{1}{5} + 8\frac{3}{4}$
4. $15\frac{1}{6} - 10\frac{3}{4}$
5. $2\frac{1}{8} + 1\frac{3}{4}$
6. $7\frac{7}{9} - 6\frac{2}{3}$
7. $14\frac{2}{3} - 5\frac{1}{2}$
8. $8\frac{7}{8} - 5\frac{3}{4}$
9. $7\frac{1}{6} + 4\frac{5}{9}$
10. There are $7\frac{7}{8}$ cups of flour in the canister. If $2\frac{1}{4}$ cups are used to make muffins, how much will be left?
11. Marilyn bought $4\frac{1}{2}$ pounds of hazelnut coffee and $2\frac{1}{4}$ pounds of chocolate-flavored coffee. How much coffee did she buy altogether?

CHAPTER 21 EXTRA PRACTICE

Set A (pp. 380–381)

Find the product.

1. $\frac{2}{5} \times 10$
2. $\frac{1}{4} \times 12$
3. $15 \times \frac{3}{5}$
4. $\frac{4}{7} \times 28$
5. $32 \times \frac{5}{8}$
6. $30 \times \frac{5}{6}$
7. $\frac{7}{8} \times 16$
8. $\frac{3}{4} \times 20$
9. $\frac{1}{2} \times 22$
10. $24 \times \frac{1}{3}$
11. There are 15 cars in a parking lot. Of the 15 cars, $\frac{1}{3}$ are red. How many of the cars in the parking lot are red?
12. Mike has 22 baseball cards. Of the 22 cards, $\frac{1}{2}$ show rookie players. How many of the cards show rookie players?

Set B (pp. 382–383)

Find the product. Write the answer in simplest form.

1. $\frac{3}{4} \times \frac{5}{6}$
2. $\frac{2}{3} \times \frac{4}{5}$
3. $\frac{3}{4} \times \frac{1}{3}$
4. $\frac{1}{3} \times \frac{3}{5}$
5. $\frac{3}{8} \times \frac{2}{3}$
6. $\frac{3}{10} \times \frac{1}{6}$
7. $\frac{1}{5} \times \frac{5}{12}$
8. $\frac{2}{3} \times \frac{5}{6}$
9. $\frac{5}{8} \times \frac{2}{5}$
10. $\frac{2}{3} \times \frac{3}{5}$
11. A muffin recipe calls for $\frac{3}{4}$ cup of flour. If you cut the recipe in half, how much flour do you need?
12. Julie bought $\frac{3}{4}$ yard of fabric. She used $\frac{1}{3}$ of it to make a doll dress. What part of a yard of fabric did she use for the dress?

Set C (pp. 384–385)

Multiply. Write the answer in simplest form.

1. $\frac{1}{2} \times 1\frac{1}{2}$
2. $2\frac{1}{3} \times \frac{1}{4}$
3. $\frac{4}{5} \times 2\frac{1}{6}$
4. $2\frac{1}{2} \times \frac{2}{3}$
5. $\frac{3}{4} \times 3\frac{1}{2}$
6. $\frac{3}{4} \times 1\frac{2}{3}$
7. $2\frac{1}{3} \times \frac{1}{5}$
8. $\frac{5}{6} \times 3\frac{1}{5}$
9. $1\frac{3}{4} \times \frac{4}{7}$
10. $2\frac{1}{2} \times \frac{5}{8}$

Set D (pp. 386–387)

Multiply. Write the answer in simplest form.

1. $2\frac{1}{4} \times 1\frac{2}{3}$
2. $3\frac{2}{3} \times 4\frac{1}{2}$
3. $1\frac{7}{8} \times 2\frac{2}{5}$
4. $2\frac{2}{3} \times 4\frac{1}{2}$
5. $4 \times 1\frac{7}{8}$
6. $3\frac{1}{2} \times 2\frac{3}{5}$
7. $1\frac{5}{7} \times 3\frac{1}{3}$
8. $4\frac{1}{3} \times 6\frac{3}{4}$
9. $5\frac{1}{4} \times 3\frac{2}{5}$
10. $2\frac{5}{8} \times 4\frac{2}{3}$
11. LaKeshia is training for a track meet. She walks $5\frac{3}{4}$ mi every day. How many miles does she walk in one week?

Set A (pp. 396–397)

Are the two numbers reciprocals? Write *yes* or *no*.

1. $\frac{2}{3}$ and 3 **2.** $\frac{3}{4}$ and $1\frac{1}{3}$ **3.** $\frac{4}{7}$ and $1\frac{4}{7}$ **4.** $1\frac{2}{3}$ and $\frac{3}{5}$ **5.** $\frac{3}{4}$ and $1\frac{1}{4}$

Write the reciprocal of each number.

6. $\frac{1}{3}$ **7.** 10 **8.** $\frac{7}{2}$ **9.** $1\frac{3}{4}$ **10.** $1\frac{1}{3}$

11. $\frac{6}{9}$ **12.** $1\frac{1}{4}$ **13.** $\frac{3}{4}$ **14.** 6 **15.** $5\frac{2}{5}$

16. Jane writes the reciprocal of $\frac{7}{10}$. What number does she write?

17. Is the reciprocal of $1\frac{2}{5}$ greater than or less than 1?

Set B (pp. 398–399)

Divide. Write the answer in simplest form.

1. $9 \div \frac{3}{4}$ **2.** $4 \div \frac{1}{3}$ **3.** $2 \div \frac{4}{5}$ **4.** $7 \div \frac{2}{3}$ **5.** $10 \div \frac{1}{4}$

6. $3 \div \frac{3}{4}$ **7.** $8 \div \frac{1}{6}$ **8.** $6 \div \frac{3}{8}$ **9.** $1 \div \frac{5}{6}$ **10.** $15 \div \frac{2}{3}$

11. How many halves are in 7?

12. How many thirds are in 6?

13. There are 8 pounds of clay for projects. Each project uses $\frac{1}{2}$ pound of clay. How many projects can be made from the clay?

14. A pastry chef has 9 pounds of butter and uses $\frac{3}{4}$ pound for each batch of cookies. How many batches of cookies can he make?

Set C (pp. 400–403)

Use a reciprocal to write a multiplication problem for the division problem.

1. $\frac{3}{4} \div \frac{1}{5}$ **2.** $\frac{3}{5} \div \frac{6}{7}$ **3.** $\frac{1}{3} \div \frac{5}{9}$ **4.** $\frac{6}{7} \div \frac{1}{8}$ **5.** $\frac{5}{8} \div \frac{2}{3}$

Divide. Write the answer in simplest form.

6. $\frac{4}{7} \div \frac{3}{4}$ **7.** $\frac{7}{8} \div \frac{5}{6}$ **8.** $\frac{5}{7} \div \frac{5}{9}$ **9.** $\frac{1}{6} \div \frac{3}{7}$ **10.** $\frac{2}{7} \div \frac{1}{4}$

11. $\frac{2}{3} \div \frac{4}{7}$ **12.** $\frac{3}{10} \div 4$ **13.** $\frac{5}{8} \div 2\frac{1}{2}$ **14.** $3\frac{1}{3} \div \frac{2}{3}$ **15.** $2\frac{2}{3} \div 1\frac{1}{6}$

16. Chris wants to build a fence $8\frac{1}{3}$ yd long. How many $1\frac{2}{3}$ yd pieces of fence will he need?

17. A gold chain is $\frac{7}{8}$ foot long. If each link in the chain is $\frac{1}{64}$ foot long, how many links are there?

CHAPTER 23

EXTRA PRACTICE

Set A (pp. 416–419)

Identify the integers graphed on the number line.

1.

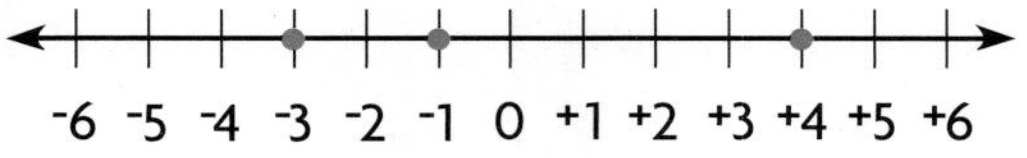

2.

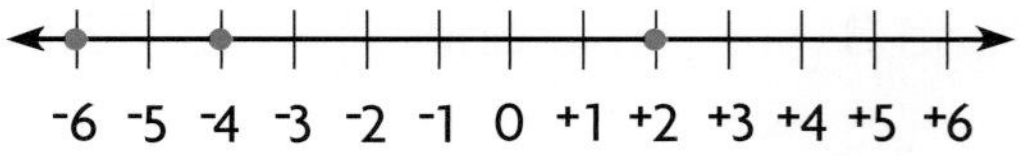

Write an integer to represent each situation.

3. spend \$100 **4.** 75° above 0° **5.** a loss of \$3 **6.** gain 12 yards

Set B (pp. 420–421)

Compare. Write <, >, or = for each ●.

1. $^{+}6$ ● $^{+}3$ **2.** $^{-}6$ ● $^{-}3$ **3.** $^{+}10$ ● $^{-}10$ **4.** $^{-}3$ ● 0 **5.** $^{-}2$ ● $^{-}4$

Order each set of integers from least to greatest.

6. $^{+}6, {}^{-}1, {}^{+}7, {}^{-}5$ **7.** $^{+}3, {}^{-}2, 0, {}^{-}4$ **8.** $^{-}9, {}^{+}8, {}^{-}11, {}^{+}4$ **9.** $^{+}2, {}^{-}2, {}^{+}1, 0$

10. The average temperatures for 3 weeks in Alaska were $^{-}21$°F, $^{-}26$°F, and $^{-}25$°F. Which is the coldest?

Set C (pp. 422–425)

Write the addition number sentence modeled.

1.

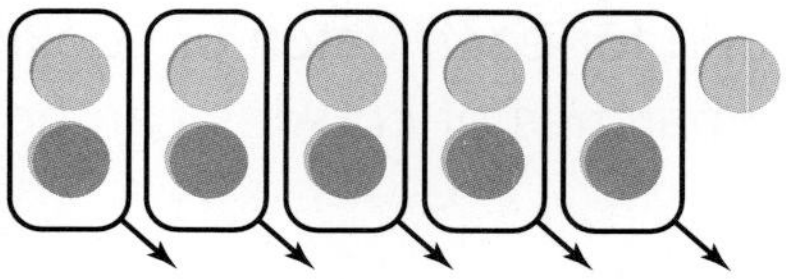

2.

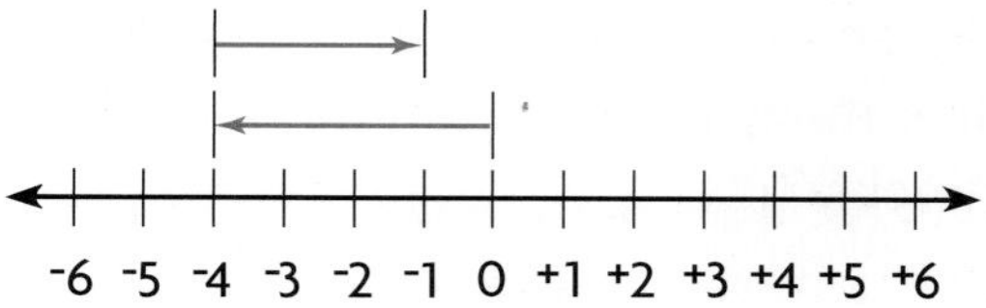

Find each sum.

3. $^{+}3 + {}^{-}4$ **4.** $^{-}5 + {}^{+}2$ **5.** $^{+}9 + {}^{-}7$ **6.** $^{+}6 + {}^{-}6$

Set D (pp. 428–431)

Draw a number line to find the difference.

1. $^{+}5 - {}^{+}1$ **2.** $^{-}5 - {}^{+}4$ **3.** $^{-}6 - {}^{+}2$ **4.** $^{-}2 - {}^{+}3$

Write the subtraction number sentence modeled.

5.

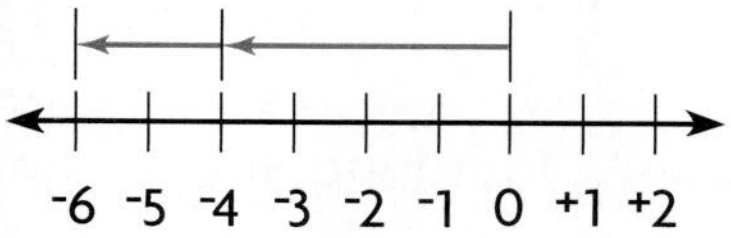

6.

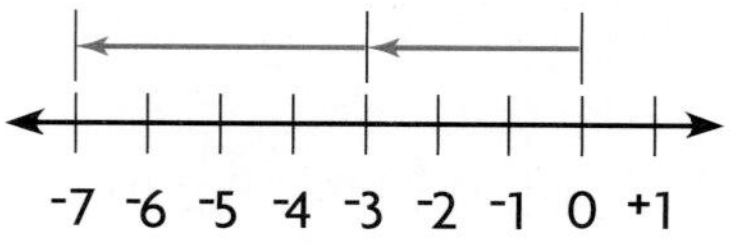

Find each difference.

7. $^{+}12 - {}^{+}4$ **8.** $^{-}5 - {}^{+}5$ **9.** $^{-}10 - 0$ **10.** $^{-}5 - {}^{+}9$

Set A (pp. 438–439)

Write the ordered pairs. Then graph them.

1.

Input, x	0	1	2	3
Output, y	4	5	6	7

2.

Input, x	5	10	15	20
Output, y	1	2	3	4

Set B (pp. 440–443)

For 1–6, identify the ordered pair for each point.

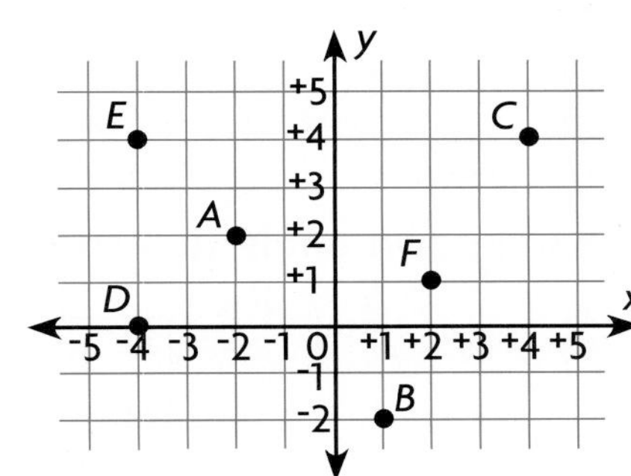

1. Point A **2.** Point B

3. Point C **4.** Point D

5. Point E **6.** Point F

Graph and label the ordered pair on a coordinate plane.

7. $A\ (^-1,^+3)$ **8.** $B\ (^-3,^-7)$ **9.** $C\ (^+6,^+7)$

10. $D\ (^+1,^-5)$ **11.** $E\ (0,^+8)$ **12.** $F\ (^+2,^+3)$

For 13–16, name the ordered pair that is described.

13. Start at the origin. Move 11 units to the left.

14. Start at the origin. Move 7 units to the left and 4 units down.

15. Start at the origin. Move 8 units to the right and 1 unit up.

16. Start at the origin. Move 6 units to the left and 6 units up.

Set C (pp. 444–447)

Use a rule to complete the table. Write ordered pairs and then make a graph.

1.

Cats, x	1	2	3	4	5
Legs, y	4	8	12		

2.

Batteries, x	2	4	6	8	10
Games, y	1	2	3		

3. Write an equation for the relationship in the table in Problem 1.

4. If you have 28 batteries, how many games do you have?

5. Look at your graph of the data in Problem 2. What does the ordered pair (6,3) represent?

6. **REASONING** How can you use the equation $y = 4x$ to find the number of legs on 50 cats?

CHAPTER 25

EXTRA PRACTICE

Set A (pp. 454–457)

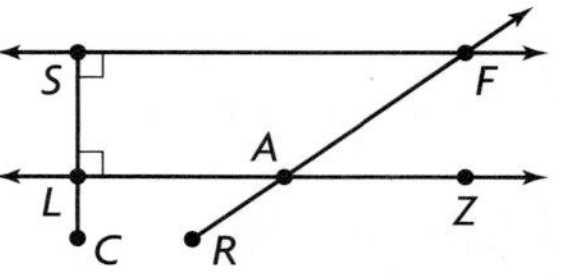

Use the figure at the right. Name an example of each term.

1. point **2.** right angle **3.** ray **4.** parallel lines **5.** line segment

Set B (pp. 460–463)

Name each polygon and tell if it is *regular* or *not regular*.

1. 

2. 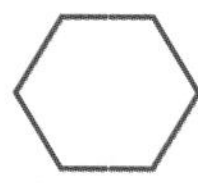

3. 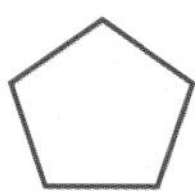

4. 

Find the unknown angle measure.

5. 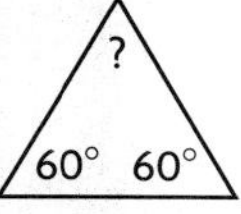

6. 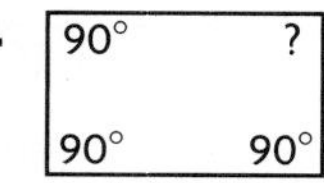

7. 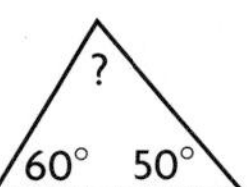

8. 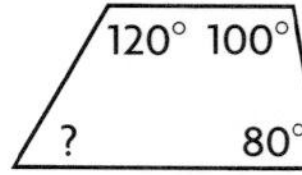

Set C (pp. 464–467)

Measure each central angle.

1. 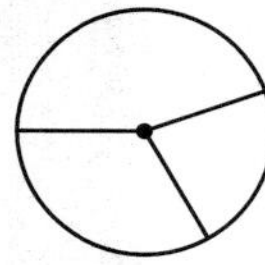

2. 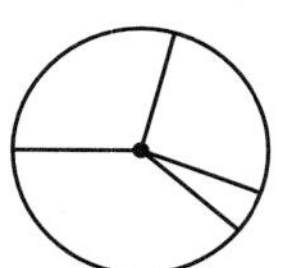

Find the unknown angle measure.

3. 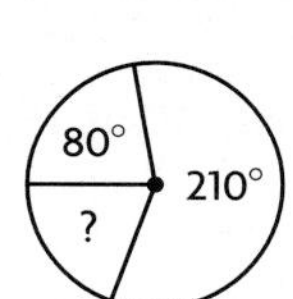

4. 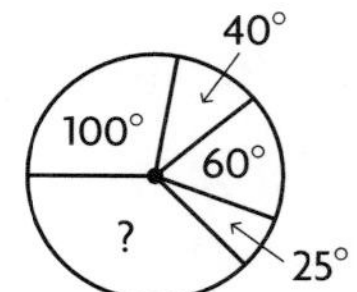

5. The pizza restaurant will deliver only within a 3-mile radius. Tasha lives 2 miles from the restaurant. Will it deliver?

Set D (pp. 468–469)

Write *similar, congruent,* or *neither* to describe each pair of figures.

1. 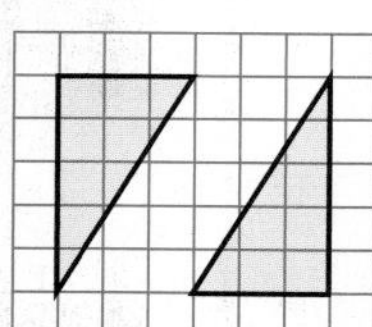

2. 

3. 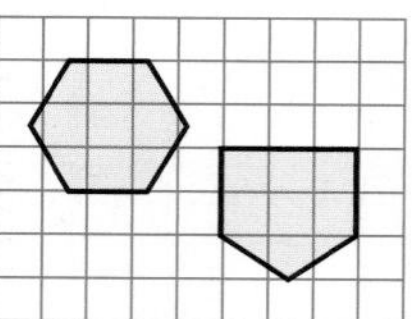

Set E (pp. 470–471)

Trace each figure. Draw lines of symmetry. Tell whether each figure has rotational symmetry. Write *yes* or *no*.

1. 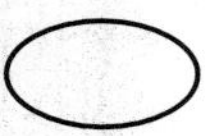

2. 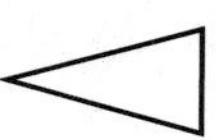

3. 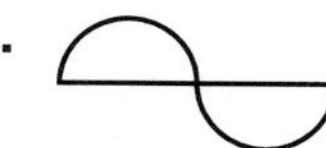

Set A (pp. 478–481)

Classify each triangle by its sides and angles.

1. 2.

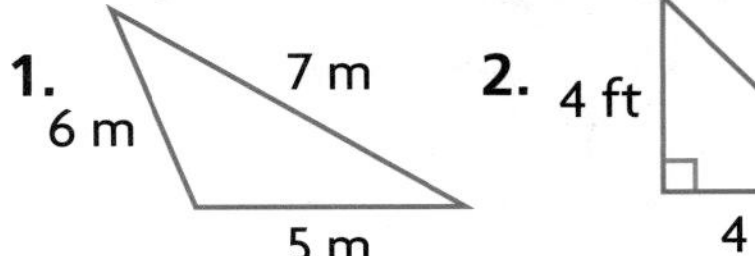

3.

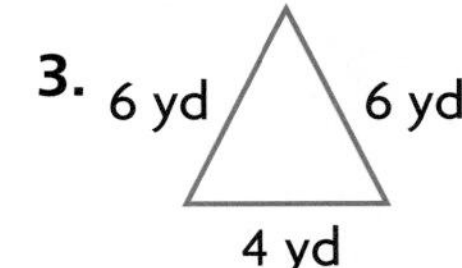

4. 5.

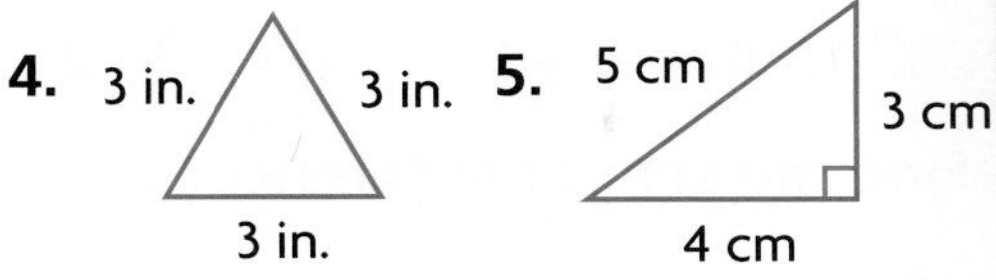

6. Alfonso biked 1 mile north, $1\frac{1}{2}$ miles southeast, and $1\frac{1}{8}$ miles back to where he started. How far did he ride? Classify his triangular path by its sides.

7. The measures of two angles of a triangle are $30°$ and $60°$. What is the measure of its third angle? Classify the triangle by its angles.

Set B (pp. 482–483)

Classify each figure in as many ways as possible. Write *quadrilateral, parallelogram, square, rectangle, rhombus,* or *trapezoid*.

1. 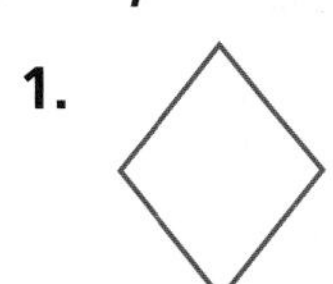

2.

3.

4.

5.

6.

7. The measures of three angles of a quadrilateral are $60°$, $120°$, and $60°$. What is the measure of the fourth angle? If all of its sides are congruent, what kind of quadrilateral is it?

Set C (pp. 486–489)

Classify each solid figure. If it is a prism or pyramid, write the number of faces, vertices, and edges.

1. 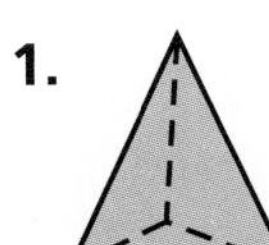

2. 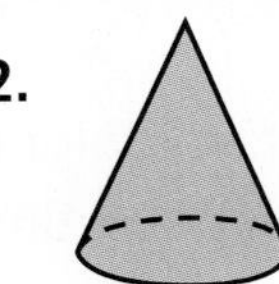

3.

4. 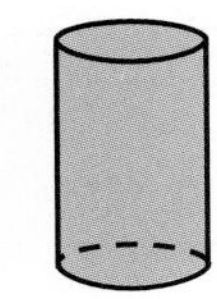

5. 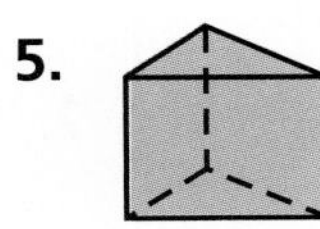

6. 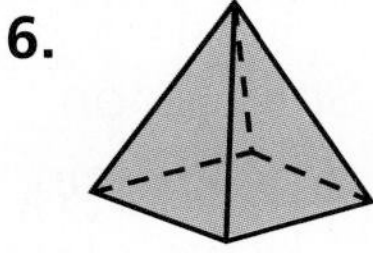

7. Karen's jewelry box is a pentagonal prism. She wants to paint each face a different color. How many different paint colors will she need?

CHAPTER 27

EXTRA PRACTICE

Set A (pp. 508–511)

Change the unit.

1. 15 ft = ■ yd
2. 72 in. = ■ ft
3. 100 mm = ■ cm

Find the sum or difference.

4. 9 ft 2 in. + 6 ft 4 in.
5. 5 ft 1 in. − 2 ft 11 in.
6. 10 ft 5 in. + 3 ft 11 in.
7. 3.9 cm + 4.8 cm
8. 6.1 cm − 5.2 cm
9. 2.3 cm + 9.6 cm

Set B (pp. 512–513)

Change the unit.

1. 8 qt = ■ gal
2. 32 pt = ■ qt
3. 3 gal = ■ c
4. 19 lb = ■ oz
5. 6 T = ■ lb
6. $\frac{1}{2}$ lb = ■ oz
7. Jane bought five 3-lb bags of apples to make apple pies. How many ounces of apples did she buy?
8. The Chungs ordered 500 gallons of juice for the breakfast. How many quarts of juice did they order?

Set C (pp. 514–515)

Change the unit.

1. 1,000 g = ■ kg
2. 3.5 kg = ■ g
3. 2,000 g = ■ kg
4. 1.5 L = ■ mL
5. 1 kL = ■ L
6. 0.5 L = ■ mL
7. Juan bought ten 1-L sodas. How many mL of soda did he buy?
8. Bob ate 10 g of fat at breakfast, 15 g at lunch, and 5 g at dinner. How many mg of fat did he eat?

Set D (pp. 516–517)

Write the time for each.

1. Start: noon
6 hr 26 min elapsed time
End: ■
2. Start: January 10, 3:05 A.M.
6 days 4 hr 16 min elapsed time
End: ■
3. Start: midnight
■ elapsed time
End: 11:45 P.M.
4. The clock shows the minutes and seconds left in a game. How much total time in seconds is left?

Set A (pp. 524–525)

Find the perimeter of each polygon.

1.

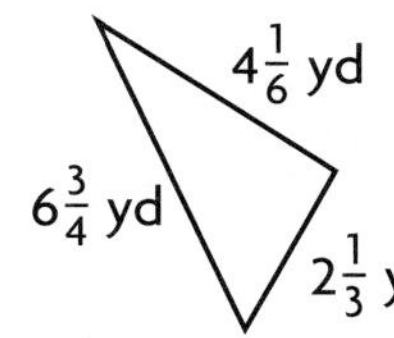

2.

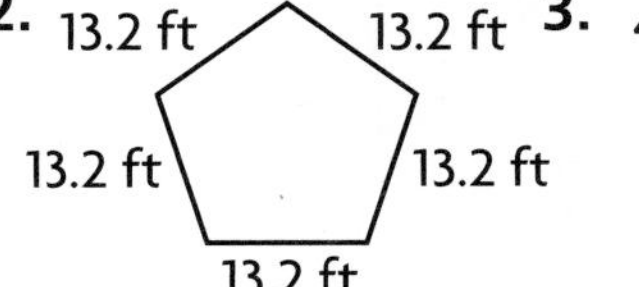

3.

4.

Set B (pp. 528–529)

Find the area of each figure.

1.

2.

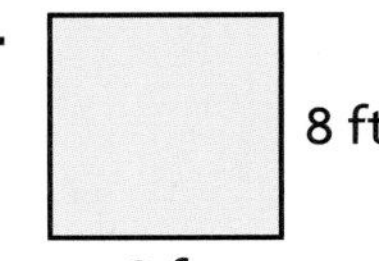

3.

4.

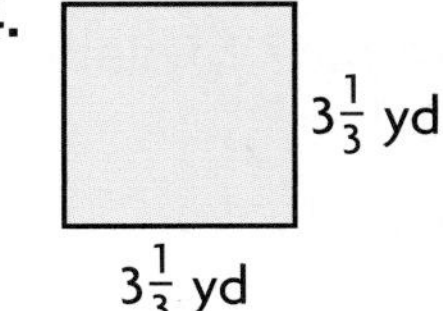

5. John has 68 ft of fence for a rectangular garden. What dimensions will make the greatest garden area?

Set C (pp. 532–535)

Find the area of each triangle.

1.

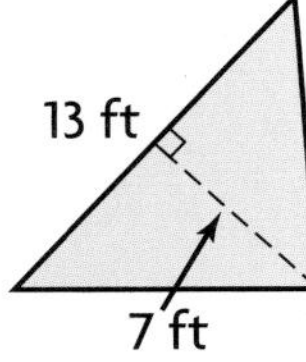

2.

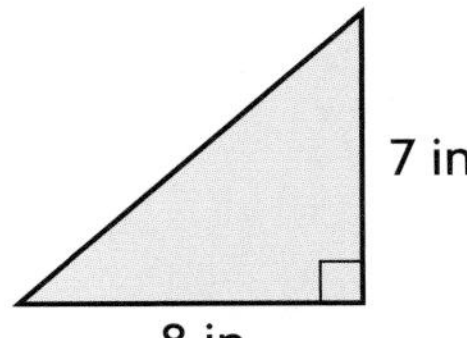

3.

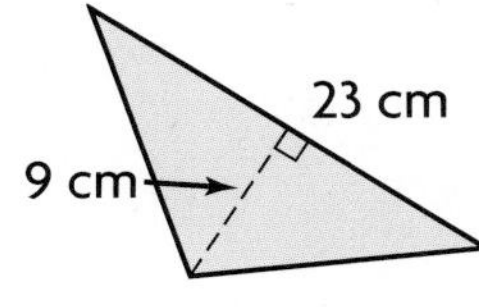

4. A road sign is a scalene triangle. The base is 2.8 meters and the height is 1.8 meters. What is the area of the sign?

Set D (pp. 536–539)

Find the area of each parallelogram.

1.

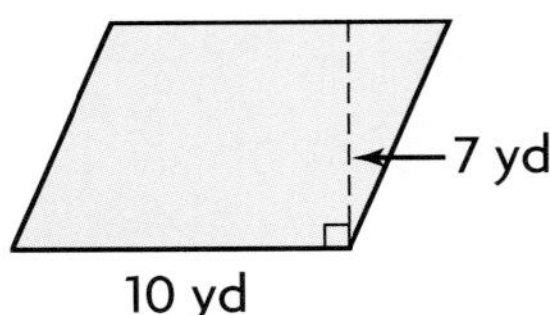

2.

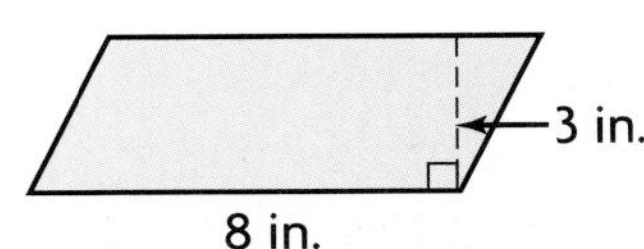

3.

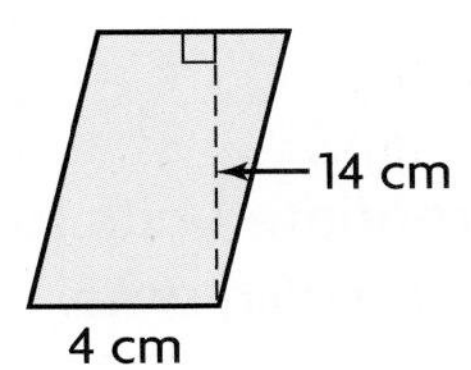

4. The area of a banner in the shape of a parallelogram is 120 ft^2. If its base is 8 ft, what is its height?

CHAPTER 29

EXTRA PRACTICE

Set A (pp. 546–547)

Match each solid figure with its net. Write *a, b, c,* or *d*.

1. 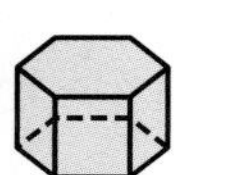

2.

3. 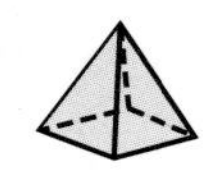

4. 

a. 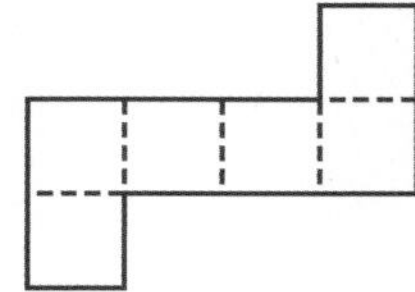

b. 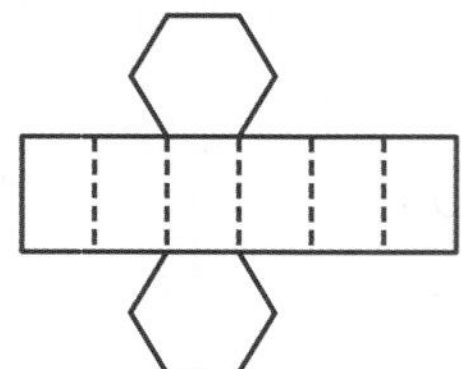

c. 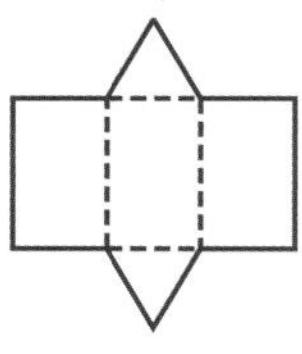

d. 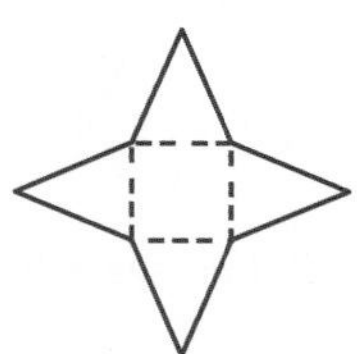

5. Which solid has more faces: a triangular pyramid or a square pyramid? Explain.

Set B (pp. 548–551)

Find the surface area of each prism.

1. 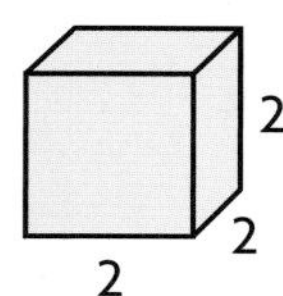

2. 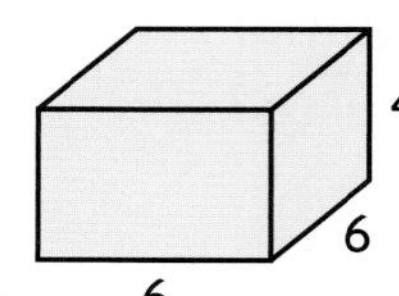

3. 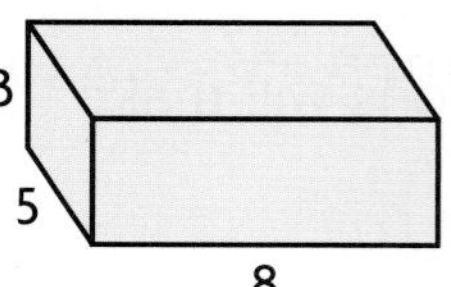

4. How many square inches of paper are needed to cover a box that is 4 inches long, 10 inches wide, and 3 inches high?

Set C (pp. 552–555)

Find the unknown measure.

1. length = 8 ft
width = 2 ft
height = 3 ft
volume = ■

2. length = ■
width = 10 ft
height = 10 ft
volume = 1,000 ft^3

3. length = 5.5 in.
width = 2 in.
height = 3 in.
volume = ■

4. Rick's fish tank has 10,640 cubic cm of water in it. If the length is 38 cm and the width is 28 cm, what is the height of the water?

Set D (pp. 556–557)

Tell the appropriate units for measuring each. Write *units, square units,* or *cubic units*.

1. room inside a space shuttle

2. fence for a play yard

3. carpet for a room

4. paper to cover a box

EXTRA PRACTICE CHAPTER 30

Set A (pp. 572–573)

For 1–2, make a tree diagram to show the possible choices.

1. **Lunch Menu**

   Entree: sandwich, hot dog, or hamburger

   Side: pretzel, apple, or pudding

2. **School Trips**

   Place: zoo, museum, or planetarium

   Day: Monday, Tuesday, Wednesday, Thursday, or Friday

3. Sarah needs a new television. She can buy one with a 13, 20, or 25 inch screen. She can buy one with or without a VCR. How many choices does she have?

4. Maurice has to go to the store. He can go through the park or around the park. He can either walk or run. How many choices does he have?

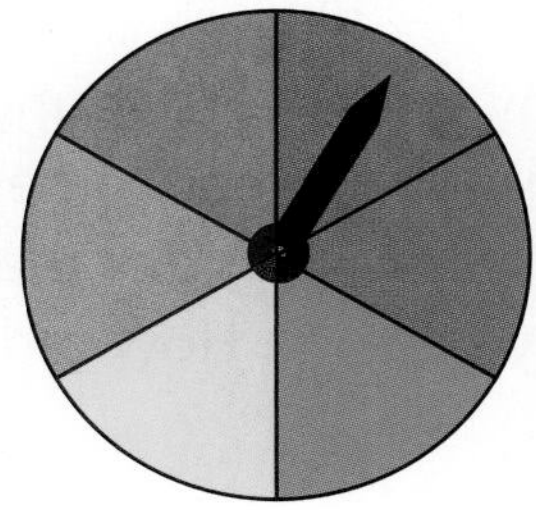

Set B (pp. 574–575)

Write the probability of the pointer landing on each color.

1. green
2. purple or yellow
3. white
4. purple, green, or yellow
5. Mark has a bag of fruit candy. Of the candies, 5 are strawberry, 4 are grape, and 1 is apple. What is the probability that Mark will pick grape?
6. If you toss one coin and roll one number cube, how many outcomes are possible? What is the probability that the result will include heads?

Set C (pp. 576–577)

There are 12 marbles in a bag. 1 is blue, 4 are red, 2 are orange, and 5 are yellow. Write each probability as a fraction. Tell which event is more likely.

1. You pull a red marble; you pull a yellow marble.
2. You pull a blue marble; you pull an orange marble.
3. You pull a marble that is not yellow; you pull a marble that is not orange.
4. You pull a marble that is not red; you pull a marble that is not blue.
5. You pull a red marble; you pull a marble that is not orange.
6. You pull a blue marble; you pull a marble that is not yellow.

Tips for Taking Math Tests

Being a good test-taker is like being a good problem solver. When you answer test questions, you are solving problems. Remember to **UNDERSTAND, PLAN, SOLVE,** and **CHECK.**

Read the problem.

- Look for math terms and recall their meanings.
- Reread the problem and think about the question.
- Use the details in the problem and the question.

1. Tony is 10 years younger than Bill. Amber is 5 years older than Tony. Iris is 3 years younger than Amber. Bill is 15 years old. How old is Iris?

A 7 **C** 13

B 12 **D** 27

TIP! Understand the problem.
The critical information is that Bill is 15 years old. List all the given relationships and express them as equations. Use Bill's age to begin solving the equations. Be sure to answer the question asked. The answer is A.

- Each word is important. Missing a word or reading it incorrectly could cause you to get the wrong answer.
- Is there information given that you do not need?
- Pay attention to words that are in **bold** type, all CAPITAL letters, or *italics*.
- Some other words to look for are *round, about, only, best,* or *least to greatest*.

2. In 1995, Mariko ran the 100-yard dash in 12.40 seconds. In 1996, she ran it in 12.343 seconds. In 1997, she ran it in 12.6 seconds. In 1998, she ran it in 12.502 seconds. In what year was Mariko's fastest time?

F 1995 **H** 1997

G 1996 **J** 1998

TIP! Look for important words.
The word <u>fastest</u> is important. The fastest runner takes the least amount of time. So, the least number of seconds is her fastest time. The answer is G.

Plan

Think about how you can solve the problem.

- Can you solve the problem with the information given?
- Pictures, charts, tables, and graphs may have the information you need.
- You may need to recall information not given.
- The answer choices may have information you need.

3. What is the greatest common factor of 27, 33, and 63?

A 1 **C** 3

B 2 **D** 9

TIP! **Get the information you need.**

The greatest common factor is the greatest number that is a factor of all of the given numbers. Make a list of the factors of each number and look for the common factors. The greatest of these common factors is the answer. The answer is C.

- You may need to write a number sentence and solve it to answer the question.
- Some problems have two steps or more.
- In some problems you need to look at relationships instead of computing an answer.
- If the path to the solution isn't clear, choose a problem solving strategy and use it to solve the problem.

4. A gravel path surrounds a rectangular grass field. The field is 60 feet long and 40 feet wide. The path is 3 feet wide. What is the area covered by the path?

F 300 sq ft **H** 600 sq ft

G 309 sq ft **J** 636 sq ft

TIP! **Decide on a plan.**

Using the strategy *draw a diagram* will help you find the dimensions of the larger rectangle so you can find the area. Match the description in the problem and label it. First find the area of the field and the path. Then subtract the area of the field. The answer is J.

Solve

Follow your plan, working logically and carefully.

- Estimate your answer. Are any answer choices unreasonable?
- Use reasoning to find the most likely choices.
- Make sure you solved all steps needed to answer the problem.
- If your answer does not match any of the answer choices, check the numbers you used. Then check your computation.

5. A point located at (2,4) on a coordinate plane was translated to (5,10). Which statement describes how the point was translated?

A 3 units up and 6 units to the right

B 3 units up and 6 units to the left

C 3 units to the right and 6 units up

D 3 units to the right and 6 units down

TIP! Eliminate choices.

Thinking about the location of the point (5,10), you know it is to the right and up from the point (2,4). You can eliminate choices B and D because they say left or down. Now decide which of choices A or C is correct. When you subtract 2 from 5 and 4 from 10, you see that the point (5,10) is 3 units to the right and 6 units up from (2,4). The answer is C.

- If your answer still does not match one of the choices, look for another form of the number, such as a decimal instead of a fraction.
- If answer choices are given as pictures, look at each one by itself while you cover the other three.
- If you do not see your answer and the answer choices include **Not here,** make sure your work is correct and then mark **Not here.**
- If answer choices are statements, relate each one to the problem.
- Change your plan if it isn't working. Try a different strategy.

6. Eduardo needs to find the total weight of a package he is mailing in order to calculate the postage needed. The package contains 15 copies of a document that weighs 2.5 ounces. The mailing envelope weighs 0.25 ounces. What is the total weight of the package?

F 17.75 oz

G 37.5 oz

H 41.25 oz

J Not here

TIP! Choose the answer.

You need to calculate the weight of all the copies of the document (15×2.5) and then add the weight of the envelope. If your answer doesn't match one of the answer choices, check your computation and the placement of the decimal point. If you know your work is correct, and your answer (37.75) is not one of the choices, mark the letter for Not here. The answer is J.

Check

Take time to catch your mistakes.

- Be sure you answered the question asked.
- Check that your answer fits the information in the problem.
- Check for important words you might have missed.
- Be sure you used all the information you needed.
- Check your computation by using a different method.
- Draw a picture when you are unsure of your answer.

7. The distance from Todd's home to the soccer field is $\frac{3}{8}$ mile. The distance from Mia's home to the field is $\frac{2}{3}$ as great as Todd's distance. How far is it from Mia's home to the soccer field?

A $\frac{5}{11}$ mi **C** $\frac{7}{24}$ mi

B $\frac{1}{3}$ mi **D** $\frac{1}{4}$ mi

TIP!

Check your work.

Since Mia's distance is $\frac{2}{3}$ of Todd's distance, it must be less than Todd's. To check your multiplication, draw a square. Divide the square into 8 equal columns and 3 equal rows. Shade the amounts given. Count the overlapping parts that are shaded and write the fraction that names that part. Be sure your fraction is in simplest form. The answer is D.

Don't Forget!

Before the test

- Listen to the teacher's directions and read the instructions.
- Write down the ending time if the test is timed.
- Know where and how to mark your answers.
- Know whether you should write on the test page or use scratch paper.
- Ask any questions you may have before the test begins.

During the test

- Work quickly but carefully. If you are unsure how to answer a question, leave it blank and return to it later.
- If you cannot finish on time, look over the questions that are left. Answer the easiest ones first. Then go back to answer the others.
- Fill in each answer space carefully. Erase completely if you change an answer. Erase any stray marks.
- Check that the answer number matches the question number, especially if you skip a question.

ADDITION AND SUBTRACTION FACTS TEST

	K	L	M	N	O	P	Q	R
A	5 +6	9 −3	8 +6	10 − 7	18 − 9	7 +9	11 − 4	0 +9
B	12 − 4	9 +9	14 − 5	8 −5	9 +4	11 − 6	5 +3	16 − 9
C	6 +5	0 +8	9 −6	4 +4	13 − 6	3 +6	12 − 3	7 +4
D	6 +3	14 − 6	8 +8	7 −7	13 − 5	5 +8	9 +7	11 − 8
E	4 +6	17 − 9	10 − 5	6 +6	8 −4	1 +9	8 +7	12 − 9
F	8 +5	4 +7	11 − 3	3 +7	10 − 2	9 +0	12 − 8	7 +2
G	5 +7	13 − 4	6 +8	20 −10	8 −0	6 +9	14 − 7	4 +8
H	6 +7	11 − 7	9 −9	9 +8	16 − 8	8 +10	17 −10	7 −3
I	13 − 7	14 − 9	5 +5	9 +10	10 − 6	3 +9	9 −7	7 +6
J	11 − 2	9 +5	15 − 7	10 +10	13 − 9	7 +8	11 − 5	8 +4

MULTIPLICATION FACTS TEST

	K	L	M	N	O	P	Q	R
A	4×8	5×4	8×9	6×2	9×5	12×8	3×9	1×5
B	7×1	6×3	5×3	3×8	5×5	3×3	0×8	8×5
C	5×7	9×6	6×4	3×4	8×6	7×3	11×7	4×2
D	3×2	9×8	1×6	7×2	4×6	7×5	12×12	4×7
E	7×8	2×4	5×9	12×10	7×6	6×9	6×0	4×4
F	1×9	7×7	12×5	9×3	8×11	6×6	2×9	11×11
G	3×7	4×3	9×9	7×4	6×10	9×7	2×5	6×8
H	12×3	11×5	8×3	12×7	10×4	4×5	10×10	9×4
I	3×5	9×12	12×11	12×6	11×10	8×8	11×3	5×12
J	7×9	4×12	9×11	3×10	9×10	12×2	8×12	6×12

DIVISION FACTS TEST

	K	L	M	N	O	P	Q	R
A	$2\overline{)8}$	$9\overline{)54}$	$7\overline{)21}$	$4\overline{)16}$	$3\overline{)18}$	$7\overline{)0}$	$4\overline{)32}$	$6\overline{)48}$
B	$4\overline{)8}$	$8\overline{)40}$	$1\overline{)4}$	$5\overline{)35}$	$2\overline{)12}$	$5\overline{)10}$	$7\overline{)56}$	$6\overline{)30}$
C	$8\overline{)64}$	$7\overline{)42}$	$6\overline{)18}$	$9\overline{)27}$	$12\overline{)96}$	$6\overline{)54}$	$4\overline{)20}$	$7\overline{)49}$
D	$8\overline{)48}$	$9\overline{)72}$	$5\overline{)60}$	$7\overline{)28}$	$4\overline{)24}$	$5\overline{)40}$	$3\overline{)27}$	$9\overline{)36}$
E	$7\overline{)14}$	$8\overline{)24}$	$3\overline{)9}$	$11\overline{)99}$	$5\overline{)20}$	$3\overline{)6}$	$2\overline{)14}$	$4\overline{)12}$
F	$6\overline{)36}$	$4\overline{)28}$	$6\overline{)72}$	$12\overline{)60}$	$2\overline{)10}$	$8\overline{)0}$	$5\overline{)45}$	$9\overline{)81}$
G	$2\overline{)16}$	$3\overline{)15}$	$3\overline{)21}$	$5\overline{)30}$	$9\overline{)63}$	$8\overline{)32}$	$9\overline{)9}$	$3\overline{)33}$
H	$9\overline{)18}$	$6\overline{)24}$	$10\overline{)70}$	$3\overline{)12}$	$12\overline{)120}$	$11\overline{)66}$	$5\overline{)25}$	$11\overline{)77}$
I	$8\overline{)72}$	$11\overline{)22}$	$12\overline{)36}$	$7\overline{)84}$	$11\overline{)55}$	$10\overline{)100}$	$4\overline{)36}$	$8\overline{)80}$
J	$11\overline{)121}$	$10\overline{)90}$	$7\overline{)63}$	$12\overline{)108}$	$2\overline{)24}$	$10\overline{)110}$	$12\overline{)144}$	$11\overline{)132}$

MULTIPLICATION AND DIVISION FACTS TEST

	K	L	M	N	O	P	Q	R
A	$2\overline{)18}$	8×4	$5\overline{)15}$	10×6	8×1	$3\overline{)24}$	$6\overline{)12}$	5×8
B	8×2	$7\overline{)77}$	$9\overline{)81}$	4×10	7×12	$1\overline{)6}$	$8\overline{)80}$	4×9
C	$12\overline{)36}$	11×5	7×7	$10\overline{)90}$	$5\overline{)45}$	6×7	$8\overline{)16}$	9×9
D	10×2	$4\overline{)32}$	$9\overline{)99}$	7×8	12×3	$9\overline{)108}$	$11\overline{)88}$	12×4
E	8×10	12×9	$12\overline{)84}$	$2\overline{)20}$	9×0	10×11	$3\overline{)36}$	$10\overline{)100}$
F	$4\overline{)44}$	$12\overline{)72}$	7×11	12×6	$7\overline{)56}$	$9\overline{)45}$	10×3	9×7
G	$12\overline{)144}$	$6\overline{)60}$	9×2	8×12	$12\overline{)108}$	$11\overline{)44}$	5×10	$7\overline{)84}$
H	6×5	8×8	$11\overline{)33}$	$5\overline{)55}$	6×11	12×5	$11\overline{)132}$	$6\overline{)42}$
I	$8\overline{)96}$	$10\overline{)120}$	11×8	10×9	$5\overline{)60}$	11×4	10×10	$10\overline{)80}$
J	11×6	12×11	$4\overline{)40}$	$7\overline{)35}$	3×6	$8\overline{)56}$	9×8	12×12

TABLE OF MEASURES

METRIC	CUSTOMARY
Length	
1 meter (m) = 1,000 millimeters (mm)	1 foot (ft) = 12 inches (in.)
1 meter = 100 centimeters (cm)	1 yard (yd) = 3 feet, or 36 inches
1 meter = 10 decimeters (dm)	1 mile (mi) = 1,760 yards, or 5,280 feet
1 kilometer (km) = 1,000 meters	
Capacity	
1 liter (L) = 1,000 milliliters (mL)	1 tablespoon (tbsp) = 3 teaspoons (tsp)
1 metric cup = 250 milliliters	1 cup (c) = 8 fluid ounces (fl oz)
	1 pint (pt) = 2 cups
	1 quart (qt) = 2 pints
	1 gallon (gal) = 4 quarts
Mass/Weight	
1 gram (g) = 1,000 milligrams (mg)	1 pound (lb) = 16 ounces (oz)
1 kilogram (kg) = 1,000 grams	1 ton (T) = 2,000 pounds

TIME

1 minute (min) = 60 seconds (sec)	1 year (yr) = 12 months (mo), or about 52 weeks
1 hour (hr) = 60 minutes	1 year = 365 days
1 day = 24 hours	1 leap year = 366 days
1 week (wk) = 7 days	

SYMBOLS

$=$	is equal to	1:3	ratio of 1 to 3	$\angle ABC$	angle *ABC*
$\neq$	is not equal to	%	percent	$\triangle ABC$	triangle *ABC*
$>$	is greater than	$\perp$	is perpendicular to	π	pi (about 3.14)
$<$	is less than	$\parallel$	is parallel to	°	degree
$\approx$	is approximately equal to	$\overleftrightarrow{AB}$	line *AB*	°C	degrees Celsius
$\lvert 4 \rvert$	the absolute value of 4	$\overrightarrow{AB}$	ray *AB*	°F	degrees Fahrenheit
		$\overline{AB}$	line segment *AB*	(2,3)	ordered pair (*x*,*y*)

FORMULAS

Perimeter of polygon = sum of lengths of sides	Area of rectangle $A = l \times w$
Perimeter of rectangle $P = (2 \times l) + (2 \times w)$	Area of parallelogram $A = bh$
Perimeter of square $P = 4 \times s$	Area of triangle $A = \frac{1}{2}bh$
Circumference $C = \pi \times d$	Volume of rectangular prism $V = l \times w \times h$

Grade Five California Mathematics Content Standards

By the end of grade five, students increase their facility with the four basic arithmetic operations applied to fractions and decimals, and learn to add and subtract positive and negative numbers. They know and use common measuring units to determine length and area and know and use formulas to determine the volume of simple geometric figures. Students know the concept of angle measurement and use a protractor and compass to solve problems. They use grids, tables, graphs, and charts to record and analyze data.

Number Sense

1.0 Students compute with very large and very small numbers, positive integers, decimals, and fractions and understand the relationship between decimals, fractions, and percents. They understand the relative magnitudes of numbers:

1.1 Estimate, round, and manipulate very large (e.g., millions) and very small (e.g., thousandths) numbers.

O-ꟷ 1.2 Interpret percents as a part of a hundred; find decimal and percent equivalents for common fractions and explain why they represent the same value; compute a given percent of a whole number.

A test had 48 problems. Joe got 42 correct.

1. What percent were correct?
2. What percent were wrong?
3. If Moe got 93.75% correct, how many problems did he get correct?

1.3 Understand and compute positive integer powers of nonnegative integers; compute examples as repeated multiplication.

Extend the tables shown below:

$2^4 = 16$ $10^4 = 10{,}000$

$2^3 = 8$ $10^3 = 1{,}000$

$2^2 = 4$ $10^2 = 100$

$2^1 = ?$ $10^1 = ?$

$2^0 = ?$ $10^0 = ?$

O-ꟷ 1.4 Determine the prime factors of all numbers through 50 and write the numbers as the product of their prime factors by using exponents to show multiples of a factor (e.g., $24 = 2 \times 2 \times 2 \times 3 = 2^3 \times 3$).

Find the prime factorization of 48 and use exponents where appropriate.

Note: The sample problems illustrate the standards and are written to help clarify them.

The symbols O-ꟷ and ⬬ identify the Key Standards for grade five.

Number Sense (Continued)

1.5 Identify and represent on a number line decimals, fractions, mixed numbers, and positive and negative integers.

2.0 Students perform calculations and solve problems involving addition, subtraction, and simple multiplication and division of fractions and decimals:

2.1 Add, subtract, multiply, and divide with decimals; add with negative integers; subtract positive integers from negative integers; and verify the reasonableness of the results.

Determine the following numbers:

1. $11 + (^{-}23)$
2. $(^{-}15) - 128$
3. $(^{-}27) + (^{-}45)$

2.2 Demonstrate proficiency with division, including division with positive decimals and long division with multidigit divisors.

Find the quotient:

6 divided by .025

2.3 Solve simple problems, including ones arising in concrete situations, involving the addition and subtraction of fractions and mixed numbers (like and unlike denominators of 20 or less), and express answers in the simplest form.

Given the following three pairs of fractions ($\frac{3}{8}$ and $\frac{1}{6}$, $5\frac{3}{4}$ and $2\frac{1}{3}$, 16 and $12\frac{7}{8}$), find for each pair its:

1. Sum
2. Difference

2.4 Understand the concept of multiplication and division of fractions.

2.5 Compute and perform simple multiplication and division of fractions and apply these procedures to solving problems.

Given the following three pairs of fractions ($\frac{3}{8}$ and $\frac{1}{6}$, $5\frac{3}{4}$ and $2\frac{1}{3}$, 16 and $12\frac{7}{8}$), find for each pair its:

1. Product
2. Quotient in simplest terms

Ericka has $3\frac{1}{2}$ yards of cloth to make shirts which take $\frac{7}{8}$ yards for each shirt. How many shirts can she make? How much cloth will she have left over?

Algebra and Functions

1.0 Students use variables in simple expressions, compute the value of the expression for specific values of the variable, and plot and interpret the results:

1.1 Use information taken from a graph or equation to answer questions about a problem situation.

1.2 Use a letter to represent an unknown number; write and evaluate simple algebraic expressions in one variable by substitution.

$3x + 2 = 14$. What is x?

1.3 Know and use the distributive property in equations and expressions with variables.

1.4 Identify and graph ordered pairs in the four quadrants of the coordinate plane.

Plot the points $(1, 2)$, $(^{-}4, ^{-}3)$, $(12, ^{-}1)$, $(0, 4)$, $(^{-}4, 0)$.

1.5 Solve problems involving linear functions with integer values; write the equation; and graph the resulting ordered pairs of integers on a grid.

Measurement and Geometry

1.0 Students understand and compute the volumes and areas of simple objects:

1.1 Derive and use the formula for the area of a triangle and of a parallelogram by comparing it with the formula for the area of a rectangle (i.e., two of the same triangles make a parallelogram with twice the area; a parallelogram is compared with a rectangle of the same area by pasting and cutting a right triangle on the parallelogram).

Find the area and perimeter.

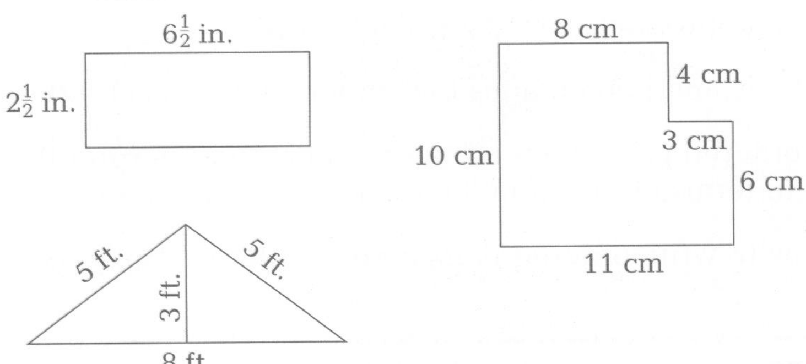

Measurement and Geometry (Continued)

🔑 **1.2** Construct a cube and rectangular box from two-dimensional patterns and use these patterns to compute the surface area for these objects.

🔑 **1.3** Understand the concept of volume and use the appropriate units in common measuring systems (i.e., cubic centimeter [cm^3], cubic meter [m^3], cubic inch [$in.^3$], cubic yard [$yd.^3$]) to compute the volume of rectangular solids.

1.4 Differentiate between, and use appropriate units of measures for, two- and three-dimensional objects (i.e., find the perimeter, area, volume).

2.0 Students identify, describe, and classify the properties of, and the relationships between, plane and solid geometric figures:

🔑 **2.1** Measure, identify, and draw angles, perpendicular and parallel lines, rectangles, and triangles by using appropriate tools (e.g., straightedge, ruler, compass, protractor, drawing software).

🔑 **2.2** Know that the sum of the angles of any triangle is 180° and the sum of the angles of any quadrilateral is 360° and use this information to solve problems.

2.3 Visualize and draw two-dimensional views of three-dimensional objects made from rectangular solids.

Statistics, Data Analysis, and Probability

1.0 Students display, analyze, compare, and interpret different data sets, including data sets of different sizes:

1.1 Know the concepts of mean, median, and mode; compute and compare simple examples to show that they may differ.

1.2 Organize and display single-variable data in appropriate graphs and representations (e.g., histogram, circle graphs) and explain which types of graphs are appropriate for various data sets.

1.3 Use fractions and percentages to compare data sets of different sizes.

🔑 **1.4** Identify ordered pairs of data from a graph and interpret the meaning of the data in terms of the situation depicted by the graph.

🔑 **1.5** Know how to write ordered pairs correctly; for example, (x, y).

Mathematical Reasoning

1.0 Students make decisions about how to approach problems:

1.1 Analyze problems by identifying relationships, distinguishing relevant from irrelevant information, sequencing and prioritizing information, and observing patterns.

1.2 Determine when and how to break a problem into simpler parts.

Mathematical Reasoning (Continued)

2.0 Students use strategies, skills, and concepts in finding solutions:

2.1 Use estimation to verify the reasonableness of calculated results.

2.2 Apply strategies and results from simpler problems to more complex problems.

2.3 Use a variety of methods, such as words, numbers, symbols, charts, graphs, tables, diagrams, and models, to explain mathematical reasoning.

2.4 Express the solution clearly and logically by using the appropriate mathematical notation and terms and clear language; support solutions with evidence in both verbal and symbolic work.

2.5 Indicate the relative advantages of exact and approximate solutions to problems and give answers to a specified degree of accuracy.

2.6 Make precise calculations and check the validity of the results from the context of the problem.

3.0 Students move beyond a particular problem by generalizing to other situations:

3.1 Evaluate the reasonableness of the solution in the context of the original situation.

3.2 Note the method of deriving the solution and demonstrate a conceptual understanding of the derivation by solving similar problems.

3.3 Develop generalizations of the results obtained and apply them in other circumstances.

GLOSSARY

Pronunciation Key

a	**a**dd, m**a**p	f	**f**it, hal**f**	n	**n**ice, ti**n**	p	**p**it, sto**p**	yo͞o	**f**use, **f**ew
ā	**a**ce, r**a**te	g	**g**o, lo**g**	ng	ri**ng**, so**ng**	r	**r**un, poo**r**	v	**v**ain, e**v**e
â(r)	c**are**, **air**	h	**h**ope, **h**ate	o	**o**dd, h**o**t	s	**s**ee, pa**ss**	w	**w**in, a**w**ay
ä	p**a**lm, f**a**ther	i	**i**t, g**i**ve	ō	**o**pen, s**o**	sh	**s**ure, ru**sh**	y	**y**et, **y**earn
b	**b**at, ru**b**	ī	**i**ce, wr**i**te	ô	**o**rder, j**aw**	t	**t**alk, si**t**	z	**z**est, mu**s**e
ch	**ch**eck, ca**tch**	j	**j**oy, le**dge**	oi	**oi**l, b**oy**	th	**th**in, bo**th**	zh	vi**s**ion, plea**s**ure
d	**d**og, ro**d**	k	**c**ool, ta**k**e	ou	p**ou**t, n**ow**	t͟h	**th**is, ba**th**e		
e	**e**nd, p**e**t	l	**l**ook, ru**l**e	o͝o	t**oo**k, f**u**ll	u	**u**p, d**o**ne		
ē	**e**qual, tr**ee**	m	**m**ove, see**m**	o͞o	p**oo**l, f**oo**d	û(r)	b**ur**n, t**er**m		

ə the schwa, an unstressed vowel representing the sound spelled *a* in **a**bove, *e* in sick**e**n, *i* in poss**i**ble, *o* in mel**o**n, *u* in circ**u**s

Other symbols:
- • separates words into syllables
- ′ indicates stress on a syllable

absolute value [ab•sə•lo͞ot′ val′yo͞o] The numerical value of a given number, or the distance of an integer from zero on a number line (p. 417)

acute angle [ə•kyo͞ot′ an′gəl] An angle that has a measure less than a right angle (less than 90°) (p. 454)
Example:

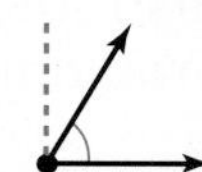

acute triangle [a•kyo͞ot′ trī′an•gəl] A triangle that has three acute angles (p. 478)

addends [ad′endz] Numbers that are added in an addition problem (p. 409)

angle [an′gəl] A figure formed by two rays that meet at a common endpoint (p. 454)
Example:

area [âr′ē•ə] The number of square units needed to cover a surface (p. 528)

array [ə•rā′] An arrangement of objects in rows and columns (p. 264)
Example:

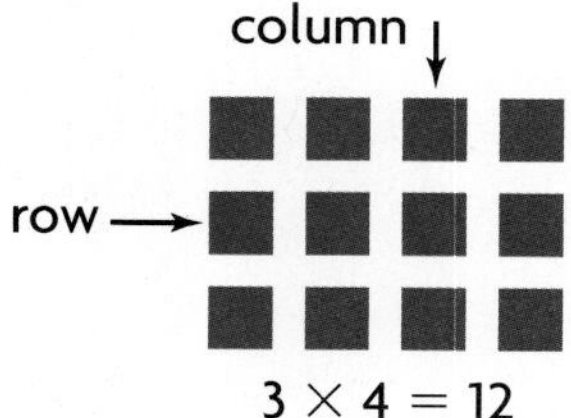

Associative Property of Addition [ə•sō′shē•ə•tiv prä′pər•tē əv ə•di′shən] The property that states that when the grouping of addends is changed, the sum is the same (p. 78)
Example: $(5 + 8) + 4 = 5 + (8 + 4)$

Associative Property of Multiplication [ə•sō′shē•ə•tiv prä′pər•tē əv mul•tə•plə•kā′shən] The property which states that the way factors are grouped does not change the product (p. 90)
Example: $(2 \times 3) \times 4 = 2 \times (3 \times 4)$

average [av′rij] See *mean.* (p. 122)

axis [ak′səs] The horizontal or vertical number line used in a graph or coordinate plane (p. 122)

bar graph [bär graf] A graph that uses horizontal or vertical bars to display countable data (p. 108)
Example:

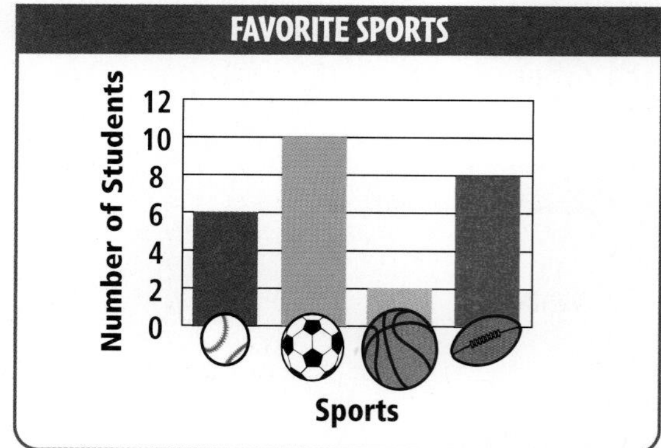

base [bās] A number used as a repeated factor (p. 266) *Example:* $8^3 = 8 \times 8 \times 8$. The base is 8.

base [bās] A polygon's side or a solid figure's face by which the figure is measured or named (pp. 486, 532)
Examples:

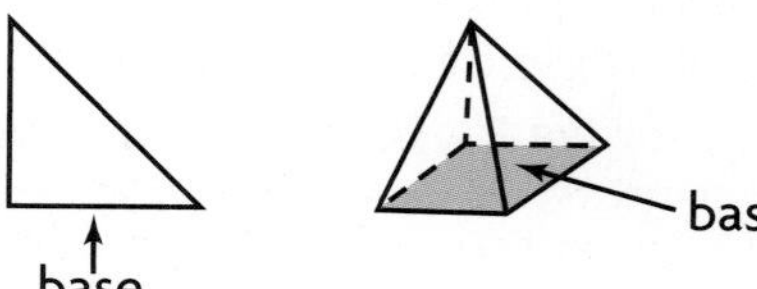

benchmark [bench′märk] A familiar number used as a point of reference

billion [bil′yən] One thousand million; written as 1,000,000,000 (p. 5)

capacity [kə•pa′sə•tē] The amount a container can hold (p. 512)

centimeter (cm) [sən′tə•mē•tər] A unit for measuring length in the metric system 0.01 meter = 1 centimeter (p. 506)

central angle [sen′trəl an′gəl] The angle formed by two radii of a circle that meet at its center (p. 465)

certain [sur′tən] Sure to happen; will always happen (p. 569)

chord [kôrd] A line segment with endpoints on a circle (p. 464)
Example:

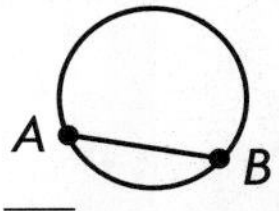

$\overline{AB}$ is a chord.

circle [sər′kəl] A closed figure with all points on the figure the same distance from the center point (p. 464)
Example:

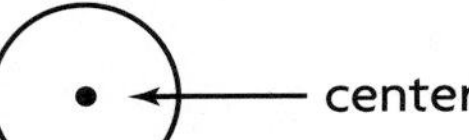

circle graph [sur′kəl graf] A graph that shows how parts of the data are related to the whole and to each other (p. 109)
Example:

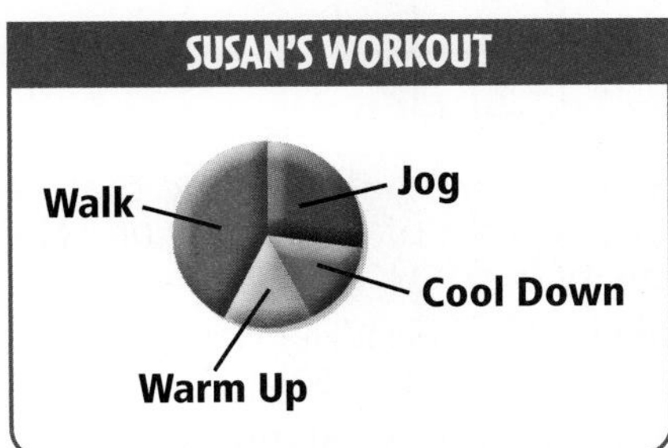

circumference [sər•kum′fər•əns] The distance around a circle (p. 526)

common factor [kä′mən fak′tər] A number that is a factor of two or more numbers (p. 258)

common multiple [kä′mən mul′tə•pəl] A number that is a multiple of two or more numbers (p. 256)

Commutative Property of Addition [kə•myōō′tə•tiv prä′pər•tē əv ə•di′shən] The property that states that when the order of two addends is changed, the sum is the same (p. 78)
Example: $4 + 5 = 5 + 4$

Commutative Property of Multiplication [kə•myōō′tə•tiv prä′pər•tē əv mul•tə•plə•kā′shən] The property that states that when the order of two factors is changed, the product is the same (p. 90)
Example: $4 \times 5 = 5 \times 4$

compass [kum′pəs] A tool used to construct circles and arcs (p. 464)

compatible numbers [kəm•pa′tə•bəl num′bərz] Numbers that are easy to compute mentally (p. 182)

composite number [käm•pä′zət num′bər] A number having more than two factors (p. 264)
Example: 6 is a composite number, since its factors are 1, 2, 3, and 6.

cone [kōn] A solid figure that has a flat, circular base and one vertex (p. 487)
Example:

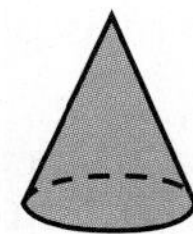

congruent [kən•grōō′ənt] Having the same size and shape (p. 468)

contrast [kən•trast′] To examine for differences (p. 289)

coordinate plane [kō•ôr′də•nət plān′] A plane formed by two intersecting and perpendicular number lines (p. 440)
Example:

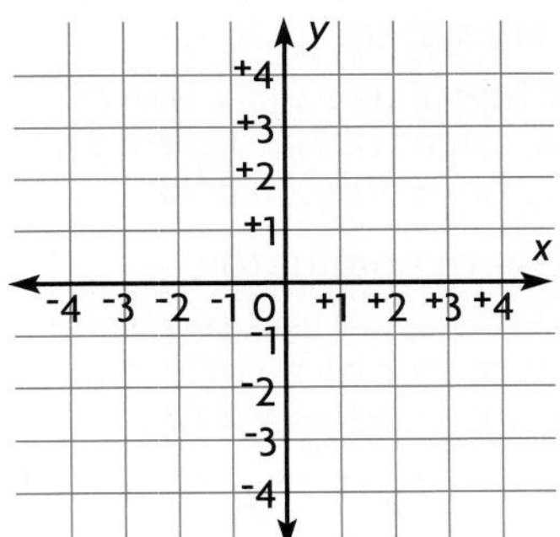

coordinates [kō•ôr′də•nəts] The numbers in an ordered pair (p. 440)

cube [kyōōb] A solid figure with six congruent square faces (p. 546)
Example:

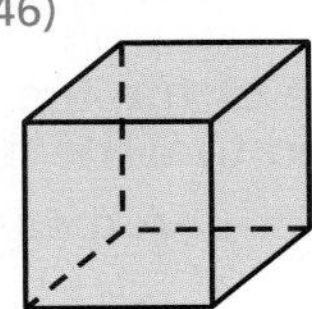

cubic unit [kyōō′bik yōō′nət] A unit of volume with dimensions 1 unit × 1 unit × 1 unit (p. 552)

cumulative frequency [kyōō′myə•lə•tiv frē′kwən•sē] A running total of data (p. 98)

cylinder [si′lən•dər] A solid figure that has two parallel bases that are congruent circles (p. 487)
Example:

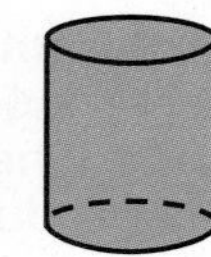

D

data [dā′tə] Information collected about people or things, often to draw conclusions about them (p. 98)

decimal number [de′sə•məl num′bər] A number with one or more digits to the right of the decimal point (p. 17)

decimal point [de′sə•məl point] A symbol used to separate dollars from cents in money, and the ones place from the tenths place in decimal numbers (p. 17)

decimal system [de′sə•məl sis′təm] A system of computation based on the number 10

decimeter (dm) [de′sə•mē•tər] A unit of length in the metric system; 10 decimeters = 1 meter (p. H70)

degree [di•grē′] A unit for measuring angles or for measuring temperature (pp. 458, 511)

degrees Celsius (°C) [di•grēz′ səl′sē•us] A standard unit for measuring temperature in the metric system (p. 511)

degrees Fahrenheit (°F) [di•grēz′ fâr′ən•hīt] A standard unit for measuring temperature in the customary system (p. 511)

denominator [di•nä′mə•nā•tər] The number below the bar in a fraction that tells how many equal parts are in the whole (p. 344)
Example: $\frac{3}{4}$ ←denominator

diameter [dī•am′ə•tər] A line segment that passes through the center of a circle and has its endpoints on the circle (p. 464)
Example:

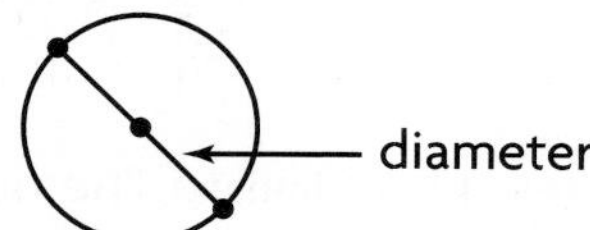

difference [dif′ər•əns] The answer to a subtraction problem (p. 31)

digit [di′jit] Any one of the ten symbols 0, 1, 2, 3, 4, 5, 6, 7, 8, 9 used to write numbers (p. 1)

dimension [də•men′shən] A measure in one direction (p. 531)

Distributive Property of Multiplication [di•strib′yə•tiv prä′pər•tē əv mul•tə•plə•kā′shən] The property that states that multiplying a sum by a number is the same as multiplying each addend in the sum by the number and then adding the products (p. 92)
Example: $3 \times (4 + 2) = (3 \times 4) + (3 \times 2)$
$3 \times 6 = 12 + 6$
$18 = 18$

dividend [di′və•dend] The number that is to be divided in a division problem (p. 181)
Example: $36 \div 6$; $6\overline{)36}$ The dividend is 36.

divisible [də•vi′zi•bəl] Capable of being divided so that the quotient is a whole number and there is a zero remainder (p. 254)
Example: 21 is divisible by 3.

division [də•vi′zhən] The process of sharing a number of items to find how many groups can be made or how many items will be in a group; the operation that is the opposite of multiplication (p. 181)

divisor [də•vī′zər] The number that divides the dividend (p. 181)
Example: $15 \div 3$; $3\overline{)15}$ The divisor is 3.

double-bar graph [du′bəl bär′graf] A graph used to compare two similar kinds of data (p. 130)
Example:

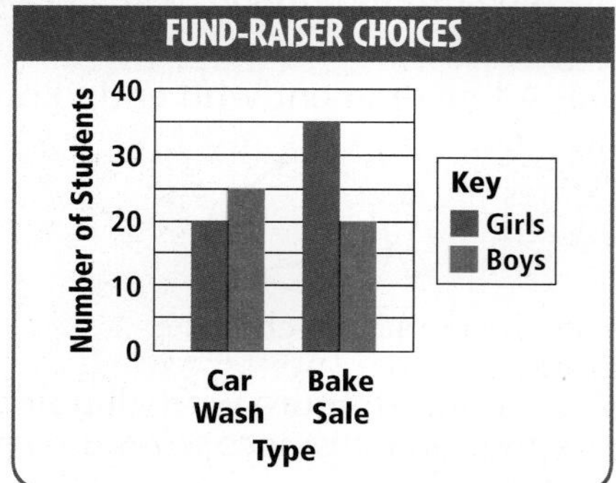

E

edge [ej] The line made where two or more faces of a solid figure meet (p. 486) *Example:*

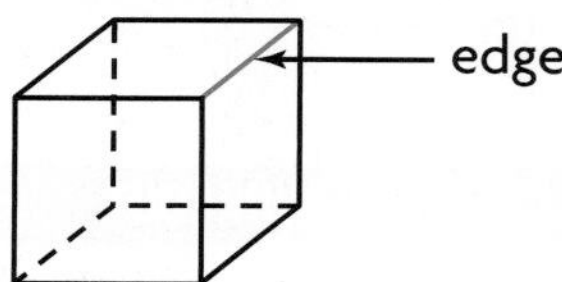

elapsed time [i•lapst′ tīm] The time that passes between the start of an activity and the end of that activity (p. 516)

equally likely [ē′kwəl•lē lī′klē] Having the same chance of occurring (p. 574)

equation [i•kwā′zhən] An algebraic or numerical sentence that shows that two quantities are equal (p. 72)

equilateral triangle [ē•kwə•la′tə•rəl trī′an•gəl] A triangle with three congruent sides (p. 478)
Example:

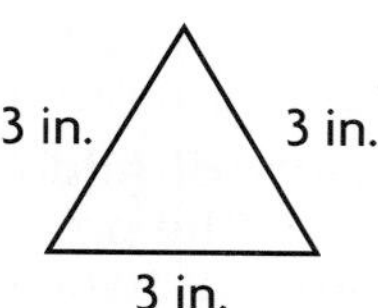

equivalent [ē•kwiv′ə•lənt] Having the same value

equivalent decimals [ē•kwiv′ə•lənt de′sə•məlz] Decimals that name the same number or amount (p. 22) *Example:* 0.4 = 0.40 = 0.400

equivalent fractions [ē•kwiv′ə•lənt frak′shənz] Fractions that name the same number or amount (p. 284) *Example:* $\frac{3}{4} = \frac{6}{8}$

equivalent ratios [ē•kwiv′ə•lənt rā′shē•ōz] Ratios that make the same comparison (p. 306)

estimate [es′tə•mət] An answer that is close to the exact answer and is found by rounding, by clustering, or by using compatible numbers (p. 56)

evaluate [i•val′yə•wāt] To find the value; in an algebraic expression, to replace a variable with a value and then perform the operation(s) (p. 86)

event [i•vent′] A set of outcomes (p. 569)

expanded form [ik•spand′id fôrm] A way to write numbers by showing the value of each digit (p. 4) *Example:* 832 = 800 + 30 + 2

exponent [ek′spō•nənt] A number that shows how many times the base is used as a factor (p. 266) *Example:* $10^3 = 10 \times 10 \times 10$; 3 is the exponent.

expression [ik•spre′shən] A mathematical phrase or the part of a number sentence that combines numbers, operation signs, and sometimes variables, but doesn't have an equal sign (p. 68)

face [fās] A polygon that is a flat surface of a solid figure (p. 486)
Example:

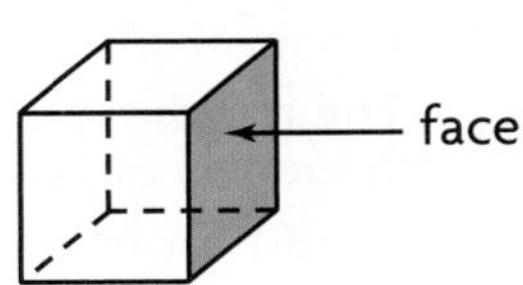

factor [fak′tər] A whole number multiplied by another whole number to find a product (p. 253)

factor tree [fak′tər trē] A diagram that shows the factors of a number, then the factors of those numbers, and so on until only prime factors are left (p. 272)

figurate numbers [fi′gyə•rət num′bərz] Numbers that can be represented by geometric figures (p. 481)
Example:

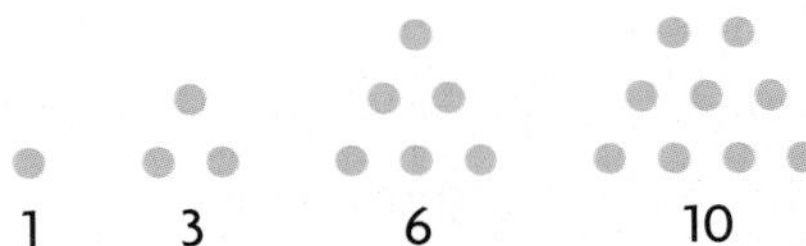

foot (ft) [fo͝ot] A unit of length in the customary system; 1 foot = 12 inches (p. 508)

formula [fôr′myə•lə] A set of symbols that expresses a mathematical rule (p. 80) *Example:* $A = l \times w$

fraction [frak′shən] A number that names a part of a whole or a part of a group (p. 343)

frequency table [frē′kwen•sē tā′bəl] A table that uses numbers to record data about how often something happens (p. 98)

function table [funk′shən tā′bəl] A table that matches each input value with an output value. The output values are determined by the function. (p. 444)

gallon (gal) [ga′lən] A customary unit for measuring capacity; 4 quarts = 1 gallon (p. 512)

gram (g) [gram] A unit of mass in the metric system; 1,000 grams = 1 kilogram (p. 514)

greater than (>) [grā′tər than] A symbol used to compare two numbers, with the greater number given first (p. 9)
Example: 6 > 4

greatest common factor (GCF) [grā′təst kä′mən fak′tər] The greatest factor that two or more numbers have in common (p. 258)
Example: 6 is the GCF of 18 and 30.

height [hīt] A measure of a solid figure, taken as the length of a perpendicular from the base to the top of a plane or solid figure (p. 532)
Example:

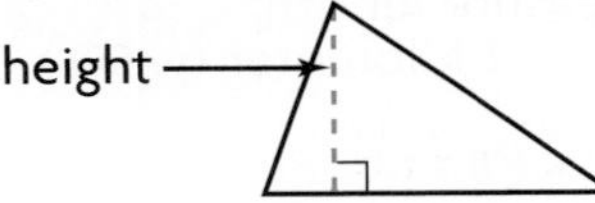

hexagon [hek′sə•gän] A polygon with six sides and six angles (p. 460)

histogram [his′tə•gram] A bar graph that shows the number of times data occur within intervals (p. 128)

hundredth [hun′drədth] A decimal or fraction that names one part of 100 equal parts (p. 18)
Examples: 0.56 fifty-six hundredths
$\frac{45}{100}$ forty-five hundredths

impossible [im•pä′sə•bəl] Not able to happen; can never happen (p. 569)

inch (in.) [inch] A customary unit used to measure length; 12 inches = 1 foot (p. 504)

inequality [in•i•kwä′lə•tē] A mathematical sentence that shows that two amounts are not equal

integers [in′ti•jərz] The set of whole numbers and their opposites (p. 416)

intersecting lines [in•tər•sek′ting līnz] Lines that cross each other at exactly one point (p. 455)
Example:

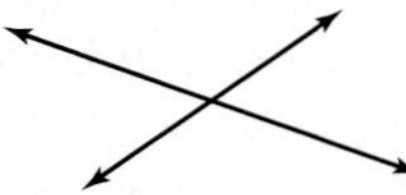

interval [in′tər•vəl] The distance between one number and the next on the scale of a graph (p. 116)

inverse operations [in′vərs ä•pə•rā′shənz] Operations that undo each other, like addition and subtraction, or multiplication and division (p. 191)

isosceles triangle [ī•sä′sə•lēz trī′an•gəl] A triangle with exactly two congruent sides (p. 478)
Example:

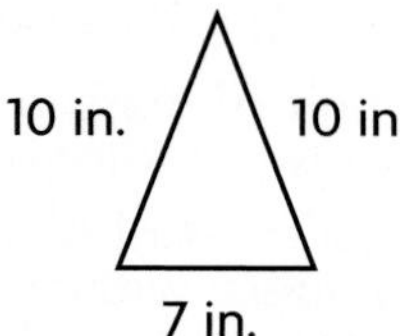

K

kilogram (kg) [ki′lə•gram] A metric unit that is used to measure mass; 1,000 grams = 1 kilogram (p. 514)

kiloliter (kL) [ki′lə•lē•tər] A metric unit that is used to measure capacity; 1,000 liters = 1 kiloliter (p. 514)

kilometer (km) [kə•lä′mə•tər] A metric unit that is used to measure length; 1,000 meters = 1 kilometer (p. 508)

L

least common denominator (LCD) [lēst kä′mən di•nä′mə•nā•tər] The least common multiple of two or more denominators (p. 352)
Example: The LCD for $\frac{1}{4}$ and $\frac{5}{6}$ is 12.

least common multiple (LCM) [lēst′ kä′mən mul′tə•pəl] The least number, other than zero, that is a common multiple of two or more numbers (p. 256)

less than (<) [less than] A symbol used to compare two numbers with the lesser number given first (p. 9)
Example: $4 < 6$

like fractions [līk frak′shənz] Fractions that have the same denominator (p. 344)
Example: $\frac{2}{5}$ and $\frac{4}{5}$ are like fractions.

line [līn] A straight path in a plane, extending in both directions with no endpoints (p. 454)
Example:

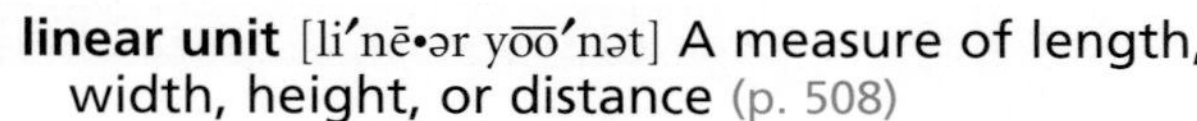

linear unit [li′nē•ər yōō′nət] A measure of length, width, height, or distance (p. 508)

line graph [līn graf] A graph that uses a line to show how data change over time (p. 109)

line plot [līn plät] A graph that shows frequency of data along a number line (p. 99)
Example:

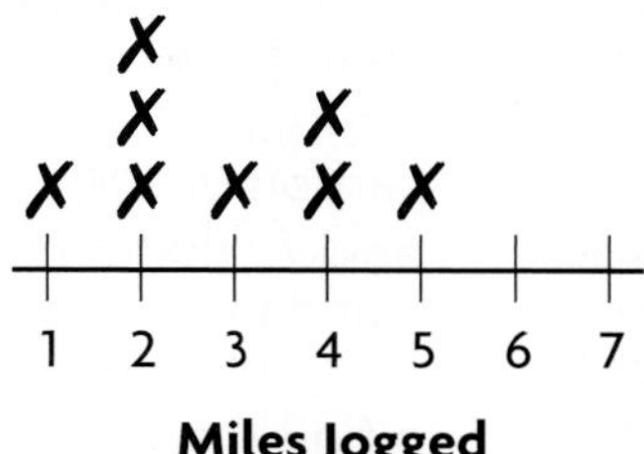

line segment [līn seg′mənt] A part of a line between two endpoints (p. 454)

line symmetry [līn si′mə•trē] A property shown by a figure if, when the figure is folded or reflected on a line, the two parts are congruent (p. 470)

liter (L) [lē′tər] A unit of capacity in the metric system; 1 liter = 1,000 milliliters (p. 514)

M

map scale [map skāl] A ratio that compares distance on a map with actual distance (p. 308)

mass [mas] The amount of matter in an object (p. 514)

mean [mēn] The average of a set of numbers, found by dividing the sum of the set by the number of addends (p. 102)

median [mē′dē•ən] The middle number in a set of data that are arranged in order (p. 104)

meter (m) [mē′tər] A unit for measuring length in the metric system; 1 meter = 100 centimeters (p. 508)

mile (mi) [mīl] A unit of length in the customary system; 5,280 feet = 1 mile (p. 508)

milligram (mg) [mil′ə•gram] A unit for measuring mass in the metric system; 1,000 milligrams = 1 gram (p. 514)

milliliter (mL) [mi′lə•lē•tər] A metric unit that is used to measure capacity; 1,000 milliliters = 1 liter (p. 514)

millimeter (mm) [mi′lə•mē•tər] A unit for measuring length in the metric system 1 millimeter = 0.001 meter (p. 506)

million [mil′yən] 1,000 thousands; written as 1,000,000 (p. 2)

mixed number [mikst num′bər] A number that is made up of a whole number and a fraction (p. 294) *Example:* $1\frac{5}{8}$

mode [mōd] The number or numbers that occur most often in a set of data (p. 104)

multiple [mul′tə•pəl] The product of a given whole number and another whole number (p. 144)

multiplication [mul•tə•plə•kā′shən] A process to find the total number of items made up of equal-sized groups, or to find the total number of items in a given number of groups. It is the opposite operation of division. (p. 144)

multistep problems [mul′ti•step prä′bləmz] Problems requiring more than one step to solve (p. 374)

negative integer [ne′gə•tiv in′ti•jər] Any integer less than 0 (p. 416)
Examples: $^-4$, $^-5$, and $^-6$ are negative integers.

net [net] A two-dimensional pattern that can be folded into a three-dimensional prism or pyramid (p. 546)
Example:

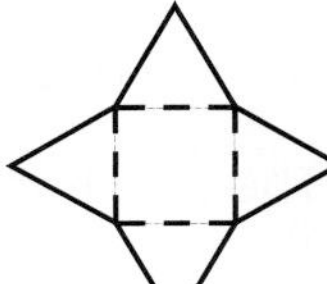
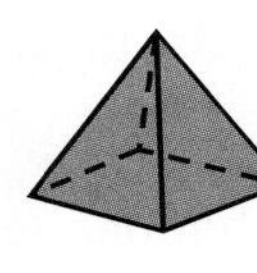

number line [num′bər līn] A line with equally spaced ticks named by numbers. A number line does not always start at 0. (p. 8)
Example:

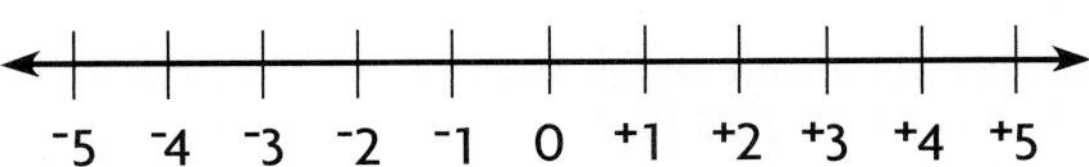

numerator [nōō′mə•rā•tər] The number above the bar in a fraction that tells how many equal parts are being considered (p. 344)
Example: $\frac{3}{4}$ ←numerator

obtuse angle [äb•tōōs′ an′gəl] An angle that has a measure greater than a right angle (greater than 90°) (p. 454)
Example:

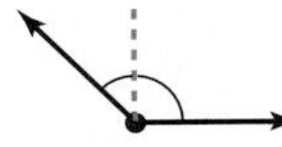

obtuse triangle [äb•tōōs′ trī′an•gəl] A triangle that has one obtuse angle (p. 478)

octagon [äk′tə•gän] A polygon with eight sides and eight angles (p. 460)
Example:

opposites [ä′pə•zəts] Integers that are the same distance, but in opposite directions, from 0 on a number line (p. 417)

ordered pair [ôr′dərd pâr] A pair of numbers used to locate a point on a grid. The first number tells the left-right position and the second number tells the up-down position. (p. 122)

origin [ôr′ə•jən] The point where the two axes of a coordinate plane intersect, (0,0) (p. 440)

ounce (oz) [ouns] A customary unit used to measure weight (p. 512)

outcome [out′kum] A possible result of an experiment (p. 570)

outlier [out′lī•ər] A value separated from the rest of the data (p. 99)

parallel lines [pâ′rə•lel līnz] Lines in a plane that do not intersect (p. 455)
Example:

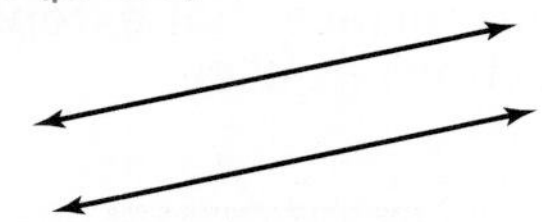

parallelogram [pâ•rə•lel′ə•gram] A quadrilateral whose opposite sides are parallel and congruent (p. 482)
Example:

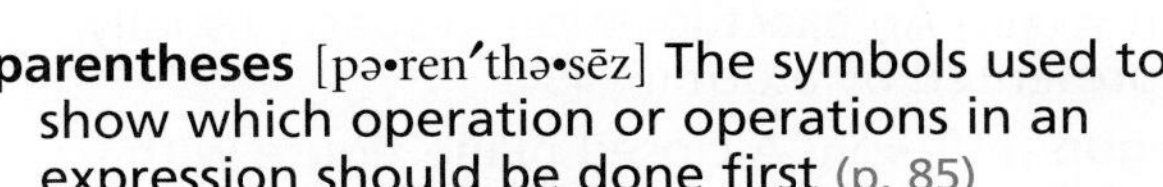

parentheses [pə•ren′thə•sēz] The symbols used to show which operation or operations in an expression should be done first (p. 85)

pentagon [pen′tə•gän] A polygon with five sides and five angles (p. 460)
Example:

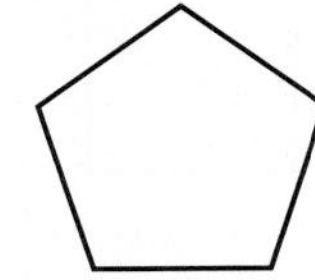

percent [pər•sent′] A ratio of a number to 100 (p. 316)

perimeter [pə•rim′ə•tər] The distance around a closed plane figure (p. 524)

period [pir′ē•əd] Each group of three digits in a large number (p. 1)

perpendicular lines [pər•pen•dik′yə•lər līnz] Two lines that intersect to form right angles (p. 455)
Example:

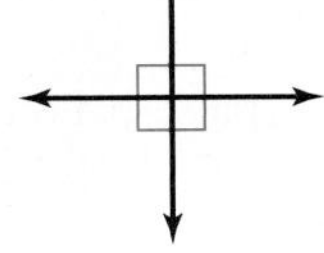

pi (π) [pī] The ratio of the circumference to the diameter of a circle; an approximate decimal value of pi is 3.14. (p. 527)

pictograph [pik′tə•graf] A graph that displays countable data with symbols or pictures (p. 108) *Example:*

HOW WE GET TO SCHOOL

Walk	⊛ ⊛ ⊛
Ride a Bike	⊛ ⊛ ⊛ ⊛
Ride a Bus	⊛ ⊛ ⊛ ⊛ ⊛ ⊛
Ride in a Car	⊛ ⊛

Key: Each ⊛ = 10 students

pint (pt) [pīnt] A customary unit for measuring capacity; 2 cups = 1 pint (p. 512)

place value [plās val′yo͞o] The value of a place, such as ones or tens, in a number (p. 4)

plane [plān] A flat surface that extends without end in all directions (p. 454)
Example:

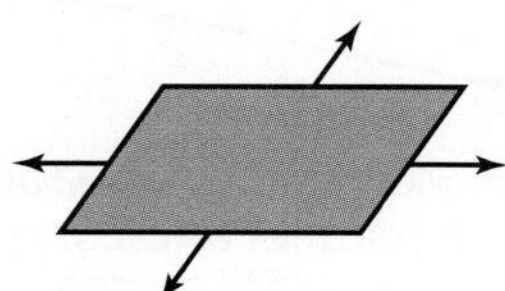

plane figure [plān fig′yər] A figure that lies on a flat surface and is all in one plane (p. 460)

point [point] An exact location in space, usually represented by a dot (p. 454)

polygon [pä′lē•gän] A closed plane figure with straight sides that is named by the number of its sides and angles (p. 460)
Examples:

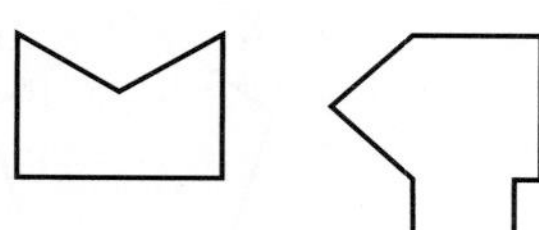

polyhedron [pä•lē•hē′drən] A solid figure with faces that are polygons (p. 486)
Examples:

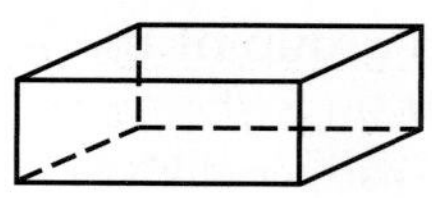

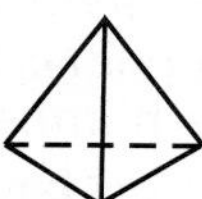

positive integer [pä′zə•tiv in′ti•jər] Any integer greater than zero (p. 416)

possible outcomes [pä′sə•bəl out′kumz] The events that have a chance of happening in an experiment (p. 570)

pound (lb) [pound] A customary unit used to measure weight (p. 512)

precision [pri•sizh′ən] A property of a measurement related to the unit of measure you choose; the smaller the unit, the more precise the measurement will be (p. 504)

prediction [pri•dik′shən] A reasonable guess as to the outcome of an event (p. 212)

prime factorization [prīm fak•tə•ri•zā′shən] The process of factoring a composite number into its prime components, as with a factor tree (p. 272)

prime number [prīm num′bər] A number that has exactly two factors: 1 and itself (p. 264)
Examples: 5, 7, 11, 13, 17, and 19 are prime numbers.

prioritize [prī•or′ə•tīz] To put events in order of importance (p. 388)

prism [priz′əm[A solid figure that has two congruent, polygon-shaped bases, and all other faces are rectangles (p. 486)
Examples:

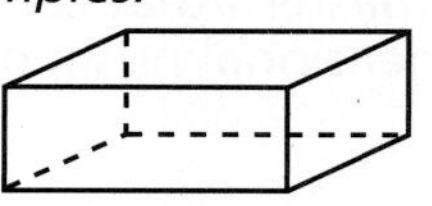

rectangular prism

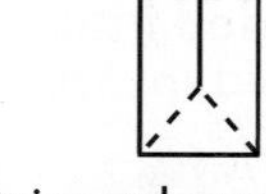

triangular prism

probability [prä•bə•bil′ə•tē] The likelihood that an event will happen (p. 574)

product [prä′dəkt] The answer to a multiplication problem (p. 31)

Property of One [prä′pər•tē əv wun] The property that states that the product of any number and 1 is that number (p. 90)
Examples: $5 \times 1 = 5$; $16 \times 1 = 16$

protractor [prō′trak•tər] A tool used for measuring or drawing angles (p. 458)

pyramid [pir′ə•mid] A solid figure with a polygon base and all other faces triangles that meet at a single vertex (p. 486)
Example:

Q

quadrilateral [kwäd•rə•lat′ə•rəl] A polygon with four sides and four angles (p. 482)
Example:

quart (qt) [kwôrt] A customary unit for measuring capacity; 2 pints = 1 quart (p. 512)

quotient [kwō′shənt] The number, not including the remainder, that results from dividing (p. 181)
Example: $8 \div 4 = 2$. The quotient is 2.

radius [rā′dē•əs] A line segment with one endpoint at the center of a circle and the other endpoint on the circle (p. 464)
Example:

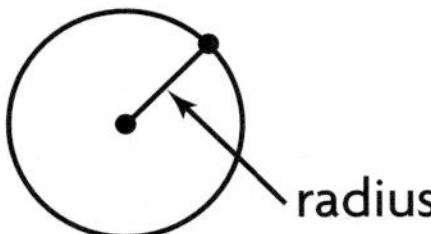

range [rānj] The difference between the greatest number and the least number in a set of data (p. 99)
Example: 2, 2, 3, 5, 7, 7, 8, 9
The range is $9 - 2 = 7$.

ratio [rā′shē•ō] The comparison of two quantities (p. 302)

ray [rā] A part of a line; it begins at one endpoint and extends forever in one direction. (p. 454)
Example:

reciprocal [ri•sip′rə•kəl] One of two numbers whose product is 1 (p. 396)
Example: 8 and $\frac{1}{8}$ are reciprocals since $8 \times \frac{1}{8} = 1$.

rectangle [rek′tan•gəl] A plane figure with opposite sides that are equal and with four right angles (p. 482)
Example:

rectangular prism [rek•tang′gyə•lər pri′zəm] A solid figure in which all six faces are rectangles (p. 486)
Example:

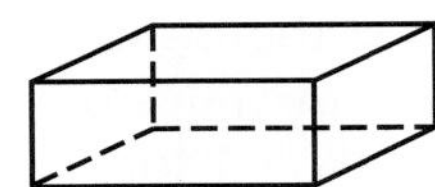

reflection [ri•flek′shən] A movement of a figure to a new position by flipping it over a line (p. 484)
Example:

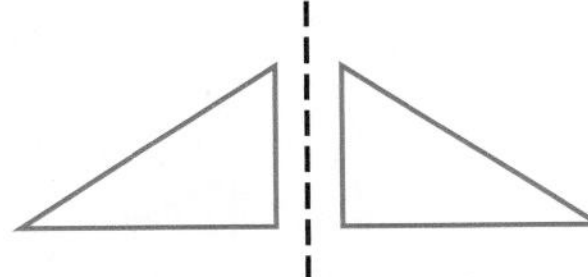

regular polygon [re′gyə•lər pä′lē•gän] A polygon in which all sides are congruent and all angles are congruent (p. 460)

rhombus [räm′bəs] A parallelogram with 4 congruent sides and with opposite angles that are congruent (p. 482)
Example:

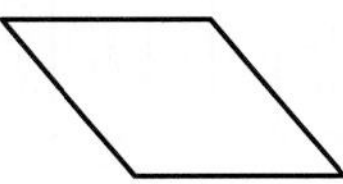

right angle [rīt an′gəl] A special angle formed by perpendicular lines and equal to 90° (p. 454)
Example:

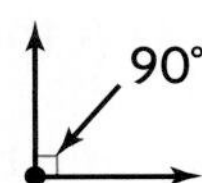

right triangle [rīt trī′an•gəl] A triangle that has a right angle (p. 478)
Example:

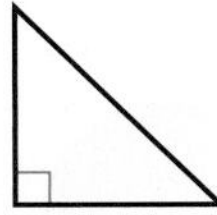

rotation [rō•tā′shən] A movement of a figure by turning it around a vertex or point of rotation (p. 484)
Example:

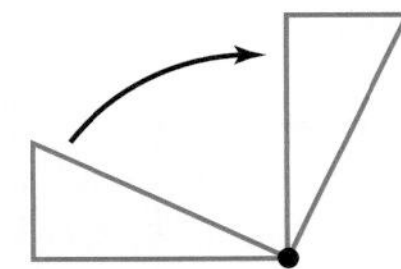

rotational symmetry [rō•tā′shən•əl si′mə•trē] A property of a figure that, when rotated less than 360° about a central point or a point of rotation, still matches the original figure (p. 470)

rounding [round′ing] Replacing a number with one that tells about how many or how much (p. 32)

scale [skāl] A series of numbers starting at 0 and placed at fixed distances on a graph to help label the graph (p. 116)

scale drawing [skāl drô′ing] A reduced or enlarged drawing whose shape is the same as an actual object and whose size is determined by the scale (p. 308)

scalene triangle [skā′lēn trī′an•gəl] A triangle with no congruent sides (p. 478)
Example:

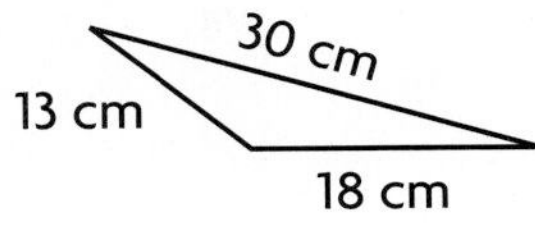

schedule [skej′o͞o•əl] A table that lists activities or events and the times they happen

similar [si′mə•lər] Having the same shape, but not necessarily the same size (p. 468)
Example:

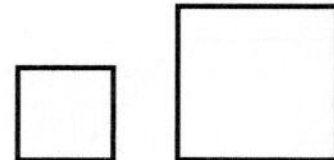

simplest form [sim′pləst fôrm] A fraction is in simplest form when the numerator and denominator have only 1 as their common factor. (p. 290)

solid figure [sä′ləd fig′yər] A three-dimensional figure (p. 486)

solution [sə•lo͞o′shən] A value that, when substituted for the variable, makes an equation true (p. 74)

sphere [sfir] A round object whose curved surface is the same distance from the center to all its points (p. 487)
Example:

square [skwâr] A polygon with 4 equal sides and 4 right angles (p. 482)
Example:

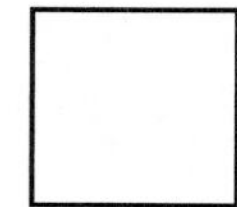

square number [skwâr num′bər] The product of a number and itself (p. 266)
Example: $4^2 = 16$; 16 is a square number.

square pyramid [skwâr pir′ə•mid] A solid figure with a square base and with four triangular faces that have a common point (p. 486)
Example:

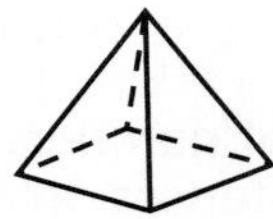

square unit [skwâr yo͞o′nət] A unit of area with dimensions 1 unit × 1 unit (p. 528)

standard form [stan′dərd fôrm] A way to write numbers by using the digits 0-9, with each digit having a place value (p. 1)
Example: 456 ←standard form

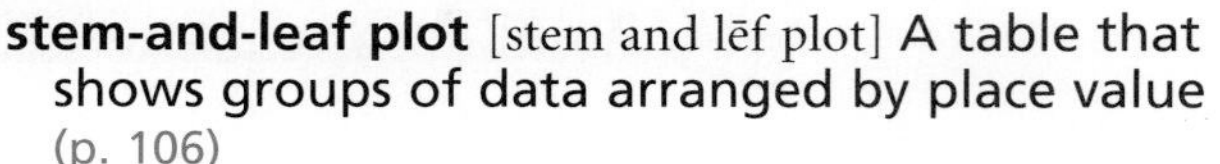

stem-and-leaf plot [stem and lēf plot] A table that shows groups of data arranged by place value (p. 106)
Example:

Stem	Leaves
1	1 2 4
2	0 3 4 5
3	4 5 7
4	0 0 1 2

Number of Tickets Sold

subtraction [səb•trak′shən] The process of finding how many are left when a number of items are taken away from a group of items; the process of finding the difference when two groups are compared; the opposite of addition (p. 36)

sum [sum] The answer to an addition problem (p. 31)

surface area [sûr′fəs âr′ē•ə] The sum of the areas of all the faces, or surfaces, of a solid figure (p. 548)

ten-thousandth [ten thou′zəndth] A decimal or fraction that names one part of 10,000 equal parts (p. 20)
Example: 0.5784 = five thousand, seven hundred eighty-four ten-thousandths

tenth [tenth] A decimal or fraction that names one part of 10 equal parts (p. 18)
Example: 0.7 = seven tenths

tessellation [tes•ə•lā′shən] A repeating pattern of closed figures that covers a surface with no gaps and no overlaps (p. 497)
Example:

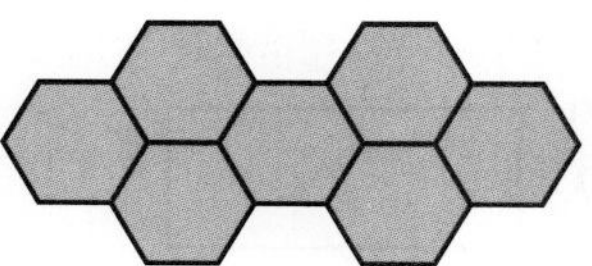

thousandth [thou′zəndth] A decimal or fraction that names one part of 1,000 equal parts (p. 20)
Example: 0.006 = six thousandths

ton (T) [tun] A customary unit used to measure weight; 2,000 pounds = 1 ton (p. 512)

transformation [trans•fər•mā′shən] The moving of a figure in a translation, reflection, or rotation (p. 484)

translation [trans•lā′shən] A movement of a geometric figure to a new position without turning it or flipping it; a slide (p. 484)
Example:

trapezoid [tra′pə•zoid] A quadrilateral with exactly one pair of parallel sides (p. 482)
Example:

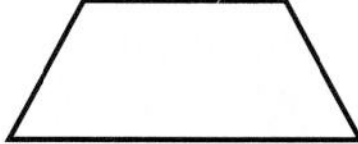

tree diagram [trē′ dī′ə•gram] An organized list that shows all possible outcomes for an event (p. 572)

triangle [trī′an•gəl] A polygon with three sides and three angles (p. 460)
Example:

triangular numbers [trī•ang′gyə•lər num′bərz] Numbers that can be represented by triangular figures (p. 481)
Example:

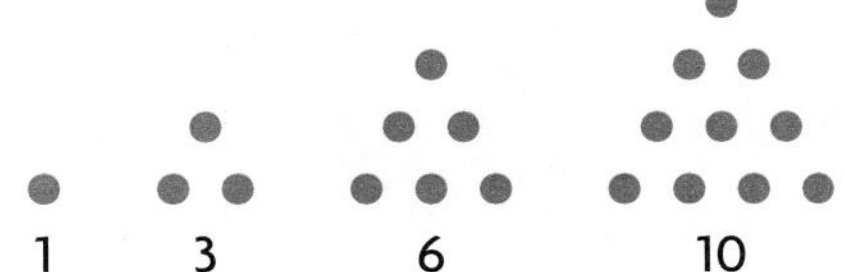

unlike fractions [un•līk′ frak′shənz] Fractions that have different denominators (p. 569)
Example: $\frac{3}{4}$ and $\frac{2}{5}$ are unlike fractions.

unlikely [un•līk′lē] Having a less-than-even chance of happening (p. 569)

variable [vâr′ē•ə•bəl] A letter or symbol that stands for one or more numbers (p. 68)

Venn diagram [ven dī′ə•gram] A diagram that shows relationships among sets of things (p. 253)
Example:

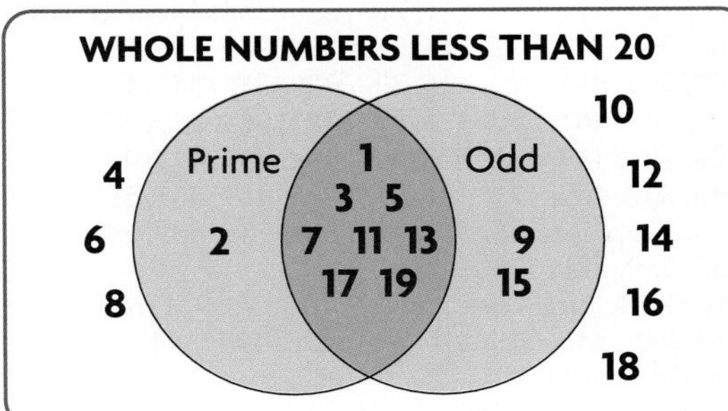

vertex [vûr′teks] The corner point made where two or more edges, rays, or sides of a polygon meet (pp. 454, 486)
Examples:

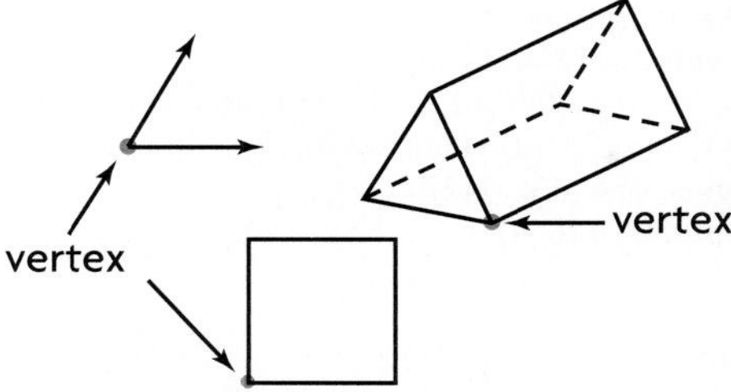

volume [väl′yəm] The measure of the space a solid figure occupies (p. 552)

whole number [hōl num′bər] One of the numbers 0, 1, 2, 3, 4, ... The set of whole numbers goes on without end. (p. 17)

word form [wurd fôrm] A way to write numbers in standard English (p. 1)
Example: 4,829 = four thousand, eight hundred twenty-nine

***x*-axis** [eks•ak′səs] The horizontal number line on a coordinate plane (p. 438)

***x*-coordinate** [eks•kō•ôr′də•nət] The first number in an ordered pair, which tells the distance to move right or left from (0,0) (p. 440)

***y*-axis** [wī•ak′səs] The vertical number line on a coordinate plane (p. 438)

***y*-coordinate** [wī•kō•ôr′də•nət] The second number in an ordered pair, which tells the distance to move up or down from (0,0) (p. 440)

Zero Property of Addition [zē′rō prä′pər•tē əv ə•di′shən] The property that states that when you add zero to a number, the result is that number (p. 78)

Zero Property of Multiplication [zē′rō prä′pər•tē əv mul•tə•plə•kā′shən] The property that states that when you multiply by zero, the product is zero (p. 90)

INDEX

A

E

S

T

PHOTO CREDITS

TOC
v (t) Scott Camazine/Photo Researchers, (br) Scott Camazine/Photo Researchers, viii Duomo Photography, xi Latent Image Photography, xi Tom & Dee Ann McCarthy/The Stock Market, xv Ron Scherl/StageImage, xv Tui De Roy/Bruce Coleman,Inc., xvii Tony Freeman/PhotoEdit.

0 Earth Imaging/Stone, 3 (l) E.R. Degginger/Photo Researchers, (r) Scott Camazine/Photo Researchers, 4 (t) Corbis Digital Stock, (b) NASA, (ti) NASA, 5 NASA, 6 Telegraph Colour Library/FPG International, 8 Mary Eleanor Browning/Photo Researchers, 10 Michael Hubrich/Dembinsky Photo Associates, 13 (t) Telegraph Colour Library/FPG International, 16 (bg) W. Metzen/H. Armstrong Roberts, (lbi) E. F. Anderson/Visuals Unlimited, (cti) Ed Reschke/Peter Arnold, Inc. (ti) Dr. Dennis Kunkel/Phototake, (lti) Grant Heilman/Grant Heilman Photography, (cbi) S Lowry/Univ. Ulster/Stone, (rti) Larry Lefever/Grant Heilman Photography, (rbi) G. Shih-R. Kessel/Visuals Unlimited, 18 Richard Shiell/Dembinsky Photo Associates, 20 Dennis O'Clair/Tony Stone Images, 22 Gloria H. Chomica/Masterfile, 23 Shin Yoshino/Minden Pictures, 24 Mike Powell/Allsport USA, 26 Jeff Foott/Bruce Coleman, Inc., 27 John Warden/Tony Stone Images, 30 Stephen J. Krasemann/DRK, 32 Todd Warshaw/Allsport USA, 36 Steve Kaufman/DRK, 38 Al Bello/Allsport Photography, 44 Julia Patriarche/www.chalkdust.net, 46 Rod Planck/Dembinsky Photo Associates, 48 Julia Patriarche/www. chalkdust.net, 49 Duomo Photography, 53 Index Stock Photography, 55 Duomo Photography, 56 Ron Kimball, 61 Superstock, 64 Tony Stone Images, 65 Mark E. Gibson,Inc., 66 S. Nielsen/Bruce Coleman, Inc., 68 Carl Roessler/Bruce Coleman, Inc., 72 R. Andrew Odum/Peter Arnold, Inc., 74 Mark J. Thomas/Dembinsky Photo Associates, 75 John Giustina/FPG International, 76 Ken Scott/Dembinsky Photo Associates, 77 Index Stock Photography, 78 Robert E. Daemmrich/Tony Stone Images, 80 Bill Losh/FPG International, 81 Peter Gridley/FPG International, 84 (i) Domino arrangement by: Scott Suko/Event organized by: EXPO Children's Museum, Gainesville, FL, 84 Henry Blackham/The Stock Market, 86 Photo courtesy of Ralf Laue, 88 AP Photo/Wide World, 93 E.R. Degginger/Dembinsky Photo Associates, 96 Mitchell Layton/Duomo Photography, 102 (b) The Granger Collection, (tl) Corbis, 106 Richard M. Smith/FPG International, 107 Duomo Photography, 110 Ron Kimball, 111 (t) Ron Kimball, (b) Frank Hecker/OKAPIA/Photo Researchers, 114 Stocktrek Corp/The Stock Market, 116 Gerry Ellis/Minden Pictures, 117 Konrad Wothe/Minden Pictures, 118 Tom McHugh/Steinhart Aquarium/Photo Researchers, 119 Robert Torrez/Tony Stone Images, 120 FPG International, 122 Dean Conger/Corbis, 124 Jimmy Rudnick/The Stock Market, 125 Chie Nishio/Omni-Photo Communications, 126 Rafael Macia/Photo Researchers, 127 Ken Cole/Animals Animals, 133 Toyohiro Yamada/FPG International, 140 Dewitt Jones/Corbis, 141 (t) Robert Holmes/Corbis, (b) Robert Holmes/Corbis, 142 David Nunuk/Science Photo Library/Photo Researchers, 144 NASA, 146 Art Wolfe/Photo Researchers, 147 David Madison/Bruce Coleman, 148 Duomo Photography, 149 Anup Shah/Animals Animals, 150 Richard J. Wainscoat/Peter Arnold, Inc., 151 Monkmeyer/Rogers, 152 Michael Hubrich/Photo Researchers, Inc., 153 Spencer Jones/FPG International, 156 Robert A. Tyrrell Photography, 160 M. Lammertink, 162 Robert A. Tyrrell Photography, 163 Rod Planck/Photo Researchers, 164 (fg) E. R. Degginger/Dembinsky Photo Associates, 165 John Mitchell/Photo Researchers, 168 Hans Pfletschinger/Peter Arnold, Inc., 169 Runk/Schoenberger/Grant Heilman Photography, 178 Tom Till, 179 Mark E. Gibson, 180 Tony Arruza/Corbis, 182 ITAR-TASS/Sovfoto, 183 Philip James (University of Toledo), Steven Lee (University of Colorado), NASA, 184 NASA, 185 NASA, 187 Kazoobie, Inc., 188 Archive Photos, 189 Ron Scherl/StageImage, 191 (t) Archive Photos, (b) The Granger Collection, New York, 192 U.S. Postal Service, 193 Gemma Dickmann, 194 Ed Bock/The Stock Market, 195 Tony Freeman/PhotoEdit, 198 Don Smetzer/Stone, 202 (br) Hulton Getty/Liaison International, 203 (c) Corbis, 204 Herb Scharfman/Sports Illustrated/Time, 206 APA /Archive Photos, 208 Associated Press/ Wide World Photos, 209 Frank Atura/Asolo Theatre Company, 210 John Cancalosi/DRK, 211 AFP/Corbis, 212 Steinouer Photography, 213 Claudia Parks/The Stock Market, 216 FoodPix, 218 Tom & Dee Ann McCarthy/The Stock Market, 222 Jay Syverson/CORBIS, 223 Tony Freeman/PhotoEdit, 225 Jose L. Pelaez/The Stock Market, 226 Latent Image Photography, 227 Mark E. Gibson, 228 Tony Freeman/PhotoEdit, 229 Mark E. Gibson, 232 Sonny Senser/Harcourt, 238 Foodpix, 242 Wesley Bocxe/Photo Researchers, 243 Sovfoto/Eastfoto, 247 Index Stock Photography, 250 Bruce Berman/The Stock Market, 251 Mark E. Gibson, Inc., 252 W.B. Hamilton/U.S.G.S. Photo Library, 254 Bill Aron/Photo Researchers, 261 (t), (b) The Granger Collection, New York, 266 AP Wirephoto/Wide World Photos, 268 Biophoto Associates/Photo Researchers, 270 A. B. Dowsett/Science Photo Library/Photo Researchers, 278 Rosenfeld Images Ltd./Science Photo Library/Photo Researchers, 282 Myreen Ferguson/PhotoEdit, 297 D. Young-Wolff/PhotoEdit, 300 William R. Sallaz/Duomo Photography, 306 Allsport Photography (USA), 311 (l) Yann Arthus-Bertrand/Corbis, 311 (r) Ron Kimball, 314 Roy Corral/Stone, 320 (tl) Davies & Starr/Tony Stone Images, 320 (tr) Bruce Coleman, Inc, 320 (bl) Kjell B. Sandved/Bruce Coleman, Inc., 320 (br) Davies & Starr/Tony Stone Images, 321 (t) Raymond A. Mendez/Animals Animals, 321 (b) J.C. Carton/Bruce Coleman, Inc., 323 Richard Reid/Earth Scenes, 325 Jeff Greenberg/PhotoEdit, 328 Carolyn A. McKeone/Photo Researchers, 329 Ron Kimball, 333 Peter Skinner/Photo Researchers, 340 StockFood America/Bonisolli, 341 Craig Lovell/AGStock USA, 342 Scott Barrow, Inc., 344 Duomo Photography, 345 David Madison, 350 Paul Harris/Tony Stone Images, 353 S. Nielsen/Bruce Coleman, Inc., 358 Bernard Boutrit/ Woodfin Camp & Associates, 362 Lori Adamski Peek/Stone, 364 Central Florida Council Boy Scouts of America, 366 (l) Federal Clipart, 366 (r) Federal Clipart, 367 Index Stock Photography, 368 Robert Holmes/Corbis, 369 Mark Greenberg/FPG International, 372 Grant Heilman,Inc., 374 Scott Barrow, Inc., 378 Kevin R. Morris/Stone, 382 Ron Scherl/StageImage, 383 Myrleen Ferguson/PhotoEdit, 384 Larry Lefever/Grant Heilman Photography, 385 Carlyn Iverson/Photo Researchers, 392 Sharon Cummings/Dembinsky Photo Associates, 396 Tony Freeman/PhotoEdit, 399 Tui De Roy/Bruce Coleman, Inc., 403 Konrad Wothe/Minden Pictures, 404 Marty Snyderman/ Waterhouse Stock Photo, 405 F. Stuart Westmorland/Photo Researchers, 409 Index Stock Photography, 412 (l) Mark E. Gibson, Inc., 412 (r) Arthur M. Greene/Bruce Coleman, Inc., 413 (l) Stephen J. Krasemann/Photo Researchers, 413 (r) Gregory Ochocki/Photo Researchers, 414 Michael Zito/SportsChrome, 416 Mark J. Thomas/Dembinsky Photo Associates, 419 Johnny Johnson/Animals Animals, 420 Julep Gillman-Bryan, N.C. Division of Archives and History, 421 Karen Kasmauski/ Woodfin Camp & Associates, 422 David Cannon /Allsport Photography, 424 Steve Hix/FPG International, 428 Danielle Hayes/Bruce Coleman, Inc., 430 WORLDSAT International, Inc./Photo Researchers, 431 Lynette Cook/Science Photo Library/Photo Researchers, 436 Lance Nelson/The Stock Market, 440 Hazel & Douglas Wolfe, 444 NASA, 448 Index Stock Imagery, 452 Chris Mooney/FPG International, 454 J. Messerschmidt/Bruce Coleman, Inc., 460 L. West/Bruce Coleman, Inc., 462 (l) Steve Solum/Bruce Coleman, Inc., 462 (r) Alan Schein/The Stock Market, 462 (cl) Ken Straiton/The Stock Market, 462 (cr) Jeff Greenberg/PhotoEdit, 464 (t) Reuters/HO/Archive Photos, 465 AFP/Corbis, 468 (bg) Richard Dunoff/The Stock Market, 470 Mark Scott/FPG International, 476 Will & Deni McIntyre/Photo Researchers, 500 Larry Allan/Bruce Coleman, Inc., 501 Mark E. Gibson, Inc., 502 Stuart Westmorland/Stone, 506 (c) E.R. Degginger/Bruce Coleman, Inc., 511 Mark E. Gibson/Dembinsky Photo Associates, 512 Ostentoski & Zoda/Envision, 515 (cr) Barbara Reed/Animals Animals, 516 B. Daemmrich/The Image Works, 519 FoodPix, 522 David Ball/The Stock Market, 524 Lee Balterman/FPG International, 532 Jamie Squire/Allsport, 536 Art Resource, 541 John W. Bova/Photo Researchers, 544 Mark Segal/Stone, 544 (i) Weronica Ankarorn, 559 Comstock, 563 Bill Foley/Bruce Coleman, Inc., 566 (t) Lee Foster/Bruce Coleman, Inc., 566 (b) David Hall/Photo Researchers, 567 Monterey Bay Aquarium, 568 (inset) Doranne Jacobson/International Images, 586 John Elk III/Bruce Coleman, Inc., 587 (l) Phillip Roullard Photography, 587 (r) Alon Reininger/The Stock Market, 587 (c) Phillip Roullard Photography, H0 Tom McHugh/Steinhart Aquarium/Photo Researchers, H1 (t) Scott Camazine/Photo Researchers.